Identification & Analysis of Organic Pollutants in Water

edited by

Lawrence H. Keith
Research Chemist
U.S. Environmental Protection Agency
Environmental Research Laboratory
Athens, Georgia

Present Address: Radian Corporation, P.O. Box 9948,
8500 Shoal Creek Boulevard, Austin, Texas 78766

ANN ARBOR SCIENCE
PUBLISHERS INC
P.O. BOX 1425 • ANN ARBOR, MICH. 48106

Third Printing, 1979
Second Printing, 1977

Copyright © 1976 by Ann Arbor Science Publishers, Inc.
P.O. Box 1425, Ann Arbor, Michigan 48106

Library of Congress Catalog Card Number 76-1730
ISBN 0-250-40131-2

Manufactured in the United States of America
All Rights Reserved

PREFACE

A new phase of environmental chemistry, the identification and analysis of specific organic pollutants in water, has begun. The chapters in this book were contributed by many authors and represent the current state-of-the-art in the United States and abroad. Only five years ago a book of this nature would not have been practical because of the scant knowledge available and the few chemists working in this area.

The most significant contributions to advances in this research area have been the development and application of computer-assisted gas chromatography/mass spectrometry (GC/MS). Environmental samples typically contain a hundred or more compounds detectable as gas chromatographic peaks. Within an hour or two an avalanche of mass spectral data can be generated from such samples. By reducing the raw data to formats that are amenable to interpretation, computer-assisted GC/MS has made analyses of these complex samples practical and cost-effective. A further refinement is computer-matching sample spectra with reference spectra in data banks that contain thousands of mass spectra.

Prior to these advances, analysis of organic pollutants was literally in the "dark ages" with respect to other areas of analytical chemistry. Organic pollution in water could only be measured by relatively crude collective parameters such as biological oxygen demand (BOD), total organic carbon (TOC), and carbon oxygen demand (COD). While these measurements can indicate that an organic pollution problem exists, they cannot provide significant information about the chemical nature of the pollutants or help to identify them.

Research in the next five to ten years will greatly increase our knowledge of the identities, quantities and distribution of specific organic chemicals in waters of all types. This information will be useful for monitoring, pollution treatment and control, and evaluation of environmental health effects. Prior to 1970 only about 100 different organic compounds had been identified in water. Today over 1500 organic compounds have been identified in all types of water—industrial wastewaters, rivers, lakes, ground water—and 400-500 compounds have been identified in drinking waters throughout the world. If only five percent of the more than two million known organic compounds eventually find their way into our waters, then an estimated 100,000 organic compounds may be identified in the waters of the world. If this is true, we have a long way to go.

Our new knowledge of specific organic compounds in water places a heavy burden on the environmental health effects sciences. Questions relating to health effects immediately arise when new compounds are identified in municipal drinking water supplies. Still unknown are the chronic (long-term) effects of most of these chemicals at the trace concentrations in which they usually occur in drinking water. Also unknown is a threshold concentration below which a carcinogen will *not* cause cancer. Research to determine chronic health effects requires huge expenditures of time and money. Some of these problems are addressed in Section I.

Other problems that remain to be solved include better methods for separating organic pollutants from water and from each other. Section II presents state-of-the-art methods now in use and being developed for separating and concentrating these complex mixtures, often at the parts per billion level, from their water matrix. These methods include gas purging, headspace analysis, resin adsorption, carbon adsorption, liquid-liquid extraction, and semipermeable membranes. Often the analyst will discover that this phase of environmental chemistry is the most difficult, especially if quantitation of the identified compounds is involved. This active area of research should result in future improvements. After the organics are removed from water, they must be separated from one another before the individual compounds can be completely identified. Currently, capillary column gas chromatography offers the brightest promise for separating the "volatile" compounds. However, typically 80-90% (by weight) of the organic components of a water sample will not (even after derivitization) pass through a gas chromatographic column. Development of improved methods to separate and detect these nonvolatile compounds is another active research area. High-performance liquid chromatography (HPLC), presently the leading technique for separating nonvolatile compounds, is also discussed in Section II.

Section III is devoted primarily to discussions of analytical results, but many of the chapters also provide further information on methods and techniques. The types of samples analyzed include drinking water, river water, lake water, rainwater, ground water, municipal wastewater, and industrial wastewaters. These analyses have made a significant contribution to information concerning the number of organic chemicals identified in water, their occurrence and distribution.

With the exception of Chapters 7, 17, 23 and 36, the material in this book came from the symposium program "Identification and Analysis of Organic Pollutants in Water" at the First Chemical Congress of the North American Continent, Mexico City, Mexico in December, 1975.

Hopefully, the symposium and this book will stir the interest of other analytical, environmental and organic chemists and will further the development of environmental chemistry as a distinct branch of the "Science of Chemistry." Establishment of Institutes of Environmental Chemistry would further hasten the maturity of this science and provide centers of expertise for research. Ultimately, environmental chemistry should be recognized by more chemistry departments and more national chemical societies.

Two conclusions from the symposium and this book are evident: this type of research can now be accomplished with rewarding results, and much still remains to be done. We have only begun to learn and to apply what we have learned to provide a better and cleaner environment.

L. H. Keith, Ph.D.
January, 1975

To

HARRIET M. KEITH

who 20 years ago planted the seeds for this book

and

VIRGINIA H. KEITH

who helped bring it to fruition

CONTENTS

SECTION I: CHEMISTRY OF POLLUTANTS   1

1   The Foundations of Organic Pollutant Analysis   3
       Aaron A. Rosen

2   The Chemistry of Some Potential Polyhalogenated Water Pollutants   15
       O. Hutzinger, G. Sundström, F. W. Karasek, S. Safe

3   Photodecomposition of Halogenated Aromatic Compounds   35
       S. Safe, N. J. Bunce, B. Chittim, O. Hutzinger, L. O. Ruzo

4   Chemical Carcinogens in the Environment   49
       Barry Commoner

SECTION II: TECHNIQUES AND METHODS OF ANALYSIS   73

5   Glass Capillary Gas Chromatography in Water Analysis: How to Initiate Use of the Method   75
       K. Grob, G. Grob

6   GC/MS Determination of Volatiles for the National Organics Reconnaissance Survey (NORS) of Drinking Water   87
       Frederick C. Kopfler, Robert G. Melton, Robert D. Lingg, W. Emile Coleman

7   A Convenient Liquid-Liquid Extraction Method for the Determination of Halomethanes in Water at the Parts-Per-Billion Level   105
       James E. Henderson, Gary R. Peyton, William H. Glaze

8   Separation of Trace Organic Compounds from Water .   113
       J. P. Mieure, G. W. Mappes, E. S. Tucker, M. W. Dietrich

9   Resin Sorption Methods for Monitoring Selected
    Contaminants in Water   135
        Gregor A. Junk, John J. Richard, James S. Fritz,
        Harry J. Svec
10 Development of Methods for Organic Analyses for
    Routine Application in Environmental Monitoring
    Laboratories   155
        William L. Budde, James W. Eichelberger
11 Development of Computerized GC/MS Techniques
    Within the U.S. EPA   177
        J. M. McGuire, M. H. Carter, A. L. Alford
12 The Use of GC/MS in the Analysis of Unusual
    Environmental Chemicals.   185
        R. E. Finnigan, J. B. Knight
13 On the Origin of Polycyclic Aromatic Hydrocarbons in
    the Aqueous Environment   205
        Anneli Hase, Ronald A. Hites
14 Separation and Analysis of Refractory Pollutants in
    Water by High-Resolution Liquid Chromatography . . .   215
        W. W. Pitt, Jr., R. L. Jolley, S. Katz
15 Determination of Chlorination Effects on Organic Con-
    stituents in Natural and Process Waters Using
    High-Pressure Liquid Chromatography   233
        Robert L. Jolley, Guy Jones, Jr., W. W. Pitt, Jr.,
        James E. Thompson
16 Analysis of New Chlorinated Organic Compounds in
    Municipal Wastewaters after Terminal Chlorination . .   247
        William H. Glaze, James E. Henderson, IV,
        Garmon Smith
17 N-Nitroso Compounds in Water   255
        David H. Fine, David P. Rounbehler
18 Determination and Monitoring of Some Organic
    Explosives in Natural and Effluent Water by
    Single-Sweep Polarography   265
        Gerald C. Whitnack
19 Organic Fractionization and Selected Trace Metal
    Content of Sludges.   281
        Leo W. Newland, John R. Ten Eyck, Viola K. Ohr
20 Detection and Isolation of Bioaccumuable Chemicals in
    Complex Effluents   297
        Gilman D. Veith, Ned M. Austin

SECTION III: IDENTIFICATION OF ORGANIC POLLUTANTS   303

21 The Occurrence of Volatile Organics in Five Drinking Water Supplies Using Gas Chromatography/Mass Spectrometry   305
      W. Emile Coleman, Robert D. Lingg, Robert G. Melton, Frederick C. Kopfler

22 Identification of Organic Compounds in Drinking Water from Thirteen U.S. Cities   329
      L. H. Keith, A. W. Garrison, F. R. Allen, M. H. Carter, T. L. Floyd, J. D. Pope, A. D. Thruston, Jr.

23 GC/MS Identification of Trace Organic Compounds in Philadelphia Waters   375
      I. H. Suffet, L. Brenner, J. V. Radziul

24 Analysis of Drinking Water for Organic Compounds . .   399
      Robert D. Kleopfer

25 Possible Factors in the Drinking Water of Laboratory Animals Causing Reproductive Failure   417
      J. D. McKinney, R. R. Maurer, J. R. Hass, R. O. Thomas

26 Analyses of Organic Constituents in Water by High-Resolution Gas Chromatography in Combination with Specific Detection and Computer-Assisted Mass Spectrometry   433
      Walter Giger, Martin Reinhard, Christian Schaffner, Fritz Zürcher

27 Isolation and Identification of Organic Contaminants in Ground Water   453
      W. J. Dunlap, D. C. Shew, M. R. Scalf, R. L. Cosby, J. M. Robertson

28 Chloroform-Extractable Organic Compounds in the International Great Lakes   479
      William M. J. Strachan

29 Uses of Wastewater Discharge Compliance Monitoring Data   499
      E. William Loy, Jr., Donald W. Brown, J. H. Myron Stephenson, John A. Little

30 GC/MS Analysis of Organic Compounds in Domestic Wastewaters   517
      A. W. Garrison, J. D. Pope, F. R. Allen

31	The Molecular Nature and Extreme Complexity of Trace Organic Constituents in Southern California Municipal Wastewater Effluents . . . . *A. L. Burlingame, B. J. Kimble, E. S. Scott, F. C. Walls, J. W. de Leeuw, B. W. de Lappe, R. W. Risebrough*	557
32	The Characterization of Trace Organic Constituents in Petroleum Refinery Wastewater by Capillary Gas Chromatography/Real-Time High-Resolution Mass Spectrometry—A Preliminary Report . . . . *A. L. Burlingame, B. J. Kimble, E. S. Scott, D. M. Wilson, M. J. Stasch, J. W. de Leeuw, L. H. Keith*	587
33	Identification of Two Chlorinated Guaiacols in Kraft Bleaching Wastewaters . . . . *I. H. Rogers, L. H. Keith*	625
34	Fate of Selected Organic Compounds in the Discharge of Kraft Paper Mills into Lake Superior . . . . . . *Michael E. Fox*	641
35	Persistent Organic Compounds from a Pulp Mill in a Near-Shore Freshwater Environment . . . . *B. Brownlee, W. M. J. Strachan*	661
36	GC/MS Analyses of Organic Compounds in Treated Kraft Paper Mill Wastewaters . . . . *Lawrence H. Keith*	671
INDEX		709

SECTION I

CHEMISTRY OF POLLUTANTS

CHAPTER 1
THE FOUNDATIONS OF ORGANIC POLLUTANT ANALYSIS

Aaron A. Rosen

    U.S. Environmental Protection Agency
    Cincinnati, Ohio

The timing of this book is remarkably significant since it marks the 25th anniversary of a landmark development in the analysis of organic pollutants in the environment. The event commemorated is the introduction in 1950 by Braus, Middleton and Walton[1] of large-scale sampling of organic pollutants, thus making possible the application of the entire range of available methodology for systematic analysis.

APPLICATION OF SYSTEMATIC ORGANIC ANALYSIS

At the beginning of this 25-year period of progress, available methods for analysis of organic pollutants in water were embodied in the ninth edition of *Standard Methods*.[2] On the whole, organics could be measured either directly as a general aggregate (BOD and COD) or indirectly (odor, color). There were a few methods for specific organic substances in waters or wastes: oil, grease, organic nitrogen, cyanide, tannin-and-lignin, phenol, and furfural. There was already a method, though based on an impractical wet oxidation procedure, for total organic carbon, thus anticipating a major later instrumental development. There were a few unusual methods, *e.g.*, albuminoid nitrogen and methane, which are rarely used today.

In 1950 the methodology for organic chemicals in air pollution was even more primitive. No reference comparable to *Standard Methods* existed. The current emphasis was on determining the concentration of total particulates and total soluble organic particulate matter. There was a method for trapping automobile exhaust gases and determining aldehydes in the condensate. At the same time there were some early experiments on applying long-path infrared analysis to detect hydrocarbons in ambient air. For several years progress in the air pollution field was made by borrowing methodology developed in water pollution research.[3]

New Direction in Sampling

In 1950 laboratory practices in qualitative organic analysis largely followed the system in the third edition of the manual by Shriner and Fuson.[4] The principal features of this system were: determination of solubility classes, application of classification tests (for functional groups and structural characteristics), and preparation of derivatives. The authors recognized that "It is very rarely possible to identify the constituents of a mixture without previous separation." In their system, separation of mixtures was based on chemical character, *i.e.*, differences in polar character affecting solubility in water and organic solvents.

The difficulty in applying the Shriner and Fuson system to water pollution analysis lay in the need for relatively massive organic samples, free of water. For example, the instructions for separation of mixtures begin: "From 25 to 50 g of the mixture is mixed with 75 ml of ether . . ." Typical weights of isolated organics for classification tests are 0.1 g (solids) to 2 ml (liquids), and for derivative preparation, 0.3 g solids to 1.0 g liquids. Even if the practical semimicro techniques of that time were substituted in the system, requirements for each test would be 0.025 to 0.10 g of each organic compound.

Braus and his co-workers decided in 1950 that it was no longer practical to base organic pollutant analysis on attempts to develop a catalog of ultrasensitive analyses for specific organics in water samples of convenient size for laboratory processing. It was obvious that an entirely different sampling approach had to be devised to supply the comparatively massive organic samples required for systematic separation and identification. The solution was to filter 5000 gallons (sometimes more) of water through a column of activated carbon. The organic pollutants were adsorbed upon the carbon, which was then air-dried and exhaustively extracted with ether. The yield of water-free organic mixture was 2-5 g, typically about 4 g. For enhanced safety, chloroform replaced ether as the extraction solvent (chloroform was not then recognized as a possible carcinogen). Later a second extraction with ethanol was added, yielding an additional 2-10 g of organic sample.[5] Carbon adsorption sampling was extended with partial success to organics in airport atmospheres,[6] and currently with great success to industrial atmospheres.

Even 5000 gallons of water sometimes yielded a sample of organic pollutants that was too small to allow extensive analysis. The carbon filter sampling system, both the filter array and the extraction apparatus, was scaled up 80-fold.[7] By this means, chloroform extracts weighing 150-1700 g ("megasamples") were obtained, allowing for much more extensive analyses. It is noteworthy that the recent discovery of a large number of organic pollutants in the New Orleans drinking water was made using organic extract samples obtained by use of the same apparatus.[8]

Specialized Analytical Methods

A respectable array of specialized methods and instruments for organic analysis was available in 1950. The best catalog of these was assembled by Weissberger.[9] They became useful in analysis of organic environmental pollution only after the carbon filter sampling method came into use, since they were best applied to isolated pure compounds and their derivatives. Some of the available methods were: chemical microscopy, X-ray diffraction, ultraviolet spectrometry, polarography, radioactive labeling (isotope dilution methods), and molecular complexes, *e.g.,* inclusion compounds. Mass spectrometry was already in use in the analysis of petroleum components, but the method was first applied to water pollution in 1953.[10]

Three methods were in the analyst's armament in 1950 that later made especially important contributions to water pollution research. Infrared spectrometry has the special virtue of being applicable to impure organic samples and even to mixtures. Its use with water pollutants began almost simultaneously with the development of carbon filter sampling. Many forms of chromatography—column adsorption and partition, paper, electro- and thin-layer—were already in common use and awaited only the availability of suitable isolates of organic pollutants. Sensory methods—measurement of odor and taste—can be considered a specialized form of bioassay, probably the most sensitive. They are applicable to water samples and thus were used importantly even before the improved sampling method.

This chapter presents examples of the progress made in the past 25 years by the timely application of currently available analytical methods. During this period there were two major scientific advances relevant to water pollution. One was the improvement and expansion of methods to assess the health effects of pollutants, particularly carcinogens. As a result, investigations of organic pollutants in water received new legal, organizational and budgetary support. The other was the continuous development of gas chromatography, a technique that provided a quantum increase in the ability to separate organic mixtures (replacing distillation systems that were never truly effective). At the same time, gas chromatography yielded data for pollutant identification- retention time (analogous to boiling point of a pure component) and specific detector response (analogous to elemental and functional group determination).

PROGRESS IN ORGANIC POLLUTANT ANALYSIS

During the 25-year period reviewed here, the most productive advances initially involved improvement in the new sampling method and expansion in the application of conventional systematic organic analysis and related specialized instrumental methods.

Adsorption on activated carbon is a complex phenomenon still not well understood. It was recognized from the start that the carbon itself is a major variable. Braus et al.[1] showed that a carbon manufactured specifically for water treatment was nearly five times as effective as another type of carbon. Later work considered the effect of pore size, pore volume, and surface functional groups on adsorption capacity, but showed that there is little difference in adsorption activity between a number of brands of carbon expressly developed for water purification. However, carbons varied several fold in their content of extractable organic matter and a brand with a consistently low blank became the standard method material. The effectiveness of different solvents in desorbing organic material from carbon also varies notably.[11] It was shown that consecutive extraction with chloroform and ethanol, later established as the standard method, was only slightly less effective than the most active combination of six solvents studied.

Experiments with a number of pure organic compounds in dilute solution showed that their adsorption in a carbon filter was 91-100% complete, but solvent extraction usually desorbed only 70-85% of the adsorbed compound.[12] At the time many conjectured that the loss was due to decomposition catalyzed by the active carbon surface. However, a definitive study of the adsorption-desorption of phenol labeled with carbon-14 showed that the unrecovered phenol remained, trapped but unchanged, in the pores of the activated carbon.[13]

Another advance was based on the carbon-14 label in organic pollutants. Since industrial chemical wastes are derived from fossil sources, they contain no radioactive carbon, whereas organics derived from sewage and run off from vegetated land will have nearly a contemporary level of radioactivity. Accordingly, an assay for carbon-14 can reveal the relative organic pollution contributions of these two principal classes of sources.[14] This work also showed that the industrial pollutants predominate in the chloroform extract while natural and municipal pollutants largely make up the ethanol extract. Accordingly, the yield of chloroform extract is an approximate measure of industrial pollutants, many of which are known to pose health risks. This rationale led to the adoption of a limit on concentration of CCE (carbon chloroform extract) in drinking water as a practical means to avert the health hazard of countless unknown industrial organic pollutants.[15] This same concept was later adopted in the European and International drinking water standards.

The superscale carbon filter system (Megasampler)[7] has been mentioned previously. Besides providing more abundant samples for expanded analysis, the system made available sufficient organic material for long-term animal testing of adverse health effects. The result of these tests was the first demonstration, in 1963, of carcinogenicity in the organic pollutants of a public drinking water supply.[16] By 1969, the apparent carcinogen in that water supply was identified as *bis*(2-chloroethyl) ether, and its tumor-producing activity was demonstrated.[17]

Other methods of concentrating organic pollutants from water were developed. Some of these were: liquid-liquid extraction (including variations to handle 500-gallon or larger samples), adsorption on polyurethane foam or polymer beads, collection on macroreticular ion-exchange resins, and volatilization with subsequent trapping in solvents or on adsorbents. Many chapters presented in this book illustrate applications of these sampling methods.

Group Separation Methods

The availability of carbon filter extracts made possible the application of the solubility-group separation system of Shriner and Fuson, yielding initially these groups of related organic pollutants: amphoteric, basic, acidic, phenolic, and neutral compounds.[1] Experience in use of the method resulted in dropping the amphoteric group and adding a water-soluble group.

The neutral group thus separated is quite heterogeneous. In addition to aliphatic and aromatic hydrocarbons, it contains a variety of functional group classes, *e.g.*, chlorinated hydrocarbons, alcohols, esters, and aldehydes and ketones. Adding a step to isolate carbonyl compounds by dissolving them in sodium bisulfite solution or combining with Girard's or other carbonyl reagents provided little new analytical information. The same fact was true when efforts were made to treat esters as a distinct group. On the other hand, complex formation served very effectively to separate and classify neutral hydrocarbons, until gas chromatography came into use. Normal aliphatic hydrocarbons were precipitated as their urea complexes, regenerated and recovered, thus providing convincing evidence of the presence of petroleum materials. Polycyclic aromatic hydrocarbons were precipitated as their complexes with 2,4,7-trinitrofluorenone. Inadequate tools for characterizing the isolated aromatic hydrocarbons were then available, but an indication of 1,2'-dinaphthyl was observed.[18] This complexing agent is currently being revived with the new attention being given to carcinogenic aromatic hydrocarbons in the environment.

Column adsorption chromatography was developed as a more useful means to subdivide the neutral group into fractions designated as aliphatics, aromatics, and oxygenated compounds. The aliphatic and aromatic fractions of petroleum pollutants gave characteristic infrared spectra more recognizable than the spectra of the weathered total oil. The method, which is the basis of many current oil spill investigation methods, was applied in 1955 to detect the presence of petroleum refinery wastes[19] and in 1959 to the identification of the sources of oil slicks in the Santa Barbara channel.[20] This latter application startlingly foresaw the oil spill problems that were to erupt a decade later.

The chromatographic separation of neutrals is an effective way to concentrate the polycyclic aromatic hydrocarbons in environmental samples for detailed analysis. It is widely used for this purpose in air pollution research.[21]

Most chlorinated hydrocarbon pesticides concentrate in the aromatic fraction, where they can be detected by infrared spectrometry. This property made possible the early successes in monitoring insecticides as unsuspected pollutants in environmental samples.[22]

Application of Instrumental Analyses

Infrared spectrometry was applied to carbon filter samples almost from the start of the filter method. The infrared technique can be applied to mixtures and impure compounds; thus, it was very informative with the still-complex fractions of organics obtained by group separation. Some noteworthy firsts in organic pollutant analysis were accomplished by infrared spectrometry. The detection of chlorinated pesticides has been described. The definitive identification of synthetic detergents (alkylbenzene sulfonates) as the cause of severe foam problems in streams, water supplies, and sewage treatment plants was also accomplished by infrared spectrometry.[23] This proof resulted in the redirection of the entire detergent industry to the manufacture of biodegradable products.

The discovery of o-nitrochlorobenzene as a Mississippi River pollutant was made in 1958 through interpretation of its infrared spectrum in carbon filter extracts[24]; the spectrum was especially clear in the aromatic fraction. This compound was traced for a thousand river miles from an industrial waste source in St. Louis to the water supply intake at New

Orleans, where its concentration was estimated at 2 µg/l. This was the first of the potentially health-threatening organic pollutants to be discovered at New Orleans, anticipating the endrin pollution discovered in 1963 and the many recently discovered toxic organics demanding concentrated attention at the present time.

Infrared spectrometry was applied, without requiring carbon filter sampling, to an unusual, unpublished case of organic pollution of ground water. The pollutant was an organic phosphorus compound produced by deactivation of nerve gas at a military facility. The aquifer was affected by infiltration from a waste lagoon. In this case, chemical analysis had an unexpected consequence—earthquakes resulted when authorities ordered the waste thenceforth to be disposed of by deep-well injection.

Ultraviolet spectrometry is so powerful a tool that it has to be valuable in many ways in environmental research. An early application was to determine phenols in aqueous wastes and polluted waters.[25] This tool has been especially productive, combined with adsorption chromatography, in the identification and determination of carcinogenic aromatic hydrocarbons in both atmospheric and aquatic environments.[26]

Wherever the classical system of organic analysis calls for the preparation of a solid derivative and identification by its melting point, X-ray diffraction provides much surer identification. There have been only a few applications to water pollution research. One of these was the identification of pyridine bases in drinking water derived from an aquifer contaminated by an industrial waste lagoon. The mixture of bases was precipitated as their corresponding chloroplatinates, which were readily identified and estimated from their X-ray diffraction patterns.[27] X-Ray diffraction can be applied to all solid pollutants or pollutant derivatives, *e.g.*, fatty acids as their silver salts, by use of available semimicro specimen preparation procedures.

The word "chromatography" refers to an exceedingly broad array of separation methods, a fact that was equally true before gas chromatography came into general use. A few applications of other types of chromatography are sketched briefly here. In unpublished work in the early 1950s, paper chromatography was used to identify phenols as their azo dyes in industrial wastes. It was often possible to determine if cases of phenolic pollution of streams were caused by steel mills (coke ovens) or petroleum refineries because the phenols in their wastes produce characteristic paper chromatograms.

Column partition chromatography has been successfully applied to separate, identify, and measure organic acids in sewage.[28] Several applications of column adsorption chromatography have been described. This method and its analog, thin-layer adsorption chromatography, are favored means for clean-up and separation of chlorinated pesticides for identification and analysis, *e.g.*, by infrared spectrometry.[29] High-performance liquid chromatography, though a relatively recent development, is now broadly applied to environmental problems. It may be the only technique to approach the problem of analyzing the estimated 90% of organic materials in water that are nonvolatile. The first such application, in 1972, separated 77 compounds from a primary sewage, of which 13 were identified.[30]

Sensory Methods

Systematic analysis of organic pollutants began 25 years ago as a campaign against taste and odor problems. These sensory properties are the most readily evident and effectively measurable kinds of biological effects. As a first step, the procedure for measuring sensory contribution through the threshold test was made consistent with modern panel test procedures, in the form which now appears in *Standard Methods*. As the analysis of pollutants advanced, it became evident that odors are often caused by the presence of more than one odorous organic substance. The question arose: how do such odorant mixtures affect the threshold test? Psychologists had conducted primitive investigations of this question as far back as 1917. New studies reported in 1962 showed definitively that mixed odors are usually additive at the threshold level and some mixtures are even synergistic.[31] These observations supported the experimental evidence that the same phenomena operate in waters containing more than one odorous pollutant.

Measurement of odor and taste and distinguishing between them was the basis of a fundamental study on the intense odor generated by chlorination of phenols. The nature of the odorous products has long been a mystery. The odor depends on the chlorine:phenol ratio. At the optimum ratio the odor is contributed by three distinct chlorophenols, each of which was identified by one or more of these methods: paper chromatography, infrared spectrometry, or derivative formation.[32]

Contributions of Gas Chromatography

Gas chromatography is a phenomenally powerful separation system resembling fractional distillation, but it is capable of separating 50 or 100 components in a 1-mg or 2-mg sample. In comparison, earlier efforts at refining the group separation system by adding microdistillation steps were hopelessly ineffective. As gas chromatography was being developed, there were many unpublished efforts to apply it to analysis of air pollutants and water pollutants, probably in that order. Sometime between 1960 and 1963 the analysis of chlorinated pesticides in water by gas chromatography was demonstrated, using the primitive columns and detectors of that time. In 1963 a report of pollution of the water supply at Hawaii Volcanoes National Park revealed that chlordane was discovered in the water and identified by gas chromatography.

The rapid introduction of new specific detectors in effect added the simultaneous capability for functional group detection to the separating power of gas chromatography. For example, electron capture and coulometric detectors were almost specific, with appropriate precautions, for chlorinated pesticides and polychlorinated biphenyls. Flame photometric detectors were used with similar specificity for phosphorus pesticides in one mode, for alkyl mercury compounds in a second mode, and for identification of petroleum pollutant sources through characteristic patterns of sulfur compounds in a third mode.[33]

The analogy between gas chromatography and fractional distillation has been drawn. Using slightly larger samples with preparative columns, the separated pollutant compounds are isolated and recovered, making them available for all the analytical procedures already enumerated: infrared spectrometry and sensory testing. By this combination, 11 organic compounds, originating in petrochemical wastes, were identified in 1961, in the Kanawha River, a tributary of the Ohio River. Their individual and collective contributions to the river odor problem were determined,[34] resulting in the first state regulation of a group of specific organic waste compounds. Interestingly, ten of the same compounds were found two years later in Ohio River water at Cincinnati, and the next year four of the same compounds were identified at Cairo, Illinois, hundreds of miles downstream.

Preparative gas chromatography was used to solve a problem of many decades, the identity of the pollutant causing earthy-musty odors in water supplies in many areas of the world. The responsible microbiological metabolite, *geosmin*, was trapped in sufficient quantity to permit its structural determination by mass spectrometry, infrared and nuclear magnetic resonance spectroscopy and synthesis.[35]

The application of mass spectrometry to identify geosmin, a gas chromatography isolate, anticipated today's most powerful analytical system, coupled gas chromatography—mass spectrometry—computerized data interpretation. The importance of this combination is apparent from its dominance in the following chapters of this book.

FUTURE PROSPECTS

This chapter is intended as a retrospective view of the development of systematic analyses of organic pollutants in the environment. It is appropriate to use this review as a base from which to anticipate the directions of future progress. It is obvious that gas chromatography and its coupling with mass spectrometry are today's principal thrust. But it is important not to fall into the "black box" trap—to concentrate effort where a convenient instrument with computerized output is easily available. We must consider that work to date has dealt almost exclusively with the 10% or less of organic pollutants that are volatile. New methods and instruments will be needed to analyze the nonvolatile remainder for which gas chromatography is inappropriate. Liquid chromatography and pyrolysis-gas chromatography (including "carbon-skeleton" analysis) come to mind first as possible tools.

Both analysts and program managers should reconsider the principles underlying all the progress that has already been made in environmental analysis. One of these was the wise application of the large body of chemical science. The environmental field needs the resurgence of contributions of soundly-trained chemists, not merely instrument specialists, although chemistry will lead to contributions from new or little-used instruments, such as carbon-13 magnetic resonance.

The other principle is wide-open programming. Past progress was made in unexpected directions through allowance for the exercise of creativity. Most of the toxic pollutants now sought to be controlled by regulation were discovered while pursuing more general scientific objectives. Many of the most productive methods resulted from the search for taste or odor causes. Yet today most of the emphasis and virtually all the support is given to research on known "health-related" pollutants. "Esthetic" pollution effects are severely downgraded. Research programs should provide for a wider range of basic scientific interest. In the future, as in the past, environmental programs should allow for the exercise of creativity and the serendipitous fallout from programs of broad scope.

REFERENCES

1. Braus, H., F. M. Middleton, and G. Walton. *Anal. Chem.* **23**, 1160 (1950).
2. American Public Health Association. *Standard Methods for the Examination of Water and Sewage*, 9th ed. (New York: APHA, 1946).
3. Chambers, L. A., E. C. Tabor, and M. J. Foter. *A. M. A. Arch. Industrial Health* **16**, 17 (1957).

4. Shriner, R. L. and R. C. Fuson. *The Systematic Identification of Organic Compounds*, 3rd ed. (New York: John Wiley & Sons, 1948).
5. U.S. Public Health Service. "Survey of the Big Sandy River in Connection with Taste and Odor Problems in Drinking Water at Catlettsburg, Kentucky," Robert A. Taft Sanitary Engineering Center, Cincinnati, Ohio (May 1956).
6. U.S. Public Health Service. "Carbon Extract Analysis for Jet Fuel Components in Air-Jet Engine Exhaust Mixtures," Robert A. Taft Sanitary Engineering Center, Cincinnati, Ohio (October 13, 1964).
7. Middleton, F. M., H. H. Pettit, and A. A. Rosen. *Proceedings 17th Industrial Waste Conference*, Purdue Univ., Ext. Ser. **122**, 454 (1963).
8. U.S. Environmental Protection Agency. "New Orleans Area Water Supply Study," (Draft Analytical Report), Lower Mississippi River Facility, Slidell, Louisiana (November 1974).
9. Weissberger, A. *Physical Methods of Organic Chemistry* (New York: Interscience Publishers, 1946).
10. Melpolder, F. W., C. W. Warfield, and C. E. Headington. *Anal. Chem.* **25**, 1453 (1953).
11. Middleton, F. M., A. A. Rosen, and R. H. Burttschell. "Manual for Recovery and Identification of Organic Chemicals in Water," Part II. U.S. Public Health Service, Robert A. Taft Sanitary Engineering Center, Cincinnati, Ohio (May 1957).
12. Rosen, A. A. *Anal. Chem.* **31**, 1729 (1959).
13. Goldin, A. S., R. C. Kroner, A. A. Rosen, and M. B. Ettinger. *Proceedings 1st International Conference on Peaceful Uses of Atomic Energy*, **15**, 47 (1957).
14. Rosen, A. A. and M. Rubin. *J. Water Poll. Control Fed.* **37**, 1302 (1965).
15. U.S. Public Health Service. "Drinking Water Standards—1962," Washington, D. C.
16. Heuper, W. C. and W. W. Payne. *Am. J. Clinical Pathology* **39**, 475 (1963).
17. Innes, J. R. M., B. M. Ulland, M. G. Valerio, L. Petrucelli, L. Fishbein, E. R. Hart, A. J. Pallotta, R. R. Bates, H. L. Falk, J. J. Gart, M. Klein, I. Mitchell, and J. Peters. *J. Nat. Cancer Inst.* **42**, 1101 (1969).
18. Middleton, F. M., W. Grant, and A. A. Rosen. *Ind. Eng. Chem.* **48**, 268 (1956).
19. Rosen, A. A. *Anal. Chem.* **27**, 790 (1955).
20. Rosen, A. A., L. R. Musgrave, and J. J. Lichtenberg. In *Waste Disposal in the Marine Environment*, E. A. Pearson, Ed. (New York: Pergamon Press, 1960), pp. 353-371.
21. Lao, R. C., R. S. Thomas, H. Oja, and L. Dubois. *Anal. Chem.* **45**, 908 (1973).
22. Middleton, F. M. and J. J. Lichtenberg. *Ind. Eng. Chem.* **52**(6), 99A (1960).
23. Rosen, A. A., F. M. Middleton, and N. W. Taylor. *J. Am. Water Works Assoc.* **48**, 1321 (1956).

24. U.S. Public Health Service. "Report on the Recovery of Orthonitrochlorobenzene from the Mississippi River," Robert A. Taft Sanitary Engineering Center, Cincinnati, Ohio (June 1959).
25. Schmauch, L. J. and H. M. Grubb. *Anal. Chem.* **26**, 308 (1954).
26. Harrison, R. M., R. Perry, and R. A. Wellings. *Water Research* **9**, 331 (1975).
27. Burttschell, R. H., A. A. Rosen, J. C. Wells, and E. S. Yunghans. Presented at 131st National Meeting of American Chemical Society (April 10, 1957).
28. Murtaugh, J. J. and R. L. Bunch. *J. Water Poll. Control Fed.* **37**, 410 (1965).
29. Boyle, H. W., R. H. Burttschell, and A. A. Rosen. In *Organic Pesticides in the Environment*, Advances in Chemistry Series No. 60, R. F. Gould, Ed. (Washington: American Chemical Society, 1966), Paper 17, p. 207.
30. Katz, S., W. W. Pitt, Jr., C. D. Scott, and A. A. Rosen. *Water Research* **6**, 1029 (1972).
31. Rosen, A. A., J. B. Peter, and F. M. Middleton. *J. Water Poll. Control Fed.* **34**, 7 (1962).
32. Burttschell, R. H., A. A. Rosen, F. M. Middleton, and M. B. Ettinger. *J. Am. Water Works Assoc.* **51**, 215 (1959).
33. Garza, M. E., Jr. and J. Muth. *Env. Sci. Technol.* **8**, 249 (1974).
34. Rosen, A. A., R. T. Skeel, and M. B. Ettinger. *J. Water Poll. Control Fed.* **35**, 777 (1963).
35. Gerber, N. N. and H. A. Lechevalier. *Appl. Microbiol.* **13**, 935 (1965).

CHAPTER 2

THE CHEMISTRY OF SOME POTENTIAL POLYHALOGENATED WATER POLLUTANTS

O. Hutzinger, G. Sundström

> Laboratory of Environmental Chemistry
> University of Amsterdam
> The Netherlands

F. W. Karasek

> Department of Chemistry
> University of Waterloo
> Canada

S. Safe

> Department of Chemistry
> University of Guelph
> Canada

INTRODUCTION

General

Persistent chemicals which accumulate in the biomass are among the most important types of water pollutants. Several groups of synthetic chemicals (industrial compounds; "xenobiotics") belong to this class since certain structural features render them resistant to natural degradation factors and favor their accumulation in fatty tissues.

One of the most striking features of many industrial compounds is the general lack of information and knowledge about their environmental properties. This becomes particularly obvious when they are compared to a certain group of compounds: agricultural pesticides and related products. Herbicides, insecticides, fungicides, etc., are directly (100%) released into the environment during their use, and most governments have therefore required extensive information on the toxicity, environmental behavior, and metabolism of these products. A number of industrial compounds are produced in considerably larger quantities than pesticides, and in many instances leakage into the environment during manufacture and use seems unavoidable. Industrial compounds should therefore be investigated for their environmental behavior similar to pesticides, particularly if undesirable environmental properties can be expected.

Choice of Compounds

Because of the number of possible compounds to be studied and the limited research capabilities, a careful and intelligent selection has to be made. The most important criteria for the choice of industrial compounds to be studied are:

1. Large production rate
2. Evidence or likelihood of entry into the environment
3. Suspicion (or indication) of toxicity (of compound or by-product)
4. Suspicion of persistence and bioaccumulation.

Unfortunately, most of the information is not easily accessible. Data on points 1 and 2 can be obtained from government reports, import-export figures on the one hand, and by analyzing known methods of production, shipment and use patterns on the other. In absence of reliable predictive parameters, points 3 and 4 can be evaluated best by comparison with known compounds and established relationships.

For an emphasis on persistent compounds (point 4), three different types of information could be considered:

From the environment. Concentration of certain compounds could be measured along the path of a river, for instance, and the relative increase of one type of compound over others should be an indication of persistence. An example is shown in Figure 2.1.

From structure-persistence relationship. Sound, generally applicable correlations between structure and persistence (*e.g.*, biodegradability, chemical and photochemical stability) are not known. Certain relationships, however, have been established by experience and some common examples are shown in Figure 2.2.

From comparison with known persistent compounds. Structures of industrial products could be compared to those of chemicals with known environmental impact (persistence). Figure 2.3 shows some surprising structural similarities of a group of industrial compounds (flame retardants) to chemicals with known undesirable environmental properties.

CHEMISTRY OF POLLUTANTS 17

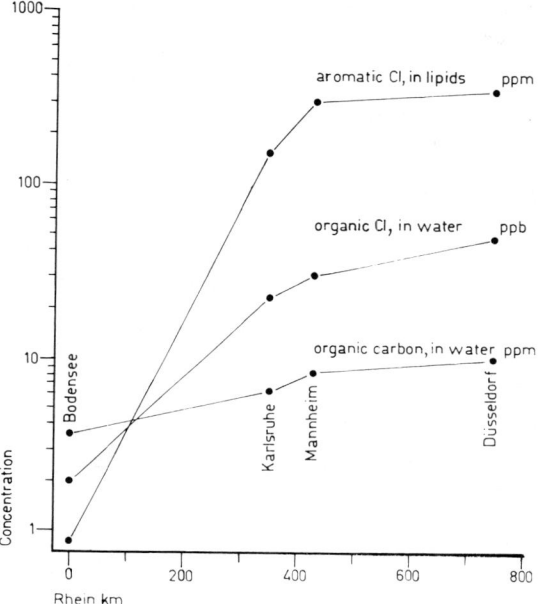

Figure 2.1 Concentration of pollution indicators along the Rhine River (Redrawn from Reference 1, with permission).

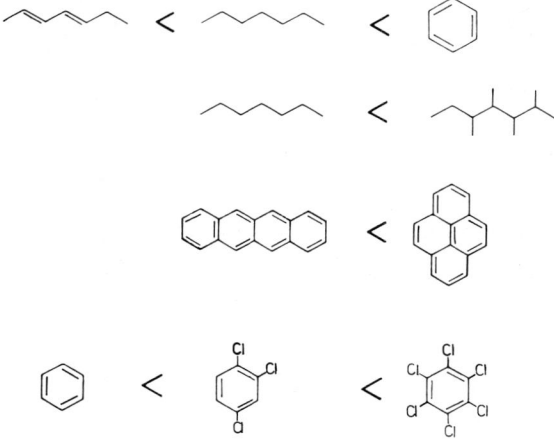

Figure 2.2 Generalized comparison of chemical and biochemical stability of some organic compounds.

A

Compound with known environmental properties.

Flame retardant.

Mirex
insecticide

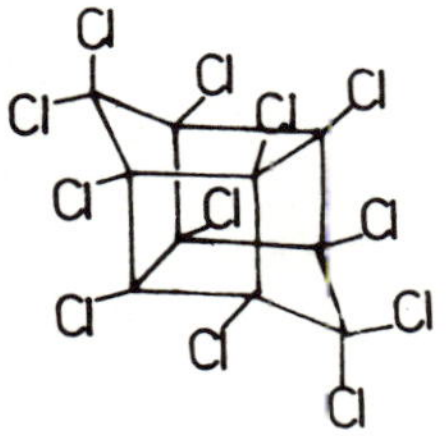

Dechlorane

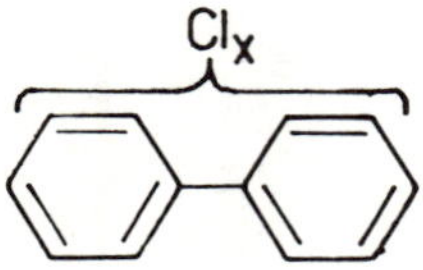

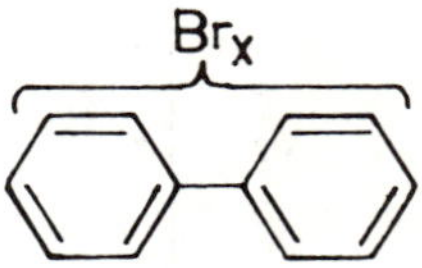

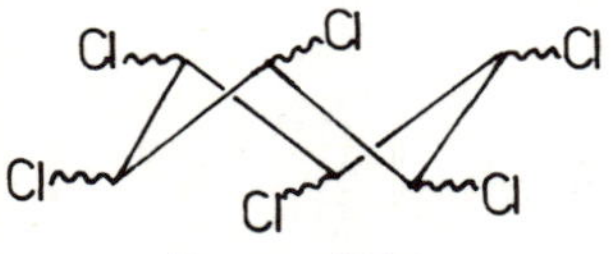

Cl_x

Br_x

insecticide

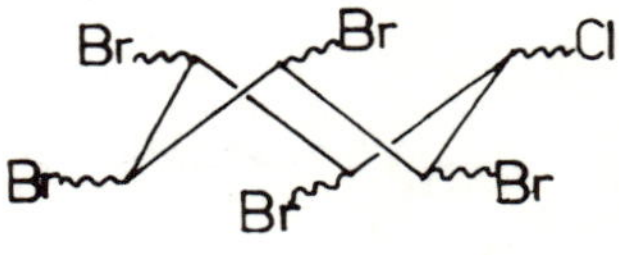

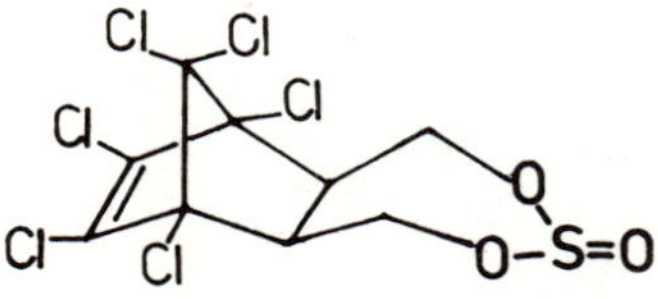

Endosulfan
insecticide

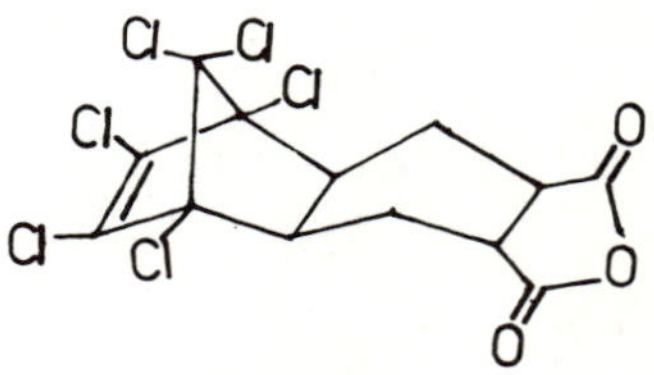

Cloran

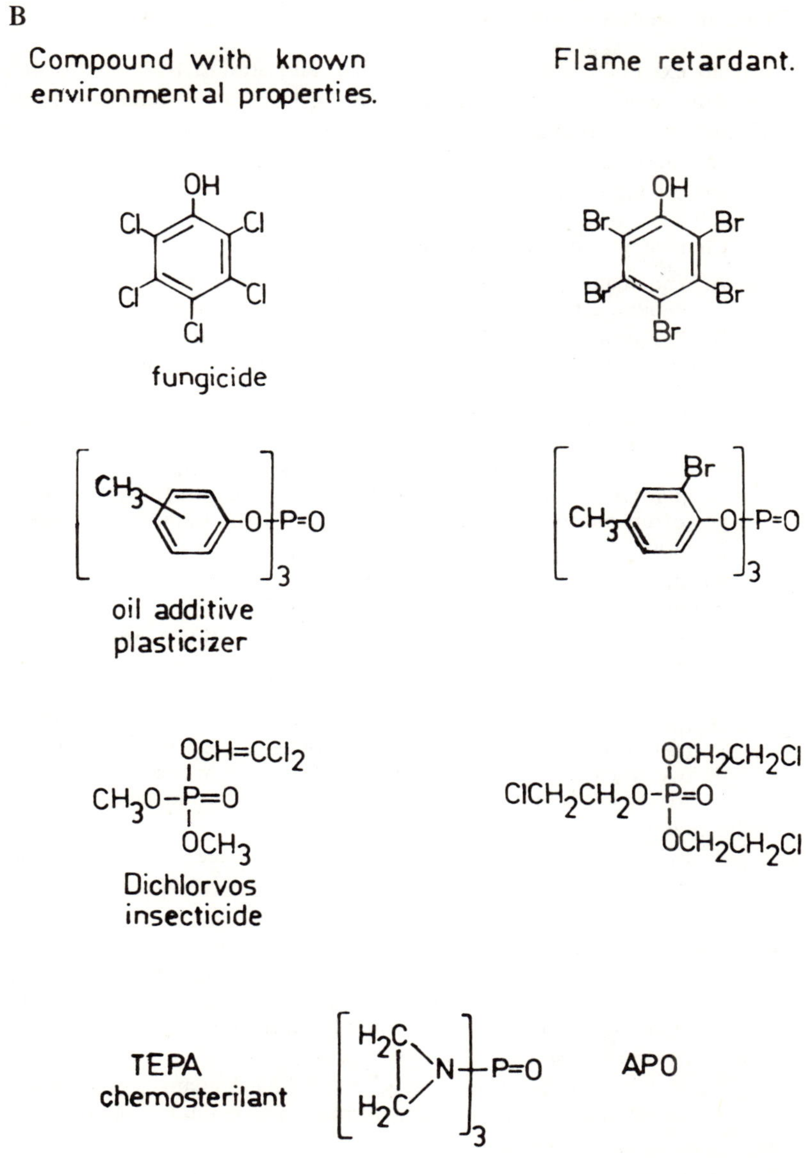

CHEMISTRY OF POLLUTANTS   19

B

| Compound with known environmental properties. | Flame retardant. |

Pentachlorophenol (OH, Cl × 5)
fungicide

Tetrabromo phenol (OH, Br × 4, Br)

$\left[CH_3\text{-}C_6H_4\text{-}O \right]_3 P=O$
oil additive
plasticizer

$\left[CH_3\text{-}C_6H_3(Br)\text{-}O \right]_3 P=O$

$$CH_3O-\underset{\underset{OCH_3}{|}}{\overset{\overset{OCH=CCl_2}{|}}{P}}=O$$
Dichlorvos
insecticide

$$ClCH_2CH_2O-\underset{\underset{OCH_2CH_2Cl}{|}}{\overset{\overset{OCH_2CH_2Cl}{|}}{P}}=O$$

TEPA
chemosterilant

$\left[\underset{H_2C}{\overset{H_2C}{\diagdown}}N \right]_3 P=O$ APO

Figure 2.3 Structural relationship between some flame retardant compounds and compounds with known environmental and toxic effects.

Types of Studies to be Undertaken

As an example of information needed, and studies to be performed in a thorough environmental chemical investigation of an industrial compound with pollution potential, an outline is presented below. The information would be necessary to assess the possible environmental impact of that hypothetical industrial compound.

I. Chemistry and analytical methods.
   A. Who makes the compound, how is it made, what is it used for; possible alternatives; are similar products being made?
   B. Chemical composition of compound, structure determination, components of mixtures; particularly important is the presence of (toxic) impurities.
   C. Analysis of products; sampling and purification (clean up) in different matrices (water, soil, animal, fat, etc.); quantitative determination (often gas chromatography with specific detectors); qualitative aspects (confirmation, identification); monitoring (often continuous); analytical methods for impurities, breakdown products and metabolites.

II. Routes of possible release, mechanisms of distribution and (after compound is being used) environmental levels.
   A. Routes into environment (sources of pollution); production, transport, use (spill, leakage); waste disposal (solid waste, sewage, incineration).
   B. Environmental distribution; transport (evaporation, co-distillation, meteorological factors, transport on particles, in air or in the ocean); adsorption (on soil particles, mud and sediments of rivers, lakes and ocean); bioaccumulation (food chains, partition factors).
   C. Environmental levels; air, water, soil, sewage; animals and their organs; processed food and materials of everyday use (*e.g.,* paper, cosmetics).

III. Environmental behavior (persistence, degradation), metabolism.
   A. Structural change in the environment (chemical and biochemical reactions of pollutants); chemical (oxidation, hydrolysis); photochemical (sunlight, λ ca. 300-350 nm); biochemical (microorganisms in sludge, soil, water, compost).
   B. Metabolism (uptake, storage, biochemical change, elimination); plant; fish and aquatic organisms; bird; mammal; man.

IV. Biological and toxicological effects.
   A. Disturbance of ecosystems, toxic effects and interference with allelo- and semiochemicals.
   B. Pharmacological and toxic effect acute and chronic toxicity, combination with other factors, possible toxic effects on man (general exposure from all sources, occupational and accidental exposure; pathological, clinical data).

CHEMISTRY OF ORGANOHALOGEN COMPOUNDS

General

To illustrate our work and our approach we have chosen four compounds of interest both from a chemical point of view as well as from a standpoint of pollution potential. In this chapter the identification, structure, elucidation and analytical aspects are considered.

One of the products (Chloralkylene 12) is a suggested PCB-replacement compound. (For structures of some PCB-replacement compounds, see Figure 2.4). Considering the environmental problems of PCB, these products should come under close scrutiny before being used.

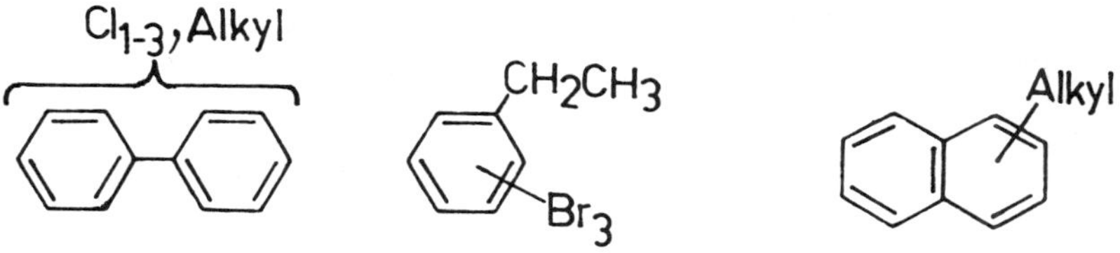

Figure 2.4 Structures of three types of compounds suggested as PCB replacements.

The other three products are flame retardants (some representative structures of flame retardants are shown in Figures 2.5 and 2.6). Two of the three compounds have general structures closely related to known problem-compounds. With the new stricter regulations on flame retardancy of synthetic fibers, the use of flame retardants is expected to increase significantly. Also, the toxicological properties of these products are receiving more attention.[2]

PCB-Replacement Compound

Alkylated Chlorobiphenyls

Mixtures of alkylated chlorobiphenyls (Chloralkylenes, Prodelec, France) have been suggested and patented as substitutes for PCB in electrical applications.[3] We have investigated the composition of one such product, Chloralkylene 12, which due to its low chlorine content (around two chlorine atoms per molecule) together with the presence of alkyl groups in the molecules has been suggested to be less persistent in the environment than PCB.[4]

In the original patent chloralkylenes are described as chlorobiphenyls (or terphenyls) with 0.2-2.5 alkyl groups per molecule containing 1-12

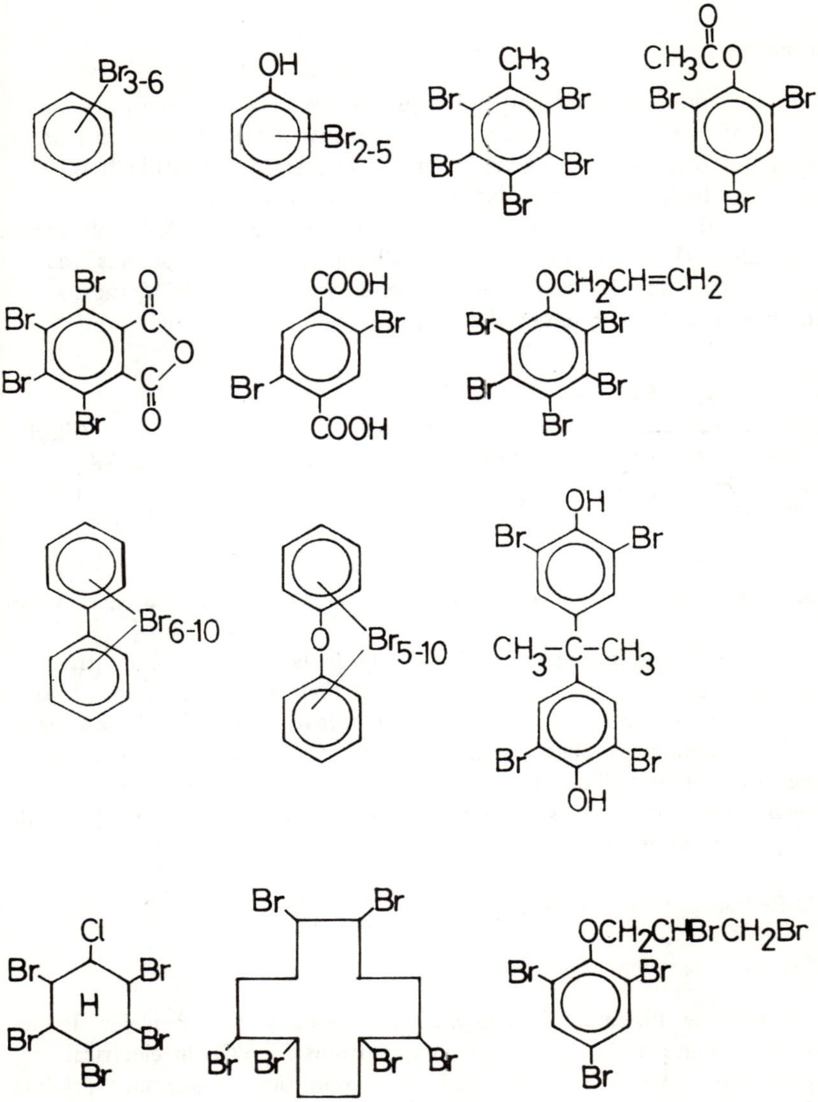

Figure 2.5 Some representatives for bromine-containing flame retardants.

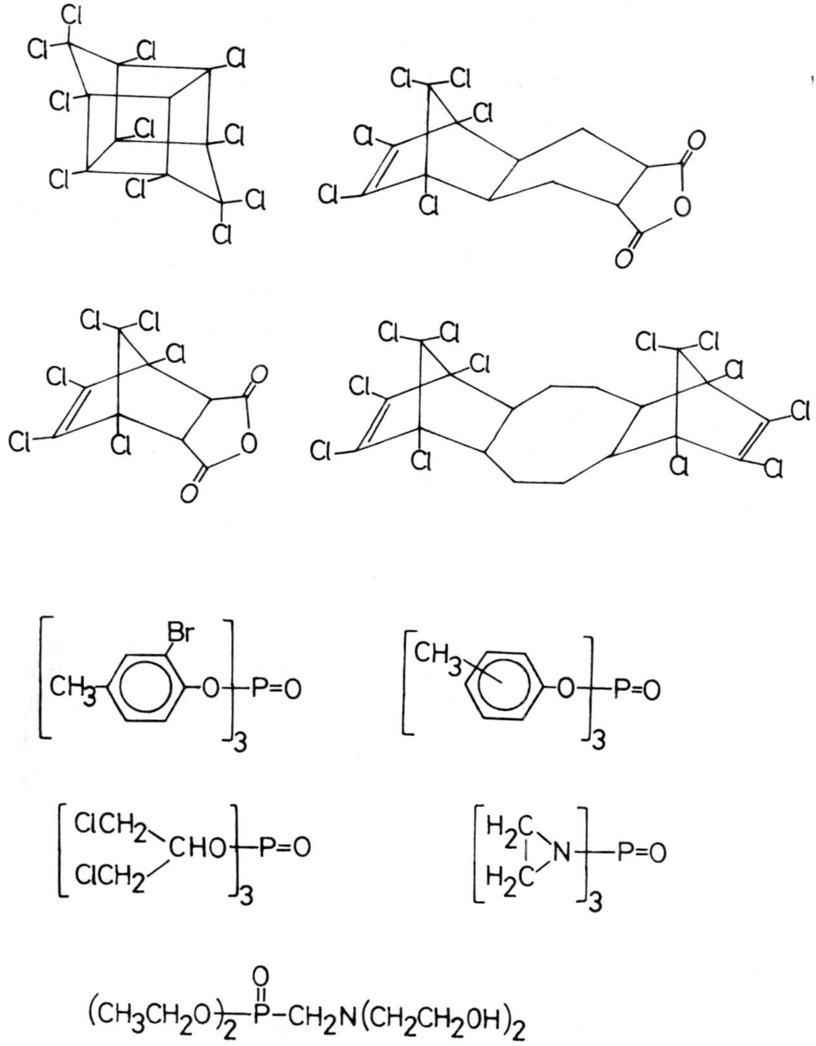

Figure 2.6 Some representatives for chlorine- and phosphorus-containing flame retardants.

carbon atoms. However, propene seems to be the reagent of choice for introduction of alkyl groups which preferentially leads to the incorporation of isopropyl groups into the aromatic nucleus. Thus the NMR spectrum of crude Chloralkylene 12 (Figure 2.7) is in its general appearance closely related to that of cumene (isopropylbenzene).

Although the NMR spectrum indicates a certain homogeneity of Chloralkylene 12, the gas chromatogram clearly shows that the reverse is true and a complex pattern comparable to that of a PCB mixture is obtained (Figure 2.8).

Preliminary investigations on the structure of the components have been performed by degradative reactions. When Chloralkylene 12 was treated with the dechlorination agent sodium bis(2-methoxy-ethoxy)-aluminum hydride (Red-Al®, Aldrich Chemical Co.), both biphenyl and a number of isopropylbiphenyl isomers were identified by comparison with synthetic material. The result is shown in Figure 2.9. As evident, ~20% of the chloralkylene mixture consists of unalkylated chlorobiphenyls, and 2,4'-dichlorobiphenyl has been identified as the major chlorobiphenyl component (ref. time 5.62 min in Figure 2.8).

In our work on the analytical properties of Chloralkylene 12 it has been shown that the compounds may be extracted from biological materials analogously to other lipophilic substances such as PCBs. Their chromatographic properties are very similar to those of PCB and DDT-substances (see Figure 2.10) and further work on separation of the different groups of components is in progress. The more destructive clean-up procedures of extracts (*e.g.*, concentrated sulfuric acid treatment used for PCBs) leaves the chloralkylenes unaffected, but the oxidation procedures sometimes used for elimination of interferences of PCB and DDT compounds in GC analyses destroy most of the chloralkylene components (Figure 2.11).

Flame Retardants

Bromobiphenyls

Polybrominated biphenyls (PBB) constitute an important group of flame retardants for synthetic fibers, polymers, etc. The structural similarity of PBB to PCB is evident (Figure 2.3), and their fate if released into the environment may be similar to that of persistent pesticides and PCB. The accidental poisoning of a large number of livestock in the U.S. shows that these compounds can create a problem.

Analogous to PCB, a proper understanding of the chemical and biological impact of PBB will require a knowledge of the composition of

CHEMISTRY OF POLLUTANTS 25

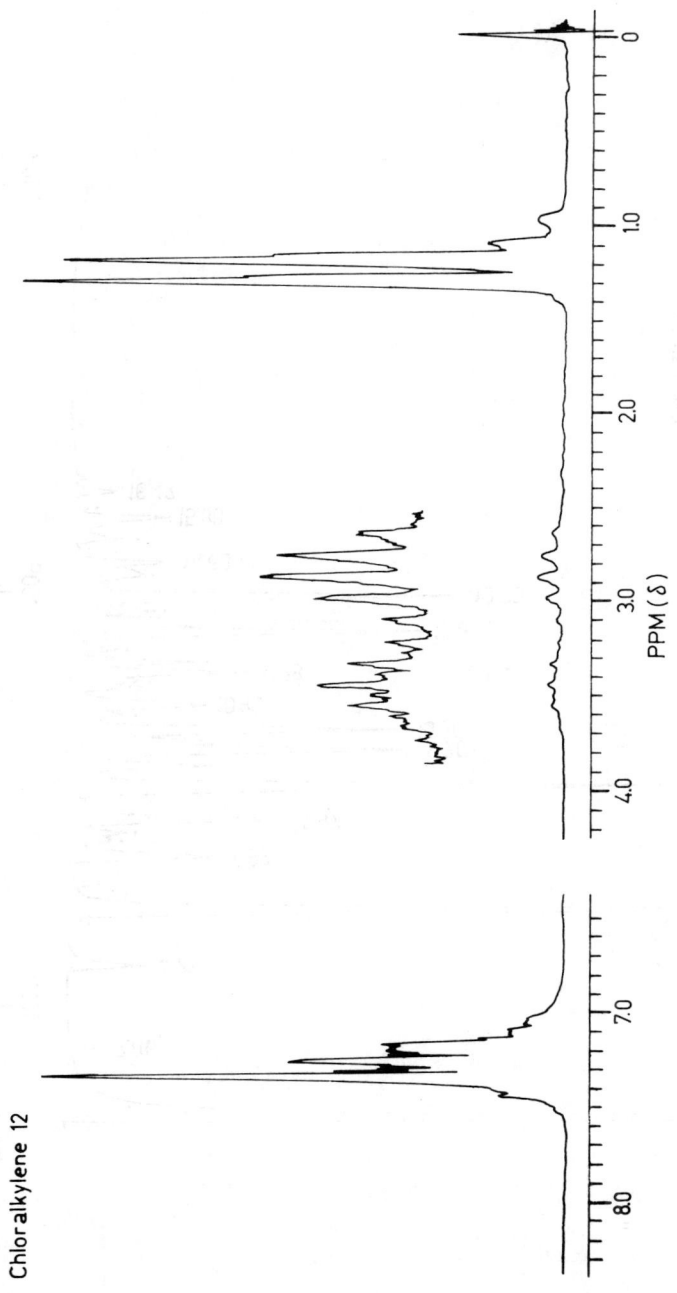

Figure 2.7 NMR spectrum (100 MHz) of crude Chloralkylene 12.

26 ORGANIC POLLUTANTS IN WATER

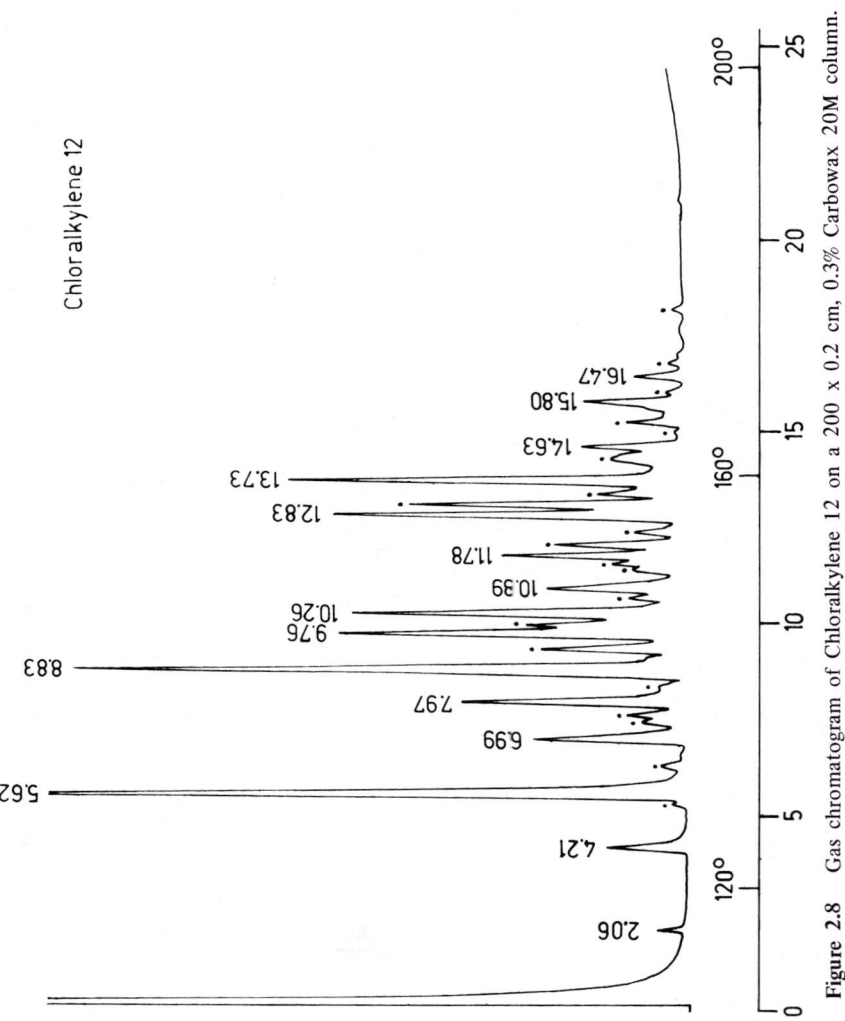

Figure 2.8 Gas chromatogram of Chloralkylene 12 on a 200 × 0.2 cm, 0.3% Carbowax 20M column.

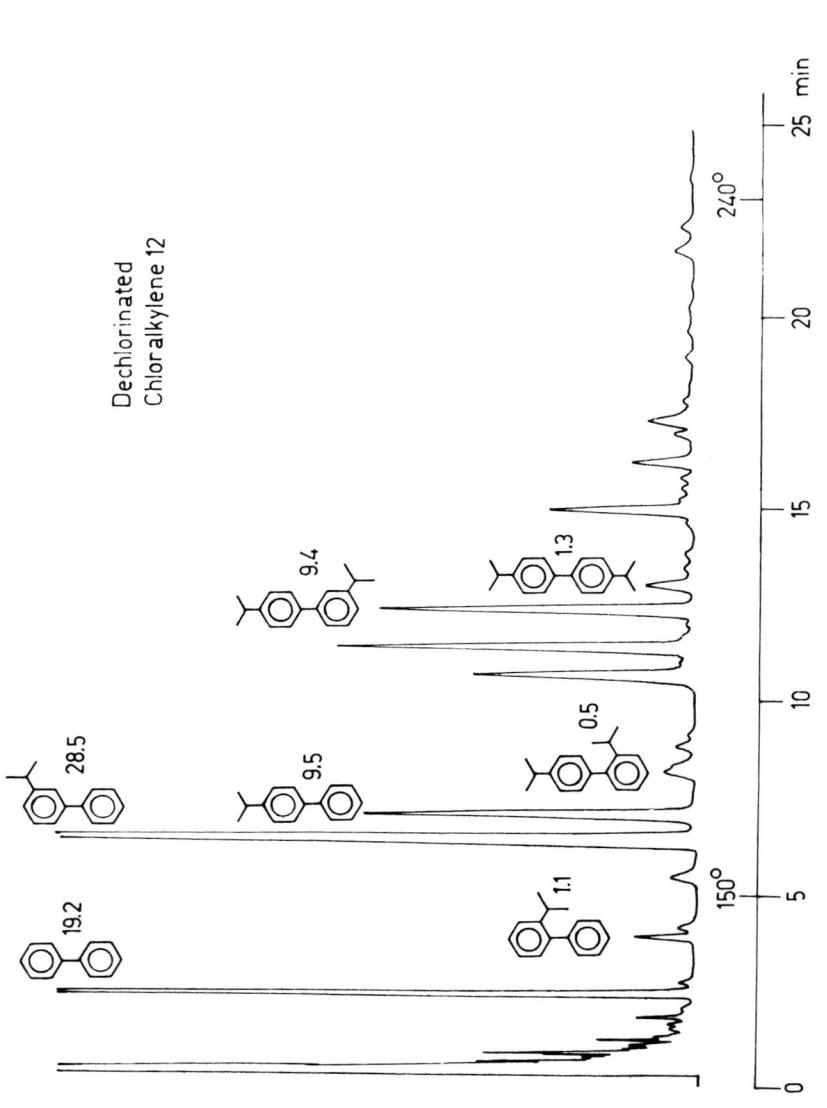

Figure 2.9 Gas chromatogram of dechlorinated Chloralkylene. Structures and amounts (%) of some components are indicated.

28 ORGANIC POLLUTANTS IN WATER

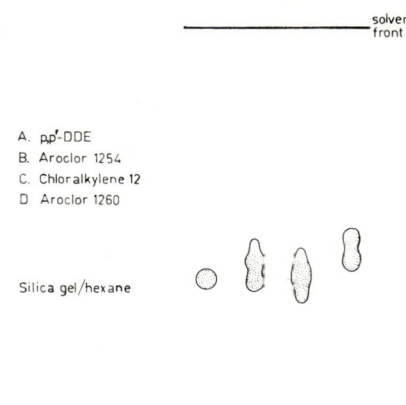

Figure 2.10 Thin-layer chromatogram of Chloralkylene and some related compounds of interest.

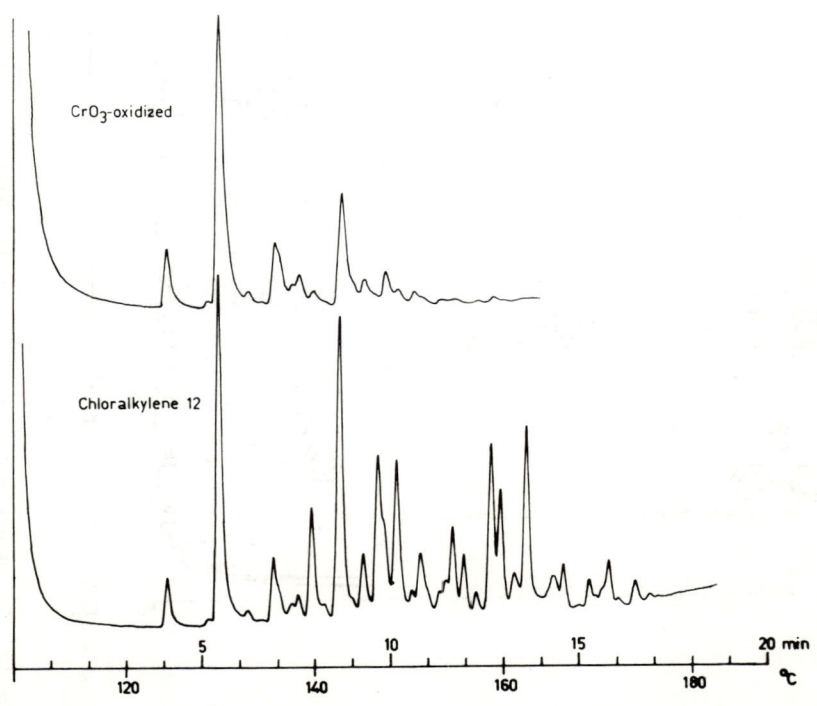

Figure 2.11 Gas chromatogram of Chloralkylene 12 and the same product after chromium oxide clean-up treatment.

CHEMISTRY OF POLLUTANTS 29

the technical mixtures and an access to pure isomeric compounds of known structure.

The bromobiphenyl preparations used as flame retardants contain from six to ten bromine atoms. Upon investigation of commercial mixtures we have found that, in very sharp contrast to the PCB mixtures, they are very simple in composition. A hexabromobiphenyl product, fireMaster® BP-6 (Michigan Chemical Co.), was found to contain only two major components (Figure 2.12), one of which constituted ~65% of the total mixture.[5] This compound (1, Figure 2.12) could be purified by recrystallization (ethanol/isopropanol) and structurally defined by NMR spectroscopy (Figure 2.13). The presence of only two singlet signals is consistent only with the 2,2',4,4',5,5'-hexabromobiphenyl structure.

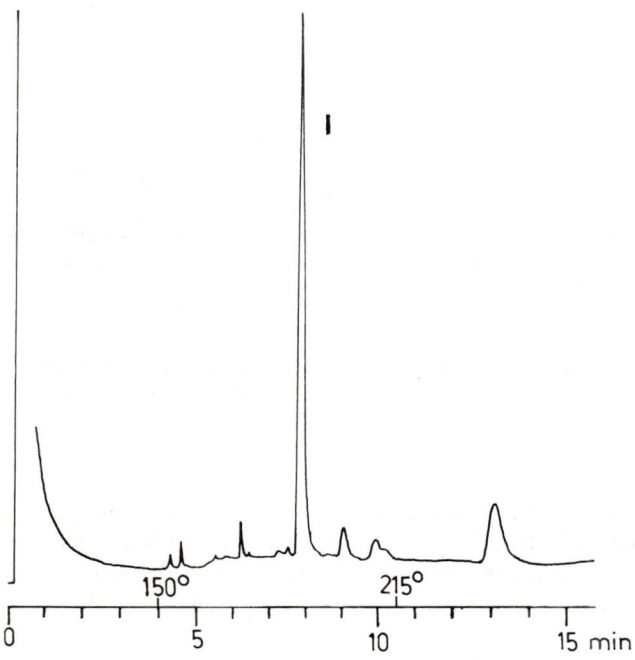

Figure 2.12 Gas chromatogram of crude fireMaster® BP-6 on a 200x0.2 cm, 0.3% Carbowax 400 column.

A more highly brominated product, FR-250 BA (Dow Chemical Co.), given as octabromobiphenyl gave a very simple NMR spectrum containing only two singlet signals (Figure 2.13). If the bromination reaction proceeds mainly *via* the above 2,2',4,4',5,5'-hexabromobiphenyl the only

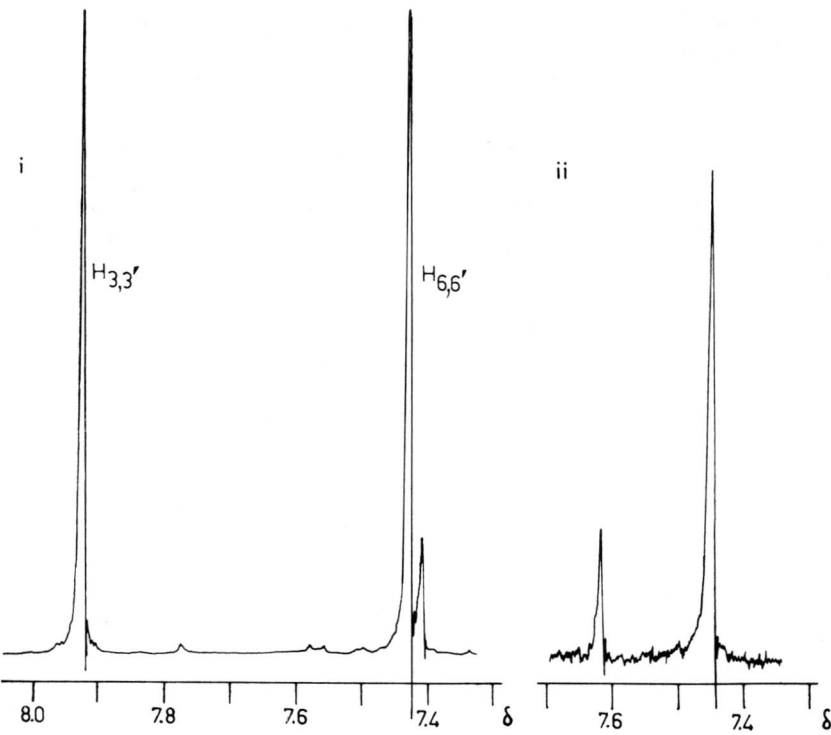

Figure 2.13 NMR spectra (220 MHz) of (1) the major product in fireMaster® BP-6, 2,2',4,4',5,5'-hexabromobiphenyl and (2) the crude octabromobiphenyl mixture FR-250 BA.

octabromobiphenyl compounds formed in any appreciable yield will be 2,2',3,3',4,4',5,5'-(II), 2,2',3,3',4,4',5,6'- and 2,2',3,3',4,4',6,6'-octabromobiphenyl. However, the NMR spectrum indicates the presence of mainly the component II giving rise to the singlet signal at δ 7.46 ($H_{6,6'}$).

Pentabromochlorocyclohexane(s)

This compound type was first synthesized by Emschwiller and Saconney in 1948 from benzene and a mixture of chlorine and bromine under photolytic conditions.[6] Many patents have been issued on the use of these compounds as flame retardant additives. The compound may, as in the case of hexachlorocyclohexane, exist in a number of different configurations, and it was suggested that the major component formed was an α-isomer (aaeeee).[6] This conclusion was based on the fact that upon chlorination of benzene the α-isomer of hexachlorocyclohexane is the most abundant isomer formed (60-70%).

The technical product (FR-651-A, Dow Chemical Co.) contains ~95% of a single isomer (see Figure 2.14) which after purification by crystallization (benzene) shows a melting point of 202-203°. This is most likely the major product described above given the m.p. 203.5° in the original literature.[6] To obtain conclusive proof for the structure of the major component we have examined the compound by spectral methods. The IR-spectrum of the compound shows absorptions in the region 1000-1500 cm^{-1} (C-H stretchings) with a pattern very similar to that of both α-hexachlorocyclohexane and α-hexabromocyclohexane. It is evident from the IR spectra of the isomeric hexachlorocyclohexanes that the C-H stretching absorptions are very useful for distinguishing between the

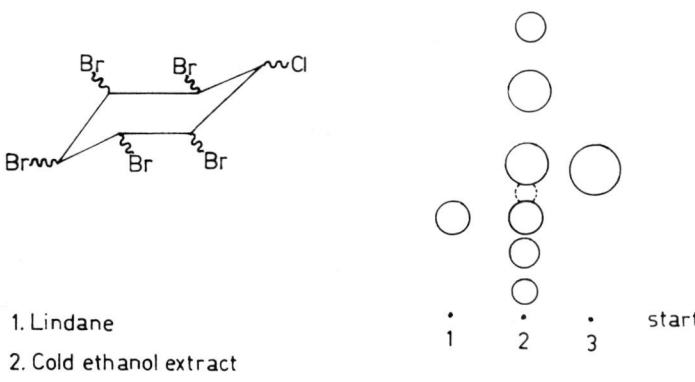

Figure 2.14 Thin-layer chromatogram of fractions from a pentabromochlorocyclohexane preparation and lindane.

different isomers. Preliminary data from X-ray analyses indicate the same structure.

The residual fraction of the pentabromochlorocyclohexane preparation (<5%) most likely consists of compounds isomeric to the major product (at least six compounds). However, the analyses of these components are hindered by their relatively low stability and they tend to isomerize or decompose upon heating in the gas chromatograph. Some products formed have tentatively been identified as tribromobenzenes. The major component in the pure state also decomposes in the gas chromatograph but can, on the other hand, be purified by sublimation around its melting point without decomposition.

Hexabromocyclododecane

The flame retardant hexabromocyclododecane has been the subject of a large number of patents. According to the manufacturer (Michigan Chemical Co.), the structure of the product is 1,2,5,6,9,10-hexabromocyclododecane. The technical product contains one major component (~80%) which can be obtained in a pure state by crystallization from chloroform (m.p. 203-7°). Neither the m.p. nor the proton-NMR spectrum of this component is in agreement with data of the two previously described hexabromocyclododecanes obtained by bromination of the *cis, trans, trans-* and *trans, trans, trans*-cyclododeca-1,5,9-trienes.[7]

The complete structure of the major component of the technical product is presently under investigation as are the chemical and biological properties. It has previously been shown that this type of hexabromocyclododecane upon treatment with base undergoes ring-closure reactions to 1,2-benzocyclo-octa-1,3,5-triene and similar products (Figure 2.15).[7]

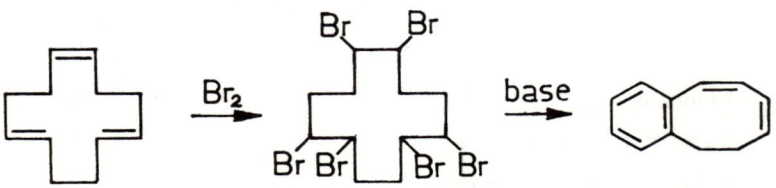

Figure 2.15   Synthesis and reaction with base of 1,2,5,6,9,10-hexabromocyclododecane.

ACKNOWLEDGMENTS

We wish to thank the following for support: NATO, Brussels, for an international research grant to O.H. and S.S.; ZWO, The Hague, Netherlands, for a visitor's grant to F.W.K.; and Environment Canada for a grant to S.S.

REFERENCES

1. Kölle, W., H. Ruf and L. Stieglitz. *Naturwissenschaften* **59**, 299 (1972).
2. Hutzinger, O., G. Sundström and S. Safe. *Chemosphere*, 3 (1976).
3. Fr. 1,603,289, 7 May 1971. *Chem. Abstr.* **76**, 105416n (1972).
4. "Substitute Impregnant for Capacitor Askarel, the Chloralkylene 12," Information Brochure, Prodelec Co., Paris (September 1972).
5. Sundström, G., O. Hutzinger and S. Safe. *Chemosphere*, 11 (1976)
6. Emschwiller, G. and J. L. Saconney. *Bull. Soc. Chim. France*, 118 (1949).
7. Eglinton, G., W. McCrae, R. A. Raphael and J. A. Zabkiewicz. *J. Chem. Soc. C*, 474 (1969).

# CHAPTER 3

PHOTODECOMPOSITION OF HALOGENATED AROMATIC COMPOUNDS

S. Safe, N. J. Bunce and B. Chittim

    Guelph-Waterloo Centre for Graduate Work in Chemistry
    University of Guelph
    Guelph, Ontario

O. Hutzinger and L. O. Ruzo

    Milieuchemie
    University of Amsterdam
    Amsterdam, The Netherlands

Halogenated aromatic compounds are among the most widespread and persistent pollutants in the global ecosystem. Chemicals which persist in the environment are typically those which have been introduced directly (*e.g.*, the pesticide DDT) or indirectly [*e.g.*, polychlorinated biphenyls (PCB), a widely used industrial chemical] into the environment and are generally resistant to environmental breakdown.[1-5] A number of commercial or industrial chemicals such as the polychlorinated terphenyls (PCT), the polybrominated biphenyls (PBB) and the polychlorinated naphthalenes (PCN) have properties and applications similar to those of PCB and are therefore potential pollutants. Both PCT[6-8] and PBB[9] have already been identified in environmental samples, and PCN as by-products in commercial PCB preparations.[10,11] Since environmental breakdown of these pollutants is important in their removal and disappearance from the environment, it is therefore of interest to determine the degradation initiated by the photochemically active part of the solar spectrum. This pathway has been shown to be an important component in the environmental breakdown and detoxication of a number of pesticides.[12,13] This chapter summarizes recent photochemical studies on

36 ORGANIC POLLUTANTS IN WATER

isomeric PCB, PBB, PCN and PCT and emphasizes the mechanism of the photochemical degradation as well as the identity of the products formed. Included in the product studies are data obtained in acetonitrile-water solutions since these results tend to approximate "natural" conditions.

The accepted route for the photochemical excitation of aromatics in the 280-320 nm region occurs by a transition of electrons in the π ground state to an excited state (π^*). From the excited state, which can be of singlet or triplet multiplicity, the carbon-halogen bond undergoes fission giving rise to an aryl and a hydrogen radical. The radicals then abstract hydrogen from the medium or dimerize and, in addition, a hydrogen halide can also be detected. Prior to bond fission an alternative reaction between the excited state and a nucleophilic species can occur giving the appropriate substitution product at the C-X bond (see Figure 3.1).

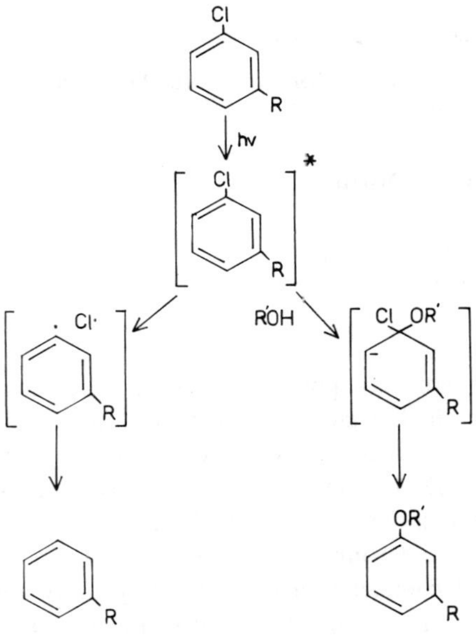

Figure 3.1 Photolysis of haloaromatics.

PHOTOLYSIS OF PCB

The photochemically-induced degradation of several PCB isomers and the commercial mixtures (Aroclors) has been reported by a number of groups.[14-22] Comparison of the reaction rates in both oxygen-saturated solutions and in degassed solutions indicated a marked reduction in rates in the presence of oxygen. This observation is consistent with the intermediacy of a triplet excited state, and this conclusion is supported by diene quenching experiments, benzophenone sensitization and intersystem crossing studies.[15] Another important observation is the preferential cleavage of the 2,2',6 or 6' *ortho* chlorine substituents. Biphenyl itself has a planar excited state geometry, and the coplanarity between the rings would increase conjugation and stabilize the excited state molecules. Clearly the *ortho* chlorine substituents sterically hinder the coplanarity between the phenyl rings. These groups are specifically cleaved in preference to substituents at the *meta* and *para* positions (Table 3.1). It is also possible that fission of the *ortho* chlorine substituents is due to stabilization of the transition state by complexation of the leaving halogen atom with the π system of the other ring. Thus the *ortho* position being closest to the ring would facilitate cleavage. Substitution by the solvent could occur *via* nucleophilic attack on the triplet or by attack on a radical cation. Photolysis in methanol thus yields 2-substituted methoxy PCB, and experiments carried out in aqueous conditions can yield chlorobiphenylols. A summary of PCB photoproducts is indicated in Figure 3.2 and Table 3.1, and the data show that dehalogenation and substitution products are also accompanied by chlorinated biphenylenes and the highly toxic chlorinated dibenzofurans. Thus photochemical degradation can both *detoxify* environmental PCB due to the formation of lower chlorinated PCB isomers and *toxify* with the formation of the more toxic chlorinated dibenzofuran and chlorobiphenylol photoproducts.

PHOTOLYSIS OF PBB

The photochemical degradation of isomeric PBB gives results which support the PCB studies. Photolysis of PBB primarily gives debrominated products with preferential elimination of the *ortho* bromine substituent also being observed[23,24] In addition, examination of the more polar nonvolatile residues by mass spectrometry indicates the formation of dimeric quaterphenyls. Parallel experiments with brominated benzenes gave similar results (Figure 3.3). The photolysis products of several isomeric bromobiphenyls in hexane are summarized in Table 3.2. The results obtained in methanol were comparable, and the corresponding methoxylated products observed in PCB experiments were not identified in the PBB photolysis. Moreover, when 2,4-dibromo- and 2,3',4',5-tetrabromobiphenyl were photolyzed in acetonitrile-water only the debrominated products were formed.

Examination of reaction rates in air-saturated solutions, in degassed solutions and benzophenone-sensitized reactions clearly confirm a triplet excited state intermediate. In addition the effect of the base, triethylamine,

Table 3.1 Photoproducts of PCB Isomers[14-21]

PCB Isomer	Photoproducts	Solvent
4,4'-dichlorobiphenyl	4-chlorobiphenyl	methanol
2,4,6-trichlorobiphenyl	2,4-dichlorobiphenyl 4-chlorobiphenyl	cyclohexane
2,4,5-trichlorobiphenyl	3,4-dichlorobiphenyl 4-chlorobiphenyl	cyclohexane
3,4,2'-trichlorobiphenyl	3,4-dichlorobiphenyl	cyclohexane
2,2',4,4'-tetrachlorobiphenyl	2,4,4'-trichlorobiphenyl 4,4'-dichlorobiphenyl 2,4,4'-trichloro-2'-methoxybiphenyl	methanol
2,2',5,5'-tetrachlorobiphenyl	2,5,3'-trichlorobiphenyl 3,3'-dichlorobiphenyl 3-chlorobiphenyl trichloromethoxybiphenyl dichlorodimethoxybiphenyl chlorohydroxybiphenyls	methanol dioxane-water
2,2',3,3'-tetrachlorobiphenyl	2,3,3'-trichlorobiphenyl 3,3'-trichlorobiphenyl trichloromethoxybiphenyl	methanol
2,2',6,6'-tetrachlorobiphenyl	2,2',6-trichlorobiphenyl 2,2'-dichlorobiphenyl trichloromethoxybiphenyl	methanol
3,3',4,4'-tetrachlorobiphenyl	3,4,4'-trichlorobiphenyl 4,4'-dichlorobiphenyl trichloromethoxybiphenyl	methanol
3,3',5,5'-tetrachlorobiphenyl	3,3',5-trichlorobiphenyl	methanol
2,3,4,5-tetrachlorobiphenyl	3,4,5-trichlorobiphenyl 3,4-dichlorobiphenyl	cyclohexane
2,3,5,6-tetrachlorobiphenyl	2,3,5-trichlorobiphenyl 3,5-dichlorobiphenyl	cyclohexane
2,2',4,4',6,6'-hexachlorobiphenyl	dechlorinated product (mono pentachlorobiphenyls quaterphenyls, biphenylene dibenzofurans, chloromethoxybiphenyls	hexane and methanol
2,2',4,4',5,5'-hexachlorobiphenyl	dechlorinated products chlorinated quaterphenyls	hexane and methanol
2,2',3,3',4,4',5,5'-octachlorobiphenyl	dechlorinated products	
Aroclor 1254	dechlorinated, hydroxylated and hydrated photoproducts	dioxane-water

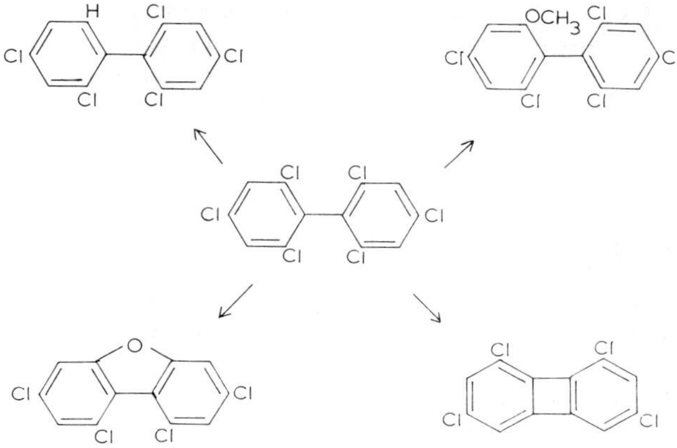

CHEMISTRY OF POLLUTANTS 39

Figure 3.2 Photodegradation of PCB.

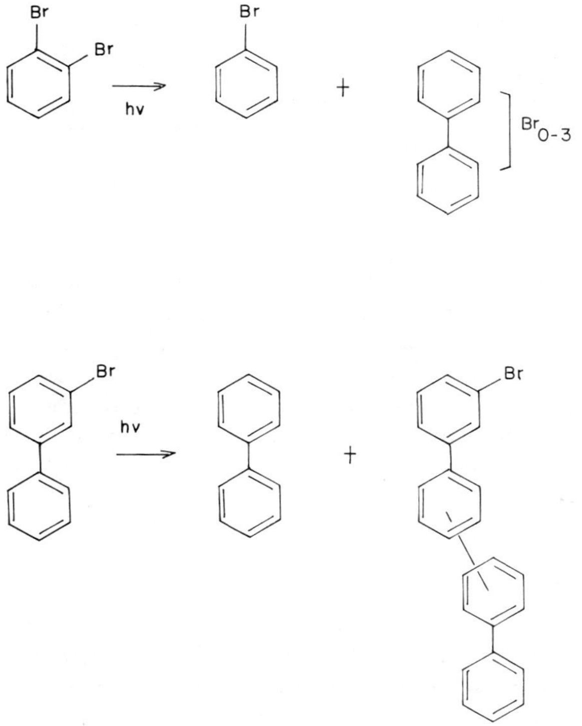

Figure 3.3 Photodecomposition of brominated aromatics in acetonitrile-water.

Table 3.2 Photoproducts of Polybrominated Biphenyls[a]

Substrate	Product[b]
2-Br	biphenyl
3-Br	biphenyl
4-Br	biphenyl
2,2'-Br$_2$	2-Br
4,4'-Br$_2$	4-Br
2,5-Br$_2$	3-Br
2,6-Br$_2$	2-Br
2,4-Br$_2$	4-Br
2,4,6-Br$_3$	2,4-Br$_2$, 4-Br
2,5,2'-Br$_3$	(Br$_2$), 3-Br
2,5,3'-Br$_3$	(Br$_2$), 3-Br
2,5,4'-Br$_3$	(Br$_2$)
2,2',5,5'-Br$_4$	(Br$_3$, Br$_2$), 3-Br
3,3',5,5'-Br$_4$	(Br$_3$)
2,5,3',4'-Br$_4$	(Br$_3$)
2,4,6,2',5'-Br$_5$	(Br-Br$_1$)
2,2',4,4',5,5'-Br$_6$	(Br$_5$-Br$_2$)
2,2',3,3',5,5',6,6'-Br$_8$	(Br$_7$-Br$_2$)

[a]n-Hexane and methanol solutions at 300 nm (0.5-2 hr).
[b]Parentheses denote products not identified beyond bromine content.

on the reaction rate and the proportion of photoreduction products was also examined. Trimethylamine-assisted photolysis of 2-, 3-, 4-bromobiphenyl and 2,2'-dibromobiphenyl are all accelerated and give a higher proportion of the photoreduction products. Photolyses in aryl halides in the presence of bases have generally been considered to proceed by way of electron transfer leading to an aryl radical which subsequently abstracts hydrogen, *e.g.*,

$$^3ArBr + Et_3N \rightarrow Et_3N^{+\cdot} + Ar^{\cdot} + Br^{-}$$

$$Ar^{\cdot} + RH \rightarrow ArH + R^{\cdot}$$

Since the addition of the amine suppresses the formation of dimeric products (*i.e.*, quaterphenyls), it is conceivable that these are formed by another route. A suggested pathway shown below might involve the reaction of the triplet

$$ArBr + {}^3ArBr \rightarrow ArBr^{\cdot +} + Ar^{\cdot} + Br^{-}$$

with a second molecule of the arylbromide thus forming an aryl radical, a bromide anion and an aryl-radical cation. Reaction of the latter species with aryl residues in a Friedl-Crafts type of reaction would then yield the observed dimeric photoproducts.

PHOTOLYSIS OF PCT

Since PCT have been identified as environmental contaminants,[6-8] a series of isomeric PCT have been synthesized and the photochemical degradation of these isomers has been investigated. Preliminary results[25] suggest a comparable photochemical pathway for the PCT and PCB. Rate studies using oxygen-saturated solutions, degassed solutions and sensitizers clearly suggest a triplet excited state which decomposes to give dehalogenated products or reacts with nucleophiles (*e.g.*, methanol) to give the appropriate substitution products (Figure 3.4, Table 3.3).

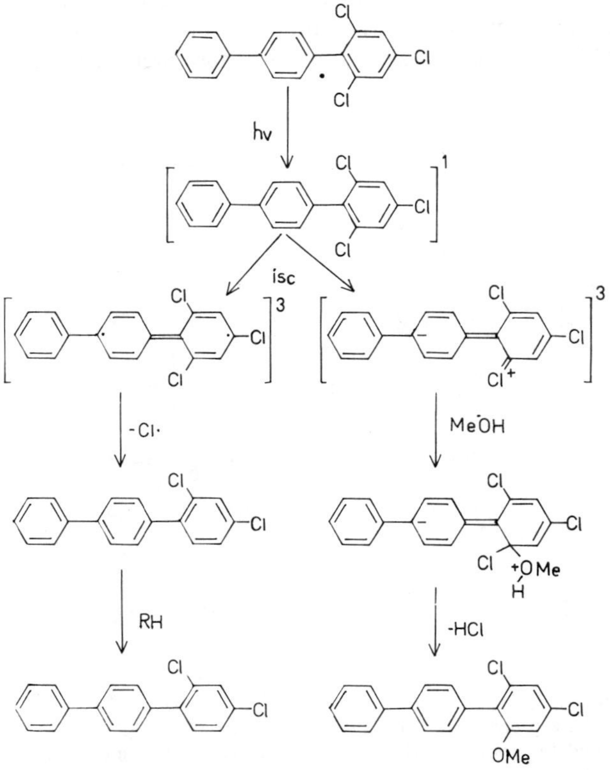

Figure 3.4 Photodecomposition of 2,4,6-trichloroterphenyl.

42   ORGANIC POLLUTANTS IN WATER

Table 3.3   Photodecomposition of Terphenyls[a]

Isomer	
2,5-dichloroterphenyl	3-chloroterphenyl
	5-chloro-2-methoxyterphenyl
2,5,4'-trichloroterphenyl	3,4'-dichloroterphenyl
	5,4'-dichloro-2-methoxyterphenyl
2,4,6-trichloroterphenyl	2,4-dichloroterphenyl
	4-chloroterphenyl
	2,4-dichloro-6-methoxyterphenyl
	4-chloro-2,6-dimethoxyterphenyl
	dichlorophenylbiphenylene

[a]Methanol solvent; the isomers were all *p*-terphenyls.

Photolysis of 2,4,6-trichloroterphenyl in hexane gave 2,4-dichloroterphenyl and 4-chloroterphenyl as the major products and trace quantities of terphenyl (Figure 3.5). These results also show preferential expulsion

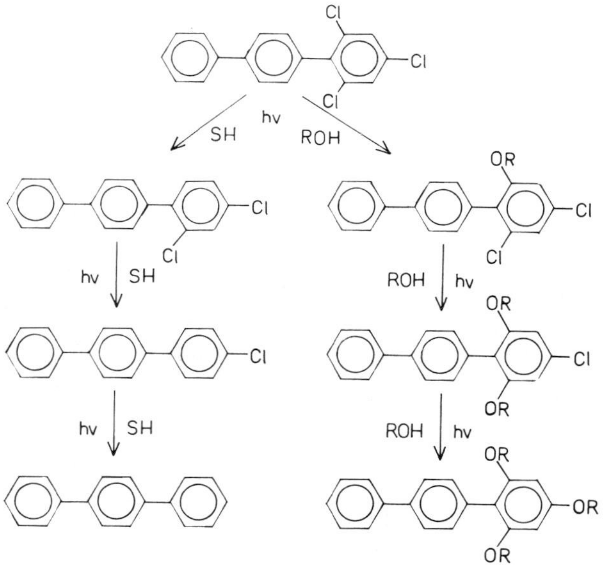

Photoproducts:
  - isolated by preparative thin layer chromatography
  - identified by gas liquid chromatography, mass spectrometry, and synthesis

Figure 3.5   Photolysis reaction scheme of 2,4,6-trichloroterphenyl.

of the sterically-hindered *ortho* chlorine substituents as observed in PCB photolysis. Photolysis of 2,4,6-trichloroterphenyl in methanol gives a complex mixture of products as shown in the gas chromatogram of the photolysis mixture (Figure 3.6). The dechlorinated products were formed

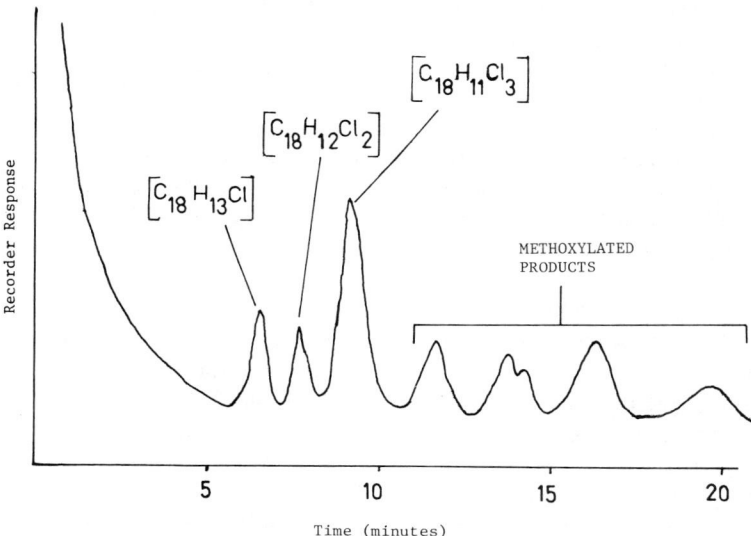

Figure 3.6 Gas-liquid chromatography analysis of 2,4,6-trichloroterphenyl photoproducts.

along with a series of methoxychloroterphenyls, and isolation of the major products by thin-layer chromatography gave two pure components which were identified spectroscopically as 2,4-dichloro-6-methoxyterphenyl and 4-chloro-2,6-dimethoxyterphenyl. As with the photolysis of PCB in methanol, the nucleophilic substitution occurs preferentially at the *ortho*-substituted position. Substitution of methanol with acetonitrile-water as solvent yielded a number of hydroxylated PCT photoproducts; this would suggest the possible formation of these hydroxylated species in sunlight-induced photochemical degradation of PCT in the environment.

PHOTOLYSIS OF PCN

Chloronaphthalenes have been found to undergo photochemical fragmentations. Chloride ion formation has been observed in the photolysis of 1-chloronaphthalene in ethanol[26] (irradiation of the same compound in benzene afforded small amounts of 1-phenylnaphthalene).[27] Kulis et al.[28] observed photoreduction of 1-chloronaphthalene to naphthalene in the presence of KOH-methanol and found the quantum yield to be dependent on the concentration of base, suggesting an electron transfer mechanism. Similar photoreductions in the presence of bases have also been observed for 1-bromonaphthalene. Mamedov et al. photolyzed[29] the isomeric monochloronaphthalenes in hexane; they observed dechlorination but also claimed that there was photoisomerization of the α to the β isomer.

We have photolyzed the isomeric chloronaphthalenes. In cyclohexane solution[15] naphthalene and binaphthyl were the major products. Photolysis in methanol gave additional products of which methoxynaphthalene and dichloronaphthalene are the most notable, together with methoxy- and chlorobinaphthyls. However, we were unable to find any evidence for isomerization of 1-chloronaphthalene to its isomer as previously claimed.[29] The quantum yield for reaction was reduced by a factor of three in the presence of atmospheric amounts of oxygen, while in the presence of benzophenone approximately the same quantum yield was observed suggesting that a triplet state is the principal precursor of both methoxylated and free radical products, as has been found in the chlorobiphenyl series.[15] The predominance of the triplet-derived products is reasonable in view of the high phosphorescence/fluorescence ratio of 5.2 which has been observed in 1-chloronaphthalene,[30] compared with 0.09 for the corresponding ratio in naphthalene itself. Such a pronounced heavy atom effect is not unusual for a π-π* transition in an aromatic system of high symmetry.

Quantum yields for the reaction of both 1- and 2-chloronaphthalene are comparable, suggesting a similar lability for the chlorine atom in both environments. This is also true for 1,2-dichloronaphthalene, photolysis of which gave approximately equal yields of both monochloronaphthalene and only small amounts of binaphthyls upon irradiation to low conversion in methanol.

The photodegradation of several isomeric PCN in methanol has been reported and the major reaction products were the dechlorination products and the dimeric binaphthyls which were accompanied in some cases by methoxylated compounds. PCN isomers included in this study were the 1- and 2-chloro, 1,8-, 1,5-, 1,2-, 2,7-, 2,3- and 1,4-dichloro,1,4,6-, 1,2,6-, 1,3,5-, 2,3,6-, 1,3,6-, 1,3,7-, and 1,2,3-trichloro and 1,3,5,8-, 1,4,6,7-, 1,3,5,7- and 1,2,3,4-tetrachloronaphthalenes.[31] The most reactive isomer was the 1,8- dichloronaphthalene and this was attributed to possible 1,8-peri-interactions which destabilize the reactive intermediates and the fact that stabilization of the triplet diradical over both chlorine atoms is not possible for this isomer.

The major organic products suggest that free radical intermediates are involved; however, simple homolytic fission seems unlikely because this reaction would be so endothermic. Triplet energies of chloronaphthalenes are ~58 kcal/mol whereas typical Ar-Cl bond dissociation energies are ~85 ~kcal/mol. By analogy with previous observations on aryl halides, some kind of electron transfer process seems likely and this postulate is supported by the observation that the quantum yield for disappearance of 1-chloronaphthalene increases eightfold in the presence of triethylamine, a known electron donor.

In aqueous acetonitrile, a solvent which can be considered a suitable model system for environmental photochemistry, the main photoproducts from 1-chloronaphthalene were chlorobinaphthyl, 1-naphthol, and a hydroxylated dimer (see Figure 3.7). In the presence of oxygen the dimers were largely suppressed. An additional product was detected in trace amounts showing m/e 104 by glc-mass spectrometry but could not be

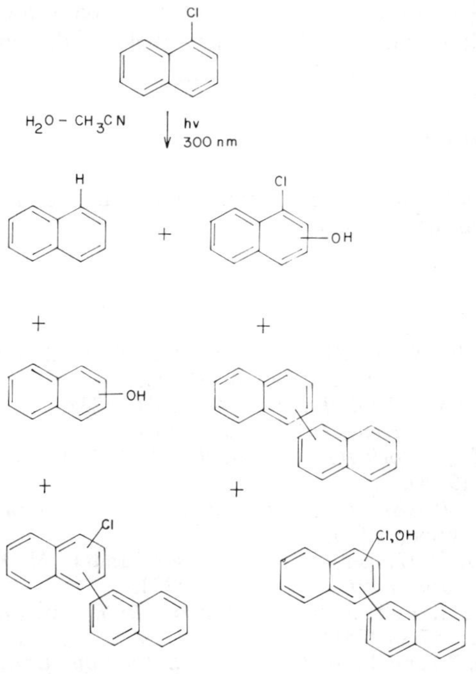

Figure 3.7 Photodecomposition of 1-chloronaphthalene in acetonitrile-water.

identified owing to its low yield and the difficulty of its separation from starting material. Polycyclic biaryls are thus formed, but in the presence of oxygen 1-naphthol is the predominant product. Recently we have photolyzed 1-chloronaphthalene with 4-chlorobiphenyl in solution and found that there is considerable sensitization. The rate of dechlorination of 1-chloronaphthalene increases as well as the amount of dimer formed. This observation is significant in that PCB and halowaxes are often found together in the ecosystem.

Since the halogenated aromatics are thermally nondegradable and chemically stable, their presence in the environment, either through misuse or accidental leakage, constitutes a long-term problem. Photochemical reactions are shown to be a minor degradation pathway since solar radiation of the wavelengths required is present in the ecosystem, down to 280 nm. The photochemical experiments have been carried out in the laboratory not only to simulate environmental condition but also to determine the precise mechanism of these reactions. The results can therefore be used to accurately predict the structures of the photoproducts. The nature of the photoproducts is environmentally significant since their toxicity must be studied in relationship to the parent compound. The hydroxy products obtained in aqueous media resemble the photoproducts that arise by substitution, and for PCB and naphthalenes the phenolic products are more toxic than the parent compounds.[32]

ACKNOWLEDGMENTS

The financial assistance of Environment Canada, Inland Waters Directorate, is gratefully acknowledged.

REFERENCES

1. Risebrough, R. W., P. Rieche, D. B. Peakall, S. G. Herman and M. N. Kirven. *Nature* **220**, 1098 (1968).
2. Fishbein, L. *Sci. Total Environ.* **4**, 305 (1973).
3. Jensen, S. *New Scientist* **32**, 612 (1966).
4. Brook, G. T. *Chlorinated Insecticides*, Vol. 1 (Cleveland, Ohio: CRC Press, 1974).
5. Edwards, C. *Persistent Pesticides in the Environment*. (Cleveland, Ohio: CRC Press, 1970).
6. Zitko, V., O. Hutzinger, W. D. Jamieson and P. M. K. Choi. *Bull. Environ. Contam. Toxicol.* **7**, 200 (1972).
7. Doguchi, M., S. Fukano and F. Ushio. *Bull. Environ. Contam. Toxicol.* **11**, 157 (1974).
8. Freudenthal, J. and P. A. Greve. *Bull. Environ. Contam. Toxicol.* **10**, 108 (1973).

9. Robertson, L. W. and D. P. Chynoweth. *Environment* **17**, 25 (1975).
10. Stalling, D. and J. N. Huckins. *J. Assoc. Offic. Anal. Chem.* **56**, 367 (1973).
11. deVos, J. G., H. H. Koeman, H. L. van der Maas, M. C. ten Noever de Brauw and R. H. deVos. *Food Cosmet. Toxicol.* **8**, 627 (1970).
12. Crosby, D. G. and N. Hamadman. *J. Agr. Food. Chem.* **19**, 1171 (1971).
13. Nakagawa, M. and D. G. Crosby. *J. Agr. Food. Chem.* **22**, 849 and 930 (1974).
14. Safe, S. and O. Hutzinger. *Nature* (London) **232**, 641 (1971).
15. Ruzo, L. O., M. J. Zabik and R. D. Schuetz. *J. Amer. Chem. Soc.* **96**, 3809 (1974).
16. Ruzo, L. O., M. J. Zabik and R. D. Schuetz. *Bull. Environ. Contam. Toxicol.* **8**, 217 (1972).
17. Ruzo, L. O., S. Safe and M. J. Zabik. *J. Agr. Food Chem.* **23** 594 (1975).
18. Ruzo, L. O., M. J. Zabik and R. D. Schuetz. *J. Agr. Food Chem.* **22**, 199 (1974).
19. Hutzinger, O., S. Safe and V. Zitko. *Environ. Health Pers.* **1**, 15 (1972).
20. Anderson, K., A. Norstrom, C. Rappe, B. Rasmuson, and H. Swahling. 166th Meeting of the Amer. Chem. Soc., Chicago, Ill (August 1973).
21. Hustert, K. and F. Korte. *Chemosphere* **1**, 7 (1972).
22. Ruzo, L. O., S. Safe and M. J. Zabik. *J. Agr. Food Chem.* **23**, 595 (1975).
23. Ruzo, L. O. and M. J. Zabik. *Bull. Environ. Contam. Toxicol.* **13**, 181 (1975).
24. Ruzo, L. O., S. Safe and N. J. Bunce. *J. Chem. Soc.* (in press).
25. Chittim, B., L. O. Ruzo, N. J. Bunce and S. Safe. Unpublished results.
26. Szychlinski, J. *Rocz. Chem.* **34**, 941 (1960).
27. Robinson, G. E. and J. M. Vernon. *J. Chem. Soc.* **(C)**, 3363 (1971).
28. Kulis, Y. Y., I. Y. Poletaeva and M. G. Kuzimin. *J. Org. Chem. USSR* (Engl. Transl.) **9**, 1242 (1973).
29. Mamedov, K. I. and I. K. Nasibov. *Zh. Prikl. Spectrosk.* **11**, 859 (1969).
30. Ermolaev, V. L. *Sov. Phys. Vspeckhi* 333 (1963).
31. Ruzo, L. O., S. Safe, N. J. Bunce and O. Hutzinger. *Bull. Environ. Contam. Toxicol.* **14**, 341 (1975).
32. Sundstrom, G., O. Hutzinger, S. Safe and D. Jones. In Proceedings of International Symposium on Sublethal Effects of Toxic Chemicals on Aquatic Animals, Wageningen, Netherlands (1975).

CHAPTER 4

CHEMICAL CARCINOGENS IN THE ENVIRONMENT

Barry Commoner

>Director, Center for the Biology
>   of Natural Systems
>Washington University
>St. Louis, Missouri

It is now recognized that cancer represents one of the most important hazards of environmental pollution. For any given type of cancer, the fraction of the total incidence attributable to environmental factors can be estimated, as a minimum, from the difference between its incidence in the population of the United States and the lowest incidence observed in any other populations. Such estimates indicate that in a highly industrialized country, such as the United States, as much as 75-80% of the cancer incidence is of environmental origin.

Laboratory and epidemiological studies indicate that the majority of known environmental carcinogens are synthetic organic chemicals—products of the huge increase in the commercial production of man-made chemicals in the last 30 years. Each year, as thousands of new substances are added to the list of synthetic compounds, new carcinogenic agents are discovered. Efforts to determine the carcinogenicity of synthetic compounds began in the 1930s, based on the incidence of tumors in experimentally-exposed laboratory animals. Figure 4.1 reports the yearly growth in the number of carcinogenic substances discovered by these means, as reported in the literature. Following a rapid increase after 1930, when a number of already suspected compounds were tested, there has been a steady rise in the number of compounds that are discovered each year to be carcinogenic toward laboratory animals (and in some cases toward people). The problem is one that is worsening with time.

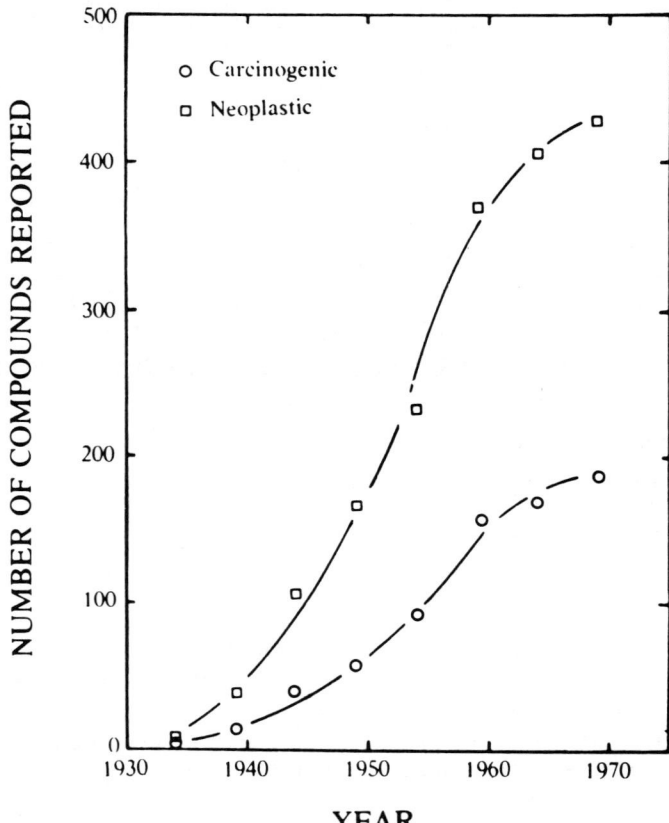

Figure 4.1 Yearly growth in the number of carcinogenic substances discovered by tests with animals.

Some of the very large amounts of synthetic organic compounds that are now manufactured enter the environment as a necessary outcome of their use (for example: pesticides and solvents used in inks and paints which evaporate into the air as these products dry). Many of them escape into the workplace environment, and may affect those who work in it. Others are disseminated into the environment as wastes, either intentionally or by accident, from industrial plants. In addition, various industrial combustion processes inadvertently produce organic compounds—some of which are carcinogenically active—that are disseminated into the environment.

These considerations suggest that the environment, in particular air and surface waters, must contain some numbers of synthetic organic compounds which are responsible for a considerable part of the incidence of cancer in the U.S. population. However, thus far only a few substances have been identified as environmental carcinogenic agents—usually by the tragic discovery of actual cases of cancer in people exposed to them. The earliest example is exposure to substances in coal tar. In 1775 Sir Percival Pott observed that cancer of the scrotum was characteristic of chimney sweeps, and it was later shown that this was due to polycyclic hydrocarbon carcinogens such as benzo(a)pyrene, present in coal tar. Coke oven workers are still, today, exposed to these same carcinogens and experience a considerably elevated cancer incidence.

Vinyl chloride is the latest example of an environmental carcinogen that remained undetected until it caused cancer in man. More general evidence of the impact of environmental agents on cancer incidence is given in a recent study by the National Cancer Institute. The study shows that cancer mortality is most pronounced in urban and industrialized areas, and particularly where the regional density of the petrochemical industry is high.[1]

There is reason, therefore, to regard the commercial production and use of synthetic organic chemicals as a major source of environmental carcinogens. The total production of synthetic organic chemicals in the U.S. has increased exponentially in the last 30 years at an average annual rate of 11% (1962-1971). Large-scale production of synthetic organic compounds is a postwar phenomenon; annual production increased tenfold between 1946 and 1966. This fact, plus the lag-time that is expected between exposure to a carcinogen and the appearance of cancer (generally of the order of 5-25 years), suggest that the actual effects of such substances on the incidence of cancer in the U.S. population may just now begin to appear. This situation emphasizes the importance of a timely effort to analyze the problem of environmental carcinogenesis and to take suitable steps toward its control.

Environmental cancer, therefore, confronts us with both a serious problem and an important opportunity. The problem, which has recently been given increasing recognition, is that of detecting, identifying and tracing the origins of the number of synthetic organic compounds that account for the environmentally-induced incidence of cancer. If we are successful in limiting human exposure to carcinogens, it should be possible to prevent an appreciable portion of the present incidence of cancer and perhaps forestall a threatened increase, as the full effect of the rising dissemination of such agents develops.

To achieve this goal there is an initial, basic prerequisite: the capability for the analysis of numerous environmental samples, each likely to contain a number of compounds, some of which are carcinogenic. The methods must be sufficiently rapid and inexpensive to enable the numerous analyses required to trace the movement (and often the chemical transformations) of substances in the environment, as they vary from place to place, with time, and with seasonal and weather conditions.

At present, available analytical methods are not suitable to the attainment of these goals. One possible strategy is to use conventional fractionation and analytical methods (*e.g.*, GC/MS) to identify all the synthetic organic compounds that occur in typical environmental samples. Then it would be necessary to determine--from existing data or by new tests where necessary—which of these compounds are carcinogens; and then, by means of suitable chemical analyses, to trace the movement of these substances in given environmental situations. Such a strategy confronts several serious difficulties:

1. The present analytical methods are far too slow and expensive to permit the detailed environmental analyses needed to trace the origins and movement of carcinogenic agents.
2. The literature is far from complete or is ambiguous in distinguishing between those synthetic organic compounds that are regarded as carcinogenic and those that are not. Consequently, the foregoing strategy would require that many compounds be studied as to their carcinogenic activity.
3. Determination of carcinogenicity by present, conventional methods is slow, cumbersome and expensive (a test of a single substance requires 2-3 years, the use of a large number of animals, and about $150,000 in costs). In any case present tests do not determine, unambiguously, that a substance is carcinogenic toward people.

Present determinations of carcinogenicity involve long-term observations of two or more species of laboratory animals. They may respond to exposure to a given carcinogen in different ways depending on the route of administration and other factors such as diet and sex of the animal. Such studies often show that a given substance is highly carcinogenic toward some species and wholly inactive toward others. Therefore, the demonstration that a given substance is an active carcinogen toward one or more laboratory species does not always signify that it is also carcinogenic toward people, but only that it *may* be. Relative to people, then, such a substance should be regarded as a *presumptive* carcinogen. (Accordingly, in this chapter the term "carcinogen" will be reserved for those substances known to cause cancer in people, and the term "presumptive carcinogen" will be used to designate a substance shown to cause cancer in one or more species of laboratory animals, but not as yet in human beings.)

An alternative strategy might be the following:

1. Carcinogenic activity would be detected by means of a simple, rapid biological screening method, initially in a mixed environmental sample.
2. By applying this detection method to a fractionated sample (*e.g.*, zones in a chromatogram), carcinogenically-active fractions would be separated and localized.
3. The latter, subjected to suitable physicochemical analysis would be identified as to molecular composition.

This strategy would, by means of rapid screening, eliminate the need for laborious analyses of samples that do not in fact contain active carcinogens; it would eliminate the need for long-term animal tests; it would be sensitive to the presence of unknown carcinogens in environmental samples and would therefore respond to the appearance of new carcinogens in the environment.

Fortunately, as a result of recent studies, it is now possible to devise such a strategy. Samples can be rapidly screened for *bacterial mutagenesis—* an activity which correlates closely with the capability of synthetic organic compounds to induce cancers in laboratory animals. Thus, the method is an effective screening technique for presumptive carcinogens. At the same time, further development of the technique also offers the possibility of determining which presumptive carcinogens are in fact carcinogenic toward people. As will be shown below, the method can be applied effectively to the analysis of pure compounds as well as heterogenous environmental samples.

This approach has its origin in recent research developments on the biochemistry of carcinogenesis. Among the considerable mass of data produced by such studies, two areas are **most** relevant to the problem under consideration here: the metabolic fate of carcinogenic agents in the body, and the relation between carcinogenesis and mutagenesis.

It is now known that most carcinogenic substances must be converted, metabolically, into an active substance before cancer is induced. (For example, 2-acetylaminofluorene, AAF, must be converted to a proximal carcinogen, N-hydroxy AAF, before carcinogenesis occurs.) Such observations have helped to resolve ambiguities regarding variations in the carcinogenicity of a given substance toward different species. In a number of instances it has been found that a substance known to be carcinogenic toward one species may fail to induce cancer in another species because the latter is unable to metabolize the original substance. Thus, when fed, AAF is a powerful hepatocarcinogen in the rat, but is totally inactive in the guinea pig; unlike the rat, the guinea pig does not metabolically convert AAF to N-hydroxy AAF. On the other hand, N-hydroxy AAF causes cancer in both animals. Metabolic activation is usually due to hydroxylase enzyme systems localized in the microsomal fraction of liver and other tissues.

On theoretical grounds there is reason to expect a close connection between carcinogenesis and mutagenesis, based on the hypothesis that both processes originate in the cell's genetic apparatus. This leads to the simple idea that any substance that is carcinogenic should also be mutagenic and *vice versa*. However, it has been found that there are many exceptions to this expectation. For example, while AAF is a powerful carcinogen, it is not mutagenic.

The foregoing results regarding the formation of active carcinogen metabolites suggest the solution to this ambiguity—that the active substance produced by metabolic conversions of the original carcinogenic substance is *both* the proximal cancer-producing agent and the mutagen. Thus, it has been found that, while AAF is not a mutagen, N-hydroxy AAF *is*. Another result of these studies is the observation that derivatives of the active metabolites may appear in the urine of carcinogen-fed animals.

In the last few years, a very sensitive system for measuring the mutagenic activity of microsome-activated carcinogens, using specially developed strains of *Salmonella typhimurium*, has been developed by Ames, et al.[2] These strains have a histidine-negative genome (and are therefore unable to grow unless histidine is supplied) and include other genetic factors that render them specially sensitive to chemical mutagenesis. One of Ames' strains (TA 1535) is designed to detect mutations due to base-pair substitutions and therefore tends to respond selectively to mutagens such as alkylating agents. Two strains (TA 1538 and TA 1537) detect frame-shift mutations; TA 1538 responds particularly well to a number of carcinogens such as 2-nitrosofluorene; TA 1537 responds to carcinogens such as 9-aminoacridine. In its genome each strain also includes mutations that greatly increase its overall sensitivity to mutagens. One of these causes loss of the DNA excision repair system, and the other the loss of the lipopolysaccharide barrier that coats the surface of the bacteria (thus enhancing the penetration of the large molecules).

Mutation causes the bacteria to revert to a histidine-positive genome (*i.e.*, the mutant bacteria can now synthesize histidine). Mutant colonies can therefore be detected by their growth on a nutrient medium that lacks histidine. In practice, nutrient plates are seeded with a suitable *Salmonella* strain; the substance to be tested is added, together with an aliquot of a microsomal preparation. Substances are usually also tested in the absence of the microsomal preparation to determine whether they are, themselves, proximal carcinogens that are active without metabolic transformation. After 48 hr of incubation the mutant colonies are counted (on duplicate plates) and compared with controls lacking the substance or sample being tested.

In the original Ames method metabolic conversion of the substance to be tested was accomplished *in situ* by incorporating a preparation of rat liver microsomes in the test culture plate. We have extended the system by testing each sample with microsome preparations from seven different rat tissues, in the expectation that some presumptive carcinogens might be effectively activated by microsomes from some tissues other than liver. It was also known from the work of Ames and others that the dose-response curve of the *Salmonella* test is not linear; the mutation rate often exhibits a maximum at a particular mutagen concentration, with higher concentrations tending to be toxic; typical dose-response curves which illustrate this effect are shown in Figure 4.2. Accordingly, it is necessary to test a presumptive carcinogen over a rather wide range of concentrations.

On the basis of these considerations we have designed an extended version of the original *Salmonella* test to distinguish optimally between presumptive carcinogens and noncarcinogens. The method has been applied systematically to 100 organic compounds, most of them synthetic. Of these, 50 were selected on the basis of observations, reported in the literature, which showed that they consistently caused a significantly elevated incidence of tumors when applied to one or more species of laboratory animals. The second group of 50 was selected on evidence in the literature, based on animal tests, that they are noncarcinogens. Each substance was tested against three different *Salmonella* strains (TA 1535, TA 1537, TA 1538) at three concentrations, with microsome preparations from seven rat tissues (and in the absence of any microsome preparation). From these data, and suitable controls, it was possible to determine how the results obtained from the two classes of substances were affected by these parameters and what combinations of conditions optimize the distinction between the two classes.

The chief results of this study are reported in Figures 4.3 and 4.4. Figure 4.3 shows the frequency distributions of the mutagenic activity ratios* of the two classes of compounds obtained with the combination of the foregoing parameters that yielded a maximum ratio. All but 1 of the 50 noncarcinogens fall into a narrow, symmetrical distribution centered around a ratio of 1—indicating that these substances do not affect the intrinsic value of mutation of the bacteria. The one exception is

---

*This is defined as $\frac{E - C}{\overline{C}}$ where E is the number of mutant colonies per plate when the compound is present, C is the corresponding value for the control, with the compound absent, obtained on the same day, and $\overline{C}$ is the average control value for all 100 tests.

56 ORGANIC POLLUTANTS IN WATER

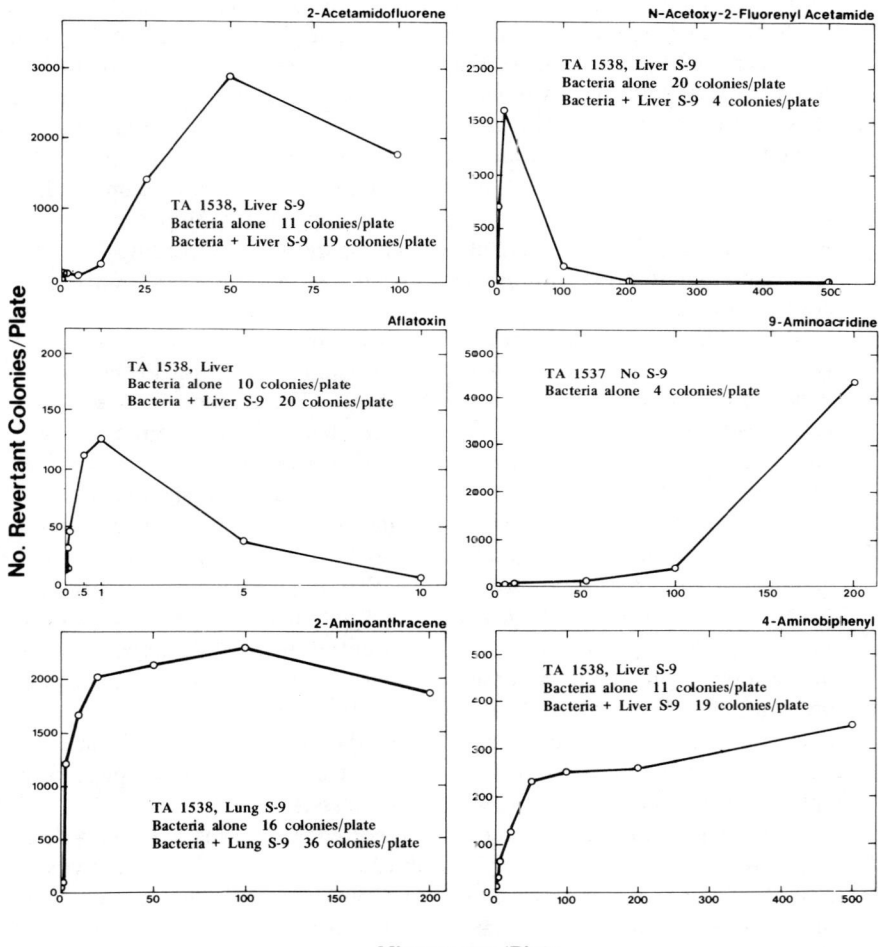

Figure 4.2 Dose-response curves of mutagenically-active compounds.

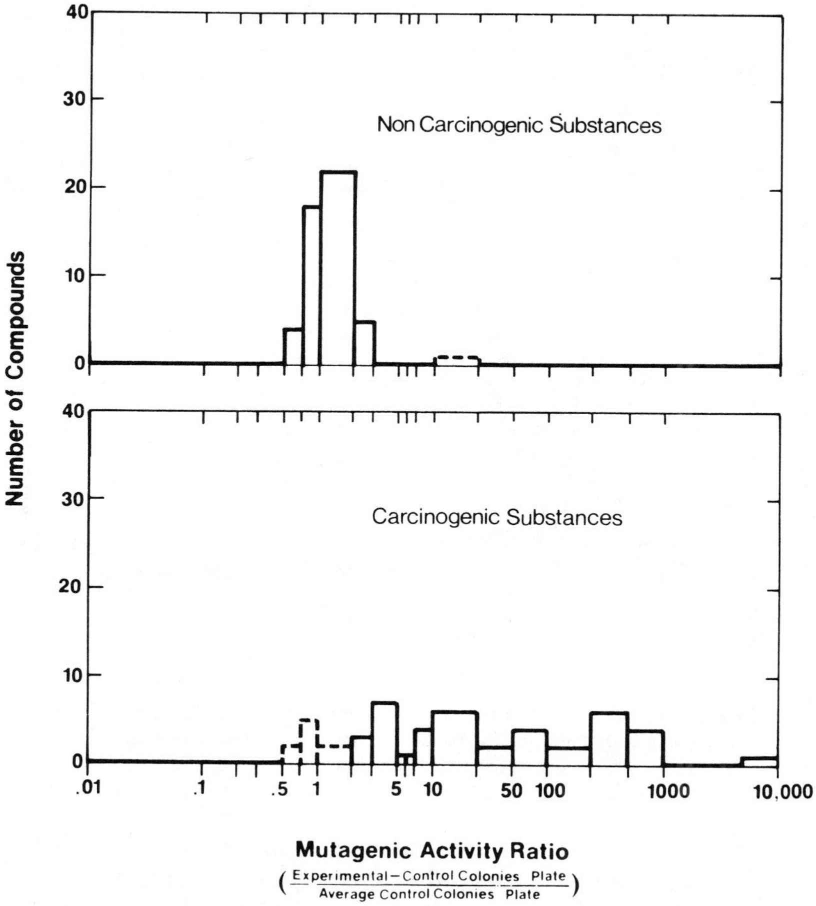

Figure 4.3 Frequency distribution of mutagenic activity ratios of presumptive carcinogens and noncarcinogens.

p-aminoazotoluene, which is clearly a "false positive," yielding an enhancement of mutation rate similar to that obtained with known presumptive carcinogens. (A closely related substance, o-aminoazotoluene, has been reported to cause tumors in test animals, and it is possible that with further tests, the p-substance will be reclassified.) Of the 50 presumptive carcinogens all but 9 exhibit significant enhancement of the *Salmonella* mutation rate.

58   ORGANIC POLLUTANTS IN WATER

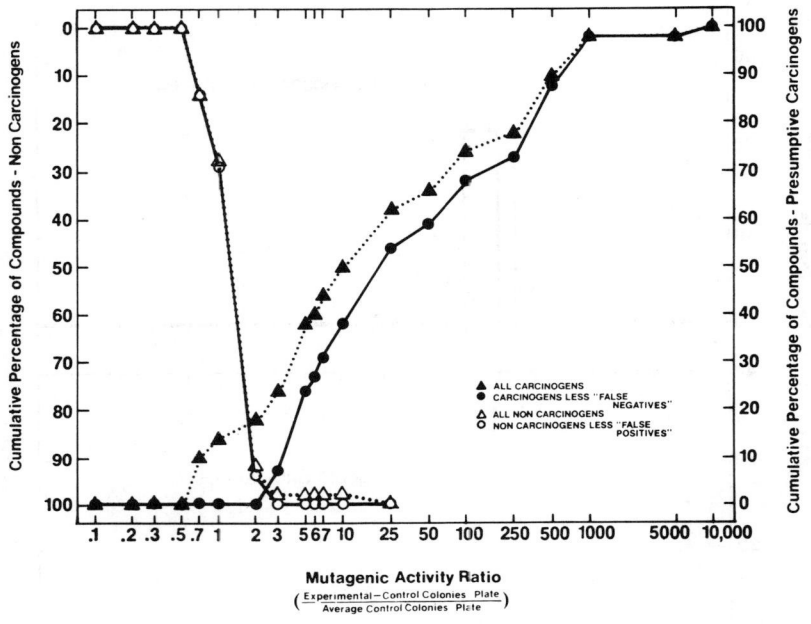

Figure 4.4 Cumulative percentage of compounds as a function of mutagenic activity ratio (for optimal strain, microsome preparation and substance concentration).

The statistical outcome of this test is summarized in Figure 4.4, which shows the variation of the cumulative percentage of compounds with mutagenic activity ratio for the two classes of compounds. Figure 4.4 shows that the probability of correctly classifying a presumptive carcinogen from the value of its mutagenic activity ratio is equal to the probability of correctly classifying a noncarcinogen when the mutagenic activity ratio is 2.0. About 83% of the presumptive carcinogens yield ratios above that value, and the same percentage of noncarcinogens yield ratios below that value. Thus, if a ratio of 2.0 is chosen as a cutoff, 83% of both noncarcinogens and presumptive carcinogens would be correctly classified by the results of mutagenesis tests. At a cutoff ratio of 2.5, 95% of the noncarcinogens and 81% of the presumptive carcinogens would be correctly classified.

The statistical reliability of the mutagenesis test system could be further improved if the occurrence of "false negatives" could be overcome.

This can be accomplished by more elaborate procedures in two of the nine cases, namely by administering the compound to a rat and testing the urine for mutagenic activity. In two additional cases it has been found that the presence of DMSO, which was used in all tests in order to solubilize water-insoluble compounds, interfered with mutagenic activity which could be observed when the DMSO concentration was reduced.

In general, in the present state of the method, one can regard a mutagenic activity ratio greater than 2.5 as indicating, with a probability of 95%, that the substance would produce tumors if administered to laboratory animals, *i.e.*, that it is a presumptive carcinogen. A mutagenic ratio of 2.0 or less indicates with a reliability of 83% that a substance is a noncarcinogen. For those few substances that fall in the range of 2.0-2.5 the results are more ambiguous.

Since most positive responses are well above this range, the method is already very useful for the analysis of environmental samples; we have begun a series of studies in which it is being applied to air and water samples. Initial results of studies of urban air particulate samples are shown in Table 4.1 and Figure 4.5. In these tests about 150 cm^2 of standard high-volume air sample filter papers were extracted in benzene/hexane, the residue taken up in DMSO and applied to *Salmonella* test plates in the usual way. Table 4.1 shows that significant enhancement of mutation rate has been noted in several samples from downtown St. Louis, but none in a sample from the residential area to the west of the city.

Table 4.1 Mutagenic Activity of St. Louis Air Particulate Samples

Sample	Date of Sample	Equivalent Volume of Air/Test Plate	Microsome Preparation	Number of Revertant Colonies/Plate			
				TA1535	TA1536	TA1537	TA1538
Control; filter paper only			−	4	2	6	24
			+	7	2	21	35
12th & Russell	9/9/72	2304 cu ft	−	7	2	15	31
			+	15	2	46	70
		5760 cu ft	+	—	—	—	112
8th & Soulard	5/17/72	1637 cu ft	−	6	1	103	126
			+	13	5	279	335
		4092 cu ft	+	—	—	—	620
Lemay Ferry & Lindbergh	6/6/72	1096 cu ft	−	6	1	14	26
			+	13	6	32	46
		3240 cu ft	+	—	—	—	57

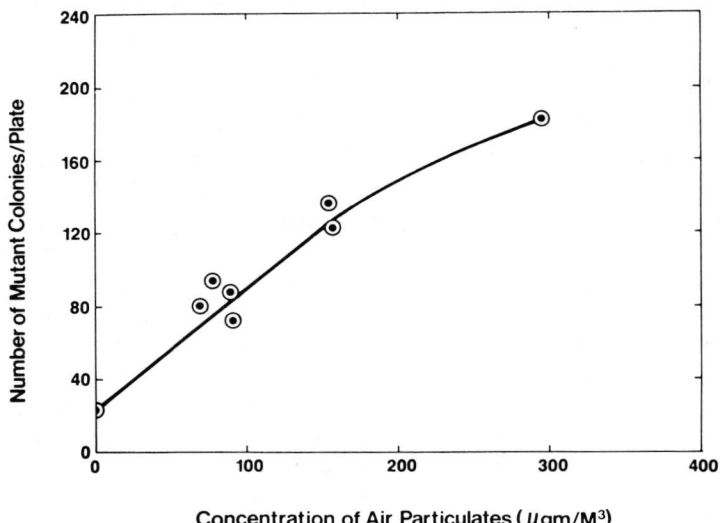

Figure 4.5 Mutagenic activity of Chicago air particulate samples (TA 1538/liver microsomes).

In collaboration with the City of Chicago Department of Environmental Control we have begun to carry out a systematic analysis of the occurrence of mutagenically-active substances in air particulate samples (4 cm^2 of filter papers) from different regions of the city. An initial set of results is shown in Figure 4.5. The amount of mutagenic activity is roughly proportional to the amount of air particulates in the sample.

That it is possible to fractionate such samples and recover the original activity in the separate fractions is shown in Table 4.2. Additional material from one of the urban air particulate samples reported in Table 4.1 was fractionated by thin-layer chromatography and a series of successive zones cut out, extracted and the extract tested in the usual way. Table 4.2 shows that fractions which differ in mutagenic activity can be obtained in this way and that the sum of their mutagenic activities is about equal to the activity of the original sample. These data, while preliminary, indicate that the mutagenic activity associated with extractable constituents of urban air particulates can be traced through fractionation procedures and eventually localized to specific compounds.

In collaboration with the Harris County, Texas Pollution Control Department we have also undertaken to apply the mutagenesis technique to

Table 4.2 TLC Fractionation of Air Particulate Sample[a]
(8th & Soulard, St. Louis)

Chromatograph Zone	Average Number of Colonies/Plate (TA 1538/Liver Microsomes)
Control	24
1	125
2	90
3	150
4	161
5	48
Sum	574
Total Extract	620

[a]Solvent system: benzene/carbon tetrachloride 85:15.

the analysis of samples from the Houston Ship Channel, which receives effluents from numerous industrial plants. We have thus far studied effluents from a pulp plant, an oil refinery, a steel mill, and two industrial waste treatment plants. Initial results are shown in Table 4.3. Of the five effluent samples tested thus far, one (black liquor from the pulp plant) appears to be unambiguously negative. Two other effluents (industrial waste treatment plant B and the oil refinery) are clearly positive and must therefore be regarded as delivering presumptive carcinogens to the receiving water. The mutagenic activity ratio of the sample from industrial waste treatment plant A (2.31) is borderline.

The results obtained with the steel mill effluent sample illustrate an interesting problem we must expect to encounter in testing for environmental samples, especially with mixed samples. It will be noted from Table 4.3 that this sample yielded a slight but not significant enhancement in mutation rate at one concentration, and that the colony count fell well below that of the control at a higher concentration. This suggests that the sample might either contain a mutagen which is toxic at higher concentrations, or a separate substance which inhibits bacterial growth and which may, therefore, reduce the apparent mutation rate. Figure 4.6 illustrates how this question may be resolved, by proceeding to the next phase in the analysis of environmental samples: fractionation and evaluation of the fractions by means of the *Salmonella* test. In this example, the steel mill effluent was separated into a series of consecutive

62 ORGANIC POLLUTANTS IN WATER

Table 4.3 Mutagenic Activity of Effluents[a] of Industrial Plants (Houston Ship Channel)

Source	Equivalent Amount of Sample per Test Plate Ml	Number of Revertant Colonies/Plate (TA 1538/Liver Microsomes)	
		Control	Experimental
Pulp Mill (Black liquor)	1	31	36
Steel Mill	25	42	41
	62.5	42	64
	125	42	17
Oil Refinery	125	26	86
	250	26	82
Industrial Waste Treatment Plant A	50	36	53
	125	36	58
	250	36	83
Industrial Waste Treatment Plant B	50	31	127
	125	31	128

[a]Aqueous samples were extracted in 8:2 benzene/isopropanol for analysis.

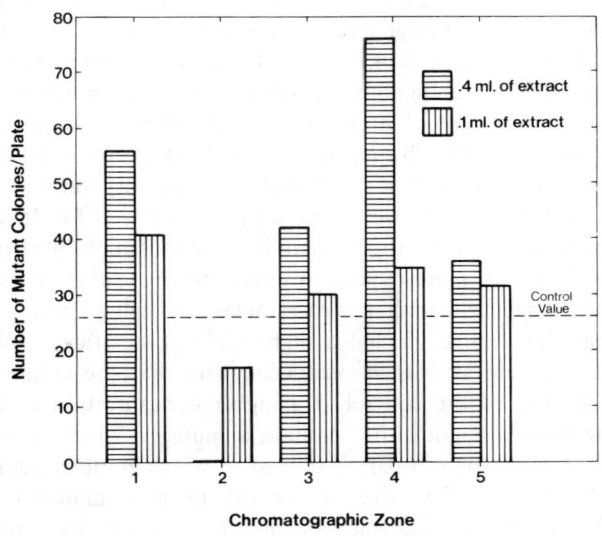

Figure 4.6 Results from fractionation of steel plant effluent.

zones by means of thin-layer chromatography (using hexane as a solvent). The zones were then extracted by means of chloroform and the samples evaporated to dryness, taken up in DMSO and tested in the usual way. Figure 4.6 shows that Zone 2 clearly contains bacteriostatic material, while Zone 4 and possibly Zones 1 and 3 contain mutagenic material.

A new potential source of environmental carcinogens which—on the basis of the government's recently announced plans—may contribute heavily to the problem in the future is the production and use of synthetic crude oil manufactured from coal by hydrogenation. Coal contains graphite-like condensed-ring structures, and the hydrogenation process is therefore likely to yield a variety of polycyclic hydrocarbons. As it happens, a number of compounds of this type are active carcinogens; the best known example is benzo(a)pyrene which has been frequently detected in the products of combustion. A coal hydrogenation plant, operating for about six years in the 1960s during which time workers were carefully guarded against undue exposure to the products, reported an incidence of skin cancer 16-37 times higher than expected from the general population.[3] Similar observations in England showed that Scottish shale oil is carcinogenic.[4] Figure 4.7 reports an initial mutagenesis test of a synthetic crude oil sample (kindly supplied by Dr. William Fulkerson of the Oak Ridge National Laboratory). It is evident that the sample contains mutagenically-active material and therefore must be regarded as containing presumptive carcinogens.

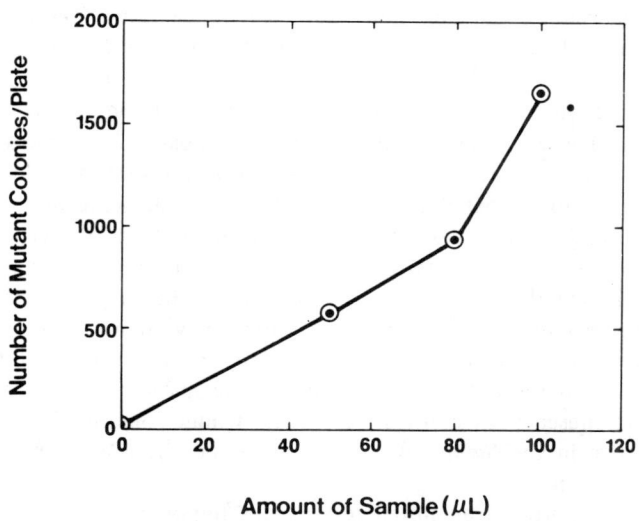

Figure 4.7 Mutagenic activity of synthetic crude oil (TA 1538/liver microsomes).

These data indicate that the *Salmonella* technique is indeed applicable to the analysis of typical environmental samples. It serves as a rapid biological indicator of the presence of substances which, on the basis of the correlation study described earlier, are with a high probability presumptive carcinogens. As shown in more detail below, together with chromatographic techniques, the *Salmonella* system can be used to isolate and ultimately identify the active compounds.

It must be re-emphasized, however, that the identification of a substance as a presumptive carcinogen does not necessarily mean that it causes cancer in *people*—that, in the terminology proposed above, it is truly a *carcinogen*. And it must be emphasized, as well, that the reported correlation is between the mutagenic activity of various organic substances in the *Salmonella* test and their carcinogenicity toward laboratory animals. In other words, a positive response in the *Salmonella* test means that presumptive carcinogens are present.

Based on some further developments it appears that the mutagenesis technique also has the potential of determining whether or not a presumptive carcinogen is also carcinogenic toward people. This derives from observations made during earlier investigations of the biochemistry of carcinogenesis, especially by the Millers[5] and the Weisburgers.[6] By means of conventional biochemical techniques, they showed that in several instances, when laboratory animals are fed carcinogen-containing diets, the proximal, active metabolites are excreted in the urine as glycosides or sulfate esters. More recently, in my own laboratory, we have shown that these active agents could be readily detected in the urine of carcinogen-fed rats by means of the *Salmonella* test[7] and similar work has been reported by Ames.[8]

As noted earlier, one observation of such investigations is that the resistance of a given species to the effect of a substance carcinogenic in some other species is a consequence of the resistant species' failure to convert the original substance to the proximal, carcinogenic metabolite. As an hypothesis this suggests an interesting and potentially very useful concept: a species which develops tumors on exposure to a presumptive carcinogen is one that is capable of metabolizing the original substance, to form the proximal, active substance—which may then appear in the urine. Thus, the original substance would not yield a significant mutagenic response in the *Salmonella* test unless the microsomal enzyme system were present; in contrast, the urine would contain substances that are active in the *Salmonella* test even in the absence of the microsomal preparation.

Table 4.4 reports on an initial test of this hypothesis, relative to the differential effects of AAF on the rat and the guinea pig. Recall that

Table 4.4 Metabolites of AAF in Rat and Guinea Pig Urine

Urine Sample	Diet	No. of Days on Diet	Equivalent Amount of Urine/Plate	Number of Colonies/Plate (TA 1538)	
				Without Liver Microsomes	With Liver Microsomes
Rat Urine I (before hydrolysis)	0.06% AAF	7 Days	4 ml	76	1,820
Rat Urine II (after β-glucuronidase hydrolysis)	0.06% AAF	7 Days	4.2 ml	696	1,624
Rat Urine III (after acid hydrolysis)	0.06% AAF	7 Days	5.25 ml	292	319
Guinea Pig Urine I	0.06% AAF	13 Days	4 ml	24	287
Guinea Pig Urine II	0.06% AAF	13 Days	4 ml	16	330
Guinea Pig Urine III	0.06% AAF	13 Days	4 ml	7	22

AAF is a powerful hepatocarcinogen in the rat but is totally inactive in the guinea pig, and that AAF is inactive in the *Salmonella* test unless a microsomal preparation is present. Table 4.4 shows that, as expected, the urine from AAF-fed rats contains substances that are highly mutagenic in the absence of a microsomal preparation but also contains material that is mutagenic only after activation by the microsomal system. In contrast, a similar analysis of urine from AAF-fed guinea pigs shows that there are no substances present which are mutagenic in the absence of the microsomal enzyme system. The guinea pig urine does, however, contain substances, both in free and conjugated forms, which are mutagenic if activated by the microsomes. This result is, therefore, in keeping with the hypothesis that a species in which a presumptive carcinogen is active is one that can convert it to a proximal, inherently active substance.

Figures 4.8, 4.9 and 4.10 represent a further development of this approach and also illustrate the power of the *Salmonella* system as a means of identifying active substances. Figure 4.8 is a photograph of the results obtained when a strip from a thin-layer chromatogram of AAF is applied to a *Salmonella* test plate seeded with the test organism (TA 1538) and a liver microsomal preparation and then incubated for 48 hr in the usual way. The location of the AAF zone on the chromatogram is clearly

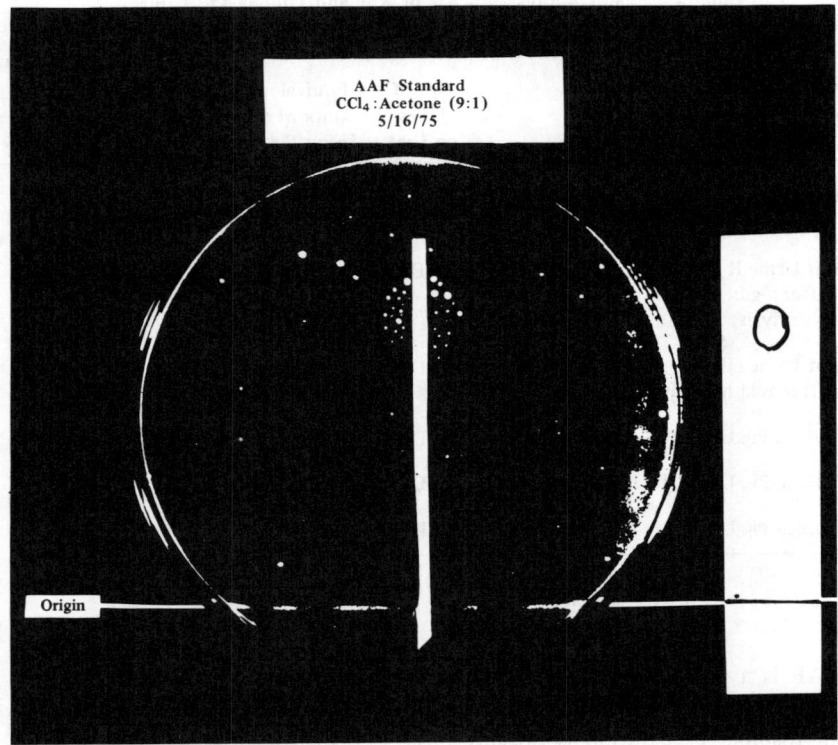

Figure 4.8 TLC paper strip of standard AAF on *Salmonella* test plate (TA 1538).

revealed by the cluster of mutant colonies. Figure 4.9, derived from this experiment, shows the number of mutant colonies as a function of distance along the chromatogram, yielding a clear indication of the peak due to AAF. Figure 4.10 illustrates the results of an experiment in which this technique was applied to the analysis of the several fractions of urine from an AAF-fed rat. It is evident that the technique can be used to localize several AAF derivatives, both those that are inherently active as mutagens (dotted lines) and those which are mutagenic only when activated by the microsome preparation (solid lines).

These results demonstrate that the rat produces a series of different metabolic products from AAF, confirming earlier observations made on the basis of conventional biochemical analyses.[5,6] More generally, they show that a sample containing a mixture of substances can readily be

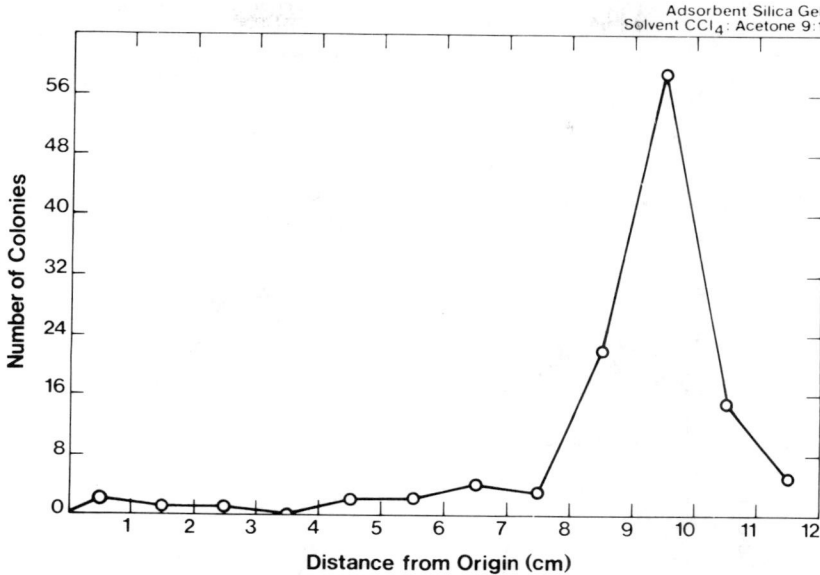

Figure 4.9 TLC analysis of standard AAF.

separated, and the separate constituents isolated by this technique. Applied to environmental samples, such as extracts of airborne particulates, this same approach should lead to the isolation and identification of the presumptive carcinogens present.

Returning to the original problem, we can now see the outlines of a strategy that should enable us to detect and identify the presumptive carcinogens present in the environment and to determine which of them are likely to be carcinogenic toward people. The general strategy would be first to determine what presumptive carcinogens are present in a population's environment by means of analytical techniques based on the *Salmonella* test. Then one would analyze urine from individuals in the population to determine whether or not the proximally active metabolites of the presumptive carcinogens detected in the environment are present. The presence of such metabolites in the urine would indicate that an individual is at risk from the substance(s). It should be noted that there is some evidence of genetic differences in human populations with respect to the activity of the liver hydroxylase system. Hence, analyses of this type might serve to identify individuals who are genetically resistant to given carcinogens because they fail to metabolize them.

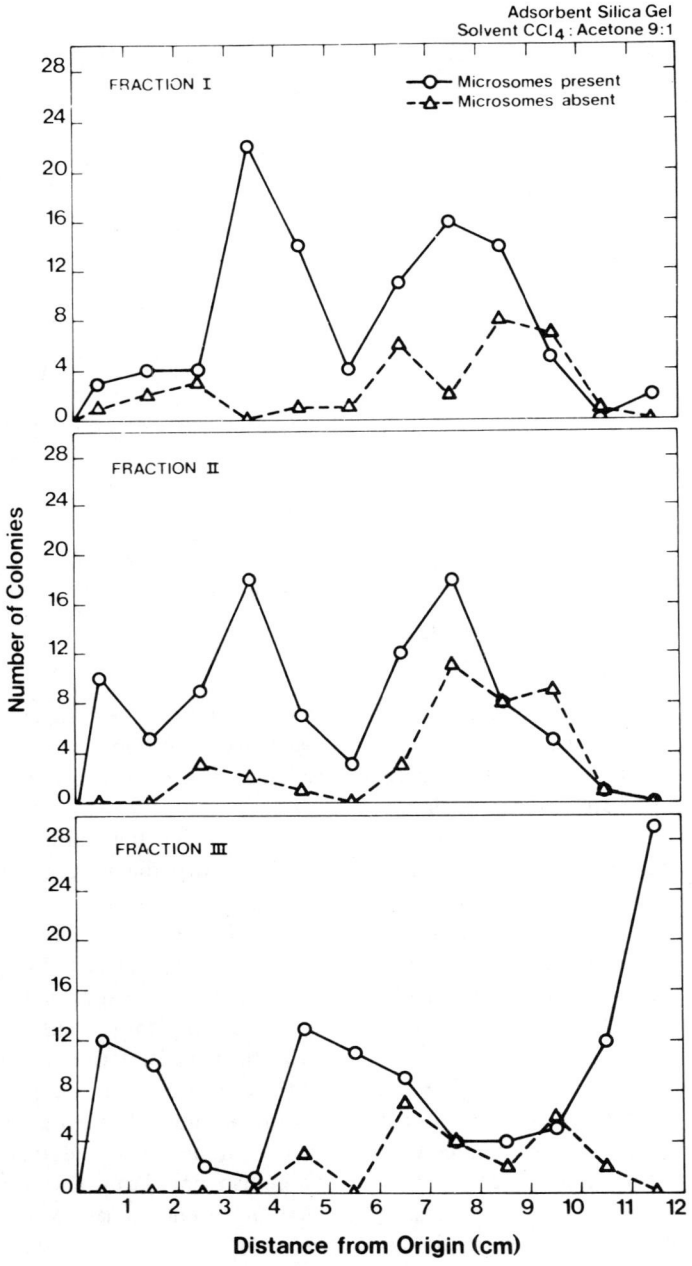

Figure 4.10 TLC separation of AAF metabolites in rat urine.

This situation is complicated by the effects of other important environmental pollutants—chlorinated hydrocarbons. The liver hydroxylase system is generally active toward foreign organic compounds, and one result of an animal's exposure to certain substances is a considerable stimulation of the activity of this enzyme system. Polychlorinated biphenyls (PCB's) are particularly powerful stimulants of hydroxylase activity and are often used to induce the enzyme activity in rats before liver microsome preparations are made for use in the *Salmonella* test. DDT and similar substances may have a corresponding effect. Thus, it is possible that individuals exposed to such substances may have an enhanced tendency to convert environmental carcinogens to their inherently-active mutagenic metabolites. By the same token, such individuals may be particularly susceptible to environmental carcinogenesis. Obviously it is of considerable importance to take this phenomenon into account in evaluating the carcinogenic hazard from DDT, PCB's and similar substances.

In any case, there are obvious advantages to mutagenic analyses of human urines as a means of evaluating exposure to potential environmental carcinogens. As an initial step in the direction of this sort of research strategy, we have begun to analyze human urine samples by means of the techniques described above. Figure 4.11 summarizes all of our results obtained thus far. The data include analyses of urine samples from a group of research laboratory workers; a group of individuals including some exposed in the workplace to a variety of synthetic organic compounds (analyzed blind); a group of workers exposed to possible carcinogens in the workplace. The urines were fractionated according to the method described earlier[7] following enzymatic and acid hydrolysis, and the mutagenic activity of materials occurring as glycosides (Fraction I) and as sulfate esters (Fraction II) were determined separately.

It should be emphasized that the results described in Figure 4.11 are not intended to provide evidence regarding the relative occurrence of urinary mutagens in chemical workers and others. The numbers involved in this preliminary population sample are clearly too small to support such a determination. Rather, what is pertinent about these results is that they show the range of values of mutagenic activity occurring in the human urine samples. It is evident that most of the values center around a ratio indicative of the absence of mutagenic activity, with only a small fraction of the individuals exhibiting ratios in the range of 2.5-5.0. This suggests that the range of "normal background" values of urinary mutagenic activity ratios is sufficiently narrow so that the occurrence of samples with values above 5.0 can be regarded as positive. Figure 4.11 shows that six urine samples fell in this elevated range; four subjects happen to be chemical workers, two nonchemical workers. We plan to

70 ORGANIC POLLUTANTS IN WATER

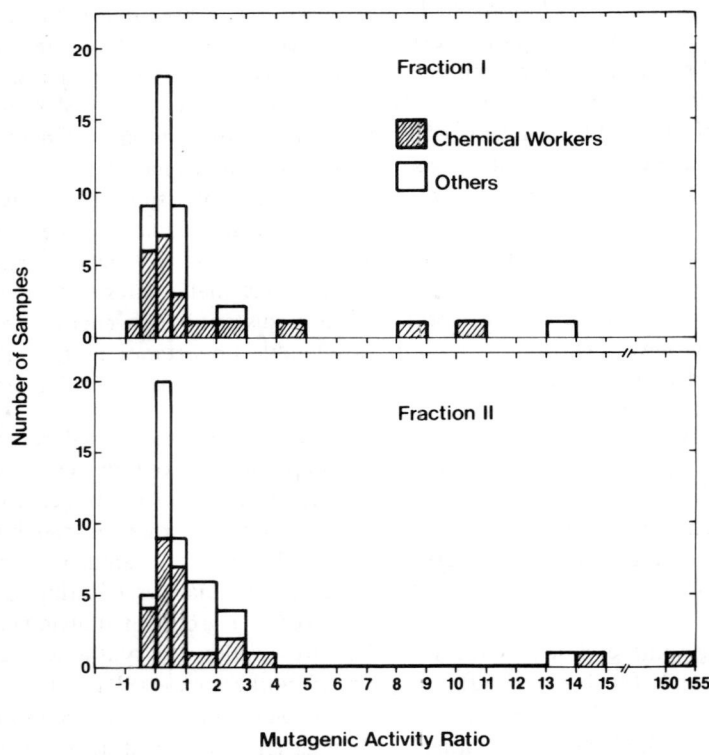

Figure 4.11 Frequency distribution of mutagenic activity of human urine samples.

extend such analyses of urine from chemical workers and others who may be heavily exposed to presumptive carcinogens, to otherwise comparable samples of the general population. Where active production of urinary mutagens is found, we hope to identify the responsible agent and to relate it to an environmental constituent.

Because of the potential carcinogens expected to occur in the general environment (for example, in urban airborne particulates), it will be of interest to determine whether individuals with urinary mutagenic activity ratios in the range of 2.5-5.0 reflect such environmental exposure, or possibly the hydroxylase-enhancing effects of PCB and similar agents. We plan to carry out urinary analyses of populations exposed to different levels of presumptive carcinogens in the general environment (for example, those exposed to different concentrations of urban airborne particulates) for this purpose.

On the basis of the foregoing data, it would appear that the opportunity now exists to develop a new strategy for controlling the growing problem of environmental cancer. It would begin with detecting presumptive carcinogens in environmental samples, identifying them and tracing their movements in the environment by means of screening based on bacterial mutagenesis. Then by determining the mutagenicity of human urine samples, it may be possible to determine which of these presumptive carcinogens represent carcinogenic risks to people. With such information in hand it would be possible to reduce this risk by tracing environmental carcinogens back to their origins, and then taking the final and most difficult step—regulating environmental emissions—that will, at last, prevent the disease.

ACKNOWLEDGMENTS

The work reported in this paper was supported in part by grants from the National Institutes of Health, The National Science Foundation/RANN, The Laras Fund and by a contract with the Environmental Protection Agency, but does not necessarily reflect their views. The project staff includes: Dr. Antony J. Vithayathil (Project Coordinator), Joyce I. Henry, Dr. Massoud Arefi-Anbarani, Jane C. Gold, and Mary Jo Reding.

REFERENCES

1. Hoover, R. and J. F. Fraumeni, Jr. *Environmental Research* **9**, 196 (1975).
2. Ames, B. N., W. E. Durston, E. Yamasaki and F. D. Lee. *Proc. Nat. Acad. Sci.* **70**, 2281 (1973).
3. Sexton, R. J. *Arch. Env. Health* **1**, 208 (1960).
4. Bell, B. *Edinb. Med. J.* **22**, 135 (1876).
5. Miller, J. A., J. W. Cramer and E. C. Miller. *Cancer Research* **20**, 950 (1960).
6. Weisburger, E. K. and J. H. Weisburger. *Adv. Cancer Research* **5**, 331 (1958).
7. Commoner, B., A. J. Vithayathil and J. I. Henry. *Nature* **249**, 850 (1974).
8. Durston, W. E. and B. N. Ames. *Proc. Nat. Acad. Sci.* **71**, 737 (1974).

SECTION II
TECHNIQUES AND METHODS OF ANALYSIS

CHAPTER 5

GLASS CAPILLARY GAS CHROMATOGRAPHY IN WATER ANALYSIS: HOW TO INITIATE USE OF THE METHOD

K. Grob and G. Grob

Swiss Federal Institute for Water Resources
and Water Pollution Control (EAWAG)
Dubendorf, Switzerland

CAPILLARY *VERSUS* PACKED COLUMNS?

Very few examples can be found in the literature comparing capillary and packed columns for the same sample and the same liquid phase. Here we present such comparisons performed under exacting conditions. Since capillary columns have exclusively been used in our laboratory for 12 years, we asked a packed column expert to concentrate his experience on our water extract. Figure 5.1 presents analyses of a mixed water extract (several water extracts combined to obtain sufficient sample) run on the stationary phase OV-1. The upper chromatogram in Figure 5.1 shows the best packed column result out of a long series of trials, although it appears relatively modest with respect to resolution. One must realize, however, that under these conditions all peaks cover several substances and appear, therefore, distorted and broadened in many ways. With ten times less sample components, the chromatogram would exhibit much better resolution.

Table 5.1 gives detailed information about the two columns used, the chromatographic conditions, and the number of peaks including reproducible shoulders. The comparison indicates that the packed column is an unsuitable tool for our particular separation problem. This statement would be less evident if a more efficient separation from a capillary column

76 ORGANIC POLLUTANTS IN WATER

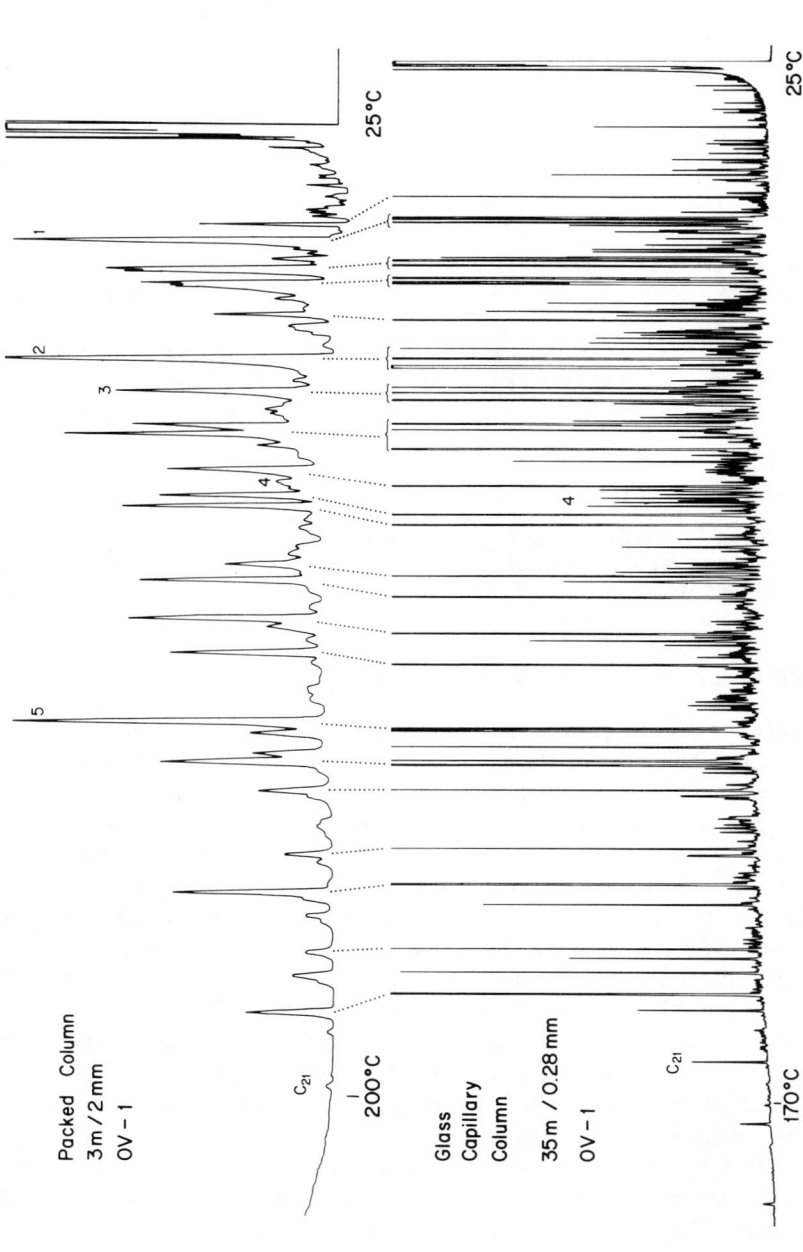

Figure 5.1 Comparison of typical separation from packed and capillary columns. Sample (extract from river water) and liquid phase identical. For characteristics of columns and chromatographic conditions, see Table 5.1. Dotted lines indicate corresponding sample components.

Table 5.1   Column Characteristics and Chromatographic Conditions

	Packed	Capillary Narrow Bore	Capillary Wide Bore
Length	3 m	35 m	60 m
Inner Diameter	2 mm	0.28 mm	0.60 mm
Coating	3% OV-1 on Gaschrom Q, 80-100 mesh	OV-1 Film $1.0 \cdot 10^{-5}$ cm	OV-1 Film $1.8 \cdot 10^{-5}$ cm
Liquid-Phase Load	120 mg	1.5 mg	14 mg
Carrier Gas Flow	He 10-22 ml/min (progr)	H_2 2.2 ml/min	H_2 14 ml/min
Temperature Range	50-200°C	25-170°C	25-200°C
Temperature Program Rate	2°C/min	3.5°C/min	2.5°C/min
Analysis Time	90 min	45 min	65 min
Sample Size	0.15 µl direct	0.6 µl splitting 1:25	0.6 µl splitting 1:8
Attenuation	x 32	x 4	x 16
Number of Peaks	118	490	320

were not available for comparison. The analyst would then be tempted to interpret the larger peaks as "major components" and might, therefore, be satisfied with his result. However, the large peaks (Nr. 1,2,3) are in fact composed of at least three important sample components, while the sharp peak (Nr. 5) contains two equivalent major substances. The unsuitability of the packed column is also evident when medium and minor components—which may be of great interest in water analysis— are considered, as exemplified by the short chromatogram section (Nr. 4).

One of the great advantages of packed over capillary columns is their much higher loading capacity (in our case about 100 times higher, according to the ratio of liquid phase load). However, the printed chromatogram has been obtained with only six times more sample than that from the capillary column. Doubling the sample size produced an intense base line "hump" caused by nonresolved material—a simple technical failure which is unfortunately often regarded as inevitable or even normal. This means that the higher loading capacity which in fact exists for single, well-resolved substances, cannot be exploited with our sample.

Finally, it may be added that the run on the capillary columns took only half the time.

We must strongly emphasize that our comparison does not imply a general judgment of packed columns which have done, and still do a perfect job in an enormous variety of analytical problems. However, even the most perfect tool fails when applied improperly. Our comparison shows that the packed column is not the appropriate tool to handle complex mixtures.

Environmental chemistry includes probably the most extreme branch of analytical chemistry as far as sample complexity is concerned. Thus, most environmental samples should be analyzed using means and methods to provide maximum separation efficiency and resolution. For the volatile part of these samples, the appropriate technique—capillary GC—is available, and has in fact routinely been used in water analysis for years, as a few selected papers[1,2] may demonstrate. The complementary technique for the nonvolatile part seems to be liquid chromatography. Here we desperately need further development and improvement.

HOW TO INITIATE USE OF CAPILLARY GC

Presumably, most environmental analysts will accept our above conclusions. Nevertheless, many environmental laboratories still run complex samples on packed columns because of the problems of introducing capillary GC. Unfortunately, for many laboratories working without expert assistance, the difficulties proved to be overwhelming. It is from such unsuccessful attempts that capillary GC has gained the reputation of being tricky, unreliable and unsuitable for routine work. Opposite opinions are given at laboratories where the introduction barrier has been broken. It is typical that in all such laboratories the broadest possible use of capillary GC is made. We do not know of any laboratory which, after a successful introduction period, has given up capillary GC.

The problems responsible for the present situation can roughly be attributed to the following sources: unsuitable equipment, unsuitable technique, and poor capillary columns. In the following pages we discuss these three sources of failure and give suggestions which might help to avoid disappointment and loss of time.

Equipment

Capillary GC separates substances in the nanogram range (10^{-8}-10^{-10} g), whereby the minute amounts should be eluted within a few seconds. From these preconditions the basic differences between capillary and packed column GC become evident, namely strongly increased importance of:

- even extremely small losses of sample material, often caused by adsorption on active surfaces (metals); or by absorption into deposits of nonvolatile sample by-products, into accumulated column bleed, or into plastic material;
- artifacts caused by impure carrier gas, by unsuitable regulating valves, by plastic parts, or by back-diffused sample material; and
- parts in the sample flow pathway causing additional diffusion.

These characteristics lead to fundamental requirements for the design of equipment. The most essential ones are cited below:

- Shortest possible pathway of sample vapor from vaporization to column inlet, as schematically shown in Figure 5.2
- Glass liner of vaporizer easily removable for replacement and cleaning.
- Vaporizer cavity with simple, axial construction to allow quick and thorough cleaning.
- Geometry of glass linear easily adaptable to sample size and injection technique (*e.g.*, splitless injection), to avoid unnecessary dilution of small samples, or diffusion of large samples into metallic parts.
- Hindered diffusion of sample vapor toward the injection septum to avoid loss of sample by absorption.

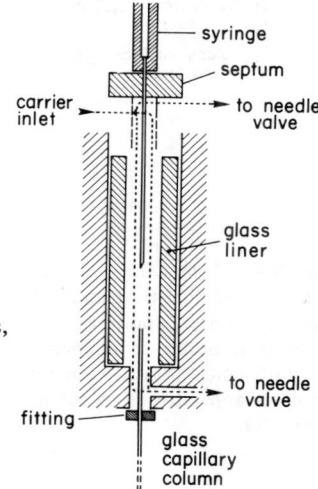

Figure 5.2 Essential parts and fundamental design of vaporizer; for schematic representation of essential functions and requirements, see text.

80   ORGANIC POLLUTANTS IN WATER

- Separate carrier flow flushing the septum to eliminate artifacts from plasticizers.
- Controllable stream-splitting device with low dead volume.
- Shortest possible distance between column outlet and flame of FID (Figure 5.3).
- Flame tip of FID easily removable for checking and cleaning.
- Easily interchangeable additional detectors with minimum dead volume.
- High quality pressure-regulating valves without plastic parts, showing minimum dependence of pressure on flow.
- Simple and flexible fittings for mounting the glass capillary columns (Figure 5.4).
- Large oven to facilitate column handling, especially for ancillary techniques such as direct sample transfer from adsorbent filters, two-dimensional chromatography and odor sensing.

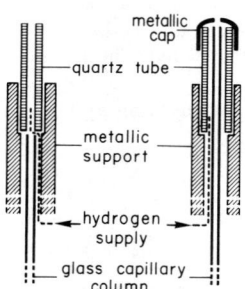

Figure 5.3   Connection of glass capillary column to flame tip of FID; two versions. Quartz capillary removable for cleaning. Metallic cap as a lower electrode of FID.

Figure 5.4   Connecting column inlet and outlet. Elastic material (*e.g.,* silicone rubber, graphite-loaded Viton) according to temperature requirements. Small washer to protect apparatus from rubber particles.

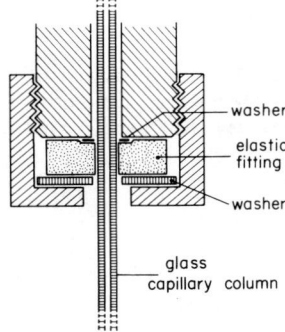

Some of these requirements imply principles which may not be commonly accepted and should therefore be explicitly mentioned. The first one contradicts the well-known injection concept according to which carrier gas and sample vapor should be thoroughly mixed before entering the column. In capillary GC this concept is not applicable for high-boiling sample material. Our requirement totally neglects vapor homogeneity with the automatic consequence that exact linear stream splitting with corresponding quantitative reproduction is not feasible. In routine practice this limitation is of no importance since quantitative work on capillary columns is best done with splitless injection. The second principle states that any kind of connecting or conveying parts splitting up the functional blocks of vaporizer/column inlet and of column outlet/detector have to be eliminated. Again, this principle becomes increasingly important with decreasing sample volatility.

Technique

A severe gap experienced by unsuccessful beginners is caused by insufficient know-how concerning the application technique. This subject is by far too broad to be covered here. Five years ago we published detailed descriptions and suggestions,[3,4] and additional indications for splitless injection.[5] Meanwhile, further intensive development has occurred. Thus, a revised draft is necessary. For the time being, the most efficient help would be practical introductory courses.

For experts on regular GC it is very tempting, and seems to be quite logical, to introduce capillary GC simply as an extension of GC with packed columns. Unfortunately, this error is probably the most widespread cause of trouble during the initiation period, since capillary GC has become an almost individual analytical field with its own equipment and methods.

Glass Capillary Columns

Laboratory-made columns have long been the basis of regular GC. Column preparation is even an essential activity of the gas chromatographer when special selectivities for different separations have to be achieved.

Making capillary columns is more delicate. The least recommendable way of introducing capillary GC is by trying to make capillary columns as a first step. The preparation work has a fair chance only when based on thorough testing of the freshly made columns to produce information about their characteristics. Exact and reliable testing, however, is only feasible when the influences of equipment and handling are perfectly known and controlled. Acquiring the necessary know-how and experience

82 ORGANIC POLLUTANTS IN WATER

is practically impossible with a column of doubtful characteristics. Thus, it is strongly recommended to begin with a column of guaranteed quality. Furthermore, it is our experience that poor results in the hands of a beginner are almost automatically attributed to the column, while in most cases they are caused by weaknesses of equipment or technique. The obvious reason for this is that an experienced gas chromatographer feels sure about his apparatus and handling; the new and suspicious thing for him is the capillary column.

The making of one's own capillary columns has been difficult for many years because the preparation procedures were poorly understood and were only successful in a few specialized laboratories. The situation is now changing. In the near future controlled procedures will be published, enabling every careful working analyst to become independent of commercial column suppliers. Practical courses to provide the corresponding knowledge will again be helpful.

After having tried out practically all solid support materials suitable for capillary columns, we must conclude that glass seems to be the best choice for the next few years. Besides its relative thermal and chemical inertness, its unequaled advantages are its easy availability and transparency.

Properly-handled glass capillary columns of good quality compare well in durability with packed columns. We have recently reported the results of an inquiry with the aim of gathering information about the fate of a large number of columns being used for routine work.[6] The main influences affecting column lifetime, cited in sequence of relative importance, have been found to be the following:

- Continuous or excessive exceeding of the upper temperature limit indicated for the given column type.
- Continuous heavy loading with polar, poorly volatile sample by-products which tend to accumulate on the solid support, thus displacing the liquid film.
- High oxygen content and, for certain types of coating, high water content of the carrier gas.
- Continuous heavy overloading as it often occurs in GC/MS-coupling when clear mass spectra from minor sample components are needed.

We have for years attempted to make columns onto which water samples may be injected routinely. Although important progress has been made, the columns still show an unsatisfactory lifetime.

AN INTERMEDIATE SOLUTION:
WIDE-BORE GLASS CAPILLARY COLUMNS

While the majority of presently-used glass capillary columns is in the diameter range of 0.2-0.4 mm, wide capillaries, 0.5-1.0 mm i.d., are successfully used in several laboratories. Out of the corresponding literature two recent papers are cited.[7,8] These columns may be regarded as intermediate between packed and capillary columns as far as carrier gas flow and sample capacity are concerned. They accept gas flow rates which are not much lower than those used with packed columns. Thus, problems with dead volume and back diffusion are greatly alleviated. In many cases, stream splitting in the sample introduction step can even be totally avoided with the result that the construction of the injection port becomes much less critical. Direct sampling also greatly facilitates the quantitative work. The increased sample size reduces the relative importance of adsorptive and absorptive sites in the entire flow path from evaporation to detection. This may allow the use of conveying lines between parts of the equipment and the column. In summary, wide-bore capillary columns may be used on the same equipment as packed columns, eventually with minor modification of certain parts.

We must emphasize, on the other hand, that the intermediate solution exists for the equipment and the chromatographic technique only, not for the column. Wide-bore capillary columns are not easier to make than narrow ones.

To demonstrate the potential of wide-bore glass capillary columns in water analysis, we prepared a 0.6-mm-i.d. column, again coated with silicone gum OV-1. Figure 5.5 shows a separation of the same water extract from which the chromatograms in Figure 5.1 had been obtained. Column characteristics and chromatographic conditions were given in Table 5.1.

A rough comparison of the three chromatograms shows that the separation accomplished by the wide-bore capillary column more closely resembles that from the narrow-bore capillary than that from the packed column. The impression is confirmed by the number of resolved peaks. The wide-bore capillary yields almost three times more peaks than the packed column, but only two-thirds as many peaks as the narrow bore.

From our comparison we may conclude the following: 20-50% of separation efficiency is lost when complex samples are analyzed on wide-bore capillary columns, even when the wide-bore column is considerably longer than the narrow one. Especially in the hands of unexperienced beginners or moderately-trained staff, this loss may be compensated for by the increased ease and safety of routine application. It may, therefore, be a *wise* decision to start work with wide-bore capillary columns as a first step when capillary GC initiation is planned.

ACKNOWLEDGMENT

The authors gratefully acknowledge the intense work by Mr. M. Hrivnac, Givaudan-Esrolko AG, Zürich, who kindly accepted the task of optimizing

84 ORGANIC POLLUTANTS IN WATER

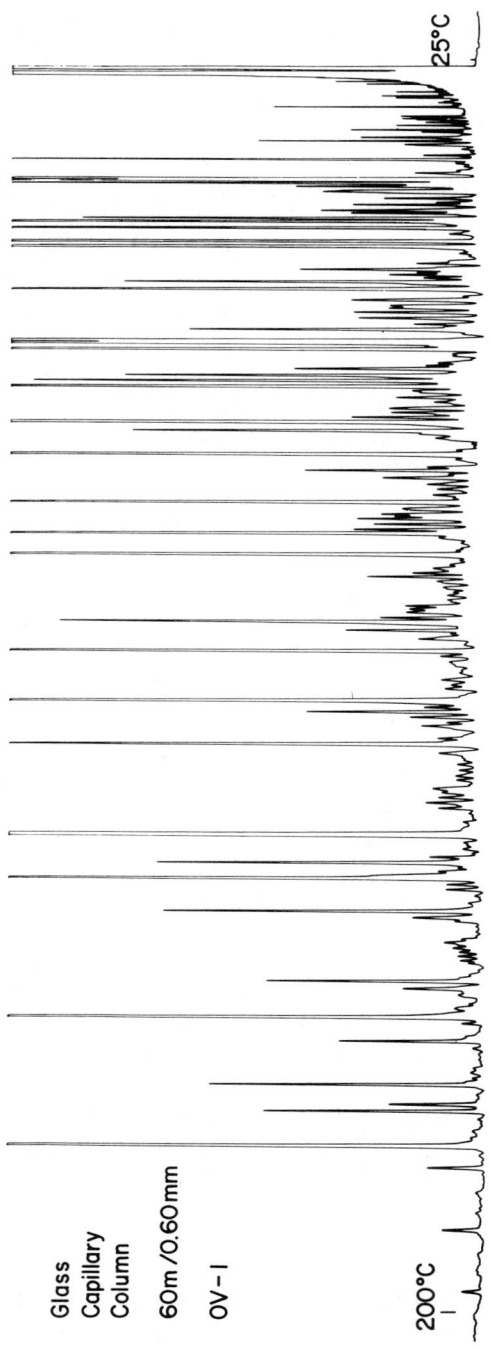

Figure 5.5 Same water extract as used for Figure 5.1, analyzed on a wide-bore glass capillary column. For column characteristics and chromatographic conditions, see Table 5.1.

the packed column analysis of our water extract. We are indebted to Mrs. J. Davis for correcting the English language.

REFERENCES

1. Giger, W., M. Reinhard and C. Schaffner. *Vom Wasser* **1974**, 343 (1974).
2. Reinhard, M., V. Drevenkar and W. Giger. *J. Chromatogr.* (in press).
3. Grob, K. and G. Grob. *Chromatographia* **5**, 3 (1972).
4. Grob, K. and H. J. Jaeggi. *Chromatographia* **5**, 382 (1972).
5. Grob, K. and K. Grob, Jr. *J. Chromatogr.* **94**, 53 (1974).
6. Grob, K. *Chromatographia* **8**, 423 (1975).
7. Badings, H. T., J. J. G. van der Pol and J. G. Wassink. *Chromatogr.* **8**, 440 (1975).
8. Sandra, P., M. Verzele and E. Vanluchene. *Chromatogr.* **8**, 499 (1975).

CHAPTER 6

GC/MS DETERMINATION OF VOLATILES FOR THE NATIONAL ORGANICS RECONNAISSANCE SURVEY (NORS) OF DRINKING WATER*

Frederick C. Kopfler, Robert G. Melton,
Robert D. Lingg, and W. Emile Coleman

>U.S. Environmental Protection Agency
>Health Effects Research Laboratory
>Water Quality Division
>Cincinnati, Ohio

INTRODUCTION

During the past several years the technique of partitioning trace organics from aqueous samples into a gas phase for analysis by chromatography has become popular. This technique is advantageous because it is suitable for the analysis of more volatile compounds which would be lost in routine extraction procedures. A decision was made to use this technique to complement methods for isolation of higher-molecular-weight organics from drinking water.[1]

At the time this decision was made several methods had been developed. McAuliffe[2] repeatedly equilibrated the aqueous sample with helium and analyzed the gas phase directly. Distribution coefficients obtained by the repeated equilibrations supplemented the retention times in identifying unknown organic compounds.

---

*Best paper presented at the Symposium on Identification and Analysis of Organic Pollutants in Water. Presented by F. C. Kopfler at the First Chemical Congress of the North American Continent, Mexico City, December 1975.

Rook[3] developed a static head-gas method in which 10 liters of water are heated at 60°C for 12 hr after which the head-gas is forced through a small amount of activated silica which traps the organics for subsequent gas chromatographic analysis. Mieure and Dietrich[4] developed a method in which the head space over an aqueous sample is continuously swept through a porous polymer trap. Zlatkis *et al.*[5] developed a similar method, but the aqueous sample is heated almost to boiling to thermally extract the volatile organics into the head space. This latter method has been applied to drinking water by Dowty *et al.*[6,7]

Grob[8] reported an elegant system in which a small volume of air is recycled through a water sample and a small charcoal filter. The organics are eluted from the charcoal with CS_2 for analysis.

While these existing methods were being considered for use in this laboratory, another method for isolating volatile organics from water was developed in another EPA Laboratory by Bellar and Lichtenberg[9] and its validity for the analysis of drinking water demonstrated.[10] Consequently, this latter method was adopted in our laboratory because this group of workers was available to give assistance in overcoming the technical difficulties encountered when an unfamiliar technique is implemented.

In this technique 5 ml of sample is introduced by syringe into the sample compartment of a specially-designed purging device. The sample is purged for 11 min with 20 ml/min of a purified gas such as helium or nitrogen. The purge gas exits through a 1/8 in. stainless steel tube filled with porous polymer or silica gel to trap the organics. The organics are then thermally desorbed and back-flushed onto a gas chromatographic column.

In August 1975 this laboratory was requested to assist in the analysis of the drinking water of New Orleans, Louisiana. This purging method was used in conjunction with a combined gas chromatograph-mass spectrometer (GC/MS) and a gas chromatograph with flame ionization detectors (GC/FID). Only 15 compounds were identified;[11] many other compounds were present in concentrations too low to produce conclusive identifications.

Soon after the completion of the New Orleans study the Administrator of the U.S. Environmental Protection Agency announced that a nationwide study, NORS, would be undertaken to determine the concentration and potential effects of certain organic chemicals in drinking water. The objective of this laboratory was to identify as comprehensively as possible the volatile organic compounds present in the tap water of five U.S. cities. Ten of the identified compounds were then to be selected for quantitative analysis.

Results of the New Orleans survey and other field studies demonstrated that achievement of these objectives required both a reliable method of sample collection and a method for collecting an increased amount of organics from the sample. The purpose of this presentation is to describe the development and use of these techniques for collecting and preparing the NORS samples for GC/MS analysis. The results of these analyses are presented by Coleman et al.[12] in Chapter 21.

SAMPLE COLLECTION AND TRANSPORTATION

From previous experience in collecting and transporting samples from the field for volatile organic analysis (VOA), it had been found that serum vials closed with Teflon-lined rubber septa held securely in place with crimped-on aluminum seals were effective in preventing leakage of the samples during transportation. Vials filled carefully to the top with water and capped so that no air space remained (Figure 6.1) developed no significant head space during storage.

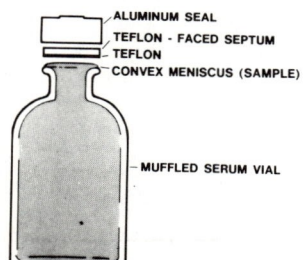

Figure 6.1 Collection of samples without headspace.

At the time the survey was announced, there was evidence indicating that chlorination of drinking water resulted in the formation of halogenated organic compounds in the water.[3,9] Since drinking water contains a chlorine residual, it was necessary to determine if the concentration of organics could change with time. The effect of storing tap water samples was investigated by filling a group of 50-ml vials with tap water. These samples were then divided into two groups, one maintained at 3°C and the other at 20°C. At selected intervals, two vials from each group were removed, placed in a 25°C water bath for 30 min and analyzed for chloroform and bromodichloromethane by the method of Bellar and Lichtenberg.[9] The results are illustrated in Figures 6.2 and 6.3. The concentrations of these compounds clearly increase even when the samples are kept cold.

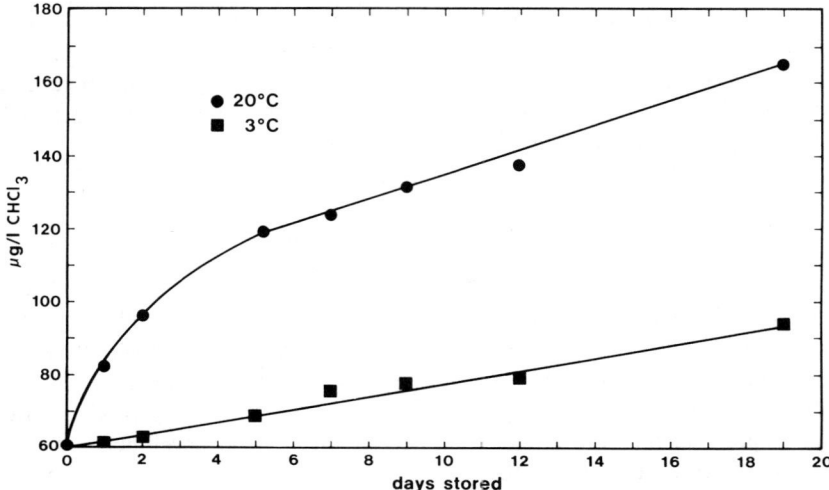

Figure 6.2 Chloroform concentrations in stored drinking water samples.

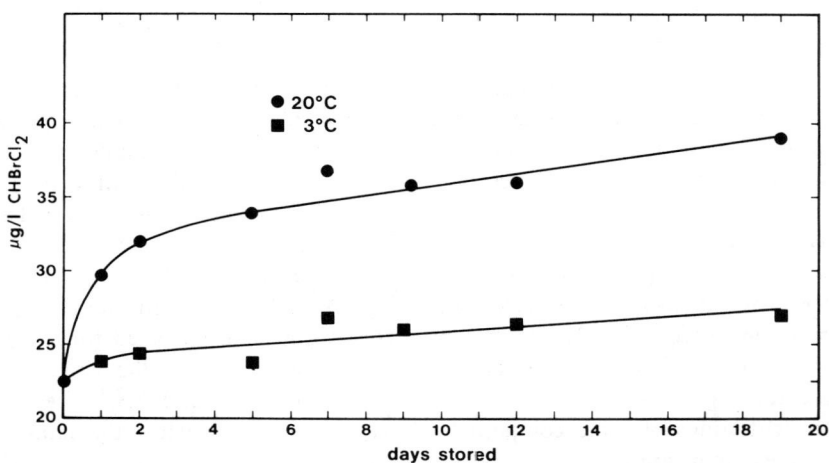

Figure 6.3 Bromodichloromethane concentrations in stored drinking water samples.

Therefore, samples for this survey were collected in 50- or 140-ml serum vials which had previously been heated at 550°C for 4 hr and then were packed in an ice chest with an ice substitute which maintained the temperature at 0 to 4°C. Because samples shipped by air mail or air freight could require more than a week to reach the laboratory, they were then hand carried by air transportation to the laboratory and stored in a refrigerator free of organic solvents. The samples from each city were analyzed between 24 to 168 hr after collection.

QUALITATIVE ANALYSIS

The Effect of Temperature

It had been demonstrated that recovery of volatile organics could be enhanced by elevating the temperature during purging.[5,8] Because formation of haloforms in the water samples were temperature-dependent, the effects of elevated temperature on some volatile organics in drinking water samples were incorporated in a broader study. The water was heated to boiling in an open container, and an aliquot transferred to a serum vial and capped as previously described. After 30 min, additional samples of the boiling water were obtained. These samples were allowed to cool and were analyzed at room temperature as were samples of the original water. The results are illustrated in columns D and E of Table 6.1 as percent change in concentration from the original tap water. It can be seen from the data in Column E that boiling the water containing a chlorine residual increased the concentration of haloforms. When continued for 30 min (column D), an efficient thermal extraction of the alkyl halides was achieved.

Table 6.1 Percentage Increase or Decrease of Five Volatile Organics Above or Below Levels Present in Cincinnati Cold Tap Water

Compound	(A) Cold Tap	(B) Carbon Filtered	(C) Carbon Filtered & Boiled	(D) Boiled 30 Min	(E) Boiled 5 Sec	(F) Hot Tap
Trichloromethane	0	(−) 99.5	(−) 99.4	(−) 94.0	(+) 96.5	(+)108.0
1,1,1-Trichloroethane	0	(−)100.0	(−)100.0	(−)100.0	(−) 50.4	(−) 14.5
Tetrachloromethane	0	(−)100.0	(−)100.0	(−) 97.3	(−) 61.6	(−) 18.0
Bromodichloromethane	0	(−)100.0	(−)100.0	(−) 97.2	(+) 72.4	(+) 78.6
Chlorodibromomethane	0	(−)100.0	(−)100.0	(−)100.0	(+)121.8	(+)141.5

This increase in haloform concentration upon heating made the direct incorporation of thermal extraction into the analysis of drinking water impossible. The reason in quantitative analysis is obvious, but the increase in chloroform content can also cause a problem in qualitative analysis: chromatograms from a computerized GC/MS are normalized on the largest peak and, since in many samples chloroform is the volatile compound present in highest concentration, an increase in chloroform concentration tends to obscure peaks of compounds present in low concentration.

The effect of chemically reducing the residual available chlorine in a tap water sample before increasing its temperature was investigated. Sodium thiosulfate was not used since thiosulfate can react with alkyl halides to produce Bunte salts (S-alkyl thiosulfates).[13] Sodium sulfite can also react with alkyl halides to produce sulfonic acids and therefore could not be used in these analyses.

The reducing agent chosen for this work was potassium ferrocyanide because of its reduction potential and because there were no known reactions with organic compounds. The ferricyanide ion produced upon oxidation of ferrocyanide can react with phenols,[14] but since neither reactants nor products are volatile under the purging conditions, no interferences were observed. A small increase in cyanogen chloride concentration may have occurred in drinking water samples to which ferrocyanide was added. At low pH, however, this problem can become severe; thus, if ferrocyanide is to be added to the sample, it should be made alkaline before addition of this reducing agent.

It is important to point out that even though residual chlorine was totally eliminated, an increase of chloroform occurred when the sample was heated to 95°C. This same increase in chloroform content was observed in samples containing ferrocyanide after standing at 25°C for 24 hr. No further increase was observed in these samples after standing for an additional 24 hr. The same increase was observed after 24 hr in samples containing a 10-fold excess of ferrocyanide. Thus, it appears that at the time of taking the samples some chlorine is already combined with an organic molecule that will ultimately undergo a conversion resulting in production of chloroform.

The increased yield of organics observed when the sample is purged at 95°C outweighed the disadvantages, and a thermal extraction step was incorporated into the final method.

Design of Larger Purging Devices

A direct 100-fold scale-up of the 5-ml purging device of Bellar and Lichtenberg was not satisfactory. The next apparatus constructed was modeled after the purging device used by Saunders *et al.*,[15] but the sample chamber was surrounded by a jacket which permitted rapid heating of the sample by circulating water from a bath maintained at 95°C. For qualitative analysis a device with a 500-ml sample chamber was constructed.

For several reasons a similar device, illustrated in Figure 6.4, containing 140 ml of sample was constructed for quantitative analysis. There was skepticism about transferring the contents of more than one serum vial quantitatively. Total removal of alkyl halides was not completed in a reasonable time at room temperature in the larger device. It was impossible to secure enough sample for several replicate quantitative determinations if the larger device was used.

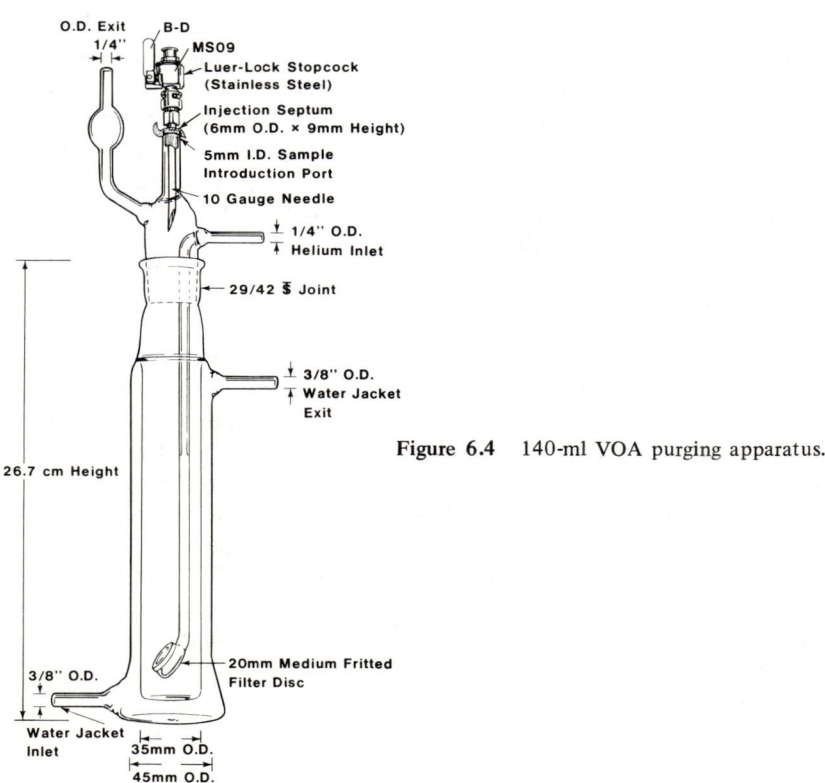

Figure 6.4 140-ml VOA purging apparatus.

The smaller purging device is similar to a gas-washing bottle equipped with a 29/42 ground glass joint on the top, a 20-mm medium fritted disc on the end of the gas tube to dispense the purge gas, a 5-mm-ID sample introduction port fitted with a silicone rubber septum, an 11-gauge by 304.8-mm (12-in.) stainless steel hypodermic needle through the septum, a stainless steel stopcock with Luer-Lock fittings on the needle, and a water jacket surrounding the sample reservoir. The design of the larger device is the same, but a 457-mm (18-in.) stainless steel needle was required for the sample introduction port. Both of these purging devices, as well as that of Bellar and Lichtenberg, are available from Paxton Woods Glass Shop of Cincinnati, Ohio.*

Traps

The traps used with the smaller device were identical to those described by Bellar and Lichtenberg.[8] The trap used with the larger purging device was similar but was constructed of 1/4-in. stainless steel tubing (4.2 mm ID) instead of 1/8-in. tubing. All traps were filled with 60/80-mesh Tenax GC and were conditioned at approximately 200°C with a helium flow of 20 ml/min for 16 to 24 hr. The traps were reconditioned for 20 min before use each day. The traps are available from Cincinnati Valve and Fitting, Cincinnati, Ohio.

Low Organic Water

Another problem encountered during the New Orleans survey was obtaining a source of water free of volatile organics to be used in development of a procedural blank. Distilled water from a number of sources was analyzed as was distilled water further treated with activated carbon and purified water produced by a Millipore Corp. Super-Q System. All were contaminated with several volatile compounds. An acceptable "low organic" water was produced from distilled water by passing it through a Millipore Corp. Super-Q system followed by purging with ultra-pure helium for 48 hr at 95°C. This water still contained traces of polar organics such as ethanol and acetaldehyde. It was then transferred to serum vials, sealed and stored in the refrigerator with the samples until analyzed.

---

*Mention of commercial sources does not necessarily imply endorsement by the U.S. government.

Fractional Purging

The initial analyses of samples conducted with the largest purging device resulted in chromatograms dominated by haloforms to such an extent as to obscure many compounds present in lower concentration. The slowness with which polar organics were purged in the preparation of the "low organic water" suggested that a fractional purging scheme could be developed to aid in the qualitative analyses of the samples.

Figure 6.5 illustrates the relative purging efficiencies of five compounds. These data were obtained by performing six successive analyses of a solution containing approximately 50 µg/l of each component except ethanol. To

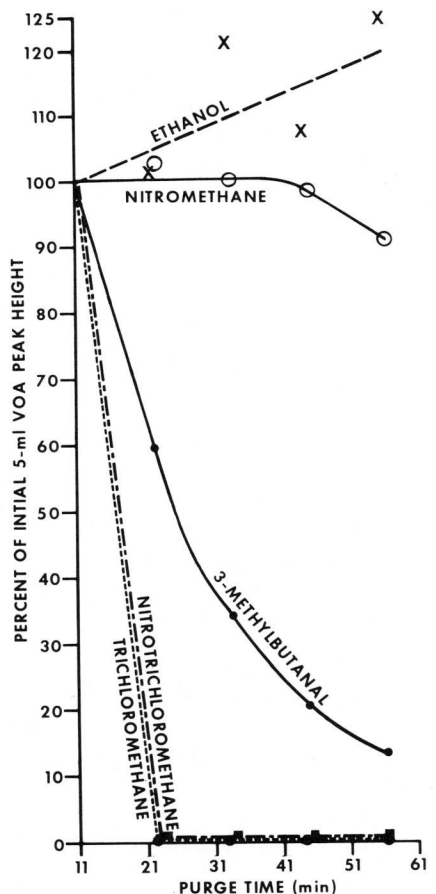

Figure 6.5 Five repetitive purges of standard solution.

96 ORGANIC POLLUTANTS IN WATER

obtain a similar response for ethanol with the GC/FID, a concentration of 2 mg/l was required. Clearly, all organics do not purge at the same rate, nor do they all purge quantitatively. It was found that purging at 95°C for 30 min removed most of the alkyl halides and other nonpolar compounds. A sufficient quantity of polar compounds remained so that a subsequent purging of the sample allowed these organics to be trapped and analyzed separately. This aided in obtaining conclusive spectra of the polar compounds.

Qualitative Procedure Used in NORS

Organics were isolated for qualitative analysis from 500- to 600-ml aliquots of each sample. The organics were fractionated by conducting three consecutive 30-min purges of the same sample. The first purge was conducted at 6°C to remove the most volatile, water-insoluble organics. A stoichiometric amount of potassium ferrocyanide was then added to reduce chlorine, and the sample was then purged at 95°C to remove most of the remaining volatile, water-insoluble organics. Finally, a third purge was conducted at 95°C to detect the presence of volatile, water-soluble organics. Details of the qualitative procedure are as follows:

1. The level of total chlorine residual in an aliquot of the drinking water sample was determined by amperometric titration.
2. The trap was attached to the purging device.
3. The content of five 125-ml sample bottles were forced with helium into the reservoir through a Teflon transfer line as illustrated in Figure 6.6
4. The sample was purged for 30 min at 6°C with the slow stirring of a magnetic stirring bar.
5. The sample purging and stirring was stopped, the trap was removed and capped, and the helium exit of the purging device was capped.
6. After the organics had been desorbed from the trap into the GC/MS, the cooled trap was replaced onto the purging device.
7. The stoichiometric amount of potassium ferrocyanide necessary to reduce the chlorine was added through the sample inlet, and the solution was stirred for 10 min at 6°C.
8. The temperature was raised to 95°C and the water was again purged for 30 min.
9. The trap was removed and the adsorbed organics analyzed while the purging of the water continued.
10. The trap was again cooled and attached.
11. The organics were collected for another 30 min and analyzed.

The "low organic water" was analyzed immediately prior to the analysis of each sample. A compound was deemed present in a sample if its

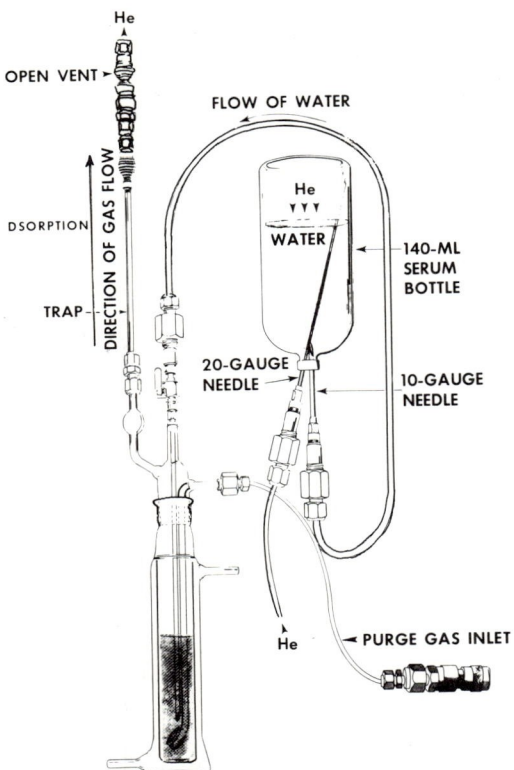

Figure 6.6 Addition of water sample to VOA purging apparatus.

concentration was greater in the sample than in the corresponding procedural blank.[12]

Figures 6.7, 6.8 and 6.9 are the reconstructed chromatograms obtained during the qualitative analysis of the Miami, Florida, sample. The benefit of the fractional purging is best illustrated in the following observations. Figure 6.7 shows the detection of extremely volatile compounds such as vinyl chloride. The chromatogram in Figure 6.8 is dominated by alkyl halides, especially the haloforms. The presence of many more water-soluble alcohols, aldehydes and ketones can easily be detected in the third analysis of the sequence, Figure 6.9.

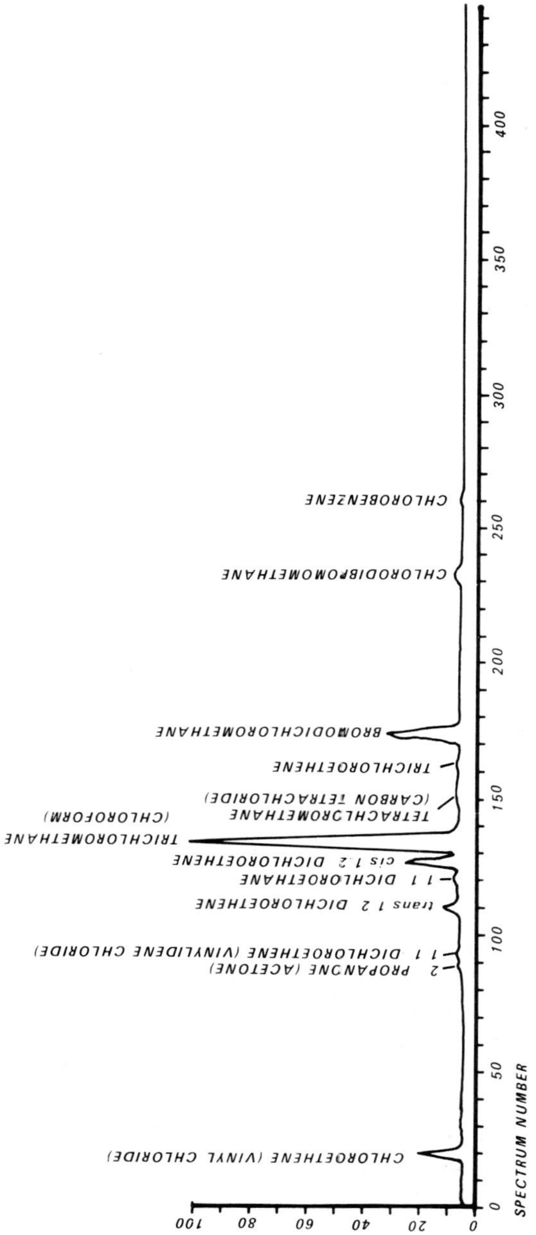

98 ORGANIC POLLUTANTS IN WATER

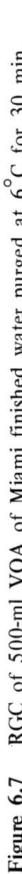

Figure 6.7 RGC of 500-ml VOA of Miami finished water purged at 6°C for 30 min.

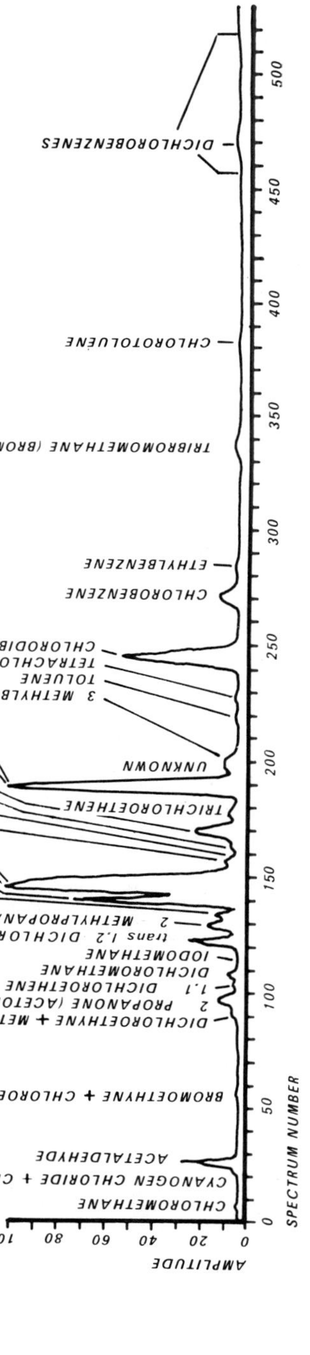

TECHNIQUES AND METHODS OF ANALYSIS 99

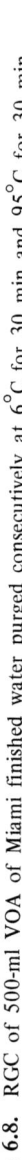

Figure 6.8. RGC of 500-ml VOA of Miami finished water purged consecutively at 6°C for 30 min and 95°C for 30 min.

100   ORGANIC POLLUTANTS IN WATER

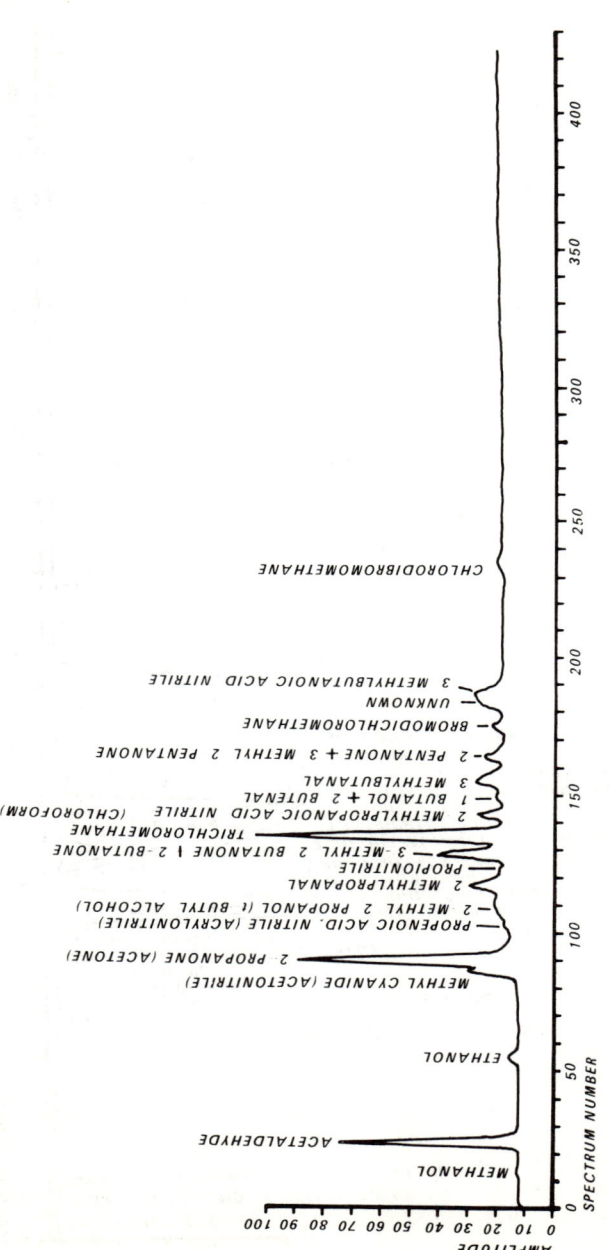

Figure 6.9 RGC of 500-ml VOA of Miami finished water purged consecutively at 6°C for 30 min and 95°C for 60 min.

QUANTITATIVE ANALYSIS

The quantitative analyses of these samples was the most complex task to be accomplished in the NORS. Before the survey began a decision was made to determine the concentration of ten volatile organics deemed significant because of their potential health effects. These compounds were chosen only after the qualitative analysis of all five samples was completed. The increase in haloform concentration in the samples precluded delaying quantitative analysis until after all qualitative analyses were completed and selection of compounds was made. Thus, it was necessary to obtain data on each sample that could later be used to quantitate any compound identified in the samples.

Given these conditions, the only approach to quantitation was to use an internal standard (IS) in all samples and blanks, to analyze them under identical conditions with computerized GC/MS, and to store the data on magnetic tapes. After the compounds to be quantified were selected, solutions containing known amounts of these compounds and the IS were analyzed in the same manner. The IS consisted of a solution of two compounds: 2-bromobutane and 2-iodopropane. These compounds were chosen because they are easily purged from water and together have a good distribution of mass peaks that are 15% or greater of the base peaks.

The internal standards were prepared from reagents supplied by Chem-Services, Inc. A 10-μl syringe was used to carefully measure 2 μl of each of the two compounds. This volume of each compound was injected into a 1-liter volumetric flask nearly filled to the mark with water. After dilution to the mark and gentle agitation to insure complete mixing, the IS solution was bottled in 50-ml serum vials and capped as described above. Fresh internal standard was prepared each day when quantitation runs were made because both compounds slowly decompose in aqueous solution.

Purging Procedure for Quantitative Analysis

To insure that a constant amount was added at each run, the IS from these capped serum bottles was added directly to the 140-ml purging apparatus. Before sample water was introduced, the sample transfer line was disconnected from the Luer-Lock valve and a 5-mm NMR sample tube septum was placed over the opening of the valve (Figure 6.4). With the trap in place, a 500-μl syringe was used to withdraw 300 μl of IS from one of the prepared vials. The NMR tube septum was then pierced with the syringe needle and the Luer-Lock valve was opened. After passing the syringe needle through the valve opening, the 300 μl of IS was discharged directly into the purging device. Upon withdrawing the syringe needle beyond the valve opening, the valve was closed and the syringe withdrawn completely. The septum was removed, the sample transfer tube was reconnected, and the sample was then transferred from a serum vial into the apparatus via the closed transfer loop previously described for qualitative analysis.

Samples were purged for 15 min at 25°C with a helium flow rate of 20 ml/min. Four replicate analyses were performed on each sample using a 1.52-m (5-ft) column of 60/80-mesh Chromasorb 101. Figure 6.10 illustrates the reproducibility of the four quantitative chromatograms obtained on one of the samples.

Analytical Standards

The compounds selected for quantitation were liquids at room temperature. Aqueous standards of these compounds were prepared by adding 2 µl of each to a 1-liter volumetric flask almost filled with distilled water. The flask was then diluted to volume with distilled water and agitated to achieve a complete solution. Dilutions of this stock solution produced a series of standards containing approximately 5, 2.5, 1, 0.5, 0.3, 0.1 and 0.05 µg/l of each compound. The exact concentration of each compound was calculated from its density. These solutions were transferred to 140-ml serum vials and capped until analyzed.

Four replicates of this set of standards were prepared. Each standard solution was fortified with the IS and analyzed as described for the samples. The intensity of the IS response in each chromatogram was used to compensate for instrument variations which existed among analyses, allowing comparison of the samples to the standards. This technique has recently been discussed by Young.[16]

Many problems were encountered in obtaining quantitative data on these samples prior to obtaining the qualitative data. The application of computerized GC/MS to this problem is to be described in a forthcoming paper by Lingg et al.[17]

TECHNIQUES AND METHODS OF ANALYSIS 103

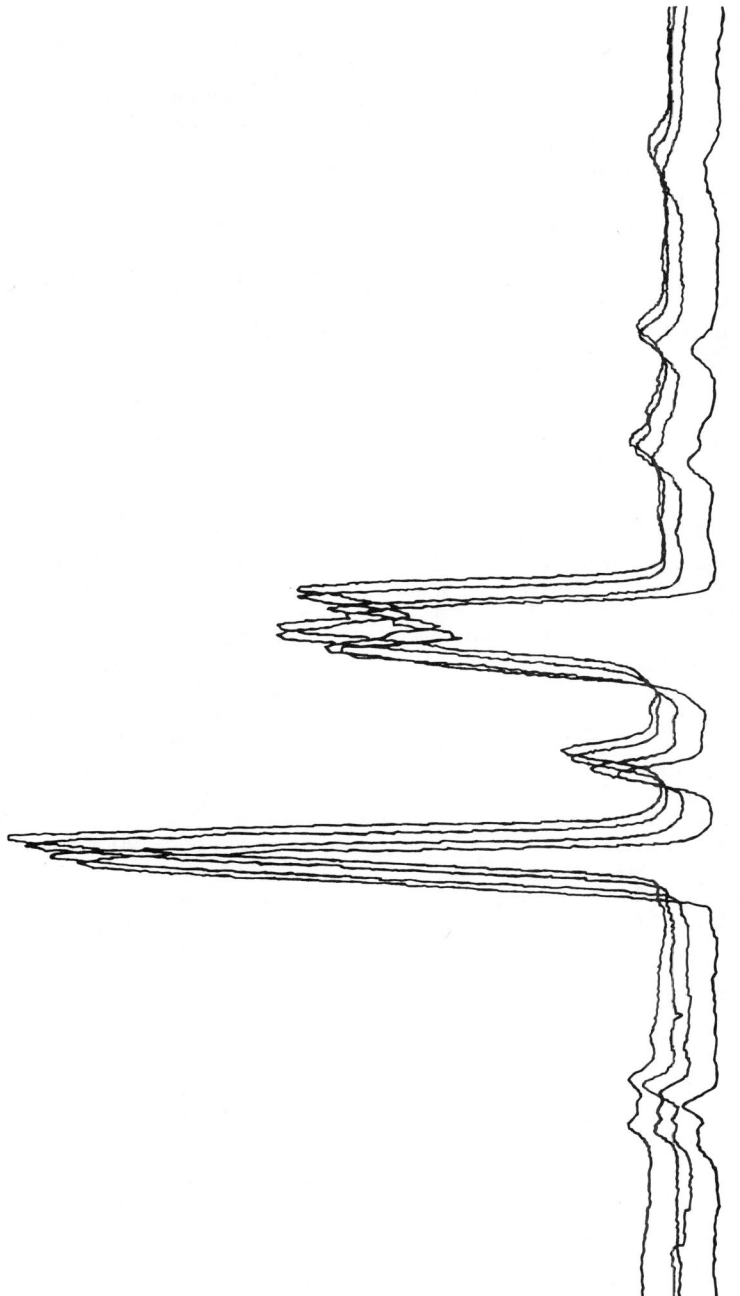

Figure 6.10 Reproducibility of four real-time 140-ml VOA chromatograms of Philadelphia finished water.

REFERENCES

1. Kopfler, F. C., R. G. Melton, J. L. Mullaney and R. G. Tardiff. "Human Exposure to Water Pollutants," presented before the Division of Environmental Chemistry, American Chemical Society, Philadelphia, Pennsylvania (April 1975).
2. McAuliffe, C. *Chem. Tech.* 1, 46 (1971).
3. Rook, J. J. *Water Treatment Exam.* 21, 259 (1972).
4. Mieure, J. P. and M. W. Dietrich. *J. Chromatogr. Sci.* 11, 559 (1973).
5. Zlatkis, A., H. A. Lichtenstein and A. Tishbee. *Chromatographia* 6, 67 (1973).
6. Dowty, B., D. Carlisle, and J. Laseter. *Science* 187, 75 (1975).
7. Dowty, B., D. Carlisle and J. Laseter. *Environ. Sci. Technol.* 9 762 (1975).
8. Grob, K. *J. Chromatogr.* 84, 255 (1973).
9. Bellar, T. A. and J. J. Lichtenberg. *J. Amer. Water Works Assoc.* 66, 739 (1974).
10. Bellar, T. A., J. J. Lichtenberg and R. C. Kroner. *J. Amer. Works Assoc.* 66, 703 (1974).
11. U.S. Environmental Protection Agency. "Analytical Report—New Orleans Area Water Supply Study," Report EPA-906/10-74-002 Region VI (Dallas, Texas) Surveillance and Analysis Division (1975).
12. Coleman, W. E., R. D. Lingg, R. G. Melton and F. C. Kopfler. "The Occurrence of Volatile Organics in Five Drinking Water Supplies Using Gas Chromatography/Mass Spectrometry (GC/MS)," presented at the First Chemical Congress of the North American Continent, Mexico City (December 1975), Chapter 21, this volume.
13. March, J. *Advanced Organic Chemistry: Reactions, Mechanisms and Structure.* (New York: McGraw-Hill Book Co., 1968), p. 330.
14. Fieser, L. F. and M. F. Fieser. *Reagents for Organic Synthesis.* (New York: John Wiley and Sons, 1974), p. 929.
15. Saunders, R. A., J. R. Griffith, and F. E. Saalfeld. *Biomed. Mass Spectrom.* 1, 192 (1974).
16. Young, I. G. *American Laboratory* 7, 11 (1975).
17. Lingg, R. D., R. G. Melton, F. C. Kopfler and W. E. Coleman. "GC/MS Techniques for the Quantification of Volatile Organics in Tap Water," in preparation (1975).

CHAPTER 7

A CONVENIENT LIQUID-LIQUID EXTRACTION METHOD FOR THE DETERMINATION OF HALOMETHANES IN WATER AT THE PARTS-PER-BILLION LEVEL

James E. Henderson, Gary R. Peyton and William H. Glaze
> Institute of Applied Sciences
> North Texas State University
> Denton, Texas

The discovery of C_1 and C_2 halogenated hydrocarbons in drinking water supplies has caused great concern among investigators studying organic water pollutants. The most common of these pollutants are chloroform, carbon tetrachloride, bromodichloromethane, chlorodibromomethane, and bromoform although other species have now been reported. These halomethanes have been shown by Rook[1] to form as a result of the chlorination process used for water disinfection. Since that work, extensive studies have been carried out to determine the extent of contamination of water supplies by these compounds. As a part of the National Organics Reconnaissance Survey (NORS) being conducted by the U.S. Environmental Protection Agency, drinking water supplies from 80 U.S. cities were sampled and assayed.[2] The results showed that halogenated hydrocarbons occurred in varying amounts in all of the cities sampled.

One analytical technique used for the NORS survey was developed by Bellar and Lichtenberg.[3] This technique employs a stripping principle using inert gas to "extract" the volatile organics, including the halocarbons, and a Tenax GC trap to absorb them. The organics can then be desorbed thermally onto the head of a gas chromatographic column where separation is effected. The halogen-specific Hall (or Coulson) electrolytic conductivity detector is used to eliminate interferences from nonhalogenated pollutants.

A major disadvantage of the system is the length of time required to complete an analysis. The limiting time factors are the purge step and the chromatographic procedure, which requires perhaps a total of about 30 min. Thus, one is severely limited in the number of samples which can be run per day. Moreover, various workers have reported difficulties in the use of Tenax for repetitive analysis of this type, which may be due in part to the thermal and/or hydrolytic degradation of the resin.

In this chapter we report on the development of a convenient and more rapid procedure for the analysis of halomethanes at the $\mu g/l$ level, using equipment commonly available in water laboratories.

EXPERIMENTAL

Samples are collected in 120-ml serum bottles. The bottles are completely filled so that no headspace is left in the bottle. They are then capped with a teflon-coated rubber septum and sealed by crimping the aluminum septum retainer over the lip of the bottle. The samples can then be conveniently transferred to the laboratory for analysis.

In the laboratory the extraction solvent (pentane in this work) is added directly to the sampling bottle. This is done using two 10-cc syringes with 22-gauge needles, one syringe containing the solvent, the other being empty. The needles of both syringes are pierced through the septum, and the (3-5 ml) solvent is injected into the bottles as indicated in Figure 7.1. The solvent displaces the water, which is collected in the second syringe and discarded. The partitioning is then brought to equilibrium by shaking the bottles for 15 min at 500 rpm with a gyrotory platform shaker. Then, an aliquot (5 μl) of the organic layer is removed with a microliter syringe for chromatographic analysis.

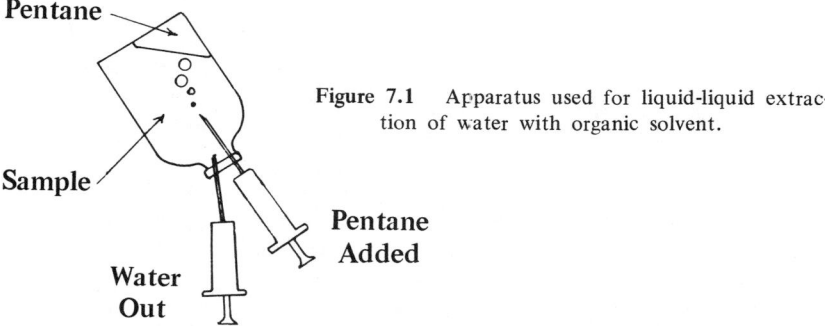

Figure 7.1 Apparatus used for liquid-liquid extraction of water with organic solvent.

The chromatographic procedure utilizes a Tracor 560 gas chromatograph with linearized ^{63}Ni electron capture detector. The column is 6 ft x 2-mm i.d. glass packed with 10% squalane on Chromasorb W/AW (80-100 mesh). The analysis is run isothermally at 67°C with argon/methane as the carrier gas (flow rate of 20 ml/min through the column and 60 ml/min added as make-up to the effluent). The attenuation is nominally set at x 16. A Varian Model 435 digital electronic integrator is used for peak area measurements.

Quantitative measurements were made using the internal standard (IS) technique with 1,2-dibromoethane as the IS. Response factors for each compound were determined using the usual procedure (see Figure 7.3). The internal standard was added to the pentane before extraction, and its level was found to fall approximately 10% during the extraction process, presumably due to water solubility. Standard solutions of the five compounds of interest in ethanol were used to prepare water standards which were run at the beginning and end of a day's runs. See Table 7.1 for results on a typical standard solution.

RESULTS AND DISCUSSION

This new method of analysis offers the following advantages over the VOA procedure of Bellar and Lichtenberg:

1. Many samples can be extracted simultaneously using a platform shaker, as opposed to the sequential method employed in the Bellar-Lichtenberg procedure.
2. Conventional injection techniques are employed, eliminating the time required for thermal desorption of the organics into the chromatograph.
3. The isothermal chromatographic procedure uses conventionally coated packings, cutting the instrumental analysis time significantly.

In addition, this method eliminates the need for the special equipment required for the stripping, adsorption and desorption procedures. In fact, little extra equipment will be required for those laboratories already equipped to do pesticide analyses. Finally, the method conceivably may be used to determine substances whose properties make them unsuitable for determination by the Bellar-Lichtenberg VOA method.*

It should be noted that the method we use is similar in principle to standard LLE techniques which are routine in the analysis of organic materials in water. The extraction method is also similar to one developed

---

*Several peaks other than those due to the five compounds reported in Table 7.1 have been observed in raw and treated water samples. Some of these have been identified tentatively as other halomethanes, including three iodine-containing compounds.[4]

Table 7.1 Concentrations of Five Halomethanes in Raw and Treated Municipal Water from One Southwestern City

Sample	Concentrations Determined ($\mu g/l$)[b,c]					Total Halomethanes
	$CHCl_3$	CCl_4	$CHCl_2Br$	$CHClBr_2$	$CHBr_3$	
Standard[a]	6.39	6.34	6.52	6.21	6.50	31.85
Municipal Water (Ground Water Source)						
(7-16-75) Raw	0.14 (0.03)	t	a	a	a	0.17
(7-16-75) Treated	0.96 (0.06)	t	6.11 (0.04)	42.8 (2.4)	74.2 (3.8)	123
(7-18-65) Raw	1.03 (0.11)	0.14	t	t	t	1.2
(7-18-75) Treated	1.23 (0.10)	t	1.23 (0.08)	15.8 (0.7)	58.5 (2.5)	77
(7-21-75) Raw	0.12	t	a	t	t	0.12
(7-21-75) Treated	0.84 (0.01)	t	0.69 (0.02)	7.53 (0.13)	43.1 (1.5)	51
(7-23-75) Raw	0.34 (0.01)	t	t	a	a	0.34
(7-23-75) Treated	1.38 (0.07)	t	7.76 (0.49)	42.1 (1.8)	53.2 (2.2)	103

[a]Concentration of standard as prepared differed by less than 1.0% from concentrations determined.
[b]t = <0.1 $\mu g/l$; a = absent.
[c]Numbers in parentheses refer to average deviation between two successive chromatographic runs on the same sample.

by Grob et al.[5] except that the volume of pentane solvent used for the extraction is much larger (x 10^1). While this decreases our concentration factor, the reproducibility of the method is increased. Moreover, the use of the very sensitive electron capture detector allows us to obtain sufficiently low limits of detection ($\sim$ 1 μg/l).

Figure 7.2 is a temperature-programmed gas chromatogram of the five-component mixture of interest with each compound spiked in 140 ml of water at the 10-mg/l level. ("Organic free" water was spiked with an appropriate amount of a standard solution of the components dissolved in ethanol). On the basis of this result, a column temperature of 67°C was chosen for the analysis. At this temperature the five peaks of interest are resolved to near-baseline (R > 1.0 for adjoining peaks) and the analysis is complete in 12 min. Retention times for the compounds of interest are as follows: $CHCl_3$ (91 s); CCl_4 (132 s); $CHCl_2Br$ (156 s); $CHClBr_2$ (292 s); IS (342 s); $CHBr_3$ (586 s). Maximum deviation in each case is 4%.

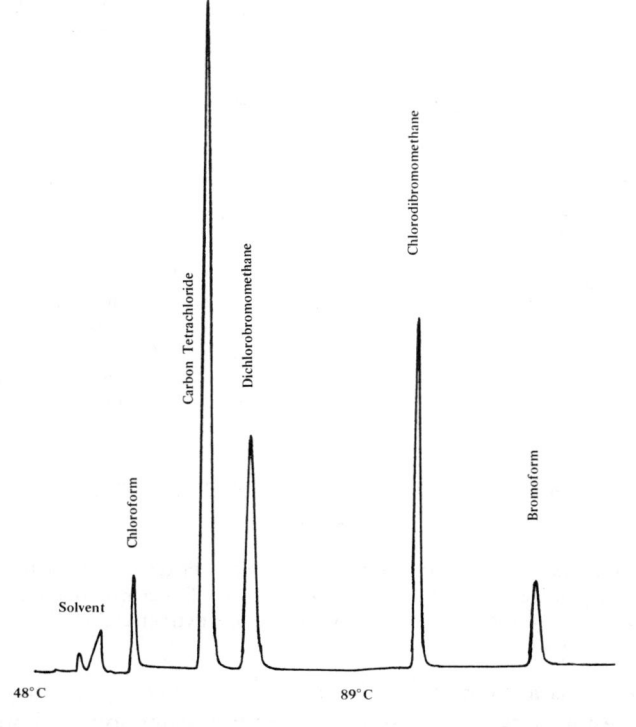

Figure 7.2 Temperature-programmed gas chromatogram of halomethanes. GC conditions described in text.

Figure 7.3 shows the response of the ^{63}Ni electron capture detector to the five compounds of interest in this work. As indicated in Figure 7.3, good linearity is obtained over a four- or five-decade concentration range. The lower limit of detectability for the method appears to be approximately 0.1 μg/l for the most easily determined species (CCl_4) and approximately 1.0 μg/l for $CHBr_3$, the least favorable case.

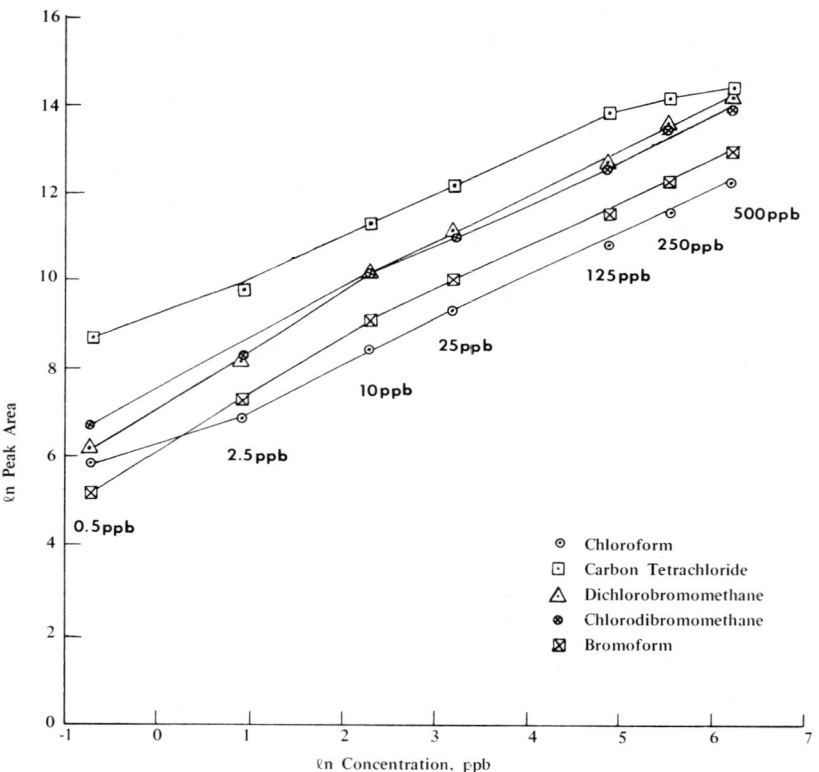

Figure 7.3 Plot of response of the ^{63}Ni electron capture detector to various concentrations of halomethanes. Concentrations refer to those in water before extraction.

Table 7.1 is a compilation of some results obtained by the method on a southwestern city's water supply, a ground water source, both before and after treatment. This particular city utilizes approximately 20 lb Cl_2 per million gallons of water (2.3 mg/l). As indicated by the data in Table 7.1, only a small fraction of the chlorine ends up as halomethanes.

Of much interest, however, is the predominance of brominated methanes in the treated waters from the test city. Presumably these compounds are formed by the action of bromine (or HOBr) formed as a result of the reaction of natural bromide ion with the added chlorine:[6]

$$Cl_2 + 2Br^- \rightarrow 2Cl^- + Br_2$$

$$Br_2 + H_2O \rightleftharpoons HOBr + Br^- + H^+$$

These results suggest that chloroform monitoring alone may be an inconclusive indicator of halomethane concentrations under certain conditions.

CONCLUSION

The new liquid-liquid extraction procedure described in this chapter appears to represent a convenient method for VOA monitoring, especially when coupled with the very sensitive electron capture detector. The method is now being utilized to monitor several southwestern water supplies before and after treatment.

ACKNOWLEDGMENTS

We are grateful for a grant from the NTSU Faculty Research Fund for support of this work, and to David Mapel and Frederick Spies for assistance in the analytical work.

REFERENCES

1. Rook, J. J. *Water Treatment Exam.* **23**, 234 (1974).
2. Kopfler, F. C., R. G. Melton, R. D. Lingg and W. E. Coleman. "GC/MS Determination of Volatiles for the National Organics Reconnaissance Survey (NORS) on Drinking Water," presented at the First Chemical Congress of the North American Continent, Mexico City (December 1975), Chapter 6, this volume.
3. Bellar, T. A. and J. J. Lichtenberg. *J. Amer. Water Works Assoc.* **66**, 739 (1974).
4. Glaze, W. H., J. E. Henderson, IV, and G. Smith. Proceedings of the Conference on the Environmental Impact of Water Chlorination, Oak Ridge, Tenn. (1975), to be published as a report of the U.S. Energy Research and Development Administration.
5. Grob, K., K. Grob, Jr. and G. Grob. *J. Chromatography* **106**, 299 (1975).
6. Kleopfer, R. D. "Analysis of Drinking Water for Organic Compounds," presented at the First Chemical Congress of the North American Continent, Mexico City (December 1975), Chapter 24, this volume.

CHAPTER 8

SEPARATION OF TRACE ORGANIC COMPOUNDS FROM WATER

J. P. Mieure, G. W. Mappes,
E. S. Tucker and M. W. Dietrich

> Monsanto Company
> St. Louis, Missouri

INTRODUCTION

Most analyses of water for trace organic compounds require prior separation of the components of interest from the aqueous matrix. One of the most useful and widely used separation procedures is liquid-liquid extraction using an organic solvent. Sensitivity is limited, however, unless the materials being analyzed are sufficiently low in volatility that further concentration via solvent evaporation is possible. In certain cases the extraction solvent may also interfere with measurement. Similar limitations apply to procedures involving collection on solid adsorbants followed by organic solvent desorption and concentration.

Two alternate approaches which we have found effective for isolating and concentrating organic solutes are headspace analyses and membrane separations. Principles and applications of these techniques are the subject of this chapter. We will also show how these techniques complement other isolation/concentration schemes.

HEADSPACE ANALYSIS

In its simplest form headspace analysis involves direct sampling and analysis of the equilibrium (static) atmosphere over a solution in a closed container. Detection limits are restricted by the equilibrium gas-liquid partition of the solutes (dissolved organics) as well as the limited amount of headspace gas which can be conveniently sampled and analyzed. An improved technique using dynamic instead of static sampling consists of sparging organics from the water phase with an inert gas. The partitioned organics are continuously trapped from the gas stream in a short chromatographic column for later analysis by gas chromatography.

This technique has gained acceptance within the last few years[1-3] and received widespread publicity because of recent use to measure volatile organics in drinking water supplies around the United States.[4] Studies in our laboratories have shown that the sparging apparatus developed by Bellar and Lichtenberg[3] is more efficient than other devices for removing organics from water. This section presents a summary of these studies along with a discussion of two variables, temperature and ionic strength, which can make the sampling more effective.

We compared the Bellar apparatus with a conventional gas scrubber, a modified Bellar apparatus and a continuous headspace sampler. The comparison was carried out by charging each device with distilled water spiked with 41 ppm 2-octanone and passing 500 ml of N_2 gas at 20 ml/min through the device. These conditions were chosen as typical of those required for highly volatile materials such as vinyl chloride which might break through the trap with larger collection volumes. The percentage of 2-octanone removed from the water was determined by direct-injection, flame-ionization gas chromatography. The results are shown in Table 8.1.

Table 8.1 Comparison of Sparging Efficiency Using 2-Octanone

Apparatus	Water Volume (ml)	% 2-Octanone Removed
Conventional Scrubber	20	28
Continuous Headspace Sampler	20	24
Bellar Apparatus	5	81
Modified Bellar Apparatus	5	62

The conventional scrubber had a lower sparging flux per unit volume of water which may have contributed to its poor removal efficiency. The modified Bellar apparatus differed from the conventional Bellar by being expanded in cross section in the region of the glass frit as shown in Figure 8.1. It was hoped that this increased surface area would facilitate the sparging action. However, we found that at 20 ml/min sparge rate,

TECHNIQUES AND METHODS OF ANALYSIS   115

Figure 8.1   Modified Bellar apparatus for sparging organics from water.

diffusion of the sparge gas through the frit was localized instead of dispersed, leaving portions of the water undisturbed. Table 8.2 summarizes 2-octanone removal under other conditions.

The continuous headspace sampler shown disassembled in Figure 8.2 swept inert gas over the surface of vigorously-stirred water. Sweep gas entered the top, diffused around the circular baffle and exited through the tubing in the center of the plate. The low value reported in Table 8.1 indicates this process is not as efficient as bubbling gas through the liquid.

The effects of ionic strength and temperature on the stripping efficiency of the Bellar apparatus were evaluated using distilled water spiked

116 ORGANIC POLLUTANTS IN WATER

Table 8.2 Comparison of Modified Bellar Apparatus Sparging Efficiency Under Varied Conditions

Water Volume (ml)	Flow Rate (ml/min)	Volume Air (ml)	Percent 2-Octane Removed
5	20	500	62
20	20	500	29
20	80	2000	69
20	513	12,800	99.5

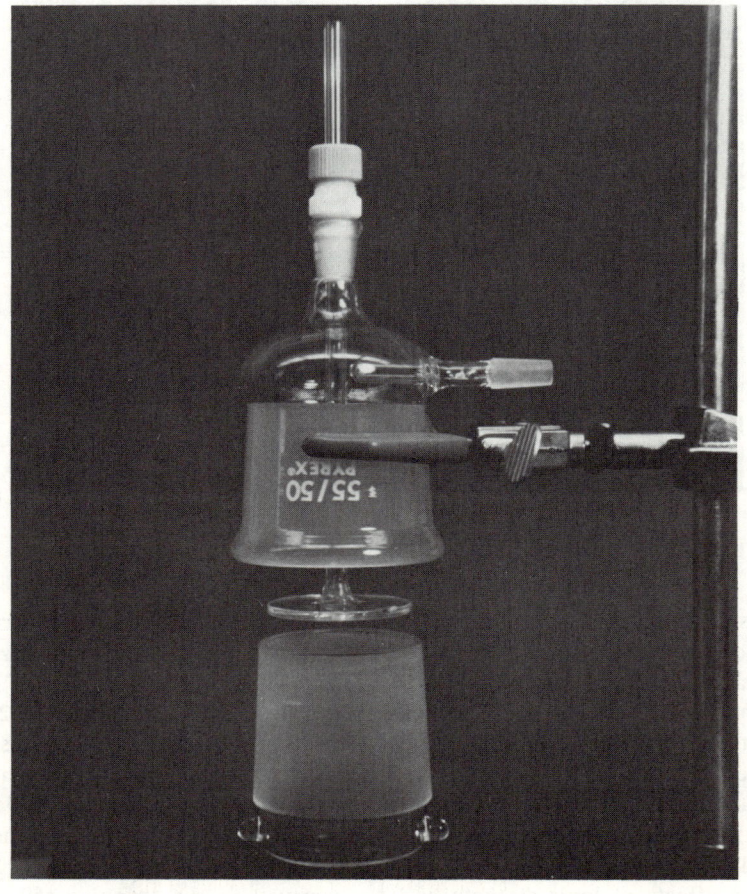

Figure 8.2 Device for headspace analysis of organics in water.

TECHNIQUES AND METHODS OF ANALYSIS

with 40 ppm each of four ketones: acetone, 2-butanone, 2-pentanone and 2-hexanone.

The effect of the addition of 0-23% (w/v) NaCl on the rate at which these ketones are released from solution is shown in Table 8.3. The percentage removed after sparging with 480 ml nitrogen increased markedly as the salt level increased.

Table 8.3 Effect of Adding NaCl on Sparging Efficiency of Four Ketones

	Percent Removed by 480 ml-Sparge			
	0[a]	12.5[a]	18.4[a]	23[a]
Acetone	12	21	29	32
2-Butanone	17	30	50	58
2-Pentanone	25	39	67	87
2-Hexanone	28	50	93	97

[a]Percent NaCl (w/v).

Similar experiments in which the temperature was increased from ambient (23°C) to 50°C also demonstrated that the stripping efficiency of this apparatus could be increased significantly at elevated temperatures. Results for 480 ml stripping gas volume are shown in Table 8.4. After a 1.5-liter sparge 96% of acetone, a highly water-soluble compound, was removed.

An attempt was made to increase stripping efficiencies by increasing both ionic strength and temperature. Unfortunately, at 50°C the frit quickly plugged with salt which eventually stopped the flow of stripping

Table 8.4 Effect of Temperature on Sparging Efficiency of Four Ketones

	Percent Removed by 480-ml Sparge	
	23°C	50°C
Acetone	12	53
2-Butanone	17	69
2-Pentanone	25	78
2-Hexanone	28	91

gas. However, it appears that this procedure can quantitatively isolate and concentrate a number of compounds with water solubilities in excess of 2% by the addition of salt or by increasing the stripping temperature.

A key feature in the use of the Bellar approach is how effectively the resin trap will collect and desorb organic chemicals.[2] In particular, to obtain high sensitivity the traps must be capable of retaining trapped materials while liters of air pass through the resin. We investigated retention for several materials with differing boiling points, polarities and water solubilities. Results are given in Table 8.5. For each compound, a known amount was introduced onto the column from a flowing air stream. From 2-19 liters of air were passed through the column. After collection, the trap was analyzed by GC techniques utilizing thermal desorption. In all cases, within the overall accuracy of the technique we used, quantitative recovery was obtained.

Table 8.5 Recovery from Collection Column

Compound	Percent Recovery[a]	Volume of Air Sampled (liters)	Collection Column
Acetic Acid	94	10	Porapak N
Benzene	92	8	Chromosorb 105
Benzyl Chloride	91	7.5	Tenax-GC
Dimethyl Acetamide	117	6	Tenax-GC
Phenol	90	10	Tenax-GC
Styrene	90	19	Tenax-GC
Triethylamine	105	2	Chromosorb 103

[a]0.3-2.5 µg on column.

MEMBRANE SEPARATIONS

Membrane separations are not new to analytical chemistry. Such membrane techniques as reverse osmosis, dialysis and ultrafiltration are used routinely. Typical applications include purification of water, desalination of protein solutions, separation of macromolecules from solvents, and microbiological analysis of water. Some of these applications, along with an excellent discussion of the theory and principles of membrane separations, are covered in a recent text.[5]

The membrane separations described in this chapter can be used to isolate ppb levels of organics for measurement. The analyte is placed on one side of a semipermeable membrane. The membrane is chosen to be permeable to the components of interest but not the undesirable matrix components. In one case the permeating molecules enter a mass spectrometer for analysis. Alternatively, a membrane dialysis configuration utilizing a convenient solvent has been developed so that other analytical techniques can be used for identification or measurement.

Membrane/Mass Spectrometry (Pervaporation)

Pervaporation through a hollow fiber probe was used by Westover *et al.* to introduce volatile contaminants from water into a mass spectrometer.[6] Our Hewlett-Packard 5982 GC/MS system uses a membrane separator to interface the chromatograph to the mass spectrometer. In this separator, as shown in Figure 8.3, the effluent from a GC column passes over the surface of a mounted, supported silicone membrane and then to the atmosphere. The organic compounds eluting from the GC permeate the membrane into the mass spectrometer. The silicone membrane excludes the helium carrier gas.

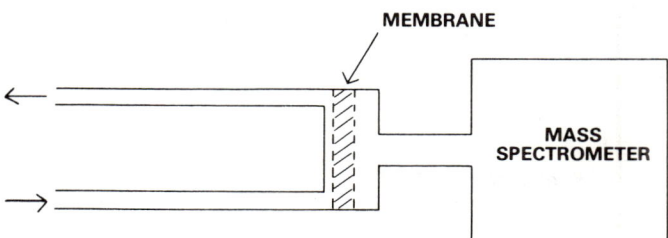

Figure 8.3 Block diagram of membrane/mass spectrometer configuration.

We used this same membrane to exclude water from the mass spectrometer while allowing organic solutes in the water to pass through the membrane into the mass spectrometer. A 2-ft length of 1/8-in. copper tubing was attached to the inlet of the separator. The other end of the tubing was raised about 12 in. above the level of the separator. This 12-in. head of water provided the driving force to cause water to flow through the separator at a rate of 4 ml/min. A 3-ml disposable syringe forced onto the copper tubing provided a reservoir for introducing water samples to the flow system.

The back surface of the membrane was open to the ion source of the mass spectrometer through the normal GC/MS connection, a 10-in. glass-lined 1/8-in. tube and an isolation valve. The temperature of the valve and tubing were kept at 250°C.

The mass spectrometer was operated in the selected ion monitoring (SIM) mode with electron impact ionization to detect permeating species. In this mode of operation the mass spectrometer monitors one to four ions. Using SIM for one or several ions characteristic of each compound to be measured, we could detect very low levels of organic species in water flowing through the system.

Several milliliters of distilled water were first passed through the system to establish a background baseline. This was followed by a 3- to 9-ml aliquot of analyte and then distilled water to reestablish the baseline. A typical experiment is shown in Figure 8.4, monitoring the mass 78 peak for a 6-ml aliquot of 0.04 ppm benzene. The benzene solution was introduced to the membrane surface at about 2.1 min of the experiment. The signal immediately increased sharply, reaching an equilibrium value in about 20 sec. The height of the signal is proportional to the concentration of benzene in the water. Introduction of distilled water at about 3.8 min caused the signal to decay to the original baseline within about 1 min.

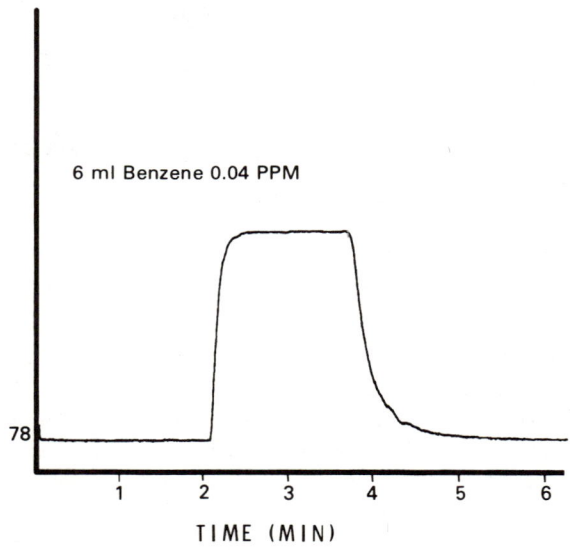

Figure 8.4 Membrane/mass spectrometric response to 0.04 ppm benzene in water.

Figure 8.5 shows the response for a 3-ml aliquot of 4 ppb benzene in water. Benzene at this low level is still easily measurable.

Similar experiments were carried out with dilute solutions of methyl salicylate, chloroform, methylene chloride, acetone, isopropanol,

TECHNIQUES AND METHODS OF ANALYSIS 121

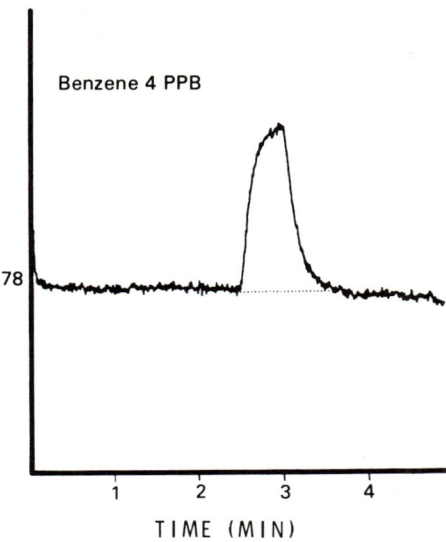

Figure 8.5 Membrane/mass spectrometric response to 4 ppb benzene in water.

and phenol in water. These compounds represent a wide range of water solubility, polarity and boiling point. Table 8.6 shows the approximate detection limits for these compounds in water and the corresponding ion used to measure each. These limits are based on actual measurements made within a factor of 10 of the detection limit and assume a signal-to-noise ratio of 2.

Table 8.6 Lower Detection Limits of Membrane/Mass Spectrometric Responses

Compound	Detection Limit (ppb)	Ion Monitored
Benzene	0.5	78
Methyl Salicylate	2	120
Chloroform	4	83
Methylene Chloride	4	49
Acetone	50	58
Isopropanol	50	45
Phenol	600	94

Figure 8.6 shows plots of mass spectrometer response in arbitrary units versus concentration in ppm for benzene, methylene chloride and isopropanol. The responses are linear over two orders of magnitude and the slopes of the three plots are the same, indicating that no complex calibration procedure is needed and quantitative calculations are straightforward.

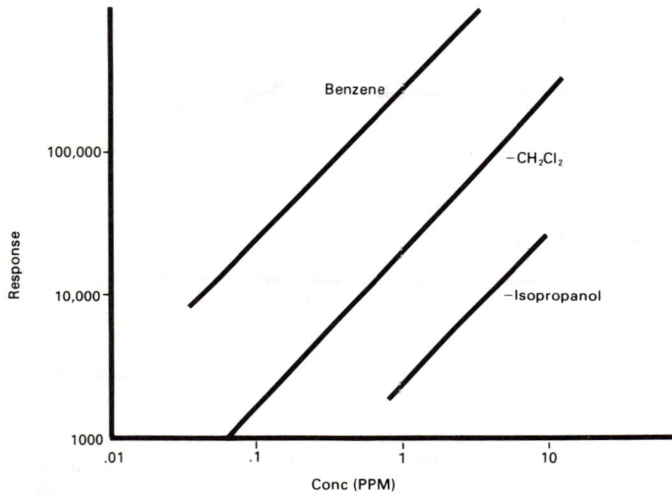

Figure 8.6   Membrane/mass spectrometric response for various concentrations of three solutes.

For most of our measurements we allowed the ion current to reach a constant value. In Figure 8.4 we saw that the equilibrium value for benzene was reached in about 20 sec. This rapid response was also observed for chloroform, acetone, isopropanol, methylene chloride and phenol. For methyl salicylate the time to reach a plateau was about 3 min. This response time is determined by the time needed to establish constant migration through the membrane and transfer line.

For the above measurements the membrane was at room temperature (25°C). To test the effect of temperature on permeation, the membrane temperature was increased to 50°C. At the higher temperature, the signal for methyl salicylate was twice as great as the signal at 25°C. However, the pressure in the ion source also increased because the water/organic selectivity for silicone membranes decreases with increasing temperature. This means that sensitivity can be increased by raising the temperature, but at the expense of decreased filament lifetime due to higher water pressure. The proper membrane temperatures must be determined by balancing the sensitivity needs against reasonable filament lifetimes.

The recent interest in determining volatile organohalides in drinking water prompted us to attempt this analysis by membrane/mass spectrometry. Figure 8.7 shows the signal observed when St. Louis County, Missouri, drinking water was analyzed for chloroform. Chloroform and the bromo-analogs all contribute to mass 83 ions, but the major contribution is from chloroform. The peak height corresponded to approximately 150 ppb chloroform.

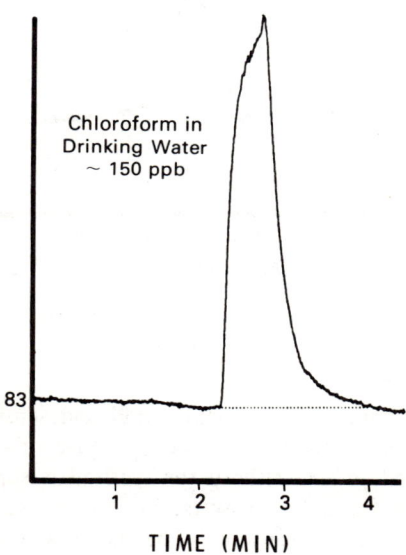

Figure 8.7   Membrane/mass spectrometric response to chloroform in St. Louis County, Missouri, drinking water.

Figure 8.8 shows analysis of the same water for bromodichloromethane. The mass 127 and 129 ions characteristic of this compound were monitored. The ratio of these signals is the same as for ions containing one chlorine and one bromine atom. This agreement is additional evidence that we measured bromodichloromethane and not an interfering compound. Since up to four characteristic ions can be continuously monitored, this method is relatively specific and free of interferences. Up to four compounds can be monitored provided interference-free ions can be found for each.

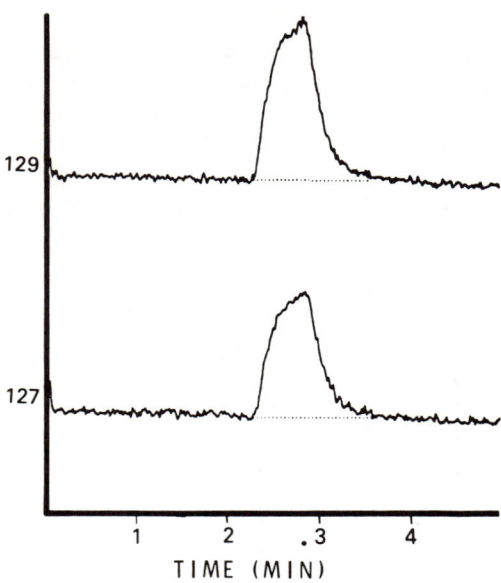

Figure 8.8 Membrane/mass spectrometric response to bromodichloromethane in drinking water.

This technique provides a convenient, rapid and sensitive method for monitoring many organic species in water. Because of its speed (minutes/sample), simplicity and ppb sensitivity the method is superior to gas sparging techniques or liquid-liquid extraction for measuring volatile organics in water for many applications. The membrane separator unit can be purchased independently and connected to other mass spectrometers.

We are currently investigating the use of a membrane probe which inserts through the mass spectrometer solids probe vacuum lock. Water to be analyzed flows inside a capillary to the probe tip, passes across a membrane surface and exits the probe through another capillary. One advantage of this approach is that components permeate directly into the ionizing region of the mass spectrometer without having to traverse a transfer line. This should permit analysis of compounds with low vapor pressures and thermally unstable compounds. The second advantage is that a membrane probe can be fabricated for any mass spectrometer equipped with a solids probe vacuum lock. If this probe is successful, we plan to use it to monitor effluent from a liquid chromatograph for aqueous solvent applications.

Dialysis into Solvent

The second general type of membrane technique uses dialysis of the components of interest from water into another solvent. The solvent volume is 1-3 orders of magnitude less than the water volume. Thus, after dialysis, the solute is not only isolated into a convenient matrix but also concentrated to facilitate analysis.

In some cases dialysis can offer unique features not obtainable with liquid-liquid extraction so that dialysis becomes the method of choice. At least four such cases can be identified:

1. Dialysis using a water miscible solvent;
2. Dialysis using membrane selectivity to prevent removal of otherwise extractable components;
3. Analysis of solutions with high emulsion potential; and
4. Automated sampling and analysis.

A device which is simple to construct yet very useful for separating organics from water is shown in Figure 8.9. This consists of a 500-ml round bottom flask with a 25-mm O-ring joint and a similar O-ring joint

Figure 8.9   Device for isolating trace organics from water by dialysis into solvent.

capped to form a 10-ml volume. The O-ring joints are clamped together with a polymer film membrane separating the chambers, using a Viton O-ring (Parker Seals 2-217, Compound V747-75). The 500-ml end is flattened slightly to permit the use of a magnetic stirrer and stirring bar to agitate the water during dialysis. Stirring was effected from the flattened end so the vortex provided maximum agitation at the membrane.

For dialysis, 500 ml of water is placed in the large volume, 10 ml of solvent in the small volume and stirring is initiated. Aliquots of solvent can be removed and analyzed periodically to monitor the extent of dialysis.

Effects of different operating parameters were investigated for three different types of solutes, represented by dibutyl phthalate, carbon tetrachloride and Aroclor* 1016, a mixture of polychlorinated biphenyls. There is significant interest in monitoring trace levels of these classes of organic compounds in water.

Unexpectedly, we found that each of the organic compounds tested at trace levels in water dialyzed at the same rate. Additional experiments showed that mass transfer of solute to the membrane surface is the rate-controlling step for dialysis at a given temperature. The effect of stirring speed on relative permeation rate is listed in Table 8.7. Even at 4400

Table 8.7 Effect of Stirring Speed

RPM	Relative Response
1000	1
2200	1.4
3300	1.8
4400	2.5

rpm, the maximum rotation of the stirrer used, mass transfer is the rate-limiting parameter. Apparently for dilute solutions (< 100 ppb) polymeric membranes have a much higher permeation potential for organic solutes than can be realized by mass transfer at normal stirring speeds.

This conclusion was verified by determining dialysis rates for different synthetic polymer membranes with different permeation characteristics. Identical dialysis rates were obtained at a constant stirring rate, using Aroclor 1016 as the solute. Polymers studied included polydimethylsiloxane (silicone) and polycarbonate-silicone copolymer MEM 213 (General Electric Co.), polyethylene and butyl rubber.

*Registered Trademark of Monsanto Company.

Permeation through a given membrane material is generally held to vary inversely with membrane thickness.[5] Polyethylene membranes 0.001-0.003 in. thick were investigated for dialysis of Aroclor 1016 in water. Measured dialysis rates were constant at ppb concentrations because of the mass transfer limitation. Varying the solvent from hexane to benzene to mineral oil also did not alter the permeation rate. Increasing temperature increases permeation rate, so this variable must be maintained constant throughout an experiment. The conclusion from these studies is that, at a given temperature, solute transfer to the membrane is the dialysis rate-limiting factor for ppb and lower concentrations.

For all subsequent studies reported in this chapter, polyethylene membranes 0.002 in. thick were used at a stirring rate of 4400 rpm, with hexane solvent.

Using these conditions solute transfer by dialysis was determined for low ppb concentrations of dibutyl phthalate, carbon tetrachloride and Aroclor 1016. Extent of dialysis was monitored by analyzing the hexane phase by electron-capture gas chromatography. For these solutes, the distribution coefficient between water and hexane greatly favored the organic phase, so solute transfer was virtually quantitative. The membrane functioned as a transfer medium permitting attainment of the final distribution.

From Figure 8.10 it can be seen that dialysis is approximately 50% complete after 3-1/2 hr and >95% complete after 24 hr. The dialysis

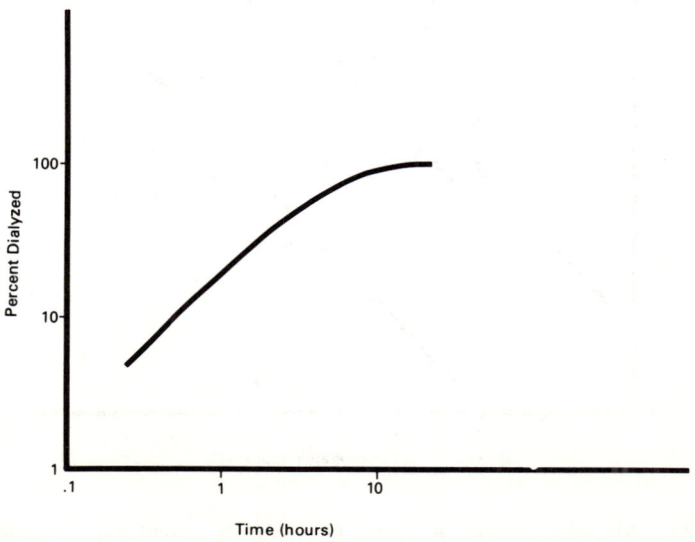

Figure 8.10 Removal of trace organic components from water by dialysis into solvent.

128 ORGANIC POLLUTANTS IN WATER

rate decreases as a function of time because the solute concentration in the aqueous phase is continuously decreasing.

As mentioned above, the dialysis cell used for these studies had a water-to-solvent volume ratio of 50 to 1 so that complete dialysis effected a 50-fold solute concentration. Other volume ratios could be used, depending on the application. For quantitation, two types of measurements were used. One approach utilized dialysis to virtual completion followed by measurement. This yielded the greatest accuracy and minimum detectability with long (24-hr) analysis times. The other approach utilized dialysis for a shorter, carefully measured interval and comparison to standards dialyzed and analyzed under equivalent conditions.

Working curves prepared for 1-hr dialysis intervals are shown in Figure 8.11. These plots are linear and parallel. The useful concentration range for this procedure is determined by the sensitivity of the measurement technique. For the measurements shown in Figure 8.11, gas chromatography with electron capture detection was used. Lower detection limits of 0.01 ppb for carbon tetrachloride and 0.1 ppb for dibutyl phthalate and Aroclor 1016 were obtained. At dibutyl phthalate concentrations greater than about 10 ppb, the working curve was nonlinear because the linear range of the detector was exceeded.

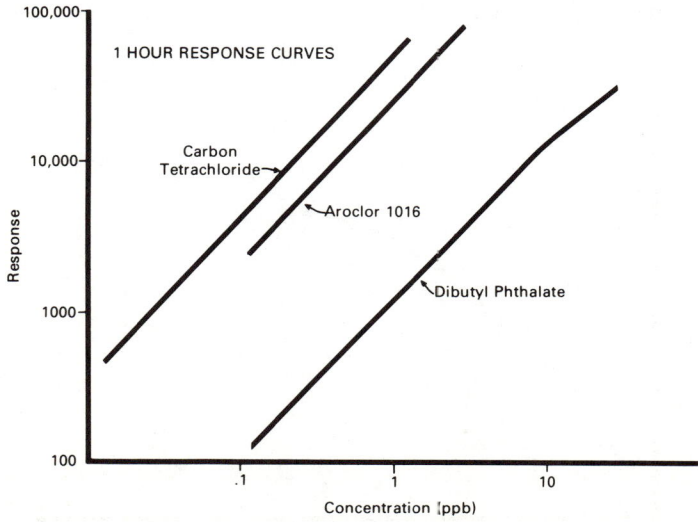

Figure 8.11 Response of electron capture GC analyzing solvent after 1-hr dialysis for various concentrations of three solutes.

Reproducibility of measurements was investigated by making replicate analyses of 7.6-ppb Aroclor 1016 solutions using 1-hr dialysis intervals. The observed values for three determinations were 6.9, 7.5 and 7.6 ppb, indicating good accuracy and precision.

We have used this technique to isolate and concentrate dissolved phthalate ester and PCB components from environmental waters. Chromatograms for a mixture of PCBs (Aroclor 1016) after dialysis into hexane or liquid-liquid extraction into hexane indicate no significant alteration of composition by preferential permeation of components as shown in Figure 8.12. The chromatograms are for qualitative comparison only and do not represent the same initial PCB concentration. A chromatogram of dialyzed phthalate esters from river water at concentrations about 5 ppb is shown in Figure 8.13.

Another type of dialysis cell for isolating organics from water is pictured in disassembled form in Figure 8.14. This device permits flow of analyte past the membrane and is useful for continuous monitoring. Continuous dialysis of a flowing stream provides a time-averaged sample. When the cell is assembled with the membrane connecting the two chambers, the analyte flows through the left chamber and organic components dialyze from the flowing stream into solvent in the right chamber. A port is provided for introduction of solvent or removal of aliquots for analysis.

This dialysis cell has been used to monitor organohalides in drinking water. Chromatograms after dialysis for 2 hr are given in Figure 8.15 for distilled water and tap water from St. Louis County, Missouri. At least eleven organohalide components were detected, including five which were identified as labeled in Figure 8.15.

This flowing dialysis principle may have considerable potential for water monitoring. Sampling and analysis can be automated for unattended operation. Since only permeated components reach the analyzer, contamination of the detector system is minimized. Large concentration factors can be achieved. Use of hollow-fiber membranes in a gas chromatographic sample loop as sample concentrators/injectors is being investigated. This may represent the ultimate in sensitivity with existing measurement techniques because the entire sample collected over a long time interval can be injected into the chromatograph. High surface area-to-internal volume ratio makes hollow fiber membranes especially attractive for this application.

One limitation of the dialysis technique should be mentioned. It is useful only for dissolved components. If the compounds of interest are adsorbed onto particulate matter or exist as a separate phase, the technique gives low results not representative of the total composition.

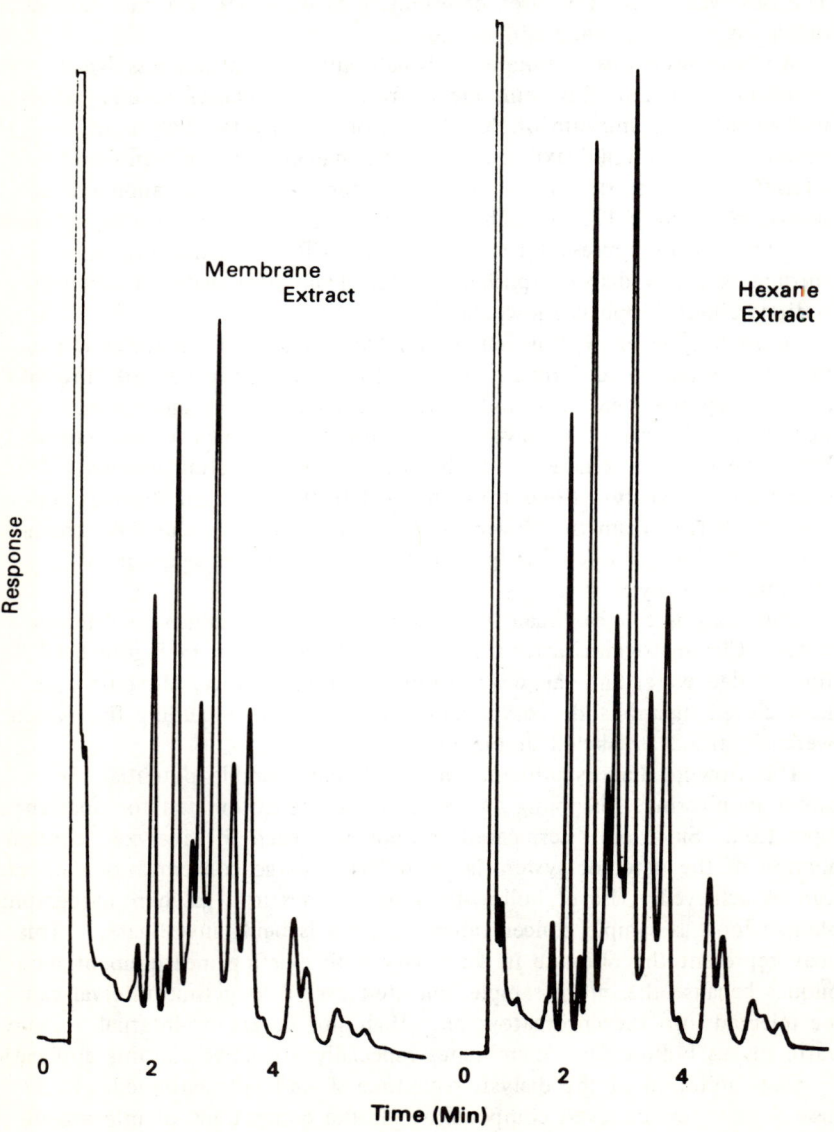

Figure 8.12 Comparison of Aroclor 1016 composition after dialysis into hexane or liquid-liquid extraction into hexane.

TECHNIQUES AND METHODS OF ANALYSIS 131

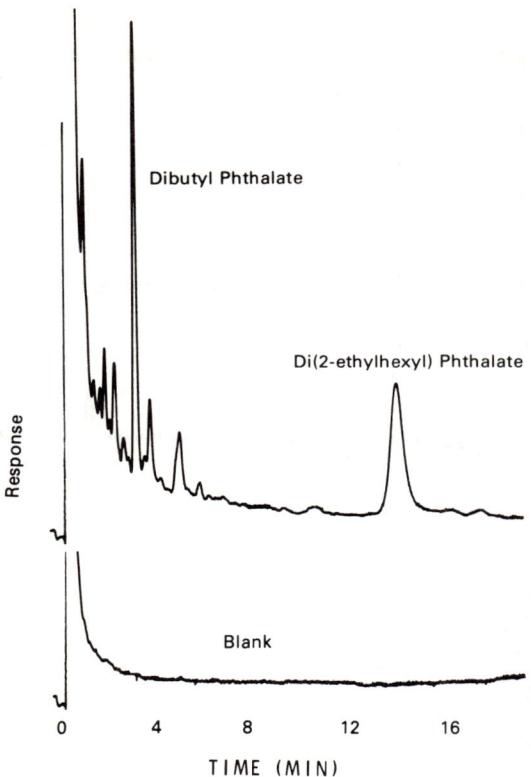

Figure 8.13 Electron capture GC determination of impurities dialyzed from river water.

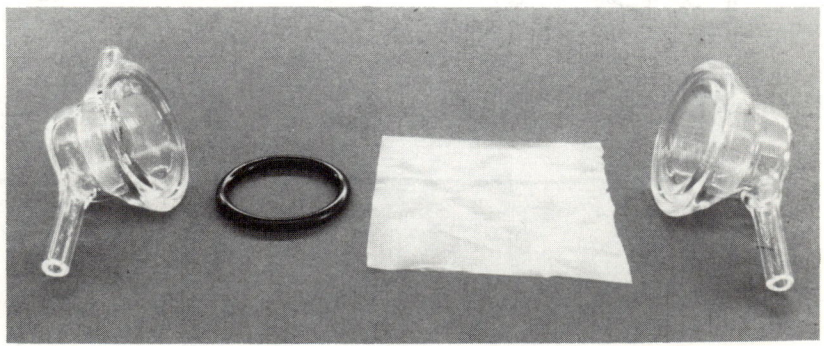

Figure 8.14 Device for continuous dialysis of flowing stream.

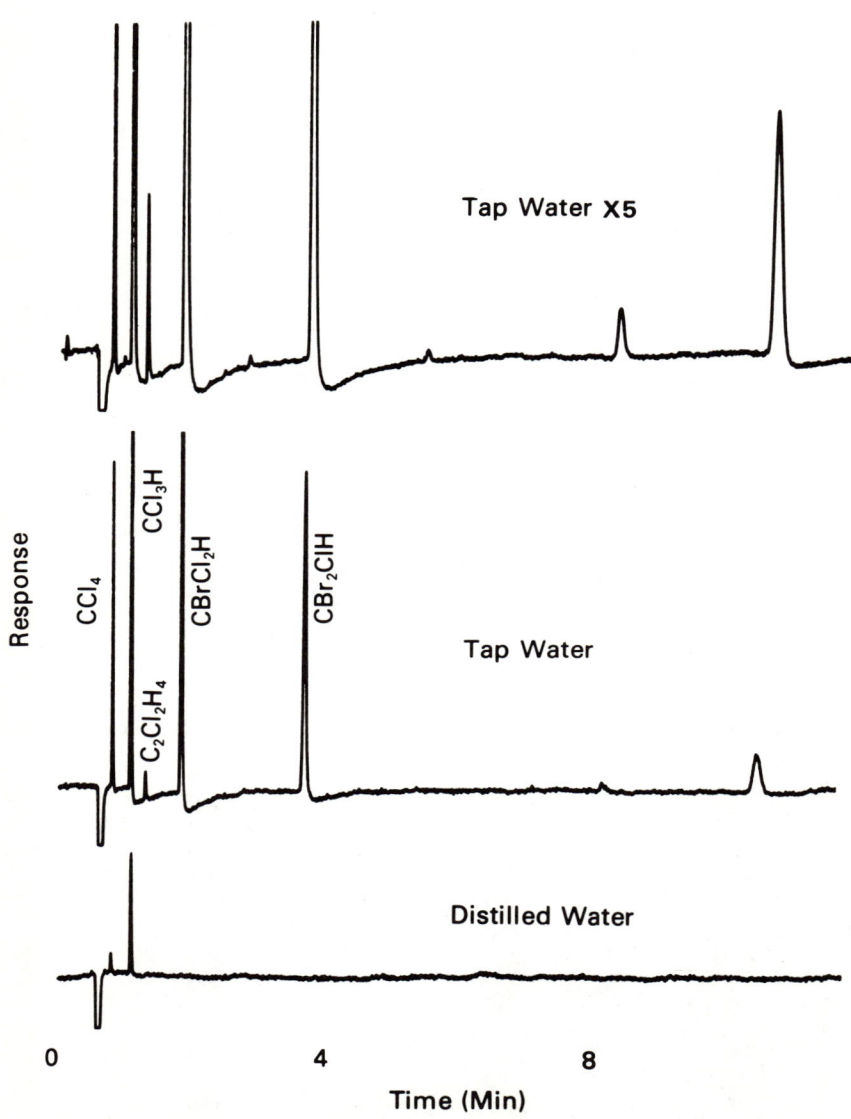

Figure 8.15 Electron capture GC determination of organohalides dialyzed from drinking water.

CONCLUSIONS

Each of these isolation procedures has unique features which can be utilized to facilitate analysis of water for trace organic components. Headspace analysis provides a means for isolating relatively volatile components without adding a solvent which would dilute the components and might mask certain compounds. Membrane/mass spectrometry combines the isolation, identification and measurement steps into one rapid and convenient technique. Dialysis into a solvent applied the full potential of membrane selectivity to time-averaged and automated sampling for measurement by any of several analytical techniques.

REFERENCES

1. Zlatkis, A., H. A. Lichtenstein, A. Tishbee, W. Bertsch, F. Shunbo and H. M. Liebich. *J. Chromatog. Sci.* **11**, 299 (1973).
2. Mieure, J. P. and M. W. Dietrich. *J. Chromatog. Sci.* **11**, 559 (1973).
3. Bellar, T. A. and J. J. Lichtenberg. *J. Amer. Water Works Assoc.* **66**, 739 (1974).
4. Bellar, T. A., J. J. Lichtenberg and R. C. Kroner. *J. Amer. Water Works Assoc.* **66**, 703 (1974).
5. Hwang, S. and K. Kammermeyer. *Membranes in Separations.* (New York: Wiley-Interscience, 1975).
6. Westover, L. B., J. C. Tou and J. H. Mark. *Anal. Chem.* **46**, 568 (1974).

CHAPTER 9

RESIN SORPTION METHODS FOR MONITORING SELECTED CONTAMINANTS IN WATER

Gregor A. Junk, John J. Richard,
James S. Fritz and Harry J. Svec

    Ames Laboratory—ERDA and
      Department of Chemistry
    Iowa State University
    Ames, Iowa

INTRODUCTION

    The evolution of analytical schemes for estimating the quality of water has resulted in rather routine methods for measuring inorganic constituents and microorganisms. Treatment processes such as ion exchange, coagulation, flocculation and disinfection are used to eliminate or at least control inorganic salts and pathogenic organisms that can cause adverse health effects. The success of these treatment processes is regularly monitored by well-documented and inexpensive procedures.
    This analytic success with inorganic material and microorganisms is contrasted with the general failure of schemes for measuring and identifying organic chemicals in water. Up to 1970, only 66 organic chemicals had been positively identified in the fresh waters of the world.[1] Of these, only 10 had been shown to be present in drinking waters. Recent investigations have raised the total number of identified chemicals in drinking water to over 300.[2] Thus considerable progress has been made in the last five years in establishing the profile of organic contaminants. This progress was achieved by using a variety of extraction and concentration techniques—charcoal adsorption, solvent extraction, freeze drying, steam distillation and, more recently, reverse osmosis,[3] inert gas stripping[4]

and resin sorption.[5] Individually, these techniques present technical problems and their combination still doesn't give a complete profile due to some remaining separation and detection difficulties.

While the methodology for the complete characterization of organic contaminants is being developed, it is advisable to also develop inexpensive and accurate schemes for the rapid measurement of selected contaminants. Not only the profile, but also the amounts and distributions must be established before the risk[6] associated with the ingestion of contaminated water can be adequately assessed.

An example of the monitoring of selected contaminants is the U.S. Environmental Protection Agency (EPA) reconnaissance survey[7] of 80 water supplies for the presence of six volatile C_1 and C_2 halocarbons. The amount and distribution data from this survey can now be used to formulate sound toxicological and epidemiological protocols for estimating the health hazards.

The purpose of this report is to describe a simple and accurate method for measuring small amounts of other less volatile and lipophilic contaminants in water. The method lends itself to large-scale sampling, both in numbers and sizes of samples being analyzed. In each case the lipophilic contaminants are extracted from the water sample by sorption on a macroreticular resin XAD-2 (Rohm and Haas, Philadelphia, Pa.).

This XAD-2 resin has the characteristics necessary for high sorption capacity. Some properties of the resin are listed in Table 9.1. On a

Table 9.1 XAD-2 Resin Characteristics

Functionality	styrene-divinylbenzene
Surface Area	300 m^2/g
Average Pore Size	90 Å
Shape	smooth spheres
Diameter	40-60 mesh (~0.5-0.25 mm)
Color	white

crude polarity scale, the resin is classified as having low polarity. It has an effective surface area for sorption of 300 m^2/g. The average pore size is 90 Å. The shape, size and opaque white color of the resin is shown in Figure 9.1.

The efficiency of XAD-2 in extracting a wide range of organic chemicals from water has been reported previously,[5] and a summary of those test results for the entire analytical scheme is listed in Table 9.2. Thirteen different chemical classes were tested with four to fifteen chemicals

TECHNIQUES AND METHODS OF ANALYSIS 137

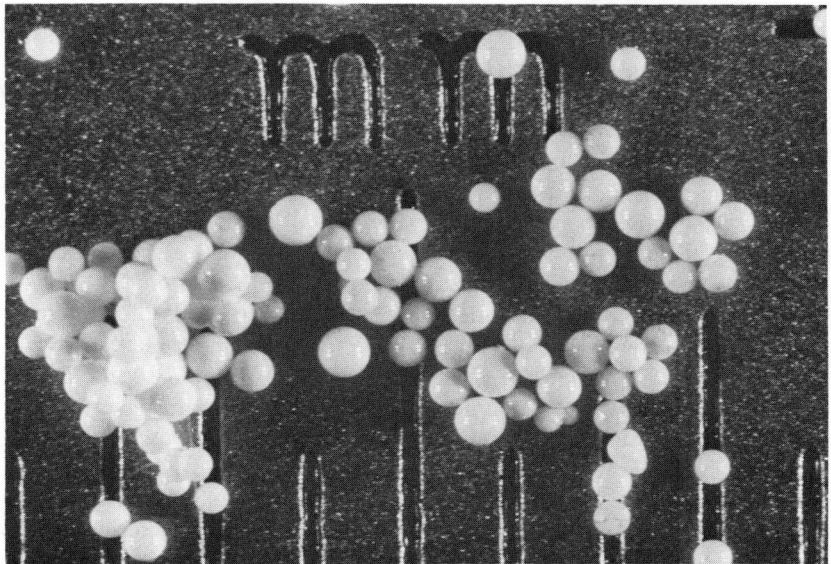

Figure 9.1 Photograph of XAD-2 beads on a millimeter-scaled ruler.

Table 9.2 Recovery Efficiency for the XAD-2 Sorption Method

Compound Type	No. Tested	Average % Recovery
Alcohols	8	94
Aldehydes + Ketones	7	95
Esters	15	93
Acids[a]	5	101
Phenols[a]	6	89
Ethers	5	90
Halogen Compounds	10	87
Polynuclear Aromatics	8	89
Alkylbenzenes	4	90
N,S Compounds	10	89
Pesticides + Herbicides[b]	5	90
Total =	83	Wt. Ave. = 91

[a]Test water acidified with 5 ml conc. HCl per liter.
[b]Test at 20 ng/l level; all others at 10 - 100 µg/l.

per class. The weighted average of the recoveries from water solutions spiked with the model chemicals was 91% for the 83 test chemicals.

The recovery efficiency for low levels of lipophilic pesticides prompted the development of a convenient scheme for measuring pesticides of interest in the waters of Iowa. Surface, subsurface and finished (processed) waters were assayed for trace levels of atrazine, DDE and dieldrin. Atrazine was chosen because it is a widely used herbicide which had been detected in runoff water from small test plots.[8-14] Dieldrin and DDE were chosen because they had previously been reported[15-18] to be present in several rivers. Other pesticides and chemicals of interest were included later in the survey to show how the simple methodology might be extended to include more chemicals, but no extensive measurements indicative of valid amounts and distributions were made.

The list of organic chemicals for which simple procedures have been shown to be applicable, with the extensiveness of their study, is shown in Table 9.3. This list includes only the chemicals studied and is not suggestive of the potential of the methodology. By employing proper separation and detection techniques the method is applicable to most lipophilic contaminants.

Table 9.3 Contaminants Discussed

Contaminants	How Studied?
Atrazine	extensive
DDE	extensive
Dieldrin	extensive
Alachlor (Lasso)	limited
Propachlor (Ramrod)	limited
Sutan (Carbamate)	very limited
2,4-D and Esters	very limited
2,4,5-T and Esters	very limited
Tetrachlorodibenzodioxin	tested only
Polynuclear Aromatics	tested only

EXPERIMENTAL

The general resin sorption scheme for measuring organic compounds in water is shown in the flow chart in Figure 9.2. The applications of this scheme to test waters,[5] municipal drinking waters[19-21] and water contaminated by plastic tubes[22] demonstrate the general utility of the method. A brief description of the total scheme is presented here as background for explaining the simple modification.

TECHNIQUES AND METHODS OF ANALYSIS   139

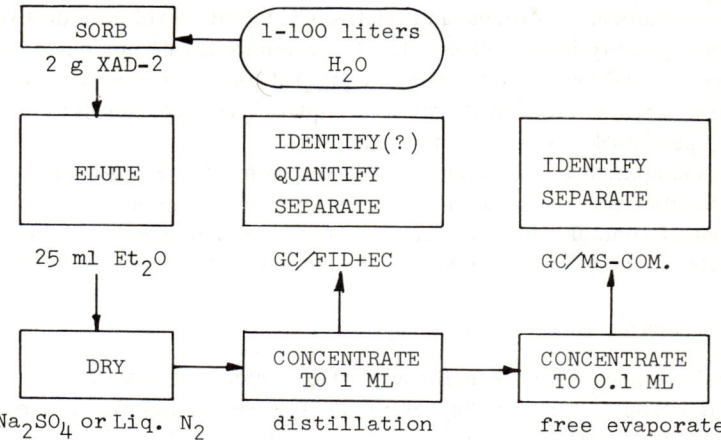

Figure 9.2 Flow chart of general resin sorption scheme.

The first step in the resin sorption scheme is passage of from 1 to 100 liters of water through a glass column containing 2 g of 40-60 mesh XAD-2. The sorbed organic compounds are then eluted from the XAD-2 with 25 ml of diethyl ether (Et_2O). Residual water is removed and most of the ether is distilled through a three-ball Snyder column. The extracted mixture of organic compounds is then separated by temperature-programmed gas chromatography using packed columns. The output data of the gas chromatographic detector are used to calculate the amounts of each of the separated components. The components are identified by subjecting an aliquot of the concentrated mixture extract to combination gas chromatograph-mass spectrometer-computer (GC/MS-Com) analysis.

This overall scheme is useful for measuring the profile of gas chromatographable constituents in water but is time-consuming, somewhat complicated and costly. Yet up to the separation, identification and quantitation it is simple. No complicated techniques are employed and only common, inexpensive laboratory equipment is used. Even the total scheme is simplified once the gas chromatographic separation conditions have been established and the identifications have been confirmed using the GC/MS-Com system.

Identifications for subsequent samples can be made using authentic standards, isothermal conditions, and relative GC retention volumes on two different polarity liquid phases. The GC columns can be purchased for ~ $40. A $6000 gas chromatograph equipped for flame ionization and electron capture detection is entirely adequate. Trained, but not highly-skilled personnel can do the entire analyses.

This simplified and inexpensive resin sorption scheme was employed to monitor the selected contaminants reported in this chapter. In effect, the number of contaminants measured in a small water sample has been greatly restricted. The GC/MS-Com system is employed only for periodic checks.

Grab Samples

Grab samples were taken for all surface waters using clean amber 4-liter solvent bottles. After settling overnight, the clear water was decanted into the 5-liter reservoir shown at the top of Figure 9.3. The stopcock (E) was adjusted to deliver a flow rate of 25-50 ml/min. When the water

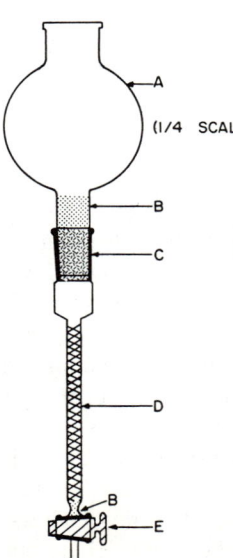

Figure 9.3 Scale drawing of grab sample extraction device. (A) 5-liter reservoir scaled by ~1/4; (B) glass wool plugs; (C) 24/40 ground glass joint with teflon sleeve; (D) 8x140-mm glass tube packed with ~2 g, 40-60 mesh XAD-2 resin; (E) teflon stopcock.

TECHNIQUES AND METHODS OF ANALYSIS 141

level reached the upper glass wool plug, the sediment from the bottle was transferred to the reservoir using several rinses with organic-free water. After all the water had passed through the resin, the stopcock was closed, the reservoir removed and 15 ml of diethyl ether was added to the resin. About 5 ml was allowed to flow through the resin and collected in a 60-ml separatory funnel. The stopcock was then closed for 15-30 min after which the remaining 10 ml of ether was collected in the separatory funnel. This elution procedure was repeated with a second 15-ml portion of ether which was combined with the first. The water layer was drained from the separatory funnel, and the final traces of water were removed from the eluate by adding 10-15 ml of petroleum ether and 2-3 g anhydrous sodium sulfate. The mixture was shaken for $\sim$ 30 sec and the liquid extract was transferred quantitatively to a concentration flask. The extract was concentrated to 1 ml using the microdistillation procedures described previously.[5] A 1- to 5-μl aliquot of the concentrated sample was gas chromatographed without further treatment.

Composite Samples

The apparatus shown in Figure 9.4 was used for composite sampling over a $\sim$ 24 hr period. The standard garden hose coupling was attached to a suitable faucet and the water flow adjusted to deliver $\sim$ 150 ml/min.

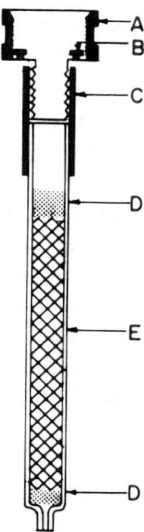

Figure 9.4 Scale drawing of composite sample extraction device. (A) standard garden hose coupling; (B) teflon washer; (C) 12.7-mm i.d. teflon tubing; (D) glass wool plugs; (E) 12.7-mm o.d. x 9 cm long glass tube packed with $\sim$2 g, 40-60 mesh XAD-2 resin.

After 24 hr the XAD-2 column was removed from the coupling and teflon sleeves were used to connect a 25-ml reservoir and a teflon stopcock to appropriate ends of the column. The column was then eluted with diethyl ether and the eluate was treated as described above for the grab samples. Either sampling procedure may be employed depending on whether one is interested in instantaneous (grab) or average (composite) concentrations over an extended time period.

Gas Chromatography

All chromatographic separations were made using 2- to 3-m packed columns. Electron capture was used for the measurement of all pesticides and 2,3,7,8-tetrachlorodibenzo-p-dioxin (TCDD) which were separated by partitioning on 5% OV-210 or 10% DC-200, or 1.5% OV-17/1.95% QF-1 or 4% SE-30/6% OV-210 liquid phases. Flame ionization was used for the measurement of all other compounds which were separated on 3% dexsil or 5% OV-17 liquid phases.

RESULTS AND DISCUSSION

Pesticides (Surface Waters)

Low levels of some pesticides were detected in the water from several Iowa cities in 1972 when a general survey of organic pollutants in finished waters was initiated. Subsequently, the simplified methodology described in the introduction was applied to a variety of surface, subsurface and finished waters during the growing seasons of 1974 and 1975. Additional more detailed data for the 1974 growing season are available in a previous publication.[23]

For rivers located in an agricultural area, the 1974 results for the South Skunk River of Ames, Iowa are representative. These data are shown in Table 9.4. As expected, the amounts varied seasonally as well as in dramatic short-term fluctuations due mainly to climatic conditions. Seasonally, the amounts decrease from spring through the summer and into the fall. In early spring dramatic increases were observed following heavy rains that caused considerable soil erosion. This pattern for the South Skunk River drainage area of 315 sq mi is the same as that observed for small test plots[9-11] and for simulated rainfall on very small fallow plots[8,13] which had been treated with pesticides. Although extensive data were not accumulated for all Iowa watersheds, spot tests of other rivers suggested that the same pattern was operative as observed for the Skunk River.

Table 9.4   Pesticide Amounts in the South Skunk River (1974)

Sampling Date	Concentration, ng/l		
	Atrazine	DDE	Dieldrin
6/9	12000	1820	33
6/11	3900	475	6
6/16	420	45	3
6/19	2300	688	36
6/22	2000	688	76
6/27	1575	87	10
7/2	540	16	13
7/9	230	10	10
7/16	260	14	6
7/22 to 10/13	steady at ~200	steady at ~5	steady at ~4

These data are used to establish the contamination cycle for the rivers. It begins in early spring when the pesticides are applied to prepared soil containing little or no vegetation. Heavy spring rains cause appreciable erosion with runoff of dissolved pesticides and pesticides associated with soil particles. A quasi-equilibrium is established in the rivers between the sorbed and dissolved pesticides. As vegetation increases and the amount and intensity of the rainfall decreases, the amounts of pesticides in the water decrease. This decrease is due to a combination of reduced supply, less runoff and much less erosion. In the fall, the land is plowed. In the spring, it is again prepared for planting and a new supply of pesticides is applied. The cycle repeats. The source of the pesticide contamination is then agricultural runoff. The amounts vary with rainfall events and the season of the year.

Monitor information similar to that for the Skunk River are tabulated in Table 9.5 for a smaller stream and in Table 9.6 for an agricultural drainage ditch. Similar concentration patterns exist.

A summary of the survey data for some other surface waters is given in Table 9.7. The individual amounts are dependent on the season and rainfall events alluded to earlier. The Des Moines and Raccoon Rivers are the raw water sources for several cities. The Red Rock and Rathbun impoundments are potential raw water sources expected to be tapped in future years.

Some selected values for other surface waters are given in Table 9.8. No attempt was made to sample these waters under identical climatic

Table 9.5   Pesticide Amounts in Indian Creek (1974)

Sampling Date	Concentration, ng/l		
	Atrazine	DDE	Dieldrin
6/9	42000	3920	25
6/11	3400	600	7
6/16	870	80	4
6/19	2000	910	30
6/22	2400	435	71
6/27	1075	100	17
7/2	510	6	9
7/9 to 8/25	steady at ∼250	steady at ∼5	steady at ∼6

Table 9.6   Pesticide Amounts in Drainage Ditch near Fernald, Iowa (1974)

Sampling Date	Concentration, ng/l		
	Atrazine	DDE	Dieldrin
6/9	9000	1150	20
6/11	1800	244	10
6/16	440	76	3
6/19	1500	407	23
6/22	700	200	72
6/27	625	235	8
7/2	190	32	11
7/9 to 8/25	steady at ∼230	steady at ∼14	steady at ∼8

and time conditions. Therefore the amounts are *not* indicative of the relative purities of these waters. Rather these results are included here to show the universal nature of the pesticide contamination of surface waters. The New Orleans result suggests that the contamination is undoubtedly not peculiar to Iowa.

During the 1975 growing season, the pesticide monitoring program for the South Skunk River was repeated and included a few refinements.

Table 9.7 Pesticide Amounts in Surface Waters Used as Raw Water Sources of Drinking Water

Water Type Location	Sampling Frequency	Concentration, ng/l		
		Atrazine (Range)	DDE (Range)	Dieldrin (Range)
Red Rock Reservoir Central Iowa	12	921 (100-1900)	212 (8-350)	18 (5-36)
Rathbun Reservoir Southern Iowa	10	1285 (165-3750)	92 (7-325)	3 (2-6)
Raccoon River Van Meter, Iowa	7	814 (120-3300)	59 (2-250)	7 (1-12)
Des Moines River Boone, Iowa	7	211 (50-800)	68 (1-248)	7 (2-14)

Table 9.8 Pesticide Amounts in Other Surface Waters (1974)

Location	Date Sampled	Concentration, ng/l		
		Atrazine	DDE	Dieldrin
Cedar River Cedar Rapids, Iowa	6/24	6350	480	42
Iowa River Iowa City, Iowa	6/24	3000	350	22
Skunk River Oskaloosa, Iowa	7/29	50	7	1
Mississippi River McGregor, Iowa	8/12	100	2	< 1
Gremore Lake McGregor, Iowa	8/12	190	4	< 1
Mississippi River Davenport, Iowa	7/30	331	< 0.5	< 0.5
Mississippi River New Orleans, La.	7/30	1200	48	7
Missouri River Council Bluffs, Iowa	8/15	80	0.5	0.6
Farm Pond southern Iowa	7/1	900	20	46
Des Moines River Ottumwa, Iowa	7/29	368	74	3

146   ORGANIC POLLUTANTS IN WATER

The 1975 results verified all the conclusions drawn from the data collected during the 1974 growing season. The refinements involved duplicate analyses on critically-filtered and simple glass wool-filtered water as well as the inclusion of the other pesticides, alachlor and propachlor.

The duplicate analyses proved that the very simple settling followed by glass wool filtration employed in the 1974 study was accurate for measuring the amounts of dissolved pesticides. These results are shown in Table 9.9 where the average amounts of atrazine, dieldrin, DDE, alachlor and propachlor are listed for 18 to 23 duplicate samples taken during the months of May, June and July of 1975. Comparison of

Table 9.9   Comparison of 10-Micron Sintered Glass Filtered Water to Simple Glass Wool Filtered Water (1975)

Pesticide	Average Concentration, ng/l		No. Comparisons
	Glass Wool	Sintered Glass	
Atrazine	1634	1648	19
Dieldrin	36	32	23
DDE	125	111	21
Alachlor	688	679	18
Propachlor	80	80	19

individual results for the two filtering procedures always agreed within 20%. The sediment load for the comparisons varied from 1 to 0.1 g/l. Sutan was also detected during the 1975 monitoring program but was not repetitiously analyzed. Obviously the simple scheme of filtration through glass wool as shown in Figure 9.3 is adequate for measuring the amounts of dissolved pesticides.

Pesticides (Ground Water)

A logical question is whether the selected pesticides, which are widely distributed in surface waters, have percolated to the ground water supplies. Part of the data in Table 9.10 answers this question. When the underground aquifer is within the alluvial plain of a contaminated river, the pesticides are present in the ground water. The natural percolation which exists in these cases does little to remove the contamination. When the well systems are outside the alluvial plain, the pesticides are generally not present or at a level which approaches the detection limit of 0.5 ng/l. This conclusion is based on the data shown on the last line of the table.

Table 9.10 Pesticides in Drinking Waters Derived from Various Raw Water Sources (1974)

Location	Type Raw Water	Concentration, ng/l		
		Atrazine	DDE	Dieldrin
Iowa City, Iowa	River	200	3	5
Des Moines, Iowa	River	60	2	1
Davenport, Iowa	River	405	5	2
New Orleans, La.	River	5000	–	50
Cedar Rapids, Iowa	Wells[a]	483	28	–
Marshalltown, Iowa	Wells[a]	60	–	–
Oskaloosa, Iowa	Wells[a]	14	< 0.5	< 0.5
Waterloo, Iowa	Wells[a]	4	< 0.5	< 0.5
Four cities	Wells[b]	–	< 0.5	–

[a]Wells located within the alluvial plain of contaminated rivers.
[b]Wells located outside the alluvial plain. Representative values for four cities which were tested—Sioux City, Fort Dodge, Iowa Falls, and Ames.

Pesticides (Drinking Waters)

An important consideration is whether pesticides applied to agricultural land eventually end up in drinking waters either unchanged or as some stable degradation product. Some typical data are shown in Table 9.10. When contaminated surface waters (rivers) are the raw water sources, the contamination is also present in the drinking waters. Highest amounts will be present in the spring and pesticides will always be present at some level above the detection limit of the analytical method. When certain wells, located within the alluvial plain of a river, are the raw water source, the pesticides are again present in the drinking waters. Only when the raw waters are from wells outside the alluvial plain of a stream will the drinking waters be free of pesticide contamination.

Water Treatment Processes

The effectiveness of current water treatment procedures in changing the pesticide contamination is reflected by the data in Table 9.11. City A draws its raw water from a river and from an infiltration gallery parallel to the river. About half comes from each raw water source. The water from these two sources was sampled and analyzed on the

148 ORGANIC POLLUTANTS IN WATER

Table 9.11 Pesticides in Raw Water Versus Finished Water (1974)

Water Type	Concentration, ng/l		
	Atrazine	DDE	Dieldrin
Raw River A[a]	25	6	2
Raw Gallery A[a]	82	5	0.5
Prefilter A	47	4	0.5
Finished A	29	2	0.4
Raw River[b]	331	0.5	0.5
Finished B[b]	469	2	1

[a]Pre-filter water is ~50/50 mixture of raw river and adjacent infiltration gallery water.
[b]Water treatment included activated carbon filtration beds.

same day. A sample of the mixed raw water after it had passed through the water plant prefilter was also taken and analyzed. The results of these analyses show that the prefilter was ineffective in reducing the pesticide contamination. The next most convenient sampling point at water plant A was the finished water as it went out to the consumers. At best, the complete treatment process reduced the amount of pesticides by less than 50%. Unknown mixing variables, which exist at all large water treatment plants, can bias results such as those given in Table 9.11. However, subsequent repeat analyses of the river water and the finished water at City A, suggested that the treatment process was much less than 50% effective in removing the pesticides.

The only convenient sampling points at City B were the raw river water and the completed finished water. It is evident that no significant improvement occurred during processing, even though City B had a modern water treatment plant which included an activated carbon filtration bed. Although this filtration bed was still effective for odor and taste control, it apparently was adding rather than removing the pesticides atrazine, DDE and dieldrin to the finished water.

The explanation for the increased pesticide concentration in the finished water is probably related to the late July sampling time. In the spring the filtration beds could have become partially saturated with respect to pesticides and other organic compounds. Entrapment of soil particles containing sorbed pesticides could also occur. By late July, these previously-sorbed pesticides were being rinsed from the filtration beds by the raw water.

A fairly positive conclusion can be drawn from these comparisons of City A and City B. Current treatment processes are ineffective in removing small amounts of many dissolved organic chemicals. Even charcoal filtration is ineffective after a relatively short time.

CHLOROPHENOXYACETIC ACIDS AND ESTERS

Three problems exist in the analysis of water for chlorophenoxyacetic herbicides used for weed and brush control. First, both free acids [2,4-dichlorophenoxyacetic acid (2,4-D) and 2,4,5-trichlorophenoxyacetic acid (2,4,5-T)] and their esters are used. Second, the esters might hydrolyze under ambient water conditions causing erroneous conclusions if the analysis measures only the esters. Third, the separations and standardizations are difficult because a variety of ester formulations are being used. Preliminary results suggest that a modified resin sorption procedure is a useful and convenient approach which circumvents these difficulties.

The acidity of the water to be tested is adjusted to ~pH 2 using conc. H_2SO_4. At this pH both the esters and acids are effectively sorbed on the XAD-2 resin. The sorbed materials are then eluted with 40 ml of diethyl ether. The ether eluate is split into two equal fractions. The first fraction is evaporated to dryness and the residue is esterified using diazomethane. Gas chromatographic separation of an aliquot of the esterification products and electron capture detection yields the amount of free acids present in the water. The residue after evaporation of the second fraction is hydrolyzed prior to the diazomethane esterification. Analysis of an aliquot from this reaction gives the total amount of chlorophenoxyacetic herbicides measured as the methyl esters. No separation of all the probable esters of 2,4-D and 2,4,5-T is required. Only sufficient GC resolution to separate the methyl esters of 2,4-D and 2,4,5-T is necessary. Additionally, only two standard solutions are necessary for quantitation.

The above procedure was employed for water taken from the Skunk River at a time when attempts to measure the esters directly were fruitless. The results are shown in Table 9.12. It appears as if the simple methodology described here is convenient for measuring the total amount of chlorophenoxyacetic herbicide contamination in water. The free acid results are distorted if the methyl esters of 2,4-D and 2,4,5-T are present in the water but these esters are rarely used. If suspected, the distortion may be checked by direct analysis of the water omitting the esterification step.

Table 9.12 2,4-D, 2,4,5-T and Their Esters in South Skunk River Water (1975)

Pesticide	ng/l
2,4,D (acid)	249
2,4-D esters[a]	< 5
2,4,5-T (acid)	152
2,4,5-T esters[a]	48

[a]Measured as methyl esters.

TCDD and PNA

The simplified resin sorption method was tested for applicability to 2,3,7,8-tetrachloro-p-dibenzodioxin (TCDD) and polynuclear aromatics (PNA). Organic-free water was spiked with these materials and carried through the sorption, elution and detection scheme. Recovery results are given in Table 9.13 where average percent recoveries are listed for the recorded concentration ranges. Because of the very effective sorption of these compounds on XAD-2, large volumes of water may be sampled to arrive at detection limits of less than 1 ng/l. Refinements in the methodology, particularly the separation and detection schemes, and applications to environmental waters are currently in progress.

Table 9.13 Efficiency of Resin Sorption for TCDD and PNA

Compound	Concentration Range, ng/l	Average % Recovery
TCDD	40-100	92
PNA[a]	20-500	89

[a]Acenaphthylene, phenanthrene, fluoranthene, pyrene, anthracene, chrysene, perylene, benzopyrenes.

Aquatic Life Samples

Three 40-cm channel catfish were taken from the Skunk River and analyzed for selected pesticides. The fish were skinned and only the consumable portion representing the muscle tissue along the spine was homogenized, extracted and analyzed. The results of these tissue analyses and the concentrations of the pesticides in the water on the same day when the fish were taken from the river are given in Table 9.14. A dramatic biomagnification of $\sim 10^4$ is evident. Even though these fish matured in water which at times contained significantly higher concentrations than those listed in the table, the water concentration was always several orders of magnitude below that observed in the fish.

Table 9.14 Selected Pesticides in Catfish and Water from Skunk River (August 4, 1975)

Sample	Concentration, ng/1000 g				
	DDE	Dieldrin	DDD	DDT	HCE
Water	12	5	< 1	< 1	< 1
Catfish	45×10^4	35×10^4	12×10^4	34×10^4	21×10^3

Comparative Exposures

Recent publicity about organic contamination of drinking water has resulted in serious concern over the associated health effects. Results presented earlier in this chapter show that many natural waters are contaminated and that current treatment processes are ineffective in removing some organic chemicals of known or suspected toxicity. Should the technology for processing water be improved to remove all traces of objectionable organic matter, little progress will have been made in reducing man's exposure to certain chemicals.

For example, the highest level observed for dieldrin in drinking water was 14 ng/l. At times the amount of dieldrin may be somewhat higher than this, but 14 ng/l is considered an upper limit for average daily exposure based on existing data. At an ingestion rate of two liters per day of water contaminated at this level, a man would be exposed to 1×10^{-5} g of dieldrin in one year.

Biomagnification effects account for a far greater probable exposure as suggested by the fish analyses given in Table 9.14. Eating a 0.5-kg catfish that has matured in contaminated water would result in a dieldrin ingestion of 1.7×10^{-4} g or an amount equivalent to that ingested from drinking water at the rate of two liters per day for over 17 years.

Another example of comparative ingestions is the probable exposure to dieldrin from drinking milk. The regulatory limit for dieldrin in milk is 3×10^{-4} g/l. Thus a man can ingest from consuming but a single liter of milk, an amount of dieldrin equivalent to that from drinking water for over 30 years. Similar examples of other exposures and other organic chemicals could be cited to place the concern about contaminated drinking water into a more valid perspective.

CONCLUSIONS

Specific conclusions from the results of the use of simplified resin sorption for the monitoring of selected contaminants are dispersed throughout this chapter. Only a few general conclusions are emphasized here.

—Simplified resin sorption is a convenient and inexpensive procedure for the monitoring of lipophilic contaminants in order to establish the amounts and distribution of selected chemicals of known or suspected toxicity. The procedure is adaptable to rapid and inexpensive routine monitoring of surface, subsurface and drinking waters.

—Pesticide contamination of water is widespread, although the amounts ingested by drinking contaminated waters are negligible compared to other probable ingestions, especially from aquatic foods and dairy products.

—The contamination of surface waters is caused by agricultural runoff and is most severe after pesticide applications in early spring because heavy rains cause appreciable erosion.

—Significant advances in water treatment technology may be necessary if removal of ng/l amounts of certain potentially toxic chemicals is eventually judged desirable.

ACKNOWLEDGMENTS

Appreciation is expressed here to several people, Ann Konermann, Lewis Naylor, Larry Wing, Nancy Nehring, Sherri Stanley, Ray Vick and Mike Avery for their contributions in sample collection, testing and instrumental analyses. The success of this research was dependent on the excellent cooperation of water plant supervisors and superintendents. Support by the National Science Foundation under contract numbers GP33526X and GP42252 is gratefully acknowledged.

REFERENCES

1. U.S. Environmental Protection Agency. "Water Quality Criteria Data Book, Vol. 1—Organic Chemical Pollution of Freshwater," Water Quality Office, Project No. 18010DPV (December 1970).
2. Junk, G. A. and S. E. Stanley. "Organics in Drinking Water. Part I. Listing of Identified Chemicals," IS-3671, NTIS, P.O. Box 1553, Springfield, VA 22161 (July 1975).

TECHNIQUES AND METHODS OF ANALYSIS   153

3. Kremen, S. S. "Reverse Osmosis: An Answer for Clean Water," *Environ. Sci. Technol.* **9**(4), 314 (1975).
4. Bellar, T. A. and J. J. Lichtenberg. "Determining Volatile Organics at Microgram-per-Liter Levels by Gas Chromatography," *J. Amer. Water Works Assoc.* **66**, 739 (1974).
5. Junk, G. A. *et al.* "Use of Macroreticular Resins in the Analysis of Water for Organic Contaminants," *J. Chromatog.* **99**, 745 (1974).
6. U.S. Environmental Protection Agency. "Assessment of Health Risk from Organics in Drinking Water," Ad Hoc Committee Report to HMAC, Science Advisory Board (April 1975).
7. Train, R. E. "Environmental News," OPA (A-107), press release by EPA Administrator, Washington, D.C. (April 1975).
8. Bailey, G. W., *et al.* "Herbicide Runoff from Four Coastal Plain Soil Types," EPA Report No. 660/2-74-017 (1974).
9. Hall, J. K. *et al.* "Losses of Atrazine in Runoff Water and Soil Sediment," *J. Environ. Quality* **1**, 172 (1972).
10. Hall, J. K. "Erosional Losses of s-Triazine Herbicide," *J. Environ. Quality* **3**, 174 (1974).
11. Ritter, W. F. *et al.* "Atrazine, Propachlor and Diazinon Residues on Small Agricultural Watersheds—Runoff Losses, Persistence, and Movement," *Environ. Sci. Technol.* **8**, 38 (1967).
12. Trichell, D. W. *et al.* "Loss of Herbicides in Runoff Water," *Weed Sci.* **16**, 447 (1968).
13. White, A. W. *et al.* "Atrazine Losses from Fallow Land Caused by Runoff and Erosion," *Environ. Sci. Technol.* **1**, 740 (1967).
14. Ritter, W. F. "Environmental Factors Affecting the Movement of Atrazine, Propachlor and Diazinon in Ida Silt Loam," Unpublished Ph.D. Thesis, Iowa State University, Ames, Iowa (1967).
15. Johnson, L. G. and R. L. Morris. "Chlorinated Hydrocarbons and Pesticides in Iowa Rivers," *Pestic. Monit. J.* **4**, 216 (1971).
16. Lichtenberg, J. L. *et al.* "Pesticides in Surface Waters of the United States—A 5-Year Summary, 1964-1968," *Pestic. Monit. J.* **4**, 71 (1970).
17. Richard, J. J. and J. S. Fritz. "Adsorption of Chlorinated Pesticides from River Water with XAD-2 Resin," *Talanta* **21**, 91 (1974).
18. Kellogg, R. L. "Dieldrin Contamination of Channel Catfish, Invertebrates and Minnows from the Des Moines River," Unpublished M.S. Thesis, Iowa State University, Ames, Iowa (1974).
19. Burnham, A. K. *et al.* "Identification and Estimation of Neutral Organic Compounds in Potable Water," *Anal. Chem.* **44**, 139 (1972).
20. Burnham, A. K. *et al.* "Trace Organics in Water: Their Isolation and Identification," *J. Amer. Water Works Assoc.* **65**, 722 (1973).
21. Junk, G. A. and H. J. Svec. "The Use of Macroreticular Resins in the Analysis of Water for Trace Organic Contaminants," 21st Annual Conference on Mass Spectrometry and Allied Topics, Paper No. P1, San Francisco (May 1973).
22. Junk, G. A. *et al.* "Contamination of Water by Synthetic Polymer Tubes," *Environ. Sci. Technol.* **8**, 1100 (1974).
23. Richard, J. J. *et al.* "Analysis of Various Waters for Selected Pesticides—Atrazine, DDE and Dieldrin," *Pestic. Monit. J.* (in press).

CHAPTER 10

DEVELOPMENT OF METHODS FOR ORGANIC ANALYSES FOR ROUTINE APPLICATION IN ENVIRONMENTAL MONITORING LABORATORIES

William L. Budde and James W. Eichelberger

U.S. Environmental Protection Agency
Environmental Monitoring and Support Laboratory
Cincinnati, Ohio

The major premise of this chapter is that the most meaningful and significant measurements ought to be the ones emphasized and routinely applied in environmental monitoring laboratories. For many years routine monitoring was accepted to mean the application of simple, easy-to-use, and inexpensive analytical methodology. Unfortunately the result of this attitude was the wide application of measurements that gathered much data, but provided very little real insight into the problem of environmental pollution.

In the field of organic analysis the prevailing philosophy was revealed in the wide application of such tests as the biochemical oxygen demand (BOD), the chemical oxygen demand (COD), and the nonspecific 4-aminoantipyrine derivative test for phenols. These crude measurements of organic pollution are of little value because they do not and cannot provide specific and reliable information about individual pollutants. Very reliable and specific information is needed to support many important environmental monitoring activities. Firm identifications and measurements are required to assess:

- The nature of organic pollution generated as a result of the development of new sources of energy, *e.g.,* increased coal mining, offshore drilling, and oil shale extraction.
- The causes of taste and odor in drinking water.

- The relationship of drinking water pollutants to the frequency of incidence of cancer and other diseases.
- The distribution of toxic compounds in all media.
- The causes of fish kills and wildlife population trends.
- The effectiveness of treatment facilities in removing classes and specific types of compounds.
- The sources of specific compounds and their fate in the environment.
- The degree of compliance with effluent and ocean dumping standards for toxic compounds.

It is often argued that the application of the complex analytical methodology needed to accurately identify and measure organic pollutants is not possible in the environmental monitoring laboratory because of the large sample load. It is indeed a challenge to conduct complex organic analyses on a routine basis. However, a number of recent developments in methods, instrumentation, and quality control have created an atmosphere that is favorable for a significant improvement in organic environmental analyses.

There are four general requirements that need to be met in order to establish a routine monitoring program for organic pollutants:

1. The development of well-defined and well-documented analytical methods.
2. The development of an overall plan for the application of these methods.
3. The large-scale utilization of automation including networks of laboratory minicomputers and large database search systems.
4. The rigid application of a fully-integrated quality control program.

Each of these requirements is discussed in subsequent sections and our progress in each area is reviewed.

ANALYTICAL METHODS

There are two fundamentally different approaches to organic analyses. Traditionally chemists have attempted to develop methods to "analyze for" compounds of interest. In these methods, a chemical or physicochemical procedure is developed to isolate and concentrate the specific compound of interest. Standards are employed to establish unequivocally that isolation indeed occurs and that all obvious interferences are eliminated. The final measurement of concentration is made with a relatively simple and inexpensive device. With this approach the isolation and concentration procedure becomes, in effect, the identification or qualitative analysis.

The "analyze for" approach has been widely applied in the environmental field and all other fields of organic analysis. The gas chromatography/electron capture detector procedures for chlorinated hydrocarbon pesticides are examples of this approach. However, in the environmental field the "analyze for" concept has many serious limitations. In order to include all environmentally-significant compounds, literally thousands of different detailed procedures would have to be developed, tested and documented. The implementation of all these procedures in an environmental monitoring laboratory would be extremely slow and costly, and would be simply impossible under the conditions of a large sample load. In addition this approach includes no provision for finding new pollutants except by an unexpected interference in a particular method. Because the relatively simple and inexpensive measurement device generally produces little or no qualitative information, serious errors from unexpected interferences are not infrequent.

The "analyze for" approach will continue to be used widely in certain areas where well-developed methodology exists, *e.g.,* chlorinated hydrocarbon pesticide analyses. However, in an environmental monitoring laboratory a more general "survey" type approach is also clearly indicated.

In order for analytical methodology to be acceptable for routine survey analyses of organic environmental pollutants, several important requirements must be satisfied. Since environmental samples invariably contain mixtures of compounds, some kind of efficient separation process must be involved. In practice this requires the use of some form of chromatography. The detector used must produce information of sufficient quality to permit very reliable qualitative analyses. The detector must operate in a continuous fashion to acquire data from each component of the mixture as it emerges from the chromatograph. Because of the volume of data generated, some kind of automated data acquisition and data reduction is required.

There are very few well-developed analytical techniques that meet these criteria. Computerized gas chromatography/mass spectrometry (GC/MS) is the lone significant example. Gas chromatography/infrared spectrometry is not as well developed, and infrared spectra are less useful than mass spectra for unequivocal identifications. High-pressure liquid chromatography has excellent separation power for many compounds not amenable to vapor phase analysis. However, no detector has been developed that gives qualitative information comparable to mass spectrometry or infrared spectrometry. Therefore, it is not surprising that essentially all the discoveries of organic pollutants in environmental samples in the last 15 years have been based on GC/MS.

158 ORGANIC POLLUTANTS IN WATER

In Figure 10.1 all organic compounds are crudely divided into three broad classes. This classification is according to the general behavior of compounds with respect to a group of widely-used isolation and concentration procedures. In the left box are compounds that range in vapor pressure from gases at room temperature and pressure (STP) up to the volatility of common laboratory solvents, *e.g.,* chloroform. This class of compounds is ideally suited to GC/MS analysis and several GC/MS-oriented sample preparation methods have been developed in recent years.

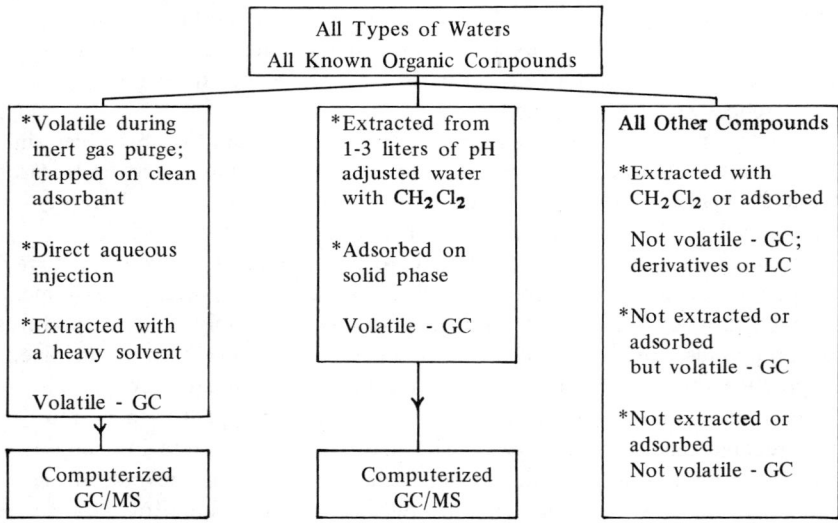

Figure 10.1 Classification of organic compounds according to analytical behavior with survey methods.

The procedure developed by Bellar and Lichtenberg[1] defines many of the compounds in this class, *i.e.,* relatively insoluble compounds such as chloroform. The detection limit of the method is about 1 μg/l. Direct aqueous injection GC/MS using a cross-linked polystyrene-type column is an auxiliary method for this class of compounds.[2] Its principal advantage is "instant analysis," and it is capable of detecting many water-soluble and relatively insoluble compounds with a detection limit of about 1 mg/l. Extraction with a heavy (*i.e.,* high boiling, late eluting) solvent such as hexadecane is another auxiliary procedure that is less thoroughly evaluated.

TECHNIQUES AND METHODS OF ANALYSIS 159

It is capable of efficient isolation and concentration of relatively nonpolar compounds to the low microgram/liter range.

The compounds represented in the center box are sufficiently volatile and stable to make them amenable to gas chromatography. However, their vapor pressures at STP and somewhat elevated temperatures are too low to permit their isolation by the inert gas entrainment procedure. This group of compounds includes many of great environmental significance including the chlorinated hydrocarbon pesticides, polycyclic aromatic hydrocarbons, chlorinated biphenyls, and a host of others. For this class of compounds GC/MS is clearly the measurement technique of choice.

Solvent extraction is a very well-developed and time-honored technique with a detection limit of about 10 ng/l. Adsorption methods have been widely used and have been significantly improved in recent years with the development of new types of adsorbants.[3] The advantage of adsorption methods is the acquisition of time-integrated samples and extremely low detection limits. It should be noted that the volatile compounds defined in the left box of Figure 10.1 may be extracted or adsorbed, but they are usually lost during extract concentration or masked by the solvent during gas chromatography.

The compounds represented in the left and center boxes of Figure 10.1 include the vast majority of environmentally-significant compounds and probably encompass several hundred thousand individual entities. Each of the sample preparation procedures is well defined and documented in the Environmental Protection Agency's GC/MS Procedural Manual that is scheduled for publication next year. Each individual procedure is a module with a specific scope and limitation, and each is suitable for application in the environmental monitoring laboratory on a routine basis. Individual laboratories may wish to select some or all modules for actual routine use depending on the time, personnel and equipment available.

The box on the far right of Figure 10.1 includes all the remaining organic compounds. Unlike the previous two groups, there is no general survey approach for this class of compounds. Numerous specific "analyze for" methods have been developed, but these are not suitable for routine application. Research is required to develop survey methods, and as these appear the box on the right will be subdivided and reorganized.

Among the compounds that are extracted with methylene chloride or adsorbed on a solid phase, but are not amenable to gas chromatography, are many of environmental significance, *e.g.,* some carbamate pesticides. A great deal of work has been devoted to the development of techniques to convert compounds in this subgroup into volatile derivatives. The difficulty with chemical derivatization is the great uncertainty about the exact nature of the reaction in a complex mixture. In addition, conversion yields vary considerably under any given set of conditions and this aggravates the problem of quantitative measurement. It appears that an approach based on high-pressure liquid chromatography using a continuous spectrometric detector would be more promising in the long run. The interfacing of a mass spectrometer with a liquid chromatograph has been reported by several authors.[4] While this development may extend the range of compounds amenable to mass spectrometry somewhat, the requirement of volatility for mass spectrometry suggests that other methods should be sought.

Compounds in the second subgroup are volatile but simply too watersoluble for extraction, adsorption or inert gas entrainment. If present in sufficiently high concentration, these types of compounds may be observed by direct aqueous injection GC/MS. Again more sensitive survey methods are required for this and the last subgroup of compounds. Approaches using liquid chromatography will require compatibility with the aqueous phase. Many of the compounds in the last subgroup are common natural products including amino acids, carbohydrates, nucleotides, etc., and their water-soluble degradation products. These can hardly be considered harmful except in the BOD sense, but conversions may occur during chlorination of effluents or purification of drinking water. Considerable research is required to clarify the nature of these problems.

OVERALL PLAN

Figure 10.2 is a flow chart of an overall plan for application of the well-defined survey methods based on GC/MS. In the general survey approach the conventional, broad band GC detector has an important place as a screening tool for the busy GC/MS laboratory. Alternatively, if the current sample load is not high, it is reasonable to go directly to the most powerful tool available to maximize its utilization.

As recently as the late 1960s mass spectra interpretation was a highly-specialized skill requiring a significant background in organic ion decomposition mechanisms and a generous portion of experience. However, with the development of the computerized mass spectral search system (MSSS) and a large database of mass spectra (currently about 42,000), routine identifications are a common fact. The MSSS is operational on a commercial time-sharing system. Many search options are available including interactive keyboard entry of MS data and direct, computer-to-computer automatic transmission.

Unequivocal identifications are not always possible with integer accuracy electron impact mass spectrometry. Experience indicates that a success

TECHNIQUES AND METHODS OF ANALYSIS 161

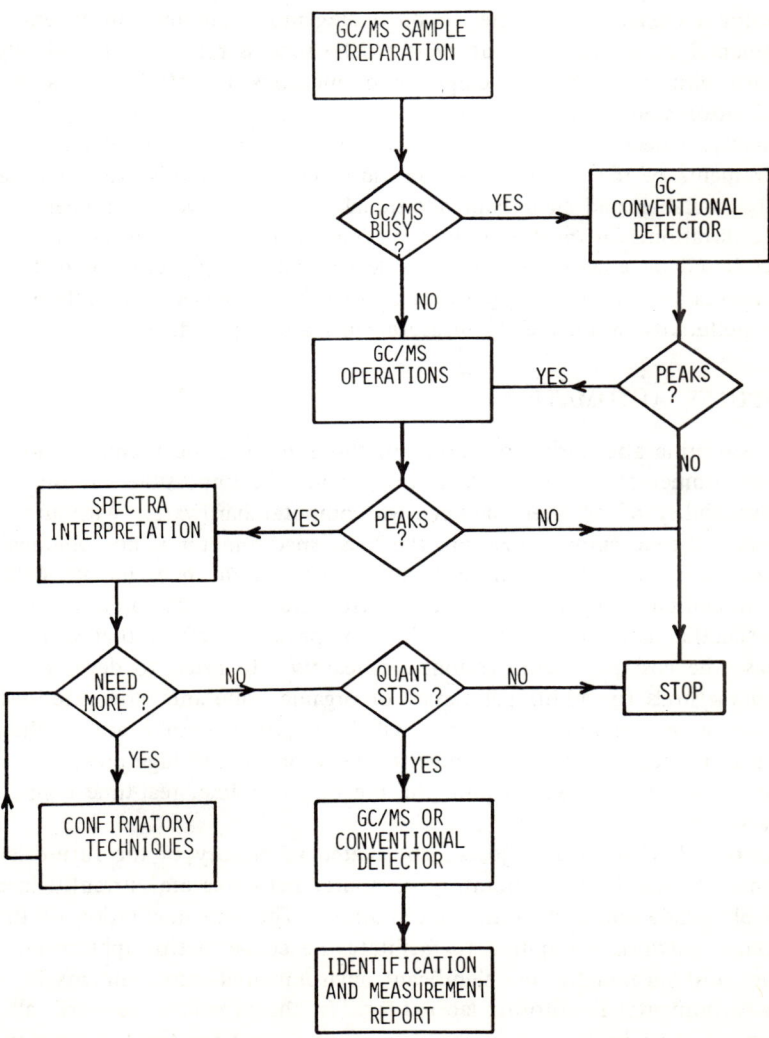

Figure 10.2 Flow chart for survey analyses of organic compounds using GC/MS.

rate of about 70% is possible on a routine basis with this technique. Other supporting and confirmatory techniques may be applied to improve this percentage, but some of these are not suitable for application on a routine basis. A choice must be made between gathering more data and achieving a higher success rate, or processing more samples. In an environmental monitoring laboratory, the 70% success rate should gradually improve with the further development of methods, the MSSS, and additional automated instrumentation.

Finally, a decision must be made concerning the accuracy and precision requirements of the concentration measurement. Standards are not always readily available, and significant additional cost is involved in concentration calibration. In most routine surveys, estimates of concentration ranges based on conservative recovery levels and average response factors are acceptable. In special situations of very high levels or particularly toxic pollutants, more precise measurements are required.

COMPUTER AUTOMATION

The routine application of survey methods for organic pollutants in the environmental monitoring laboratory would be impossible without the availability of relatively inexpensive computer hardware. Without on-line, real-time computerization, the mass spectrometer is not practical for continuous analysis of the effluent from a gas chromatograph. Without the computerized mass spectral search system, the interpretation of mass spectral data would require a level of personnel effort that would increase the cost per identification significantly. In order to develop survey methods for additional classes of organic pollutants, the extensive application of computer technology will be required. For example, the practical application of a spectrometric detector for the high-pressure liquid chromatograph will require the use of an on-line, real-time computer system.

An overall view of a projected automated laboratory of the future is shown in Figure 10.3. Laboratory computer networks are currently in a relatively crude and early state of evolution. The implementation of the computer network is required to facilitate the cost-effective application of the most meaningful and significant environmental measurements in the environmental monitoring laboratory. In the projected network, all computers must be located in the laboratory except for the large system that is used for management of data from finished samples. This large system is a centrally located facility equipped for remote input from many laboratories. The center of the projected laboratory network is a medium-scale computer that is a laboratory data management tool to

TECHNIQUES AND METHODS OF ANALYSIS 163

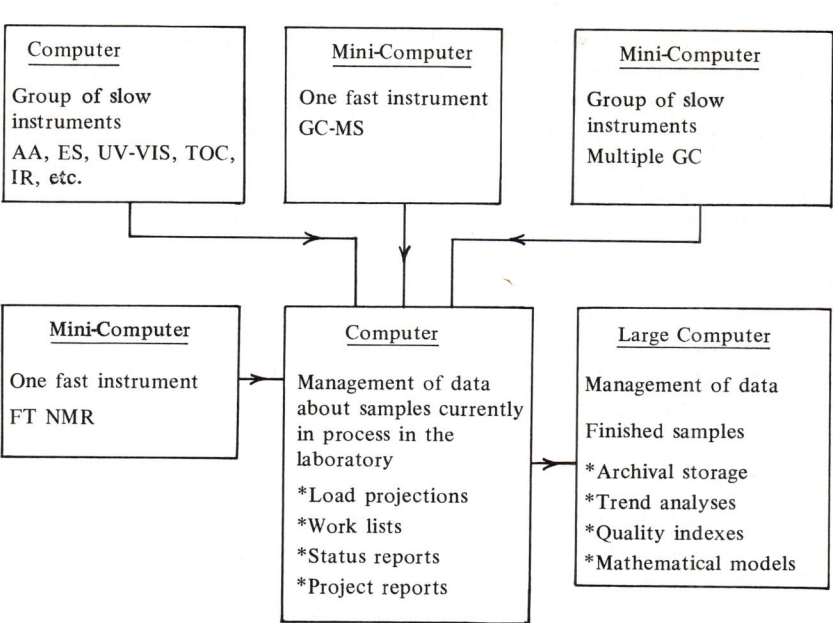

Figure 10.3 A projected laboratory computer network.

sort, merge and display information about samples currently in process. After output of a final comprehensive report on a group of samples, all subsequent data management is carried out in the large centralized system.

The laboratory computer network is a cost-effective automation approach. The application of a single large laboratory computer to handle all instrument and data management functions is significantly more costly for most laboratories. The network concept includes the advantages of incorporation of manufacturers' turnkey systems that are available for a fraction of the development costs of a comparable single system. Another advantage of a distributed computer system is the reliability of the total system compared to the reliability of a system based on a single large computer.

ANALYTICAL QUALITY ASSURANCE

The importance of analytical quality assurance in organic pollutant analysis cannot be overestimated. Data generated in surveys are being used to set standards for drinking water, surface water quality, and effluents. Possible correlations between the presence of organic contaminants in drinking water and human health effects are under widespread study. In the past many carefully-conducted measurements were not documented with sufficient data to support their reliability. This caused doubt about the validity of the measurements and concern for the correctness of correlations and proposed standards.

The analytical quality assurance that applies to the well-defined survey methods based on GC/MS for qualitative organic analysis may be conveniently divided into four categories: reagent and glassware control, instrumentation control, supporting experiments, and data evaluation.

Reagent and Glassware Control

Reagent and glassware control is required to minimize the introduction of contamination from the materials used in the sample preparation procedures. Glassware cleaning procedures have been developed and they are effective. High-quality commercial reagents and solvents are available, but quality is variable and usually unpredictable. In solvents that are used for extractions, impurities are amplified by a factor of about 2000 during extract concentration. Clearly, if background contaminants that are introduced from reagents or solvents seriously obscure compounds in the sample, purification of these materials is required.

Instrumentation Control

Instrumentation control is required to assure that the total operating instrumentation system is calibrated and is in proper working order. If a computerized GC/MS system is used to collect data, the computer data system must be included in the performance evaluation. The recommended instrumentation control procedure employs a standard reference compound and a set of reference criteria to evaluate the performance of the overall system.[5] This evaluation should be performed on each day the GC/MS system is used to acquire data from samples or reagent blanks. The records from the performance evaluations should be maintained with the sample and reagent blank records as permanent documentation supporting the validity of the data. Figure 10.4 shows the mass spectrum of the recommended reference compound.

TECHNIQUES AND METHODS OF ANALYSIS 165

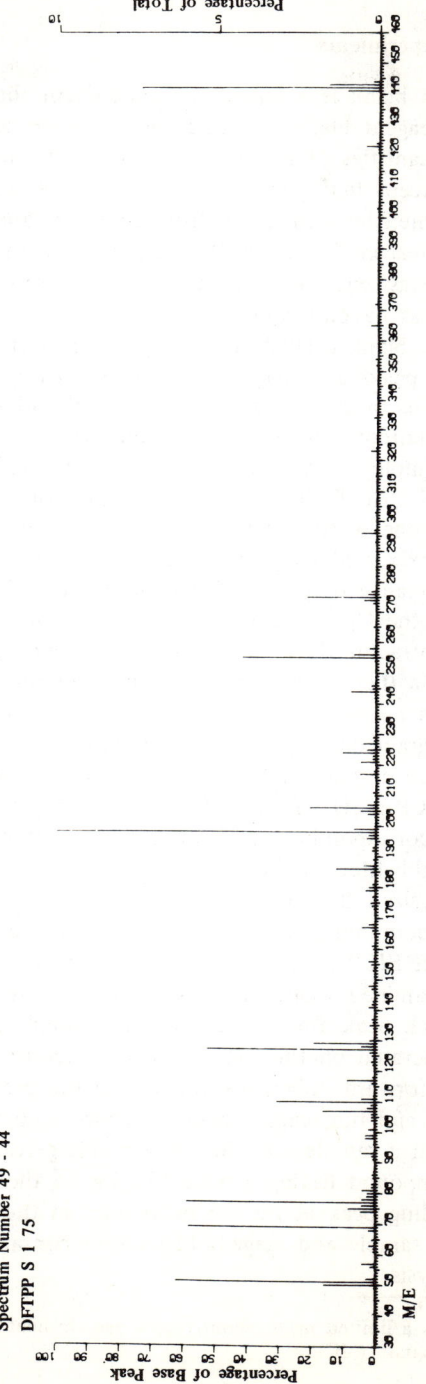

Figure 10.4   The mass spectrum of a GC/MS reference compound, decafluorotriphenylphosphine.

Supporting Experiments

The reagent blank is a supporting experiment that is required for all samples. A reagent blank is defined as an experiment that employs all procedures, quantities of materials, glassware, etc. used in the sample preparation except that no water or aqueous solution is used. It is required even when contamination from glassware and reagents is well controlled. The reagent blank result is the documentation that proves that good control was exercised, and it defines precisely the level of background that was beyond control.

The reagent blank evaluation may be a straightforward comparison of corresponding peaks and mass spectra in the reagent blank and sample. A more rigorous procedure is required to make objective judgments in ambiguous situations. An effective technique for comparison of blanks and samples employs the extracted ion current profile (EICP)* of one or several ions. An EICP is defined as a plot of the change in relative abundance of one or several ions as a function of time.[6] The data for this plot are extracted from all the ion abundance measurements made over the mass range observed during the elution of the separated components from the GC. The EICP produces an apparent increase in sensitivity by subtracting from the total ion current profile all the ion abundance data that is contributed from background, unresolved components, and other irrelevant ions. The EICP generator is a standard data reduction program on all modern computerized GC/MS systems. A fast graphics display device (CRT) is required to facilitate reviewing a large number of EICP plots. Figure 10.5 is an example of an EICP for mass 149 and the corresponding total ion current profile. The phthalate esters are clearly highlighted in the EICP.

It is emphasized that it is not necessary to have even a tentative identification of a compound to apply this technique to reagent blank evaluation. To conduct an EICP comparison, the mass spectra of all peaks in the sample are examined. One or several ions that are prominent in a spectrum from each peak are selected, and the sample and reagent blank EICPs are generated on the CRT. Their comparison permits, in most cases, straightforward judgments concerning the presence of compounds in the sample and the reagent blank. Figure 10.6 shows the EICP for mass 171 from a sample and the corresponding reagent blank. Clearly there is a compound having a mass 171 ion in the sample, but there is no corresponding peak above the noise level in the reagent blank. Figure 10.7 shows a sample and reagent blank that contain the same compounds by EICP analysis.

---

*Also known as a limited mass reconstructed gas chromatogram (LMRGC) and a mass chromatogram (MC).

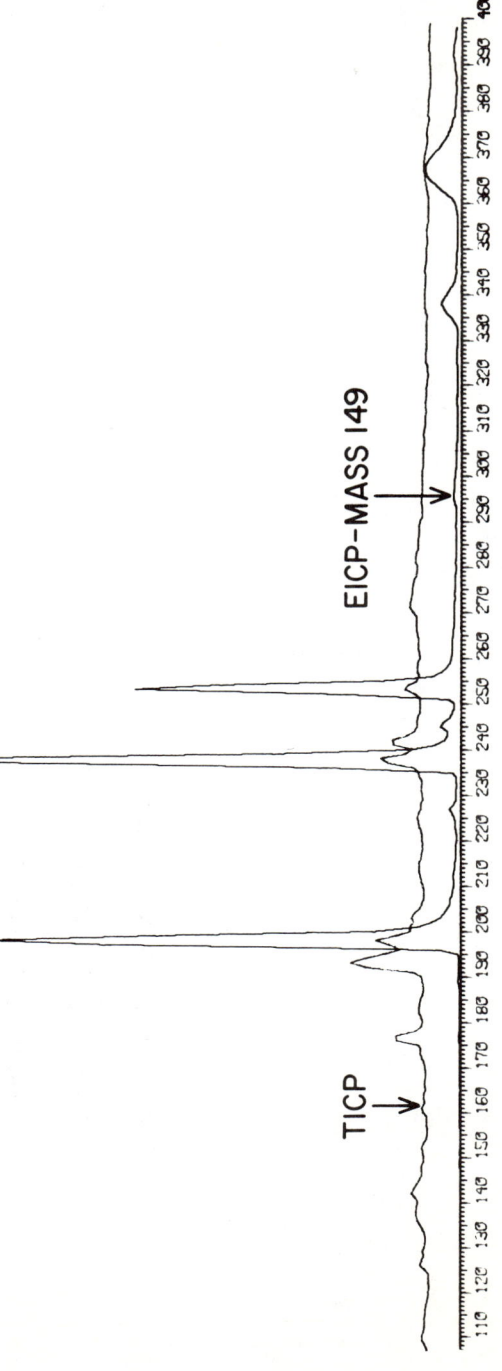

Figure 10.5 The extracted ion current profile for mass 149 and the corresponding total ion current profile.

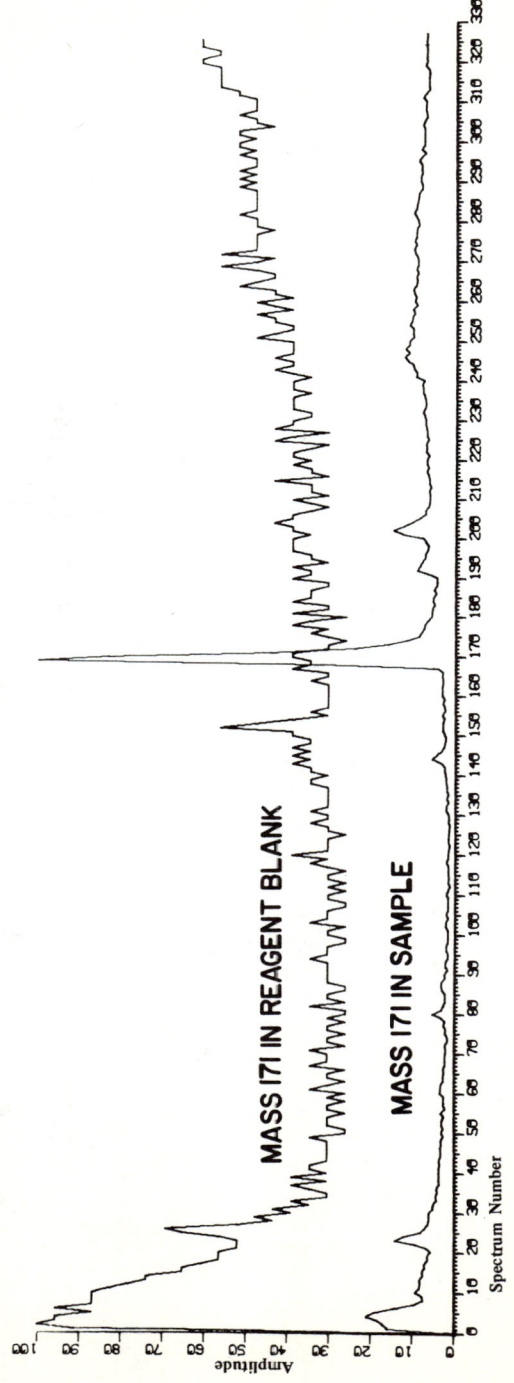

Figure 10.6 The extracted ion current profiles for mass 171 from a sample and the corresponding reagent blank.

TECHNIQUES AND METHODS OF ANALYSIS 169

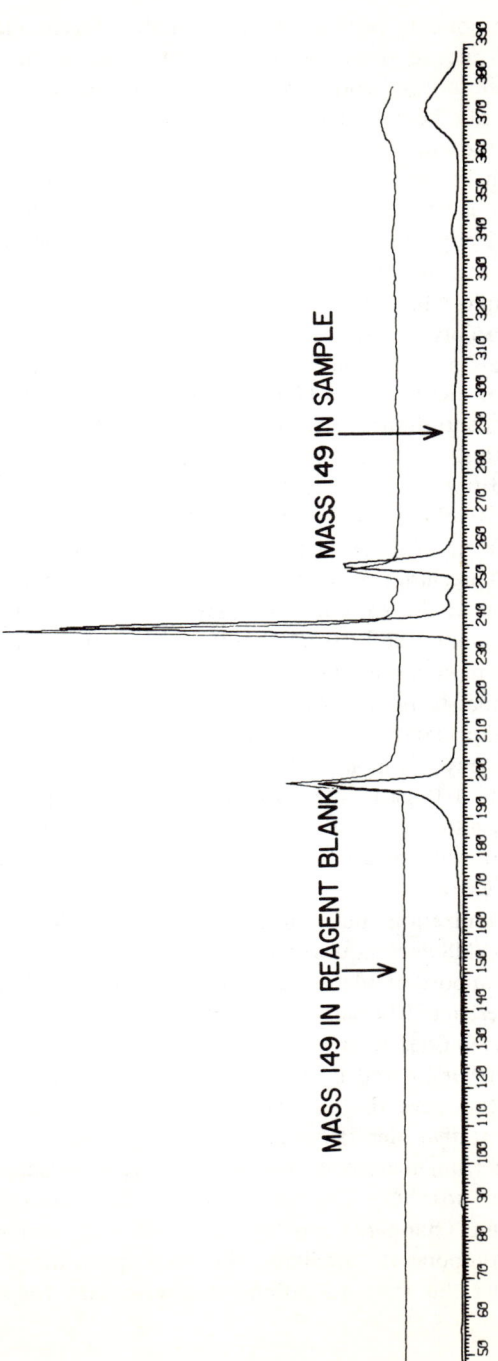

Figure 10.7 The extracted ion current profiles for mass 149 from a sample and the corresponding reagent blank.

If a corresponding peak is observed in the reagent blank, and its concentration as judged from the peak height is approximately the same as or exceeds the sample concentration, the decision is clear and the compound must not be reported. A far more difficult judgment must be made when the concentration of a component in the sample exceeds its concentration in the reagent blank. The material in the sample could, of course, be a true sample component. Alternatively, it has been observed empirically that compounds in the blank sometimes merely appear to be at lower concentrations than the same compounds in the corresponding sample. Figure 10.8 shows total ion current profiles from a sample and a corresponding reagent blank. Careful comparison of the profiles reveals a very similar pattern of peaks and valleys in certain areas, *e.g.*, spectrum numbers 170-190 and 235-245, yet a significantly lower apparent concentration in the reagent blank. There are several possible reasons for this. One rationalization used, in the case of solvent extracts, is that impurities in the solvent of the reagent blank are adsorbed more efficiently onto the drying agent and other surfaces, which creates a purifying effect. With extracts containing some water, the wetting effect of the water precludes efficient adsorption on surfaces and the impurities are carried on in the solvent. Alternatively, certain solvent impurities may be lost more readily from the blank than from the sample extract during the concentration step. This may be caused by the general organic background matrix in the sample extract which retains the solvent impurities. Both explanations are reasonable but unproved. In view of the uncertainties, any compound that is observed in the sample should not be reported if it is part of an overall pattern of peaks that is repeated in the blank, although at a lower apparent concentration. This same overall pattern will usually persist in the acid, neutral and basic fractions of a solvent extract.

Chemical ionization, field ionization, and high-accuracy mass measurements are GC/MS techniques that are capable of generating very strong evidence in support of identifications. However, the production of this evidence is restricted because only a relatively few laboratories have developed capabilities with these techniques. High-accuracy mass measurements are further limited by sample size, since some sacrifice in sensitivity is required to achieve the high accuracy.

After a tentative identification is made, by either interpretation or empirical spectrum matching, several other types of supporting experiments become possible. The retention time data from the GC/MS of a pure compound (standard) may be compared with analogous data from the sample component. Similarly, the mass spectrum of the standard, obtained under the same conditions that were used for the sample, may

TECHNIQUES AND METHODS OF ANALYSIS 171

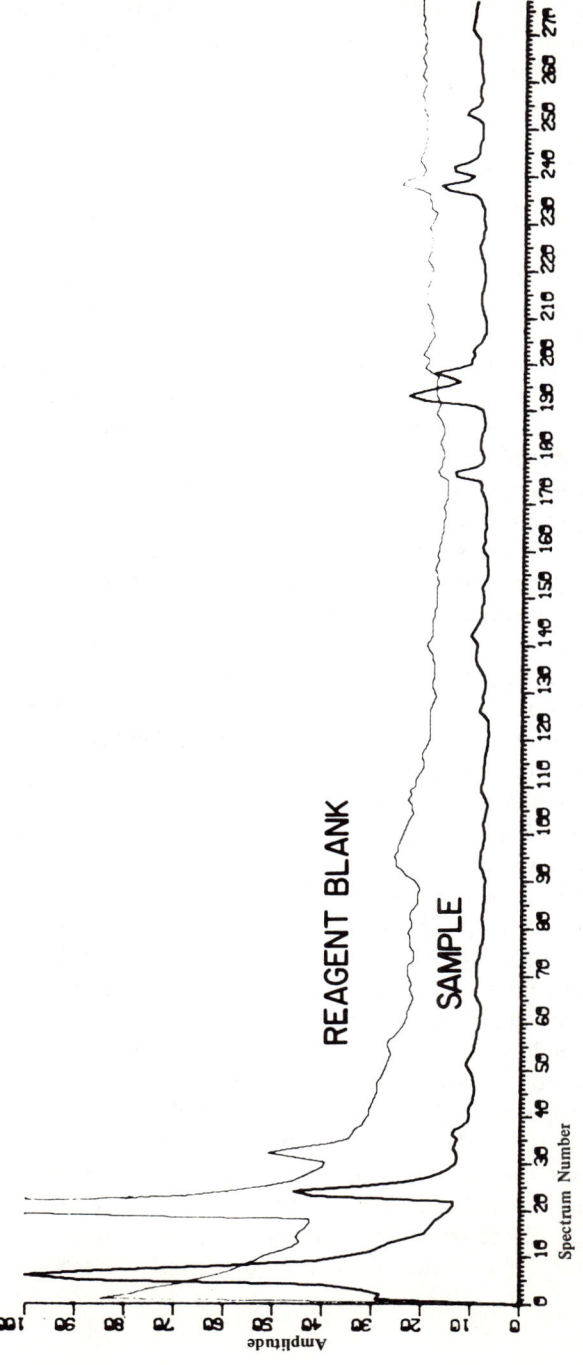

Figure 10.8 The total ion current profiles from a sample and the corresponding reagent blank.

be compared with the sample component spectrum. The standard may be dissolved in water at an appropriate concentration, isolated and measured. The recovery of this spike within the same fraction in which the suspected component appeared, and the observation of the same mass spectrum as the sample component gave, is a strong confirmation of the correctness of the identification.

Data Evaluation

The evaluation of the data must weigh the available evidence in terms of its reliability and must determine the cost and benefits to be gained by gathering additional information.

Clearly the most convincing evidence for an identification is obtained by the examination of pure standards corresponding to suspected sample components. However, the existence of this evidence is constrained by the availability of the pure standard and the additional cost and time required to examine it. Because it is not usually possible to predict which compounds will be found, some standards will not be available immediately. There are many very practical limitations imposed on the development and maintenance of a large library of pure authentic standards. Many compounds are obtainable from fine chemical supply houses, but procurement time is variable and may extend to weeks or months. Some compounds are not available from any supplier, frequently because they are by-products of industrial processes rather than manufactured products. This same limitation of standard availability also precludes calibrated concentration measurements in many cases.

Because of the problem of standard availability, it is worthwhile to determine whether conditions could exist that would lead to reliable identifications without standards. One possible criterion for reliable identification is a quantitative measure of the goodness of match between an experimental mass spectrum and a spectrum from the printed literature or a computer-readable database. A similarity index (SI), calculated on a scale from zero to one, has been described[7] and used in one computerized mass spectrum matching system.[8] Experience with this SI indicates that, in general, a value greater than about four-tenths corresponds to the existence of a reasonable match between two mass spectra.

A reasonable match often implies an identification, but sometimes it does not. It is well known that position isomers and members of homologous series of compounds often give very similar mass spectra. There is another undetermined number of compounds that are simply not uniquely characterized by their mass spectra. Figure 10.9 shows the mass spectrum of an unidentified compound and the spectrum of the

TECHNIQUES AND METHODS OF ANALYSIS 173

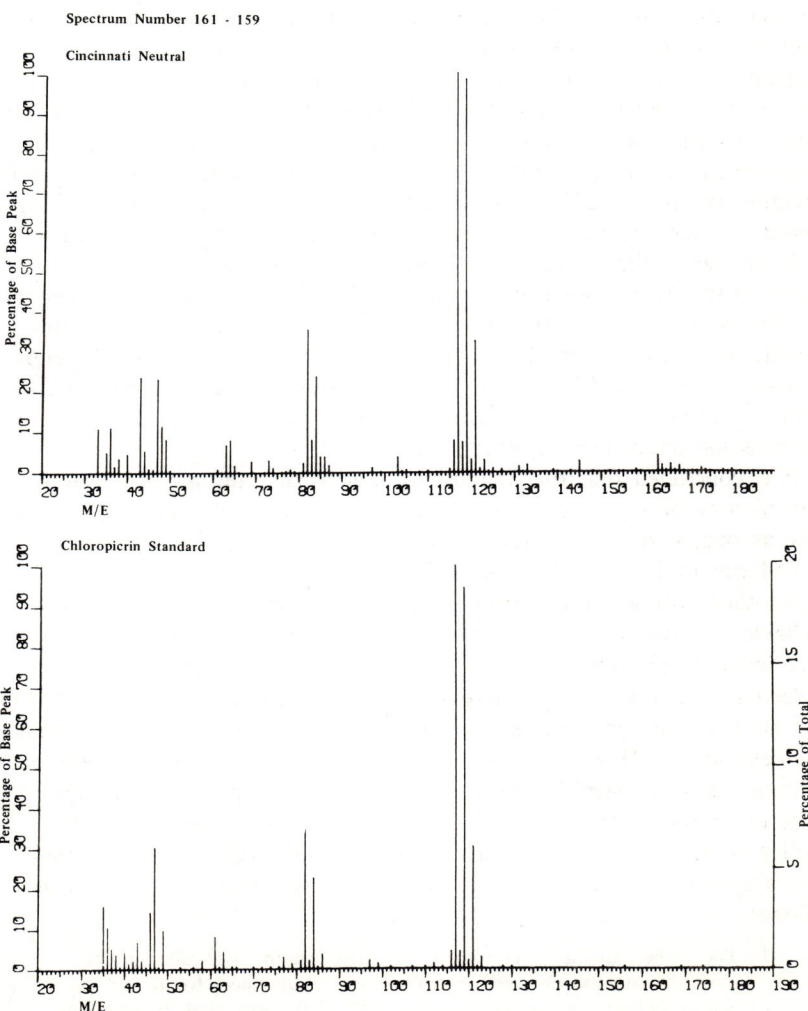

Figure 10.9 The mass spectrum of an unidentified compound and the mass spectrum of chloropicrin.

compound chloropicrin, Cl_3CNO_2. The match is clearly good by inspection and a SI of 0.453 was calculated. Nevertheless, the compound whose spectrum is illustrated in Figure 10.9 is not chloropicrin as determined by the gas chromatographic behavior of the unknown and pure chloropicrin.

Another problem with identifications based on empirical spectrum matching is that significant differences in ion abundance measurements are sometimes observed when the mass spectrum of a compound is measured on two different spectrometers. Most of these differences are probably caused by nonuniform calibration procedures or a failure to use an ion abundance calibration procedure. In addition, it is well known that different types of inlet systems may have significant effects on relative abundance measurements. With a GC or batch inlet system, generally operated in the 100-250° temperature range, temperature-dependent fragmentations are promoted with frequent reductions in the abundances of molecular and other higher mass ions. With a well-designed direct inlet system, these temperature effects may be largely precluded. As a result of these factors, it is quite common for two spectra of the same compound, that were measured with different inlet systems or spectrometers, to give a rather low SI. A low SI may also be caused by unresolved or partly resolved components which generate mass spectra containing extraneous ion abundance measurements.

It is concluded that the SI must be used with caution. A relatively high SI may be regarded as an indication of a reasonable match, but only as suggestive of the probability of an identification. A relatively low SI cannot be regarded as complete rejection.

Another criterion for reliable identifications when standards are not available is based on an assessment of the quality of the ions in the experimental and reference mass spectra. In the SI calculation,[7] molecular ions (M^+) having naturally-occurring isotopes (typically M + 1), and all key fragment ions are weighted the same as many very common fragment ions. However, the M^+, for example, is unique in every mass spectrum and has significance to an identification that far outweighs most other ions. Mass spectra may be categorized according to the quality of the ions observed and a quality index (QI) calculated that is a weighting factor for the SI. Several categories of quality are as follows:

1. The highest quality spectrum is one in which the molecular ion is observed and the observed distribution of abundances for it and its isotope-containing species is within 10% of the expected distribution. For this spectrum the QI is 1.0 and the full SI value is retained with considerably enhanced significance.
2. If a molecular ion is observed but the isotope data is not within 10% of the expected value, lower confidence is assigned by a QI of 0.75.

3. Failure to observe a molecular ion but the observation of key fragments that account for all the atoms of the molecular ion suggests a QI of 0.5. This index may be raised or lowered between 0.4-0.6 depending on the observation of consistent isotope data in the key fragment ions.
4. Finally, the lowest confidence is placed on spectra which do not contain adequate fragment ions to account for all the atoms of the molecular ion. A QI of 0.1 is assigned to these spectra.

It is recognized that position isomers may not be distinguishable under any circumstances, but this is often true even when pure standards are available.

The quality index is amenable to additional positive adjustments by 0.1-0.2 QI units if all major fragment ions are scrutinized and found to be reasonable and compatible with the assigned structure. Reasonableness should be based on compatibility with the accepted principles of fragmentation or organic ions in the gas phase. With magnetic deflection spectrometers, additional fractional quality points may be added if fragmentations are supported by the observation of ions from the decomposition of metastable ions.

Good spectrum matches having a QI of 0.75-1.0 are considered reliable identifications when pure standards are unavailable.

CONCLUSIONS

Significant progress has been made in the development of survey methods for organic analysis for routine application in the environmental monitoring laboratory. Survey methods should be applied widely and new research devoted to the development of more comprehensive methods.

The rapid decrease in the cost of computer and related hardware permits the wide application of computer automation and the development of laboratory computer networks. Analytical quality assurance techniques must be applied rigorously in all methods as invalid environmental data are worse than no data at all.

REFERENCES

1. Bellar, T. A. and J. J. Lichtenberg. *J. Amer. Water Works Assoc.* **66**, 739 (1974).
2. Harris, L. E., W. L. Budde and J. W. Eichelberger. *Anal. Chem.* **46**, 1912 (1974).
3. Junk, G. A., H. J. Svec, R. D. Vick and M. J. Avery. *Environ. Sci. Technol.* **8**, 1100 (1974).
4. Arpino, P., M. A. Baldwin and F. W. McLafferty. *Biomedical Mass Spectrometry* **1**, 80 (1974); Lovins, R. E., S. R. Ellis, G. D. Tolbert and C. R. McKinney. *Anal. Chem.* **45**, 1553 (1973).

5. Eichelberger, J. W., L. E. Harris and W. L. Budde. *Anal. Chem.* **47**, 995 (1975).
6. Hites, R. A. and K. Biemann. *Anal. Chem.* **42**, 855 (1970).
7. Hertz, H. S., R. A. Hites and K. Biemann. *Anal. Chem.* **43**, 681 (1971).
8. Heller, S. R., J. M. McGuire and W. L. Budde. *Environ. Sci. Technol.* **9**, 210 (1975).

CHAPTER 11

DEVELOPMENT OF COMPUTERIZED GC/MS TECHNIQUES WITHIN THE U.S. EPA

J. M. McGuire, M. H. Carter and A. L. Alford

U.S. Environmental Protection Agency
Southeast Environmental Research Laboratory
Athens, Georgia

In establishing water quality criteria and assessing environmental impact, methods are needed for specific identifications of organic compounds present at trace concentrations. Health and environmental effects due to organic contamination of water range from taste, odor, or color to possible toxicity or carcinogenicity. Because of differences in environmental effects among organic compounds, a nonspecific method of rather low sensitivity such as the tentative standard method for organic carbon is often inadequate as a means of pollution control. This method has a detection limit of 1 mg/l. In practice, a concentration of 1 mg/l can be far in excess of an environmentally-acceptable limit or can be totally insignificant. For example, 1 mg/l of the pesticide Endrin is completely unacceptable since it is toxic to fish at 1/1000 of this concentration. A level of 1 mg/l of geosmin in water is 10,000 times the odor threshold.

Specific identifications are also essential for enforcement procedures in order to relate pollutants to their sources. Monitoring of industrial effluents by EPA has shown that many companies do not know the composition of their effluents. Table 11.1 is a comparison of the materials one petroleum company expected in their effluent with the compounds actually identified. There is little resemblance between the 17 items predicted present and the 37 compounds actually found.

Table 11.1 Comparisons of Compounds Reported by Discharger and Compounds Identified by EPA in an Industrial Discharge

Products or Raw Materials Reported	Compounds Identified	
Propylene	all xylenes[a]	1,5-cyclooctadiene
Ethylene	isopropylbenzene	styrene[a]
Butadiene	o-ethyltoluene	o-methylstyrene[a]
Butane	diacetone alcohol	indan[a]
Octane	2-butoxyethanol	β-methylstyrene
Ethylene glycol	indene[a]	dimethylfuran isomer
Ethylene oxide	n-pentadecane	1-methyl indene[a]
Polyglycols	3-methylindene	acetophenone
Ammonia	n-hexadecane	α-terpineol
Raw gas	naphthalene[a]	both methyl
Ethane	benzyl alcohol	naphthalene isomers[a]
Refinery gases	ethylnaphthalene	α-methylbenzyl alcohol
Refinery C$_2$ Stream	2,6-dimethylnaphthalene[a]	phenol[a]
Refinery C$_3$ Stream	cresol isomer	methylethylnaphthalene
Propane	acenaphthalene	acenaphthene
Hydroformer gas	fluorene	methylbiphenyl isomer
Platformer gas	3,3-diphenylpropanol	two phthalate diesters

[a]Identification was confirmed with a standard.

Specific compound identifications are also required in the study of degradation pathways of pollutants and in assessment of the effects of experimental water treatment procedures. A process that will degrade or remove one pollutant may do so by introducing or forming another that is at least as undesirable.

During the past decade the EPA and its predecessors have developed and evaluated techniques for the identification of specific pollutants. Gas chromatography has been used to separate complex mixtures of moderately volatile compounds. However, although gas chromatography is a good method for separating many mixtures, GC retention times alone do not provide reliable identifications. The first approach to making GC identifications of pollutants more reliable was to isolate individual materials by micropreparative GC and then examine them by procedures that yield information-rich spectra or fingerprints. The first mass spectral identification of a water pollutant utilized this approach when Teasley[2] in 1967 identified a pesticide in an industrial effluent by the combined use of mass, nuclear magnetic resonance, and infrared spectrometries in a study at the Athens Environmental Research Laboratory. Continued investigations at this laboratory showed that the technique of combined gas chromatography/mass spectrometry (GC/MS) could separate complex mixtures of water pollutants and provide a mass spectrum of each resolved material without the isolation of any individual compound. The success of this method was demonstrated with a manually-operated GC/MS system by Keith and his co-workers[3] in 1969. Using this system, they were able to identify 14 components of a treated Kraft paper mill effluent. Such results showed that GC/MS was practical as a means of identifying many organic water pollutants; however, manual assignment of masses, measurement of abundances, and background correction made the system labor-intensive and subject to operator error. A faster, less error-prone system was needed. Users and manufacturers simultaneously arrived at the conclusion that these limitations could be overcome by use of a computer.

To evaluate the feasibility of interfacing the GC/MS system to a computer for EPA laboratories, an on-site survey was made in 1971 of all available computerized GC/MS systems. Based on results obtained at that time, identical systems, which were centered around the Finnigan electron impact quadrupole mass spectrometer, were purchased by the Environmental Research Laboratory in Athens and the Environmental Monitoring and Support Laboratory in Cincinnati. Since that time, more than two dozen EPA laboratories have chosen essentially the same system. Six of these laboratories have both electron impact and chemical ionization capabilities. A PDP-8 minicomputer in the system controls the operation of a quadrupole mass spectrometer and associated output devices. With the computerized system, the operator time required to gather data and prepare it for interpretation was decreased by more than an order of magnitude compared to manual GC/MS systems.

When the first computerized systems were purchased, work was begun on improving performance of both the equipment and software. This improvement was obtained through close meshing of in-house evaluations[4] with the vendors' development groups. As a result of this cooperative effort, data system speed, accuracy and reliability have been improved.

The Athens laboratory simultaneously began an extramural development program to increase the sensitivity of the computerized GC/MS system, and to provide better and faster spectral identifications.

The most obvious way to improve the sensitivity of a GC/MS system based on a quadrupole mass spectrometer is to control mass scanning so that integration times on peaks of interest are longer than on less pertinent ones. This approach, which increases sensitivity by improving the signal-to-noise ratio, was the basis for the development of a minicomputer specific ion monitoring program.[5,6] The program permits the user to choose three sets of up to eight masses each to be monitored during a

GC/MS run. He has the option of switching among these sets throughout the run. All other masses are ignored. Through suitable selection of masses, it is possible to obtain semiquantitative analyses of many materials when less than 1 ng is actually injected into the GC/MS. The program serves as the basis for a method developed for the quantification of toxaphene in water by means of chemical ionization mass spectrometry.[7]

A computerized system can also be employed to shorten the time required for spectral interpretation and compound identification. Several ways for doing this have been suggested. In 1971 an EPA research grant was made to Battelle Columbus Laboratory to develop one[8] of these methods into a computerized semi-automatic spectra matching program (EPA/Battelle program) and to develop a reference library of mass spectra. The successful program[9] was first made available to EPA laboratories in 1972. The first part of this program uses the minicomputer that controls the GC/MS system to prepare a set of spectra to be identified. Each spectrum is automatically corrected for background and abbreviated to consist of mass and abundance values for the two most intense peaks in each 14-mass unit group. The second portion of the program joins the minicomputer to a central computer and library by means of telephone lines. This portion automatically transmits spectra from the minicomputer to the central computer and returns identifications to the minicomputer. The third portion of the program involves the matching by the central computer of a spectrum from the minicomputer against a large reference library. Matches are ranked by the central computer based on the similarity between the unknown spectrum and the library spectra. The five best matches are then returned to the minicomputer for checking by the analyst. An example of this program is given in Figure 11.1. The 4th-18th lines are the abbreviated unknown spectrum. They are generated and transmitted by the minicomputer. The matches found by the central computer are then listed.

As development of the EPA/Battelle matching program was started, the National Heart and Lung Institute of NIH began the development of a Chemical Information System that included a group of mass spectral retrieval routines. This program was evaluated along with the EPA/Battelle program by EPA personnel and it was concluded that both should be combined into a single Mass Spectral Search System.[10,11]

During the initial phases of setting up this system, the Management Information and Data Systems Division of EPA, together with NIH, began collection of what has now become the world's largest collection of mass spectra.[12] Arrangements were made for Professor Fred McLafferty of Cornell University to develop quality criteria and to check and refine these 40,000 spectra to assure that the final data base would be of high

S OR P? S
OPTION
INPUT CTRL-P, CTRL-P, CTRL-A
FN-PHIL12, SP-266 -261
PHILADELPHIA CCE#1, RUN 2, 4-16-75
47,161
45,138
48,429
50,140
65,69
73,64
83,999
85,648
93,28
111,202
113,175
141,31
0,0
EDIT? N
FILE KEY = 8569
   0 1 AA-1344-2   1,1,3,3-TETRACHLORO-2-PROPANONE
SI = 0.678

FILE KEY = 4388
   0 1 AA-813-2   DICHLOROACETYL CHLORIDE
SI = 0.378

FILE KEY = 21379
   0 1 AA-813-3   DICHLOROACETYL CHLORIDE
SI = 0.325

FILE KEY = 21867
   0 1 AA-1026-2   BROMODICHLOROMETHANE
SI = 0.148

FILE KEY = 37854
   0 1 EP-0-11380   IODODICHLOROMETHANE
SI = 0.117

Figure 11.1 Computerized spectra matching program dialog.

quality. This checking is complete. At present the Mass Spectral Search System is available to all users on a commercial computer. The present cost for a search is between $2 and $6 depending on the options used.

Unfortunately, the matching systems are not 100% effective. In these cases the investigator can profitably turn to McLafferty's Probability Based Matching program and the Self-Training Interpretive and Retrieval System. The development of these programs for a PDP 11/45 computer system was sponsored by an EPA research grant to provide identifications of poorly resolved GC peaks and to provide structural information when the reference library does not contain the spectrum of the unknown.

Application of computerized GC/MS systems to environmental analyses has resulted in the identification by the EPA GC/MS System of several hundred organic compounds in water samples. More than 200 of these were tabulated in 1973 by workers in Athens;[13] Dr. Lawrence Keith is now bringing this tabulation up to date. In parallel with this, work has begun on compilation of a file of spectra of the most common water pollutants. It is expected that this file will include well over 1000 spectra within the next year. Searches of this file as a screen before searching the entire database are expected to provide the user with major reductions in both search time and cost.

Computerized GC/MS techniques provide the environmental analyst with a powerful means of determining the identity of many organic pollutants. They do, however, have an important limitation: they apply only for those compounds that can be gas chromatographed without decomposition. Many highly polar, nonvolatile compounds (often of biological origin) degrade in the GC. Such materials require a more gentle separation method such as liquid or thin-layer chromatography. Application of these techniques combined with various detectors, including MS, to environmental samples is just beginning. As suitable interfaces are developed between liquid chromatographs and mass spectrometers, computerized liquid chromatography/mass spectrometry combined with computerized GC/MS will permit the analysis of a substantial fraction of the organic pollutants in the environment.

REFERENCES

1. Keith, L. H. "Chemical Characterization of Industrial Wastewaters by Gas Chromatography-Mass Spectrometry," *Sci. of Total Environ.* **3**, 87 (1974).
2. Teasley, J. I. "Identification of a Cholinesterase-Inhibiting Compound from an Industrial Effluent," *Environ. Sci. Technol.* **1**, 411 (1967).
3. Keith, L. H. "Identification of Organic Contaminants Remaining in a Treated Kraft Pulp Mill Effluent," presented at the 157th National Meeting of the American Chemical Society, Minneapolis, Minn. (April 1969).
4. McGuire, J. M., A. L. Alford and M. H. Carter. "Organic Pollutant Identification Utilizing Mass Spectrometry," EPA Research Report EPA-R2-73-234 (1973).
5. Neher, M. B. and J. R. Hoyland. "Specific Ion Mass Spectrometric Detection for Gas Chromatographic Pesticide Analysis," EPA Research Report EPA-660/2-74-004 (1974).

6. Alford, A. L. "Evaluation of a Computer Program for GC-MS Specific Ion Monitoring," EPA Research Report EPA-660/2-74-002 (1974).
7. Thruston, A. D. "A Quantitative Method for Toxaphene by GC-CI-MS by Specific Ion Monitoring," EPA Research Report, in press.
8. Hertz, H. S., R. A. Hites and K. Biemann. "Identification of Mass Spectra by Computer-Searching a File of Known Spectra," *Anal. Chem.* **43**, 681 (1971).
9. Hoyland, J. R. and M. B. Neher. "Implementation of a Computer-Based Information System for Mass Spectral Identification," EPA Research Report EPA-660/2-74-048 (1974).
10. Heller, S. T., J. M. McGuire, and W. L. Budde. "Trace Organics by GC/MS," *Environ. Sci. Technol.* **9**, 210 (1975).
11. McGuire, J. M. "Mass Spectral Search Systems," Proceedings of EPA Office of Research and Development's First Automatic Data Processing Workshop, Bethany, West Virginia (October 1974), p. 236.
12. Heller, S. R., H. M. Fales and G. W. A. Milne. "EPA/NIH Mass Spectral Data Base—1975 Edition," National Technical Information Service, NBS Magnetic Tape 8 (1975).
13. Webb, R. G., A. W. Garrison, L. H. Keith and J. M. McGuire. "Current Practice in GC-MS Analysis of Organics in Water," EPA Research Report EPA-R2-73-277 (1973).

CHAPTER 12

THE USE OF GC/MS IN THE ANALYSIS OF UNUSUAL ENVIRONMENTAL CHEMICALS

R. E. Finnigan and J. B. Knight

Finnigan Corporation
Sunnyvale, California

INTRODUCTION

Gas chromatography-mass spectrometry (GC/MS) has become the instrumental technique of choice in most environmental laboratories today, especially where detection and identification of organic pollutants in water, often at trace levels, are required. Most often the GC/MS is used in combination with an interactive data system which greatly enhances the power of this instrumental technique. Most of the laboratories using GC/MS have employed packed columns—either U-tube or coil columns—to provide the separation of the complex organic mixtures. The quadrupole analyzer has generally been used as the mass spectrometer; in most cases it has been used with electron impact ionization. Often the GC/MS analyses have been carried out with the aid of computerized search systems which have large libraries (up to 40,000 compounds) of known spectra of organic compounds.

This chapter presents data which show the value of high-resolution glass capillary column GC used in combination with both EI and CI quadrupole mass spectrometry for analysis of some less common organic compounds which are of interest to chemists working on environmental analyses. Often, these techniques provide, at very low levels, detection and identification, of compounds which do not lend themselves to analysis by conventional packed column GC/MS using electron impact ionization alone.

POLYCYCLIC AROMATIC HYDROCARBONS IN CRUDE OIL

As an example of the power of high-resolution capillary columns in GC/MS, we show the analysis of an Alaskan crude oil sample. One of the major objectives was to determine if polycyclic aromatic hydrocarbons (PAH) were present in the crude oil samples, and their level. It is suspected that the PAH compounds are concentrated in the refining process and could possibly reach dangerous levels.

Because of the rapid elution of many of the hydrocarbon peaks from the GC, it is essential that the mass spectrometer be scanned very rapidly (*i.e.*, 1 sec or less) when analyzing samples of this type. This is necessary to retain the resolution achieved by the GC column.

Figure 12.1 shows the difference in the total ion chromatogram obtained when using 1-sec and 3-sec scans in the analysis of an oil sample by capillary column. As can be seen, the GC resolution is seriously degraded when scanning at the 3 sec scan rate. The ability of the quadrupole mass spectrometer to scan at high speed, repeatedly, is a necessary feature for capillary GC/MS analysis.

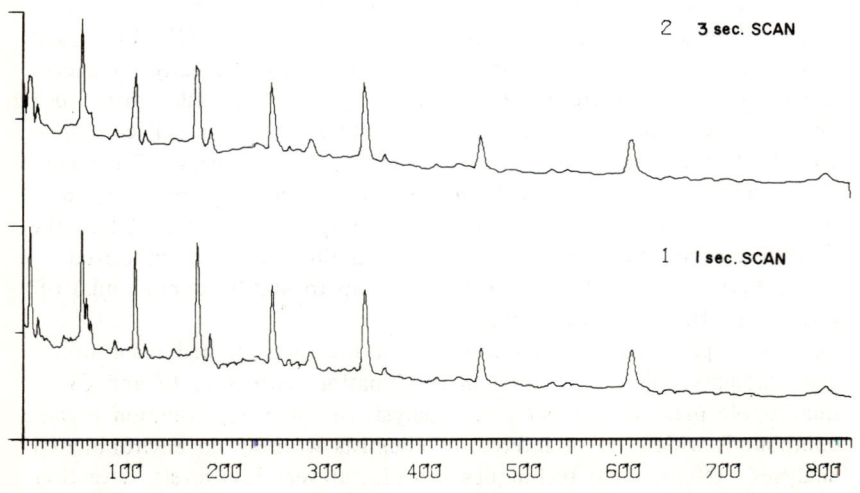

Figure 12.1 Total ion chromatograms of oil sample by glass capillary GC and and EI-MS, when scanning at 1-sec mass scan and 3-sec mass scan.

We show in Table 12.1 the conditions used for the glass capillary column EI-MS analysis of the Alaskan oil sample. In this example EI-MS is used rather than CI-MS, since the PAH compounds lend themselves to EI interpretation.

Table 12.1 Conditions Used for Analysis of Alaska Oil Sample by Glass Capillary GC - EI - MS

Column:	20 m x 0.02 in. i.d. SE-30 glass capillary
Column Temperature:	Ambient to 250°C at 4°C/min
Carrier Gas:	Helium at 2 cc/min flow
Injector Temperature:	250°C
Sample Size:	0.2 μl (Splitless injection)
Mass Range Scanned:	40-350 amu
Scan Time:	1.227 sec

The results of the analysis are shown in the following figures. Figure 12.2 shows the total ion chromatogram indicating the location of components of interest. The m/e values marked in the chromatogram correspond to spectra to where specific ion searches indicated these masses present. The m/e values searched are specific for PAH compounds. A total of 1825 scans were taken over a 45-min period. Note also that compounds with MW at m/e = 252, which very likely indicate where the most carcinogenic PAH might occur, are present at very low levels as compared to other components.

Table 12.2 shows some polynuclear aromatic hydrocarbons arranged by molecular weight. The ones labeled (*) are known carcinogens (reference Merck Index).

Figure 12.3 shows a specific ion search for m/e = 178 (possibly anthracene and phenanthrene) overlapped with the total ion chromatogram. Figure 12.4 shows the mass spectrum for either anthracene or phenanthrene (MW = 178); both give essentially identical mass spectra. Either standards or exact relative retention times (RRT) are needed to determine which compound is being identified. Figure 12.5 shows the mass spectrum #999, a PAH with MW = 228. Again, either a standard or accurate RRT is needed to determine exact identity. Similarly, Figures 12.6 and 12.7 show mass spectra of scan numbers 1163 and 1233, respectively. Both show PAH compounds with MW = 252, those

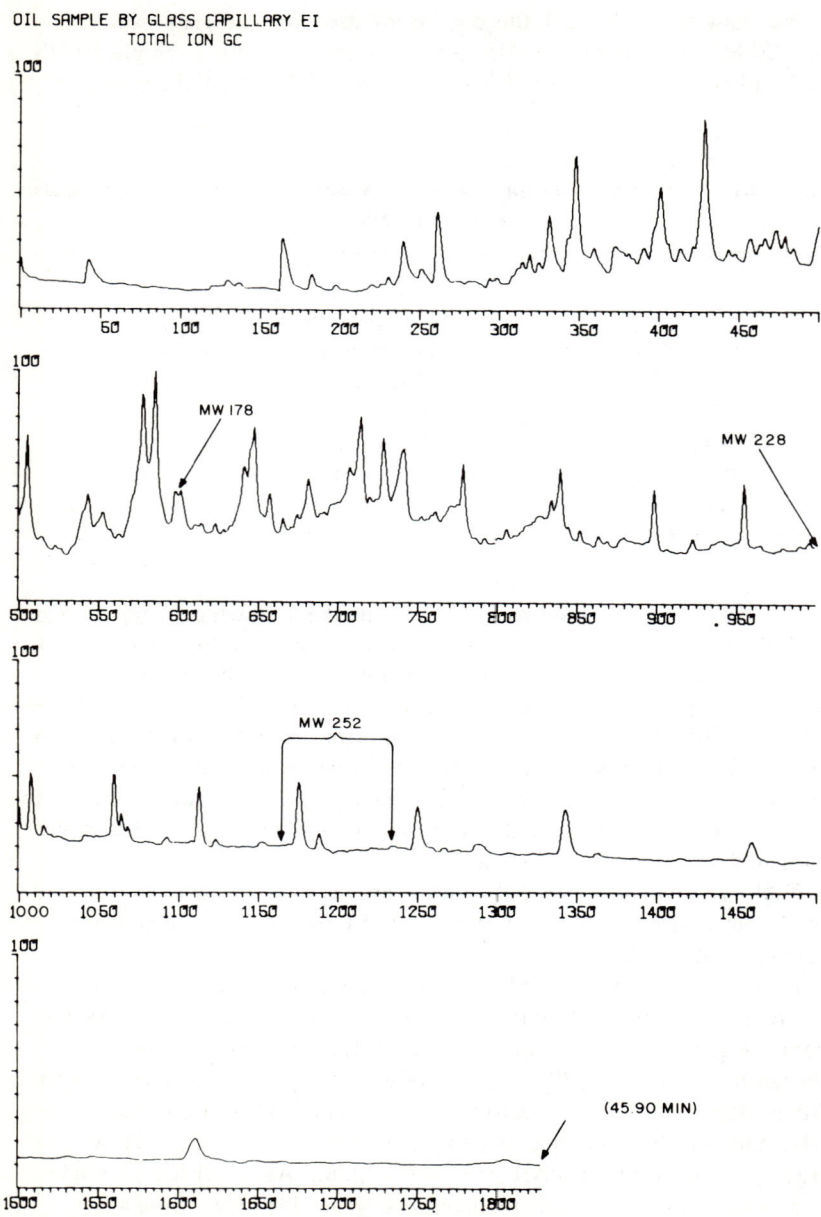

Figure 12.2 Total ion chromatogram of Alaskan oil sample. The m/e values marked on the chromatograms show where specific PAH compounds were found by limited mass search.

Table 12.2 Polyaromatic Hydrocarbons that Produce Very Similar EI Mass Spectra but are Identifiable by Capillary GC/MS

$C_{20}H_{12}$ MW 252

1. 3,4-benzfluoranthene
2. 3,4-benzpyrene*
3. 1,2-benzpyrene*
4. 10,11-benzfluoranthene
5. 11,12-benzfluoranthene
6. Perylene*

$C_{18}H_{12}$ MW 228

1. Triphenylene
2. 3,4-benzophenanthrene
3. 1,2-benzanthracene
4. Chrysene
5. Naphthacene

$C_{14}H_{10}$ MW 178

1. Phenanthrene*
2. Anthracene

* = Known carcinogens.

PAH most often associated with carcinogenic activity. Note the presence of a strong doubly-charged ion in the region of m/e = 126.

When analyzing and identifying normal paraffins in an oil sample such as this, methane CI provides significant molecular weight information. The EI spectrum of a normal paraffin is shown in Figure 12.8 whereas the CI spectrum of another homologous compound is shown in Figure 12.9. The M-1 ion is very strong in the methane CI spectrum as compared to the very small molecular ion at m/e = 170 in the EI spectrum.

STEROLS BY CAPILLARY COLUMN GC/EI/MS

The sterols considered here are often encountered when natural products are being investigated. The methyl ether derivatives make these compounds more suitable for GC analysis than the corresponding free alcohols. Often these components are present at very low levels in highly complex mixtures.

Figure 12.10 shows the structure of five of the sterols considered here: cholesterol methyl ether (MW = 400); 24-methyl-cholesterol methyl ether (MW = 414), 24-methylene cholesterol methyl ether (MW = 398(, stigmasterol methyl ether (MW = 426) and beta-sitosterol methyl ether (MW = 428).

190 ORGANIC POLLUTANTS IN WATER

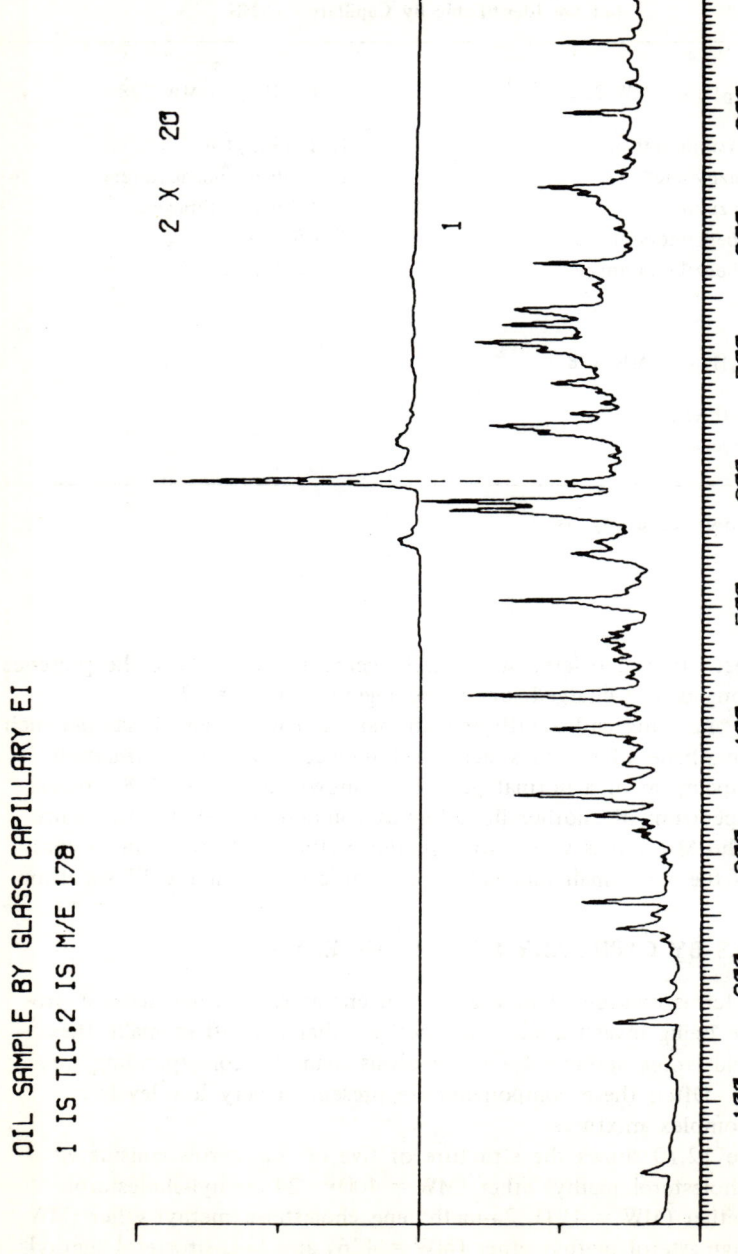

Figure 12.3 Total ion chromatogram overlapped with specific ion search for m/e = 178.

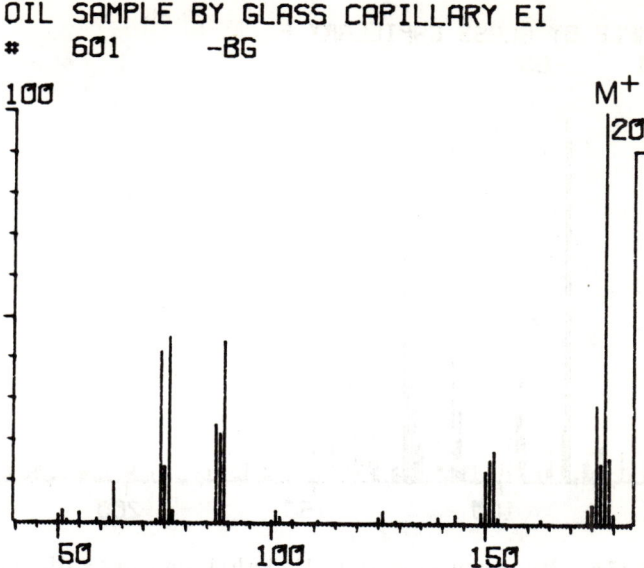

Figure 12.4 Mass spectrum for either anthracene or phenanthrene.

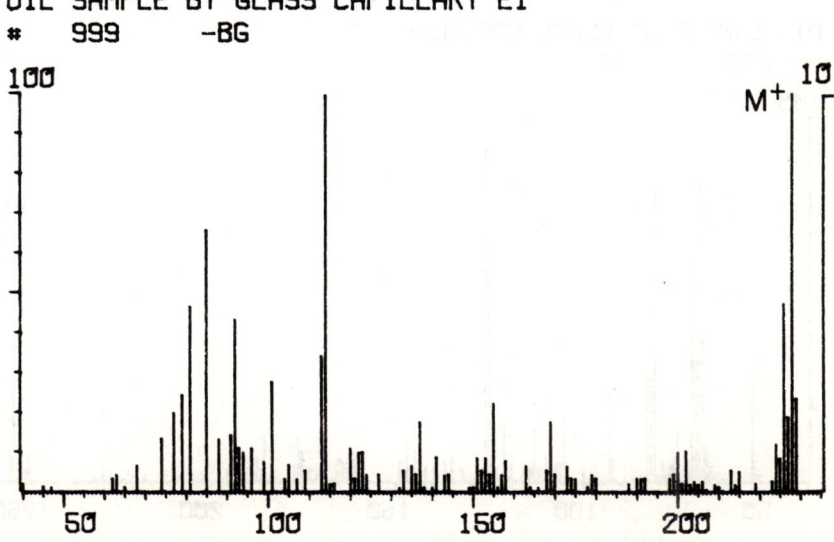

Figure 12.5 Mass spectrum of scan #999 which represents PAH compound with MW = 228.

192 ORGANIC POLLUTANTS IN WATER

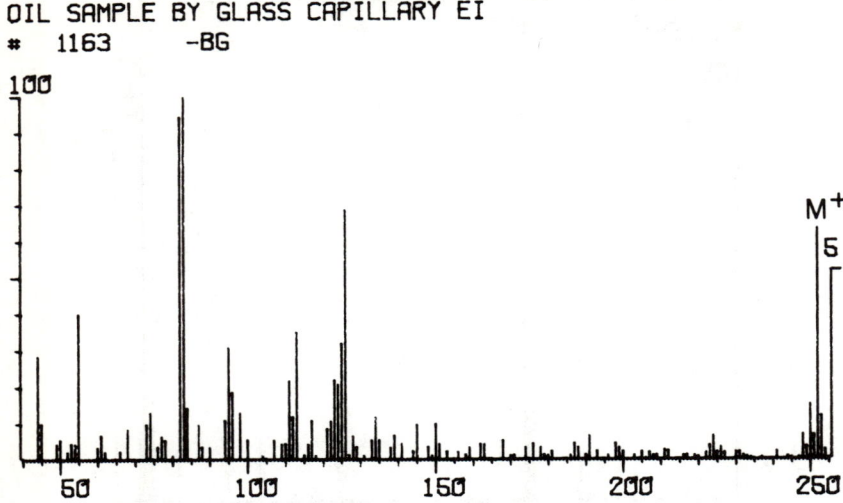

Figure 12.6   Mass spectrum of scan #1163 which represents PAH compound with MW = 252.

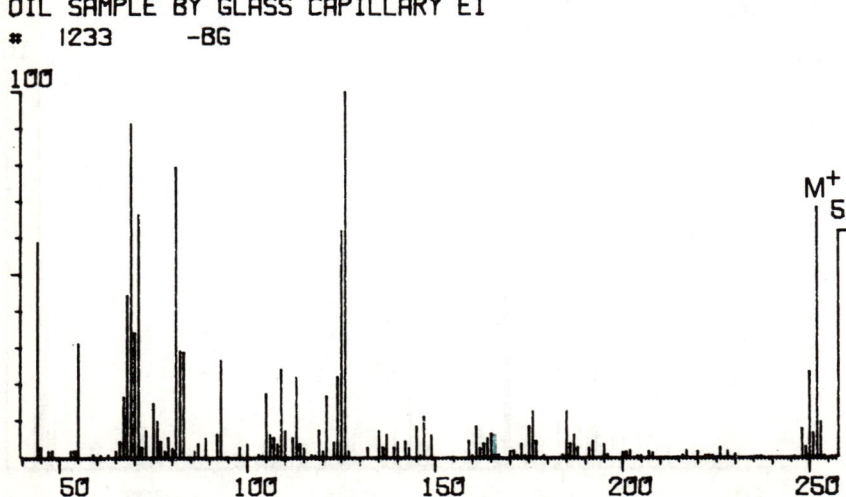

Figure 12.7   Mass spectrum of scan #1233 which represents PAH compound with MW = 252.

TECHNIQUES AND METHODS OF ANALYSIS 193

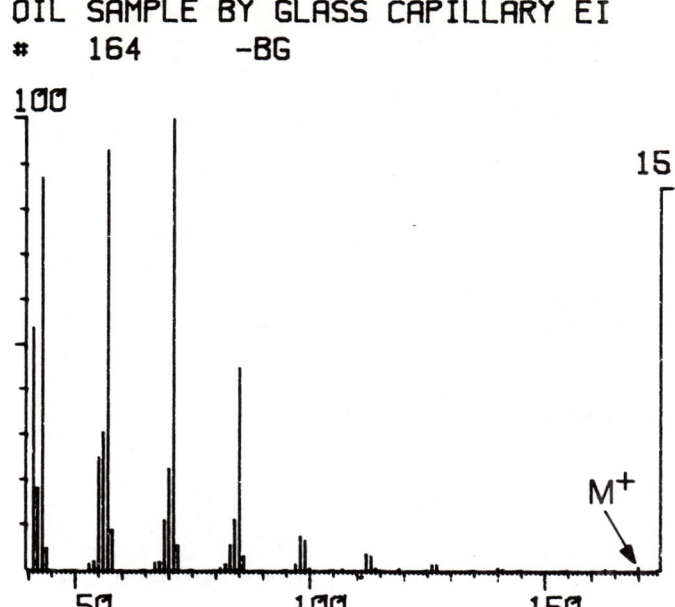

Figure 12.8 EI spectrum of normal paraffin with molecular ion at m/e = 170.

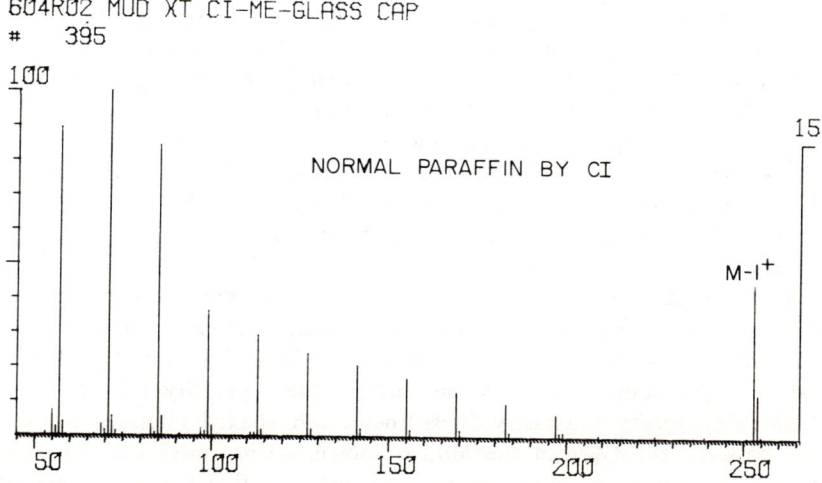

Figure 12.9 Methane CI spectrum of normal paraffin.

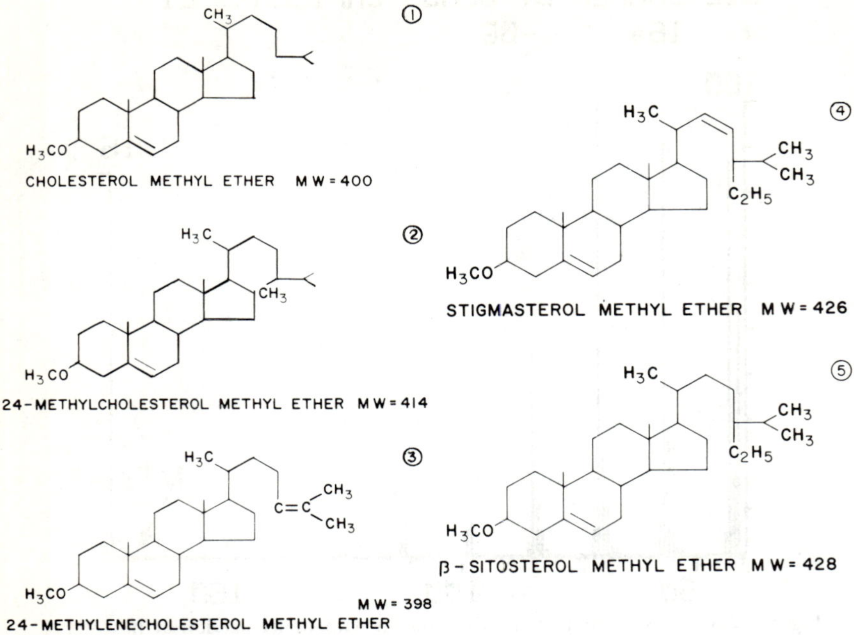

194 ORGANIC POLLUTANTS IN WATER

Figure 12.10 Molecular structure of five sterol methyl ethers analyzed by capillary column GC/MS.

① CHOLESTEROL METHYL ETHER MW = 400
② 24-METHYLCHOLESTEROL METHYL ETHER MW = 414
③ 24-METHYLENECHOLESTEROL METHYL ETHER MW = 398
④ STIGMASTEROL METHYL ETHER MW = 426
⑤ β-SITOSTEROL METHYL ETHER MW = 428

We analyzed a water sample for sterols sent to us by the Canada Centre for Inland Waters using packed column GC/MS. It is interesting to compare the packed column results with those obtained by capillary column GC/MS. Figure 12.11 shows the total ion chromatogram obtained by GC/MS of an extract taken from the Canadian waters where packed column GC techniques were used to attempt to identify sterols in water. It is believed that the sterol analysis was to determine if raw sewage was being dumped into the water since these are known fecal sterols.[1] There is no baseline separation between stigmasterol and beta-sitosterol (peaks 6 and 8).

Next we analyzed a known synthetic mixture of the sterol methyl ethers (shown in Figure 12.10) by high-resolution capillary column GC together with high-sensitivity, fast-scanning MS (Finnigan Model 3200). The sample was provided through the generosity of the Stanford University Chemistry Department. A coaxial interface was used between the GC and the MS, as shown in Figure 12.12. This provides EI operation when the reactant gas is shut off and CI operation

TECHNIQUES AND METHODS OF ANALYSIS 195

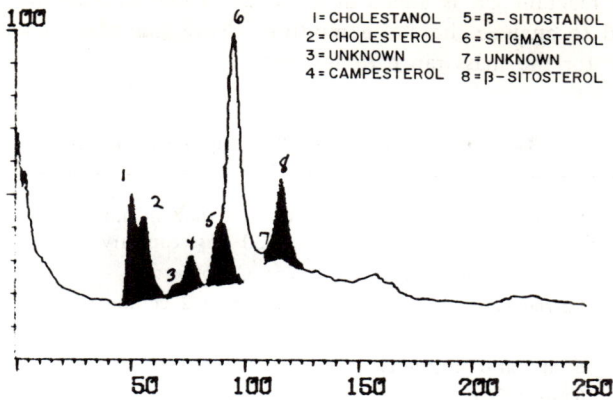

Figure 12.11 Total ion chromatogram of fecal sterol sample taken from Canadian waters. Analysis was performed using packed-column GC/MS.

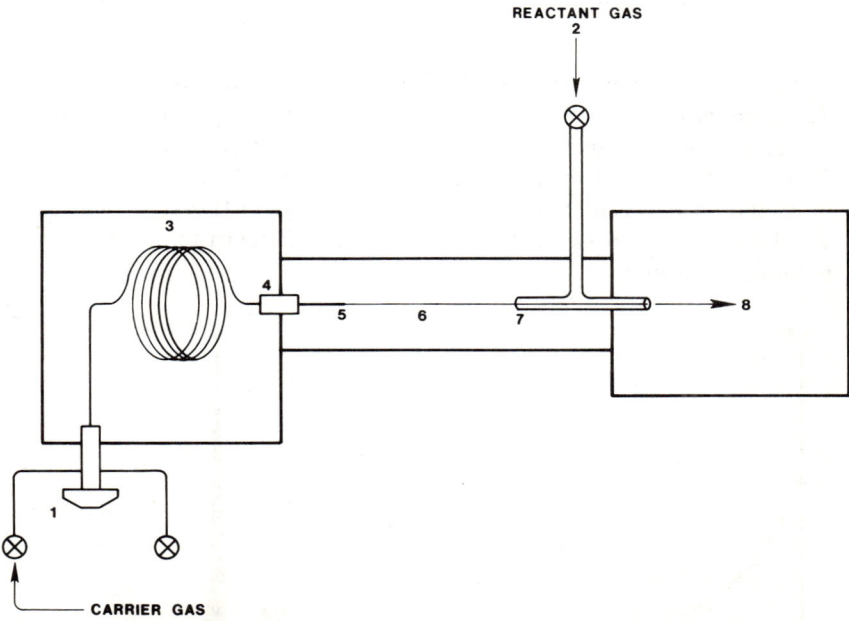

Figure 12.12 Coaxial dual-gas interface used to interface glass capillary column to MS for EI and CI-MS. 1. Gas chromatograph with Grob-type injector. 2. Reactant gas line (stainless steel, o.d. 1.6 mm. i.d. 0.5 mm). 3. Glass capillary column (H. and G. Jaeggi, Trogen, Switzerland). 4. Polyimid swagelok (Vespel SP-1, Dupont) on Teflon shrink-tube. 5. Glass/platinum seal. 6. Interface capillary (platinum, o.d. 0.3 mm, i.d. 0.1 mm). 7. Coaxial dual-gas interface. 8. Finnigan model 3200 quadrupole mass spectrometer.

196 ORGANIC POLLUTANTS IN WATER

when the reactant gas is added at point 2. Table 12.3 shows the conditions used to analyze the sterol mixture. Note that the scan time used was 2 sec for the mass range of interest.

Table 12.3 Sterol GC/MS Analysis Conditions

Column:	10 m x 0.02 in. i.d. SE-30 glass capillary
Column Temperature:	265°C isothermal
Carrier Gas:	Helium at 2 cc/min flow
Injector Temperature:	265°C
Split Ratio (Injector):	10 to 1
Sample Size:	200 ng
Ionization:	Electron impact
Mass Range:	50-550
Scan Time:	2.0 sec

Figure 12.13 shows the resulting capillary total ion chromatogram. The analysis took less than 20 min (as compared to approximately 6 min in the previous example using a packed column). Component 4 represents 20 ng of stigmasterol methyl ether. There is a significant improvement in resolution between stigmasterol and beta-sitosterol (peaks 4 and 5). The background subtracted EI mass spectrum of stigmasterol (MW = 426) is shown in Figure 12.14.

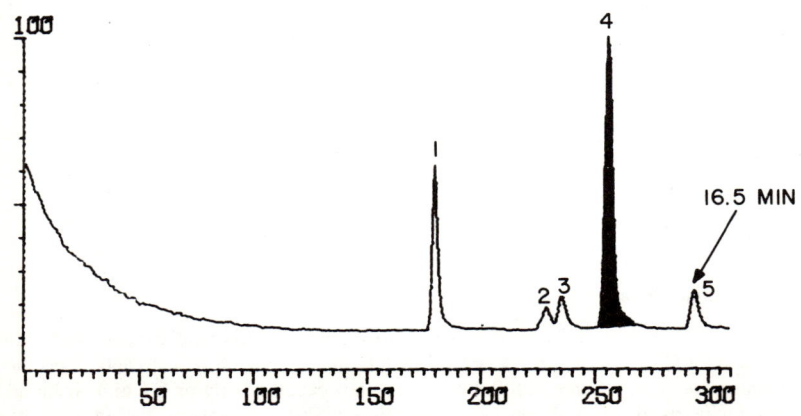

Figure 12.13 Total ion chromatogram of sterol methyl ether mixture analyzed by capillary column EI-MS.

TECHNIQUES AND METHODS OF ANALYSIS 197

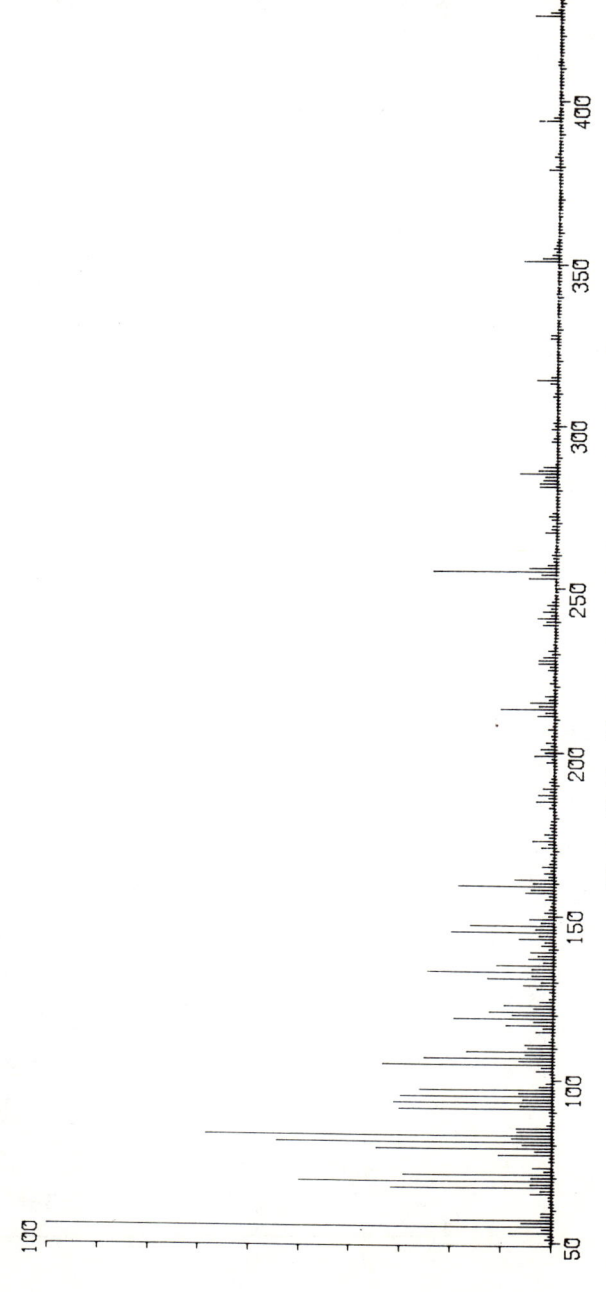

Figure 12.14 Mass spectrum of stigmasterol methyl ether.

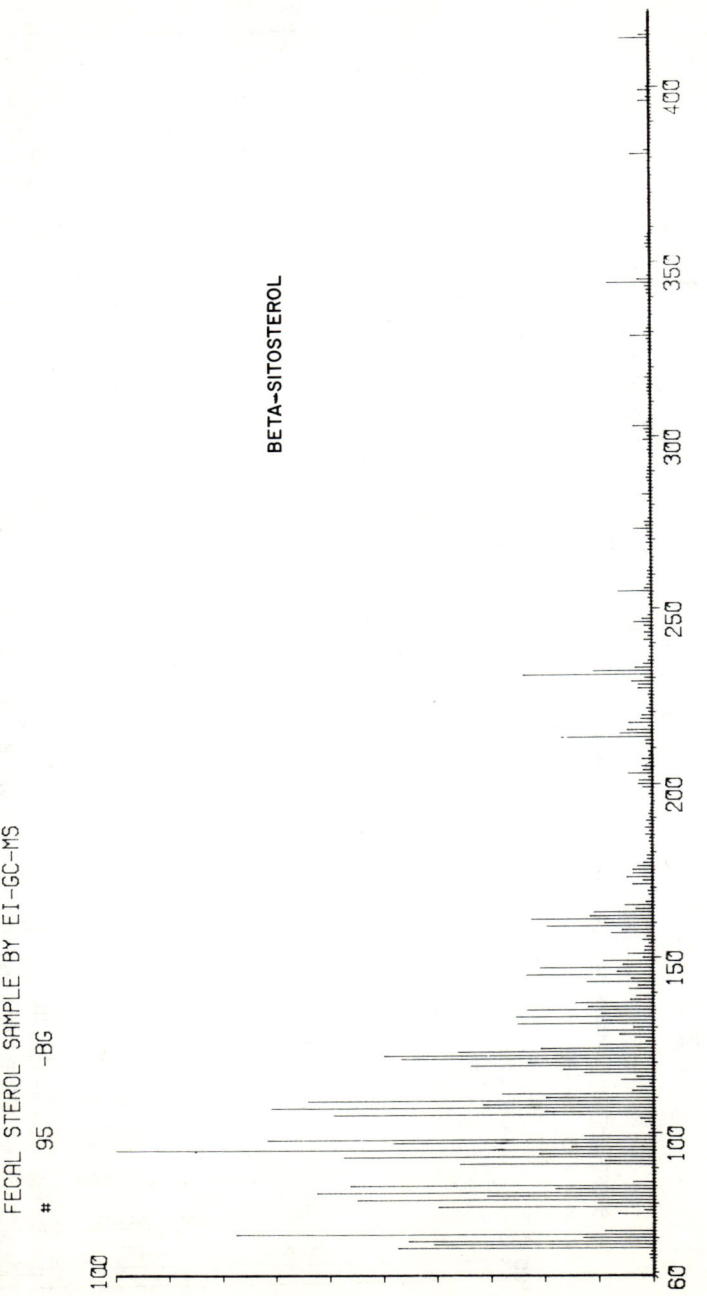

Figure 12.15 EI mass spectrum of beta-sitosterol in fecal sterol sample taken from Canadian waters.

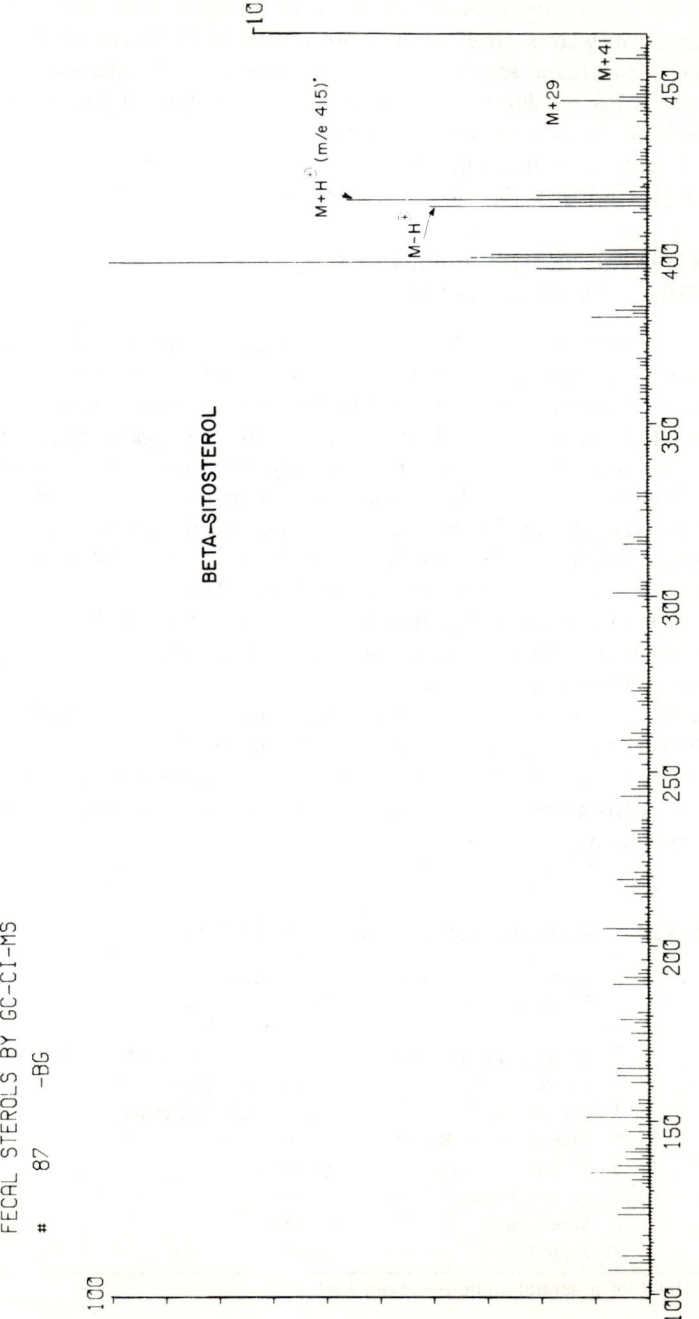

Figure 12.16 CI mass spectrum of beta-sitosterol in fecal sterol sample taken from Canadian waters.

200 ORGANIC POLLUTANTS IN WATER

Figure 12.15 shows the EI spectrum of beta-sitosterol obtained from the Canadian waters fecal sterol sample. Figure 12.16 shows the corresponding CI-methane spectrum of beta-sitosterol for comparison. The CI spectrum adds considerable information in the region of the molecular ion, helpful for unique identification.

It is concluded that high-resolution capillary column GC in combination with high-sensitivity, fast-scanning MS is ideal for analyses of this type.

IDENTIFYING ISOMERS THROUGH DIFFERENCES IN CI SPECTRA

The next example involves the use of high-resolution glass capillary column GC in conjunction with CI-MS to identify isomers through differences in CI spectra. Schnoes and Burlingame[2] have shown that GC/MS is one of the most powerful tools available to the organic geochemist. Often, because of the limitations of sample size and complexity of mixtures, high-resolution capillary columns used in conjunction with mass spectrometry furnish the most reliable structural information on isolated extracts. Because of the complexity of this mixture, solid probe MS is impossible using either low- or high-resolution MS.

In this analysis, our objective was to determine if CI-MS could be used to differentiate between sterane isomers. These isomers can be separated by high-resolution capillary column GC.

Table 12.4 shows the conditions employed in the analysis of a several million-year-old rock sample identified as marble slate fraction. It is well known that samples of this type can be first extracted and isolated into various hydrocarbon classes by techniques such as thin-layer or column chromatography.

Table 12.4 Conditions Used for Geological Sample[a] Analysis by Glass Capillary GC - CI - MS

1. Column:	50 m x 0.02 in. i.d. OV-1 glass capillary
2. Column Temperature:	Ambient to 280°C at 4°C/min
3. Carrier Gas:	Helium at a flow of 2 cc/min
4. Sample Size:	2 µg (total) in hexane
5. Injector Split Ratio:	10 to 1
6. CI Reactant Gas:	Methane
7. Source Pressure:	1 torr
8. Mass Range:	100-650
9. Scan Time:	3 sec

[a] Extract of a several-million-year-old rock.

TECHNIQUES AND METHODS OF ANALYSIS 201

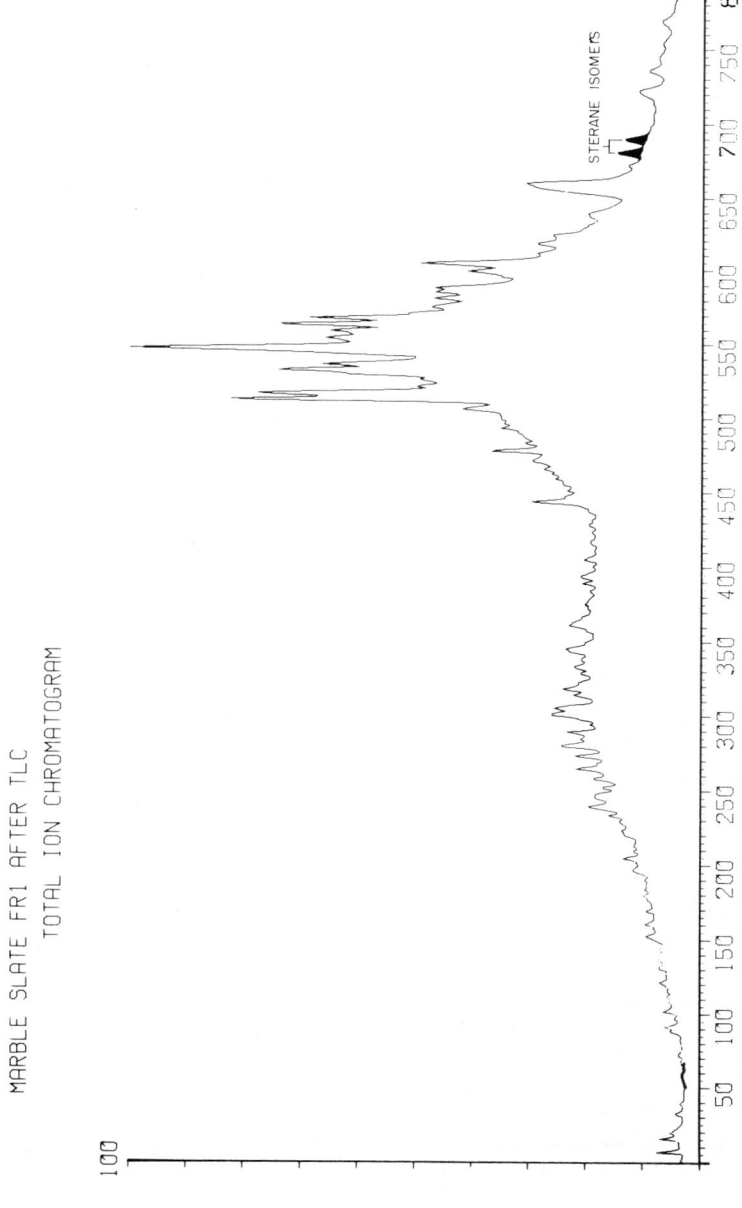

Figure 12.17 Total ion chromatogram of marble slate fraction after separation by thin-layer chromatography. Shaded portion represents sterane isomers.

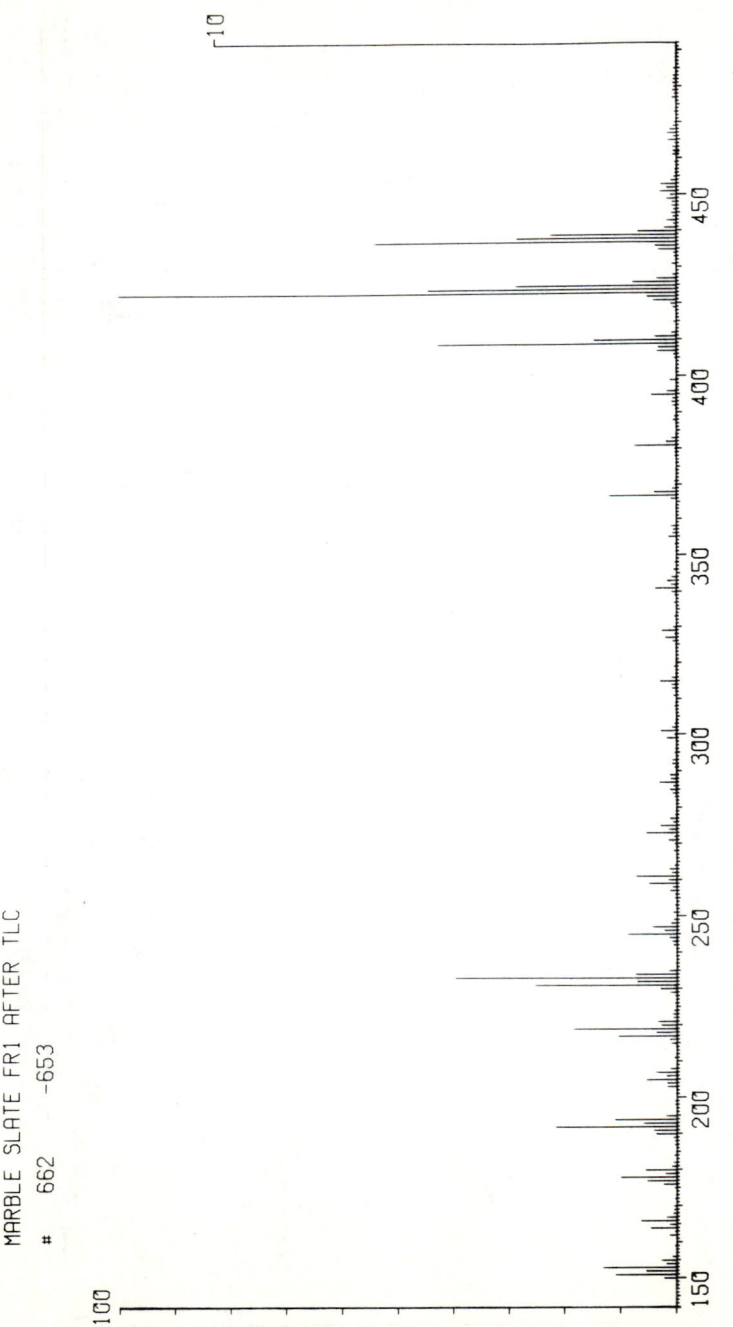

Figure 12.18 Mass spectrum by CI of scan #662 with background subtracted (scan #653).

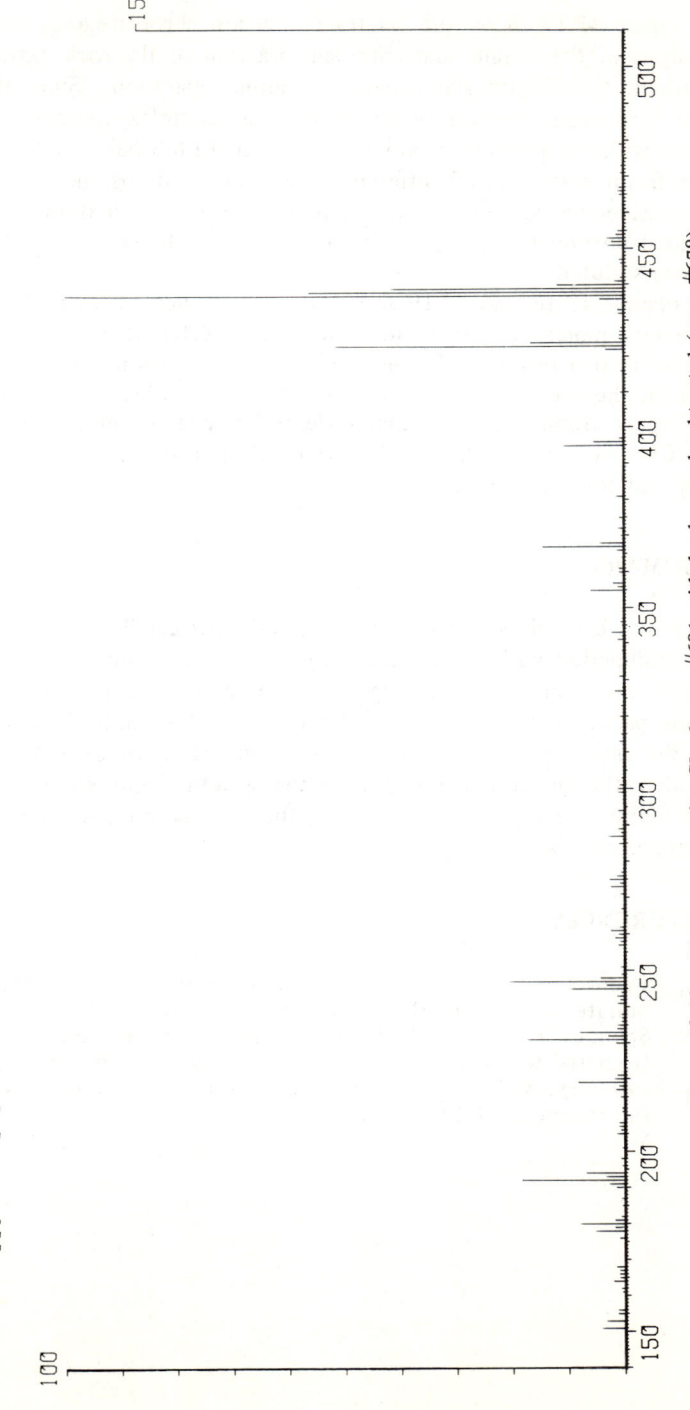

Figure 12.19 Mass spectrum by CI of scan #681 with background subtracted (scan #678).

Figure 12.17 shows the resulting total ion chromatogram obtained by analysis of the sterane and triterpane fraction of the rock extract using methane CI-MS with glass capillary column separation. Since the steranes and triterpanes represent some of the most complex compounds isolated from rocks or petroleum samples and since EI-MS has failed to show significant mass spectral differences in many of the isomers present, there is considerable interest in the CI mass spectra of such isomers. The shaded portion of the total ion chromatogram shows where two suspected isomers eluted.

Figures 12.18 and 12.19 show two compounds which have the same apparent molecular weight but significantly different m/e ratios in the 390-450 area region. Although exact structure has not yet been determined, the preliminary CI results on these and other steranes and triterpane isomers indicate that differentiation and identification is possible by CI. We are indebted to Professor Eglinton at the University of Bristol, England for this sample.

SUMMARY

It has been shown that high-resolution glass capillary column GC used in conjunction with a high-sensitivity, fast-scanning quadrupole mass spectrometer provides advantages over the conventional GC/MS system using packed columns in the analysis of complex mixtures of interest to the environmental chemist. It is found that both EI and CI-MS are needed, the choice depending upon the structural interaction desired and the compound type. In most cases, the two methods of ionization are complementary.

REFERENCES

1. Eneroth, P. and E. Nystrom. "Sterol Composition of the Steryl Sulfate Fraction in Human Feces," *Steroids* (1968).
2. Schnoes, H. K. and A. L. Burlingame. "Applications of Mass Spectrometry to Organic Geochemistry," *Topics in Organic Mass Spectrometry*, Vol. 8, A. L. Burlingame, Ed. (New York: Wiley-Interscience, 1970).

CHAPTER 13

ON THE ORIGIN OF POLYCYCLIC AROMATIC HYDROCARBONS IN THE AQUEOUS ENVIRONMENT

Anneli Hase* and Ronald A. Hites

    Department of Chemical Engineering
    Massachusetts Institute of Technology
    Cambridge, Massachusetts

INTRODUCTION

    It has been estimated that four out of five occurrences of cancer in man are due to environmental effects such as chemical carcinogens in air and water.[1] Among the most well-known chemical carcinogens are the polycyclic aromatic hydrocarbons (PAH) many of which are also potent toxins, mutagens and teratogens.[2-6] Because of their severe health effects, this class of compounds has been monitored and investigated extensively. The National Academy of Sciences has addressed the question of the occurrence of PAH in air and has summarized studies detailing both the chemical composition and the sources of PAH in air.[6]

    The aqueous environment (water and its associated sediment) has only recently received attention, but it is now recognized that PAH are ubiquitous water pollutants.[7-9] Unlike the occurrence of PAH in air, however, there is no concensus concerning their sources and identities in the aqueous environment. This controversy has recently been aggravated by the discovery that the PAH mixtures in various recent New England sediments are extremely complex.[10-12] For example, sediment from the Charles River Basin contains PAH with at least 11 different aromatic ring structures and with abundant alkyl-substituted derivatives of these compounds.[10] This chapter extends these studies[10,13-15] of

---

*Present address: Helsinki University of Technology, Otaniemi, Finland.

the Charles River Basin system in an attempt to generally define the sources of PAH in the aqueous environment.

The techniques involved in our studies have centered around mass spectrometry. The identification of the components in the sediment of the Charles River Basin was achieved by the use of combined gas chromatographic mass spectrometry (GC/MS) with on-line computer processing of the data.[10,16] Other studies of the PAH mixtures produced in combustion systems have also been carried out using GC/MS.[14] Because the details of these GC/MS procedures have been published elsewhere,[10,13-16] this chapter will not include such information but rather will concentrate on information gained by high-resolution mass spectrometry (HRMS) of the total, unseparated extracts.

Using high-resolution mass spectrometry, the elemental composition of the various components of a mixture can be determined assuming that their molecular ion abundances are significant. Because the molecular ion is usually the most intense peak in the mass spectra of PAH, HRMS is a very powerful tool for the analysis of these compounds. As opposed to GC/MS, however, one loses some information on the structures of the various isomers which are present. However, HRMS data supply detailed quantitative information on the distribution of alkyl homologs in PAH mixtures without extensive preseparation. This is the type of information with which this chapter will deal and on which conclusions will be based.

RESULTS AND DISCUSSION

The distribution of alkyl homologs in a recent sediment and a soil is exemplified by the data shown in Figure 13.1 which represents the homolog abundances of the $C_{18}H_{12}$ isomers (Z = -24*). It has also been shown that other ring systems give very similar distributions.[10] The line labeled "river sediment" represents the distribution of alkyl homologs in the Charles River Basin sediment; and the line labeled "Maine soil," which is replotted from Blumer and Youngblood,[12] represents the distribution of alkyl homologs in an "unpolluted" Maine soil. These distributions are similar to those of many different sediment and soil samples.[12] It is especially significant that the slopes of the lines shown in Figure 13.1 agree even though the data for the river sediment were obtained by high-resolution mass spectrometry of the total extract, whereas the data for the Maine soil were obtained by low resolution mass spectrometry after several preliminary separations. The agreement of these two lines is an independent verification of the two techniques and of the generality of this alkyl homolog distribution in various recent sediments.

*The value of Z in C_nH_{2n+z}.

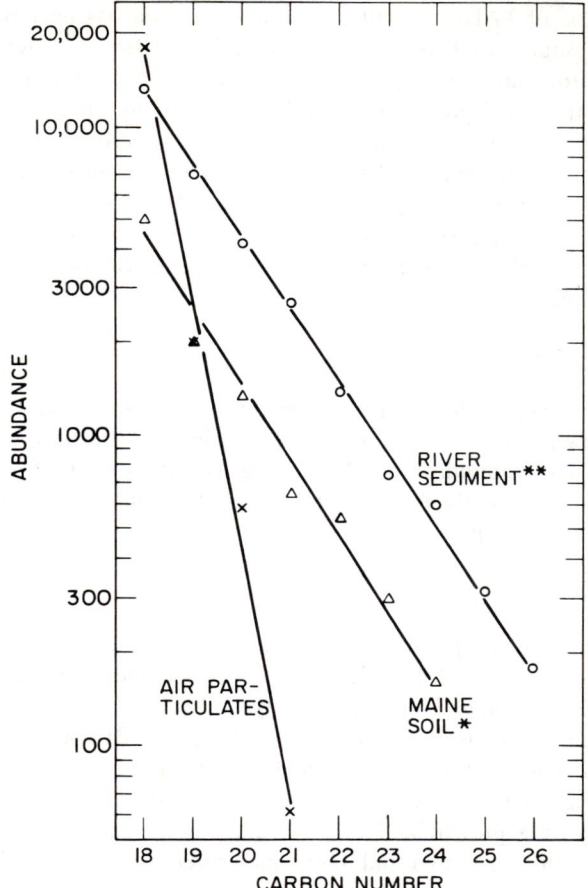

Figure 13.1 Alkyl homolog distribution for the chrysene-type series (Z = -24) in different environmental samples. Arbitrary vertical displacement for clarity.
*Replotted from Blumer and Youngblood.[12]
**Replotted from Hites and W. Biemann.[10]

Different sources of PAH in these sediments have been suggested.[10,12] These include (a) biosynthesis by plants, algae and bacteria; (b) petroleum spillage; and (c) combustion from mobile and stationary sources, refuse burning, forest fires and industrial activity (all of which lead to air particulate formation). We will address these possibilities.

Biosynthesis

Biosynthesis of PAH by plants, algae and bacteria has been suggested as a possible source of PAH (see the review by ZoBell[17]). Recent work in our laboratory and elsewhere, however, has not been able to demonstrate biosynthesis of any PAH either by plants or by anaerobic bacteria when precautions have been taken to exclude or monitor outside contamination.[15,18] Instead we have shown that the bioaccumulation effect is quite strong. It seems likely, therefore, that biological systems are not an *a priori* source of PAH but rather they could modify (by differential bioaccumulation) the distribution of compounds which were originally present in a sediment.

Petroleum Spillage

Polycyclic aromatic hydrocarbons may also originate from petroleum. PAH mixtures from petroleum are quite deficient in the unsubstituted species; the most abundant alkyl homolog usually contains three or four carbon atoms.[12] Clearly the PAH in sediments could not originate directly from petroleum.

Combustion

Because combustion-produced PAH transported by air are one of the most likely sources of PAH in the aqueous environment, we investigated the distribution of alkyl homologs in a typical sample of urban air particulates. For comparison, the Z = -24 series for PAH found in this air particulate sample is plotted in Figure 13.1; the other series are shown in Figure 13.2. These data show that pyrene-, chrysene-, benzopyrene-, etc.-type series are present and that the alkyl homologs of these compounds containing one, two and sometimes three carbon atoms also occur. The significant information to be drawn from Figure 13.2 is the low relative abundance of the alkyl homologs within a given series. For example, the abundance of the C_1-substituted pyrene-type species is approximately a factor of seven less than the unsubstituted species. In general, this result agrees with GC/MS results which indicate that alkyl-substituted PAH are in low abundance in urban air particulates[19] and in combustion-generated PAH mixtures in general.[14] Higher homologs are probably present, but since their abundances decrease very rapidly with carbon number, these compounds simply would not be detected.

Returning to Figure 13.1, we note that there is a considerable difference in relative homolog abundance between the air particulate sample and the sediment samples. Because of this difference, Blumer and

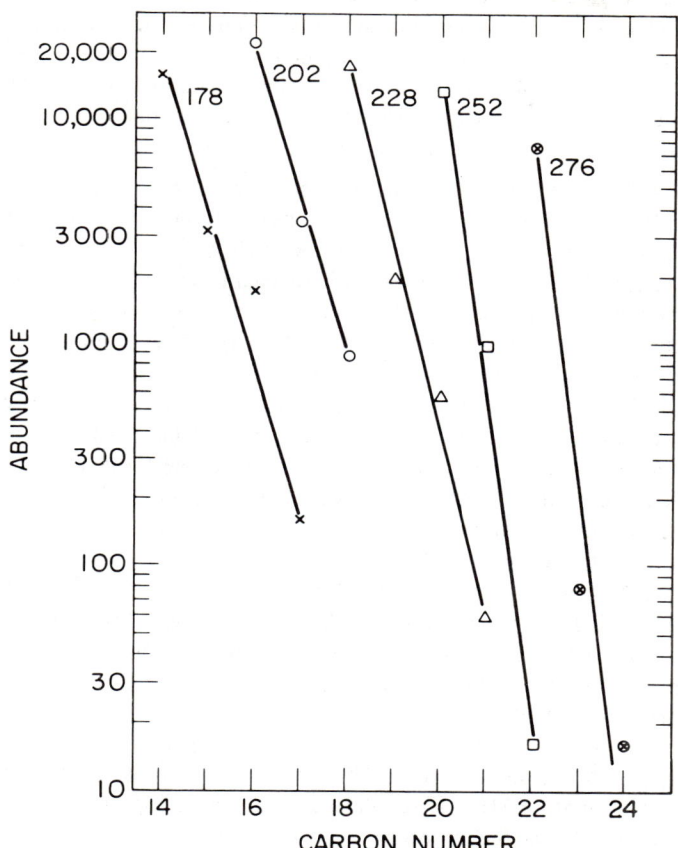

Figure 13.2 Alkyl homolog distribution for several PAH series in urban air particulates. The lines are labeled by the molecular weight of the unsubstituted species. Example isomers are: 178 (Z = -18) phenanthrene, 202 (Z = -22) pyrene, 228 (Z = -24) chrysene, 252 (Z = -28) benzo[a]pyrene, and 276 (Z = -32) benzo[ghi]perylene.

Youngblood[12] have concluded that the PAH in these recent sediments could not have come from the simple deposition of such anthropogenic airborne particulates but instead originate from forest fire particulates. We feel this conclusion is not correct for two reasons:

1. The general formation mechanism of PAH in combustion systems has been thoroughly investigated and there is now agreement that

qualitatively similar PAH mixtures are produced almost regardless of the fuel type and the combustion conditions.[14,20] One feature of PAH mixtures produced by all combustion systems is the very low abundance of alkyl homologs. Wood combustion (as in a forest fire) is no exception; experiments in our laboratory show that the burning of wood gives a distribution of PAH alkyl homologs more similar to that of urban air particulates than to that of recent sediments.

2. We feel that there are natural mechanisms by which the steep slope of the alkyl homolog distribution observed for urban air particulates can be changed into a more gradual slope such as that observed for the sediments (see Figure 13.1). One such mechanism could be differential water solubility of the higher alkyl homologs *vs* the unsubstituted species. McAuliffe[21,22] has shown that the logarithm of water solubility is a linear function of the number of carbon atoms for several alkyl benzenes and other hydrocarbons. Extrapolating somewhat, we would expect that the various alkyl homologs of PAH systems would have water solubilities distributed in a similar fashion. We, therefore, suggest the following mechanism for transition of the air particulate homolog distribution into that observed in the sediments: After airborne particle deposition on the soil or in the water, the lowest homologs (including the unsubstituted species) continuously fractionate into the water phase to an extent proportional to their carbon number as described above. The remaining species, which accumulate on the particulate matter and in the sediment, are therefore devoid of the lowest homologs, thus increasing the relative abundance of the higher homologs.

As a test of this hypothesis, we examined PAH in water *and* sediment samples from the Charles River Basin system and compared them to urban airborne particulate PAH. These data are shown in Figures 13.3 and 13.4. Figure 13.3 is the homolog distribution of the anthracene and phenanthrene series ($Z = -18$), and Figure 13.4 is the homolog distribution of the pyrene and fluoranthene series ($Z = -22$). In both cases the slope of the air particulate homolog plot is intermediate between that found in the water and in the sediment. This indicates that the water is enriched in the lowest homologs, while the sediment is enriched in the higher homologs relative to the air particulates. These data support our hypothesis that the main source of polycyclic aromatic hydrocarbons in the aqueous environment is anthropogenic, airborne, combustion-produced, polycyclic aromatic hydrocarbons.

TECHNIQUES AND METHODS OF ANALYSIS 211

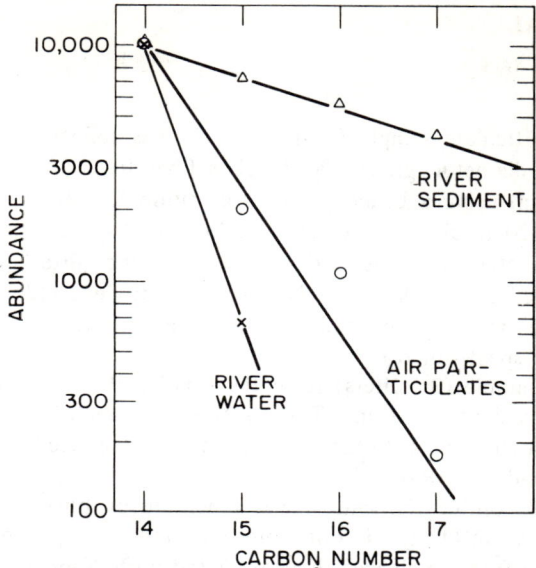

Figure 13.3 Alkyl homolog distribution for the phenanthrene-type series (Z = -18) in Charles River water and sediment and in urban air particulates. Data are normalized such that the unsubstituted species is 10,000 units in each case.

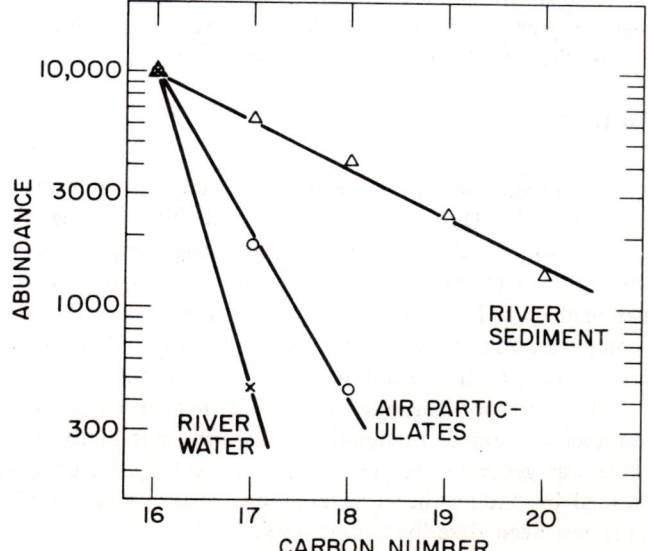

Figure 13.4 Alkyl homolog distribution for the pyrene-type series (Z = -22) in Charles River water and sediment and in urban air particulates. Data are normalized such that the unsubstituted species is 10,000 units in each case.

EXPERIMENTAL

Samples

1. An air particulate sample from urban air was collected on a pre-cleaned, 10-cm diameter, grade 900 AI glass fiber filter (Arthur H. Thomas Co., Philadelphia) using a Unico gas suction pump for 50 hr. The collecting point was 20 m above street level in Boston, Massachusetts and the sample was collected in August 1975. The glass fiber filter was then Soxhlet extracted using 200 ml of Nanograde methylene chloride (Mallinckrodt). The extract was reduced to 500 μl and analyzed using high-resolution mass spectrometry.

2. A water sample (3.5 liters) from the Charles River, Boston, was collected from a depth of 1 m. The water was extracted with 200 ml of Nanograde methylene chloride; the extract was reduced to 100 μl and was analyzed by HRMS.

3. The wood combustion sample was generated by burning pine wood in the laboratory; 200 mg of soot were collected on a precleaned, cooled stainless steel surface which was then extracted with Nanograde methylene chloride in an ultrasonic bath. The extract was centrifuged, reduced to 500 μl and analyzed by GC/MS.

4. Some previously unpublished data from the HRMS analysis of Charles River sediment is presented here; the details of sediment sampling and analysis are described by Hites and W. Biemann.[10]

Analytical Methods

The samples were analyzed using HRMS by directly inserting the sample into the ion source of a DuPont Instruments 21-110B mass spectrometer. The samples were vaporized continually at increasing temperature while several exposures on a photographic plate were made. After development, the plate was read on a D.W. Mann comparator which is on-line to an IBM 1802 computer, and the exact masses were converted to elemental compositions. Details of this technique have been described by K. Biemann.[23] The corresponding intensities were then arranged as tables of carbon number *vs* number of double bonds and rings in the molecule. One such table was generated for each exposure and then a composite table was formed by adding the corresponding entries of each table. This technique has been described previously.[10,15]

ACKNOWLEDGMENTS

We are grateful to the National Science Foundation for support (grant number DES-7515044); the instrumentation was supported in part by National Institutes of Health Research Grant RR00317 from the Division of Research Facilities and Resources.

REFERENCES

1. Clayson, D. B. *Eur. J. Cancer* 3, 405 (1967).
2. Gerarde, H. W. *Toxicology and Biochemistry of Aromatic Hydrocarbons*. (London: Elsevier Publishing, 1960).
3. Daly, J. W., D. M. Jerina and B. Witkop. *Experimentia* 28, 1129 (1972).
4. Steere, N. V. *J. Chem. Ed.* 49, A19 (1972).
5. Arcos, J. C. and M. F. Argus. *Chemical Induction of Cancer. Structural Bases and Biological Mechanisms*, Volume II A (New York: Academic Press, 1974).
6. National Academy of Sciences. *Particulate Polycyclic Organic Matter*, in the series on *Biologic Effects of Atmospheric Pollutants*, Washington, D. C. (1972).
7. Martin, A. and M. Blumer. *Polycyclic Aromatic Hydrocarbons: Occurrence and Analysis—A Partial Bibliography*, unpublished manuscript, Woods Hole Oceanographic Institution, Massachusetts (1975).
8. Andelman, J. B. and J. E. Snodgrass. *Critical Reviews in Environmental Control* 4, 69 (1974).
9. Harrison, R. M., R. Perry and R. A. Wellings. *Water Research* 9, 331 (1975).
10. Hites, R. A. and W. G. Biemann. *Advances in Chemistry* 147, 188 (1975).
11. Giger, W. and M. Blumer. *Anal. Chem.* 46, 1663 (1974).
12. Blumer, M. and W. W. Youngblood. *Science* 188, 53 (1975); Youngblood, W. W. and M. Blumer. *Geochim. Cosmochim. Acta* 39, 1303 (1975).
13. Hites, R. A. and K. Biemann. *Science* 178, 158 (1972).
14. Hase, A., P. H. Lin and R. A. Hites. *Proceedings of Symposium on Polycyclic Aromatic Hydrocarbons*, Battelle Institute, October 1975, Raven Press (in press).
15. Hase, A. and R. A. Hites. *Geochim. Cosmochim. Acta* (in press).
16. Hites, R. A. and K. Biemann. *Anal. Chem.* 40, 1217 (1968).
17. Zobell, C. E. *Proceedings of Joint Conference on Prevention and Control of Oil Spills*, Washington, D. C., June 15-17, 1971, American Petroleum Institute, Washington, D. C. (1971), pp. 441-451.
18. Grimmer, G. and D. Duevel. *Z. Naturforsch.* 25B, 1171 (1970).
19. Lao, R. C., R. S. Thomas, H. Oja and L. Dubois. *Anal. Chem.* 45, 908 (1973).

20. Lindsey, A. J. *Combust. and Flame* **4**, 261 (1960).
21. McAuliffe, C. *Chem. Geol.* **4**, 225 (1969).
22. McAuliffe, C. *J. Phys. Chem.* **70**, 1267 (1966).
23. Biemann, K. *Applications of Computer Techniques in Chemical Research.* (London: Institute of Petroleum, 1972), pp. 5-22.

CHAPTER 14

SEPARATION AND ANALYSIS OF REFRACTORY POLLUTANTS IN WATER BY HIGH-RESOLUTION LIQUID CHROMATOGRAPHY

W. W. Pitt, Jr., R. L. Jolley, and S. Katz

Oak Ridge National Laboratory
Oak Ridge, Tennessee

INTRODUCTION

The continuing discharge of stable organic compounds from various sources to surface waters and the buildup of organic constituents by recycling constitute a serious threat to our water quality.[1-3] Furthermore, as such water contamination becomes excessive, additional processing steps may become necessary to permit reuse of the water. Because most present analytical techniques for dissolved refractory organics are nonspecific and often indirect, they are not adequate for use in evaluating new developments of advanced treatment processes.[4-5] These analytical techniques do not provide sufficient information to determine the chemical forms or concentrations of the stable organic compounds that resist degradation. Thus, we do not know the effectiveness of various treatment steps for specific refractory compounds; nor do we know what harmful effects might result if these refractory compounds continue to build up in the water supply.

To better define the pollution problem, an effort is being made to determine the specific organic compounds present in various waste effluents and natural waters using high-resolution liquid chromatography (HRLC). An important first step in providing the analytical information to answer pertinent questions and permit rational development of additional waste treatments was the demonstration that HRLC can provide reliable measurement of the refractory organic compounds at microgram-per-liter levels in sewage effluents.[6] It was demonstrated that more than 150 specific refractory organic compounds are present in effluents from municipal sewage treatment plants at microgram-per-liter levels and that many more refractory compounds may be present at higher concentrations in effluents from industrial waste treatment plants.[7,8]

By presenting the results of the analyses achieved by HRLC on a diverse set of aqueous samples, we intend to show the wide ranging capability of the method. Although HRLC does not supplant other methods (*i.e.*, gas chromatography) it does fill many of the gaps left by them. No attempt is made here to fully describe any specific application of the method. In some cases, the details are reported elsewhere,[8] and in other cases the method has only recently been applied.

METHOD

Sample Preparation

Since the lower limit of detection for HRLC with a uv photometer is 100 µg/liter to 100 mg/liter (depending on the uv absorptivity of the individual compounds), and the concentrations of specific contaminants in water samples may be 1 µg/liter or less, concentration of samples by factors up to 10,000 may be necessary prior to analysis. Low-temperature evaporation is a reasonably convenient method of concentration and provides adequate recovery of stable, nonvolatile organic compounds. For concentration by factors greater than 500, we chose a two-step procedure. A concentration of 5- to 100-fold was first obtained in a vacuum still or rotary evaporator; the resulting volume (~ 100 ml) of concentrate was then further reduced by lyophilization. Prior to concentration, the water samples were filtered through a 0.45-µm membrane to remove suspended matter; the samples were then acidified by percolation through a bed of weak cation exchanger beads in the H^+ form to reduce the quantity of precipitated inorganic salts in the concentrate. The remaining solids were removed by centrifugation. As determined by carbon analyses before and after concentration, the recovery of noncarbonate organic compounds in the final concentrate was generally greater than 85%. Only when concentrating natural water containing a large amount of humus was the recovery less than 85%. Some of the aqueous waste samples, particularly those from coal liquefaction processes, had sufficient concentration of organics and required little or no additional concentration; these were either lyophilized or injected onto the chromatographic column without preparation.

High-Resolution Liquid Chromatography

Because of its sensitivity and capability for detecting and quantifying many individual organic compounds in complex aqueous samples, high-resolution anion exchange chromatography[9,10] was used to analyze for individual aquatic pollutants. Two different systems, the UV-Analyzer (for uv-absorbing compounds) and the carbohydrate analyzer, were used. Each consisted of a heated, high-pressure anion exchange column; a sample injection valve; a concentration-gradient generating and pumping system; a detector; and a strip-chart recorder. The ion exchange column for each system was a 150-cm length of type 316 seamless stainless steel tubing (0.22- to 0.62-cm i.d.) packed with strongly basic anion exchange resin. To minimize peak broadening, a 0.05- to 2.5-ml sample (the volume depends on the inside diameter of the ion exchange column and the nature of the sample) was injected into the column by a six-port injection valve mounted as near to the top of the column as possible.

The UV-Analyzer chromatograms were developed by eluting the sample constituents with an ammonium acetate-acetic acid buffer solution (pH 4.4) whose acetate concentration gradually increased from 0.015 to 6.0 M. The eluent was pumped through the ion exchange column at the rate of about 250 ml cm^{-2} hr^{-1} with a pressure drop of 100 to 200 atm. The absorbances of the column effluent at 254 and 280 nm, referred (at the same wavelengths) to the stream entering the column, were monitored by a two-wavelength, dual-beam flow photometer and recorded on a strip chart. An additional detection method, cerate oxidimetry,[11] was used with many samples.

The carbohydrate analyzer utilized a sodium borate-boric acid buffer (pH 8.9) as the eluent. This eluent, whose boron concentration increased from 0.085 to 0.845 M during an analysis, was pumped through the column at the rate of about 350 ml cm^{-2} hr^{-1}, also with a pressure drop of 100 to 200 atm. A modified version of the analyzer was used in which the separated carbohydrate compounds were monitored with a cerate oxidimetric system that mixed the column effluent with 2.5 x 10^{-4} N Ce(IV) in 2.5 N H_2SO_4. The Ce(III) produced by reaction at 100°C for 10 min with oxidizable compounds (carbohydrates) is monitored fluorimetrically (Ex, 254 nm; Em, 360 nm). This detector provided a capability of detecting sugars at the nanomole level on the carbohydrate analyzer and phenolic compounds at 100 µg/liter on the UV-Analyzer.

Identification of Pollutants

Tentative identification of organic pollutants represented by a chromatographic peak may be made from the elution volume (or retention position) of the peak and the ratio of uv absorbances at the two wavelengths, and/or from the cerate oxidimetric response relative to uv absorbance.

More positive identification requires corroborative evidence which we obtain routinely by collection of separated fractions from the chromatographic run, removal of the eluent by freeze-drying, dissolution of the organic compound in methanol, and subsequent examination by uv spectrometry, gas chromatography, mass spectrometry, and other techniques.[12,13] Peak identification for routine chromatographic separations on sample types for which the peak identities have been previously established may also be accomplished by comparison of elution volumes and absorbance ratios.

RESULTS

Effluents from Municipal Sewage Treatment Plants

Both chlorinated and unchlorinated primary and secondary effluents from municipal sewage treatment plants have been analyzed by high-pressure liquid chromatography. Using the UV-Analyzer, over 100 uv-absorbing and/or cerate-oxidizable compounds have been separated and detected in samples of concentrated (100- to 1000-fold) primary effluents. To date, 56 soluble organic compounds have been identified in the primary effluent, 25 of these quantified at the parts per billion level.[14] In addition, 103 unknown constituents were characterized with respect to gas chromatographic and mass spectrometric properties. As many as 50 or more uv-absorbing and/or cerate-oxidizable compounds have been separated and detected in samples of concentrated (500- to 3000-fold) secondary effluents. To date, 13 of these have been identified and quantified at the ppb level,[14] and an additional 20 have been characterized with respect to gas chromatographic and mass spectrometric properties.[14]

Natural Waters

The identification and quantification of trace organic compounds present in various natural waters have also been studied using high-resolution, anion exchange chromatography. The sites chosen for obtaining natural waters for the study, and the basis of selection, are listed in Table 14.1. Samples from these sites were concentrated from 500- to 10,000-fold depending on the sample. The concentrated samples were analyzed on the anion exchange chromatograph, in which the cerate-oxidative detector was in series with the uv detector, and on the carbohydrate analyzer. Shown in Figure 14.1 are UV-Analyzer chromatograms for the water samples from Watts Bar Lake and Lake Marion, and Figure 14.2 shows

Table 14.1 Initial Sampling Sites for the Measurement of the Molecular Organic Contaminants in Natural Waters

Geographic Location	Physical Location	Reasons for Selection
Oak Ridge, Tennessee	Walker Branch Watershed	Well-characterized ecologically and relatively undisturbed
Watts Bar Lake Kingston, Tennessee	Cooling water inlet[a] to TVA steam plant	Reasonably unpolluted water routinely chlorinated for antifouling purposes
Fort Loudoun Lake Knoxville, Tennessee	Midstream across from USN & USMC Reserve Training Center	Downstream from major city sewage treatment plant
Lake Marion, South Carolina	Santee Dam	Example of waters with high natural organic content
Mississippi River Memphis, Tennessee	Cooling water intake[a] of Allen steam plant	Nation's major watershed and site of samples taken for other studies
South Fork of Holston River Kingsport, Tennessee	Just above junction with North Fork	Downstream from large industrial organic chemical plant

[a] Inlets chosen to provide baseline for later studies of effects of chlorination.

220 ORGANIC POLLUTANTS IN WATER

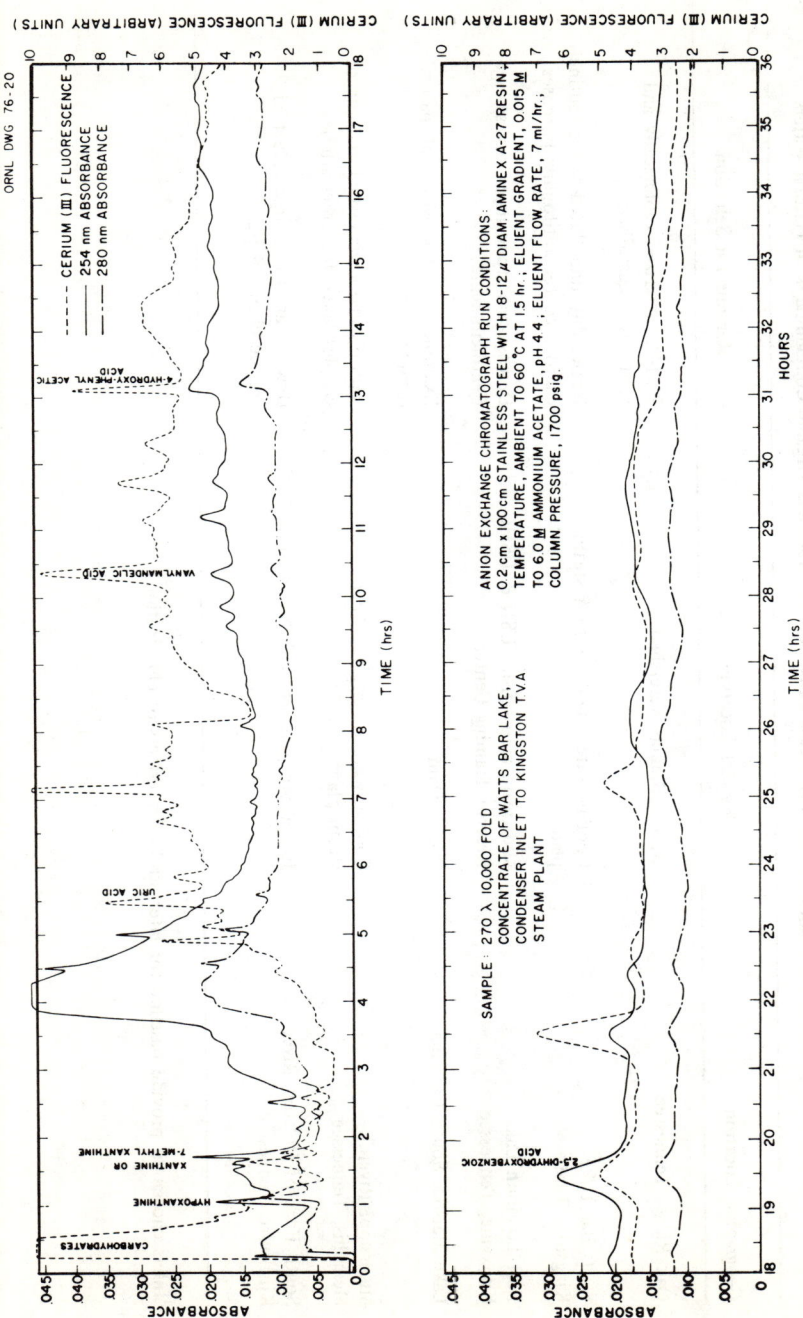

TECHNIQUES AND METHODS OF ANALYSIS 221

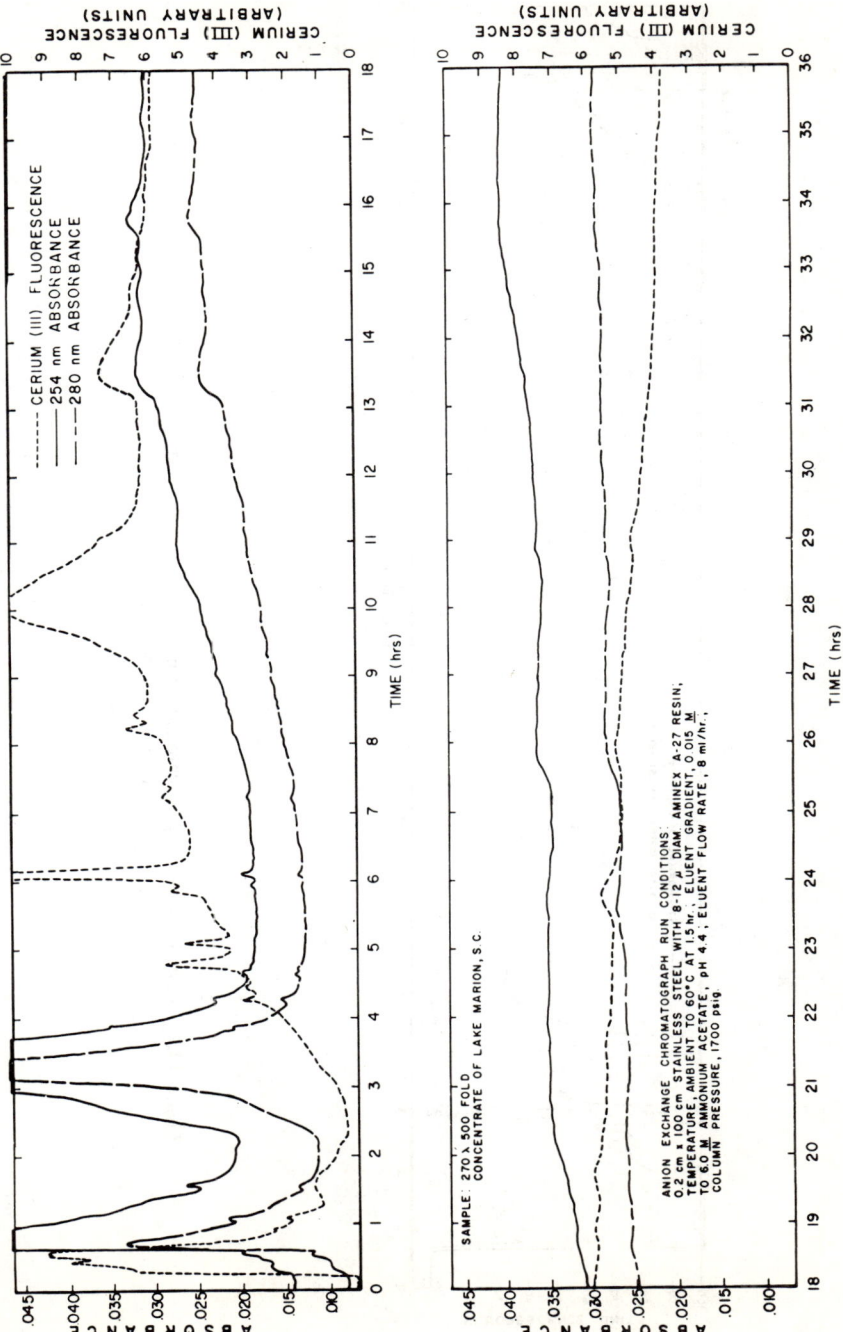

Figure 14.1 UV-Analyzer chromatograms of trace organics in natural waters.

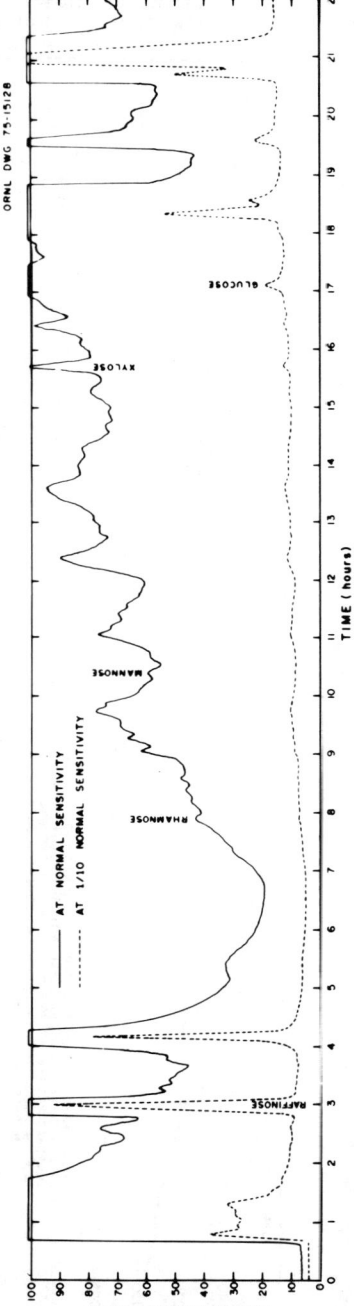

Figure 14.2 Carbohydrate analyzer chromatogram of natural water sample (1.2 ml 1000-fold concentrate of Fort Loudoun Lake water).

a carbohydrate analyzer chromatogram of a water sample from Fort Loudoun Lake. These indicate the diversity of sample types and concentrations to be found in natural waters. The 16 organic compounds listed in Table 14.2 have been identified in one or more of the samples of natural waters.

Table 14.2 Molecular Constituents Identified in Natural Water Samples

Constituent	Sample Site[a]	Identification Method[b]
p-Cresol	5	AC, CC, MS
Diethylene glycol	1	AC, MS
Glycerine	1,2,3,4,5	AC, GC, MS
Glycine	3	AC, GC, MS
Mannitol	1	AC, GC, MS
Methyl-α-d-glucopyranoside	4	AC, GC, MS
Methyl-β-d-glucopyranoside	4	AC, GC, MS
Sucrose	1,3	AC, GC, MS
Xylitol	1	AC, GC, MS
Urea	2,3	AC, GC, MS
Inositol	1,2,3,4,5	GC, MS
o-Methylinositol	1,2,3,4,5	GC, MS
Linoleic acid	1,3	MS
Oleic acid	1,3	MS
Palmitic acid	1,3	MS
Stearic acid	3	MS

[a] 1 - Watts Bar Lake; 2 - Fort Loudoun Lake; 3 - Lake Marion; 4 - Mississippi River; 5 - Holston River.

[b] AC - anion exchange chromatography; GC - cation exchange chromatography; GC - gas chromatography; MS - mass spectroscopy.

Potable Waters

Approximately 100 liters of potable water from a metropolitan water supply was obtained at the water treatment plant and returned to Oak Ridge National Laboratory (ORNL) for concentration and analysis by high-resolution liquid chromatography. The water sample was obtained after sand filtration and contained a 2.1 mg/liter chlorine residual. Prior to sampling, the water had been chlorinated (chlorine concentration of 12 mg/liter) at the treatment plant intake and passed through an open settling basin (3-day residence time). The sample was stored at ice water temperatures for the several days required for transportation to ORNL and subsequent low-temperature ($< 35°C$) vacuum concentration. Prior to concentration, the chlorine residual was destroyed by the addition of a stoichiometric amount of thiosulfate solution. This potable water sample, which contained a 1.6 mg/liter concentration of organic carbon, was then concentrated 3600-fold by vacuum distillation in preparation for analysis by HRLC.

Only 8 major and about 20 small uv-absorbing peaks and/or cerate-oxidizable chromatographic peaks were detected during chromatography. The eluate fractions containing the chromatographic peaks were collected and processed through the routine multicomponent identification procedure. The following four organic constituents were identified and quantified in the fractions: inositol, 0.5 ppb; o-methylinositol, 11 ppb; mannitol, 0.7 ppb; and succinic acid, 8 ppb. Six other compounds were characterized with respect to the gas chromatographic and mass spectral characteristics of their trimethylsilyl derivatives. In addition, a methylene chloride extraction of the solids precipitated during the concentration procedure (principally, inorganic phosphate and sulfate salts) contained dioctylphthalate, dibutylphthalate, and glycol-l-palmitate. Thirty small chromatographic peaks were also detected in the concentrated sample with the carbohydrate analyzer.

Aqueous Streams from Coal Pyrolysis Experiments

Because of the high probability of coal becoming the major energy source in the near future, and the consequent utilization of coal for the production of gaseous and liquid fuels, coal-related waste effluents may become environmentally significant. High-pressure liquid chromatography is being applied to the determination of nonvolatile organic constituents in aqueous effluents from coal liquefaction processes. Aqueous stream samples from coal pyrolysis experiments have been analyzed by high-resolution liquid chromatography. All samples were chromatographed without prior concentration, and each chromatogram showed many uv-absorbing and cerate-oxidizable constituents (Figures 14.3 and 14.4). Using the multicomponent analytical identification and quantitation procedure, six phenolic compounds were identified and quantified (Table 14.3). The lower values obtained from the flame ionization detector response reflect the fact that the HRLC peaks were not single components. Twelve other compounds were characterized with respect to their gas chromatographic and mass spectral properties.

Since the predominating compounds in coal process aqueous wastes are phenolic, the chromatographic conditions are being optimized for separation of phenol-type compounds. A chromatogram of eight phenolic

TECHNIQUES AND METHODS OF ANALYSIS 225

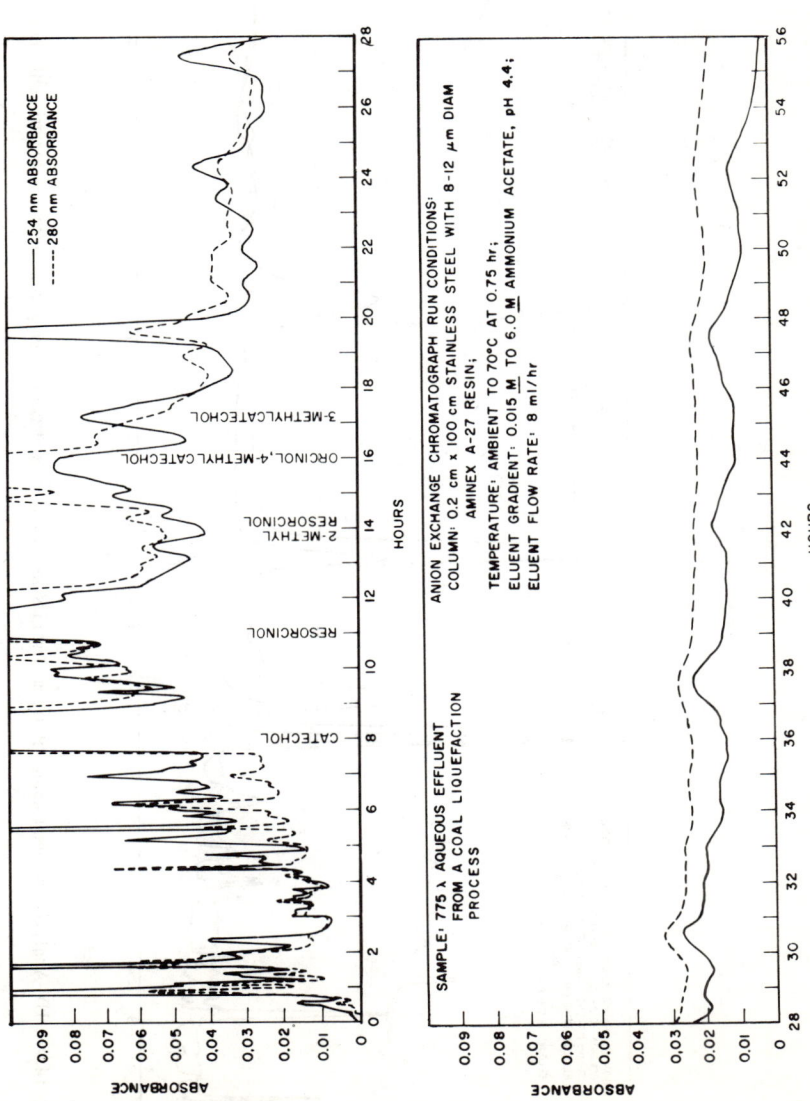

Figure 14.3 UV-Analyzer chromatogram of trace organics in an aqueous stream from a coal pyrolysis process (not concentrated).

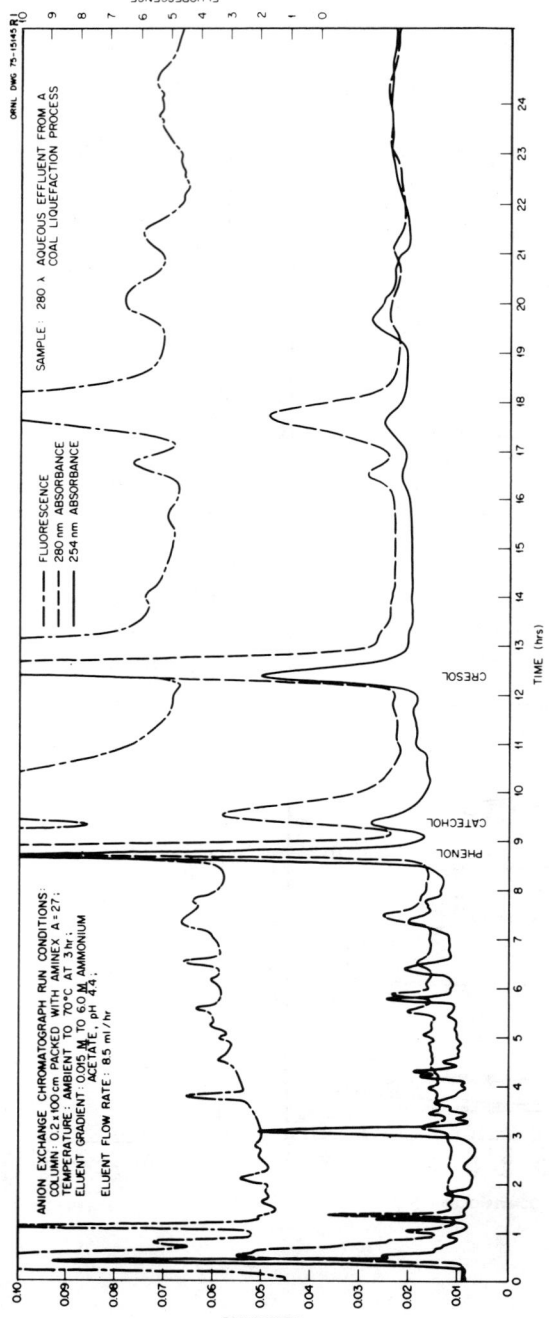

Figure 14.4 UV-Analyzer chromatogram of trace organics in an aqueous stream from a coal pyrolysis experiment (not concentrated).

Table 14.3 Soluble Organic Constituents in an Aqueous Sample from a Coal Pyrolysis Experiment

Constituent	Concentration (mg/liter)			
	FID[a]	Enzyme[b]	Prep. AC[c]	Anal. AC[d]
Catechol	660	560	3,600	2,000
3-Methylcatechol	170	–	–	–
4-Methylcatechol	110	–	–	–
Orcinol	120	–	380	500
Resorcinol	220	–	790	1,000
2-Methylresorcinol	14	–	100	10-50

[a]Based on flame ionization detector response during gas chromatography of HRLC fractions.
[b]Analyzed by B. Z. Egan using a catechol specific enzyme technique.
[c]Based on the uv absorbance of the preparative-scale chromatographic peak.
[d]Based on the uv absorbance of the analytical-scale chromatographic peak.

compounds, indicating the capability of the system, is shown in Figure 14.5. A slightly better separation than that shown was achieved using isocratic elution with 2 M ammonium acetate at 60°C.

Information concerning phenolic and other compounds in coal-processing waste streams is of great importance in assessing the possible environmental impact of coal-conversion plants. The data on chemical composition of such effluents will also provide informational feedback for modification of engineering and plant flowsheets. For example, the high-resolution liquid chromatography system was used in analyzing the efficacy of biological treatment of the phenolic wastes from coal processing plants. Shown in Figures 14.6 and 14.7 are chromatograms of an aqueous waste stream from a coal pyrolysis experiment before and after biological treatment.

DISCUSSION

The results demonstrate the capability of high-resolution liquid chromatography, with sensitive uv photometric and cerate oxidimetric detection, to separate and detect many different classes of organic compounds from aquatic environmental samples at the low levels existing in the environment. Many of the compounds could not have been detected in aqueous samples by extraction-gas chromatographic methods, and many others would have been quite difficult to detect.

228 ORGANIC POLLUTANTS IN WATER

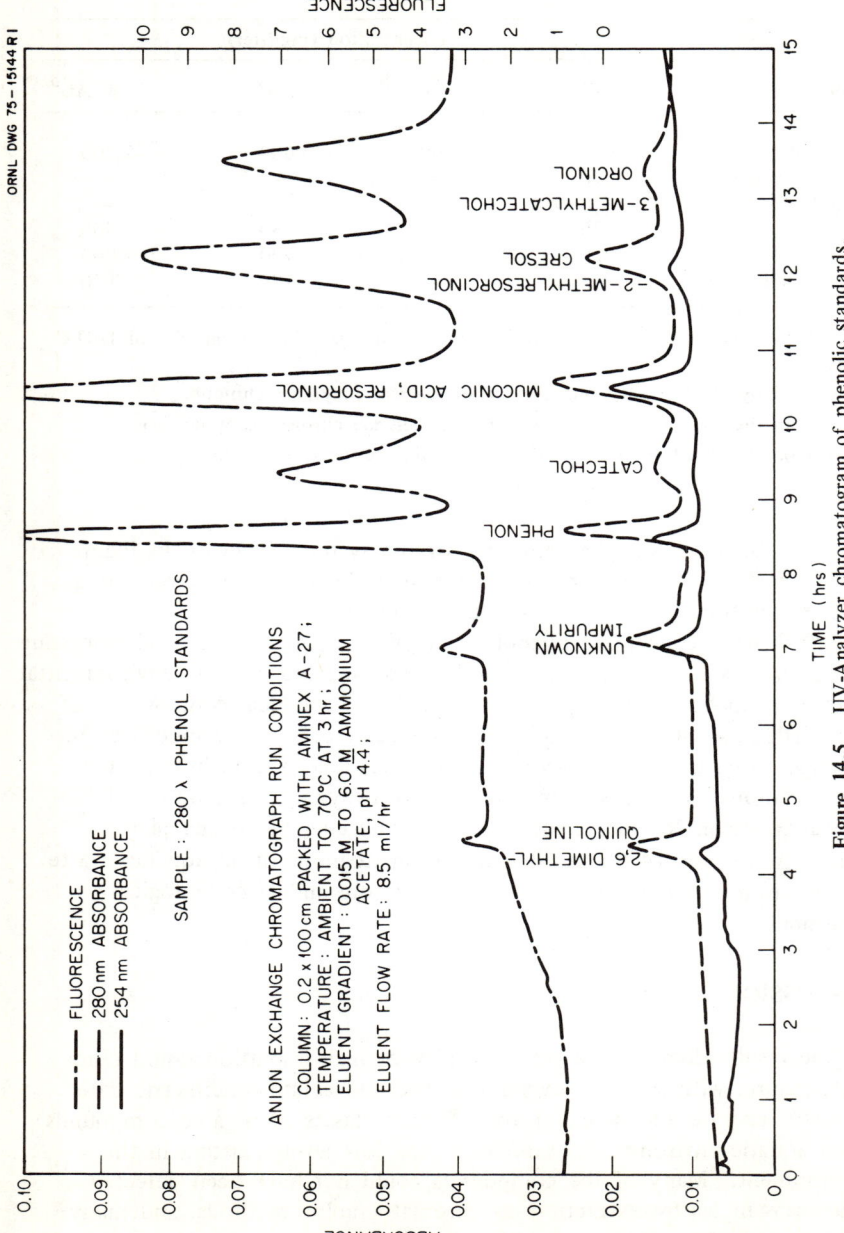

Figure 14.5 UV-Analyzer chromatogram of phenolic standards.

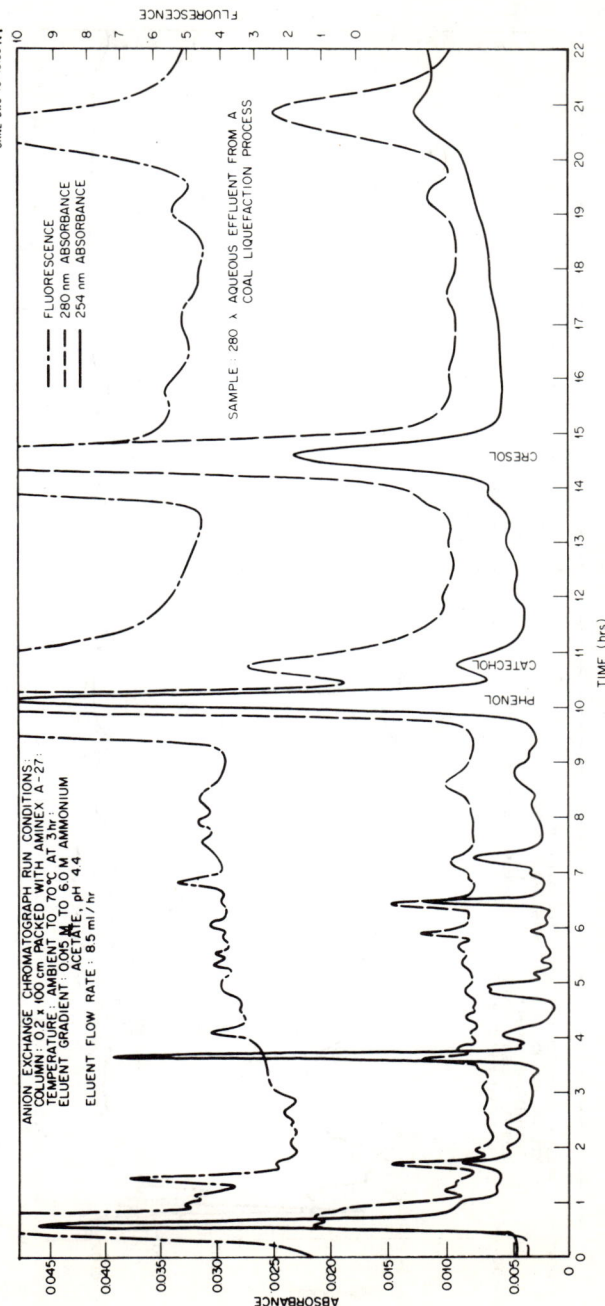

Figure 14.6 UV-Analyzer chromatogram of trace organics in an aqueous stream from a coal pyrolysis experiment before biochemical treatment.

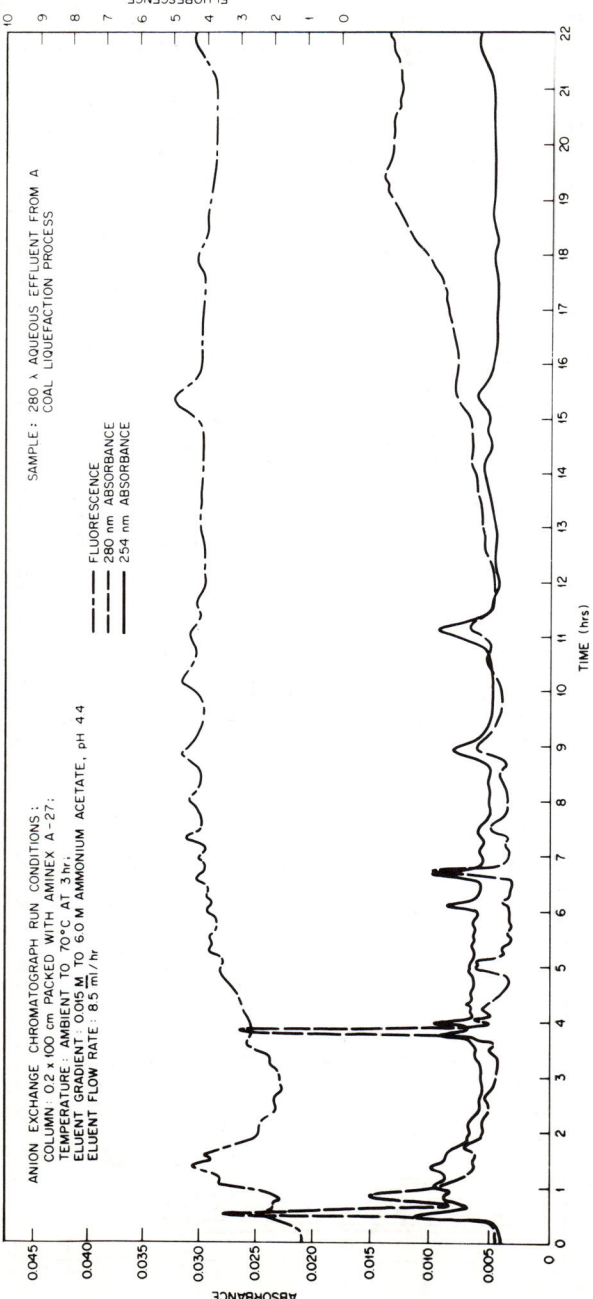

Figure 14.7 UV-Analyzer chromatogram of trace organics in an aqueous stream from a coal pyrolysis experiment after biochemical treatment.

The likelihood of altering the nature of the compounds in the samples, as occasionally occurs with other methods, is reduced by use of HRLC. The samples are analyzed in the same media in which they exist in the environment (water), and the treatment prior to analysis is either gentle (vacuum evaporation) or nonexistent. The detection of sources of pollution, the testing of the effectiveness of treatment steps, and the determination of the ultimate environmental effects of process effluents are obvious desired uses of HRLC. Use of HRLC for these purposes should occur with greater frequency in the near future.

ACKNOWLEDGMENTS

Research sponsored by the U.S. Energy Research and Development Administration, the U.S. Environmental Protection Agency, and the National Science Foundation RANN.

REFERENCES

1. Bunch, R. L., E. F. Barth, M. B. Ettinger. *J. Water Pollution Control Fed.* **33**, 122 (1961).
2. Middleton, F. M. and A. A. Rosen. *Pub. Health Rep.* **71**, 1125 (1956).
3. Hunter, J. V. In *Organic Compounds in Aquatic Environments*, S. D. Faust and J. V. Hunter, Eds. (New York: Marcel Dekker, 1971), pp. 51-94.
4. Painter, H. A. In *Water and Water Pollution Handbook*, Vol. 1, L. L. Ciaccio, Ed. (New York: Marcel Dekker, 1971), pp. 329-64.
5. Tyckman, D. W., F. W. Irvin and R. H. F. Young. *J. Water Pollution Control Fed.* **39**, 458 (1967).
6. Katz, S., W. W. Pitt, Jr., C. D. Scott and A. A. Rosen. *Water Res.* **6**, 1029 (1972).
7. Jolley, R. L., W. W. Pitt, Jr. and C. D. Scott. In *Realism in Environmental Testing and Control,* 1973 Proceedings of the Institute of Environmental Sciences, 19th Annual Technical Meeting, Mt. Prospect, Ill., pp. 247-52.
8. Pitt, W. W., Jr., R. L. Jolley, and S. Katz. EPA-660/2-74-076 Washington, D.C. (August 1974).
9. Scott, C. D. *Clin. Chem.* **14**, 521 (1968).
10. Scott, C. D., R. L. Jolley, W. F. Johnson, and W. W. Pitt, Jr. *Clin. Chem.* **535**, 701 (1970).
11. Katz, S. and W. W. Pitt, Jr. *Anal. Lett.* **5**(3), 177 (1972).
12. Mrochek, J. E., W. C. Butts, W. T. Rainey, Jr. and C. A. Burtis. *Clin. Chem.* **17**, 72 (1971).
13. Jolley, R. L., S. Katz, J. E. Mrochek, W. W. Pitt, Jr. and W. T. Rainey. *Chem. Technol.* 312 (1975).
14. Pitt, W. W., Jr., R. L. Jolley, and C. D. Scott. *Environ. Sci. Technol.* **9**, 1068 (1975).

CHAPTER 15

DETERMINATION OF CHLORINATION EFFECTS ON
ORGANIC CONSTITUENTS IN NATURAL AND
PROCESS WATERS USING HIGH-PRESSURE
LIQUID CHROMATOGRAPHY

Robert L. Jolley, Guy Jones, Jr.,
W. W. Pitt, Jr., and James E. Thompson

    Oak Ridge National Laboratory
    Oak Ridge, Tennessee

INTRODUCTION

Because of the comparatively high biocidal effectiveness and relatively simple application, chlorine has become the principal agent for sterilizing potable waters, disinfecting effluents from sewage treatment plants, and removing biological surface films in cooling systems of electric power-generating plants. Although the side effects of chlorination were questioned almost a decade ago,[1] only recently has credible evidence been presented to show that chlorine-containing organic compounds are formed during the chlorination of drinking water,[2-4] sewage effluents,[5-8] and cooling waters.[9,10]

An understanding of the chlorination effects on organic constituents in natural and process waters and an assessment of their possible environmental effects must be founded on detailed information concerning concentrations and compositions of organics in the waters of environmental concern and knowledge of their aqueous chlorination reactions. Several compilations of the identities and concentrations of organics in a variety of waters have been made within the past two years.[11-14] Details of the aqueous chlorination reactions of organics have been previously summarized.[7,10,15-17] Thus, from the known wide variety of organic compounds in waters and from the large variety of possible chlorination reactions it should be expected that a large number of chloro-organic products exist in chlorinated waters.[18] The total environmental impact of these chloro-organics, although not yet established, is a subject of active investigation.[19-25]

This chapter summarizes experimental results obtained by analyzing chlorinated sewage effluents and cooling waters using ^{36}Cl to tag chloro-organics and high-pressure liquid chromatography to separate the ^{36}Cl-tagged constituents. The results are discussed with respect to the effects of physical and chemical parameters.

METHODS

Details of the ^{36}Cl-tracer-high-pressure liquid chromatographic procedures have been previously described.[7,26] The basic procedure consists of several discrete steps, namely: chlorination of the water sample with ^{36}Cl-tagged chlorine gas or hypochlorite solution; 500- to 1500-fold concentration of the chlorinated sample by low-temperature vacuum distillation; separation of the chlorine-containing constituents by high-pressure, high-resolution anion exchange chromatography (HPLC); and detection and quantitation of the ^{36}Cl-tagged chloro-organics by liquid scintillation counting.

The water aliquots analyzed in these experiments were grab samples obtained from the secondary treatment settling basin outlet of the East Wastewater Treatment Plant of Oak Ridge, Tennessee, and from the cooling water inlets of the Allen Steam Plant on the Mississippi River at Memphis, Tennessee, and the Kingston Steam Plant on the Watts Bar Lake at Kingston, Tennessee. Both electric power-generating plants are operated by the Tennessee Valley Authority. The water samples were stored in polyethylene containers at -60°C and thawed just prior to use. The chlorine requirement (demand) for each water sample was determined by using standard methods.[27]

The 0.5- to 1.5-liter aliquots of water collected from the sites of interest were chlorinated in the laboratory with 3- to 8-mg quantities of chlorine, as chlorine gas or hypochlorite solution containing 0.03 to 0.1 mCi of ^{36}Cl, under conditions similar to those used for disinfection of sewage or antifoulant treatment of cooling waters. In each chlorination, a sufficient quantity of the chlorinating agent was used to obtain a total chlorine residual (*ortho*tolidine) of about 1 mg/liter after 5 min, as determined by the chlorine requirement of the water sample.[27] After the desired chlorination contact time (usually 15 to 75 min), the chlorine residual was measured by the *ortho*tolidine method and the reaction was terminated by adding a slight stoichiometric excess of thiosulfate. The chlorinated sample was then concentrated in preparation for subsequent chromatography. The chromatographic system used in our study has been discussed elsewhere.[28] This system, the dual-column UV-Analyzer (Figure 15.1), uses parallel strongly basic anion exchange resin columns for simultaneous separation of the constituents from two different 280-μl samples and is, therefore, useful for chromatographic comparison of multiple samples. For these experiments, the ^{36}Cl-tagged sample was chromatographed on one column and a reference or unchlorinated water sample on the parallel column. The separated constituents were detected via uv photometry, cerate oxidimetry, or via liquid scintillation counting for radioactivity. When detection was by ^{36}Cl analysis, one of the volumetric syphons was removed and the eluate collected in fractions for radioactivity measurement.

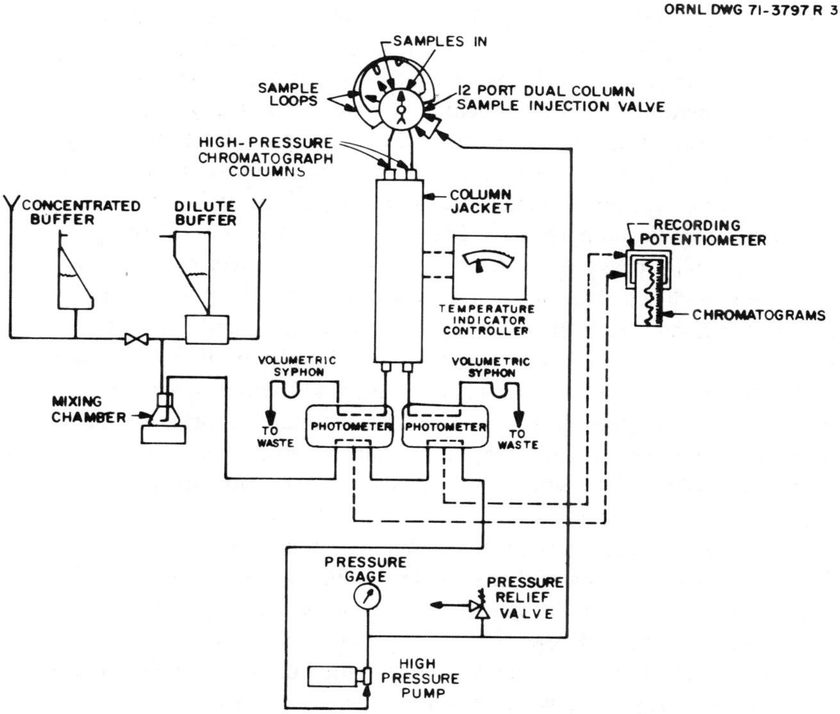

Figure 15.1 Dual-column UV-Analyzer for analytical chromatographic separations of uv-absorbing and ^{36}Cl-tagged chlorine-containing constituents in concentrates of water samples.

CHROMATOGRAPHIC RESULTS

A series of HPLC separations was made with 500- to 1500-fold concentrates of ^{36}Cl-tagged chlorinated secondary effluents from domestic sewage treatment plants,[7,8] a 1500-fold concentrate of ^{36}Cl-tagged chlorinated cooling water from Watts Bar Lake collected from the cooling water inlet of the Kingston Steam Plant,[9] and a 1470-fold concentrate of ^{36}Cl-tagged chlorinated cooling water from the Mississippi River collected near the cooling water inlet of the Allen Steam Plant.[10] Over 50 chlorine-containing organic constituents were separated from each of the water samples. Comparison of the chromatograms (Figures 15.2-15.4) reveals a definite similarity in general chromatographic profile but also many differences.

In each HPLC separation, approximately 99%, or greater, of the ^{36}Cl activity was found to be associated with the chloride ion. Based on two chlorinations of sewage effluent in which the chlorination reaction was permitted to continue through the concentration step (*i.e.*, was not terminated with thiosulfate) and in which greater than 99% of the ^{36}Cl activity was associated with the chloride peak, it was concluded that a major chlorination effect on organics in water is probably oxidation.[7,8] However, a significant fraction of the activity, up to 1.3%, in each of the HPLC separations discussed in this chapter was found to be associated with the other chromatographic peaks that represent stable chlorine-containing organic compounds. The conclusion that the ^{36}Cl-tagged chromatographic peaks other than chloride are relatively stable chloro-organics is deduced from an extensive study of chlorinated secondary sewage treatment plant effluents which contained a similar concentration and composition of inorganic constituents. Results of that study showed that less than 1% of the ^{36}Cl activity associated with chromatographic peaks other than chloride was contributed by metal-chloride or metal-inorganic chloramine complexes, or by isotopic exchange with chlorine-containing constituents which were present in the effluents prior to chlorination.[7,8]

Tentative identifications of 17 chromatographic peaks were established[7,8] for the sewage effluents and are shown in Figure 15.2. Identification was established in each case by comparing the anion exchange elution volume determined for the unknown chromatographic peak with that determined for a reference standard. The reference standards used for comparison were commercially-available, chlorinated analogs of organic compounds identified in wastewater.[11-14,29]

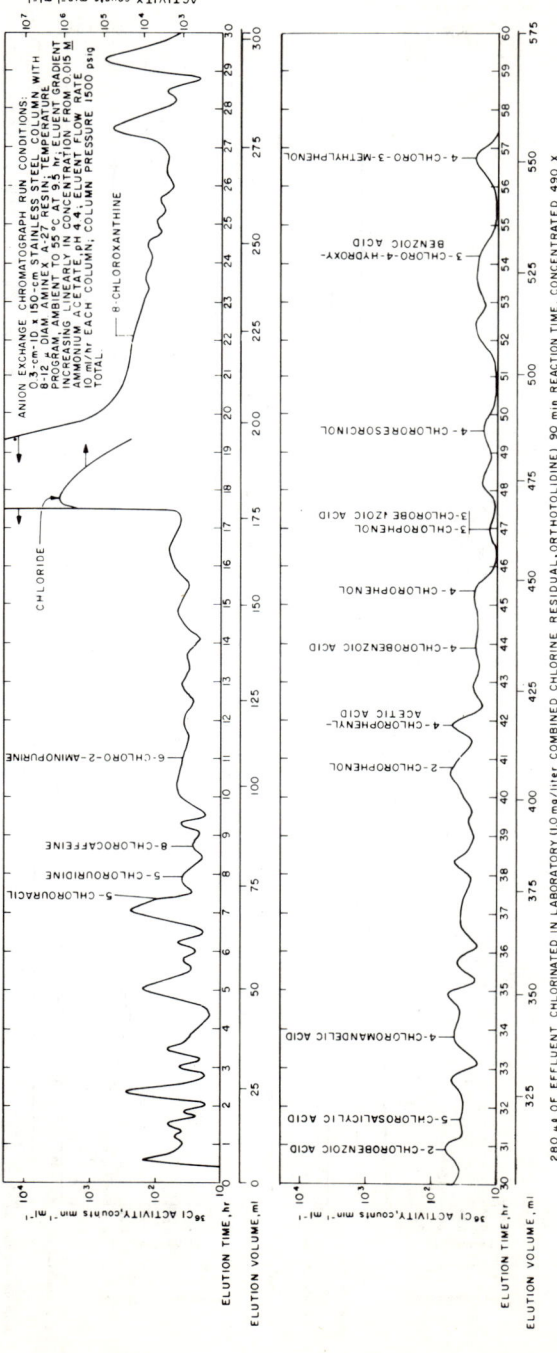

Figure 15.2 Chromatogram showing the ^{36}Cl-tagged chlorine-containing constituents in a sample of chlorinated secondary effluent from a domestic sanitary sewage treatment plant.

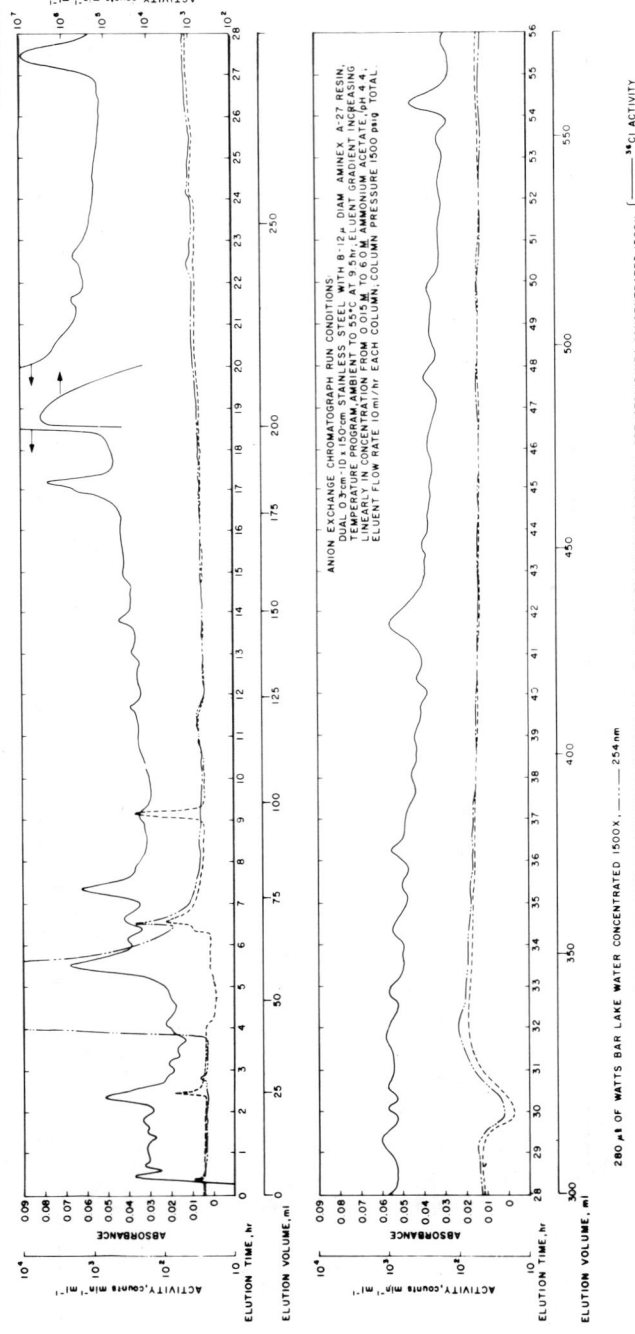

Figure 15.3 Chromatogram showing both ^{36}Cl-tagged chlorine-containing and uv-absorbing constituents in a sample of Watts Bar Lake water collected from the Kingston Steam Plant cooling water inlet and chlorinated in the laboratory. A chromatogram of the uv-absorbing constituents in the unchlorinated cooling water is included for comparison. Both samples were chromatographed simultaneously on the dual-column UV-Analyzer.

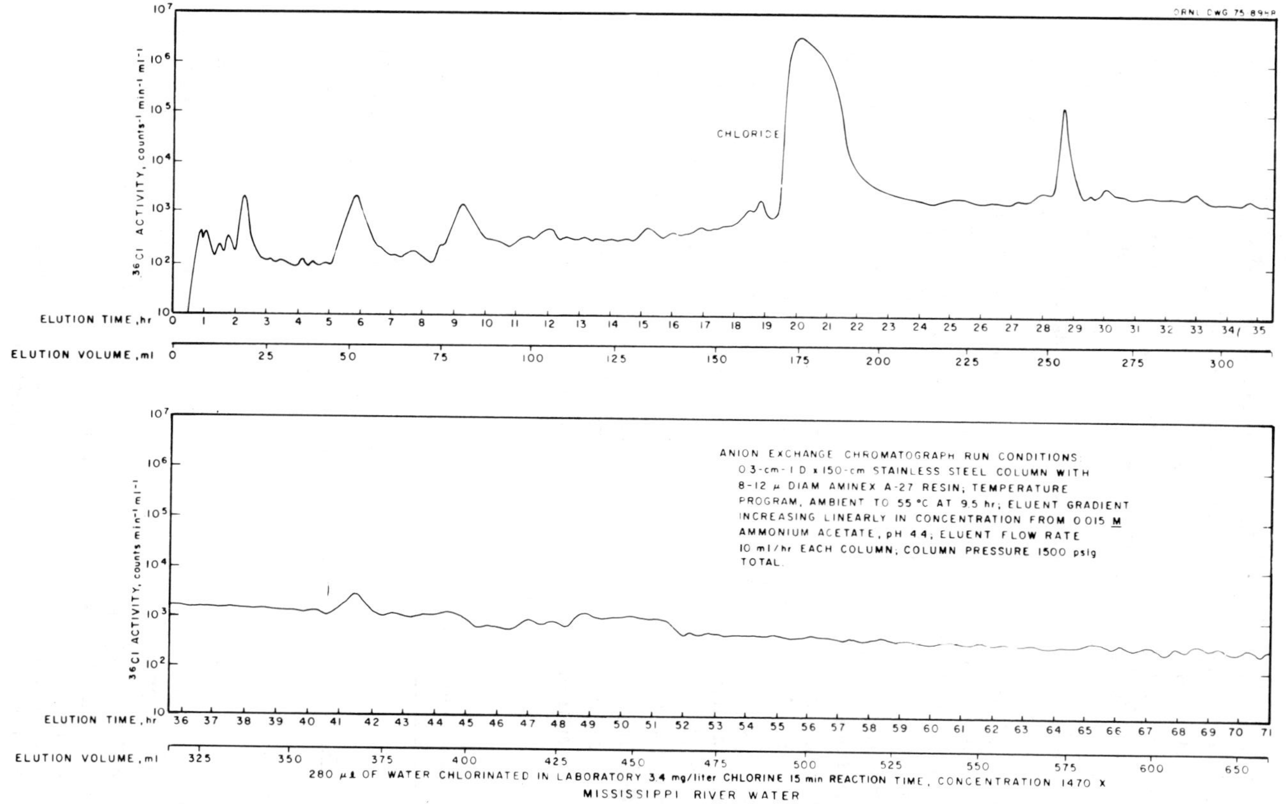

Figure 15.4 Chromatogram showing the ^{36}Cl-tagged chlorine-containing constituents in a sample of chlorinated Mississippi River water collected from the Allen Steam Plant cooling water inlet and chlorinated in the laboratory.

240 ORGANIC POLLUTANTS IN WATER

DISCUSSION

The chlorine available for chemical reactions at near-neutral pH values, 5 to 9, in ammonia-containing aqueous solutions, as in most natural and polluted waters, is present principally as the following species: hypochlorous acid (HOCl), hypochlorite ion (OCl^-), monochloramine (NH_2Cl), and dichloramine ($NHCl_2$). The HOCl-OCl^- system is an effective chlorinating reagent in aqueous solutions.[29] However, the preponderance of evidence indicates that NH_2Cl is chiefly an aminating agent in aqueous solution,[7] whereas no literature references were found relative to the effectiveness of $NHCl_2$. Hence, it is a reasonable assumption that HOCl is the effective chlorinating agent for organics in aqueous solution.[7,30] Equilibrium conditions are established rapidly for HOCl, OCl^-, and NH_2Cl, but somewhat more slowly for $NHCl_2$. The equilibrium concentrations of the reactive species are dependent on the initial chlorine and ammonia concentrations, the pH and temperature. Because HOCl reacts extremely rapidly with ammonia[29] and the equilibrium concentration of HOCl is inversely related to the ammonia concentration,[9] less formation of chloroorganics would be anticipated in those waters containing higher concentrations of ammonia when all other factors are equal. This conclusion is qualitatively supported by the analytical data presented in Table 15.1 for the series of chlorinations discussed in this chapter. The chlorination yield, that fraction of the chlorine dosage associated with stable chlorine-containing organic compounds at the end of the chlorination reaction period, is relatively much lower for the sewage effluent sample, which

Table 15.1 Selected Analytical Data and Experimental Results

	Secondary Sewage Effluent[7,8]	Watts Bar Lake[9] Sample	Mississippi River[10] Sample
pH	7.4	7.5	7.3
Chloride, ppm	22	0.5	8
Organic Carbon, ppm	12	2.7	9
Organic Nitrogen, ppm	5.8	2.7	< 0.05
Ammonia (as N), ppm	11	0.5	0.15
Chlorine Dosage, ppm	3.2	2.5	3.4
Chlorine Residual (OT), ppm	1	1	1.2
Chlorination Yield of Cl as chloro-organics	1.0%/45 min	0.8%/75 min (0.5%/15 min)[a]	3.1%/15 min

[a]Estimated value after 15-min chlorination contact time.

contains 11 ppm of ammonia-N, than for the Mississippi River water sample, which contains only 0.15 ppm. The chlorination yields are calculated based on the assumption that the kinetics of isotopic exchange of the ^{36}Cl-tagged chlorinating agent with the inert chloride in the water sample is much more rapid than the kinetics of the organic chlorination reactions; therefore, essentially complete isotopic dilution is achieved.

The significant effect of ammonia on equilibrium concentrations of the reactive chlorine-containing species is shown in Table 15.2. If

Table 15.2 Equilibrium Concentrations of Selected Constituents (ppm)

	Sewage Effluent[7,8]	Watts Bar Lake[9] Sample	Mississippi River[10] Sample
Chlorine Residual (OT)	1.0	1.0	1.2
NH_3 (as N)	11	0.5	0.15
$HOCl^a$	<0.0001	0.0016	0.2886
OCl^{-a}	<0.0001	0.0016	0.1843
NH_2Cl^a	0.9979	0.8907	0.0321
$NHCl_2{}^a$	0.0020	0.1061	0.6950
Reaction Yield after 15 min	0.9	~0.5	3.1

[a]Equilibrium values were calculated by using the computer program developed by J. E. Draley, Argonne National Laboratory, and modified by R. H. Rainey, Oak Ridge National Laboratory.

HOCl is the effective chlorinating agent for organic compounds in aqueous solutions, as we suspect, then a relatively greater reaction yield would be anticipated for the cooling water samples, *i.e.*, the Watts Bar Lake and Mississippi River water samples, than for the sewage effluent. The higher yield in the sewage effluent as compared with that obtained for Watts Bar Lake water, although somewhat surprising, is probably attributable to a much higher concentration of dissolved organic carbon (Table 15.1). Indeed, the higher yield in the Mississippi River water sample as compared with the Watts Bar Lake water sample is probably attributable both to a lesser ammonia concentration, with a resulting higher HOCl concentration, and also to a higher concentration of dissolved organics. The progressive decrease of NH_2Cl and increase of $NHCl_2$ as the ammonia content of the water samples decreased is of interest. The relatively high concentration of $NHCl_2$ in the chlorinated sample of Mississippi River water may be significant, and the chemistry of dichloramine with respect to aqueous reactions with organic compounds should be studied.

Chlorination of sewage effluents with either chlorine gas or hypochlorite solution gave essentially the same results.[7,8] The percentage yield of chloro-organics was approximately constant with respect to chlorine dosage, but increased with chlorination contact time.[7,8] The effects of two important parameters, light and temperature, were not studied in this investigation. All of the laboratory chlorinations were conducted within Pyrex vessels under normal conditions of laboratory lighting (*i.e.,* fluorescent light fixtures in the ceiling); thus the light intensity should not have been sufficient to initiate reactions. The chlorinations were conducted at 25°C, whereas a temperature range of 10 to 30°C might be anticipated for natural and process waters. Evaluation of the effects of both parameters appears to be warranted.

Most of the organic constituents detected and separated by the concentration technique and HPLC methodology used in this investigation had relatively low volatilities and molecular weights of probably less than several thousand. The techniques used did not detect or measure the formation of volatile chloro-organics, such as chloroform, which might be anticipated as reaction products.[2-4,9] Nor did the methods detect or quantify the chlorination effect on large natural polymers (*e.g.,* nucleic acids and humic acids). Disinfection of bacteria with chlorine is known to result in the formation of 5-chlorouracil and 5-chlorocytosine in the nucleic acid polymer.[31] Consequently, such polymers chlorinated at the site of the pyrimidine bases uracil and cytosine might be anticipated as chloro-organic products of cooling water chlorination and sewage effluent disinfection with chlorine. About 50% of the soluble organic carbon in sewage effluents[32] and about 90% of the soluble organic carbon in natural waters[33] consists of humic materials. These materials might be expected to chlorinate at active sites on the aromatic nuclei.

The similarities in the complex chromatographic profiles of the chlorinated waters studied in this investigation suggest that many of the same chloro-organic products are formed in each. This could reasonably be expected since natural waters include a myriad of microorganisms, animal excreta, plant and animal metabolites, as well as dead and decaying plant and animal material, not totally dissimilar from dilute sewage effluents. Based on comparable elution positions, the chromatographic peaks for the Watts Bar Lake and Mississippi River water samples corresponding to the peaks identified in the sewage effluent sample were quantified. These data are presented in Table 15.3. Although corroborative mass spectral

Table 15.3 Chloro-Organics in Cooling Waters and Process Effluents (ppb)[a]

	Sewage Effluent	Watts Bar Lake Sample	Mississippi River Sample
Nucleoside			
5-chlorouridine	1.7	0.6	7
Purine			
8-chlorocaffeine	1.7	1.1	6
6-chloro-2-aminopurine	0.9	1.0	3
8-chloroxanthine	1.5	3	–
Pyrimidine			
5-chlorouracil	4	0.6	3
Aromatic acid			
2-chlorobenzoic acid	0.3	1.1	10
3-chlorobenzoic acid	0.6	0.2	8
4-chlorobenzoic acid	1.1	0.3	8
3-chloro-4-hydroxybenzoic acid	1.3	0.8	3
4-chloromandelic acid	1.1	1.8	6
4-chlorophenylacetic acid	0.4	3	20
5-chlorosalicylic acid	0.2	3	18
Phenol			
4-chloro-3-methylphenol	1.5	0.2	0.7
2-chlorophenol	1.7	0.2	4
3-chlorophenol	0.5	0.2	6
4-chlorophenol	0.7	0.2	2
4-chlororesorcinol	1.2	0.5	7

[a]Concentrations were calculated by assuming complete isotopic dilution of the ^{36}Cl-tagged chlorinating agent with the inert chloride in the water sample.

evidence is desirable, it is significant that estimated concentrations of chlorinated purines and pyrimidines are comparable for the chlorinated cooling waters and sewage effluents. The concentrations of chlorinated aromatic acids, which may be related to humic materials, are considerably higher for the Mississippi River water sample. Although individually the compounds are present at parts-per-billion concentrations, collectively they amount to several thousand tons of chloro-organics released to the nation's surface waters annually[10] and hence represent an area needing thorough investigation.

ACKNOWLEDGMENTS

The authors wish to express their gratitude to S. Katz and C. D. Scott for their assistance, and to W. D. Bostick, S. Katz, M. Stewart, and H. E. Zittel for critically reading this manuscript.

This research was sponsored by the Energy Research and Development Administration, U.S. Environmental Protection Agency, and the National Science Foundation.

REFERENCES

1. Ingols, R. S., P. E. Gaffney and P. C. Stevenson. *J. Water Pollution Control Fed.* **38**, 629 (1966).
2. Rook, J. J. *Water Treat. Exam.* **23**(2), 234 (1974).
3. Bellar, T. A., J. J. Lichtenberg and R. C. Kroner. *J. Amer. Water Works Assoc.* **66**, 703 (1974).
4. Stevens, A. A. and G. G. Robeck. "Chlorination of Organics in Drinking Water," Proceedings of the Conference on the Environmental Impact of Water Chlorination, Oak Ridge, Tennessee, October 22-24, 1975, R. L. Jolley, R. B. Cumming and C. W. Gehrs, Eds. (in press).
5. Glaze, W. H., J. E. Henderson, J. E. Bell and V. A. Wheeler. *J. Chromatogr. Sci.* **11**, 580 (1973).
6. Glaze, W. H., J. E. Henderson, G. Smith and O. D. Sparkman. "Analysis of New Chlorinated Organic Compounds Formed by Chlorination of Municipal Wastewater," Proceedings of the Conference on the Environmental Impact of Water Chlorination, Oak Ridge, Tennessee, October 22-24, 1975, R. L. Jolley, R. B. Cumming, and C. W. Gehrs, Eds. (in press).
7. Jolley, R. L. Ph.D. Dissertation, The University of Tennessee, Knoxville, Tennessee (1973).
8. Jolley, R. L. *J. Water Pollution Control Fed.* **47**, 601 (1975).
9. Jolley, R. L., C. W. Gehrs and W. W. Pitt. "Chlorination of Cooling Water: A Source of Chlorine-Containing Organic Compounds with Possible Environmental Significance," Proceedings of the Fourth National Symposium on Radioecology, Corvallis, Oregon, May 12-14, 1975, C. E. Cushing, Ed. (in press).
10. Jolley, R. L., G. Jones, W. W. Pitt and J. E. Thompson. "Chlorination of Organics in Cooling Water and Process Effluents," Proceedings of the Conference on the Environmental Impact of Water Chlorination, Oak Ridge, Tennessee, October 22-24, 1975, R. L. Jolley, R. B. Cumming and C. W. Gehrs, Eds. (in press).
11. Pitt, W. W., R. L. Jolley and S. Katz. U.S. Environmental Protection Agency Report EPA 660/2-74-076 (August 1974).
12. World Health Organization. "Current Knowledge of Concentrations of Organic Contaminants in Waste Water, River Water, and Drinking Water," Working Document No. 4, WHO International Reference Centre for Community Water Supply, Working Meeting on Health

Effects Relating to the Re-Use of Waste Water for Human Consumption, Amsterdam (January 13-16, 1975).
13. U.S. Environmental Protection Agency. "Suspect Carcinogens in Water Supplies," Office of Research and Development Interim Report (April 1975).
14. Junk, G. A. and S. E. Stanley. U.S. Energy Research and Development Administration Report IS-3671.
15. Morris, J. C. U.S. Environmental Protection Agency Report EPA 600/1-75-002.
16. Morris, J. C. "Chemistry of Aqueous Chlorine in Relation to Water Chlorination," Proceedings of the Conference on the Environmental Impact of Water Chlorination, Oak Ridge, Tennessee, October 22-24, 1975, R. L. Jolley, R. B. Cumming and C. W. Gehrs, Eds. (in press).
17. Carlson, R. M. and R. Caple. "Organo-Chemical Implications of Water Chlorination," Proceedings of the Conference on the Environmental Impact of Water Chlorination, Oak Ridge, Tennessee, October 22-24, 1975, R. L. Jolley, R. B. Cumming and C. W. Gehrs, Eds. (in press).
18. Carlson, R. M., R. E. Carlson, H. L. Kopperman and R. Caple. *Environ. Sci. Technol.* **9**, 674 (1975).
19. Gehrs, C. W., L. D. Eyman, R. L. Jolley and J. E. Thompson. *Nature* **249**, 675 (1974).
20. Gehrs, C. W. and R. L. Jolley. "Chlorine-Containing Stable Organics: New Compounds of Environmental Concern," Proceedings of the 19th Congress of the International Association of Limnology, Winnipeg, Manitoba, August 22-29, 1974 (in press).
21. Eyman, L. D., C. W. Gehrs and J. J. Beauchamp. *J. Fisheries Res. Board Can.* (in press).
22. Tardiff, R. G. "Halogenated Organics in Tap Water: A Toxicological Evaluation," Proceedings of the Conference on the Environmental Impact of Water Chlorination, Oak Ridge, Tennessee, October 22-24, 1975, R. L. Jolley, R. B. Cumming and C. W. Gehrs, Eds. (in press).
23. Kraybill, H. F. "Origin, Classification and Distribution of Chemicals in Drinking Water Assessed for Their Carcinogenic Potential," Proceedings of the Conference on the Environmental Impact of Water Chlorination, Oak Ridge, Tennessee, October 22-24, 1975, R. L. Jolley, R. B. Cumming and C. W. Gehrs, Eds. (in press).
24. Cumming, R. B. "The Potential for Increased Mutagenic Risks to the Human Population Due to the Products of Water Chlorination," Proceedings of the Conference on the Environmental Impact of Water Chlorination, Oak Ridge, Tennessee, October 22-24, 1975, R. L. Jolley, R. B. Cumming and C. W. Gehrs, Eds. (in press).
25. Cantor, K. P. and R. N. Hoover. "The Epidemiological Approach to the Evaluation of Water-borne Carcinogens," Proceedings of the Conference on the Environmental Impact of Water Chlorination, Oak Ridge, Tennessee, October 22-24, 1975, R. L. Jolley, R. B. Cumming and C. W. Gehrs, Eds. (in press).

26. Jolley, R. L. *Environ. Lett.* **7**, 321 (1974).
27. American Public Health Association. *Standard Methods for the Examination of Water and Wastewater*, 13th ed. (Washington, D.C.: American Public Health Association, 1971).
28. Pitt, W. W., C. D. Scott and G. Jones. *Clin. Chem.* **18**, 767 (1972).
29. Morris, J. C. In *Principles and Applications of Water Chemistry*. S. D. Faust and J. V. Hunter, Eds. (New York: John Wiley and Sons, Inc., 1967), pp. 22-53.
30. Stasiuk, W. N. Ph.D. Thesis, Rensselaer Polytechnic Institute, Troy, New York (1974).
31. Prat, R., C. Nofre and A. Cier. *Compt. Rend.* **260**, 4859 (1965).
32. Rebhun, M. and J. Manka. *Environ. Sci. Technol.* **5**, 606 (1971).
33. Christman, R. F. and R. A. Minear. In *Organic Compounds in Aquatic Environments*. S. D. Faust and J. V. Hunter, Eds. (New York: Marcel Dekker, Inc., 1971), pp. 119-143.

CHAPTER 16

ANALYSIS OF NEW CHLORINATED ORGANIC COMPOUNDS IN MUNICIPAL WASTEWATERS AFTER TERMINAL CHLORINATION

William H. Glaze, James E. Henderson, IV
and Garmon Smith

>Institute of Applied Sciences and
> Department of Chemistry
>North Texas State University
>Denton, Texas

INTRODUCTION

Macroreticular resins of the XAD type have been applied to the extraction of organic compounds from ground water,[1] sea water,[2] and other natural and anthropogenic sources.[3] On the basis of the evaluations by Junk and co-workers,[1] it appears that the method is comparable to liquid-liquid extraction for the isolation of neutral organics and, providing pH control is utilized, for basic and acidic substances. In this study we utilize XAD extractions for the concentration of organics from a typical municipal waste treatment plant effluent before and after chlorination. We are particularly interested in the possible formation of new chlorinated organics as a result of the chlorination process, on the structures of these new organic substances, and their concentrations in the chlorinated effluent.

Previous works by Jolley and co-workers,[4] which are reported in more detail in Chapters 14 and 15 of this book, have utilized high-pressure liquid ion exchange chromatography (HPLC) to monitor chlorinated effluents. While their method has the advantage of avoiding a prior extraction process, it suffers from the well known deficiencies of HPLC detector technology, and by the fact that the ion exchange HPLC technique is most suitable for ionic and highly polar materials. Moreover, specific compound identification must be accomplished by the comparison of retention times or by the very limiting procedure of collecting HPLC fractions for subsequent spectroscopic study. Thus, the list of specific compounds identified by Jolley et al. is brief and contains primarily organic acids and other polar organics.

The use of the XAD-2 resin extraction technique in our work[5] may be viewed as complementary to the HPLC method. As expected, the method yields larger efficiencies for the isolation of neutral, moderately polar organics. Also, the organic solvent extracts of the compounds captured by the resin are conveniently analyzed by combined gas chromatography/mass spectrometry (GC/MS), a method of significant power.

It should be noted, however, that both HPLC and GC separations methods have limitations when applied to survey studies of this type. Moreover, the identification/quantitation device employed as an adjunct to the chromatograph will also be limited, particularly with respect to the diversity of species which may be analyzed as well as to the level of information obtainable on a given species. Suffice it to say then, that the present investigation, as well as all others which have been carried out until this time, will provide only a limited view of the nature of organic materials in wastewater effluents. The XAD-GC/MS method we employ reveals qualitative and quantitative information on volatile, neutral organics. Since chlorinated hydrocarbons are typical examples of such compounds, the method would seem to be particularly suited to the study of chlorination effects. However, one should in no sense interpret our results to mean that other species are not present for which our method is not well suited.

For just this reason, we have judged it to be advisable to develop a gross parameter similar to total organic carbon (TOC), which will measure organic-bound chlorine in environmental samples. While not without limitations themselves, gross parameters do provide a convenient method for judging the overall quality of a sample. Thus, we report here on the use of a XAD method for the determination of total organic-bound chlorine (TOCl) in water samples. The method appears to be applicable to the 1-ppb TOCl level, and may eventually be regarded as a useful parameter for water quality.

EXPERIMENTAL

Details of the XAD extraction scheme have been reported earlier.[5] In brief, the method consists of passing the water sample (~ 1 liter) over an 8 cm x 1 cm-i.d.-column of Rohm and Haas macroreticular resin which has been scrupulously cleaned.[1] A flow rate of approximately 20 ml/min of water through the resin is maintained. Following this, the resin is washed with approximately 25 ml of carefully purified diethyl ether. The ether is collected in a modified Kuderna-Danish apparatus and either concentrated or analyzed as is.

Water samples are taken from an activated sludge plant after final clarification. Chlorinations are carried out in the laboratory using solutions of molecular chlorine. The sample is divided before chlorination into two equal-sized samples, one of which serves as a control (*before*) and one of which is chlorinated at the desired Cl_2 dose level (*after*). After chlorination both *before* and *after* samples are treated with a slight excess of Na_2SO_3 to quench excess hypochlorite in the *after* samples.

The pyrolysis/microcoulometer system is a commercial unit manufactured by Dohrmann Envirotech (Model S-300/E-300). Aliquot injections are made with a microliter syringe into a platinum sample boat accessory. The furnace temperature is routinely 800°C, but inlet temperatures of 200°C were used in some early work. Readout is either in nanograms chlorine on a digital meter or in the form of a differential signal displayed on a strip-chart recorder.

The GC/MS system is a Finnigan Model 9500 GC with all-glass system, interfaced by a glass jet-type separator to the Model 3200 quadrupole MS system. The spectrometer is controlled by a Model 6100 data system which collects, stores and displays chromatographic and mass spectral data. GC conditions were described earlier.[5]

RESULTS AND DISCUSSION

Figure 16.1 is a reconstructed gas chromatogram of an XAD-2/diethyl ether extract of Denton, Texas municipal wastewater. The original sample was taken after final clarification and subjected to laboratory chlorination using a chlorine dose of 2000 ppm. Contact time was 1 hr. The vertical markers in the figure show the location of peaks whose mass spectra contain chlorine isotope clusters. A total of approximately 40 such spectra have been accumulated on this sample, which is representative of all "superchlorinated" wastewater extracts we have studied. Many of these compounds have been tentatively identified as shown in Table 16.1 using either manual interpretive methods or the Mass Spectral Search System available in our laboratory through Cyphernetics Corporation. A number of identifications have also been confirmed by comparison with standard spectra.

The quantitative estimates shown in Table 16.1 are in the same concentration range determined by Jolley *et al.*[4] for other new chlorinated

250 ORGANIC POLLUTANTS IN WATER

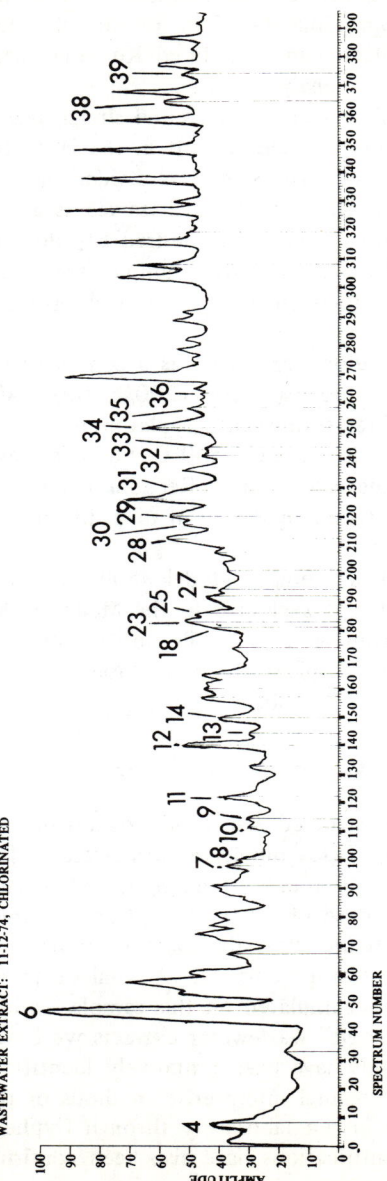

Figure 16.1 Reconstructed gas chromatogram of Denton, Texas, municipal wastewater after 2000 ppm chlorine treatment for one hour. Vertical line markers show the positions of new chlorinated organics not present in control samples. Peak numbers correspond to compound identifications in Table 16.1.

Table 16.1 Summary of New Chlorinated Organics Found in "Superchlorinated" Municipal Wastewater

Compound Number	Compound Name[a]	Identification Status	Concentration[b] μg/l
1	Chloroform	f,g	–
2	Dibromochloromethane	f,g	–
3	Dichlorobutane	d,g	27
4	3-chloro-2-methylbut-1-ene	f	285
5	Chlorocyclohexane (118)	d,g	20
6	Chloroalkyl acetate	d	–
7	o-Dichlorobenzene	f	10
8	p-Dichlorobenzene	f	10
9	Chloroethylbenzene	e	21
10	Tetrachloroacetone	e	11
11	Pentachloroacetone	f	30
12	Hexachloroacetone	f	30
13	Trichlorobenzene	f	–
14	Dichloroethylbenzene	f	20
15	Chlorocumene (154)	d,g	–
16	N-methyl-trichloroaniline (209)	d,g	10
17	Dichlorotoluene	e,g	–
18	Trichlorophenol	e	–
19	Chloro-α-methyl benzyl alcohol	e,g	–
20	Dichloromethoxytoluene	e,g	32
21	Trichloromethylstyrene (220)	d,g	10
22	Trichloroethylbenzene (208)	d,g	12
23	Dichloro-α-methyl benzyl alcohol (190)	d	10
24	Dichloro-bis(ethoxy)benzene (220)	d,g	30
25	Dichloro-α-methyl benzyl alcohol (190)	d	–
26	Trichloro-N-methylanisole	e,g	–
27	Trichloro-α-methyl benzyl alcohol	e	25
28	Trichloro-α-methyl benzyl alcohol	e	25
29	Tetrachlorophenol	f	30
30	Trichloro-α-methyl benzyl alcohol	e	50
31	Trichlorocumene (222)	d	–
32	Tetrachloroethylstyrene (268)	d	–
33	Trichlorodimethoxybenzene (240)	d	–
34	Tetrachloromethoxytoluene (258)	d	4-
35	Dichloroaniline derivative (205)	c	13
36	Dichloroaromatic derivative (249)	c	15
37	Dichloroacetate derivative (203)	c,g	20
38	Trichlorophthalate derivative (296)	c	–
39	Tetrachlorophthalate derivative (340)	c	–

[a] Compounds may be listed more than once if GC retention times indicate distinct positional isomers.

[b] Quantitative values are shown for runs other than 11-12-74 in some cases and represent typical values. Moreover, these values should only be considered as estimates since explicit recovery data is not available for our extraction system.

[c] Mass spectral information too incomplete to propose a structure; probable molecular weight indicated in parentheses.

[d] Fragmentation pattern tentatively suggests proposed compound; probable molecular weight indicated in parentheses.

[e] Probable identification based on mass spectral interpretation.

[f] Completed identification based on MS interpretation and confirmed by comparison with a reference spectrum.

[g] Identified in runs other than 11-12-74 (Figure 16.1).

compounds. It should be noted, however, that the values in Table 16.1 are only approximate; for example, a large amount of volatiles such as chloroform, probably the most abundant new chlorinated compound, are lost during concentration of the sample. The total concentration of chlorinated organics in samples of this type as determined by our TOCl procedure is shown in Table 16.2

Table 16.2 Total Organic-Bound Chlorine
Denton, Texas, Wastewater Effluents

Pyrolysis Temp (°C)	Chlorine Dose (mg/l)	TOCl (µg/l)		
		Before Chlorination	After Chlorination	Blank
800	2000	10.9 ± 0.2	906.3 ± 73.6	4.4 ± 0.2
800	1200	22.4 ± 0.3	608.9 ± 5.3	
800	3500	35.4 ± 0.4	163.8 ± 2.9	21.8 ± 0.2
800	1200	36.9 ± 0.8	337.7 ± 20.2	
800	1400	23.0 ± 0.2	402.8 ± 1.0	
200	2000	13.0 ± 0.1	479.7 ± 6.8	1.7 ± 0.1
800	25	15.0 ± 0.2	80.0 ± 0.8	

The compounds listed in Table 16.1 are of two general types: aromatic and nonaromatic derivatives. The former predominate, as expected from the relative rates of chloro-substitution reactions of aromatic and non-aromatic derivatives. However, it is important to note that under superchlorination conditions one observes mono- and polychlorinated derivatives of the most simple aromatics such as benzene, toluene, ethyl benzene, *etc.* Moreover, among the nonaromatic derivatives observed, one notes the chloroderivatives of acetone. Whether these species are formed from the direct substitution of acetone is not clear; however, it is possible that an acid-catalyzed mechanism for such a process may occur under the low pH conditions used in the superchlorination process.

TOCl Values of Chlorinated Municipal Effluents

Table 16.2 is a compilation of TOCl values obtained by our method on a series of wastewater samples before and after terminal chlorination.*

---

*The term total organic *chlorine* is a misnomer, of course, since the Dohrmann microcoulometer determines total *halide* in the sample.

As indicated in a previous section, the method as presently used suffers from the loss of some volatile halides such as chloroform. However, the method has recently been modified to avoid this problem, and we shall report shortly on the use of the modified method for TOCl in drinking waters as well as wastewaters.

It is clear from the results shown in Table 16.1 however, that superchlorination significantly increases the concentration of chlorinated organics in a typical municipal effluent. These results are particularly significant in view of the proposed use of superchlorination conditions as a waste treatment method.

The last entry in Table 16.2 shows the increase in TOCl level in a municipal effluent upon mild chlorination (25 ppm). Again, this value represents TOCl minus some volatiles such as chloroform, and the TOCl value of 80 μg/l in the chlorinated effluent represents a minimum value. More information on this matter will be forthcoming from our laboratory soon.

CONCLUSION

The results presented here confirm earlier studies[4,5] which show that chlorination of a typical secondary municipal effluent results in the formation of parts-per-billion concentrations of many new chlorinated organic compounds. The use of Amberlite XAD resin is shown to be a convenient method for the concentration of these substances which then may be identified by GC/MS techniques or determined by the gross parameter TOCl (total organic chlorine). Both GC/MS and TOCl methods confirm also that superchlorination (2000-4000 ppm dose) significantly increases the level of chlorinated organics in the effluent.

ACKNOWLEDGMENTS

We are grateful for grants from the U.S. Environmental Protection Agency (R-803007) and from the NTSU Faculty Research Fund for support of this work.

REFERENCES

1. Burnham, A. K., G. V. Calder, J. S. Fritz, G. A. Junk, H. J. Svec, and R. Willis. *Anal. Chem.* **44**, 139 (1972); Junk, G. A., J. J. Richard, M. D. Grieser, D. Witiak, J. L. Witiak, M. D. Arguello, R. Vick, H. J. Svec, J. S. Fritz and G. V. Calder. *J. Chromatography* **99**, 745 (1974).
2. Harvey, G. R., W. G. Steinhauer and J. M. Teal. *Science* **180**, 643 (1973).

3. Jung, K., *et al.* "Resin Sorption Methods for Monitoring Selected Contaminants in Water," presented at the First Chemical Congress of the North American Continent, Mexico City (December 1975), Chapter 9, this volume.
4. Jolley, R. L. Ph.D. Dissertation, University of Tennessee, Knoxville, Tennessee, 1973; *Environ. Lett.* **7**, 321 (1974); Pitt, W. W., Jr., R. L. Jolley and C. D. Scott. *Environ. Sci. Technol.* **9**, 1068 (1975).
5. Glaze, W. H., J. E. Henderson IV, J. E. Bell and V. A. Wheeler. *J. Chromatographic Sci.* **11**, 580 (1973); Glaze, W. H. and J. E. Henderson IV. *J. Water Pollution Control Fed.* **47**, 2511 (1975); Glaze, W. H., J. E. Henderson IV and G. Smith. Proceedings of the Conference on the Environmental Impact of Water Chlorination, Oak Ridge, Tennessee, U.S. Energy Research and Development Administration Report (in press).

CHAPTER 17

N-NITROSO COMPOUNDS IN WATER

David H. Fine and David P. Rounbehler

    Thermo Electron Research Center
    Waltham, Massachusetts

INTRODUCTION

N-nitroso compounds have been identified as a major candidate class of carcinogens that are likely to be causally related to human cancer.[1] Of approximately 100 N-nitroso compounds tested more than 75% have been shown to be carcinogenic.[2] Dimethylnitrosamine (DMN) is structurally the simplest N-nitroso compound and is carcinogenic in a wide range of species including the mouse, hamster, mink, rat, guinea pig, rabbit and rainbow trout. DMN is carcinogenic in mink at the lowest dose tested in rodents, 50 μg/kg body weight, given in the diet two times per week.[3] Most of the animals succumbed to tumors after a total uptake of 25-70 mg DMN/kg body weight.

N-nitroso compounds, such as DMN, N-nitroso pyrrolidine and N-nitroso amino acids, may be present in water where they may be formed from both natural and man-made precursors, namely amines and nitrite. Amines occur naturally in sewage, and dimethylamine and diethylamine are found in fish products.[4] Furthermore, many drugs, agricultural pesticides and industrial chemicals are secondary or tertiary amines. Nitrates are widely distributed in nature, particularly in plants, and nitrites are readily formed from them, especially in the presence of certain bacteria. Nitrates occur in significant concentrations in water, particularly in the runoff water from sewage, agricultural land and feedlots.

Standard analytical methods for N-nitroso compounds are based on gas-liquid chromatography (GLC) and are sensitive at the part-per-billion level, but are available only for the more volatile compounds.[5] The lengthy and time-consuming procedures involve extensive cleanup and concentration by a factor of at least 1000 followed by GLC and detection by either the Coulson conductivity or the alkali flame ionization GLC detectors. Because the GLC detectors are nonspecific to N-nitroso compounds, confirmation by GLC mass spectrometry is mandatory. However, although most N-nitroso compounds are expected to be derivatives of complex amines of high molecular weight, because of their thermal instability and high molecular weight they are not amenable to gas chromatographic procedures. Thus, although over 100 papers have been published on the measurement of the environmental distribution of N-nitroso compounds, all the data have been restricted to about 14 compounds; nothing is currently known about the distribution of the majority of this important class of compounds.

In an attempt to overcome the limitations of previous techniques, we have recently developed[6] a new specific method of detection for N-nitroso compounds, called Thermal Energy Analysis (TEA). Using a selective catalyst, the TEA pyrolyzer selectively cleaves N-nitroso compounds, splitting off the nitrosyl radical:

$$\begin{array}{c} A \\ \diagdown \\ Z \end{array} \! N - NO \rightarrow \begin{array}{c} A \\ \diagdown \\ Z \end{array} \! N \cdot + \cdot NO$$

where A and Z may be any organic radical. The pyrolysis products then pass through a cold trap held at -150°C, which freezes out the solvent together with the organic fragments. The nitrosyl radical passes through the cold trap unchanged, after which it is allowed to react with ozone, giving electronically-excited nitrogen dioxide. The excited nitrogen dioxide rapidly relaxes to its ground state with the emission of light in the near infrared region of the spectrum. The emitted light is monitored by means of an infrared-sensitive photomultiplier tube, the intensity of the emission being proportional to the number of nitrosyl radicals present. The TEA has been shown to be uniquely selective to the N-nitroso functional group in both volatile and nonvolatile N-nitroso compounds. In addition, TEA is sensitive to picogram quantities[7] with a linear response extending over five orders of magnitude. The TEA has been interfaced to both a gas chromatograph, TEA-GLC[8] and more recently to a high-performance liquid chromatograph, TEA-HPLC.[9] Because of the selectivity and sensitivity characteristics of TEA, prior concentration and clean-up steps are unnecessary.[10]

In a recent study[11] the TEA-GLC and TEA-HPLC techniques were applied to the analysis of community air in several U.S. cities. DMN was shown to be present at the 0.95 $\mu g/m^3$ concentration level in the air of

Baltimore, Maryland. Possible sources of DMN may have been a nearby chemical plant which manufactures unsymmetrical dimethyl hydrazine for which DMN is used as a precursor, or a nearby sewage treatment facility. In this chapter water from Baltimore is examined for the possible presence of DMN and other N-nitroso compounds.

EXPERIMENTAL

A single-column gas chromatograph was used, equipped with a N-nitroso compound-specific TEA detector (Thermo Electron, Model 502). The temperature of the TEA cold trap was kept at -150°C. The gas chromatographic column was prepared from a 6.5 m x 2-mm i.d. stainless steel tube packed with 15% FFAP on Chromosorb W (acid-washed, DCMS-treated, 80-100 mesh) and conditioned for 36 hr at 220°C, with a carrier gas flow rate of 10-30 ml/min. Argon was used as the carrier gas.

The high-performance liquid chromatograph was constructed by combining a high-pressure pump (Waters, Model 6000) with an injection valve (Waters, Model UK6). A 30-cm-long, 4-mm i.d. μ Porasil column (Waters) was used, interfaced directly to the TEA detector (Thermo Electron, Model 502). The TEA-HPLC was operated isocratically, with 2 ml/min of a mixture containing 5% acetone and 95% isooctane.

Surface water samples were collected in brown bottles and stored at 4°C. Water (500 ml) was extracted three times with 50 ml of doubly-distilled dichloromethane (Baker). The combined extracts were dried over 75 g of sodium sulfate. The sodium sulfate was extracted twice with 35 ml of dichloromethane and the combined dried extracts plus washings concentrated on a Kuderna-Danish evaporator at 52-53°C to a final volume of approximately 0.5 ml.

RESULTS

The efficiency of the extraction process was determined by recovery of N-nitroso compounds from water which had been spiked at the 0.10-, 0.05- and 0.01- μg/l concentration level. The recovery efficiency ranged from about 68% for DMN to 100% for N-nitroso-sarcosinate (see Table 17.1). Figure 17.1 is the TEA-GLC chromatogram for 25 μl of the dichloromethane extract from a water sample which had been spiked at the 0.010 μg/l level with DMN, diethylnitrosamine (DEN), dipropylnitrosamine (DPN), dibutylnitrosamine (DBN), and N-nitrososarcosinate (SARCOSN). For most compounds the peaks are clearly discernable, the detection limit for DMN and DEN being about 0.002 μg/l.

Table 17.1 Recovery Efficiency of N-Nitroso Compounds from Water (0.01 to 0.10 μg/l level)

Compound	Recovery Efficiency, %
Dimethylnitrosamine	68
Diethylnitrosamine	88
Dipropylnitrosamine	94
Dibutylnitrosamine	98
N-nitrosopiperidine	100
N-nitroso sarcosinate	100
N-nitrosoatrazine	95

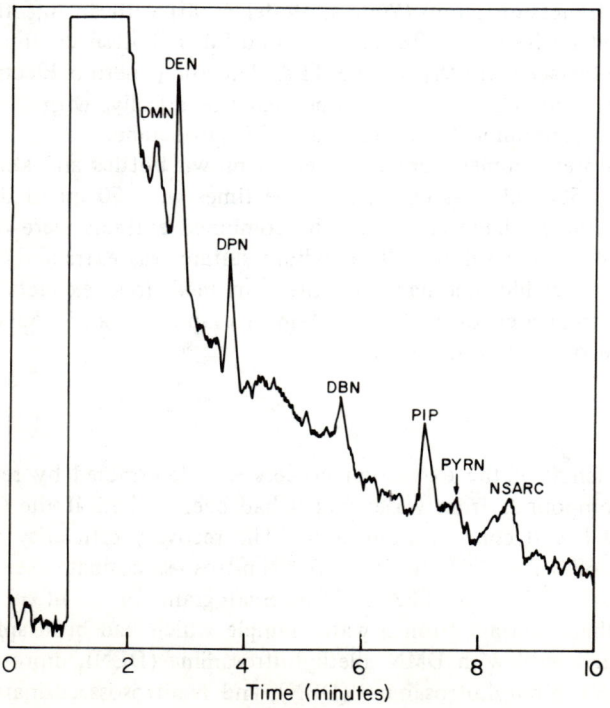

Figure 17.1 TEA-GLC chromatogram of 10 μl of dichloromethane extract following recovery of 0.010 μg/l each of a mixture of seven N-nitroso compounds.

The absence of possible artifacts in the extraction process was established by the fact that DMN could not be detected when analyzing frozen water to which both sodium nitrite and dimethylamine hydrochloride had each been added at the 10-15 μg/l concentration level. Thus, even if free dimethylamine and nitrite had been present in the water, DMN could not have been formed during the extraction, concentration or chromatography steps.

Figure 17.2 is a typical TEA-GLC trace for a water extract from the ocean in Curtis Bay, Baltimore. Two peaks are present, one corresponding

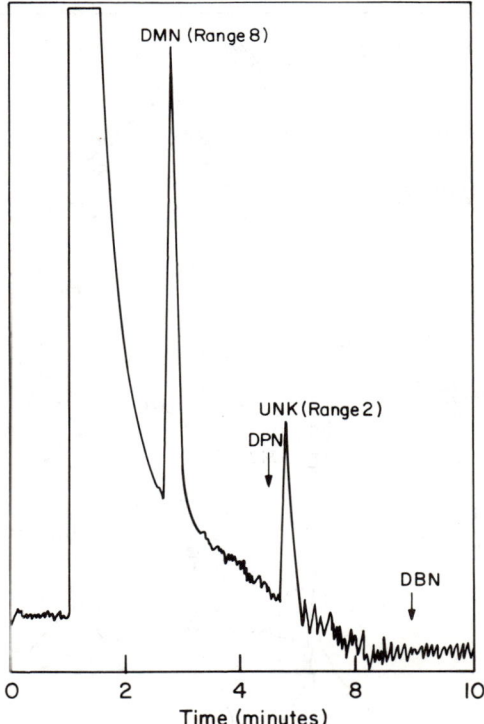

Figure 17.2 TEA-GLC chromatogram of 10 μl of a dichloromethane extract of water from the Curtis Bay region of Baltimore. DMN is present at the 0.24 μg/l level and an unknown N-nitroso compound (NXI) at the 0.02 μg/l level.

in retention time to DMN at the 0.24 μg/l concentration level, and the other to the presence of an as yet unidentified N-nitroso compound, NXI, at the 0.02 μg/l level (assuming 1 mole N-NO per molecule and a molecular

260 ORGANIC POLLUTANTS IN WATER

weight of 34). Similar chromatograms were obtained throughout the Curtis Bay area. Figure 17.3 is a map of Curtis Bay summarizing the observed DMN levels. The largest concentration which was measured was 3.0 µg/l of DMN and 0.019 µg/l of NXI which was found in water taken directly from a sewage treatment facility in South Baltimore. The same unidentified N-nitroso compound, NXI, had previously been found in the nearby community air. Ocean water from other parts of Baltimore, and also from Boston were analyzed and found to contain no traces of N-nitroso compounds, even at the 0.002 µg/l level.

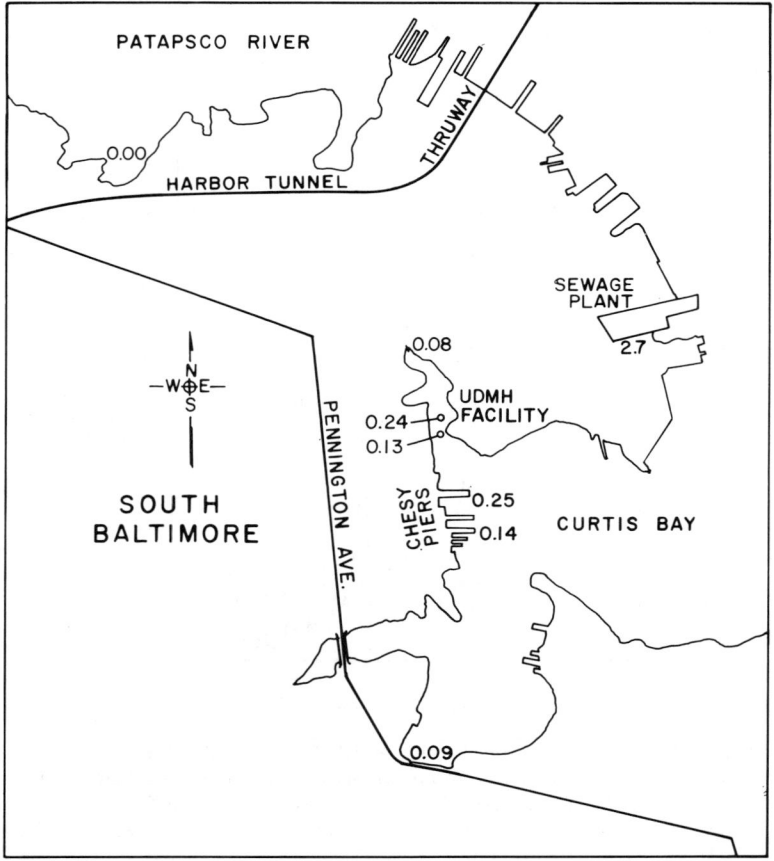

Figure 17.3 Map of Curtis Bay region of South Baltimore showing DMN levels in the water.

DISCUSSION

In this study identification was made by both coincidence of retention time and by comparison of the quantitative amount present in both gas chromatography and high-performance liquid chromatography. Coincidence of retention time using both gas-liquid chromatography (where the elution order is solubility/vapor pressure-dependent) and high-performance liquid chromatography (where the elution order is polarity-dependent) may be taken as confirmation of identity, particularly because the TEA detector that was used has been shown to be extraordinarily selective to N-nitroso compounds.[10,12] In all cases only one major peak was found to be present which corresponds in retention time to DMN; furthermore, the amount of DMN present was identical by both the TEA-HPLC and TEA-GC techniques.

Although it has long been suspected that N-nitroso compounds may be present in water, adequate analytical methods have only recently become available to test the hypothesis. In a previous investigation[13] it had been shown that volatile N-nitroso compounds were absent even at the 0.002 µg/l concentration level in water from the Mississippi River in New Orleans, Louisiana, and from drinking water in Waltham, Massachusetts. More recently,[11] several nonvolatile N-nitroso compounds, including the N-nitroso derivatives of Captam and Atrazine, have been tentatively identified as being present in New Orleans drinking water; however, because the N-nitroso pesticide derivatives are nonvolatile and therefore not amenable to analysis by TEA-GC the latter finding was by TEA-HPLC alone and cannot be taken as confirmatory. The finding here of DMN in sewage effluent and in the ocean near Baltimore is the first confirmed report of the presence of a volatile N-nitroso compound in water. The result is significant because it establishes for the first time that N-nitroso compounds can exist for considerable periods of time in nature.

Although DMN has now been found in both the ambient community air[11] and the ocean water of South Baltimore, the source of the DMN is not established. The highest level reported was 3 µg/l in the final effluent water from a sewage treatment facility, suggesting that the sewage effluent and/or runoff water from the factory are the prime sources, with the air levels being due to subsequent vaporization of DMN from the adjacent ocean. On the other hand, DMN is involved in a nearby chemical plant and the prime potential source must be considered a leak into the air, followed by absorption of DMN in the ocean. Extensive further measurements in other cities and regions are required before the source of the DMN can be properly ascertained.

The DMN levels reported here are for sea water and sewage effluent and do not therefore pose a direct health threat in Baltimore. However, if the results are considered indicative of other areas, then the final treatment water from other sewage treatment centers is suspect and needs to be examined, particularly in regions where the final effluent eventually finds its way back into the drinking water cycle.

ACKNOWLEDGMENTS

We are grateful to Eugene Sawicki and Kenneth J. Krost of EPA, Research Triangle Park, North Carolina, for many valuable technical discussions. We thank Nancy Belcher and Morris Tobin for assistance with the analysis.

Work was supported by Contract No. 68-02-2312 with the U.S. Environmental Protection Agency. The TEA-HPLC interface was developed under Contract NO1 CP 45623 with the U.S. National Cancer Institute.

REFERENCES

1. Lijinsky, W. and S. S. Epstein. "Nitrosamines as Environmental Carcinogens," *Nature* **225**, 21 (1970).
2. Magee, P. N. and J. M. Barnes. "Carcinogenic N-nitroso Compounds," *Advanc. Cancer Res.* **10**, (1967); Druckrey, H., R. Preussmann, S. Ivankovic and D. Schmahl. "Organotrope Carcinogene Wirkung bei 65 Verschiedenen N-nitroso Verbindugen an BD-rotten," *Z. Krebsforsch.* **69**, 103 (1967).
3. Koppang, N. and H. Rimslatten. "The Toxic and Carcinogenic Effect of Dimethylnitrosamine in Mink," Fourth Meeting on the Analysis and Formation of N-Nitroso Compounds, International Agency for Research on Cancer, Lyon, France. Meeting held in Tallinn, USSR, October 1975 (in press).
4. Miyahra, S. *Nippon Kagaku Zasshi* **81**, 19 (1966); Preusser, E. *Biol. Zentr.* **85**, 19 (1966).
5. Telling, G. M., T. A. Bryce and J. Althorpe. "Use of Vacuum Distillation and Gas Chromatography-Mass Spectrometry for Determination of Low Levels of Volatile Nitrosamines in Meat Products," *J. Agric. Fd. Chem.* **19**, 937 (1971); Gough, T. A. and K. S. Webb. "A Method for the Determination of Traces of Nitrosamines Using Combined GC/MS," *J. Chromatog.* **79**, 57 (1973); *N-Nitroso Compounds in the Environment*, Bogovski, P., E. A. Walker, and W. Davis, Eds. International Agency for Research on Cancer, Lyon, France. IARC Publication No. 3 (1972) and IARC Publication No. 9 (1974).
6. Fine, D. H., F. Rufeh and D. Lieb. "Group Analysis of Volatile and Nonvolatile N-Nitroso Compounds," *Nature* **247**, 309 (1974).

7. Fine, D. H., F. Rufeh, D. Lieb and D. P. Rounbehler. "Description of the Thermal Energy Analyzer (TEA) for Trace Determination of Volatile and Non-Volatile N-Nitroso Compounds," *Anal. Chem.* **47**, 1188 (1975).
8. Fine. D. H. and D. P. Rounbehler. "Trace Analysis of Volatile N-Nitroso Compounds by Combined Gas Chromatography and Thermal Energy Analysis," *J. Chromat.* **109**, 271 (1975).
9. Oettinger, P. E., F. Huffman, D. H. Fine, and D. Lieb. "Liquid Chromatograph Detector for Trace Analysis of Nonvolatile N-Nitroso Compounds," *Anal. Letters* **8**, 411 (1975); Fine, D. H., F. Huffman, D. P. Rounbehler, and N. M. Belcher. "Analysis of N-Nitroso Compounds by Combined High Performance Liquid Chromatography and Thermal Energy Analysis," Fourth Meeting on the Analysis and Formation of N-Nitroso Compounds, International Agency for Research on Cancer, Lyon, France. Meeting held in Tallinn, USSR, October 1975. (in press).
10. Fine, D. H., D. P. Rounbehler and P. E. Oettinger. "A Rapid Method for the Determination of Subpart Per Billion Amounts of N-Nitroso Compounds in Foodstuffs," *Anal. Chemica Acta.* **78**, 383 (1975).
11. Fine, D. H., D. P. Rounbehler, N. M. Blecher and S. S. Epstein. "N-Nitroso Compounds in Air and Water," International Conference on Environmental Sensing and Assessment, Las Vegas, October 1975 (in press).
12. Walker, E. A. and M. Castegnaro. "New Data on Collaborative Studies on Analysis of Volatile Nitrosamines," Fourth Meeting on the Analysis and Formation of N-Nitroso Compounds, International Agency for Research on Cancer, Lyon, France. Meeting held in Tallinn, USSR, October 1975. (in press).
13. Fine, D. H., D. P. Rounbehler, F. Huffman, A. W. Garrison, N. L. Wolfe and S. S. Epstein. "Analysis of Volatile N-Nitroso Compounds in Drinking Water at the Part Per Trillion Level," *Bull. Environ. Contam. Toxicology* **14**, 404 (1975).

CHAPTER 18

DETERMINATION AND MONITORING OF SOME ORGANIC EXPLOSIVES IN NATURAL AND EFFLUENT WATER BY SINGLE-SWEEP POLAROGRAPHY

Gerald C. Whitnack
    Code 6052
    Naval Weapons Center
    China Lake, California

INTRODUCTION

    The manufacturing, loading, storage and disposal of ammunition, propellants, fuels and other ordnance material containing organic explosives has created severe environmental problems by the emission of these explosives to the air and water. An accurate determination and reliable monitoring of parts-per-billion of many such organic explosives in effluent and natural water presents a difficult problem in analysis, particularly with regard to environmental effects of both military and industrial ordnance-related facilities. At present, there are no satisfactory analytical methods and inexpensive instruments that can be used in the field to monitor organic explosives in water at the low parts-per-billion level required. The analytical methods generally used today in the detection and determination of organic explosives in water require a preconcentration and/or removal of the explosive from the water prior to performing the analysis by chromatography, spectroscopy, or polarography.[1-9] In addition, there are no EPA- or ASTM-approved standard methods of analysis for organic explosives in water such as those available for inorganic toxic metals in water.
    Single-sweep polarographic techniques and other sensitive polarographic methods such as pulse and AC polarography offer several unique advantages

for both laboratory and field determinations of low concentrations of explosives in water that are necessary to establish environmental pollution standards for ordnance wastes.[8,9]

It is the purpose of this chapter to present and discuss some useful single-sweep polarographic techniques and two reliable and inexpensive field polarographs and accessories that have been developed recently in this laboratory for application to the direct analysis of some organic explosives in natural and effluent water. Many of the water samples have been connected with pollution abatement problems currently under investigation by military and industrial ordnance-related facilities. These polarographic methods require no preconcentration or removal of the explosive from the water and little or no chemical treatment of the sample prior to analysis. The apparatus can be easily adapted to field use for continuous or periodic monitoring.

Organic explosives investigated with the single-sweep polarographic techniques described herein are 1,2-propyleneglycoldinitrate (PGDN), 2,4,6-trinitrotoluene (TNT), 1,3,5-trinitro-1,3,5-hexahydrotriazine (RDX), 1,3,5,7-tetranitro-1,3,5,7-tetraazacyclooctane (HMX), and nitroglycerine.

A digital field polarograph and polarographic cell with a constant temperature device for field use were designed and built in this laboratory to monitor pollutants in water. An application of this digital polarographic monitoring system to the analysis of a specific explosive, PGDN, in effluent water is made. The effluent was obtained from a carbon adsorption process used to remove the explosive to an acceptable environmental level (0.1 ppm) from an ordnance torpedo fuel wastewater that contained large amounts of the explosive.

EXPERIMENTAL

Apparatus and Materials

A single-sweep polarograph known as the A-1660 Davis differential cathode-ray *Polarotrace* (Southern Analytical Instruments Company, England) was used to establish the optimum conditions for analysis of the explosives. A Moseley 2D X-Y recorder was used to record the data obtained with the A-1660 polarograph and also with each of the field polarographs described in this work. The graph paper used in recording the polarograms was 10 x 10 to 1/2 in.

The polarographic measurements were made at $25° \pm 0.10°C$ on 2 ml of water placed in a 5-ml capacity quartz cell for most of the laboratory studies. The field polarographic cell holds larger volumes of water and when used in a continuous mode, about 15 ml of effluent water is retained in the cell during the operation of the monitoring system. The small stainless steel constant-temperature device keeps the temperature of the water sample being analyzed to ± 0.2°C in the field.

The dropping mercury electrodes used throughout this study had drop times of 7 sec in distilled water and m = 5 to 7 mg per drop of mercury. Dissolved oxygen was removed from all solutions, prior to recording the polarograms, with oxygen-free nitrogen obtained by passing nitrogen gas over copper metal turnings at 450°C. Redistilled mercury was used for the anode and for the dropping mercury electrode. Current peak potentials are referred to a mercury pool unless stated otherwise. All pH measurements were made with a Beckman expanded-scale pH meter. The organic explosives studied in this work all produced well-defined single-sweep polarograms in most of the water samples investigated. The polarograms used for the quantitative data reported here were proportional to explosive concentration over a range of 5×10^{-6} to 5×10^{-9} g/ml.

The organic explosives used in this study were all 95 to 99% pure. Standard solutions of the explosive were prepared in spectroquality acetone, obtained from Matheson Coleman & Bell. The solutions were prepared in the 10^{-4} to 10^{-5}-g/ml range and microaliquots (0.01 to 0.05 ml) were added directly to the water sample in the polarographic cell. Although the solubility of HMX and RDX are very low in water (less than 0.005 g/100 g at 30°C), the addition to the water sample of these microaliquots of standard solutions of the explosive in acetone gave linear response in the low concentration ranges studied in this work.

Polarography (Single-Sweep Procedure)

A 2-ml aliquot of a natural or effluent water sample is placed in a 5-ml capacity quartz cell to which a small volume of c-p mercury has been added as the anode; the dropping mercury electrode (DME) serving as the cathode is then lowered into the water sample in the cell. The solution in the cell is now flushed with oxygen-free nitrogen for 3 to 4 min to remove the dissolved oxygen from the water. The start potential of the polarograph is set to first scan from 0 to -0.50 V, and then from -0.40 to -0.80 V, -0.80 to -1.20 V, -1.20 to -1.60 V, and finally from -1.60 to -2.00 V, respectively. Single-sweep polarograms for all the organic explosives studied here are observed over this voltage range. Several polarograms may be seen with most of the explosives and the one or two that are the best defined in a particular water sample are used to determine the concentration of the explosive in that media. For example, the polarogram that appeared the best for measurement of PGDN in the effluent water obtained from a carbon adsorption cleanup process is shown in Figure 18.1, curve A.

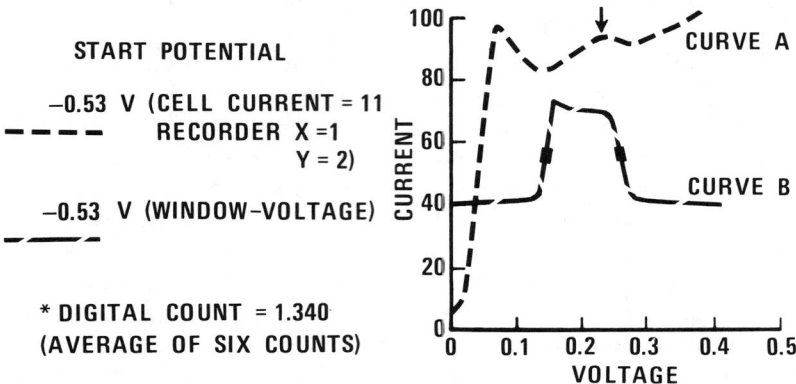

Figure 18.1   1,2-Propyleneglycoldinitrate in effluent water, 0.22 ppm (digital polarograph analysis).

Typical polarograms for several other organic explosives in different water samples are shown in Figures 18.2-18.7. Nitroglycerine gave two well-defined polarograms in a sample of river water (Figure 18.2). The wave producing a peak potential of about -0.85 V appears the best for quantitative measurement in this water. The smaller wave with a peak potential of about -0.55 V might be useful where the presence of interfering contaminants in the water prohibits the use of the better wave at -0.85 V. RDX, at an added concentration of 0.10 millimolar in a sample of lake water, produced several useful polarograms (Figure 18.3). All but

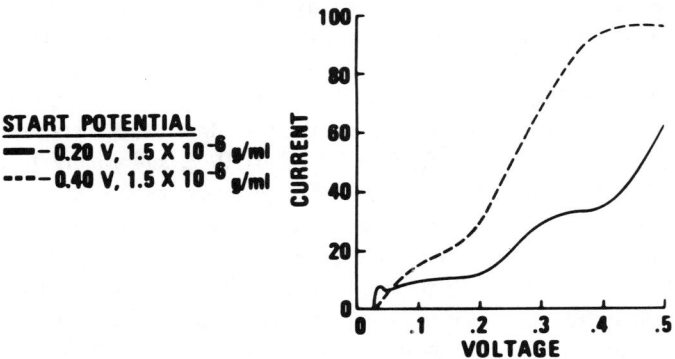

Figure 18.2   Nitroglycerine in river water.

TECHNIQUES AND METHODS OF ANALYSIS 269

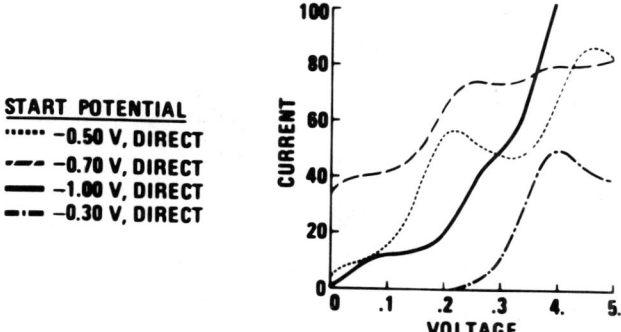

Figure 18.3   RDX in lake water, 0.10 millimolar.

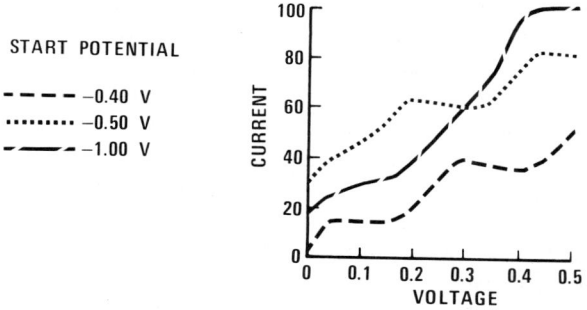

Figure 18.4   RDX in effluent water, 4.02 ppm.

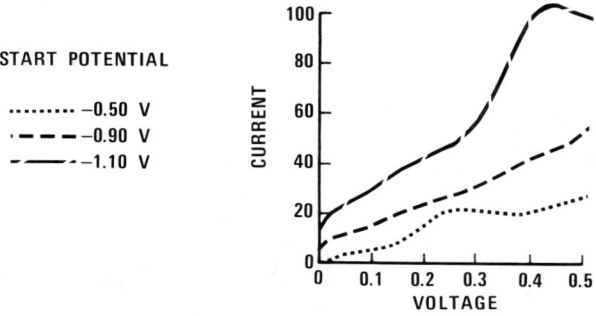

Figure 18.5   HMX in effluent water, 2.84 ppm.

270   ORGANIC POLLUTANTS IN WATER

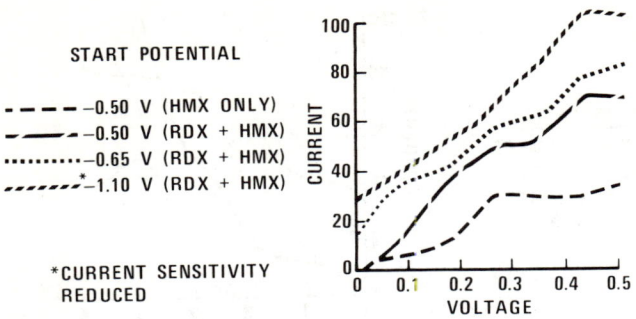

Figure 18.6   RDX (4.02 ppm) and HMX (4.26 ppm) in effluent water.

the small wave, with a peak potential of -1.10 V, are due to RDX and appear useful in determining RDX alone or in admixture in this type of water. The wave with a peak potential of about -1.30 V would seem to be useful in the determination of RDX in the presence of NG and/or TNT which appear at more positive potentials (Figures 18.2 and 18.7). RDX and HMX produce several useful polarograms in a typical effluent water from the carbon adsorption process used in the PGDN work (Figures 18.4 and 18.5). A mixture of RDX and HMX at the parts-per billion level in the same type of effluent water is shown in Figure 18.6. It can be seen that HMX gives a wave at -1.52 V, which is more negative

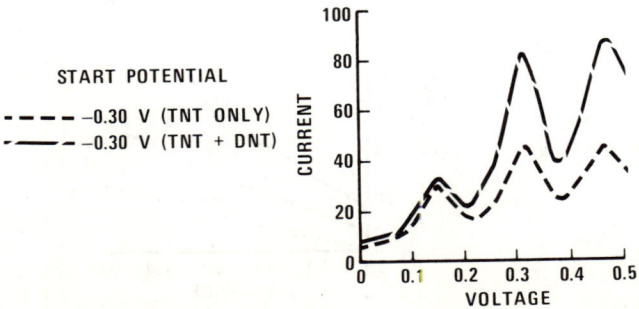

Figure 18.7   2,4,6-TNT (2.08 ppm) and 2,4-DNT (2.02 ppm) in effluent water.

than any waves produced by RDX in this media, and thus should be useful in the determination of HMX in the presence of RDX. A mixture of 2,4-DNT and 2,4,6-TNT in the effluent water is shown in Figure 18.7. It can be seen that TNT can be determined in the presence of 2,4-DNT from the most positive wave seen at -0.45 V. 2,4-DNT does not appear to affect this TNT wave.

After the voltage range to provide the best polarogram for quantitative data has been established, the polarograph is set to record only that voltage range for each analysis. A polarogram can then be obtained every 7 sec with the A-1660-type single-sweep polarograph or field polarographs used in this study. The new digital field instrument records an average current taken from four successive 7-sec polarogram readings made by the instrument from four successive 7-sec drops of mercury from the DME. Thus, with the digital polarograph a number representing a concentration is seen in the window of the instrument every 28 sec. After the polarogram and/or number for the unknown is obtained, a known concentration of the explosive being determined is added to the water sample in the polarographic cell, a few bubbles of oxygen-free nitrogen are passed into the solution, and a second polarogram and/or number is obtained over the same voltage range as before. This standard addition technique allows one to rapidly obtain the concentration of the unknown explosive in the same media from a ratio of the two wave heights of the polarograms or comparison of digital counts. Only microaliquots (0.01 to 0.05 ml) of standard solutions of the respective explosive being determined are added to the water sample in the polarographic cell to provide the data for quantitative analysis.

If there are interfering cations such as copper or zinc in the water, their interference may sometimes be removed with the use of chelating agents such as EDTA. An interfering anion such as iodide or iodate may be reduced or oxidized chemically to a species that does not interfere.[9] For example, if iodide interferes it can be converted to iodate, which may not interfere, by the addition of a drop of chlorine water to the water sample in the polarographic cell before analyzing for the explosive. In the case of effluent water coming from an efficient charcoal adsorption process, the water is already free of most inorganic interference, the pH is near 7, and the conductivity of the water is generally sufficient to give well-defined polarograms for quantitative measurements. The peak potential for PGDN was constant within ±60 mV between the pH range of 6-8 found in this water.

When interferences cannot be removed by the suggested methods or others, the organic explosive must be removed from the water with pure organic solvents such as benzene or n-hexane before analysis can be made.

The solvent is then evaporated off and the remaining material is taken up in a suitable supporting electrolyte such as 0.1 N LiCl in 75% ethanol or acetone for polarographic analysis.[4] Hetman[8] recently published work listing six suitable supporting electrolytes for the polarographic analysis of several explosives. These solutions contained pyridine and alcohol with different salts such as KCl, LiClO$_4$ and NH$_4$NO$_3$. In these media detection limits of 1 to 5 µg/ml were reported by cathode ray and alternating current polarography, respectively.

Field Polarograph and Monitoring System

The need for simple and inexpensive equipment for monitoring parts-per-billion concentrations of PGDN in effluent water, emanating from a carbon adsorption process to remove this explosive from a torpedo fuel wastewater containing large amounts of the explosive, led to the design and construction in this laboratory of both a digital and an analog *Field Polarograph*. The polarograph models weigh nearly 6 pounds each and are completely solid-state. They can be operated with an external battery pack. They were designed for both laboratory and field use, and the digital version has a built-in alarm to warn if a pollutant exceeds a concentration that has been previously determined as dangerous in the water. The pollutant concentration is displayed every 28 sec by digital numbers in an easy-view window on the instrument. A permanent record can be obtained with a strip-chart or x-y recorder and by digital printout with actual time. A specially-designed polarographic cell and constant-temperature device for field use are used to complete the monitoring system. Figure 18.8 shows the digital polarograph (left) with constant-temperature stand and DME assembly for laboratory use. In Figure 18.8 (far right) is shown the new constant-temperature stainless steel cylinder for field use and the quartz polarographic cell (right foreground) which is inserted into the center hole of the cylinder. The DME is inserted into the center hole of the polarographic cell. Three solid (carbon-type) electrodes are shown in Figure 18.8 (left foreground) as potential replacements for the DME in the field where a dropping mercury electrode might be prohibitive. The use of these electrodes in the determination and monitoring of some organic explosives in water is the subject of research now being carried out in this laboratory.

Monitoring Operation of Digital Polarograph

When the digital polarograph is used to monitor the concentration of PGDN in the effluent water released in the carbon cleanup process for the torpedo wastewater, the start potential of the instrument is set at

TECHNIQUES AND METHODS OF ANALYSIS 273

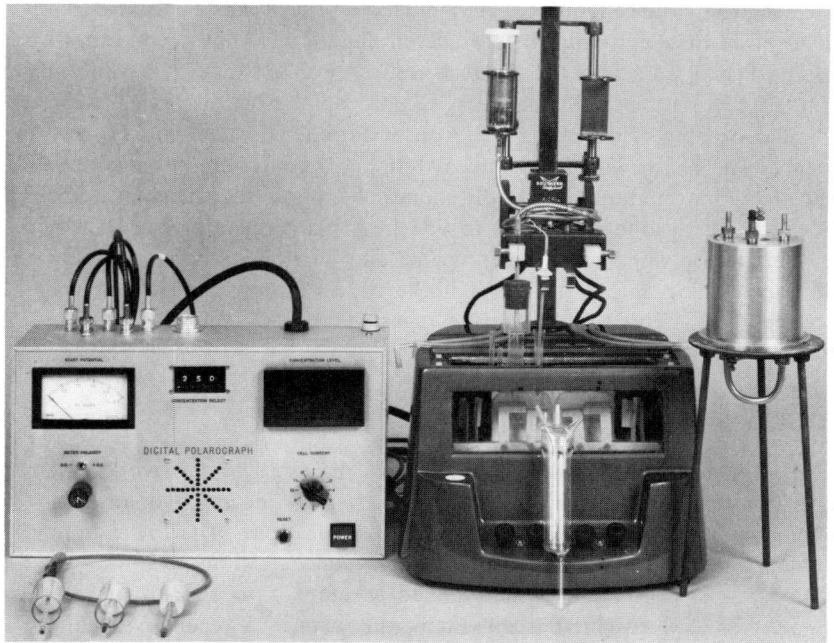

Figure 18.8 Digital polarograph and accessories.

-0.53 V *vs* the Hg pool. A typical polarogram for PGDN produced by the instrument at this start potential is shown in Figure 18.1, curve A. The window setting for the digital polarograph operation is also shown in this figure (curve B). Thus, it can be seen that the digital polarograph records a number for the largest current in the voltage range from -0.67 to -0.78 V, with the peak potential of the PGDN occurring at -0.75 V. This is well within the window setting for the analysis of the low concentration range of PGDN in this water. The 11.0 division (graph paper) wave height (Figure 18.1, curve A) represents a PGDN concentration in the effluent water, of 0.25 ppm. The digital number representing this concentration of PGDN is shown in Figure 18.1.

The alarm built into the digital polarograph will be set off when a slight positive change over the number previously dialed into the concentration select window is registered in the concentration level easy-view window of the instrument. The digital count of the polarograph is repetitive to within ± 1.5% on the 28-sec readings it registers (current average from four 7-sec drops of mercury). The number dialed into the concentration select window is predetermined from known concentrations of the PGDN added to the effluent water. For example, if a concentration of 150 parts-per-billion PGDN is unacceptable in the water, then a number of 750, representing that concentration, is dialed into the instrument. The alarm will not sound as long as the effluent water contains less than that amount but it will sound as soon as the dialed-in number is exceeded, by one digit (750 to 751). A plant operator can then push a reset button on the instrument panel and every 28 sec get another number reading and alarm sound if the concentration of PGDN is really unacceptable. This process can be repeated as many times as necessary for confirmation of water purity.

RESULTS AND DISCUSSION

The polarographic analytical procedures discussed in this work have been used successfully in this laboratory over a wide range of organic explosive concentration with very little modification. The lower limit of detection is about 20 parts-per-billion and concentrations as low as 50 to 100 parts-per-billion can be determined within ± 5 to 10% of the actual amount of explosive that is present in the water. A recent publication[9] from this laboratory pointed out the main advantages of single-sweep polarography in determining many inorganic and organic pollutants in ground and surface waters. The procedures given in this chapter for the analysis of some organic explosives in water are similar to those reported previously in the above paper; however, the procedures described here include the use of new instrumentation with a technique that does not require a prior concentration or removal of the explosive from the water before doing the polarographic analysis. This is particularly valuable when *in situ* analysis is required or analysis on a continuous basis, as described in this chapter for the determination of PGDN in effluent water.

Data obtained for PGDN concentration in typical effluent water from a carbon adsorption process, with the digital polarograph and monitoring system described, are shown in Table 18.1. The data compare well with data obtained by the thin-layer and vapor-phase chromatographic method.[3] A range of 0.10 to 1000 ppm PGDN concentration was found in the samples of effluent water and original torpedo fuel wastewater, respectively, studied in this carbon cleanup process. Only the samples of water containing over 150 ppm PGDN needed to be diluted for analysis. Well-defined polarograms were observed in all cases with the digital polarograph and auxiliary equipment described in this chapter. The individual analysis

TECHNIQUES AND METHODS OF ANALYSIS 275

Table 18.1 Polarographic Analysis of Effluent and Wastewater for 1,2-Propylene Glycol Dinitrate (Data Obtained with Field Polarograph)

Sample	pH	1,2-Propylene Glycol Dinitrate, ppm
Effluent No. 1	7.50	0.12
Effluent No. 2	8.05	0.37
Effluent No. 3	6.35	0.35
Column No. 1	6.90	2.06
Column No. 2	7.10	1.91
Column No. 1A, 880 gal	8.40	1.54
Feed, Pretreated	6.50	167.00
Feed, Original Wastewater	6.20	1062.00

can be made directly in the respective water sample or the analysis can be made on a continuous basis. Figure 18.9 shows the analog field polarograph with polarographic cell, constant-temperature device, and a carrying case for portability. This instrument has several controls on the

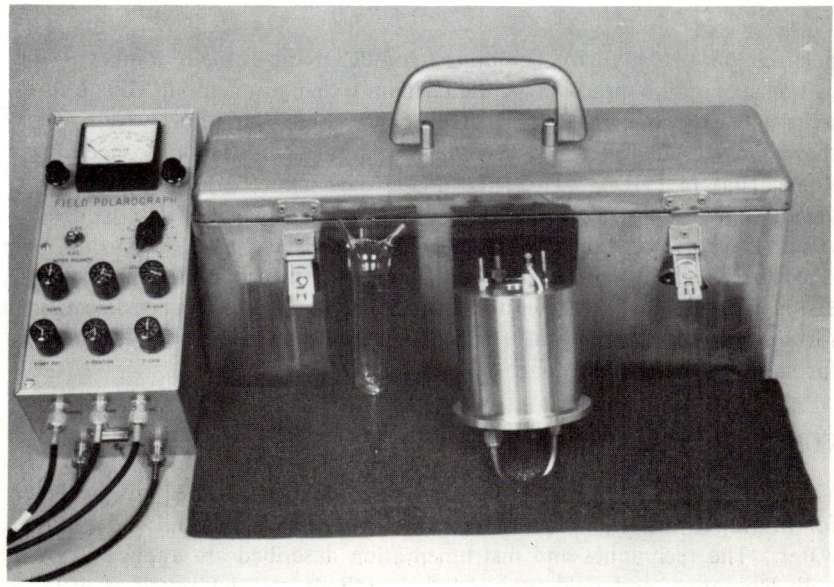

Figure 18.9 Field polarograph and accessories.

outside of the instrument case to allow greater versatility in the analysis of different pollutants in the field. In the digital polarograph the analytical parameters are preset inside the instrument case to analyze for a particular pollutant. This ensures that no external controls can be changed to alter the output data when the polarograph is in field use. Table 18.2 shows typical analog and digital data obtained with the digital

Table 18.2 PGDN in Effluent Water (Analog and Digital Polarographic Data)

x-y Recorder (Graph Paper Divisions) 10 x 10 to 1/2 in.	Digital Count[a] (Cell Current = 11)	PGDN Concentration[b] (ppm)
8.0	1.205	0.19
9.0	1.312	0.21
10.0	1.407	0.24
12.0	1.628	0.29

[a]Average of six counts.
[b]Determined by standard addition technique

polarograph and auxiliary equipment for PGDN in effluent water from a carbon adsorption process. A digital count change of about 100 represents a PGDN concentration change of 0.02 ppm, and the x-y recorder shows only a change of 1.0 division on graph paper for this concentration change of PGDN. Thus, the digital data are considerably more sensitive in response to a slight concentration change of PGDN in the effluent water. A simple block diagram of the field polarograph is shown in Figure 18.10. A complete description of both the analog and the digital field polarographs, with evaluation of their performance in analyzing for certain pollutants in natural and effluent water, is the subject of further study.

CONCLUSIONS

Simple and reliable single-sweep polarographic procedures are presented for the rapid analysis of several organic explosives in natural and effluent water. The techniques and instrumentation described are applicable to both laboratory and field use, and the methods can determine as little as 50 parts-per-billion of explosive without concentration or removal of the explosive from the water prior to the analysis. An individual analysis

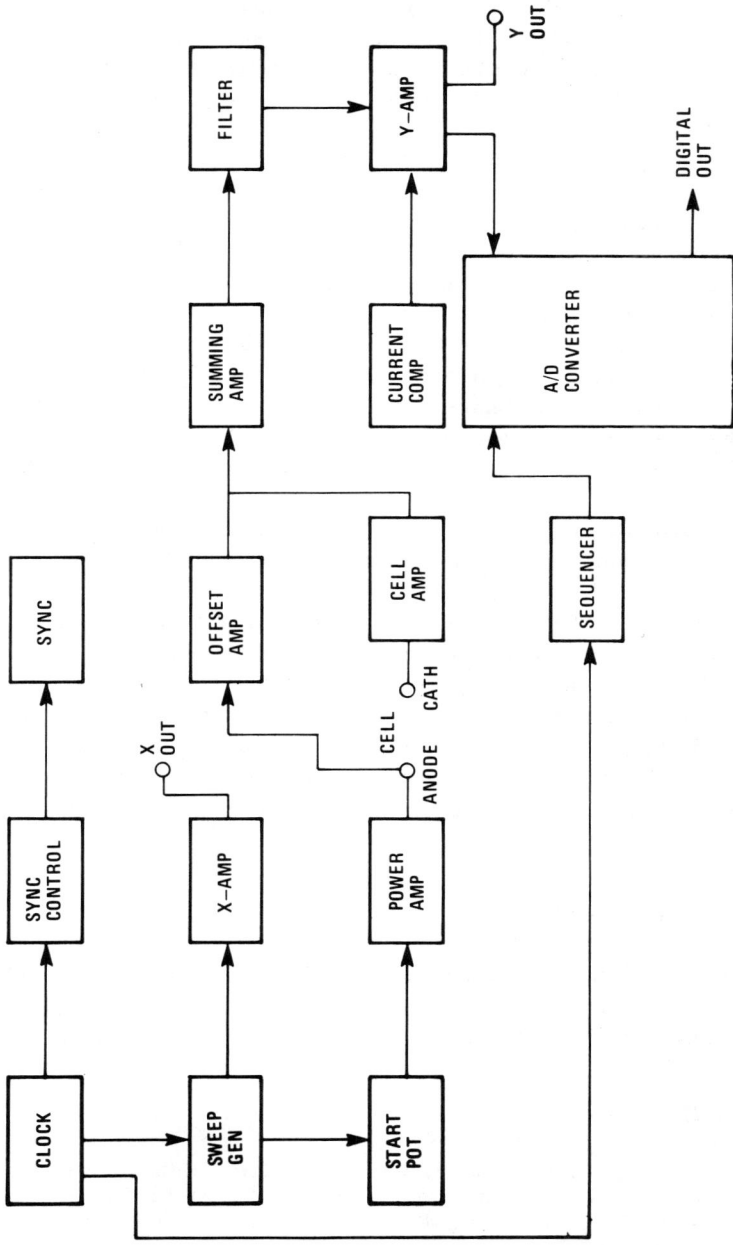

Figure 18.10 Field polarograph (block diagram).

can be performed directly on the water sample within 5 min or the analysis can be made on a continuous basis every 28 sec with the digital polarograph. Concentrations of explosives can be determined in the low parts-per-billion range within ± 5 to 10% of the actual amount present.

An analog and digital version of a field polarograph are presented together with auxiliary equipment for monitoring explosives and other pollutants directly in the water. Data on the use of these polarographs and monitoring system with the described polarographic analytical procedures are presented for PGDN analysis in some effluent water samples from a carbon cleanup process.

The polarographic techniques and field instrumentation described can be used to monitor many other types of both organic and inorganic pollutants in water. It is suggested that the polarographic methods described in this manuscript be tried and evaluated as EPA- and ASTM-approved methods of analysis for organic explosives in both natural and effluent water.

Recently, an article by R. W. Andrew[10] described an inexpensive solid-state modification of the Heathkit Polarograph for use in the determination of low-level environmental contaminants. This instrument is now marketed by Heath International, Inc., under the name of Digi-Scan. A model of the Digi-Scan was recently received and tested by the author as a potential inexpensive device for the determination of organic explosives in water. The detection and lower limit of analysis for this polarograph was found to be about 1 to 3 parts-per-million of organic explosive, not sensitive enough for the low parts-per-billion levels needed in the ordnance related pollution problems discussed here.

ACKNOWLEDGMENTS

The author gratefully acknowledges the technical assistance of Walter J. Becktel in the design and construction of the field polarographs and the support of the work by the U.S. Naval Sea Systems Command.

REFERENCES

1. Hoffsommer, J. C. and J. M. Rosen. Naval Ordnance Laboratory, NOLTR 71-151, White Oak, Maryland (August 12, 1971).
2. Glover, D. J. and J. C. Hoffsommer. *Bull. Environ. Contam. Toxicol.* **10**(5), 302 (1973).

3. Hoffsommer, J. C. *J. Chromatog.* **51**, 243 (1970).
4. Whitnack, G. C. *Anal. Chem.* **35**, 970 (1963).
5. Freer, C. S. Technical Reports 4667 and 4668. Picatinny Arsenal, Dover, New Jersey (March 1974 and July 1974).
6. Hetman, J. S. *Advan. Polarog., Proc. Intern. Congr., 2nd, Cambridge* **2**, 640 (1960).
7. Hetman, J. S. *Anal. Chem.* **32**, 1699 (1960).
8. Hetman, J. S. *Z. Anal. Chem.* **264**, 159 (1973).
9. Whitnack, G. C. *Anal. Chem.* **47**, 618 (1975).
10. Andrew, R. W. *Chemical Instrumentation* **6**, 163 (1975).

CHAPTER 19

ORGANIC FRACTIONIZATION AND SELECTED TRACE METAL CONTENT OF SLUDGES

Leo W. Newland, John R. Ten Eyck,
and Viola K. Ohr

> Environmental Sciences Program
> Texas Christian University
> Fort Worth, Texas

INTRODUCTION

Heavy metal concentrations in natural systems may be derived from two primary sources: through nature and through man's activities. The direct leaching of metal-bearing rocks and soils constitutes the natural sources but, except for a few cases, the amounts added by natural means are less than those added by man's activities.

Man's activities are many, but the most prevalent metal contamination comes from mining and the processing of ores. The petroleum and metallurgical industries contribute considerable arsenic, zinc, mercury and chromium; the paint industry barium, chromium, titanium, zinc and lead; the photography industry cadmium, silver, lead, molybdenum and selenium. The textile industry may contribute silver, barium, copper, tellurium, mercury, nickel and lead; and the chlorine-alkali industry is the major contributor of mercury to the environment.[1]

Schmitt and Hall[2] analyzed the metal concentration of settled solids from sedimentation basins and backwash solids from drinking water treatment plants. These solids are normally concentrated and discharged into the sanitary sewer system. Their findings show that Mn, Cu, Zn and Ni were among the highest concentrations found in the basin and backwash solids.

Trace element concentrations in sewage are related to the kinds and amounts of urban and industrial discharge into the sewage treatment system. The composition of urban and industrial wastes changes as the products of the industrial processes and activities change. Consequently, the sewage influent will vary in chemical composition from time to time. In addition, heavy metals accumulate in the sewage treatment plant with primary concentrations occurring in the primary settling basins, activated sludge and digested sludge. The incoming heavy metals exist in two forms, dissolved and suspended.

Chen and co-workers[3] extensively investigated the characterization of trace metals in the dissolved and particulate phases, and the size distribution of the particulate-borne fractions in wastewater effluents and digested sludge. The study indicated that most of the cadmium, chromium, copper, mercury, zinc and iron in the primary effluent was associated with particulates. This indicated that a large portion of these metals would settle out in the primary settling basins and accumulate in the digesters. Nickel, lead and manganese were mainly in the dissolved state and were less apt to settle out in the primary settling basins.

The secondary treatment, conventional activated sludge process and sedimentation, removes suspended particulates from sewage with 90-95% removal efficiency. The treatment process is less efficient in removing trace metals; however, the results indicate that approximately 70-85% of the total metals and 40-75% of the dissolved chromium, copper, iron and mercury are removed by secondary treatment. The process removes less than 30-60% of the total and less than 40% of the dissolved cadmium, nickel, lead, zinc and manganese. The secondary effluent contains 50-200% more dissolved cadmium and zinc than the dissolved fraction in the primary effluent.

Chen[3] also showed that about 14-40% of the trace metals in the bulk-digested sludge sample remained in the concentrate. More than 90% of the cadmium, chromium, copper, mercury, zinc and iron and between 70-80% of the nickel, lead and manganese in the concentrate were retained by 0.2-micron filters. Among these metals, approximately 70% of the cadmium, chromium, copper, mercury, zinc and iron and 50% of the nickel, lead and manganese were associated with particulates in the range of 0.8-8 microns. Digested sludge contains about 2.3% by weight of particulate matter.

The data indicate that on a dry weight basis, the particulates in effluents and in digested sludge have roughly the same amount of trace metals attached to them. The only exception may be the content of iron borne by the particulates in the digested sludge (at least two to three times more iron than those in effluents).[3]

Several other investigations have been conducted on the dissolved and particulate trace metals in sewage plants. In the book *Interaction of Heavy Metals and Biological Treatment Processes 1965*[4] it was stated that nickel passes through the plant almost entirely in the soluble form. Chromium and copper exhibit erratic solubility behavior in the primary effluent when compared in various plants.

The analysis of wastewater effluents has been the subject of much investigation. Trace element content has received particular interest as a result of the potential toxicity of these elements to aquatic and terrestrial life. Comprehensive analyses of sewage sludges for total heavy metal content have been conducted by a number of researchers.[5-8] Beech[9] has also reported an interesting analysis in which both total and acetic acid-extractable concentrations of selected heavy metals were examined. The results of many analyses in sewage sludges have been routinely reported in studies concerned with the effects of heavy metals found in sludges applied to soils.[10-15] A survey of these analyses reveals that the heavy metal content of sludges may vary considerably depending upon the nature of the wastewater source and the type of treatment conducted.

Painter et al.[16] and Bunch et al.[17] investigated the soluble organic content of secondary effluents in an attempt to classify the constituents. They identified about 35% of the organics as ether extractable proteins, carbohydrates, tannins and lignins, and detergents. A noted similarity was found in the sorption behavior of the organics in secondary effluents and humic substances.[18,19]

Rebhun and Manka[20] fractionated the effluent from high-rate trickling filters in an effort to classify the total organic content. Their fractionation procedure combined the methods of Painter et al.[16] and Bunch et al.,[17] with methods previously developed for the fractionation of humic substances in soil and water. Samples of effluent were treated with an acid and the organic fractions extracted with organic solvents. About 40-50% of the organics were classified as humic substances (humic, fulvic and hymathomelanic acids). Of the total humic substances, fulvic acid was reported as the major fraction (over 50%).

Manka et al.[21] fractionated the effluents from a stabilization pond and an extended-aeration activated sludge plant using the procedure developed by Rebhun and Manka.[20] The composition and average distribution of the organic fractions obtained were found to correspond closely to the results reported previously by Rebhun and Manka.[20] An investigation of the humic substances by gel permeation chromatography further revealed that the majority of fulvic and hymathomelanic acids had molecular weights in the range 500-1000, with the majority of the humic acids being in the range of 10,000-50,000. Some of the fulvic, humic and hymathomelanic acids were also found to have molecular weights of < 500, 1000-5000, and 5000-10,000.

Large applications of sewage sludges have been noted to be injurious to a number of plants as a result of the toxic effects of trace elements. Bingham and his workers[22] noted injuries to assorted vegetables (spinach, soybeans and lettuce) as a result of Cd concentrations of 4-13 µg/g in soil. Turner[23] reported injuries to vegetables grown in solution cultures with as little as 0.1 µg Cd/ml. Lunt[24] noted a reduction in the growth of oats and beans grown on an acid soil with sludge applications as a result of Cu and Zn toxicities. A reduction in the growth of rye due to high levels of Cu and Zn has also been reported.[11] Severe inhibition of germination of sorghum, sudangrass and millet planted in sludge-amended soil has been noted by Sabey and Hart.[25]

EXPERIMENTAL

The sludge samples used in this study were collected from four locations (Figure 19.1) at the Village Creek Sewage Treatment plant, Fort Worth, Texas. The plant uses the activated sludge process to treat raw wastewater which is approximately 70% residential and 30% industrial in origin. Sludge samples were collected in clean, acid-washed bottles at four stages in the treatment process:

1. primary sludge from the primary settling basins
2. activated sludge from the bottom of the final settling basins
3. digested sludge from the anaerobic digestors
4. dried sludge from the drying beds.

An organic fractionization procedure used for the analysis of soil and peat samples was employed to separate the sludge into four major fractions: humic acid, soluble fulvic acid, insoluble fulvic acid called β-humus, and an alkali-insoluble residue called humin.[26] Prior to the fractionization process, the samples were dried at 40°C, ground to pass through a 30-mesh screen, and acid-washed with 0.1 N HCl to remove adsorbed cations and increase the quantity of organic material extracted.

The prepared sample was fractionated according to the scheme given by Black.[26] The sample was first treated with 0.5 N NaOH and shaken 12 hr on a mechanical shaker. The sample was centrifuged and the supernatant decanted through glass wool to remove suspended plant materials. The residue was treated again with the alkali, shaken for 1 hr, centrifuged and the supernatant decanted and added to the previous extract. A repetition of the procedure with deionized water was then made to yield the final alkali-insoluble and -soluble fractions.

TECHNIQUES AND METHODS OF ANALYSIS 285

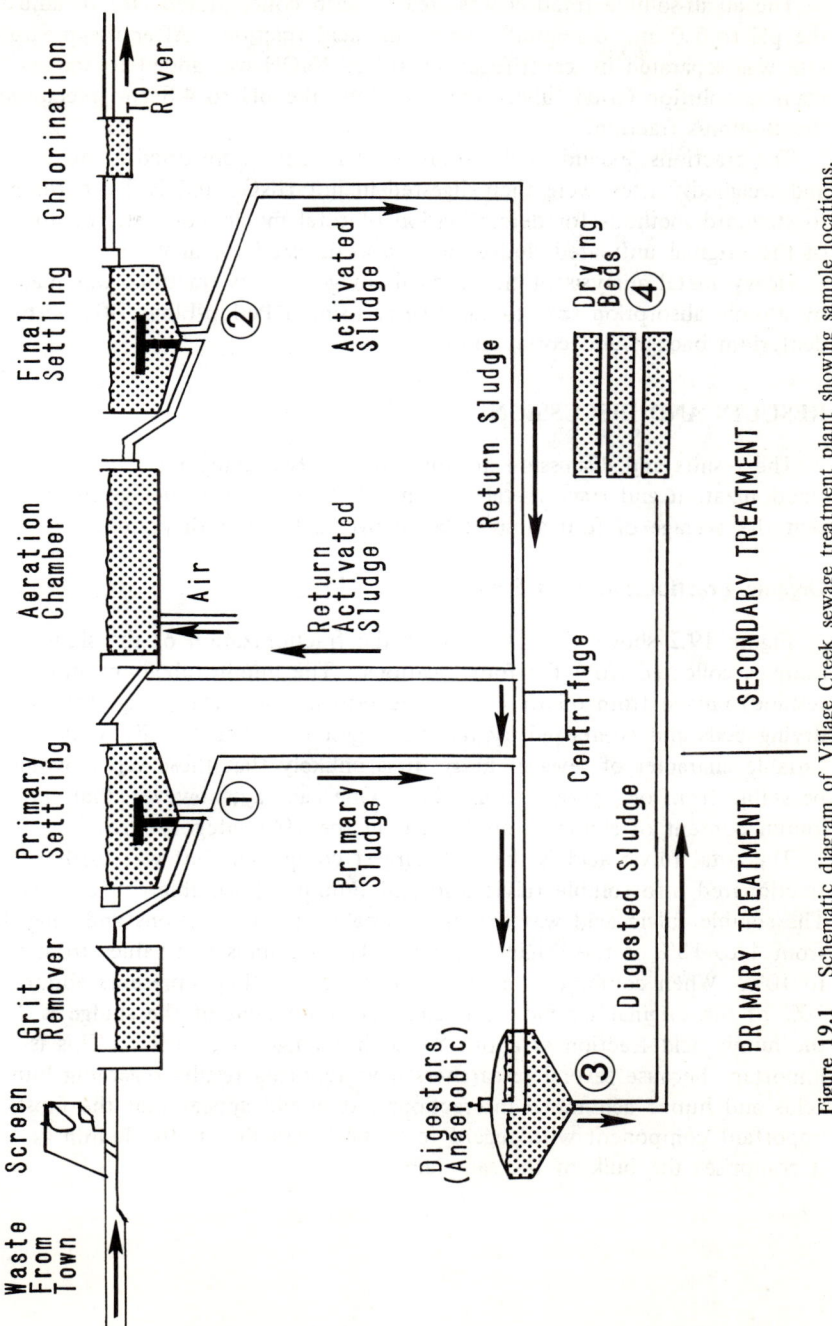

Figure 19.1 Schematic diagram of Village Creek sewage treatment plant showing sample locations.

The alkali-soluble fraction was treated with concentrated HCl to adjust the pH to 1.0 and precipitate the humic acid fraction. After the precipitate was separated by centrifugation, 0.1 N NaOH was added to the remaining solution (total fulvic acid) to adjust the pH to 4.8 and precipitate the B-humus fraction.

The fractions, excluding the soluble fulvic acid, were dried at 50°C and weighed. They were then digested in hot HNO_3 and HCl according to standard methods for determination of total metals content. Samples of the original untreated sludges were also digested for analysis.

Heavy metal analysis of the original sludge and its fractions was made by atomic absorption spectrophotometry using a Perkin-Elmer 403 with deuterium background correction.

RESULTS AND DISCUSSION

The results and discussion are divided into two categories: organic fractionization and trace metal content of sludge. All values given represent the average of four samples taken over a two-month period.

Organic Fractionization of Sludge

Figure 19.2 shows the results from the fractionization of the sludge samples collected from the four locations. The humin (alkali-insoluble residue) ranged from 65 to 77% in the primary, secondary, digester and drying beds and comprised by far the largest component. Given the variable character of sewage sludge, it is unlikely that these values would be stable from one period to another. One can say, however, that the humin content of sludges would occur in the 70% category.

The total fulvic acid is the next largest component but was further fractionated into soluble fulvic acid and β-humus (insoluble fulvic acid). The soluble fulvic acid was the most variable of the fractions and ranged from 4 to 18% of the original sample. The β-humus had values from 6 to 10%. When combined the total fulvic acid portion represents about 20% of the original sample. The smallest component of the sludge is the humic acid fraction at about 8% with a range of 5 to 9%. This is important because several researchers have reported results regarding humic acids and humic acid-metal interactions. It would appear that the most important component with regard to metal interaction is the humin as it comprises the bulk of the sample.

TECHNIQUES AND METHODS OF ANALYSIS 287

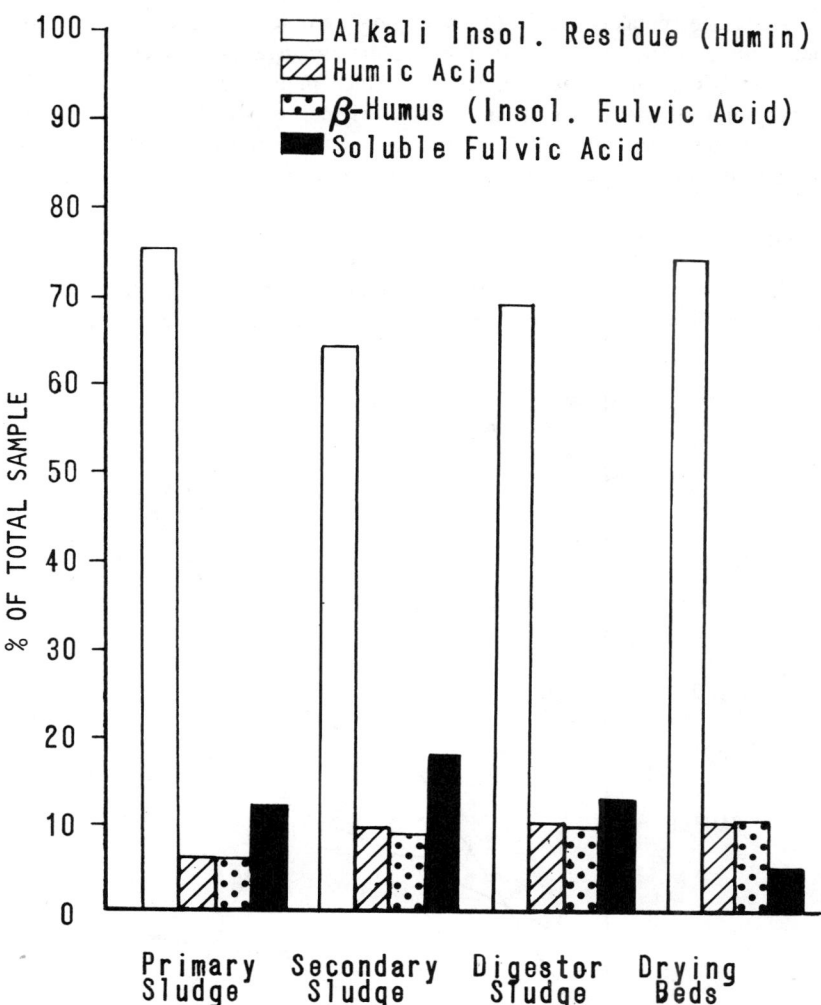

Figure 19.2 Humin, humic acid, β-humus, and soluble fulvic acid content of sludges (primary, secondary, digester, and drying bed).

Trace Metal Contents

The trace metal contents of an original sample and four fractions were determined. Five metals—cadmium, copper, manganese, nickel and zinc—were analyzed for in the samples. All trace metal values are expressed as micrograms of metal per gram of total original sample ($\mu g/g$), *i.e.*, all metal values were adjusted to represent parts per million of the original weight of sample from the various locations (primary, secondary, etc.).

Humic Acid

The results of the trace metal content of the humic acid fraction are shown in Figure 19.3. Note the y-axis varies from trace to 18 $\mu g/g$ (ppm).

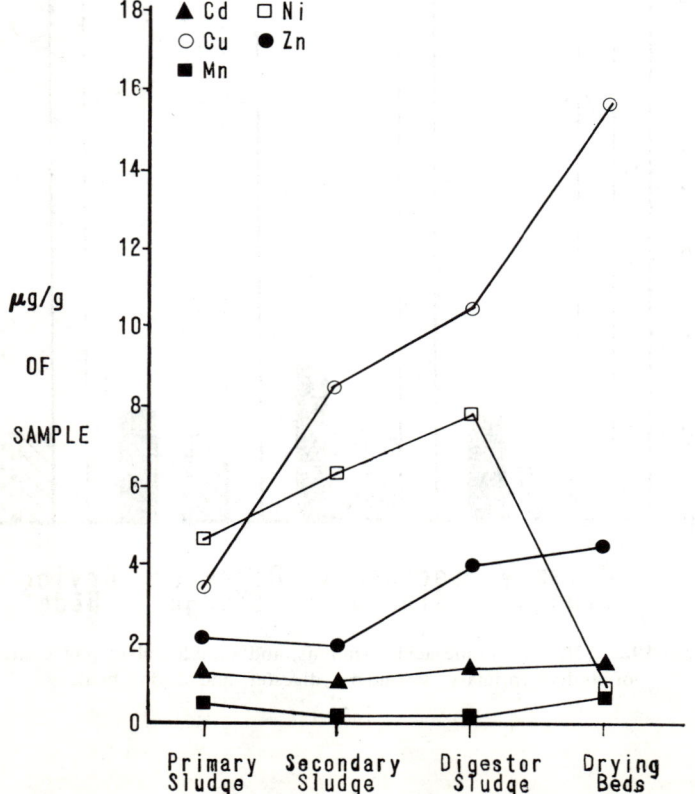

Figure 19.3 Trace metal (Cd, Cu, Mn, Ni and Zn) concentrations for the humic acid fraction of the primary, secondary, digester and drying bed sludge samples.

Copper represents the highest concentration and increases steadily as it moves through the plant, *i.e.*, from primary to secondary to digester to the drying beds. Nickel occurs in next highest concentration with a steady increase as it moves through the plant with a sudden drop from the digester to the drying beds. The remaining three metals—zinc, cadmium and manganese—were all significantly less ranging generally from trace to 4 ppm.

The humic acid fraction contained no more than 3% of any given metal when compared to the original sample. The data for this fraction are complementary to the fractionization values which deny that the humic acid is or is likely to be an extensive reservoir for copper, cadmium, manganese, nickel or zinc.

β-Humus (Insoluble Fulvic Acid)

Analysis of the β-humus fractions revealed metal concentrations in the trace to 70-ppm range (Figure 19.4). Zinc concentrations were highest with copper values second. Copper showed large **increases** as the effluent moved through the plant. Manganese, nickel and **cadmium** values were generally less than 10 ppm, but nickel did show an increase to 13 ppm in the drying beds. The increase in nickel concentration for the drying beds is exactly the opposite of the nickel values for the humic acid fraction. The explanation for this is unknown.

Soluble Fulvic Acid

The soluble fulvic acid fraction showed that copper concentrations were highest with zinc values second (Figure 19.5). These values ranged from 60 to 120 ppm. Manganese and nickel values were in the medium range (30 to 60 ppm) and cadmium was the lowest (10 to 20 ppm). These values are quite different from the β-humus values both in magnitude and graphical appearance. The values show a gradual decrease in organically-bound copper, zinc and manganese from the secondary (final) settling to the digester and to the drying beds. Nickel and cadmium show slight increases.

Total Fulvic Acid (β-Humus and Soluble Fulvic Acid)

When the values for the fulvic acid fractions (β-humus and solubles) are combined, it becomes apparent that zinc and copper are in the high concentration category, nickel and manganese are of medium value, and cadmium is in the low concentration range. Three metals—copper, nickel and cadmium—have values that steadily increase as the sludge moves

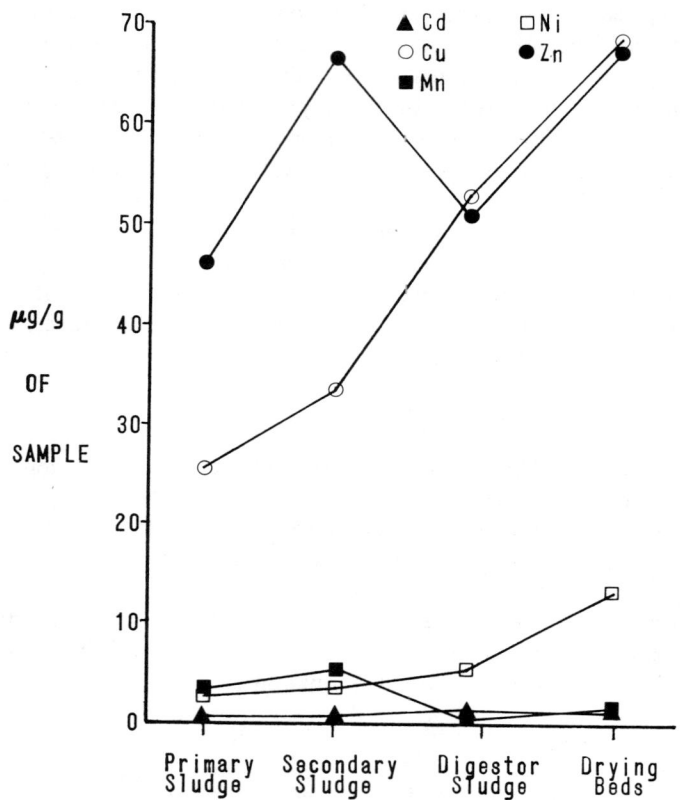

Figure 19.4 Trace metal (Cd, Cu, Mn, Ni and Zn) concentrations for the β-humus fraction of the primary, secondary, digester and drying bed sludge samples.

through the treatment plant while zinc and manganese values rise from primary to secondary, but decrease throughout the remainder of the treatment process. This can be explained for zinc and manganese when one compares the total fulvic acid values (Figure 19.2) of the primary, secondary, digester and drying bed sludges. The values are 18%, 26.8%, 21.8% and 15%, respectively. This does not explain the behavior of copper, nickel and cadmium however. These metal concentrations increase even though the fulvic acid fraction is decreasing. It is important to note that copper and zinc are the dominant trace metals of the five tested for in the samples.

The total fulvic acid fraction contained approximately 10-12% of the metals when compared to the original fraction. The implication of this

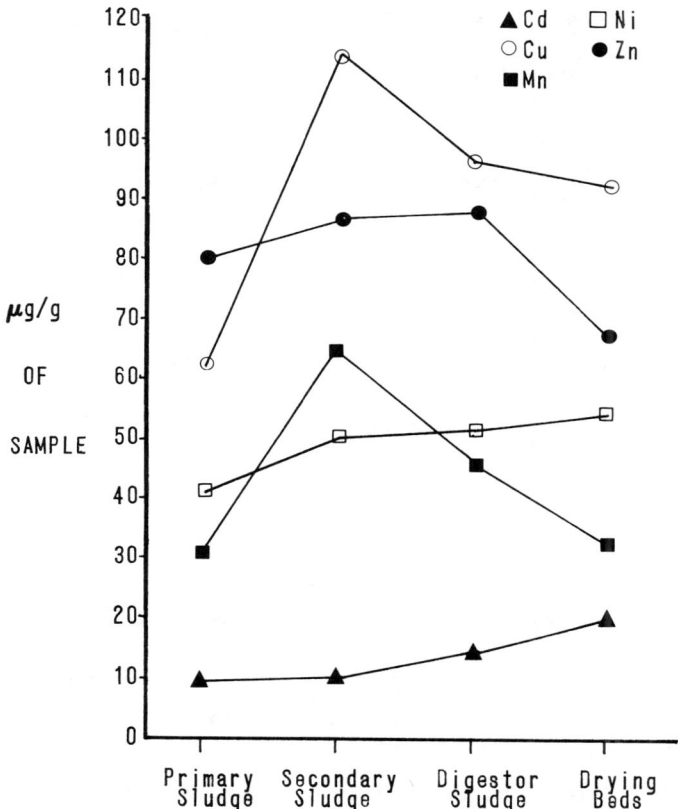

Figure 19.5 Trace metal (Cd, Cu, Mn, Ni and Zn) concentrations for the soluble fulvic acid fraction of the primary, secondary, digester and drying bed sludge samples.

data is that the fulvic acid fraction is 3 to 4 times more important with regard to cadmium, copper, nickel, manganese or zinc than the humic acid fraction.

Humin (Alkali-Insoluble Residue)

The humin, as one would predict from the fractionization values, should contain the highest concentration of trace metals. The results are shown in Figure 19.6 and indicate that metal concentrations range

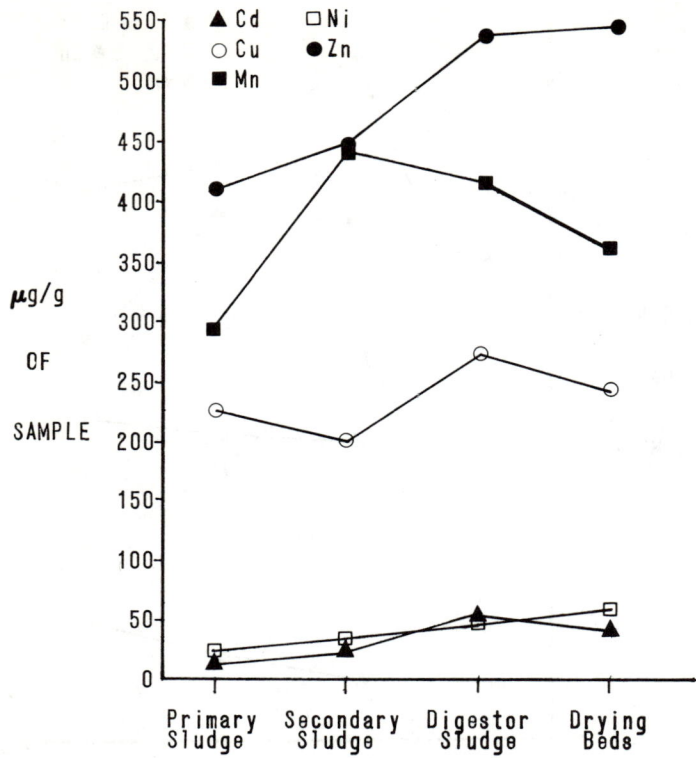

Figure 19.6 Trace metal (Cd, Cu, Mn, Ni and Zn) concentrations for the humin fraction of the primary, secondary, digester and dry bed sludge samples.

from 20 to 550 ppm for the five metals. Zinc showed the highest concentrations for all sludge samples, with manganese second and copper third in value. Nickel and cadmium were significantly lower and ranged from 17 to 64 ppm. Trends of increase and decrease in the trace metals analyzed for in the humin fraction are apparent but less prominent; *e.g.*, copper increases from primary to the drying beds while manganese was highest in the secondary sludge and decreased in either direction.

Most of the trace metals are found in this fraction—as much as 85% in some cases (*e.g.*, manganese in primary sludge). Therefore, if one were analyzing for trace metals in sludge (especially cadmium, copper, manganese, nickel or zinc), the humin fraction should be looked at first.

Original Sludge

The results for the metal analyses of the original sludge samples are shown in Figure 19.7. Some immediate observations again classify the metals into three categories—low, medium and high concentrations. Zinc stands alone in the high-concentration category and varies from 900 to 1190 ppm. Copper and manganese are both in the medium range and vary from 340 to 670 ppm, generally averaging about 500 $\mu g/g$. Again the cadmium and nickel were in the low-concentration category ranging

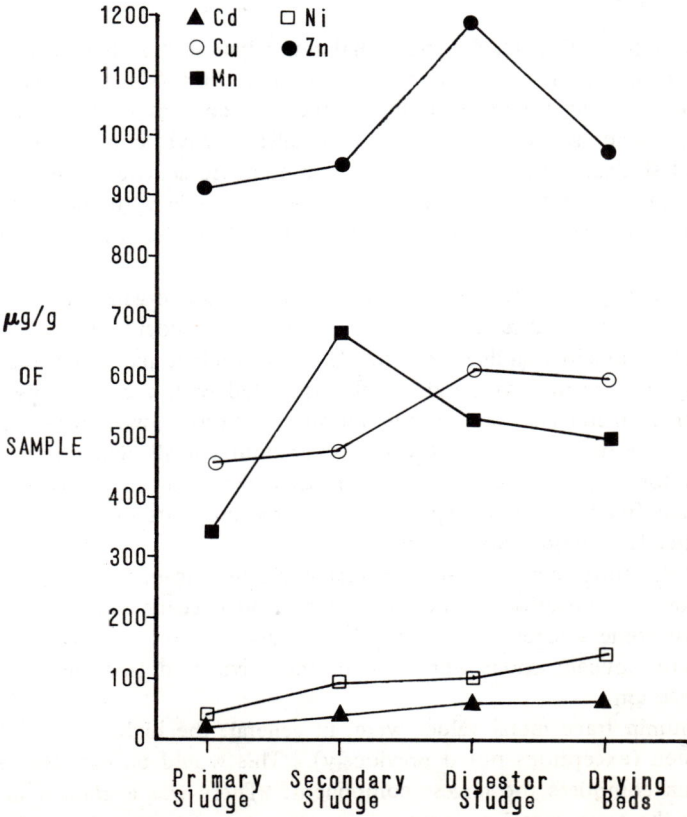

Figure 19.7 Trace metal (Cd, Cu, Mn, Ni and Zn) concentrations for the intact samples from the primary, secondary, digester and drying bed sludges.

from 25 to 140 ppm, averaging about 75. Three of the metals (copper, nickel and cadmium) show a definite increase as the sludge moves through the plant. One trace metal concentration, zinc, peaks in the digester and the remaining element, manganese, had its highest concentration in the secondary sludge material.

Total trace metal concentrations for these original samples were within the concentration ranges found by others for municipal sludges. The original samples contain trace metal concentrations sufficiently high to warrant monitoring when this material is to be used for other purposes, *e.g.*, as fertilizer. Some purposes, as in a highway road fill, would not be affected.

CONCLUSIONS

The fractionization results show that municipal sludges have a humin content of approximately 70%, total fulvic acid, 20% (soluble fulvic acid, 12%; insoluble fulvic acid, 8%), and humic acid content of about 8%. The humic acid fraction represents the smallest organic component of municipal sludge. The values for the humin, fulvic acid, and humic acid fractions vary from the primary sludge to the secondary to the digester to the drying beds, but variation is reasonable given the fluctuation in sludge source.

It is also apparent from this research that the soil organic matter fractionization procedure used is a reasonable and prudent method to fractionate municipal sludge. It provides a convenient method for determining the organically-bound metals associated with the fractions.

The trace metal analyses for the various fractions showed that the humic acid portion contained the lowest amounts of all elements analyzed for (cadmium, copper, manganese, nickel and zinc). The humic acid fraction is not responsible for large amounts of organically-bound metals in municipal sludges.

The total fulvic acid contained reasonable amounts of zinc and copper. The nickel concentration in the total fulvic acid exceeds that of the humin for some sludges, *e.g.*, drying bed sludge, but in most samples the other metal concentrations were less for the fulvic acid fraction than the humin fraction.

The humin trace metal values were, in general, the highest of all fractions tested (exceptions noted previously). This would be expected since the humin comprises the largest component. There was a gradual increase in the trace metal concentrations, except for manganese, as the sludge progressed through the plant. Similar trends were noted for some metals in other fractions.

The original sludge samples contained three categories of metal concentrations—low, medium and high concentrations. Nickel and cadmium were in the low category; manganese and copper were in the medium range; and zinc was in the high-concentration category. All concentrations were in the range of values previously reported in the literature.

The soil organic matter fractionization procedure is an effective method for characterizing the organically-bound trace metals (cadmium, copper, manganese, nickel and zinc). Additional trace metal investigations coupled with existing and future information about these soil organic fractions may provide an insight into the bonding mechanisms of trace metals with the complex organic molecules in sludge.

ACKNOWLEDGMENTS

The authors express appreciation to the T.C.U. Research Foundation for support of this research (Grant Number TCU/RF G7562) and to the City of Ft. Worth for the use of the atomic absorption spectrophotometer.

REFERENCES

1. Ghosh, M. M. and P. D. Zupper. *J. Water Pollution Control Fed.* **45**, 424 (1973).
2. Schmitt, C. R. and J. E. Hall. *J. Amer. Water Works Assoc.* **67**, 40 (1975).
3. Chen, K. Y., C. S. Young, T. K. Jan and N. Rohatge. *J. Water Pollution Control Fed.* **46**, 2663 (1974).
4. U.S. Department of Health, Education and Welfare. "Interaction of Heavy Metals and Biological Sewage Treatment Processes," Division of Water Supply and Pollution Control, Cincinnati, Ohio (1965).
5. Berggmen, B. and S. Oden. EPA Publication (1974).
6. Blakeslee, P. A. EPA Publication, U.S. Dept. of Agriculture University Workshop, Champaign, Illinois (1973).
7. Berrow, M. L. and J. Webber. *J. Sci. Ed. Agric.* **23**, 93 (1974).
8. Preuss, E. and H. Kollmann. *Naturwissenschaften* **61**, 270-1 (1974).
9. Beech, G. *J. Chem. Ed.* **51**, 328 (1974).
10. Boxwell, F. C. *J. Environ. Qual.* **4**, 267-72 (1975).
11. Dowdy, R. H. and W. E. Larson. *J. Environ. Qual.* **4**, 229-33 (1975).
12. Dowdy, R. H. and W. E. Larson. *J. Environ. Qual.* **4**, 278-82 (1975).
13. King. L. D. and H. D. Morris. *J. Environ. Qual.* **1**, 325-9 (1975).
14. Kirkham, M. B. *Environ. Sci. Technol.* **9**, 765-8 (1975).
15. Garcia, W. J., C. W. Blessin, G. E. Inglett, and R. O. Carlson. *J. Agr. Food Chem.* **22**, 810-5 (1974).
16. Painter, H. A., M. Vincy and A. Bywaters. Presented at the Metropolitan and Southern Branch, The Institute of Sewage Purification, London (December 1960).

17. Bunch, R. L., E. F. Barth and M. B. Ettinger. *J. Water Pollution Control Fed.* **33**, 122 (1961).
18. Rebhun, M. and W. J. Kaufman. Serl. Rept. No. 67-9, University of California, Berkeley, California (December 1967).
19. Abrams, I. and R. Breslin. Presented at the 26th Annual Meeting International Water Conference of the Engineers Society of Western Pennsylvania, Pittsburgh (October 1965).
20. Rebhun, M. and J. Manka. *Environ. Sci. Technol.* **5**, 606-9 (1971).
21. Manka, J., M. Rebhun, A. Mandelbaum and A. Bortinger. *Environ. Sci. Technol.* **8**, 1017-1020 (1974).
22. Bingham, F. T., A. L. Page, R. J. Mahler and T. J. Ganje. *J. Environ. Qual.* **4**, 207-11 (1975).
23. Turner, M. A. *J. Environ. Qual.* **2**, 118-9 (1973).
24. Lunt, H. H. *Water Sewage Works* **100**, 295-301 (1973).
25. Sabey, B. R. and W. E. Hart. *J. Environ. Qual.* **4**, 252-256 (1975).
26. Black, C. A., Ed. *Methods of Soil Analysis*, Part 2, American Society of Agronomy, Madison, Wisconsin (1965), pp. 1409-28.

CHAPTER 20

DETECTION AND ISOLATION OF BIOACCUMUABLE CHEMICALS IN COMPLEX EFFLUENTS

Gilman D. Veith and Ned M. Austin

> U.S. Environmental Protection Agency
> Environmental Research Laboratory
> Duluth, Minnesota

INTRODUCTION

The advent of computerized gas chromatography/mass spectroscopy (GC/MS) has made it feasible to conduct rapid, detailed analyses of complex waste effluents being discharged into natural waters. Recent advances both in instrumentation and chemical techniques are resulting in numerous compilations of chemicals identified in waste discharges, each containing in excess of 50 to 100 organic chemicals.

The expansion of our knowledge of individual organic chemicals in wastewaters creates an increasing need to rapidly assess the potential significance and/or hazards of each chemical in the environment. Two important characteristics, in addition to the quantity being discharged, are the toxicity and the persistence of the chemicals in aquatic ecosystems. The persistence of chemicals in the environment determines, in part, the distance from the source that the discharge may produce adverse effects as well as the degree to which the various trophic levels of the ecosystem may be affected.

Although it is unlikely that detailed studies of the persistence and bioaccumulation potential can be made for all chemicals identified in waste effluents, it is possible to judge the potential behavior of organic chemicals by comparing critical structural properties of the chemical with similar properties of chemicals for which data regarding the behavior in the environment exists. Studies have shown that one of the more important characteristics of organic chemicals is the lipophilic properties as measured by the *n*-octanol/water partition coefficient or a bioconcentration factor for aquatic organisms.

Figure 20.1 presents a conceptual model for the assessment of organic chemicals in the environment. The figure shows the relationship between the initial mass of a chemical remaining (corrected for dilution) with

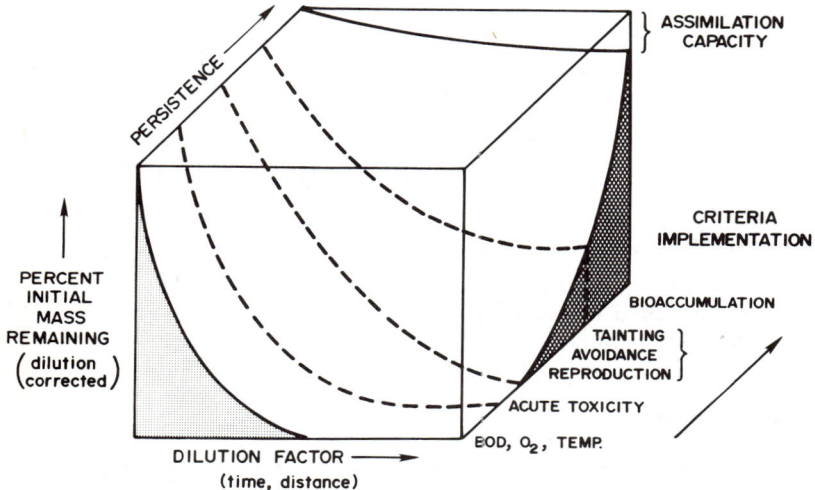

Figure 20.1 Conceptual model of the behavior of organic chemicals and implementation of water quality criteria.

distance or time from the discharge as a function of the persistence of the chemical. For rapidly-degradable chemicals, the figure indicates a definite assimilation capacity for the chemical in the environment. However, as the persistence of the chemical increases, the effects of the discharge may be exerted much greater distances from the discharge, and the assimilation capacity may be simply the sedimentation rates into the historical layer of the sediments of large lakes or the oceans. Concomitantly, the implementation of water quality criteria must shift from the protection against lethal effects of oxygen depletion and acute toxicity to chronic effects such as growth and reproduction.

Finally, there are increasing amounts of data from environmental monitoring studies which suggest that the accumulation of some toxic chemicals by fish may pose a threat to consumers of the aquatic organisms (principally fish) even though effects on the aquatic populations themselves are not discernible. During the years from 1960 to 1970, it became clear that chlorinated pesticides, chlorinated industrial chemicals, and the organic forms of some heavy metals are so persistent and bioaccumulative in the environment that quantities in the water below analytical detection limits can endanger entire ecosystems.

Neely et al.[1] have illustrated the relationship of the bioconcentration factor of selected hydrophilic chemicals in trout to the n-octanol/water partition coefficient. The data clearly indicate that chemicals with high partition coefficients have higher bioconcentration factors in aquatic organisms. Metcalf et al.[2] have found similar results using model ecosystems.

EXPERIMENTAL

The experimental approach has been to develop a separation technique using reverse-phase high-pressure liquid chromatography (HPLC) to isolate chemicals having potentially high lipid/water partition coefficients. The liquid chromatography was performed on a Varian series 4200 high-pressure liquid chromatograph (Varian Aerograph, Walnut Creek, California) equipped with a solvent programmer for the twin 5000 PSI pumps, uv and ir detectors and a constant-temperature water bath. Retention volumes were measured on a jacketed 25 cm x 2 mm (i.d.) column packed with permanently bonded octadecylsilane on 10 micro Li chromosorb (Varian, Micro Pak C-H, Palo Alto, California). The column temperature was held at 50°C. Because of the extremely wide range of retention volumes of organic chemicals, a solvent program consisting of an increase of 22% methanol in water initial to 75% methanol in water at a rate of 2%/min was used.

To evaluate the relationship between the reverse-phase HPLC retention volume and bioconcentration factors, the bioconcentration factors (C_F/C_W at steady-state) for the chemicals were determined using fathead minnows. Twenty-five fathead minnows were placed in aquaria (5 gal) receiving 400 ml/min filtered Lake Superior water heated to 25 ± 1°C. The chemicals were metered into the tanks using a metering pump (Fluid Metering Inc., Oyster Bay, N.Y.) from a continuous saturator system[3] or an acetone/water stock solution. The water concentration of the chemical (C_W) was measured at least every day and composites of five fish were removed for tissue residue analysis (C_F) after 4, 8, 16, 24 and 32 days. The steady-state concentration in fish was divided by the mean water concentration to obtain the bioconcentration factor (BF = C_F/C_W).

RESULTS AND DISCUSSION

Table 20.1 presents retention volume data for selected chemicals which represent alcohols, acids, amines, esters, ethers, hydrocarbons, phenols, pesticides and PCB's. The results indicate there is a 250-fold range in retention volume (relative to phenol) despite the solvent program. As expected, the more water-soluble chemicals such as alcohols and some phenols have much shorter retention volumes than those of the hydrophobic chemicals.

Table 20.1 Retention Data of a Variety of Chemicals using Gradient Elution Reverse-Phase HPLC

Chemical	K (Phenol = 1.0)	Log K (Phenol = 0.00)
Hydroquionone	0.27	-0.564
m-Aminophenol	0.36	-0.439
Resorcinol	0.36	-0.439
Catechol	0.73	-0.138
Phenol	1.00	0.000
o-Aminophenol	1.00	0.000
4-Methoxyphenol	1.17	0.067
Aniline	1.18	0.073
Benzyl alcohol	1.40	0.146
o-Toludine	1.82	0.260
4-Nitrophenol	1.83	0.263
o-Anisidine	2.00	0.301
Benzaldehyde	2.18	0.338
3-Methylphenol	2.25	0.352
2-Phenylethanol	2.27	0.357
4-Methylphenol	2.33	0.368
Indole	2.45	0.390
o-Chloroaniline	2.48	0.395
Benzene	2.50	0.398
n-Methylaniline	2.90	0.464
Acetophenone	3.18	0.503
Cinnamyl alcohol	3.32	0.521
4-Chlorophenol	3.73	0.571
Anisole	4.00	0.602
2,4-Dimethylphenol	4.00	0.602
Dimethyl phthalate	5.42	0.734
1-Naphthol	6.46	0.810
3-Methyl-4-chlorophenol	7.04	0.848
2,4-Dichlorophenol	8.73	0.941
Pentachlorophenol	10.42	1.018
4-Phenylphenol	11.46	1.059
Diethyl phthalate	12.45	1.095
Naphthalene	12.65	1.103

Figure 20.1, continued

Chemical	K (Phenol = 1.0)	Log K (Phenol = 0.00)
2,4,5-Trichlorophenol	14.27	1.154
Diphenylamine	14.36	1.157
Diphenyl ether	18.50	1.267
Anthracene	22.33	1.344
n,n-Dibutyl phthalate	24.45	1.388
p,p'-Methoxyclor	39.81	1.600
Endrin	39.82	1.600
p,p'-DDD	41.70	1.620
p,p'-DDT	44.69	1.650
Chlordane	50.15	1.700
Hexachlorobenzene	52.50	1.72
Aroclor 1016	35.50-51.30	1.55-1.71
Aroclor 1242	35.50-60.29	1.55-1.78
Aroclor 1248	35.50-60.28	1.55-1.78
Aroclor 1254	35.50-69.20	1.55-1.84

Much of the data in Table 20.1 have been plotted in Figure 20.2 as a function of the n-octanol/water partition coefficient reported in the literature. Figure 20.2 illustrates that the reverse-phase retention volume is a good indicator of the partition coefficient despite the diversity of molecular types of chemicals. The nonlinearity of Figure 20.2 is due largely to the increasing polarity of the solvent as the methanol content increases.

Since the bioconcentration factor has been correlated with the n-octanol/water partition coefficient, studies are being conducted to determine the suitability of the HPLC system in predicting the bioconcentration factor as well as the partition coefficient. The bioconcentration factors for Aroclor 1242, pp'-DDE, and pp'-methoxyclor are approximately 50,000, 48,000 and 22,000 respectively.

Although the determination of bioconcentration factors is on-going in the evaluation of this approach, the data can be empirically related to experience with known, persistent environmental contaminants. For example, chlordane, hexachlorobenzene, endrin, methoxyclor, DDT, DDE, Aroclor 1016, Aroclor 1242, Aroclor 1248, Aroclor 1254 and Aroclor 1260 are all chemicals or mixtures of chemicals which bioaccumulate in the environment and are commonly-observed persistent residues. The log of the retention volume of all these chemicals relative to phenol used as

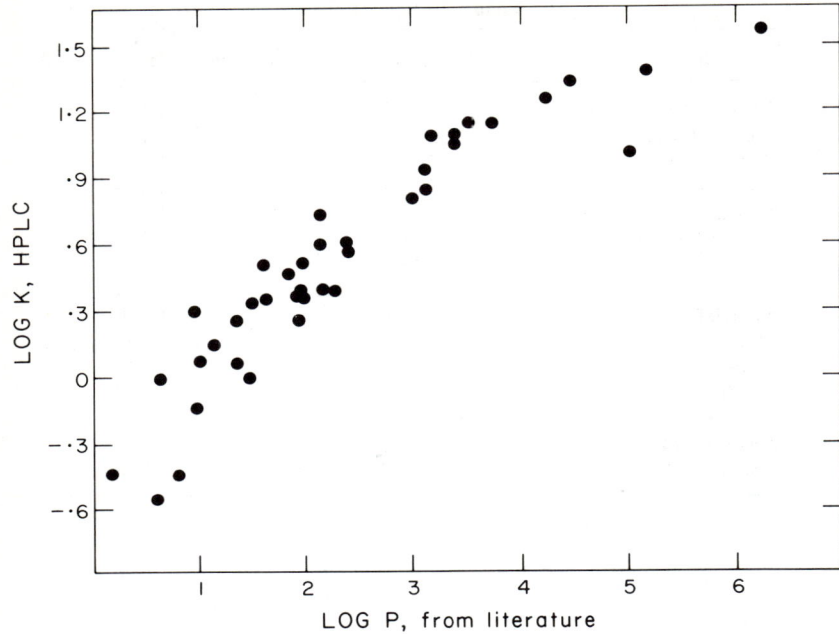

Figure 20.2 Variation of HPLC retention time and n-octanol/water partition coefficient.

an internal standard range from a log K' = 1.55 to log K' = 1.84. Consequently, if a concentrate of wastewater is injected on this reverse-phase column, those chemicals eluting with a log K' greater than 1.5 (relative to phenol) are likely to be those which have a high bioconcentration potential. Furthermore, it is proposed that the highest priority in GC/MS identification and bioconcentration evaluation be placed on those chemicals with a log K' greater than approximately 1.2 relative to phenol.

The evaluation of this proposal is continuing in both the assessment of the potential hazards of complex effluents as well as the screening of new chemicals for proposed uses.

REFERENCES

1. Neely, W. B., D. R. Bronson and G. E. Blau. *Environ. Sci. Technol.* **8**, 1113-1115 (1974).
2. Metcalf, R. L., J. R. Sanborn, P. Lu and P. Nye. *Archives Environ. Contam. Toxicol.* **3**, 151-165 (1975).
3. Veith, G. D. and V. M. Comstock. *J. Fish. Res. Bd. Canada* **32**, 1849-1851 (1975).

SECTION III
IDENTIFICATION OF ORGANIC POLLUTANTS

CHAPTER 21

THE OCCURRENCE OF VOLATILE ORGANICS IN FIVE DRINKING WATER SUPPLIES USING GAS CHROMATOGRAPHY/MASS SPECTROMETRY

W. Emile Coleman, Robert D. Lingg,
Robert G. Melton, and Frederick C. Kopfler

U.S. Environmental Protection Agency
Health Effects Research Laboratory
Water Quality Division
Cincinnati, Ohio

INTRODUCTION

After widespread publicity of the results of the study conducted by the U.S. Environmental Protection Agency (EPA) on New Orleans drinking water in November of 1974,[1-3] EPA Administrator Russell E. Train announced that he was ordering an immediate nationwide survey to determine the concentration and potential effects of certain organic chemicals in drinking water. A National Organics Reconnaissance Survey (NORS) was undertaken after President Ford signed "The Safe Drinking Water Act" in December of 1974. EPA was to conduct a comprehensive study of public water supplies and drinking water sources to determine the nature, extent, sources of, and means to control contamination by chemicals or other substances suspected of being carcinogenic.

As part of the NORS of drinking water supplies, the then Water Supply Research Laboratory in Cincinnati was charged with investigating the occurrence of volatile organics in drinking water of five selected cities: Miami, Florida; Seattle, Washington; Ottumwa, Iowa; Philadelphia, Pennsylvania; and Cincinnati, Ohio. These supplies were chosen to represent the major types of raw water sources in use in the United States today. Table 21.1

Table 21.1 Chemical Properties of Finished Water in Five Cities

City	Type of Supply	Type of Raw Water	Nonvolatile Total Organic Carbon mg/l	Conductivity μMHOS/CM	Chlorine mg/l	pH
Cincinnati	Surface	Industrial Waste	1.3	295	2.7	8.6
Miami	Ground	Natural Waste	6.5	350	2.3	8.7
Ottumwa	Surface	Agricultural Waste	2.3	500	1.4	9.2
Philadelphia (Torresdale Plant)	Surface	Municipal Waste	1.9	260	2.0	8.3
Seattle	Surface	Natural Waste	1.0	50	0	6.6

identifies the chemical properties of the finished water of the five cities along with the type supply and raw water source.

The objective of this laboratory in the NORS was to characterize as completely as possible, using existing analytical technology, the volatile organics in finished drinking water. This chapter discusses primarily the gas chromatography/mass spectrometry (GC/MS) techniques used for obtaining the results related to the above objective; that is, identifying volatile organic compounds in the drinking water supplies of these five cities. Ten selected compounds were also quantified as a part of this study. A more comprehensive report of the five-city survey is being prepared for transmittal to Congress.[4]

As part of a storage study on the effects of residual chlorine in tap water samples collected in the field and shipped to our laboratory for subsequent analyses, some compounds were identified by GC/MS that might be products of chlorination of raw water at the treatment plant. The results of ensuing experiments also are discussed in this chapter.

EXPERIMENTAL

To determine the presence of volatile organics in drinking water a computerized GC/MS system and a modified Bellar and Lichtenberg[5] trapping technique were employed. This modified technique is described in detail by Kopfler *et al.*[6]; they also describe a three-stage technique of stripping organics from a 500-ml volume of water commonly called the "500-ml VOA (volatile organics analysis)." The qualitative results of analyses in this chapter are all based on the 500-ml VOA. The method of quantification of the selected compounds is described in another paper by Lingg et al.[7]

Because of the wide variety (class and polarity) of compounds present in the water, chromatographic separation of each compound on the same GC column was not attainable. However, identification of the detected compounds was enhanced by having computer capabilities which enabled searches for specific ions (Extracted Ion Current Searches)[8] in a total ion current (TIC) reconstructed gas chromatogram (RGC).

Apparatus

Reagents

Ultra-pure helium, at a flow rate of 20 cc/min, was used to back-flush the organics from the trap onto the GC column and also as a carrier gas through the column at 30 cc/min.

Standards used to confirm relative retention times of suspect compounds were prepared with in-house distilled water. The sources of standards that we were able to obtain are listed later in Table 21.5. Preparation of standards is discussed by Lingg et al.[7]

The reagents used for the chlorination experiment were all reagent grade. The chlorine solution was prepared by bubbling gaseous chlorine (Cl_2) through nitrogen-purged distilled water to achieve a stock solution concentration in the range of 200-1000 mg/l. Appropriate dilutions were made from the stock solution (after the chlorine content was accurately measured by amperometric titration) to obtain the desired chlorine concentration. Seattle drinking water was used to make solutions containing 50 ppb each of benzene, ethanol, nitromethane, 2-pentanone, toluene and m-xylene; 1 ppm ferric chloride and 1.12 ppm chlorine in 140-ml sealed serum vials.[6,7]

Trap and Desorber

The trap and desorber for the 500-ml VOA were essentially the same as the ones designed by Bellar and Lichtenberg,[5] except the o.d. of the trap is 1/4 in. and the desorber was built to accommodate the wider diameter tubing. The desorber temperature was stabilized at 180°C.

GC Columns

Two packed, glass columns were used for the 500-ml VOA samples. One contained Chromosorb 101, 60/80 mesh and the other Tenax-GC, 60/80 mesh. The o.d. of each 3.05-m (10-ft) column was 6.35 mm; the i.d., 2 mm.

Gas Chromatograph/Mass Spectrometer

A Finnigan Model 9500 Gas Chromatograph equipped with one of the above columns was interfaced with a Finnigan Model 1015D Mass Spectrometer to obtain electron impact mass spectra.

Automated Data System

Data were acquired and analyzed with a System Industries 150 computer system. In this laboratory, all GC/MS data were acquired on a rapid access *disk* memory unit. With the use of graphic software, data were displayed on a TEKTRONIC 4010 CRT data terminal. Additional features of the automated data system are discussed in other papers.[9,10]

Mass Spectrometer Operating Parameters

Typical operating parameters for a 500-ml VOA are listed in Table 21.2. Note that the medium mass range (10-250 amu) was used because it offered additional sensitivity over the high mass range. The MS was calibrated each day on the medium range using perfluorotributylamine (FC 43).[9]

Table 21.2 Mass Spectrometer Operating Parameters for the 500-ml VOA

Parameter	Value
Ionization potential	70 eV
Emission current	500 μa
Ion energy	4V
Repeller potential	6V
Lens potential	100V
Analyzer temperature	70°C
Multiplier voltage	2KV
Analyzer pressure	5×10^{-6} TORR
Output preamplifier	10^{-7} amps/V
Mass range setting	Medium (10-250)
Separator oven temperature	≈230°C
Transfer line temperature	≈225°C

Data Acquisition System Operating Parameters

Computer software program: IFSS[11]
Mass range scanned: 14-16; 19-27; 29-31; 33-250
Maximum repeat count: 4
Integration time: 3 m sec
Repeat count before checking lower threshold: 4
Lower threshold: 4
Upper threshold: 4

The IFSS mode is a signal optimization mode used to adjust the Integration time as a Function of the Signal Strength. Using this option results in a nearly constant signal-to-noise ratio and minimizes saturation of the analog to digital converter.

The above parameters were selected so that the time required to scan a complete mass spectrum was approximately 3 sec. In scanning the mass range of 14-250 amu, m/e 17, 18, 28 and 32 were omitted to eliminate normalization of the RGC on the water or air present in the background and the sample.

Manual GC/MS Operations for 500-ml VOA

The procedure used to transfer the sample from the trap to the GC/MS system after the trap was inserted into the desorber at $180°C$ is described in Table 21.3. Time zero (0) equals the time that the needle of the desorption device was injected through the GC septum and the start of the back-flushing of the trap with helium. The computer scan could be started earlier than the time indicated in Table 21.3; however, we observed no organic components eluting from the GC column before water.

Table 21.3 Typical Manual GC/MS Operations for a 500-ml VOA

Operation	Time (minutes)	GC Oven Temperature ($°C$)
1. Purge trap	0 to 4	Room
2. Heat GC oven to $120°C$	4 to 6	Room to 120
3. Start computer scan	after water elutes (≈9 min)	120
4. Hold GC oven temperature at $120°C$	6 to 16	120
5. Start temperature program at $20°C/min$	16	120 to 200
6. Terminate run	50-60	200

N.B. - GC injector temperature = $150°C$

The ionizer in the MS was turned "on" immediately after the desorption unit was withdrawn from the GC inlet to monitor the water on the oscilloscope as the water eluted from the GC colunn.

Qualitative Analysis

Before this national survey, we qualitatively analyzed 5-ml samples of drinking water using the Bellar and Lichtenberg[5] technique. The GC/MS system seldom detected more than 10 to 15 volatile organics unless they were present in unusually high concentrations. By increasing the volume of the water 100-fold and employing three consecutive 30-min helium purges[6] of the same water, the number of volatile organics detected in the 500-ml VOA more than quadrupled. Furthermore, to facilitate identifications, the volatile organics isolated by the 500-ml VOA were actually partitioned by conducting the three consecutive purges of the same water. As stated also by Kopfler,[6] the first purge at 6°C removed the most volatile, nonwater-soluble organics (many halogenated compounds). The second purge at 95°C removed most of the remaining volatile, nonwater-soluble organics plus some water-soluble compounds not detected in the first purge. The final purge at 95°C showed the presence of volatile, water-soluble organics. Blank water, as described by Kopfler et al.,[6] was treated similarly.

GC/MS data acquisition of the volatiles isolated from the three consecutive purges of the finished water from the five cities and of the blank water were stored temporarily on disk memory units. The data from each GC/MS run in the form of RGC's, mass chromatograms (MC), and spectra were displayed on a CRT data terminal or obtained as hard copy output on a digital plotter.

Most of the identifications were made from the data acquired on the second purge because of the overall increase in concentration of the organics at the elevated temperature. Identifications of compounds isolated from the third purge were easiest because the GC peaks were well resolved and most of the compounds originally present in the water were removed by the previous purges. A compound was considered "detected" in the drinking water of the five cities when the same compound, as determined by GC/MS data, either was absent in the blank water or was above the level of that found in the blank water. The results of the second purges at 95°C were used to make these comparisons.

Identification of the volatile organics was confirmed by running standard compounds on the same GC column and under the same analytical conditions as the field samples were run. The GC relative retention times and mass spectra of these standards were compared with those of the compounds detected in the drinking water samples. It was also noted whether the mass spectrum of the standard and that of the detected compound in the drinking water agreed with a mass spectrum published in the literature.[1,2]

The expansion feature[9] of the data system allowed us to look at very small GC peaks that were normalized into the baseline in the original RGC. Without this feature, many compounds would have gone undetected and identification of them would have been thwarted by improper assignment of "eyeball" spectrum numbers.

After the mass spectra of all obvious GC peaks were obtained from each RGC, it was apparent that many spectra were contaminated by interfering ions from adjacent or underlying peaks that had not been detected initially. If two or more compounds overlap on the gas chromatogram, their mass spectra will be superimposed when this GC peak is scanned. Each chemical compound, however, gives rise to a unique combination of ions that forms its mass spectrum, and it is therefore possible to select a unique mass ion or ions (m/e value) for each compound in the overlapping spectra that is representative of only that compound. The computerized technique of extracting specific ions from a TIC RGC was used extensively in this study not only to isolate one compound but also to select classes of compounds having characteristic ions. The computer program[8] developed at Battelle Columbus Laboratories under an EPA grant is used for selected-ion monitoring of organic compounds by computer-controlled GC/MS (quadrupole) systems. It enhances detection of specific compounds in complex mixtures and alleviates background interferences. Current applications range from effluent monitoring to detecting drug residues and metabolites in overdose victims. Figure 21.1 demonstrates the practicality of this technique in drinking water analyses. Plots (A) and (B) are RGC's of Miami's finished water and of blank water, respectively. Plots (C) through (G) are mass chromatograms derived from extracted ion current searches of the RGC(A). In (C), m/e 29, 43 and 68 represent ions CHO^+, $C_2H_3O^+$ and $C_4H_6N^+$, respectively; (D) m/e 47 represents CCl^+; (E) m/e 61 and 62 represent ions $C_2H_2Cl^+$ and $C_2H_3Cl^+$; (F) m/e 77 and 91 represent ions $C_6H_5^+$ and ($C_7H_7^+$ and CBr^+); (G) m/e 79 and 81 represent ions $^{79}Br^+$ and $^{81}Br^+$.

Quantitative Analysis

The compounds selected for quantitative assessment[4,8] are listed below:

Benzene	p-Dichlorobenzene
Chlorobenzene	1,1-Dichloroethene (Vinylidene chloride)
Chloroethene (Vinyl chloride)	cis-1,2-Dichloroethene

312 ORGANIC POLLUTANTS IN WATER

trans-1,2-Dichloroethene Toluene
Tetrachloroethene Trichloroethene

Alternates:

Nitrotrichloromethane (Chloropicrin)
p-Chlorotoluene

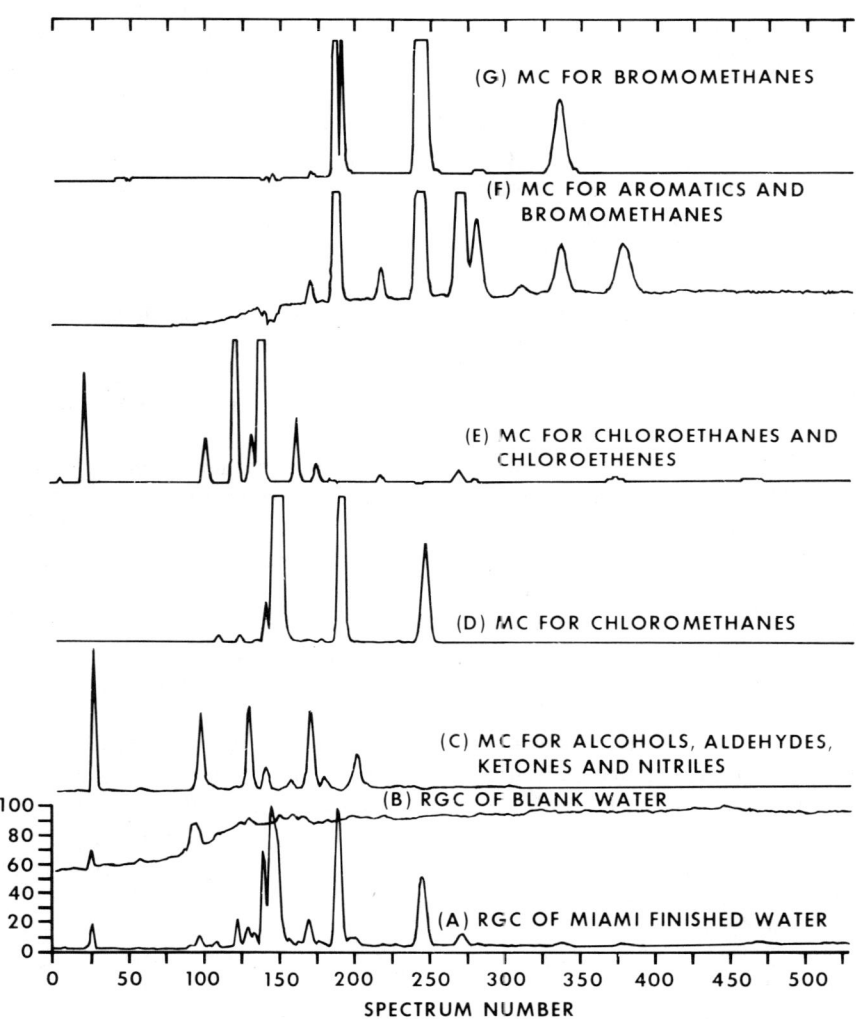

Figure 21.1 500-ml VOA (first purge at 95°C) of (A) Miami finished water and (B) blank water. C through G are mass chromatograms of (A): in (C) m/e = 29, 43 and 68; (D) m/e = 47; (E) m/e = 61 and 62; (F) m/e = 77 and 91; (G) m/e = 79 and 81.

Vinyl chloride, chlorotoluene and *p*-dichlorobenzene were not quantified in this laboratory because the standards were not received before the deadline for these analyses.

Chlorination Experiment

Kopfler *et al.* demonstrated the effects of chlorine on the increased production of chloroform in drinking water in storage and also when heat was applied to drinking water having a chlorine residual.[6] The results of their findings and the identification of compounds such as nitromethane, nitrotrichloromethane (chloropicrin), benzene, chlorobenzene, dichlorobenzenes, dichloroiodomethane, toluene, and an isomer of chlorotoluene in the five-city survey prompted this laboratory to investigate further the occurrence of these compounds in an experimental situation.

The three solutions for this experiment were allowed to react in the dark at room temperature (RT) for 4 days. (A) contained 50 ppb each of nitromethane, ethanol, 2-pentanone, benzene, toluene, and *m*-xylene and 1 ppm ferric chloride; (B) contained all reagents in (A) plus 1.12 ppm chlorine; and (C) contained all reagents in (B) except ferric chloride. Each sample was subsequently analyzed by GC/MS using a 140-ml VOA apparatus[6] purged at RT.

These solutions were made in Seattle drinking water rather than in distilled water because Seattle water had no measured chlorine residual (it was a naturally-buffered system), and the reagents added and the suspected chlorination by-products were not detected in previous analyses of Seattle water. Because of the use of iron pipes as water carriers and the presence of ferric ions in most finished water supplies, ferric chloride ($FeCl_3$) was added as a possible catalytic agent.

RESULTS AND DISCUSSION

The results of the volatile organic analyses and of the chlorination experiment using GC/MS techniques are presented in Tables 21.4 through 21.9 and in Figures 21.2 through 21.7. A total of 72 compounds were identified in drinking water supplies of the five selected cities. Table 21.4 lists the compounds according to their functional groups. It is evident from this table that aliphatic halogenated compounds accounted for the highest percentage of all compounds found in the five-city survey. Halogenated organics (aromatic and aliphatic) accounted for 53% of the total. Seventy percent of the total was attributed to compounds having elements other than carbon, hydrogen and oxygen.

Table 21.5 lists the 72 identified compounds as they eluted from a 10-ft Chromosorb 101 packed column or as noted otherwise. Their

Table 21.4 Occurrence of Functional Groups in 72 Organics

Functional Group	Present in 72 Organics	Percent of 72 Organics
Halogenated aromatic	5	7
Halogenated aliphatic	33	46
Aromatic ring	9	13
Nitro group	3	4
Sulfur atom	2	3
Alcohol group	4	6
Aldehyde or ketone group	12	17
Ester group	2	3
Nitrile group	6	8
Ether group	3	4

relative retention times (RRT) were compared with the RRT of chloroform (RRT = 1.00). Chloroform eluted from the column approximately 9 min after the start of the data acquisition.

The relative distribution of compounds per city is shown in Figure 21.2. The number of compounds found in each city is as follows: Cincinnati, 52; Miami, 60; Ottumwa, 21; Philadelphia, 50; and Seattle, 23. Figure 21.2 also indicates that the distribution of halogenated compounds in all cities except Seattle was approximately 50% of the number of compounds found in each city. The lower distribution in Seattle ($\simeq 30\%$) could be attributed to the type of raw water and the fact that a "zero" chlorine residual was measured in the finished water. (The TOC content of Seattle finished water was also the lowest of the five cities, see Table 21.1.) The compounds found in a specific city's supply along with their significance in drinking water is discussed by Tardiff.[4]

Another interesting result of this study was that only 13 of the 72 compounds were common to all five cities (see Table 21.6). One would infer, and evidence supports, that the type and number of compounds found in a drinking water supply depends heavily on environmental factors such as temperature, pollution, agricultural practices, source of water, and raw water treatment techniques.

Table 21.7 summarizes the occurrence of the 72 compounds according to carbon number, molecular weight and boiling point. After plotting the boiling points of the 72 compounds on normal probability paper, it was concluded from the linearity of the plot that the boiling points of the volatile organics were normally distributed above and below the median boiling point (77°C).

Table 21.5 Elution Order[a] of 72 Organics

	Name	Molecular Weight	Number of Carbons	Boiling Point (°C)	Ions Used For Mass Chromatograms m/e	Relative Retention Time[b]	Source of Standards[c]
1.	Chloroethyne (Chloroacetylene)	60	2	-32	60	0.04	_[d]
2.	Chloromethane	50	1	24	50	0.05	a
3.	Dimethyl ether	46	2	-24	46	0.06	a
4.	Methanol	32	1	65	31	0.10	h
5.	Cyanogen chloride	61	1	14	61	0.15	i
6.	Chloroethene (Vinyl chloride)	62	2	-14	27,62	0.16	i
7.	Acetaldehyde (Ethanal)	44	2	21	29,44	0.18	b
8.	Formic acid, methyl ester (Methyl formate)	60	2	71	31,60	0.29	c
9.	Bromomethane	94	1	5	94	0.31	i
10.	Bromoethyne	104	2	5	104	0.37	_[d]
11.	Chloroethane	64	2	12	64	0.38	a
12.	Ethanol	46	2	79	31	0.39	g
13.	Dichlorofluoromethane	102	1	9	67	0.40	j
14.	Fluorotrichloromethane	136	1	24	101	0.60	j
15.	Dichloroethyne (Dichloroacetylene)	94	2		94	0.62	_[d]
16.	Methyl cyanide (Acetonitrile)	41	2	82	41,40	0.64	c
17.	Propanal	58	3	49	29	0.64	c
18.	Pentane	72	5	36	43,72	0.65	h
19.	2-Propanone (Acetone)	58	3	57	43	0.66	h
20.	1,1-Dichloroethene (Vinylidene chloride)	96	2	36	61,96	0.71	b

316 ORGANIC POLLUTANTS IN WATER

Table 21.5, continued

	Name	Molecular Weight	Number of Carbons	Boiling Point (°C)	Ions Used For Mass Chromatograms m/e	Relative Retention Time[b]	Source of Standards[c]
21.	Diethyl ether (Ethyl ether)	74	4	35	45,74	0.71	c
22.	2-Chloropropane	78	3	37	43,78	0.72	d
23.	Dichloromethane (Methylene chloride)	84	1	40	49,84	0.74	h
24.	Formaldehyde, dimethyl acetal (Dimethoxymethane)	76	3	46	75	0.75	b
25.	Acetic acid, methyl ester (Methyl acetate)	74	3	58	43,74	0.76	c
26.	Iodomethane (Methyl iodide)	142	1	43	142,127	0.77	c
27.	Propenoic acid, nitrile (Acrylonitrile)	53	3	77	53,27	0.77	c
28.	Carbon disulfide	76	1	47	76	0.80	i
29.	2-Methyl-2-propanol (t-Butyl alcohol)	74	4	83	59	0.81	c
30.	*trans*-1,2-Dichloroethene	96	2	48	61,96	0.84	b
31.	Nitromethane	61	1	101	30,46	0.86	c
32.	2-Methylpropanal	72	4	61	43,72	0.89	b
33.	1,1-Dichloroethane	98	2	57	63,98	0.91	b
34.	2-Methylpropenal	70	4	68	70	0.91	b
35.	Propanoic acid, nitrile (Propionitrile)	55	3	97	54	0.91	c
36.	*cis*-1,2-Dichloroethene	96	2	59	61,96	0.95	c
37.	3-Methyl-2-butanone	86	5	93	43,86	0.95	b
38.	2-Butanone	72	4	80	43,72	0.96	b
39.	Bromochloromethane	128	1	67	128	0.99	b

40.	Trichloromethane (Chloroform)	118	1	83	1.00	i
41.	2-Methylpropanoic acid, nitrile (Isobutyronitrile)	69	4	108	1.07	c
42.	1-Butanol	74	4	31,56	1.10	b
43.	2-Butenal	70	4	41,70	1.10	b
44.	1,1,1-Trichloroethane	132	2	61,97	1.10	c
45.	1,2-Dichloroethane	98	2	27,62	1.11	b
46.	Tetrachloromethane (Carbon tetrachloride)	152	1	117,82	1.14	c
47.	3-Methylbutanal (Isovaleraldehyde)	86	5	44,86	1.15	b
48.	Benzene	78	6	78	1.17	c
49.	Trichloroethene (Trichloroethylene)	130	2	95,130	1.21	f
50.	2-Pentanone	86	5	43,86	1.22	c
51.	Bromodichloromethane	162	1	83	1.29	d
52.	Dibromomethane	172	1	174	1.29	d
53.	3-Methylbutanoic acid, nitrile (Isovaleronitrile)	83	5	43,68	1.37	e
54.	Dichloronitromethane	129	1	83,30	1.40	—d
55.	4-Methyl-2-pentanone	100	6	43,100	1.41	c
56.	Dimethyldisulfide (2,3-Thiabutane)	94	2	94,79	1.43	e
57.	Nitrotrichloromethane	163	1	30,117	1.48	e
58.	Toluene	92	7	91	1.49	c
59.	1,1,2-Trichloroethane	132	2	97,83	1.50	c
60.	2,6-Dimethyl-4-heptanone	142	9	57,85	1.52e	b
61.	Tetrachloroethene	164	2	166	1.54	b
62.	Chlorodibromomethane	206	1	129	1.64	d

Table 21.5, continued

	Name	Molecular Weight	Number of Carbons	Boiling Point (°C)	Ions Used For Mass Chromatograms m/e	Relative Retention Time[b]	Source of Standards[c]
63.	Chlorobenzene	112	6	132	112	1.86	c
64.	Dichloroiodomethane	210	1	132	83,175	1.90	_[d]
65.	Ethylbenzene	106	8	136	91,106	1.93	c
66.	Bromotrichloroethene	208	2		129,208	2.10	_[d]
67.	Tribromomethane (Bromoform)	250	1	150	173	2.31	c
68.	Azulene	128	10		128	2.41[e]	b
69.	Chlorotoluene, an Isomer of	126	7	159-162	91,126	2.59	c
70.	1,3-Dichlorobenzene (m-Dichlorobenzene)	146	6	172	146	3.12	c
71.	1,4-Dichlorobenzene (p-Dichlorobenzene)	146	6	174	146	3.22	c
72.	1,2-Dichlorobenzene (o-Dichlorobenzene)	146	6	176	146	3.56	c

[a]10-ft Chromosorb 101 Column.
[b]Retention time of trichloromethane is approximately 9 min. Time zero equals the beginning of data acquisition; relative retention time of chloroform = 1.00.
[c]Company: a. Air Products and Chemicals Inc.   f. Fischer Scientific Co.
              b. Aldrich Chemical Co.   g. HSMHA S. C. Co.
              c. Chemical Services Inc.   h. Burdick and Jackson Inc.
              d. Columbia Organic Chemicals Co.   i. Matheson Gas Products Co.
              e. Eastman Organic Chemicals Co.   j. PCR Inc.
[d]No Source Found
[e]10 ft Tenax GC column.

IDENTIFICATION OF ORGANIC POLLUTANTS 319

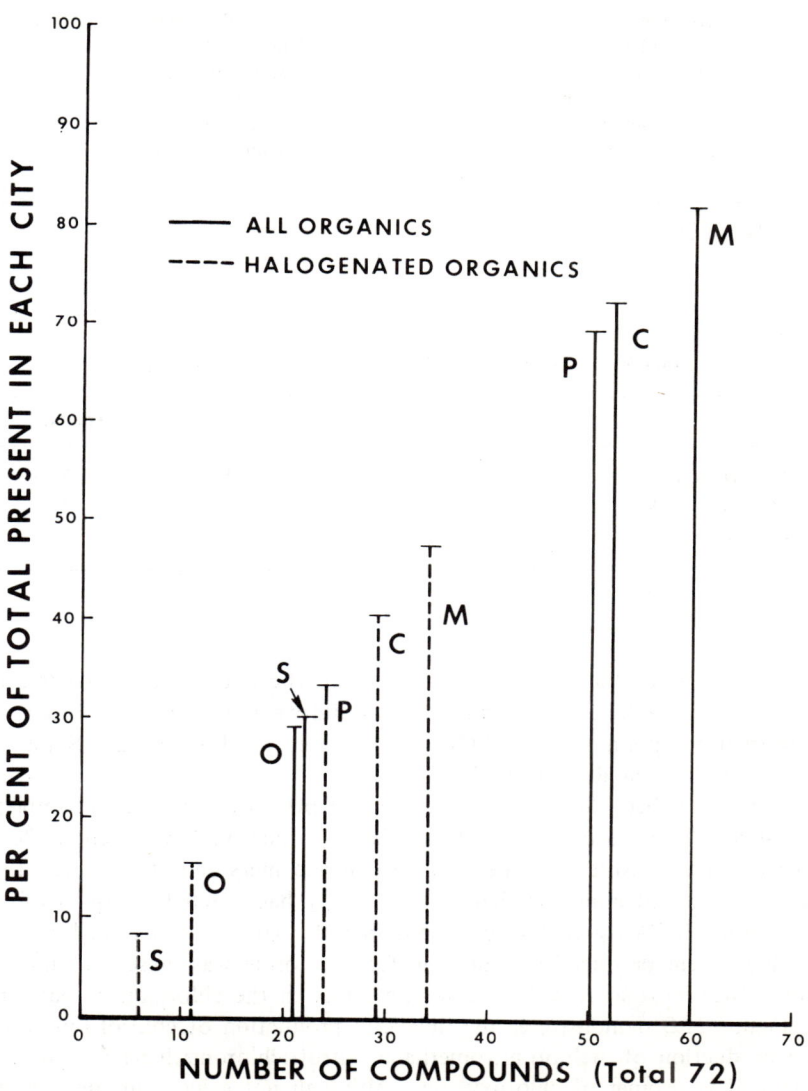

Figure 21.2 Percentage occurrence of 72 organics in finished water from Seattle (S), Ottumwa (O), Philadelphia (P), Cincinnati (C) and Miami (M).

Table 21.6   Organic Compounds Common to All Five Cities

Acetaldehyde	3-Methylbutanal
Bromodichloromethane[a]	3-Methyl-2-butanone
2-Butanone	2-Methylpropanal
Chlorodibromomethane[a]	Propanal
Chloromethane[a]	2-Propanone (Acetone)
Dichloromethane[a]	Trichloromethane[a]
Ethanol	

[a]Halogenated

Table 21.7   Summary of 72 Volatile Organic Compounds

	Range	Average	Median
Boiling Point (°C)	-32 to 176	78	77
Molecular Weight	32 to 250	101	92
Carbon Number	1 to 10	3	2

The results of the quantitative analyses of the selected compounds are presented in Table 21.8. Additional quantitative information on other identified compounds in the NORS is presented by Tardiff[4] and Symons[13] and in an Interim Report to Congress.[14]

Table 21.9 lists the residual chlorine and pH measurements of samples of "spiked" Seattle drinking water. (See Chlorination Experiment under Experimental.) Note that the residual chlorine measured at RT after 4 days (1.02 ppm) indicated that 0.10 ppm Cl_2 had reacted. Results of the 140-ml VOA by GC/MS indicated that the same chlorination by-products were produced whether the ferric chloride was present or not, thus indicating that it had no catalytic effect in the chlorination reaction.

Figure 21.3 demonstrates the increased production of chloroform and the production of dichloronitromethane, nitrotrichloromethane (chloropicrin), and an isomer of chloroxylene. Although not evident in this figure, trace amounts of chlorotoluene were detected. It is quite obvious from comparing (A) and (B) in Figure 21.3 that *m*-xylene was readily chlorinated and that chloroform was greatly increased. It is also obvious that the purging efficiency of ethanol and nitromethane from water is poor, as demonstrated by Kopfler et al.[6] The mass spectra of the chlorinated

Table 21.8 Quantitative Analysis of Selected Organics from Five-City Survey

	Concentration (µg/l) (ppb)				
	Cincinnati	Miami	Ottumwa	Philadelphia	Seattle
Benzene	0.3	< 0.1	< 0.1	0.2	*
Bromodichloromethane	15	73	+	20	4
Chlorobenzene	0.1	1	*	< 0.1	*
Chlorodibromomethane	3	32	+	5	3
1,1-Dichloroethene (Vinylidene Chloride)	+	0.1	*	< 0.1	*
cis-1,2-Dichloroethene	< 0.1	14	*	0.1	*
trans-1,2-Dichloroethene	*	1	*	*	*
Nitrotrichloromethane (Chloropicrin)	3	0.4	*	2	*
Tetrachloroethene (Tetrachloroethylene)	0.3	< 0.1	0.2	0.4	*
Toluene	< 0.1	*	*	0.7	*
Trichloroethene (Trichloroethylene)	0.1	0.3	< 0.1	0.5	*
Trichloromethane (Chloroform)	38	301	1	65	21

+ = detected but not quantified.

* = not detected by GC/MS.

Table 21.9 Residual Chlorine and pH Measurements of Spiked Seattle Water

Amount of Chlorine Initially Added mg/l	Reaction Time	Reaction Temperature °C	pH	Total Chlorine Residual mg/l
0	0 min	25	6.6	0
1.12	15 min	25	5.0	1.08
1.12	4 day	25	5.5	1.02

322 ORGANIC POLLUTANTS IN WATER

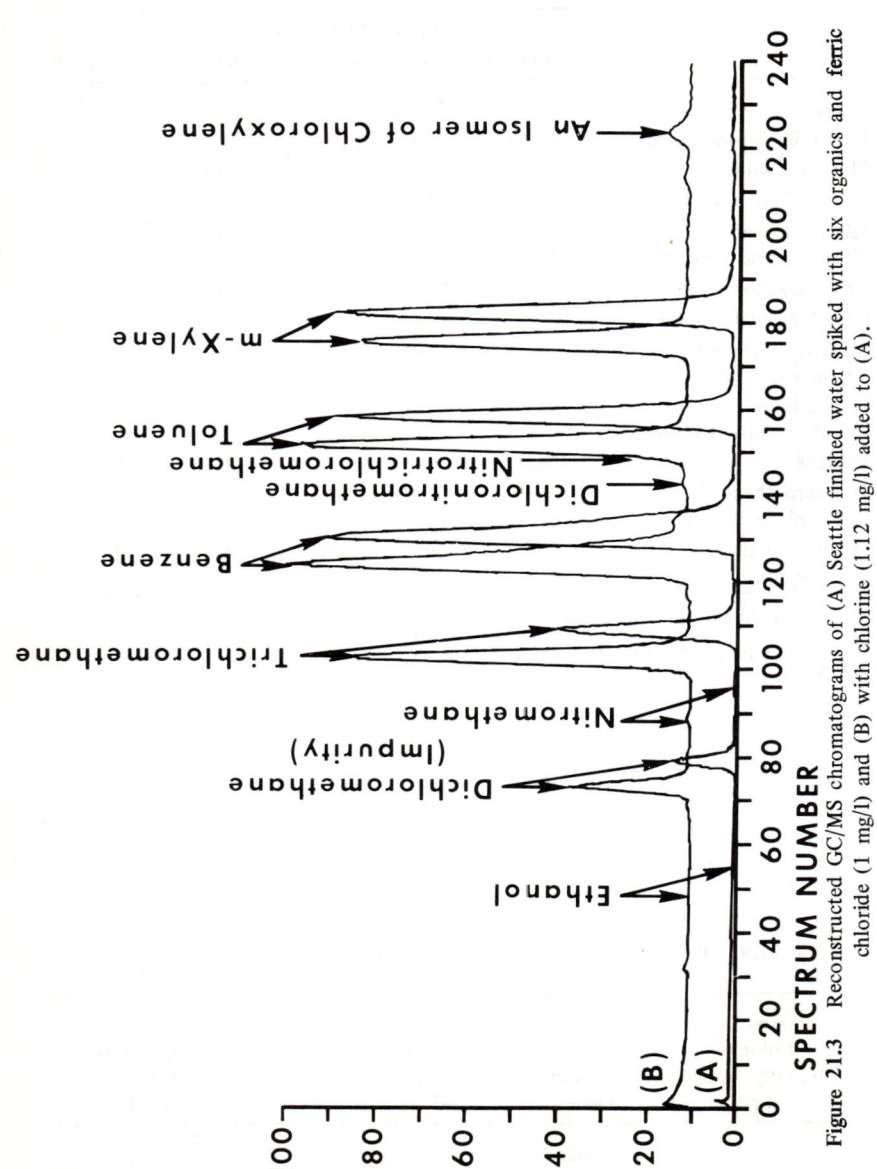

Figure 21.3 Reconstructed GC/MS chromatograms of (A) Seattle finished water spiked with six organics and ferric chloride (1 mg/l) and (B) with chlorine (1.12 mg/l) added to (A).

IDENTIFICATION OF ORGANIC POLLUTANTS   323

by-products are presented in Figure 21.4 through 21.6. In Figure 21.5 the mass spectrum of chloropicrin is somewhat obscured by major fragment ions (m/e 91, 92 and 65) of toluene. Identification of compounds in this area of the RGC was very difficult because toluene, tetrachloroethylene, chloropicrin, 1,1,2-trichloroethane, 2,6-dimethyl-4-heptanone and bromotrichloromethane all have RRT's of approximately 1.50. Bromotrichloromethane (not found in this study) has an almost identical RRT and mass spectrum as chloropicrin. The key spectral peak in differentiating the latter two compounds is m/e 30 (NO^+).

Figure 21.7 demonstrates how chloropicrin was successfully identified with the use of extracted ion current searches.

Another identified compound that may be a by-product of chlorination is dichloroiodomethane. Kleopfer[15] found this compound in several Midwest water supplies.

The general reactions listed below summarize this brief chlorination experiment.

$$\text{Nitromethane} \xrightarrow[H_2O]{Cl_2} \text{Nitrotrichloromethane (chloropicrin)} \quad \text{(reacts readily)}$$

$$\text{Benzene} \xrightarrow[H_2O]{Cl_2} \text{(does not react)}$$

$$\text{Toluene} \xrightarrow[H_2O]{Cl_2} \text{Chlorotoluene} \quad \text{(reacts slowly)}$$

$$m\text{-Xylene} \xrightarrow[H_2O]{Cl_2} \text{Chloroxylene} \quad \text{(reacts readily)}$$

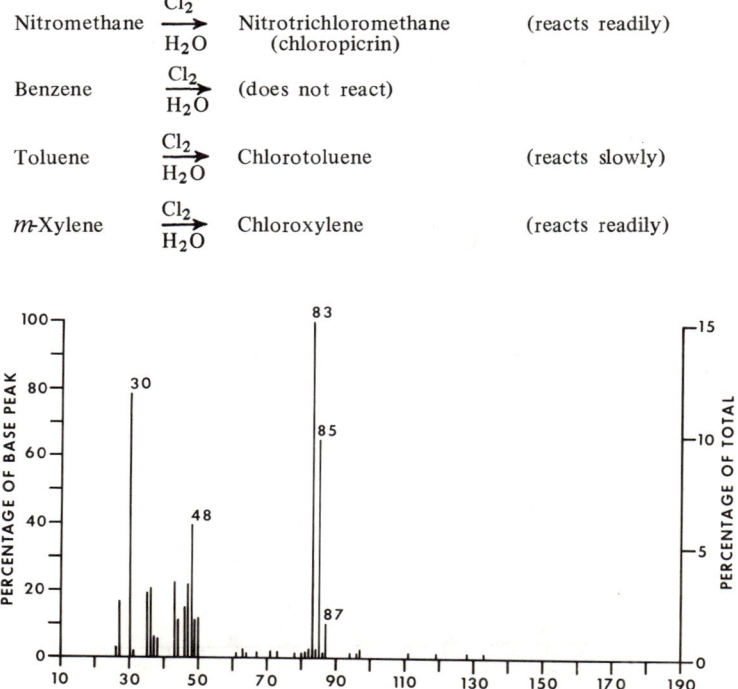

Figure 21.4   Mass spectrum (143-140) of dichloronitromethane, which was formed by the reaction of nitromethane (50 µg/l) and chlorine (1.12 mg/l) in Seattle finished water.

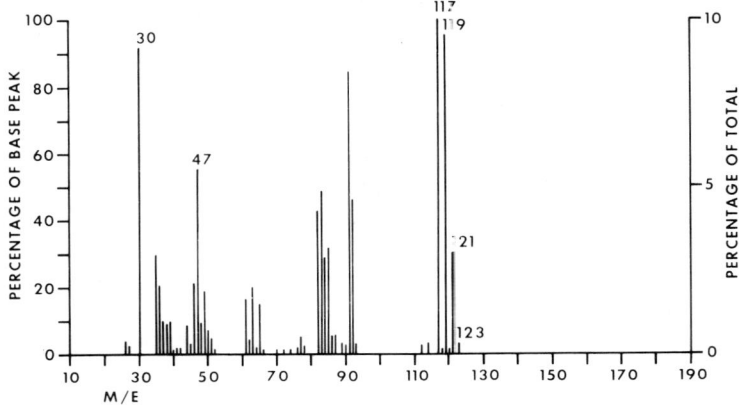

Figure 21.5 Mass spectrum (148-145) of trichloronitromethane, which was formed by the reaction of nitromethane (50 μg/l) and chlorine (1.12 mg/l) in Seattle finished water.

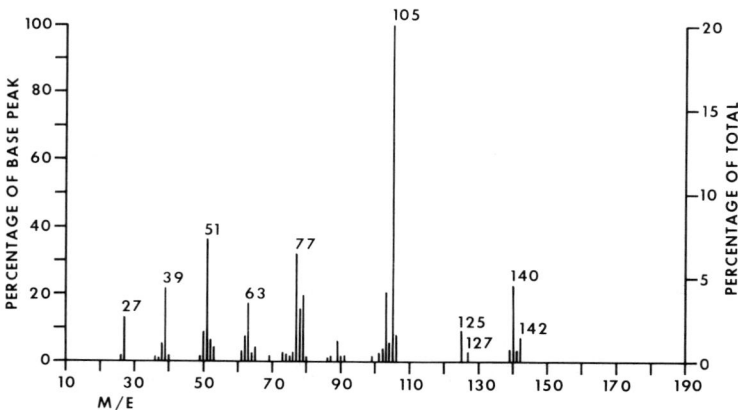

Figure 21.6 Mass spectrum (223-216) of an isomer of chloroxylene, which was formed by the reaction of m-Xylene (50 μg/l) and chlorine (1.12 mg/l) in Seattle finished water.

IDENTIFICATION OF ORGANIC POLLUTANTS 325

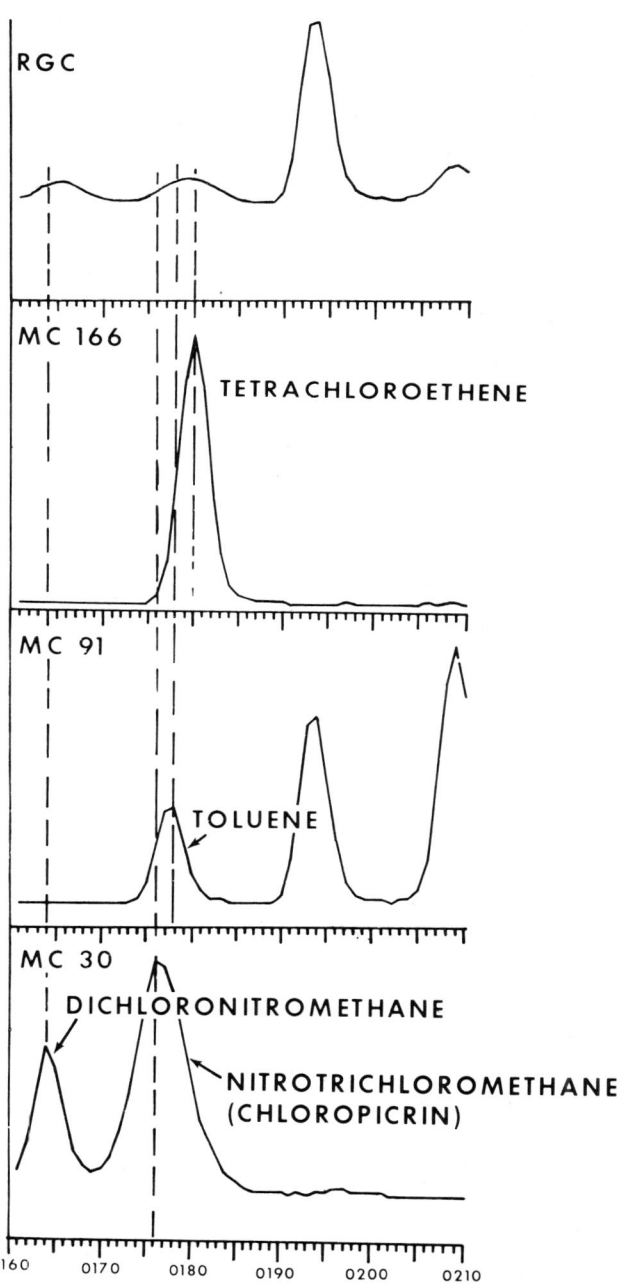

Figure 21.7 Mass chromatograms derived by extracted ion current searches.

It is highly possible and probable that the compounds mentioned in this chapter have been with us for some time. Only because of powerful tools like GC/MS and the other techniques discussed here have we been able to detect and identify low levels of volatile organics in drinking water which heretofore had been unidentified.

ACKNOWLEDGMENTS

The authors wish to thank and commend the many others who contributed to this chapter: Ms. Verna E. Tilford and Ms. Diana Routledge who typed the manuscript; Mr. Dale Dietrich and his staff who provided the art work and graphics; and Mrs. Marion G. Curry who edited the manuscript. Special commendation is due Mr. Donald Mitchell who helped with the analyses and Dr. Richard Schoenig who compiled an enormous amount of data and also helped with the analyses.

REFERENCES

1. U.S. Environmental Protection Agency. "Industrial Pollution of the Lower Mississippi River in Louisiana," Region VI, Dallas, Texas, Surveillance and Analysis Division (1972).
2. U.S. Environmental Protection Agency. "Analytical Results—New Orleans Area Water Supply Study," Report EPA-906/10-74-002, Region VI (Dallas, Texas) Surveillance and Analysis Division (1975).
3. Dowty, B., D. Carlisle and J. L. Laseter. *Science* **187**, 75 (1975).
4. Tardiff, R. G., W. L. Budde, W. E. Coleman, J. DeMarco, R. C. Dressman, J. W. Eichelberger, W. H. Kaylor, L. H. Keith, F. C. Kopfler, R. D. Lingg, L. J. McCabe, R. G. Melton and J. L. Mullaney. "Organic Compounds in Drinking Water: A Five-City Study," in preparation (1975).
5. Bellar, T. A. and J. J. Lichtenberg. *J. Amer. Water Works Assoc.* **66**, 739 (1974).
6. Kopfler, F. C., R. G. Melton, R. D. Lingg and W. E. Coleman. "GC/MS Determination of Volatiles for the National Organics Reconnaissance Survey of Drinking Water," presented at the First Chemical Congress of the North American Continent, Mexico City (December 1975), Chapter 6, this volume.
7. Lingg, R. D., R. G. Melton, F. C. Kopfler and W. E. Coleman. "GC/MS Techniques for the Quantification of Volatile Organics in Tap Water," in preparation (1975).
8. U.S. Environmental Protection Agency. "Specific Ion Mass Spectrometric Detection for Gas Chromatographic Pesticide Analysis," Report EPA-660/2-74-004. (Washington, D. C.: U.S. Government Printing Office, 1974).
9. System Industries, Inc. "System/150 GC/MS Data Processor," Sunnyvale, California (1974).

10. Lingg, R. D., R. G. Melton, W. E. Coleman and F. C. Kopfler. Proceedings AWWA Water Quality Technology Conference, Dallas, Texas (1974).
11. Eichelberger, J. W. and W. L. Budde. "Expanding the Range of Concentrations Amenable to Gas Chromatography/Mass Spectrometry in a Single Injection," presented at the 22nd Annual Conference on Mass Spectrometry and Allied Topics, Philadelphia, Pennsylvania (May 1974).
12. Stenhagen, E., S. Abrahamsson and F. W. McLafferty. *Registry of Mass Spectral Data* (New York: John Wiley and Sons, 1974).
13. Symons, J. M., T. A. Bellar, J. K. Carswell, J. DeMarco, K. L. Kropp, G. G. Robeck, D. R. Seeger, C. J. Slocum, B. L. Smith and A. A. Stevens. *J. Amer. Water Works Assoc.* **67**, 634 (1975).
14. U.S. Environmental Protection Agency. "Preliminary Assessment of Suspected Carcinogens in Drinking Water," (with Appendices) *Interim Report to Congress* (Washington, D.C., June 1975).
15. Kleopfer, R. D. "Analyses of Drinking Water for Organic Compounds," presented at the First Chemical Congress of the North American Continent, Mexico City (December 1975), Chapter 24, this volume.

CHAPTER 22

IDENTIFICATION OF ORGANIC COMPOUNDS IN DRINKING WATER FROM THIRTEEN U.S. CITIES

L. H. Keith, A. W. Garrison, F. R. Allen, M. H. Carter,
T. L. Floyd, J. D. Pope and A. D. Thruston, Jr.

>U.S. Environmental Protection Agency
>Southeast Environmental Research Laboratory
>Athens, Georgia

INTRODUCTION

The idea that it is desirable or necessary to specifically identify the individual organic compounds present in drinking water is rather recent. The embryo of this concept can be found in the records of a 1964 conference on pollution of interstate waterways of the Lower Mississippi River and its tributaries, called by the U.S. Secretary of Health, Education and Welfare. One of the recommendations of that conference was to establish a committee to direct and advise in the identification and abatement of all sources of pollution affecting the Lower Mississippi River. By 1967 consumer complaints of "oily" and "chemical" flavors in the drinking water of New Orleans and nearby communities were occurring with increasing frequency. Fish caught below Baton Rouge, Louisiana were no longer saleable because of these same bad tastes. On March 23, 1967 Dr. Leslie L. Glasgow, Chairman of the Louisiana Stream Control Commission, wrote to the Commissioner of the U.S. Federal Water Pollution Control Administration requesting technical assistance in obtaining specific information on the effects of the presence of organic chemicals on water quality and aquatic life in the Mississippi and Calcasieu Rivers. The first step was to identify the compounds present in the rivers. Work was begun in 1969 and culminated with a report[1] in 1972 in which 46 organic compounds were listed as being present in the raw or treated water supplies from three water plants (Carrollton plant of New Orleans, Jefferson #2 at Marrero, and the U.S. Public Health Service Hospital at Carville). An additional 44 compounds not detected in the three water plants were identified in wastewaters from 10 industrial plants that discharge into the Mississippi River just above New Orleans.

In July 1974, representatives of the State of Louisiana and the City of New Orleans requested the Region VI Environmental Protection Agency (EPA) administrator to undertake an analytical survey of the organic chemicals present in the finished water of three New Orleans area water plants (Carrollton, Jefferson #1 and Jefferson #2). Samples were taken in August 1974, extracted by EPA's R. S. Kerr Environmental Research Laboratory in Ada, Oklahoma, and analyzed by the Water Supply Research Laboratory, Cincinnati, Ohio and by the Analytical Chemistry Branch of the Athens, Georgia Environmental Research Laboratory. On November 8, 1974, a preliminary report[2] was released by the EPA that listed 66 organic chemicals identified in the finished water of the three New Orleans area water plants with concentrations that ranged in most cases from 0.01-5 µg/l. The USAEC—Ames Laboratory at Iowa State University added six more compounds to the list, and through additional research at the Athens Environmental Research Laboratory another 22 compounds were identified, raising the total identifications to 94.

Simultaneously with the release of the New Orleans preliminary report, the EPA announced plans for a National Organics Reconnaissance Survey (NORS) of water supplies of representative cities across the nation to determine the identity, concentrations and potential effects of chemicals in these drinking waters. Eighty cities were selected for analysis of six halocarbons in their raw and finished water supplies. Both raw and finished water were studied in order to correlate the presence of the haloforms with either the raw water source or the chlorination treatment. Ten of these same cities were chosen as sites for a more comprehensive survey of the organic content of finished drinking waters. An interim report to Congress[3,4] discusses the preliminary results of the 80-city survey and 5 of the comprehensive analyses.

The Analytical Chemistry Branch of the Athens laboratory analyzed the three New Orleans area carbon chloroform extracts (CCE's) in 1974 and the 10-city CCE's of the 1975 NORS. The results of the analyses from these 13 water plant CCE's (Table 22.1) are the subject of this chapter.

Table 22.1 Drinking Water Supply Sources

	Source Character	Water Plants
	Mississippi River at New Orleans	1. Carrollton (City of New Orleans) 2. Jefferson Parish #1 (east bank) 3. Jefferson Parish #2 (west bank)
NORS	Uncontaminated Upland Water	1. Seattle, Washington 2. New York, New York
NORS	Ground Water	1. Miami, Florida 2. Tucson, Arizona
NORS	Contaminated by Agricultural Runoff	1. Ottumwa, Iowa 2. Grand Forks, North Dakota
NORS	Contaminated by Industrial Waste	1. Cincinnati, Ohio 2. Lawrence, Massachusetts
NORS	Contaminated by Municipal Waste	1. Philadelphia, Pennsylvania 2. Terrebonne Parish, Louisiana

SAMPLING

Detailed descriptions of the sampling procedures appear elsewhere.[2,3] A variety of sampling procedures were used at New Orleans because this was the first site studied and because the best method of concentrating a wide variety of organic compounds present in trace amounts in water was not—and still is not—known.

At each New Orleans water plant, triplicate 1-liter samples of finished water were extracted in separatory funnels with 2 ml of tetralin. The extract was sealed in septum vials at the plant site. If extraction of very volatile components (*e.g.,* chloroform, benzene) was quantitative, a 500-fold concentration factor would be achieved. A small peak from chloroform was detected. In EPA laboratories tetralin extraction has since been supplanted by the Volatile Organics Analysis (VOA) technique because the latter is a superior method for concentrating trace amounts of very volatile organics from water.

VOA, accomplished by purging the water sample with helium, adsorbing the compounds on a small column of Tenax or Chromosorb 103 and then thermally desorbing them onto a gas chromatographic column, was carried out by the EPA's Cincinnati Water Supply Research Laboratory and is covered in Chapter 6.

Adsorption of the compounds in the New Orleans water supplies with XAD-2 resin was carried out by Mr. Gregor Junk, USAEC-Ames, Iowa

State University, according to his established technique.[5] The eluates from these resin columns were analyzed at both the Athens and the Ames laboratories. Thirty-six of the ninety-four compounds identified were found in the resin eluates[6] and six of these were found only by Junk.

Three types of carbon adsorption units were used in the New Orleans study but only one—the CAM sampler—was used in the 10-city NORS. The CAM sampler is a 3-in.-diameter by 18-in.-long glass cylinder that contains about 3/4 lb of granular activated carbon. In the New Orleans CAM samplers used for the "70-year" CCE's, the coarse carbon was used on each end of the column as a filter and holder for the fine mesh carbon. Two CAM samplers were always connected in series at each of the sites. The second type of carbon adsorption unit used was the mega-sampler. It consisted of two large columns connected in series; each column holds about 22 lb of carbon. The third type of unit used was the mini-sampler, a small polyvinyl chloride tube that holds about 70 g (0.15 lb) of 14 x 40-mesh activated carbon.

The mega-sampler was only used at the Carrollton water plant; 300,000 gal of water was passed through it over a 7-day period. The CAM samplers were used at all three New Orleans water plants. A so-called "70-year" sample was collected over a 7-day period (25,500 liters) and represents the amount of water a person would theoretically drink in 70 years at the rate of 1 liter/day. The mini-sampler was used at the three New Orleans water plants; a 2-month equivalent (60-liter) sample was collected over a 2-day period.

Sampling was much less extensive for the 10 cities in the NORS; 6000 liters of water were passed through two CAM samplers in series over a 7-day period.

Appropriate blanks were prepared for every method of sample concentration used, including separate adsorbent/solvent blanks for each of the carbon and resin extracts. These blanks were examined for interfering background components by both gas chromatography (GC) and gas chromatography/mass spectrometry (GC/MS).

All CAM carbons were air-dried and extracted with chloroform at the EPA's Robert S. Kerr Environmental Research Laboratory in Ada, Oklahoma. The operations were conducted in a special carbon-handling room designed to minimize potential contamination.[2] Extractions with chloroform were carried out in Soxhlets for 48 hours, followed by vacuum concentration of the chloroform at temperatures not exceeding 27°C in rotary evaporators to volumes of 30-60 ml. The extracts were transferred to 10-ml glass ampules, sealed under nitrogen and mailed to the Athens Environmental Research Laboratory. Some of the New Orleans CAM

IDENTIFICATION OF ORGANIC POLLUTANTS 333

carbons that had been first extracted with chloroform were dried and extracted with ethanol. However, these carbon alcohol extracts (CAE's) gave only a few poorly resolved GC peaks, both before and after methylation with diazomethane. CAE's were not collected from the NORS samples.

Carbon from the mega-sampler was dried and taken to EPA's Robert A. Taft Center in Cincinnati, Ohio for extraction.[2] Using a large-scale, permanently-installed extraction unit the carbon was extracted for 40 hr with 50 gal of analytical reagent grade redistilled chloroform. Excess solvent was distilled off until about 0.5 gal remained in the pot. This concentrated extract was hand-carried to the Athens laboratory in teflon bottles.

The mini-samples were dried and extracted at EPA's Region VI Laboratory Facility in Houston, Texas. One-half of these "2-month" CCE's were mailed to the Athens laboratory.

Upon receipt of each of these extracts they were concentrated further in Kuderna-Danish and Micro-Snyder column evaporators (Figures 22.1 and 22.2); the final volume was adjusted with a gentle stream of nitrogen

Figure 22.1 Kuderna-Danish evaporator.

Figure 22.2 Micro-Snyder evaporator.

so that 1 µl of the extract corresponded to 1 liter of water passed through the carbon or XAD resin filter (e.g., the NORS CCE's were concentrated to a volume of 6.0 ml since 6000 liters of water were passed through the carbon filters). The extracts from the respective carbon and resin blanks and each of the solvent blanks were all concentrated to volumes equivalent to their corresponding sample extracts. An exception to this procedure was the mini-samples. We received half of the 2-month extracts and these were each concentrated to 0.3 ml so that 1 µl corresponded to 100 ml of water passed through the carbon filter. Blank extract volumes were adjusted accordingly.

PRE-ANALYSIS FRACTIONATION

Organic analyses were performed using concentrated extracts from carbon and XAD-2 resin adsorption columns. Various attempts were made to fractionate the extracts to improve subsequent GC peak separation—most of these were unsuccessful. For example, TLC (silica gel: benzene) could be used to separate the mega-sample into three distinct fractions. However GC/MS of these separate fractions gave no identifications that were not also made by direct GC/MS of the original extract. The TLC separation had been attempted because it is particularly recommended for isolation of polynuclear aromatic hydrocarbons; in this case it was wasted effort.

Aliquots of the mega-sample CCE, the Carrollton "70-year" CCE, and their respective blanks were analyzed by the EPA organochlorine pesticide method.[7] This procedure involves fractionation by column chromatography on florsil with detection by electron capture GC. The presence of dieldrin and endrin, as determined by this technique, was confirmed by matching GC retention times with standards on two other GC columns of varying polarity. Further confirmation, as well as detection of chlordene, α-chlordane, and pentachlorophenyl methyl ether, was obtained by GC/MS analysis of the appropriate fractions. With the exception of dieldrin, these compounds were present at concentrations too low for GC/MS detection using the total extracts. Dieldrin was detected in the total extracts only after searching for characteristic peaks by limited mass range mass spectral techniques. The Jefferson #1 and Jefferson #2 70-year CCE's were successfully analyzed by electron capture GC for the detection and measurement of endrin without prior fractionation.

Other attempts were made to improve GC peak separation by various types of fractionation. Steam distillation neatly separated the mega-sample CCE into volatile distillables and nondistillables, but did little to improve actual peak separation without major adjustment of GC conditions.

Solubility extraction of the CCE into strongly and weakly acidic, basic and neutral fractions served only to show that most of the compounds are neutral. Fortunately, extract fractionation was not necessary for qualitative or even quantitative analysis. Computer controlled GC/MS analysis, with computer manipulation of acquired data, allowed identification of compounds by direct injection of CCE's.

QUALITATIVE AND QUANTITATIVE METHODS

Instrumentation

Gas chromatography of the New Orleans CCE's was performed using a Varian 1400 GC with a flame ionization detector (FID). A 10-ft x 1/8-in. i.d. glass column was packed with 3% SP-2100 on 80-100 mesh Supelcon AW/DMCS. Helium was used as a carrier gas at 20 ml/min, and the temperature was programmed from 40-280°C at 6°/min after an initial 6-min hold at 40°. Similar conditions were used to obtain mass spectra on the Finnigan 1015 GC/MS system.

The 70-year Carrollton CCE was rechromatographed on a 30-m by 0.4-mm i.d. glass capillary column coated with Supelco SP-2100. The column was prepared at the Athens Environmental Research Laboratory. Replacement of the Gohlke separator in the Finnigan 1015 GC/MS with a 9-in. stainless steel capillary tube enabled direct connection of the glass capillary column with the mass spectrometer. Improved chromatographic separation (Figure 22.3) and enhanced sensitivity of this arrangement led us to use the capillary column with all the following CCE's from the 10-city comprehensive NORS.

Sample injection size was reduced from the 2.0 μl used with the packed column to 0.4 μl on the capillary column; no injection splitter was used with the latter. Chromatographic conditions for GC/MS analyses using the capillary column were modified and included multiple-temperature and carrier gas (helium) flow programming. Injection was made with the GC oven door open, the column at room temperature (about 30°C), and the pressure in the mass spectrometer ion chamber at 1.5×10^{-5} torr. The GC oven door was closed 5 min after injection and the temperature was slowly increased to about 50°C over the next 6 min. At 11 min after injection the oven temperature controller was set at 60°C. Two minutes later temperature programming at 2°/min was started. Twenty-three minutes after injection (80°C) the temperature program rate was increased to 6°/min and carrier gas flow was increased to produce a pressure of 2.8×10^{-5} torr in the ion chamber (previously determined to correspond to a helium flow of 2 cc/min at room temperature).

336 ORGANIC POLLUTANTS IN WATER

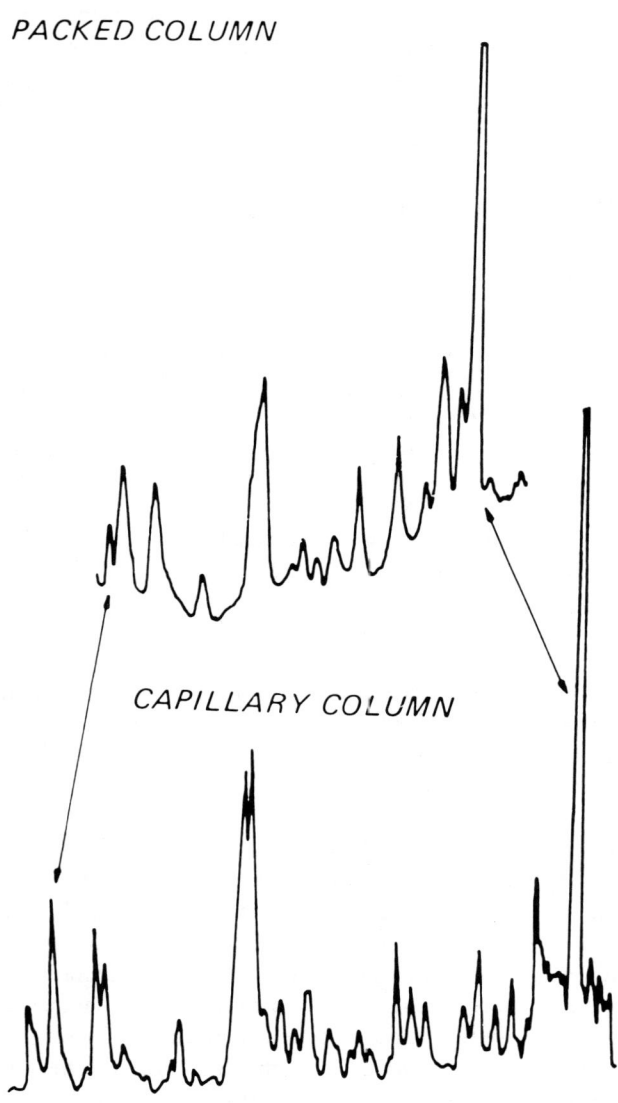

Figure 22.3 Comparison of a portion of the GC traces from the Carrollton "70-year" CCE using (Top) a conventional packed column and (Bottom) a glass capillary column. The stationary phase is SP-2100 in both examples.

Thirty-three minutes after injection (130°C) the temperature program rate was increased to 10°/min. The final temperature of 250°C was maintained for 20 min.

Computer-controlled collection of mass spectral data was begun immediately after sample injection. To prevent filament damage as solvent entered the MS, the ionization current was shut off 2.5 min after injection and turned on again 3.5 min after injection. Electron energy was maintained at 70 eV and filament current at 400 ua. The Finnigan 1015 mass spectrometer was connected by a System Industries System 150 interface to the PDP-8/e computer. A mass spectrum from m/e 41 to 350 was acquired approximately every 2.5 sec under control of the computer.

Chemical ionization mass spectrometry (CIMS) was performed on selected components in the Carrollton 70-year CCE with a separate computerized Finnigan 1015 mass spectrometer interfaced to a Finnigan 9500 GC using methane as the carrier/reactant gas. The same PDP-8/e computer was used for data acquisition.

Some initial GC/MS work was done on a Varian MAT CH5/DF system interfaced to a Varian 2740 GC via a dual Watson-Bieman separator and to a Varian SS-100 Data System. This instrument was later used in the New Orleans study for determination of empirical formulae of the major fragments in the mass spectra of alachlor and the alachlor chlorine homolog and for confirmation of the atrazine empirical formula.

A computerized Digilab FTS-14D/IR Fourier transform infrared spectrophotometer equipped with the Digilab GC-IR accessory and interfaced to a Perkin-Elmer 990 GC was used to confirm the presence of alachlor and atrazine in the Carrollton mega-sample CCE.

Method of Identification

At the end of GC/MS data acquisition a computer-reconstructed gas chromatogram (RGC) is plotted. This is a recording of the ion current summation of each mass spectrum, *i.e.*, the mass spectrometer is being used as a GC detector. The first step is to compare the RGC with the FID GC that was used to optimize chromatographic conditions. Although there is no correlation between the mass spectrometric ionization efficiency and the flame ionization response factors, correlation of the medium and large peaks in the two chromatograms is usually readily apparent. The RGC and FID chromatograms for the Carrollton 70-year CCE are shown in Figure 22.4. Peaks are numbered to correspond with the alphabetical listing of compounds given later in Table 22.4.

Programs, raw data and reduced data are stored on Diablo discs. Mass spectra of interest are plotted, with appropriate subtractions to remove

338 ORGANIC POLLUTANTS IN WATER

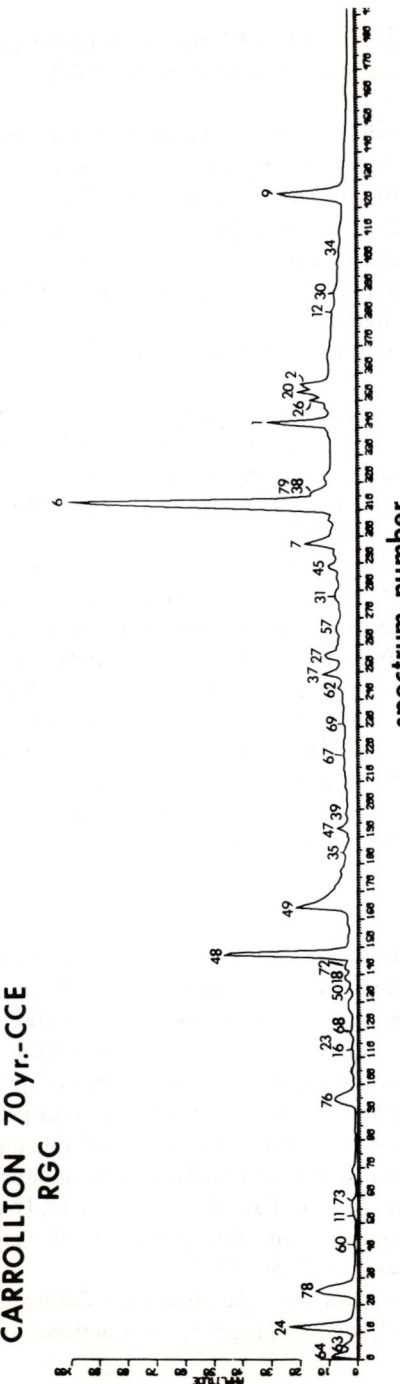

Figure 22.4a Reconstructed gas chromatogram (RGC) from the Carrollton 70-year CCE. Numbers correspond to compounds in Table 22.4.

IDENTIFICATION OF ORGANIC POLLUTANTS 339

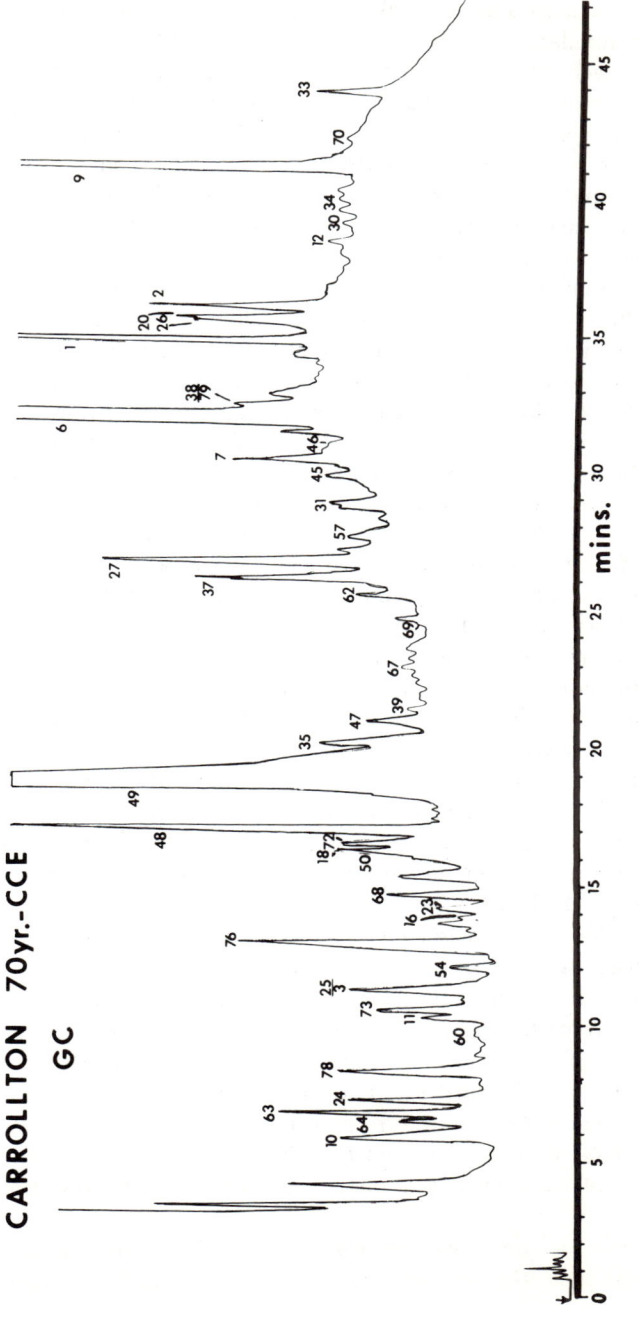

Figure 22.4b FID gas chromatogram (GC) from the Carrollton 70-year CCE. Numbers correspond to compounds in Table 22.4.

background mass spectral peaks, using either a cathode ray tube (CRT) with an auxillary hard copy unit or a Houston plotter.

Since each environmental sample usually has large numbers of chromatographic peaks (often a hundred or more) computer matching of spectra is employed as the first method of identification. Computer-abbreviated spectra are sent via direct acousticoupler-telephone connection[8] to a larger computer that presently contains the EPA library of about 40,000 abbreviated mass spectra. The list of matches is returned within minutes with the best match presented first and the others following in descending order of Similarity Index (SI), which is a measurement of the degree to which the sample spectrum matches the library spectrum.

An example of this computer-aided identification is shown in Figure 22.5. At the top is the plot of Carrollton 70-year CCE mass spectrum No. 311 with No. 308 chosen as the background spectrum and subtracted from it (refer to compound number 6 in Figure 22.4, top). In the middle of Figure 22.5 is the computer match. There were only 3 hits (matches) with corresponding Similarity Indexes of 0.555, 0.122 and 0.103. Generally, a SI of 0.4 or greater is considered to be a fairly reliable tentative identification, especially when, as in this instance, the next best match has a very low SI. The tentatively-identified compound is always visually checked against a reference mass spectrum either from the data bank (as shown at the bottom of Figure 22.5), our files, other compilations of spectra (such as the Aldermaston Eight Peak Index of Mass Spectra[9]), or the scientific literature. However, the tentative identification is never considered confirmed until it has been verified by comparison of two different physical measurements of a standard (usually the GC/MS spectrum and GC retention time) obtained on our instruments under similar operating conditions. In this chapter, compounds that have been confirmed by this definition are marked with an asterisk in tables where they are listed.

When there is no reasonable computer match (or when it has been determined that the sample compound does not correspond to any of the computer matches) the identification must be made by "manual" spectral interpretation. Often supplementary information from gas chromatography/high resolution mass spectrometry (GC/HRMS), gas chromatography/chemical ionization mass spectrometry (GC/CIMS) or gas chromatography Fourier transform-infrared spectroscopy (GC-IR) is necessary to make tentative identifications from spectral interpretations. Examples of these identifications are discussed in a later section of this chapter.

Searches for compounds by mass spectral fragments that are specific, or at least indicative of them, can be a helpful routine. Figure 22.6

IDENTIFICATION OF ORGANIC POLLUTANTS

SAMPLE SPECTRUM

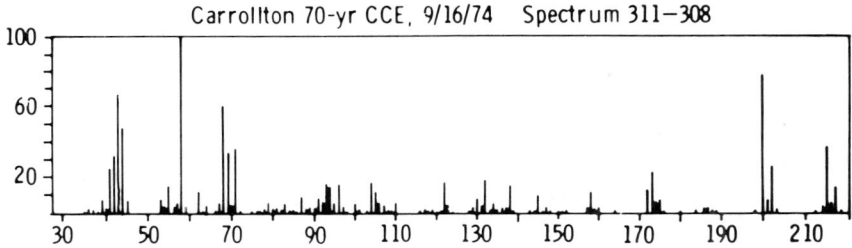

COMPUTER MATCHING

3 Hits   Methyl 3-oxooctadecanoate 312 C19 H56 O3
Atrazine (SEWL)   File key—6524
File key—8671   S I—0.122
S I—0.555

1-Methyl-4-(1,2,3,4-tetrahydro-2-naphthyl)piperazine 230 C15 H22 N2
File key—4842
S I—0.103   Cost of search: 0.36

COMPUTER REFERENCE SPECTRUM

File key? 8671
2-Chloro-4-ethylamino-6-isopropylamino-6-triazine (Atrazine) (SEWL)-T6
N CN EnJ BMY CM2 FG- 215 C8 H14 Cl N5 BSR 0186

m/e	INT	m/e	INT	m/e	INT	m/e	INT	m/e	INT
27	37	28	38	43	86	44	46	55	13
58	100	68	41	69	29	79	3	87	5
93	13	94	10	104	10	105	6	122	8
130	3	132	10	137	5	147	2	158	5
172	6	173	13	174	3	175	3	200	33
201	2	202	10	215	14	216	2	217	3

Figure 22.5 (Top) Mass spectrum of Carrollton 70-year CCE, spectrum No. 311 minus No. 308; (Middle) computer match of this data—3 hits; (Bottom) print-out of Atrazine's abbreviated spectral data that corresponds to the "best hit."

342 ORGANIC POLLUTANTS IN WATER

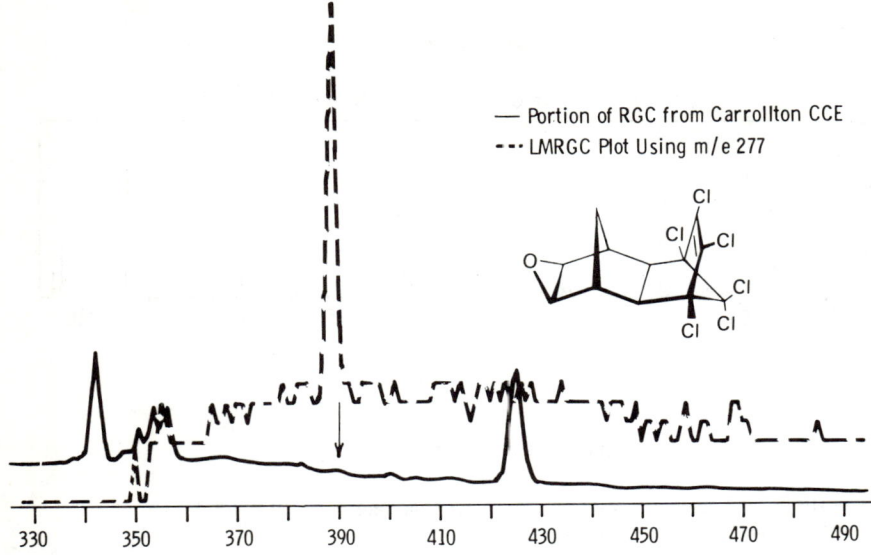

Figure 22.6 RGC of a portion of the Carrollton 70-year CCE with an LMRGC of m/e 277, one of the characteristic ion fragments of dieldrin, superimposed above it. The dieldrin peak is greatly enhanced.

shows a portion of the Carrollton 70-year RGC and a limited mass RGC (LMRGC) based on m/e 277, one of the fragments in a characteristic chlorine isotope cluster in the mass spectrum of dieldrin. The dieldrin peak in the RGC was so small that it could not be located until it was enhanced by the limited mass search.

Methods of Quantitation

Two methods of quantitation were used. The New Orleans samples were quantitated using GC peak areas, and the NORS samples were quantitated using RGC peak areas. Each has advantages and disadvantages.

A Perkin-Elmer PEP-1 Data System, interfaced to a Varian 1400 GC operated under the conditions described above, was used for computerized GC quantitation and retention time measurements. Since atrazine was present in all extracts of New Orleans samples, it was chosen as an internal standard. A stock solution of 5 parts-per-thousand of atrazine (99.7% pure) in chloroform was the internal standard for quantitation of all identified pollutants for which standards were obtained. Solutions of known amounts of pure reference compounds were prepared and mixed with a known amount of the atrazine reference stock solution. Mixtures were designed so as to obtain good GC peak resolution. The atrazine was assigned a flame response factor (RF) of 1.000 and, since its concentration was known, the computer system was able to calculate the RF, as well as the relative retention time (to atrazine), of each standard.

After tentative identification of pollutants by GC/MS, a PEP-1 computer program was written for the GC-computer analysis of each extract, allowing the computer to use the known flame responses to calculate pollutant concentrations (Figure 22.7). The concentrations of the concentrated extract are expressed in g/liter which corresponds to µg/liter in the original water samples since the concentration factor is one million times. In some cases, the RF calculated for one standard was also used for other compounds of the same chemical class. The relative retention times, calculated for all pollutants and printed out by the computer, were then manually compared with those of the available standards. The blanks had to be dosed with atrazine as an internal standard, since atrazine was absent from them.

The advantage of the above method is that a standard is not needed for every compound identified. FID response factors for compounds with similar structures and functional groups are usually similar enough that a model compound can be used to determine response factors [*e.g.,* once the response factor of 0.184 was determined for tridecane (Figure 22.7) it could be used for seven other normal and branched alkanes in the sample]. Disadvantages of the above method are that careful correlation is required between the RGC and the FID GC peaks (sometimes a difficult task with minor components of complex mixtures) and a separate PEP-1 GC computer program must be generated for each sample.

A pair of System Industries GC/MS peak quantitation routines called QNTATE and QNTSET were used for the 10 NORS samples. Data acquisition on standard compounds is made under the same conditions as data acquisition of the samples. A computer file containing information from the standard runs is established using the System Industries QNTSET program. The beginning, end and maximum of each peak (using spectrum numbers from the RGC), the quantity (expressed in nanograms) of the corresponding compound and its name are entered into the QNTSET file. Next, the QNTATE program is called into execution. The beginning, end and maximum spectrum numbers of the sample peak are specified along with information pertaining to the standard peak in the QNTSET file with which the sample peak is to be compared. For each sample peak the computer prints out the name of the compound and the amount present (expressed in nanograms). Peak areas in both QNTSET and QNTATE programs are computed with baseline subtracted. The user

344   ORGANIC POLLUTANTS IN WATER

```
RUN     1    12: 37.9    9/  9/ 74   CCE 1  CARRØLTØN
INST    4  , GC METHØD    42   , GC FILE   55 :

TIME      AREA        RRT        RF          C          NAME
 3.48    1.4457      .107,     1.000,      .8987,   !
 4.26    1.0659      .132,     1.000,      .6626,   !
 5.40     .0066      .167,     1.000,      .0041,   !
 5.95     .7198      .184,     1.000,      .4474,   !
 6.59     .2873      .204,     1.319,      .2356,   112-TRICHLØRØETHANE:
 6.89     .7331      .213,      .214,      .0975,   TØLUENE:
 7.38     .4588      .228,     2.380,      .6788,   CHLØRØDIBRØMEMETHANE:
 7.99     .0368      .247,     1.000,      .0229,   !
 8.42     .5926      .261,     2.500,      .9209,   DICHLOROIODOMETHANE
 9.01     .0438      .279,     1.000,      .0272,   !
 9.75     .0554      .302,     1.000,      .0345,   !
 9.96     .0464      .308,     1.000,      .0288,   !
10.35     .1933      .320,     4.738,      .5694,   BRØMØFØRM:
10.64     .3883      .329,      .208,      .0502,   M- ØR P-XYLENE:
11.03     .0783      .342,     1.000,      .0487,   !
11.37     .6423      .352,     1.642,      .6556,   C2H2CL2BR2, Ø-XYLENE:
12.22     .1120      .378,      .184,      .0128,   NØNANE:
13.10    1.1664      .406,     1.000,      .7250,   UNK MW 145:
13.58     .0677      .421,     1.000,      .0421,   !
13.83     .1578      .428,      .217,      .0213,   P-ET TØL ØR PRØP BZ:
14.06     .0668      .435,      .957,      .0398,   BIS-2-CL-ET ETHER:
14.38     .1710      .445,      .184,      .0196,   BRANCHED DECANE:
14.85     .2593      .460,      .240,      .0387,   TRIMETHYL BZ ISØMER:
15.49     .3321      .480,      .184,      .0380,   DECANE, UNK ARØM:
16.15     .1760      .500,      .316,      .0346,   LIMØNENE:
16.39     .5249      .508,      .557,      .1817,   BIS-2-CL-IPR ETHER:
16.69     .5580      .517,      .184,      .0638,   BRANCHED UNDECANE:
17.18    1.5324      .532,     1.635,     1.5575,   HEXACHLØRØETHANE:
17.62     .0975      .546,     1.000,      .0607,   !
17.95     .0707      .556,     1.000,      .0439,   !
18.65    8.2608      .578,      .298,     1.5302,   ISØPHØRØNE:
20.32     .5689      .630,      .278,      .0983,   DIHYDRØCARVØNE:
21.16     .3025      .656,      .866,      .1629,   HEXACHLØRØBUTADIENE:
21.59     .0696      .669,      .184,      .0080,   DØDECANE:
21.95     .0418      .680,     1.000,      .0260,   !
22.54     .0380      .698,     1.000,      .0237,   !
22.80     .0557      .706,     1.000,      .0346,   !
23.09     .0980      .715,      .184,      .0112,   TRIDECANE:
23.42     .0692      .726,     1.000,      .0431,   !
23.75     .1090      .736,     1.000,      .0678,   !
24.36     .0017      .755,     1.000,      .0011,   3-M-3-Ø-6-H-TRIAZINE:
24.84     .0895      .770,     1.000,      .0556,   !
25.72     .2123      .797,      .184,      .0243,   TETRADECANE:
26.28    1.0357      .814,      .412,      .2653,   DIMETHYLPHTHALATE:
26.90    1.1325      .834,      .310,      .2182,   DI-T-BU-BENZØ●UINØNE:
27.36     .2777      .848,     1.000,      .1726,   !
27.85     .2404      .863,      .184,      .0275,   PENTADECANE:
28.52     .0273      .884,     1.000,      .0170,   !
28.89     .1072      .895,     1.000,      .0667,   !
29.11     .1895      .902,      .314,      .0370,   DIETHYLPHTHALATE:
30.12     .2243      .933,      .416,      .0580,   CL7 NØRBØRNENE:
30.65     .7836      .950,     1.000,      .4871,   DE-ETHYL ATRAZINE:
31.65     .2206      .981,      .416,      .0571,   CL7NØRBØRNENE ISØMER:
32.25    7.7993     1.000,     1.000,     4.8480,   ATRAZINE:
32.66     .3798     1.012,      .314,      .0741,   DIPRØPYL PHTHALATE:
33.07     .3955     1.025,     1.000,      .2459,   !
33.64     .0662     1.043,     1.000,      .0412,   !
34.23     .0254     1.061,     1.000,      .0158,   !
34.54     .1663     1.071,     1.000,      .1034,   !
35.02    1.7922     1.085,      .715,      .7965,   LASSØ -- ALACHLØR:
35.76     .4150     1.108,      .314,      .0810,   DIBUTYL PHTHALATE:
35.91     .5095     1.113,     1.112,      .3522,   BLADEX:
36.34     .6254     1.126,      .715,      .2780,   LASSØ CL HØMØLØG:
36.84     .0641     1.142,     1.000,      .0398,   !
37.08     .1093     1.149,     1.000,      .0680,   !
38.31     .0644     1.187,     1.000,      .0400,   !
38.70     .1261     1.200,      .700,      .0549,   MACHETE:
39.37     .1122     1.220,      .613,      .0428,   DIELDRIN:
39.85     .1392     1.235,      .350,      .0303,   DIHEXYL PHTHALATE:
40.25     .1443     1.248,     1.000,      .0897,   !
40.56     .1976     1.257,     1.000,      .1228,   !
41.03     .0937     1.272,     1.000,      .0583,   !
41.37    3.1139     1.282,      .329,      .6368,   BENZYL BUTYL PHTHALA:
42.48     .7768     1.317,      .248,      .1198,   TRIPHENYL PHØSPHATE:
44.22     .4542     1.371,      .352,      .0994,   DI-2-ET HEXYL PHTHAL:
```

Figure 22.7   GC computer print-out for the Carrollton 70-year CCE.

specifies the spectral limits of the peak, and the computer then assumes that the baseline connects the first spectrum before the starting value and the first spectrum after the terminating value.

Advantages to the QNTSET/QNTATE method of quantitation are ease and speed in setting up the standard files and direct comparison of the raw data in the computerized sample files with the same type of data in the standard files. Comparison of peaks from an RGC with peaks from a FID gas chromatogram is eliminated. Disadvantages are that sensitivity, calibration stability and baseline may vary more over the time periods encountered between sample and standard runs (often several days) than these same parameters in a gas chromatograph. Also, the ionization efficiency of compounds of similar structure or with similar functional groups seems to vary more than the flame response factors using a FID.

Both methods require that standards and samples be run under identical conditions as far as possible. Best results are obtained when samples and standards are run immediately after one another. In practice, of course, this is not usually feasible since standard runs cannot be made until the compounds in the sample are tentatively identified. Presumably once the standard mixture is prepared and run the sample could then be rerun. Since other factors in the overall analytical scheme such as efficiency of adsorption of the sample compounds onto the carbon and their desorption by chloroform were possibly of greater magnitude than the variation in quantitation conditions, this reanalysis was not considered to be worthwhile. The concentrations of the compounds reported are therefore only approximations and could vary from their true values by as much as a factor of 2. Since the adsorption, desorption and concentration steps are not likely to be 100% efficient most of the concentration values reported are probably lower than their true values.

RESULTS

Compound Identifications

Alachlor and Its Chlorine Homolog

Alachlor (also known as Lasso and Lazo) is an acetanilide derivative introduced by the Monsanto Company in 1967 as a pre-emergence herbicide for use with soybeans, corn, cotton and peanuts. Its chemical name is 2-chloro-2′,6′-diethyl-N-(methoxymethyl) acetanilide (Structure I). It does not give an easily discernible molecular ion by electron impact mass spectrometry (EIMS), and its spectrum was not in the computer data bank, which made its identification difficult.

$$\text{C}_2\text{H}_5 \quad \text{CH}_2\text{OCH}_3$$
$$\underset{\text{C}_2\text{H}_5}{\text{N}} - \underset{\underset{\text{O}}{\|}}{\text{C}} - \text{CH}_2\text{Cl}$$

I

The identification of this compound was particularly desirable since the size of its GC peak indicated that it was present in a relatively high concentration. Also, an associated peak in the gas chromatogram had a similar mass spectrum so one identification might lead to another. Chemical ionization mass spectrometry (CIMS), however, gave a characteristic molecular-ion-plus-one peak (m/e 270) (Figure 22.8). High resolution mass spectrometry (HRMS), obtained on the fly using the GC effluent, produced possible empirical formulae of the important mass fragments, including the molecular ion. These empirical formulae were correlated with reasonable fragmentation modes of the parent molecule, so that the most probable parent ion formula was deduced (Figure 22.8). Since other herbicides had been identified in the samples, a herbicide handbook[10] was searched for compounds having the appropriate molecular formula and structural characteristics. Alachlor was the only possibility. A standard of alachlor produced the same low resolution mass spectrum and GC retention time, confirming the identification.

The identity of alachlor was further confirmed by Fourier transform gas chromatography-infrared spectroscopy (see Figure 22.8), with a spectrum obtained on the fly from the Carrollton mega-sample extract. This spectrum matched that of the standard obtained in the same mode.

The electron impact (EI) spectrum of the alachlor chlorine homolog (compound 2, Table 22.4) did not give a definitive molecular ion. Chemical ionization MS showed the molecular weight to be 303, and the chlorine isotope pattern indicated the presence of two chlorine atoms instead of one. Below m/e 188 the low resolution EI mass spectrum of this compound was similar to that of alachlor. The largest observable fragment was at m/e 272, which can be rationalized as loss of a methoxyl group as in the case of alachlor (Figure 22.9). The keys to the homolog's structure are the fragments at m/e 188 and 160. HRMS showed that they possessed the same empirical formulae as the m/e 188 and 160 fragments in alachlor's HRMS. The postulated structures giving rise to these alachlor fragments are shown in Figure 22.9. Loss of 84 mass units

IDENTIFICATION OF ORGANIC POLLUTANTS 347

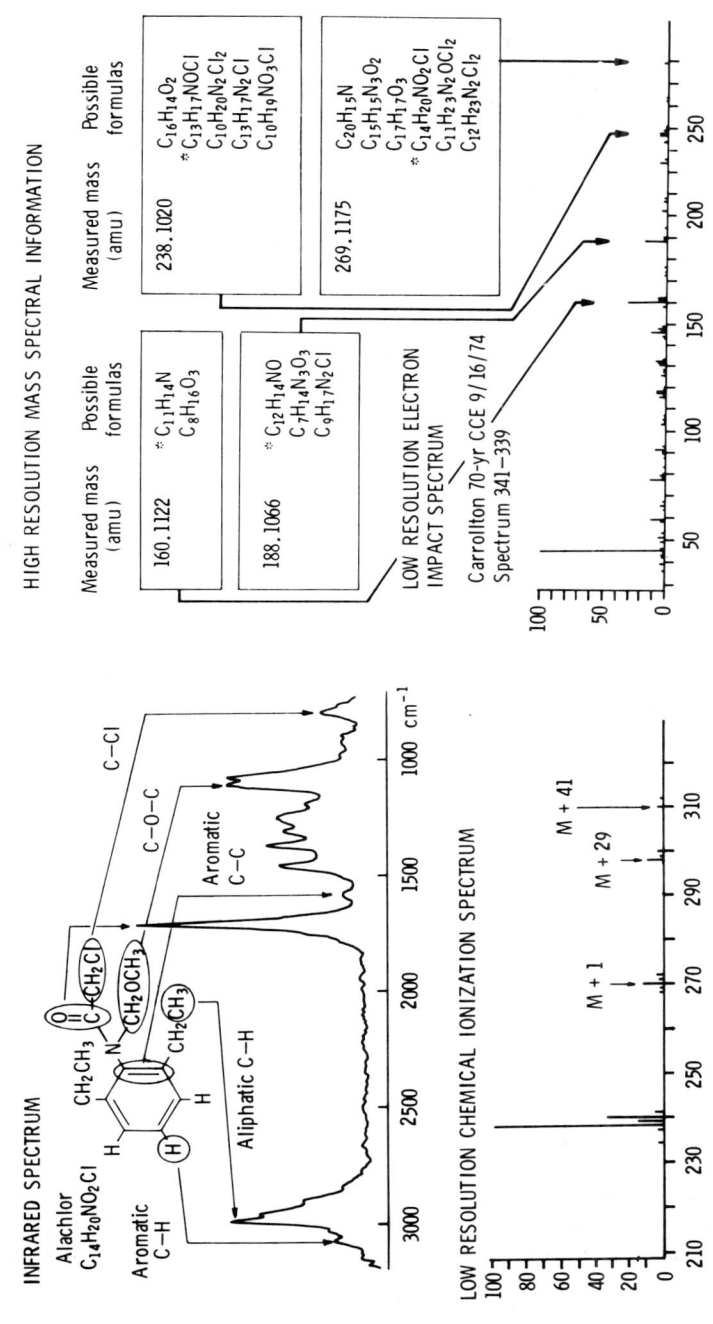

Figure 22.8 Spectra of alachlor in the Carrollton 70-year CCE. Starred (*) formulas correspond to correct alachlor ion formulas.

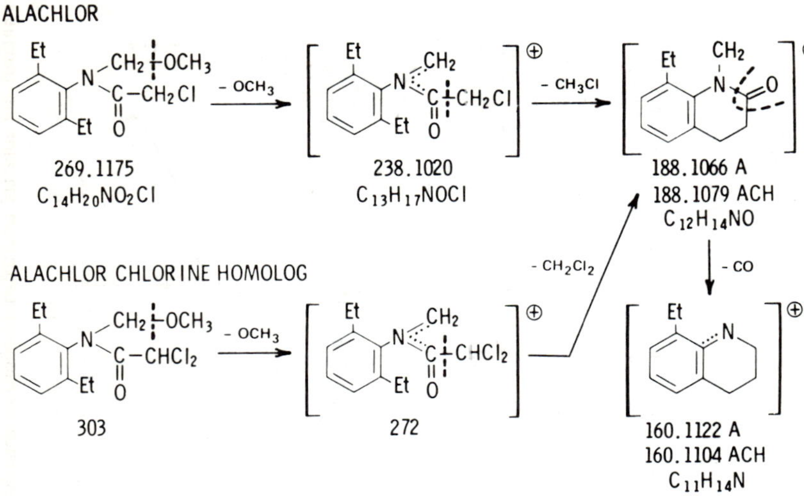

348 ORGANIC POLLUTANTS IN WATER

Figure 22.9 Proposed fragmentations of alachlor (A) and alachlor chlorine homolog (ACH).

from fragment m/e 272 is caused by loss of CH_2Cl_2 and therefore identifies the position of the second chlorine atom as being on the same carbon as the single chlorine of alachlor. The Fourier transform GC-IR spectrum of this chlorine homolog was also very much like that of alachlor. On the basis of the above information the chlorine homolog was tentatively identified as 2,2-dichloro-2',6'-diethyl-N-(methoxymethyl) acetanilide (Structure II).

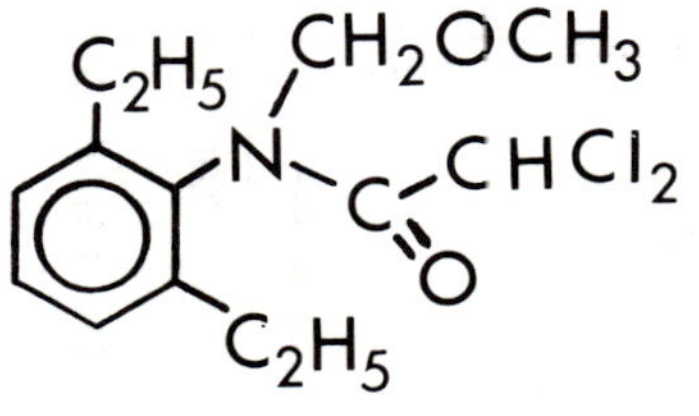

II

In an attempt to confirm the identification of the alachlor homolog and to determine whether it could have been produced during chlorination treatment of New Orleans drinking water, solutions of 5 mg/l alachlor in distilled water were treated with up to 138 mg/l of chlorine gas for 22 hr at room temperature. Two chlorinated compounds were formed, the concentrations of which were dependent upon the concentration of chlorine. The concentration of alachlor diminished correspondingly. The spectra of these compounds did not match that of the alachlor homolog found in the New Orleans water; they appeared to correspond to one and two chlorine additions to the aromatic ring of alachlor. Therefore the homolog is probably not formed by chlorination of alachlor in the New Orleans water supply; it could be a by-product of alachlor manufacture.

Deethylatrazine

Deethylatrazine (desethylatrazine) was found in New Orleans water only in samples collected by carbon adsorption, not in resin adsorption samples. Dealkylation appears to be the major mechanism involved in microbial degradation of alkyl-substituted chloro-s-triazines.[11] However, microbial degradation produces both deethyl- and deisopropylatrazine. Similarly oxidative dealkylation by free radicals produced both monoalkyl derivatives of atrazine plus the completely dealkylated products.[12,13] Limited mass range searches for the expected predominant fragments of deisopropylatrazine failed to yield any trace of this compound in the CCE's.

One possible explanation for the apparently exclusive formation of the deethyl product and none of the deisopropyl material was deduced from earlier experiments by other chemists at the Athens Environmental Research Laboratory.[14] Under synthetic reaction conditions involving atrazine (III) and sodium nitrite in water at 27°C with a low pH, only 2-chloro-4-(N-nitroso-N-ethylamino)-6-isopropylamino-s-triazine (IV, N-nitroso atrazine) was isolated. IV was stable in water at pH values greater than 4 and at 25°C. Surprisingly, it is highly photoreactive and was rapidly decomposed by uv light to deethylatrazine (V) and atrazine (III) in water. It was estimated that, near the surface of a water body, the half-life for photodecomposition of IV is less than 10 min throughout the United States. Experiments with IV under sunlight in both distilled water and water from a local river confirmed the calculated half-life.

Analysis for N-nitroso atrazine (IV) in the CCE samples by GC was unsuccessful since a standard of IV decomposed during gas chromatography under all conditions tried. Analysis for IV using limited mass range plots from the GC/MS data of the New Orleans CCE's was also unsuccessful. However, Fine *et al.* reported the possible presence of IV in the New Orleans CCE's on the basis of high-performance liquid chromatography

(HPLC) chromatograms of the CCE's using a new Thermal Energy Analyzer (TEA) as a selective detector for N-nitroso compounds.[15] A N-nitroso atrazine (IV) standard matched the retention time of one of the 24 peaks in the HPLC-TEA chromatograms of the New Orleans CCE's. Its concentration was estimated to be about 0.1 µg/l in the drinking water. An extract of water from the Mississippi River, although not concentrated as much as the CCE's, showed many of the same peaks in its HPLC-TEA chromatogram including the one with the same retention time as IV. At present these data are only tentative and must remain so until the identity of each peak is confirmed. Preliminary analysis of the New Orleans CCE's by other chemists using liquid chromatography with a dual-wavelength ultraviolet (uv) detection system showed no detectable N-nitrosoatrazine (IV) down to the lower limit of 0.1 µg/l. It is possible that this is not contradictory to Fine's tentative identification of IV since the estimated concentration was essentially at the threshold detection limit of the UV detection system. However, until the peaks observed with HPLC-TEA are confirmed by unambiguous means (such as HPLC-MS) the presence of N-nitrosoatrazine and other possible N-nitrosoamines should be considered with caution.

Trimethyl Isocyanurate

A good computer spectral match (SI = 0.509) for 1,3,5-trimethyl-2,4, 6-trioxohexahydrotriazine (trimethyl isocyanurate, VI) was obtained for a small peak in the Carrollton CCE's. This same compound was also matched in the Cincinnati CCE. No commercial source of VI was found so it was synthesized from cyanuric chloride by a published procedure.[16] The melting point, IR and mass spectra of the product matched reported values. The GC retention time and mass spectrum of VI matched the retention times and mass spectra of the tentatively identified compounds in the Carrollton and Cincinnati samples, thus providing confirmation of the identifications. At present the source of this compound in the two drinking water samples remains a mystery.

Chloral

Chloral (trichloroacetaldehyde) was found in the drinking water CCE's of six of the ten cities in the NORS (Table 22.2). This frequent occurrence leads to the postulation that its formation, like the haloforms, is from chlorination of other organic compounds during the addition of chlorine to the raw and/or finished drinking water. The reaction mechanism of chloral formation may be completely different from the *predominant* reaction mechanism of haloform production. An indication that chloral is a by-product of treatment rather than a contaminant in the raw water from industrial sources (it is a widely-used monomer for copolymerizations) is its presence in the drinking water from the two cities that obtain their water supplies from uncontaminated uplane water (New York City and Seattle). Chloral in water forms chloral hydrate.[17]

Table 22.2 Concentration of Chloral Found in Drinking Water Supplies

Drinking Water	Concentration ($\mu g/l$)
Philadelphia, Pennsylvania	5
Seattle, Washington	3.5
Cincinnati, Ohio	2
Terrebonne Parish, Louisiana	1
New York City, New York	0.02
Grand Forks, North Dakota	0.01

Chloral hydrate ("knockout drops") is a hypnotic drug sometimes used as a sedative and is considered highly toxic.[18] The LD50 for oral injection by rats is 285 mg/kg.[19] Chloral hydrate applied to the skin of mice (4-5% solutions in acetone) resulted in skin tumors in 4 of 20 of the animals.[20] The highest concentration reported here for chloral in drinking waters is only 5 $\mu g/l$ (Table 22.2).

Chloral can be synthesized by chlorinating ethanol, acetaldehyde, polyethylene glycol, ethylene chlorohydrin, chloroacetaldehyde and *bis*(2-chloroethyl) ether.[17] Bellar *et al.* postulated the formation of chloral (VII) and chloral hydrate (VIII) as intermediates in the reaction mechanism involving conversion of ethanol to chloroform.[21]

$$CH_3CH_2OH \longrightarrow CH_3C{\overset{O}{\underset{H}{\diagup}}} \longrightarrow Cl_3C-C{\overset{O}{\underset{H}{\diagup}}} \longrightarrow Cl_3C-CH{\overset{OH}{\underset{OH}{\diagup}}} \longrightarrow CHCl_3$$

Ethanol　　Acetaldehyde　　Chloral　　Chloral Hydrate　　Chloroform

　　　　　　　　　　　　　　　　　VII　　　　VIII

Rook[22] has since demonstrated that chlorination of natural humic substances in raw waters (specifically structures containing the polyhydroxybenzene moiety) is the primary source of haloforms in drinking waters although the reaction mechanism remains to be explained.

Chloral was not identified in any of the NORS samples analyzed by the inert gas stripping technique referred to as Volatile Organics Analysis (VOA). To determine if chloral (which exists as chloral hydrate in water) could be stripped from water under conditions similar to those used for VOA,[23] chloral hydrate was prepared at a concentration of 1 µg/ml in Ultra Pure Reagent Grade Water (New England Reagent Laboratory, East Providence, Rhode Island). Five milliliters of this solution was used to spike 500 ml of water previously stripped with nitrogen for 20 min at a temperature of 95°C. The spiked solution, containing chloral hydrate at 10 µg/l (twice the highest concentration of chloral found in the 13-city survey), was stripped again at 95° and the volatiles were trapped on Tenax resin.[23] Thermal desorption and analysis by GC/MS failed to reveal any trace of chloral, although chloroform was identified. Therefore, inert gas stripping is not a suitable technique for isolating and concentrating chloral from water prior to GC or GC/MS analysis.

Chloral hydrate is known to decompose in neutral, acidic and basic solutions to produce chloroform.[17] At a pH of 8 and temperature of 35°C its half-life is 2 days.[17] To determine if the failure to detect chloral by the VOA technique was due to its rapid decomposition rather than its high polarity, we conducted a kinetic study of chloroform formation from chloral hydrate in Ultra Pure Reagent Grade Water at 90°C. The results using a direct aqueous injection/flame ionization detector are shown in Figure 22.10 and summarized in Table 22.3. After 24 hr only

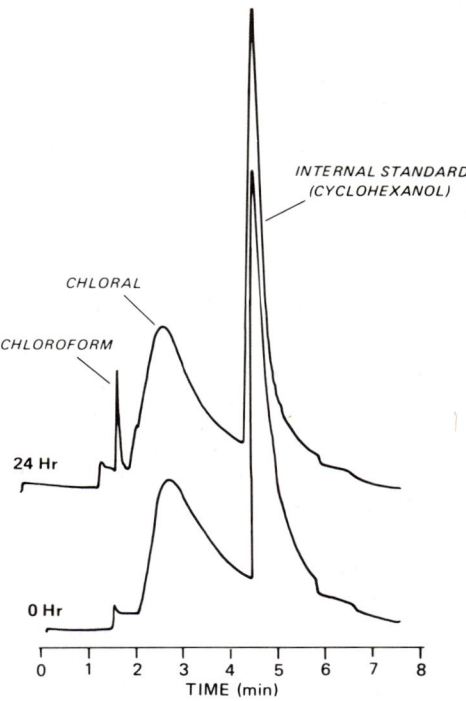

Figure 22.10 GC of aqueous chloral hydrate at 0 hr and after 24 hr at 90°C. GC conditions: direct aqueous injection, isothermal at 65°C, 30 m x 0.4 mm glass capillary column coated with SP-2100.

Table 22.3 Kinetic Study of the Formation of Chloroform from Chloral Hydrate

Time (hr)	Chloroform (mg/l)	Chloral (mg/l)
0	0.02	5.1
1-2	0.06	5.1
3-5	0.07	5.1
9-10	0.10	4.5
23-24	0.13	4.8

about 2% of the chloral hydrate had decomposed to chloroform. Apparently chloral hydrate is so polar it simply doesn't appreciably strip out of aqueous solutions even at elevated temperatures.

Although these studies indicate a possible widespread distribution of chloral hydrate in drinking waters, a data base of 13 samples is not sufficient to statistically confirm this distribution. Additional studies on the efficiency of chloral hydrate adsorption and desorption on both activated carbon and XAD resins, as well as its frequency of occurrence and its suspected production from chlorination treatment are needed.

Summary of Organic Compounds Found in CCE's of Thirteen Finished Drinking Water Supplies

Table 22.4 provides detailed information on the compounds identified in the three New Orleans area finished drinking waters. Table 22.5 lists the compounds identified in the 10-city NORS CCE's. One hundred nine different compounds were identified in the CCE's of these thirteen water supplies and eight more compounds were partially identified (*e.g.*, alkylbenzene -C_4). Seventy percent of the identified compounds have been confirmed by comparison with standards.

A wide variety of chemical classes were present in these CCE's. Table 22.6 summarizes some of these.

No polycyclic aromatic hydrocarbons (PAH's) or polychlorinated biphenyls (PCB's) were identified in any of these extracts. However, many of the compounds listed in Tables 22.4 and 22.5 have been identified previously in various industrial wastewaters.[24,25] It is our opinion that the majority of these chemicals come from industrial or municipal waste discharges although some of them may occur naturally in the water (*e.g.*, acetone, dihydrocarvone, β-santalene); some may be produced by reaction of chlorine with other chemicals (*e.g.*, chloroform, dichlorobromomethane, bromodichloromethane, bromoform, dichloroiodomethane and, perhaps, chloral hydrate); and some may originate from agricultural runoff (*e.g.*, dieldrin, endrin, lindane, atrazine, alachlor etc.).

Recovery Studies

Recoveries from Carbon Filters in Series

Quantitative results in Table 22.4 for the 70-year CCE compounds are sums of the amounts recovered from two CAM carbon filters in series—filter #1 and filter #2. Table 22.7 gives for each compound that was present at a concentration ≥ 0.04 μg/l the percent of the total concentration that was recovered in each of the two filters for the Carrollton 70-year sample.

IDENTIFICATION OF ORGANIC POLLUTANTS 355

Table 22.4 Compounds Identified in Carbon Chloroform and XAD Resin Extracts from Three New Orleans Drinking Water Plants
(All concentrations are in µg/l)

Compound	RRT+	Carrollton Water Plant				Jefferson #1 Water Plant			Jefferson #2 Water Plant			Notes
		70-Year++ CCE	2-Month CCE	Mega-Sample CCE	Resin Extract	70-Year++ CCE	2-Month CCE	Resin Extract	70-Year++ CCE	2-Month CCE	Resin Extract	
1 Alachlor* (Lasso®) [2-chloro-2',6'-diethyl-N-(methoxymethyl) acetanilide]	1.09	0.82	1.9	0.44	0.67	2.3	2.9	0.35	2.1	1.4	0.17	c,d,f,g,h,i,j
2 Alachlor, one additional chlorine homolog [2,2-dichloro-2',6'-diethyl-N-(methoxymethyl) acetanilide]	1.14	0.28	1.7	0.18	0.21		P		0.05	P		d,f,g,h,i
3 Alkylbenzene-C_2 (m-xylene or p-xylene)	0.35	0.03	7.5	0.05			5.6			6.2		
4 Alkylbenzene-C_3 (p-ethyltoluene or n-propylbenzene)	0.42		2.4	0.01			1.9		0.03	2.2		
5 Alkylbenzene-C_4	~0.52	<0.1			J							g
6 Atrazine* (2-chloro-4-ethyl-amino-6-isopropylamino-s-triazine)	1.00	4.9	3.7	2.7	1.0	5.2	4.8	0.64	5.4	3.2	0.18	c,d,e,f,g,h,i,j
7 Atrazine, deethyl* (2-chloro-4-amino-6-isopropylamino-s-triazine) (desethylatrazine)	0.96	0.51	0.78	0.22		0.27	0.80		0.27	0.75		c,g

Table 22.4, continued

Compound	RRT+	Carrollton Water Plant 70-Year++ CCE	2-Month CCE	Mega-Sample CCE	Resin Extract	Jefferson #1 Water Plant 70-Year++ CCE	2-Month CCE	Resin Extract	Jefferson #2 Water Plant 70-Year++ CCE	2-Month CCE	Resin Extract	Notes
8 Benzaldehyde*	0.44			0.03								h
9 Benzyl butyl phthalate*	1.29	0.64	1.4	0.24		0.83	1.8		0.75	1.6	0.08	g,h
10 Bromodichloromethane	0.18	P			<0.1			<0.1				b
11 Bromoform*	0.32	0.57										b,g
12 Butachlor* (Machete®) [2-chloro-2',6'-diethyl-N-(butoxymethyl)-acetanilide]	1.21	0.05		0.02	<0.1	0.06			0.05			g,h
13 Butyl octyl maleate*					J			J				h
14 α-Chlordane				<0.1 (T)								h
15 Chlordene*				<0.1 (T)								h
16 bis (2-Chloroethyl) ether*	0.44	0.04				0.16			0.12			g
17 Chloroform*				P	J							a,b
18 bis (2-Chloroisopropyl) ether*	0.51	0.18				0.08		J	0.03			g,h
19 m-Chloronitrobenzene*												
20 Cyanazine* (Bladex®) [2-(4-chloro-6-ethylamino-s-triazin-3-ylamino)-2-methylpropionitrile]	1.13	0.35		<0.01		0.21			0.31			g,h

IDENTIFICATION OF ORGANIC POLLUTANTS

#	Compound												Notes
21	DDE* [2,2-bis(p-chlorophenyl)-1,1-dichloroethylene]	0.48				0.05(J)			0.05(J)	2.0			d
22	n-Decane*	0.44	0.02	2.4	0.06		2.0			5.2			
23	Decane, branched	0.23	1.1	5.8	0.03		5.4						
24	Dibromochloromethane*	0.35	<10	0.6	<0.1		0.4	<0.1		0.06	P		b,d,g
25	Dibromodichloroethane isomer	0.33		0.16						0.63			g
26	Dibutyl phthalate*	1.12	0.10		0.09	0.05	0.36	0.01		0.23		0.03	e,f,g
27	2,6-Di-t-butyl-p-benzoquinone*	0.84	0.22				0.21			0.25			
28	m-Dichlorobenzene*	0.46		<3				<0.1					f
29	Dicyclopentadiene*							J					
30	Dieldrin*	1.23	0.04		0.01	0.01(J)	0.07	0.01(J)		0.05			g,h
31	Diethyl phthalate*	0.91	0.03	0.24	0.02	J	0.03	0.03		0.01	0.18		e,f,g,h
32	Di-(2-ethylhexyl) adipate*	1.32			0.10								
33	Di-(2-ethylhexyl) phthalate*	1.38	0.10	0.40	11	0.05	0.46	0.50	J	0.27	1.2	0.16	d,f,g
34	Dihexyl phthalate	1.24	0.03			0.05						0.16	g
35	Dihydrocarvone	0.63	0.14				0.06			0.07			
36	Diisobutyl phthalate*	1.06		<1	0.59	<0.05							f
37	Dimethyl phthalate*	0.83	0.27	0.60	P		0.13	0.82		0.18	0.74		c,f,g,h
38	Dipropyl phthalate*	1.02	0.07		0.01		0.13			0.14			g
39	n-Dodecane*	0.67	0.01	0.10				0.40			0.37		
40	Endrin*	1.27 (1st peak)			0.004 (T)		0.008 (T)			0.006 (T)			

Table 22.4, continued

Compound	RRT+	Carrollton Water Plant				Jefferson #1 Water Plant				Jefferson #2 Water Plant				Notes
		70-Year++ CCE	2-Month CCE	Mega-Sample CCE	Resin Extract	70-Year++ CCE	2-Month CCE	Resin Extract	70-Year++ CCE	2-Month CCE	Resin Extract	
41 Ethyl acetate	0.32				P						P	
42 Ethylbenzene	0.45		2.3	0.02			1.6			1.8		
43 o-Ethyltoluene*	0.42					0.04			0.02			
44 m-Ethyltoluene*	0.94					0.05			0.02			
45 1,2,3,4,5,7,7-Heptachloro-norbornene*	0.98	0.06		0.03	0.02	0.07		0.04	0.07		0.01	d,f,g,h
46 Heptachloronorbornene isomer	0.65	0.06		0.02		0.04			0.04			g,h
47 Hexachloro-1,3-butadiene*	0.53	0.16	0.70	0.07	0.04	0.27		0.12	0.21		<0.01	d,e,f,g,h
48 Hexachloroethane*	0.60	4.3		0.39	0.03	0.19		<0.1	0.30			f,g,h
49 Isophorone*	0.50	1.6	2.9			2.8	<11		2.9	9.5		c,g
50 Limonene*	0.58	0.03										
51 Methyl benzoate*								<0.01				e
52 Methylnaphthalene	0.38				J			J				
53 Naphthalene*	~0.38	0.03	2.4				2.4	J		2.1		
54 n-Nonane*					<0.1							
55 Pentachloroethane*												

#	Compound										Notes
56	Pentachlorophenyl methyl ether	0.86									
57	n-Pentadecane*	0.03		<0.1(T)			0.10	0.01	0.10	<0.01	
58	Propazine* [2-chloro-4,6-bis-(isopropylamino)-s-triazine]	~1.0		<0.1			<0.1		<0.1		
59	Simazine* [2-chloro-4,6-bis-(ethylamino)-s-triazine]	~1.0	<0.1				<0.1		<0.1		g
60	1,1,1,2-Tetrachloroethane*	0.30	0.04	0.11		J					
61	Tetrachloroethylene*	0.26		<1		0.1	0.20	<5	<0.1	0.20	<5	b,g
62	n-Tetradecane*	0.80	0.02	0.10			0.10		0.12		
63	Toluene*	0.22		11	0.08		0.10	7.1	<0.01	12	b
64	1,1,2-Trichloroethane*	0.21	0.35	6.2	<0.2	<0.1	0.45	8.5	<0.1	0.41	6.4	
65	1,1,1-Trichloropropane	~0.07				<0.1					
66	1,2,3-Trichloropropane	0.36			<0.2						
67	n-Tridecane*	0.72	0.01	0.30			0.17		0.20		
68	Trimethylbenzene isomer	0.46	0.04	6.1	0.02	0.01	5.1		0.02	5.3	
69	Trimethyl Isocyanurate*	0.73	0.01	0.01							c,g
70	Triphenyl phosphate*	1.31	0.12	0.03							h
71	n-Undecane*	0.58		2.5	0.03		<10		2.1		
72	Undecane, branched	0.52	0.06	5.3	0.04						
73	o-Xylene*	0.33	0.33	4.1	0.12		2.8		3.4		
74	Unknown chlorinated fluorinated hydrocarbon	0.78	<0.01		<0.1		<0.1		<0.1		

Figure 22.4, continued

Compound	RRT+	Carrollton Water Plant				Jefferson #1 Water Plant			Jefferson #2 Water Plant			Notes
		70-Year CCE++	2-Month CCE	Mega-Sample CCE	Resin Extract	70-Year CCE++	2-Month CCE	Resin Extract	70-Year CCE++	2-Month CCE	Resin Extract	
75 Unknown chlorinated fluorinated hydrocarbon	0.82	<0.01			<0.1			<0.1			<0.1	
76 Unknown compound, apparent MW 145	~0.40	<0.7		<0.3		<0.9			<0.9			g
77 Unknown dichlorinated compound MW 200	0.82			0.05								
78 Dichloroiodomethane	0.26	1.1		0.84		1.3			1.6			g,k
79 Unknown dichlorinated compound MW 249	~1.0	<0.04		<0.02		P			P			
80 Unknown phthalate	1.22			0.01								
81 Unknown phthalate	1.24			0.01								
82 Unknown phthalate	1.36			0.12								

Blank spaces indicate that compound was not detected in that specific sample.

*Confirmed by matching GC retention time and, in most cases, mass spectrum with that of a standard.

+Relative retention time (to atrazine); mostly determined by the GC computer system.

++Concentration values for 70-year CCE compounds are sums of concentrations in CCE #1 (filter #1) and CCE #2 (filter #2).

IDENTIFICATION OF ORGANIC POLLUTANTS 361

J Detected only by Junk, USAEC-Ames Laboratory. Identified in the Carrollton or Jefferson No. 1 resin extracts by GC/MS, but not quantitated except for DDE and dieldrin. These compounds were not detected at the Athens EPA Laboratory in the resin extracts.

P Present, usually in very low concentrations (<0.05 μg/l), but not quantitated due to interferences with other GC peaks or lack of a discernible GC peak (detected only by mass spectrometry).

T Detected only by Thruston at the Athens EPA Laboratory by GC/MS after fractionation by column chromatography.

NOTES:

aChloroform was identified at the Athens EPA Laboratory only in the tetralin extract of the Carrollton water.

bThese compounds were also identified at the Water Supply Research Laboratory, EPA, Cincinnati, by VOA analysis of the Carrollton water.

cAlso identified in the Carrollton CAE.

dAlso identified by Junk, USAEC-Ames Laboratory, in the Carrollton resin extract.

eAlso identified by Junk, USAEC-Ames Laboratory, in the Jefferson #1 resin extract.

fAlso identified by Melton at the Water Supply Research Laboratory, EPA, Cincinnati, by GC/MS analysis of a solvent extract of a 1-liter grab sample of the Carrollton water.

gAlso identified by duplicate analysis of the Carrollton 70-year CCE using the Varian CH5/DF system.

hAlso identified by GC/MS (Finnigan) analysis of the Carrollton mega-sample CCE after fractionation by TLC.

i Confirmed by high-resolution mass spectrometry (only a structural indication in the case of compound #2).

j Confirmed by **GC-FT-IR**.

kIdentification based on a report received after this table was prepared (EPA Mass Spectrometer User's Group Newsletter No. 15, April 1975)—therefore not in alphabetical order.

®Registered trade names are given only as an aid to compound recognition, and do not indicate the source of the compound.

Table 22.5 Compounds Identified in Carbon Chloroform Extracts from Drinking Waters of Ten U.S. Cities (All concentrations are in $\mu g/l$)

Compound	Cincinnati	Miami	Ottumwa	Philadelphia	Seattle	Grand Forks	Lawrence	N.Y. City	Terrebonne Parrish	Tucson
1. Acetaldehyde*				0.1	0.1					
2. Acetone*					1.0					
3. Acetophenone*				1.0						
4. Atrazine*			0.1							
5. Bromodichloromethane*	1.0	4.5		1.0	0.1	0.6	0.6	1.3	2.0	
6. Bromoform*		1.5								3.0
7. t-Butyltoluene				0.01						
8. Camphor*	0.1	0.5	0.1		0.5					
9. Chloral* (trichloroacetaldehyde)	2.0			5.0	3.5					
10. Chlorobenzene*		1.0								
11. Chlorodibromomethane*	0.5	15		0.5		0.01	0.01	0.02	1.0	0.01
12. Chloropicrin* (trichloronitromethane)			0.05							
13. p-Chlorotoluene*		1.5				0.1		0.4	1.0	
14. Cumene* (isopropylbenzene)									0.01	
15. Cyclohexanone*		0.1	0.1							
16. Cymene isomer		0.1								
17. 2,6-Di-t-butyl-p-benzoquinone*										
18. Di-t-butyl ketone							0.02			
19. Di-n-butyl phthalate*		5.0	0.1	0.05	0.01		0.01		0.02	
20. o-Dichlorobenzene*		1.0								
21. m-Dichlorobenzene*		0.5								
22. p-Dichlorobenzene*		0.5	0.1							
23. Diethyl malonate*	0.01									
24. Diethyl phthalate*	0.1	1.0		0.01	0.01		0.04	0.01		
25. Di(2-ethylhexyl) phthalate*		30		0.5			0.08		0.04	
26. 1,4-Dioxane*							0.01			

IDENTIFICATION OF ORGANIC POLLUTANTS 363

#	Compound						
27.	Di-n-propyl phthalate	0.5					
28.	2-Ethylbutanal						
29.	p-Ethyltoluene			0.05	0.02	0.04	0.05	0.01
30.	Hexachloro-1,3-butadiene*				<0.01			
31.	Hexachloroethane*	0.5						
32.	Isoamyl chloride						0.01	
33.	Lindane* (γ-BHC)	0.01						0.01
34.	2-Methyl-5-ethylheptane							
35.	Methyl ethyl maleimide						0.02	
36.	5-Methylhexa-3-ene-2-one						0.07	
37.	3-Methyl-3-pentanal		1.0					
38.	n-Nonane*						0.02	
39.	n-Pentanal		0.5					
40.	2-Pentanone*		0.1					
41.	n-Propylbenzene*	0.01	0.05					
42.	n-Propylcyclohexane	0.2						
43.	β-Santalene			0.01				
44.	α-Terpineol*		0.5					
45.	Tetrachloroethylene*	0.1	0.1		0.2	0.07	0.05	<0.01
46.	1,1,3,3-Tetrachloro-2-propanone* (tetrachloroacetone)	0.5		1.0				
47.	Tetramethylbenzene isomer	0.2						
48.	Tetramethyltetrahydrofuran isomer	0.2	0.5					
49.	Tri-n-butyl phosphate*	0.05						
50.	Trimethyl isocyanurate* (1,3,5-tri-methyl-2,4,6-trioxohexahydrotriazine)	0.5						

*Identifications confirmed by matching mass spectrum and GC retention time of a standard with conditions identical to those under which data from the samples was obtained.

Table 22.6 Chemical Classification of Organics in the CCE's[a]

Chemical Family or Use	Compounds Present
Pesticides and Associated Compounds	8[b]
Herbicides and Associated Compounds	8[c]
Halogenated Aliphatics	20
Chlorinated Aromatics	7
Aliphatic Hydrocarbons	10
Aromatic Hydrocarbons	17
Plasticizers	14
Misc. Compounds of Industrial Origin	21
Misc. Compounds of Possible Natural Origin	7
Partially-Identified Compounds	5
Total	117

[a] Also included in these data are the six compounds identified only by Gregor Junk and co-workers from XAD resin extracts of the New Orleans drinking water (DDE, butyl octyl maleate, m-chloronitrobenzene, dicyclopentadiene, naphthalene and methylnaphthalene).
[b] α-Chlordane, chlordene, DDE, dieldrin, endrin, lindane, heptachloronorbornene and its isomer.
[c] Alachlor, alachlor chlorine homolog, atrazine, deethylatrazine, butachlor, cyanazine, propazine and simazine.

Of the 29 compounds in Table 22.7, 24 were collected on filter #1 to the extent of 96% ore more, indicating quantitative adsorption on carbon (but not necessarily quantitative recovery). These 24 include a wide variety of compounds, indicating a generally high degree of adsorption. However, there are exceptions. Three trihalogenated methanes or ethanes were adsorbed to the extent of only 64 to 85% on filter #1, and more hexachloroethane was found on filter #2 than #1.

Recoveries From Carbon Filters of Different Sizes

The two-month equivalent (mini-sample) CCE's contained much higher concentrations—up to 10-fold—of lower boiling compounds than did the 70-year and mega-sample CCE's, but about the same concentrations of the less volatile compounds. This may be seen by comparing the FID chromatogram of the Carrollton mini-sample CCE with that of the mega-sample, both in Figure 22.11, and with that of the 70-year sample in

Table 22.7 Percent of Compounds[a] Recovered from Carrollton 70-Year Filters No. 1 and No. 2 in Series

Compound	Total Concentration (μg/l)	Percent from Filter No. 1	Percent from Filter No. 2
Alachlor	0.82	97	3
Alachlor homolog	0.28	100	0
Atrazine	4.9	98	2
Atrazine, de-ethyl-	0.51	96	4
Benzyl butyl phthalate	0.64	100	0
Bromoform	0.57	100	0
Butachlor	0.05	100	0
Cyanazine	0.35	100	0
Chlorodibromomethane	1.1	64	36
Bis(2-chloroethyl) ether	0.04	100	0
Bis(2-chloroisopropyl) ether	0.18	100	0
Dibromodichloroethane isomer	0.33	100	0
Dibutyl phthalate	0.10	80	20
2,6-Di-t-butyl-p-benzoquinone	0.22	100	0
Dichloroiodomethane	1.1	85	15
Dieldrin	0.04	100	0
Di-(2-ethylhexyl) phthalate	0.10	100	0
Dimethyl phthalate	0.27	100	0
Dipropyl phthalate	0.07	100	0
1,2,3,4,5,7,7-Heptachloronorbornene	0.06	100	0
Heptachloronorbornene isomer	0.06	100	0
Hexachloro-1,3-butadiene	0.16	100	0
Hexachloroethane	4.3	36	64
Isophorone	1.6	98	2
1,1,2-Trichloroethane	0.35	69	31
Trimethylbenzene isomer	0.04	100	0
Triphenyl phosphate	0.12	100	0
Undecane, branched	0.06	100	0
o-Xylene	0.33	100	0

[a]Including only compounds present in total concentration of 0.04 μg/l or greater.

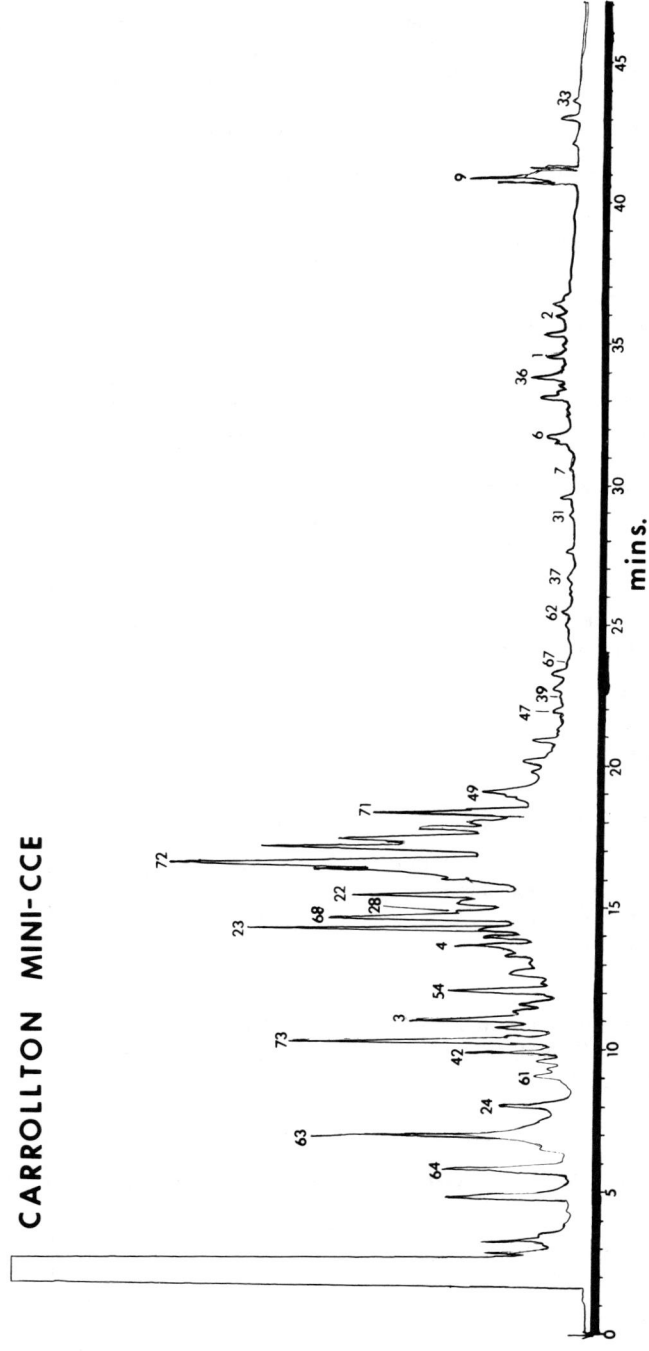

IDENTIFICATION OF ORGANIC POLLUTANTS 367

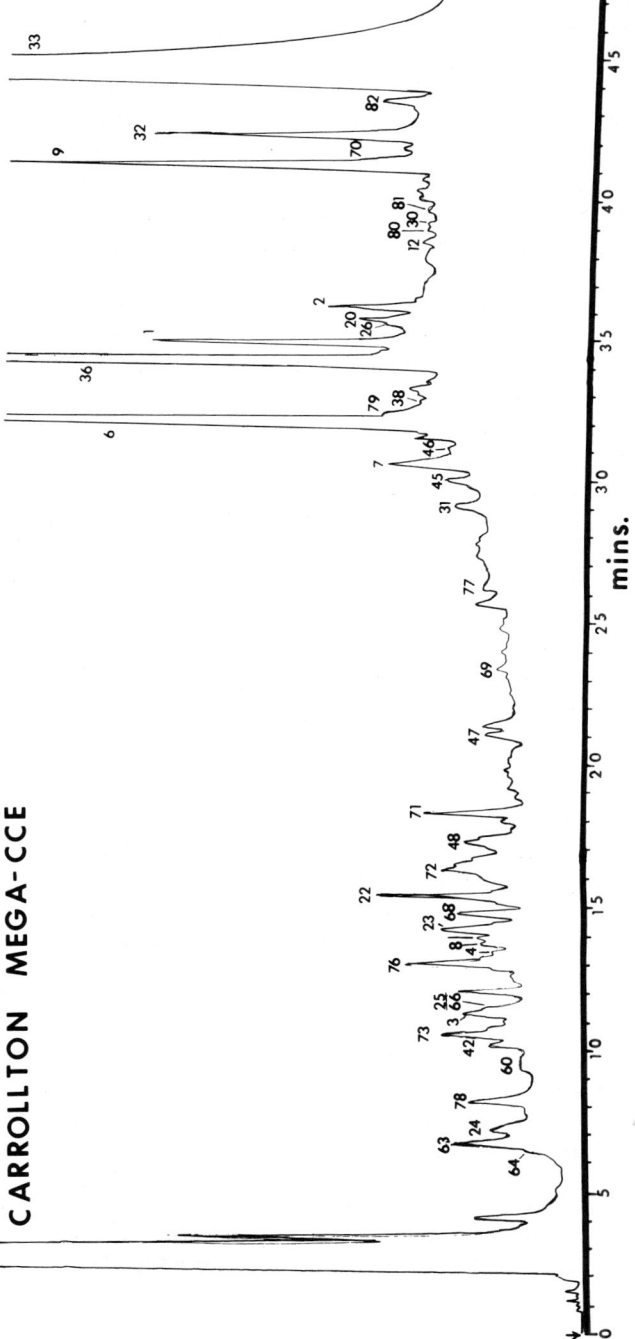

Figure 22.11 FID gas chromatograms of the Carrollton mini- and mega-CCE's. Numbers correspond to compounds in Table 22.4.

Figure 22.4. Although the 70-year sample was collected about 10 days before the mini-sample, during which time the compound quantities in the river may have changed, the concentrations should not have changed as a function of volatility. These results may be an indication of organic overloading on the 70-year carbon filter, although this explanation is contrary to most of the results shown in Table 22.7. Seventy grams of carbon were used to extract 60 liters of water in the mini-sample filters (0.9 l/g), while 230 grams were used for 25,500 liters in the 70-year filters (75 l/g). This is an 83-fold difference in the water/carbon ratio.

Concentrations in the mega-sample CCE's were low relative to the 70-year CCE's. Of 30 compounds quantitated in the Carrollton drinking water by both methods, the concentrations of 21 in the mega-sample were between 25 and 76% of that found in the 70-year sample. The water/carbon ratio for the mega-sample was 57 l/g.

Several operational factors were biased against the mega-sample quantitations relative to those for the 70-year sample. The mega-sample carbon was dried for 10 days before extraction, while the 70-year sample carbon was dried for only 2 days. The first-step concentration of the mega-sample CCE's was not as carefully controlled as it was with the 70-year CCE's in that direct distillation of the chloroform was used for the mega-sample whereas vacuum evaporation at room temperature was used for the 70-year CCE's. However, recent quantitative studies by Webb[26] indicate that the first-step concentration is not as critical as the final-step concentration, which was the same with all of these samples.

Recovery from Resin

Not enough data are available from the New Orleans study to make a good comparison of XAD-resin *vs* carbon extraction efficiencies. More compounds and higher concentrations were obtained from the carbon, but sampling circumstances were biased against the resin technique. A well-designed comparison of these two accumulator techniques is needed.

Recovery Studies With Atrazine

Since several *s*-triazine herbicides were identified in the New Orleans CCE's the recovery of atrazine from water was studied to determine approximate recoveries that might be expected for the atrazine-type herbicides. Four extraction methods were tested using spiked samples:

(a) Liquid-Liquid Extraction. Two 1-liter tap water samples were spiked with 100 and 200 μg, respectively, of atrazine, allowed to stand 5 days in laboratory light, and extracted 3 times with 100-ml portions

of methylene chloride. The extracts were evaporated to 1 ml using a Kuderna-Danish evaporator followed by final concentration with a slow stream of dry nitrogen, and then were analyzed by GC.

(b) Activated Cocoanut Charcoal (an incidental laboratory supply). Five liters of tap water were spiked with 100 µg of atrazine and passed by gravity flow through a column of 11 g (20-ml volume) of the charcoal (6-14 mesh). A control consisted of a similar charcoal column and 5 liters of unspiked water. For a direct recovery test, 100 µg of atrazine in 2 ml of chloroform was added directly to 11 g of the same charcoal. The three charcoal samples were air-dried at room temperature overnight, then extracted with methylene chloride in a Soxhlet extractor for 6 hr. The extracts were evaporated to 1 ml as in experiment (a) and analyzed by GC.

(c) "Fine" Mesh Carbon (used at New Orleans). Experiment (b) was repeated with 3.5 g (20-ml volume) of the "fine" mesh Nuchar 190 carbon used in the New Orleans study.

(d) "Coarse" Mesh Carbon (used at New Orleans). Experiment (b) was repeated with 10 g (20-ml volume) of the "coarse" mesh carbon used for the New Orleans study.

The results are summarized in Table 22.8.

Table 22.8 Recovery Studies Using Atrazine-Spiked Solutions

	Experiment Description	% Recovery
(a)	Methylene chloride extraction, 100 µg/l	98
	Methylene chloride extraction, 200 µg/l	95
(b)	Cocoanut charcoal/atrazine in water	29
	Cocoanut charcoal/atrazine in $CHCl_3$	28
(c)	Nuchar 190 fine mesh/atrazine in water	92
	Nuchar 190 fine mesh/atrazine in $CHCl_3$	103
(d)	Nuchar 190 coarse/atrazine in water	59
	Duplicate of above	71
	Nuchar 190 coarse/atrazine in $CHCl_3$	102

From Table 22.8 it is seen that Nuchar 190 is an acceptable carbon for adsorption/desorption of atrazine—especially the fine mesh. Liquid-liquid extraction with methylene chloride would also be a good method of recovery for these herbicides. However, the cocoanut charcoal gave very poor recoveries, probably because it was not desorbed efficiently by methylene chloride.

Carbon Variance

One of the disadvantages of using activated carbon for concentrating trace organics from water is the large variance in both its "activity" (adsorption/desorption characteristics) and its cleanliness. Table 22.8 provides an insight into how much various types of carbon can differ in their activities towards adsorbing/desorbing organic compounds.

The Nuchar 190 used for the New Orleans study was relatively free of background organic contaminants. Total organic peak area summation as measured on the PEP-1 system was 5.1 for the blank CCE. In contrast, the first carbon tried for the NORS gave a peak area summation of 61.6. The peak area summation for the chloroform solvent blank used in the NORS was only 0.25 so the organics were clearly eluting from the carbon. This carbon could not be used, but satisfactory results were obtained with Filtersorb 300 (although Nuchar C-190 was cleaner there was not a sufficient quantity on hand for the entire 10-city survey). Table 22.9 summarizes the results of these analyses.

Table 22.9 Peak Area Summations of Various CCE Blanks

CCE Source	Peak Area Summation
First NORS CCE blank	61.1
New Orleans Nuchar 190	5.1
Nuchar C-190	0.3
Filtersorb 300	1.2
Westvaco (sealed bag)	13.6
Westvaco (opened bag)	30.5

Probably the Westvaco open bag adsorbed organics from the air although this is not known for certain because the history of the bag is not known; the carbon was not from our laboratory.

A procedure needs to be developed for precleaning carbon. Dunlap and Shew have experienced some success by boiling the carbon in organic-free water with 5% hydrochloric acid followed by washing with organic-free water and drying in a clean oven at 150°C for about 4 hr.[27]

CONCLUSIONS

Although we know much more now about trace organic contaminants in our drinking waters than we did one or two years ago, we still have a long way to go before we can feel confident about their quantities and distribution. Computerized GC/MS has provided an efficient and cost-effective method of identifying these compounds, but we are still perfecting methods for concentrating and separating trace organics quantitatively and efficiently from water. Concurrently, we are just beginning to gather and organize data on trace organics so it can be computerized for output and assessment with useful correlations such as geographic location, type of water, etc. Concomitant with the availability of information pertaining to the identification of thousands of organic compounds in water there will be a surge of research efforts relating to the health and ecological effects of these compounds. At present there is little evidence either for or against significant health effects of most of these organic comounds at microgram per liter concentrations.

The results described in this and other chapters of this book are clearly the initial efforts of a relatively new trend in environmental chemistry—the qualitative and quantitative analysis of trace organic compounds in environmental samples. Drinking water is but one of many kinds of samples where this information is sought. Awareness is finally increasing that we need to know what and how much of various chemicals are in our drinking waters in order to intelligently assess their significance. This same logic applies to our rivers, lakes and ground waters which are the sources of our drinking waters. We expect to learn much more about the occurrence and distribution of these trace organic compounds—both natural and synthetic—in our environment over the next 10 years.

ACKNOWLEDGMENTS

The authors thank Ronald G. Webb for the synthesis of chloropicrin and trimethyl isocyanurate used as standards for the confirmation of these two identifications. Thanks are also due Ann Alford for providing some of the GC/MS data and to Leo Azarraga and George Yager for GC/IR analysis of alachlor and its chlorine homolog. Advice and help from William T. Donaldson and John M. McGuire are gratefully appreciated.

DISCLAIMER

Mention of commercial products and trade names is for informational purposes only and does not imply endorsement by the U.S. Environmental Protection Agency or the Athens Environmental Research Laboratory.

REFERENCES

1. U.S. Environmental Protection Agency. "Industrial Pollution of the Lower Mississippi River in Louisiana," Region VI, Surveillance and Analysis Division (April 1972).
2. U.S. Environmental Protection Agency. "Draft Analytical Report, New Orleans Area Water Supply Study," Region VI, Surveillance and Analysis Division, Report EPA 906/10-74-002 (November 1974).
3. U.S. Environmental Protection Agency. "Preliminary Assessment of Suspected Carcinogens in Drinking Water–Appendix," Interim Report to Congress (June 1975).
4. U.S. Environmental Protection Agency. "Preliminary Assessment of Suspected Carcinogens in Drinking Water," Interim Report to Congress (June 1975).
5. Junk, G. A., J. J. Richard, M. D. Grieser, D. Witiak, J. L. Witiak, M. D. Arguillo, R. Vick, H. J. Svec, J. S. Fritz and G. V. Calder. "Use of Macroreticular Resins in the Analysis of Water for Trace Organic Contaminants," *J. Chromatogr.* **99**, 745-762 (1974).
6. U.S. Environmental Protection Agency. "Analytical Report, New Orleans Area Water Supply Study," Region VI, Surveillance and Analysis Division, Report EPA 906/9-75-003 (September 1975).
7. National Pollutant Discharge Elimination System, Appendix A. *Federal Register* **38**, #75, Part II (November 28, 1973).
8. Heller, S. R., J. M. McGuire and W. L. Budde. "Trace Organics by GC/MS," *Environ. Sci. Technol.* **9**, 210-213 (1975).
9. Mass Spectrometry Data Centre. *Eight Peak Index of Mass Spectra*, 1st Ed., AWRE, Aldermaston, Reading, England (1970).
10. Weed Science Society of America. *Herbicide Handbook*, 3rd ed. Champaign, Ill. (1974), pp. 8-11.
11. Kaufman, D. D. and P. C. Kearney. In *Residue Reviews*, Vol. 32, F. A. Gunther and J. D. Gunther, Eds. (New York: Springer-Verlag Co., 1970), pp. 235-266.
12. Plimmer, J. R. and P. C. Kearney. "Free Radical Oxidation of Pesticides," *Weed Sci. Soc. Amer. Abstr.*, 20 (1968).
13. Plimmer, J. R., P. C. Kearney and U. Klingebiel. "s-Triazine Herbicide Dealkylation by Free-Radical Generating Systems," *J. Agr. Food Chem.* **19**, 572-573 (1971).
14. Wolfe, N. L., R. C. Zepp, J. A. Gordon and R. C. Fincher. *Bull. Environ. Contam. Tox.* (in press).
15. Fine, D. H., D. P. Rounbehler, N. M. Belcher and S. S. Epstein. "N-Nitroso Compounds in the Environment," presented at the International Conference on Environmental Sensing and Assessment, Las Vegas, Nevada (September 18, 1975) and at the Fourth IARC Meeting, Tallinn, Estonia, USSR (October 16, 1975).

16. Cignitti, M. and L. Paoloni. "Tautomeric Forms of Oxy Derivatives of 1,3,5-Triazine. I. Infrared Spectra," *Rend. 1st Super. Sanit.* **23**, 1037-1047 (1960); *Chem. Abstr.* **56**, 3037 (1962).
17. Luknitskii, F. I. "The Chemistry of Chloral," *Chem. Rev.* **75**(3), 259-289 (1975).
18. Hawley, G. G. *The Condensed Chemical Dictionary*, 8th ed. (New York: Van Nostrand Reinhold Co., 1971), p. 195.
19. Christensen, H. C., T. T. Luginhyhl and B. S. Carroll. "The Toxic Substances List" 1974 Edition, U. S. Dept. of Health, Education and Welfare, PHS, Nat. Inst. for Occupational Safety and Health, Rockville, Md.(1974), p. 194.
20. Shubik, P. and J. L. Hartwell. "Survey of Compounds Which Have Been Tested for Carcinogenic Activity. Supplement 2," U.S. Dept. of Health, Education and Welfare (1969), p. 42.
21. Bellar, T. A., J. J. Lichtenberg and R. C. Kroner. "The Occurrence of Organohalides in Chlorinated Drinking Waters," U.S. Environmental Protection Agency Report EPA-670/4-74-008 (November 1974).
22. Rook, J. J. "Formation of Haloforms During Chlorination of Natural Waters," *Water Treatment Exam.* **23**, 234-243 (1974).
23. Bellar, T. A. and J. J. Lichtenberg. "Determining Volatile Organics at the μg/l Level in Water by Gas Chromatography," *J. Amer. Water Works Assoc.* **66**, 739-744 (1974).
24. Keith, L. H. "Chemical Characterization of Industrial Wastewaters by Gas Chromatography/Mass Spectrometry," *Sci. Total Environ.* **3**, 87-102 (1974).
25. Keith, L. H. "Analysis of Organic Compounds in Two Kraft Mill Wastewaters," U.S. Environmental Protection Agency, Report No. EPA-660/4-75-005 (June 1975).
26. Webb, R. G. "Isolating Organic Water Pollutants: XAD Resins, Urethane Foams, Solvent Extraction," U.S. Environmental Protection Agency Report No. EPA-660/4-75-003 (June 1975).
27. Dunlap, William J. and D. Craig Shew. EPA, R. S. Kerr Environmental Research Laboratory, Ada, Oklahoma, Private Communication (November 3, 1975).

CHAPTER 23

GC/MS IDENTIFICATION OF TRACE ORGANIC COMPOUNDS IN PHILADELPHIA WATERS

I. H. Suffet

    Department of Chemistry
    Environmental Engineering and Science
    Drexel University
    Philadelphia, Pennsylvania

L. Brenner

    Police Department
    City of Philadelphia
    Philadelphia, Pennsylvania

J. V. Radziul

    Research and Development
    Water Department
    City of Philadelphia
    Philadelphia, Pennsylvania

INTRODUCTION

    The widespread distribution of both natural and industrial organic compounds in surface waters and drinking waters derived therefrom is being revealed by many investigators. Recently, attention has been directed to the special category of organics produced during water and wastewater treatment processes. These studies have been reviewed for the year 1974 by Suffet et al.[1] Widespread interest has been focused on the area of organic contaminants in drinking water as a result of the 1974 reports of Rook,[2] Bellar et al.[3] and the EPA analytical report identifying 66 organics in New Orleans tap water.[4] Editorial comment of the EPA report[4] by the lay press generated undue adverse criticism about the quality of the drinking water which had previously met all existing federal and state water quality regulations and of the regulatory agencies responsible for monitoring the safety of such drinking water. The possible carcinogenic insult of these trace organic contaminants has been debated and is of issue. The reports, however, highlighted the lack of real knowledge required to make effective decisions concerning trace organics in drinking water in order to properly safeguard the public health.

The Philadelphia Water Department, in concert with Drexel University, initiated research in 1969 into the nature of trace organics causing tastes and odors in drinking water. The research involved innovative development of analytical equipment methodologies for trace organics, as well as the study of treatment kinetics and ancillary subjects. In 1975, GC/MS was utilized to augment the previous work which utilized "gas chromatographic profiles." The purpose was to identify by GC/MS the organics indigenous to the Philadelphia water supply and wastewater treatment plant effluents. This chapter presents the identification observed to date on Philadelphia water supply and the methods employed to obtain these identifications. A previous report by Suffet and Radziul outlines the general area of study and initial results of this work.[5]

The Delaware River Estuary and Schylkill River provide the City of Philadelphia with its raw water supply. Delaware River water is treated at the Torresdale Water Treatment Plant and Schuylkill River water is treated at the Queen Lane and Belmont Water Treatment Plants (rated average capacity of 282,120 and 78 million gallons per day, respectively). Treatment consists of prechlorination and presedimentation followed by coagulation, sedimentation and rapid sand filtration. Post treatment includes chlorination and fluoridation. Water treated at Belmont and Torresdale (since April 1, 1975) is augmented by ammoniation for taste and odor control. A description of the unit processes at the Torresdale Water Treatment Plant detailing the chlorination process has been presented elsewhere.[6] Drinking water from these plants is under study as well as the influent Delaware and Schuylkill River water to the respective Torresdale and Belmont Water Treatment Plants. Data will be presented from studies of Torresdale and Belmont drinking water and Schuylkill River water entering the Belmont Plant.

EXPERIMENTAL

Figure 23.1 depicts the analytical screening procedure used in this study to identify organics by GC/MS.[5] Three isolation methods were applied to isolate trace organics for GC/MS analysis. These methods were direct head gas analysis (HGA), continuous liquid-liquid extraction

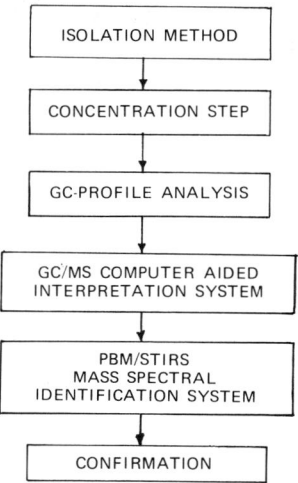

Figure 23.1 General GC/MS identification procedure.

(CLLE), and XAD-2 Macroreticular Resin Columns (MRR). Isolation was followed by an evaporative concentration step for solvents (Kuderna-Danish), GC-Profile analysis, and GC/MS analysis on three different GC columns. The MS data were interpreted when necessary by the PBM/STIRS mass spectra identification system. Confirmation of a compound's identity was completed by direct comparison between the mass spectrum of a pure reference compound and the unknown GC/MS spectra and GC retention time.

Table 23.1 summarizes the samples studied by GC/MS analysis. Samples include qualitative analyses obtained during quantitative analytical runs for *bis*(2-chloroethyl) ether by the same GC/MS methodology. Table 23.2 shows the sampling information for each analytical run. The pH of the MRR samples is adjusted to near pH 4 and 255 μmhos conductivity. A 1-liter sample for HGA is set to pH 7.0 with 0.2 M orthophosphate buffer.[7] Sodium sulfate was added to the HGA sample to adjust the total ionic strength to 1.65 M.

Drinking Water Analysis

Two new automated continuous isolation methodologies were established to isolate organics from samples of tap water for GC/MS analysis. The latest analysis (8/20/75) of Belmont drinking water was completed by simultaneous collection of organics by both methods. The water sample

Table 23.1 Sample Studies by GC/MS Analysis

	1975 Sample Date	Type Sample	SE-30 (BCEE Window)	SE-30	Bentone-Didecyl Phthalate	Carbowax
Torresdale Plant Drinking Water	2/4	M	X			
	2/5	M	X			
	2/6	M		X		
	3/24-25	M		X		
	3/24-25	L		X		
	3/25-26	M	X			
	3/25-26	L	X			
	4/17	H		X		
	4/17	M		X		X
	4/17	L		X	X	X
	4/17-19	M		X	X	
	4/19-20	M		X		
Belmont Plant Drinking Water	8/20-21	M		X	X	
	8/20-21	L		X	X	
Belmont River Influent	8/21	M		X	X	

M = MRR Sample   L = CLLE Sample   H = HGA Sample

BCEE Window - *Bis*(2-chloroethyl) ether analysis - GC/MS data for only 10-min programmed temperature run.[15]

was split between an XAD-2 macroreticular resin collection apparatus (15 l/hr) and a continuous liquid-liquid extractor (1.5 l/hr), as shown in Figure 23.2. The water sample is split by a Union Cross attached to a standard laboratory tap by a 3/8-in. normal pipe thread to a 1/4-in. Swagelok fitting. The bottom of the Union Cross was fitted with a pressure regulator to maintain line pressure at 50 lb/in.^{2} Each arm of the Union Cross was fitted with a metering valve to independently control the water flow rates to the XAD-2 MRR collection apparatus and the CLLE 9-liter water reservoir, respectively. During the analysis of Torresdale water, the MRR apparatus was operated on line, as above. The CLLE reservoir was changed after 9-liter volumes for subsequent independent isolation of the organics by the CLLE method.

Table 23.2 Summary of Sampling Information

Sample Date	Type	Volume (l)	Flow Rate (l/min)	Initial Conditions pH	Initial Conditions μmhos[b]	Sampling Conditions pH	Sampling Conditions μmhos[b]
Torresdale Plant							
Drinking Water							
2/4[a]	M	90	18	8.6	—	3.9	—
2/5[a]	M	90	18	8.6	—	4.1	—
2/6[a]	M	90	18	8.3	220	4.1	260
3/24-25[a]	M	215	15	8.5	230	4.0	250
3/24-25[a]	L	44	0.75	8.5	230	7.0	—
3/25-26[a]	M	190	15	8.6	235	4.1	255
3/25-26[a]	L	44	0.75	8.6	235	7.0	—
4/17	H	1	—	8.7	—	7.0	c
4/17	M	70	18	8.7	225	4.0	260
4/17.	L	9	0.75	8.7	225	7.0	—
4/17-19	M	420	15	8.6	230	3.9	255
4/19-20	M	150	15	8.5	—	4.2	—
Belmont Plant							
Drinking Water							
8/20-21	M	500	14	7.2	—	4.2	—
8/20-21	L	88	1.5	7.2	—	7.2	—
Belmont River Influent							
8/21	M	54	15	7.7	—	4.0	—

M = MRR Sample   L = CLLE Sample   H = HGA Sample

[a]Free chlorination practiced at water treatment plant; sample dechlorinated with Na_2SO_3.
[b]Conductivity in μmhos.
[c]Sodium sulfate added to adjust ionic strength to 1.65 M.

380   ORGANIC POLLUTANTS IN WATER

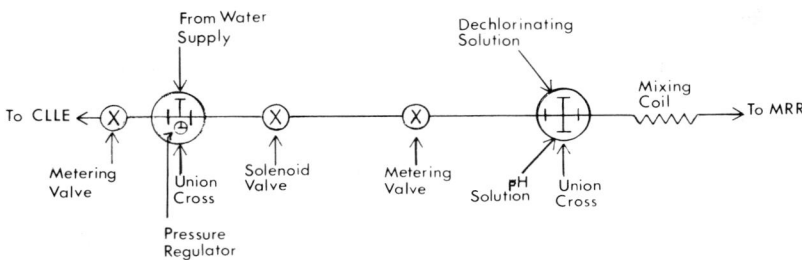

Figure 23.2   Schematic diagram of simultaneous aqueous sampling method for the subsequent isolation of organics from tap water.

Continuous Liquid-Liquid Extraction (CLLE)

The continuous liquid-liquid extraction apparatus is made of glass and teflon to prevent contamination. The apparatus is a modification of a previously designed apparatus for continuous liquid-liquid analysis of organophosphate pesticides in water.[8,9] Quantitative recovery of nanogram-to-microgram-per-liter concentrations of organophosphate pesticides were determined in 6-liter volumes with the original apparatus.

The heart of the CLLE apparatus is a 32-ft teflon helix mixing coil, 0.125 in. o.d., 0.063 in. i.d. wrapped around a 1-in. diameter pipe.[8] The sample of tap water and solvent is pumped through the coil at a water: solvent ratio of 15:1 from a dual-channel micropump at approximately 0.75 liter per hour per coil. Two parallel coils were used in the present design. The original apparatus was modified to use a heavier-than-water solvent, $CHCl_3$.

The original CLLE apparatus was also modified to enable greater solvent evaporation capacity and solvent recycling by incorporation of a distillation column with an internal cartridge heater, a water-cooled condenser, and a solvent reservoir. The system was automated for unattended operation. When a mixture of the $CHCl_3$ and water enters the phase separation column, the denser $CHCl_3$ migrates to the lower portion of the chamber while the water phase migrates to the upper portion. The waste outlet for the extracted water allows the water to exit while maintaining a constant pressure head. This head ensures a nonfluctuating interface and forces the flowing $CHCl_3$ into the concentration chamber. A level controller on the solvent reservoir controlled an on-off switch for the solvent evaporation system to prevent overheating of the solvent.

The CLLE apparatus draws water from a 9-liter reservoir filled on-line (Belmont drinking water) or manually (Torresdale drinking water and

Belmont River water influent). Emulsions were controlled by adding to the separator column loosely-packed glass wool and utilizing the upward migrating water phase to break up suspensions which can form.

Redistilled chloroform was used for the CLLE apparatus. Two batches of chloroform solvent were used for this work. Torresdale water was extracted with "Spectra Grade," Lot #2XA, Mallincroft Chemical Co., St. Louis, Missouri. Belmont water was extracted with "Laboratory Grade," Lot #752005, Fisher Scientific, Pittsburgh, Pennsylvania. Blanks were run before each analysis under operational conditions. At Torresdale (on 3/24 and 4/17) blanks were run by passing through the system 3 liters of $CHCl_3$-extracted water (16 X). At Belmont (8/20) a blank was run by passing air instead of recycled water through the system for the length of a run. The air was pre-extracted by bubbling it through a $CHCl_3$ trap.

XAD-2 Macroreticular Resin Collection Apparatus

Drinking water samples from the Union Cross (Figure 23.2) pass through a solenoid valve which is controlled by a Matheson "Lab Stat" Level Controller. The water sample is then metered by a valve which feeds the MRR apparatus. The water from the metering valve is passed through a second Union Cross enabling the addition of solutions to adjust sample pH and to dechlorinate the sample. A 25-ft x 0.25-in. o.d. teflon mixing coil which is wrapped around a 3-in. diameter pipe is utilized to assure homogeneous mixing of the sample before it flows through the MRR column.

The pH and the chlorine content of the water are controlled by the addition of solutions from two reservoirs at the Union Cross at the head of the teflon mixing coil. These reservoirs feed the solutions to the teflon mixing coil by a dual-channel micropump (Model 2-6004, Buchler Instruments, Fort Lee, New Jersey) at a flow rate of 1 ml/min. The first reservoir contains 4 ml of 85% H_3PO_4 per liter of water to adjust the pH to 4.0. The second reservoir contains 10 g Na_2SO_3 per liter of water to dechlorinate the sample. Dechlorination was completed in the mixing coil when free chlorination was practiced at the water treatment plant, Table 23.1. No dechlorination was observed when ammoniation was practiced for taste and odor control. A 5- x 24-cm glass column (Fischer and Porter Company, Warrington, Pennsylvania) equipped with a sintered glass disk and teflon stopcock was filled with 50/150 mesh XAD-2 macroreticular resin to an 11-cm bed height. The XAD-2 resin was obtained from Rohm & Haas Co. (Philadelphia, Pennsylvania).

After the XAD-2 column run is completed, the water is allowed to drain to the top of the resin bed. Then the bed is eluted with 200 ml of ether. The ether is allowed to remain in contact with the resin column for 10 min before elution and 110 to 125 ml of eluant is collected. The eluant is dried by passing it through a 2-cm i.d. column containing a 5- to 7-cm bed of anhydrous Na_2SO_4. The drying column is rinsed with ether to adjust the volume back to the original collected volume and the eluant is concentrated 100 X in a Kuderna-Danish evaporator.

All solvents used were pesticidal quality. Solvents were cleaned by passing them through a 2-cm i.d. column containing a glass wool plug atop a 2.5-in. bed of XAD-2 resin and a 2.5-in. bed of Filtrasorb-200 activated carbon (Calgon Co., Pittsburgh, Pennsylvania). Two columns were used—one for ether only, and the other for methanol, ethanol and distilled water from basic permanganate.

The XAD-2 isolation column was cleaned between runs by successive elution with 5 bed volumes each of ethanol, filtered water and methanol with contact times of 1-, several-, and 5-min/bed volumes. The methanol eluant is drained to the top of the resin bed, and the column is stored under methanol as recommended by Junk et al.[10] Ether blanks were run before each analysis. After the blank was completed, the column was recleaned before sampling. A true method blank with water was not run because of trace organics present in the distilled water.

River water was collected in 9-liter bottles. The pH was immediately adjusted to 4 with 85% H_3PO_4 (≈ 1.5 ml). After settling for 4 hr the supernatant was decanted through a glass column filled with 8 in. of loosely-packed, ether-extracted glass wool and into a 15-liter reservoir attached to the MRR column by a pentaflex connection system (Fischer-Porter Co.). The top of the MRR column contained a 2-in. layer of loosely-packed, ether-extracted glass wool to assure turbidity removal. The reservoir was refilled to maintain a flow rate of 15 l/hr.

GC/MS Analysis System

Samples were analyzed by injection of up to 10 μl liquid or 10 ml head gas into a Varian Model 1400 Gas Chromatograph coupled to a Finnigan Model 1015 S/L Quadrapole Mass Spectrometer via a glass jet separator. A splitter at the GC column effluent was adjusted to obtain 5×10^{-6} torr vacuum in the mass spectrometer's manifold that was maintained throughout GC/MS analysis. A Systems Industries Model 150 Data System was used to examine spectra with hard copies available from a Houston Complot or Model 4631 Hard Copy Unit on a Tektronix 4010 Display Unit. Data was acquired in the IFSS mode (Integration time as a Function of Signal Strength).

Instrument conditions were: manifold temperature, 100°C; ionizing voltage, 70 eV; filament current, 250 μ amps; electron multiplier, 3000 volts; mass range, 25-300 amu; sensitivity, 10^{-7} amps/volt; base integration time, 1 millisec/amu. The delay time between scans is 5 sec. Helium flow was adjusted to produce an absolute retention time of 9.0 min and 2.5 min for 2-ethyl-1-hexanol on SE-30 and Carbowax 20M columns, respectively, under the conditions listed below. Flow through a GC column of bentone-didecyl phthalate was adjusted to produce a retention time of 4 min for toluene.

The GC inlet temperature was 250°C for the SE-30 column, 200°C for the Carbowax column and 125°C for the bentone-didecyl phthalate column. The gas chromatographic columns and conditions are outlined as follows:

1. 6 ft x 0.085 in. i.d. stainless steel; 20% SE-30 on 80/100 Gas-Chrom Q. Programmed temperature: 80° initial, 200° final, 6°/min.

2. 6 ft x 0.085 in. i.d. stainless steel; 5% bentone 41-5% didecyl phthalate on 80/100 Gas-Chrom Q. Programmed temperature: 70° initial (10 min), 120° final, 6°/min.

3. 6 ft x 0.085 in. i.d. stainless steel 20% Carbowax 20M on 80/100 Gas-Chrom Q. Programmed temperature: 80° initial, 200° final, 6°/min.

Mass Spectra Identification

Computer-assisted **interpretation of mass** spectral data was accomplished using Cornell University's Mass Spectral Identification System (PBM and STIRS).[11] PBM (Probability-Based Matching System) and STIRS (Self-Training Interpretive and Retrieval System) together are complementary techniques for identification of unknown mass spectra.

PBM is a reverse search technique and examines the uniqueness of each m/e value in the reference file. It attempts to "fit" the reference spectra to the unknown, and arrives at a confidence level, K, which is an indication of the probability that the reference spectrum is identical with the unknown spectrum. Reference spectra are abbreviated lists of 15 to 26 peaks per compound. In practice, K values range from 20 to 100 (the larger the K, the better the match). Delta K reflects the deviation of the obtained K value from the maximum K value that would be expected if the unknown spectrum was identical to the reference spectrum. Therefore, the best match is for a high K value coupled with a low Delta K. PBM is useful for unresolved or mixed spectra as it is a reverse search.

STIRS uses a more complex algorithm for correlation of unknown and reference spectra. A series of match factors (13 total) are calculated, and the best matches for each match factor (MF) are available for output. Each match factor is sensitive to certain substructures. In practice, it was found that MF 11 (overall match factor) was the most useful. Numerical values for a MF ranges from 0 to 1000. MF 5 through MF 9 are zero unless a molecular weight is given by the user. STIRS may not necessarily output the true compound, but the substructure consistent with the spectrum should be represented in the results.

Two versions of the system were used. Initial results were obtained directly from Cornell (April-May 1975). An updated version was instituted June 1, 1975 on TYMNET. In the initial system STIRS was obtained directly. PBM results were obtained indirectly from data stored on STIRS. This step was completed with the help of Mr. Henry Dayringer, under the supervision of Dr. F. W. McLafferty of Cornell University. In the TYMNET System, PBM and STIRS are obtained sequentially.

The PBM/STIRS output was examined using the traditional tools of carbonium ion chemistry, isotopic ratios and chemical intuition. The aqueous solubility, stability and compatibility with the isolation and chromatographic method were also considered. Mass spectra of the most probable choices from the evaluation were compared to their mass spectra as compiled in the "Registry of Mass Spectral Data," 1974[12] to assure that matches were sensible.

RESULTS AND DISCUSSION

Results of the GC/MS analyses are presented in Tables 23.3 through 23.5. Figures 23.3 and 23.4 are computer plots showing identified compounds (full spikes) and unidentified compounds (half spikes) on the SE-30 GC column. At this time the total number of compounds identified in the drinking water from the Torresdale and Belmont Water Treatment Plants are 33 and 21, respectively (Tables 23.3 and 23.4). Twenty-one compounds were identified in the influent to the water treatment plant at Belmont (Table 23.5). Confirmation of a compound is completed by comparison of its GC retention time on the SE-30 or bentone column and its mass spectrum with the mass spectrum of a pure reference compound on the same GC/MS system. The sugar, L-sorbofuranose, listed in Table 23.3 as compound IV.5, was not commercially available. It was identified by a mass spectrum reported by De Jongh and Biemann.[13]

The identified compounds are divided into four classes: halogenated, aromatics, hydrocarbons and miscellaneous. Aliphatic and aromatic halogenated organics account for the largest group of organics identified in the drinking water. Some compounds are listed with the term isomer

IDENTIFICATION OF ORGANIC POLLUTANTS 385

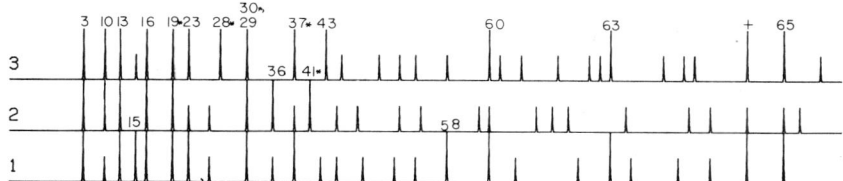

Figure 23.3 GC/MS analysis of Torresdale drinking water. Computer plot of organic compounds identified (full-spike) and unidentified (half-spike) on the SE-30 GC column. MRR samples from (1) 2/6/75, (2) 3/24/75, and (3) 4/17-19/75. See Table 23.7 for RRT data.

[*represents an isomer (actual isomer unidentified);

+compound is not commercially available (1,2:4,6-di-O-isopropylidene-L-sorbofuranose)].

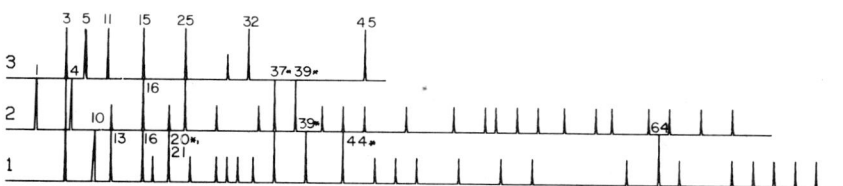

Figure 23.4 GC/MS analysis of Belmont drinking water. Computer plot of organic compounds identified (full-spike) and unidentified (half-spike) on the SE-30 GC column. (1) Macroreticular resin column, (2) continuous liquid-liquid extraction and (3) blank continuous liquid-liquid-extraction. See Table 23.7 for RRT data.

[*represents an isomer (actual isomer unidentified)].

Table 23.3 Organic Compounds Identified by GC/MS in the Drinking Water at the Torresdale Water Treatment Plant

	2/4	2/5	2/6	3/24-3/25	3/25-3/26	4/17	4/17-4/19	4/19-4/20
I. Halogenated								
(1) Chloroform			M	M	M	M,H	M	M
(2) Dibromochloromethane			M	M		M,H	M	M
(3) Dichlorobromomethane			M	M	M	M,L,H	M	M
(4) Bromoform							M	
(5) Bis(2-chloroethyl) ether[a]	M	M	M	M	M	M,L	M	M
(6) 1,2-Bis(2-chloroethoxy)ethane			M			M	M	M
(7) 1,1,2,2-Tetrachloroethane				L		H		
(8) Tetrachloroethylene			M				M	
(9) Hexachloroethane							M	
(10) Dichlorobenzene isomer			M				M	
(11) 1,1,1-Trichloropropanone (Trichloroacetone)			M	M	M	M	M	M
(12) 2-Chloroethyl chloroformate					L			
II. Aromatic								
(1) Toluene	M		M	M		M	M	M
(2) o-Xylene	M	M	M	M		H,M	M	M
(3) Xylene isomer[c] (m or p) or ethyl benzene		M				H,M	M	
(4) m-Xylene						L		
(5) p-Xylene						L		
(6) C3-benzene isomer[d]	M	M	M			M[b]	M	M
(7) C3-benzene isomer[c]	M	M					M	M
(8) 1,2,3-Trimethylbenzene				M				
(9) Benzoic acid								
(10) Acetophenone							M	M
(11) Methyl m-toluate			M					
(12) Tolunitrile isomer (m or p)[c]								M

III. Hydrocarbons
   (1) Decane   M
   (2) Branched Hydrocarbon (C7/8)   M
   (3) Branched Hydrocarbon (C9/10)   L   M
   (4) Branched Hydrocarbon (C10/11)   M

IV. Miscellaneous
   (1) Diethyl carbonate   M
   (2) Ethyl ether   L
   (3) Tributyl phosphate   M
   (4) 1,2:3,5-Di-O-isopropylidene-D-xylofuranose   M
   (5) 1,2:4,6-Di-O-isopropylidene-L-sorbofuranose   M

M = MRR Sample   L = CLLE Sample   H = HGA Sample

[a]Identified in river influent to Torresdale Water Treatment Plant by 1-liter batch LLE on 5/2/75.
[b]Identified in river influent to Torresdale Water Treatment Plant by a 9-liter CLLE on 4/17/75.
[c]The exact isomer could not be determined on the SE-30 column. Mass spectra of these isomers are indistinguishable.
[d]C3-benzenes (trimethyl, methyl, -ethyl, or propyl benzene).

Table 23.4 Organic Compounds Identified by GC/MS in Drinking Water at the Belmont Water Treatment Plant

		8/20/75
I.	Halogenated	
	(1) Chloroform	M
	(2) Dibromochloromethane	M,L
	(3) Dichlorobromomethane	M
	(4) 1,1,2,2-Tetrachloroethane	M
	(5) Bromoform	M,L
	(6) 1,1,1-Trichloropropanone (Trichloroacetone)	M,L
	(7) Tetrachloroethylene	M
	(8) Dichlorobenzene isomer	M,L
	(9) Dichlorobenzene isomer	M,L
	(10) 1,1,1-Trichloroethane	L
	(11) 1,2-Dichloroethylene	L
II.	Aromatic	
	(1) Xylene isomer (*o, m* or *p*)	M
	(2) Tolunitrile isomer (*m* or *p*)	L
	(3) Benzaldehyde	M
III.	Hydrocarbons	
	(1) Branched hydrocarbon (C7-11)	M
	(2) Branched hydrocarbon (C7-11)	M
	(3) Branched hydrocarbon (C7-11)	M
	(4) Branched hydrocarbon (C7-11)	M
	(5) Decane	M
IV.	Miscellaneous	
	(1) Dimethyl phthalate	M
	(2) Ethanol	L

M = MRR Sample L = CLLE Sample

or branched hydrocarbon, *e.g.,* (C7-11). This means that the manual or computerized interpretation of the mass spectral data and the GC retention times did not enable precise determination of the isomer(s) and compound(s) present. Figures 23.3 and 23.4 indicate that many other compounds are also present that have not yet been identified. Some peaks are unresolved mixtures; other peaks are too low in concentration to obtain good mass spectra. Further investigation is continuing on these tentatively-identified compounds.

Table 23.5 Organic Compounds Identified by GC/MS in River Water Influent to Water Treatment Plant at Belmont (8/20/75) by MRR

I. Halogenated	III. Hydrocarbons
(1) Chloroform	(1) Nonane
(2) Trichlorophenol isomer	(2) Decane
(3) o-Dichlorobenzene	(3) Undecane
(4) Tetrachloroethylene	(4) Dodecane
	(5) Branched hydrocarbon (C7)
II. Aromatic	(6) Branched hydrocarbon (C8)
(1) Toluene	(7) Branched hydrocarbon (C10/11)
(2) o-Xylene	(8) Branched hydrocarbon (C10/11)
(3) m-Xylene	(9) Branched hydrocarbon (C10/11)
(4) p-Xylene	
(5) Ethyl benzene	IV. Miscellaneous
(6) C3-benzene isomer	(1) Methyl isobutyrate
(Trimethyl, methyl -ethyl, or propyl benzene)	(2) 2-Benzothiazole

MRR = Macroreticular Resins

Blanks were run for both the XAD-2 MRR and the CLLE collection apparatus. GC/MS analysis of the ether solvent after elution from a cleaned MRR column showed ethyl acetate to be the only impurity. Table 23.6 shows the identified impurities in the CLLE blanks run prior to sampling. Apparently the change of supplier of $CHCl_3$ affected solvent quality.

Table 23.7 lists 66 compounds of interest as they elute from a 6-ft, 20% SE-30 column. Relative retention times are reported as compared to 2-ethyl-1-hexanol whose absolute retention time under the GC/MS conditions was adjusted to 9 min. A nonpolar GC liquid phase (SE-30) and polar phase (Carbowax 20M) were chosen for chromatographic work. SE-30 was an excellent choice as compounds eluted from the column in the approximate order of their boiling point. Most identified compounds were in the boiling point range 100-250°C. The compounds have a wide range of polarities and the use of the Carbowax 20M column at an initial temperature of 80° was unsuccessful. Only *bis*(2-chloroethyl) ether (BCEE) was identified on this column. Retention time data showed that the identified compounds eluted at low temperature and were unresolved.

The SE-30 column did not resolve alkyl benzene isomers, so a more specific phase, bentone-didecyl phthalate, was used.[14] Bentone-didecyl phthalate resolved the C2-benzene isomers (Tables 23.3 and 23.5), 1,2,3-trimethyl benzene (Table 23.3) and a series of hydrocarbons (Tables

Table 23.6 Blanks for Continuous Liquid-Liquid Extraction Apparatus

	3/24	4/17
A. Torresdale Drinking Water		
Tetrachloroethylene	X	X
Xylene isomer (*m* or *p*) or ethyl benzene	X	X
o-Xylene	X	X
Ethanol		X
Toluene		X
Unidentified GC Peak	X	

Water Blank: 3 liters of pre-chloroform extracted distilled water recycled 16 X at a water:solvent ratio of 15:1.

B. Belmont Drinking and River Water Influent (8/20)

Diethyl carbonate
Tetrachloroethylene
1,1,2,2-Tetrachloroethane
Pentachloroethane
Hexachloroethane
Tetrachloropene isomer (Tentative identification)

Air Blank: 44 liters of pre-chloroform extracted air at air:solvent ratio of 15:1.

23.4 and 23.5) and confirmed other compounds found using the SE-30 column.

There is no substitute for efficient GC separation for interpretation of mass spectra by a computer-based mass spectral identification system. In some instances though, mass chromatography can separate the components of a mixture of compounds present under one chromatographic peak. A particularly important application of this occurred during our work related to *bis*(2-chloroethyl) ether. Table 23.7 indicates that on the SE-30 column, BCEE cannot be resolved from C3-benzene isomers. Figure 23.5 shows a mass chromatogram for the base peak of BCEE at m/e 93 superimposed on a total ion current chromatogram for the water sample of 2/6 at the Torresdale Water Treatment Plant.

The isolation method determines the specific type of organics measured. No isolation method appears inclusive for all organics;[5] each method has its unique ability. Two general "screening" procedures for organics

Table 23.7 The Elution Order of Organic Compounds on 20% SE-30 Liquid Phase

Compound No.	Relative Retention Time	Name	Molecular Weight	Boiling Point °C[19,20]
1	0.09	Ethanol	46	79
2	0.11	Ethyl ether	74	34
3	0.18	Chloroform	118	61
4	0.20	1,1,1-Trichloroethane	132	74
5	0.25	1,2-Dichloroethylene	97	60
6	0.25	Carbon tetrachloride	154	76
7	0.25	Benzene	78	80
8	0.27	2,3-Dimethylpentane	100	89
9	0.28	Methyl isobutyrate	102	92
10	0.30	Dichlorobromomethane	164	90
11	0.41	Diethyl carbonate	118	126
12	0.43	Toluene	92	111
13	0.44	Dibromochloromethane	208	119
14	0.46	2,2,4,4-Tetramethylpentane	128	122
15	0.52	Tetrachloroethylene	165	121
16	0.56	1,1,1-Trichloroacetone	161	134
17	0.58	2,2,5,5-Tetramethylpentane	128	150
18	0.63	Ethyl benzene	106	136
19	0.64	*m*-Xylene	106	139
20	0.65	*p*-Xylene	106	138
21	0.66	Bromoform	253	150
22	0.70	2-Chloroethyl chloroformate	143	156
23	0.70	*o*-Xylene	106	144
24	0.70	Nonene	126	146
25	0.70	1,1,2,2-Tetrachloroethane	167	146
26	0.72	Nonane	128	151
27	0.83	Benzaldehyde	106	178
28	0.86	1-Methyl-2-ethylbenzene	120	165
29	0.86	1-Methyl-3-ethylbenzene	120	161
30	0.86	*Bis*(2-chloroethyl) ether	143	178
31	0.88	1,3,5-Trimethylbenzene	120	165
32	0.88	Pentachloroethane	202	162
33	0.91	1-Methyl-4-ethylbenzene	120	162
34	0.94	Decene	140	171
35	0.95	1,2,4-Trimethylbenzene	120	169
36	0.97	Decane	142	174
37	0.98	*p*-Dichlorobenzene	146	174
38	1.00	2,2,4,6,6-Pentamethylheptane	170	170
39	1.00	*m*-Dichlorobenzene	147	173
40	1.00	2-Ethyl-1-hexanol	130	184
41	1.02	1,2,3-Trimethylbenzene	120	176
42	1.04	*o*-Dichlorobenzene	147	181
43	1.08	Acetophenone	120	202
44	1.13	*m*-Tolunitrile	117	213
45	1.16	Hexachloroethane	237	185

Table 23.7, continued

Compound No.	Relative Retention Ti Time	Name	Molecular Weight	Boiling Point °C[19,20]
46	1.16	p-Tolunitrile	117	218
47	1.17	2,3-Xylenol	122	217
48	1.18	Undecene	154	193
49	1.18	2,6-Xylenol	122	203
50	1.22	Undecane	156	196
51	1.26	2,4-Xylenol	122	211
52	1.27	2,5-Xylenol	122	212
53	1.31	3,5-Xylenol	122	215
54	1.36	3,4-Xylenol	122	225
55	1.38	Benzoic acid	122	249
56	1.42	Dodecene	168	216
57	1.46	Dodecane	170	216
58	1.46	Methyl m-toluate	150	217
59	1.48	2-Benzothiazole	135	231
60	1.52	1,2-Bis(2-chloroethoxy) ethane	125	230
61	1.67	2-Methylnaphthalene	142	241
62	1.83	2,4,6-Trichlorophenol	197	246
63	1.89	1,2:3,5-di-0-isopropylidene D-xylofuranose	230	–
64	1.92	Dimethyl phthalate	194	282
65	2.31	Tributyl phosphate	266	289
66	3.26	Dibutyl phthalate	278	340

Relative retention reference compound: 2-ethyl-1-hexanol of absolute retention time set to 9 min.

were used in the present study: continuous liquid-liquid extraction and XAD-2 macroreticular resins. These methods were chosen because they could be automated, and previous GC work had shown that a sufficient quantity of organics could be collected in a reproducible manner.

Both isolation methods were operated simultaneously to try to use their inherent selectivities to collect different organics. MRR was used at pH 4, and CLLE at pH 7. It was hoped that the selectivity of the process might also help to separate GC interferences, e.g., by removing one of two compounds under one chromatographic peak. Figure 23.4 shows that there are different GC peaks from these isolation methods. The use of two independent isolation methods minimizes the likelihood that if a compound is identified from both isolation methods it is altered during sampling and analysis and adds a measure of assurance to the confirmation of a compound.

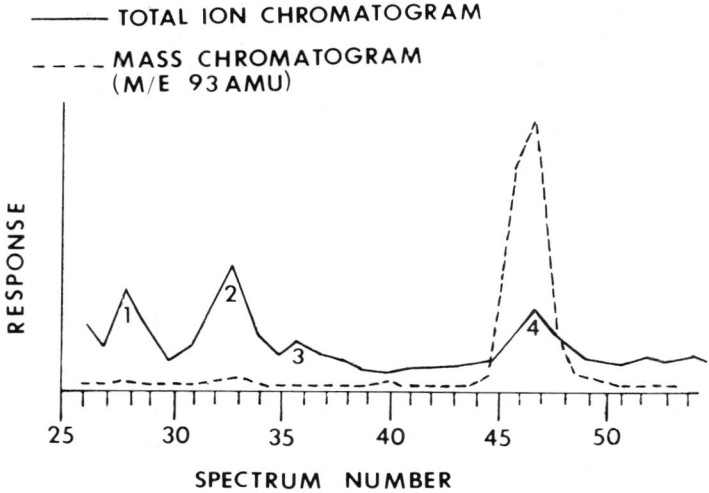

Figure 23.5 Total ion chromatogram and mass chromatogram on SE-30 of Torresdale drinking water collected by XAD-2 macroreticular resin column on 2/6/75. Identified compounds are: (1) 1,1,1-Trichloroacetone, (2) Xylene isomer (*m* or *p*) or Ethyl benzene, (3) *o*-Xylene, (4) *Bis*(2-chloroethyl) ether and a C3-benzene.

Table 23.2 shows that for GC/MS identification, an MRR sample of a minimum of 90 liters of drinking water is necessary. It was found that a convenient sample of about 500 liters could be automatically collected, eluted with ether, concentrated and analyzed by GC/MS within two working days. During this same period a CLLE composite of 50 liters of the same water (Figure 23.1) could be processed through the GC/MS analysis step. The total time required for the simultaneous processing of MRR and CLLE samples is less than 16 man-hours.

Table 23.3 shows that in Torresdale drinking water, 26 of the compounds were identified in samples isolated by the MRR method. Seven compounds were found by CLLE, exclusively. Also, CLLE and HGA confirmed compounds isolated by the MRR method. It appears that more than 44 liters of drinking water must be sampled to obtain sufficient quantities of organics by the CLLE method as used.

Table 23.4 shows that in Belmont drinking water 17 of the compounds were identified in samples isolated by the MRR method. Four compounds were identified from CLLE extracts, exclusively. Also, the CLLE method confirmed five of the compounds isolated by MRR. Figure 23.4 shows that there are many unidentified peaks in the CLLE sample. Sufficient sample size again appears to be the problem. Benzaldehyde, dimethyl phthalate and one of the dichlorobenzene isomers were isolated by MRR and identified in the drinking water. These compounds were not identified in the river water entering the water treatment process. Unidentified mixtures of organic compounds of higher concentrations are found in this GC region and the difference between the volumes of drinking water and river water samples (500:54) probably explain this discrepancy.

Of particular note, 1,1,1-trichloroacetone (TCA) has been isolated from drinking water by the XAD-2 MRR and CLLE methods. TCA is a known precursor in the formation of chloroform via the haloform reaction. This indicates that this reaction sequence is occurring within the water treatment process. A complete report on this aspect of the work has been prepared.[6]

Also, possible precursors to haloforms and TCA were found in the drinking water at Torresdale. These are the isopropylidene sugars in Table 23.3 (compounds IV.4 and IV.5). These two sugars have been shown to undergo a haloform reaction at pH > 7.5 with 10 ppm chlorine.[15] One implication of these results is that haloforms may be forming throughout the distribution system after water is released from the water treatment plant.

In an EPA investigation of the Torresdale Water Supply, 59 compounds were identified.[16-18] The isolation methods used were head gas analysis[19] and the carbon adsorption method (CAM)[16] at pH 7-8: Less than 10 compounds were identified only by the CAM.[18] Our present investigation found an additional 21 compounds that were not identified by the EPA and confirmed the identities of 12 compounds that were identified. It must be noted that three differences exist between our analysis and that of the EPA: 1) isolation methods, 2) sampling pH (MRR pH 4) and 3) GC columns.

Head gas analysis best isolates trace organic compounds with low boiling points (e.g., 60 of the 72 compounds reported by Coleman et al.[17] in the EPA study of five cities by HGA had boiling points less than 125°C. Table 23.7 shows that most compounds identified from XAD-2 resin and CLLE have boiling points between 100-250°C. This indicates the influence of the isolation method on the compounds found in the sample.

The GC/MS results of the drinking water at the Torresdale Water Treatment Plant (Table 23.3 and Figure 23.3) show a constant background of organics in February, March and April. Twenty of the thirty-three compounds were identified during at least two of these three months. There are indications (Figure 23.3) from GC peaks that other as yet unidentified compounds may also be part of a constant background. Suspected sources of the petroleum-related products (aromatics and hydrocarbons) in this background are industrial effluents, wastes from shipping, and street runoff. The chlorination process within the water treatment plant appears to add chlorinated compounds. It must be noted that not all the halogenated compounds can be implicated in the chlorination process as some apparently are from industrial sources, *e.g.*, tetrachloroethylene, 1,2-*bis*(2-chloroethoxy) ethane and *bis*(2-chloroethyl) ether.

Obvious future use of these continuous-collection isolation methods are "screening" the source of pollution (*e.g.*, precursors of the haloform reaction) and testing the effectiveness of water treatment operations. The quantitative reliability of the HGA and the CLLE methods[8,9] will be used to monitor specific pollutant problems.

ACKNOWLEDGMENTS

This innovative research project would not have been possible without the encouragement and leadership of Water Commissioner Carmen F. Guarino, for which the authors express their gratitude and appreciation. The authors acknowledge the financial support and laboratory assistance furnished by the Philadelphia Water Department and Alan Hess, Chief of Water Treatment, who reviewed the manuscript, and Richard Grochowski and associates of the Belmont laboratory who helped develop and operate the CLLE apparatus.

Special commendation is due Mr. Barry Silver of the Torresdale Laboratory of the Water Department, who operated the MRR apparatus and participated in the GC/MS work.

The authors are also indebted to the Rohm & Haas Corporation, Philadelphia, Pennsylvania, for the use of their GC/MS System, as well as to Dr. F. W. McLafferty, Department of Chemistry, Cornell University, for the complimentary use of the initial PBM/STIRS System and for his technical review of that aspect of this manuscript.

REFERENCES

1. Suffet, I. H., S. Friant, C. Marcinkiewicz, M. J. McGuire and D. T.-L. Wong. "Organics, Nature and Analysis of Chemical Species," *J. Water Poll. Control Fed.* **47**, 1169 (1975).
2. Rook, J. J. "Formation of Haloforms during Chlorination of Natural Water," *Water Treatment Exam.* **23**, 234 (1974).
3. Bellar, T. A., J. J. Lichtenberg and R. C. Kroner. "The Occurrence of Organohalides in Chlorinated Drinking Water," *J. Amer. Water Works Assoc.* **66**, 703 (1974).
4. U.S. Environmental Protection Agency. "Draft Analytical Report— New Orleans Area Water Supply Study," Lower Mississippi River

Facility, Slidell, La., Surveillance and Analysis Division, Region VI, Dallas, Texas EPA-906/10-74-002 (1974).
5. Suffet, I. H. and J. V. Radziul. "Analysis of Organic Pollutants in Drinking Water," Proceedings of the International Conference on Environmental Sensing and Assessment, Las Vegas, Nevada (September 14-19, 1975), Volume 2, Session 30-1, (New York: Institute of Electrical and Electronic Engineers, Inc., 1976).
6. Suffet, I. H., L. Brenner and B. Silver. "The Identification of 1,1,1-Trichloroacetone (1,1,1-Trichloropropanone) in Drinking Water—A Known Precursor in the Haloform Reaction," submitted for publication.
7. Friant, S. Ph.D. Thesis, Drexel University, Philadelphia, Pennsylvania (1976).
8. Wu, C. and I. H. Suffet. "Continuous Liquid-Liquid Extraction of Organic Pesticides from Aqueous Solutions," *ASTM, Special Technical Publication* **582**, 90 (1975).
9. Wu, C. and I. H. Suffet. "Design and Optimization of a New Continuous Liquid-Liquid Extraction Apparatus—Application: Analysis of Organophosphate Pesticides in Water," presented at the Environmental Chemistry Division, American Chemical Society Bicentennial Meeting, New York (April 1976).
10. Junk, G. A., J. J. Richard, M. D. Grieser, D. Witiak, J. L. Witiak, M. D. Arguello, R. Vick, H. J. Svec, J. S. Fritz and G. V. Calder. "Use of Macroreticular Resins in the Analysis of Water for Trace Organic Contaminants," *J. Chromatogr.* **99**, 745 (1974).
11. Kwok, K-S, R. Venkataraghavan and F. W. McLafferty. "Computer-Aided Interpretation of Mass Spectra, III. A Self-Training Interpretive and Retrieval System," *J. Amer. Chem. Soc.* **95**, 4185 (1973); McLafferty, F. W., R. H. Hertel and R. D. Villwock. "Probability Based Matching of Mass Spectra. Rapid Identification of Specific Compounds in Mixtures," *Org. Mass Spectrom.* **9**, 690 (1974). More information on the Cornell PBM/STIRS system can be obtained from Dr. John Aikin, Office of Computer Services, Cornell University, Ithaca, New York 14850.
12. Stenhagen, E., S. Abrahamsson and F. W. McLafferty. *Registry of Mass Spectral Data* (New York: Wiley-Interscience, 1974).
13. De Jongh, D. C. and K. Biemann. "Mass Spectra of O-Isopropylidene Derivatives of Pentoses and Hexoses," *J. Amer. Chem. Soc.* **86**, 67 (1964).
14. Ottenstein, D. M., D. A. Bartley and W. R. Supina. "Gas Chromatographic Separation of Styrene in the Presence of Xylenes and Propylbenzenes," *Anal. Chem.* **46**, 2225 (1974).
15. Suffet, I. H. Unpublished data, Drexel University (1975).
16. U.S. Environmental Protection Agency. "Preliminary Assessment of Suspected Carcinogens in Drinking Water," EPA Interim Report to Congress, Washington, D. C. (June 1975).
17. Coleman, W. E., R. D. Lingg, R. G. Melton and F. C. Kopfler. "The Occurrence of Volatile Organics in Five Drinking Water Supplies Using GC/MS," presented at the First Chemical Congress of the North American Continent, Mexico City (December 1975), Chapter 21, this volume.

18. Keith, L. H., A. W. Garrison, F. R. Allen, M. H. Carter, T. L. Floyd, J. D. Pope and A. D. Thruston. "Identification of Organic Compounds in Drinking Water from Thirteen U.S. Cities," presented at the First Chemical Congress of the North American Continent, Mexico City (December 1975), Chapter 22, this volume.
19. Kopfler, F. C., R. G. Melton, R. D. Lingg and W. E. Coleman. "GC/MS Determination of Volatiles for the National Organics Reconnaissance Survey (NORS) Drinking Water," presented at the First Chemical Congress of the North American Continent, Mexico City (December 1975), Chapter 6, this volume.
20. Chemical Rubber Co. *CRC Handbook of Chemistry and Physics*, 50th edition. (Cleveland, Ohio: Chemical Rubber Co., 1969-70).
21. Fieser, L. F. and M. Fieser. *Advanced Organic Chemistry* (New York: Reinhold Publ. Co., 1961).

CHAPTER 24

ANALYSIS OF DRINKING WATER FOR ORGANIC COMPOUNDS

Robert D. Kleopfer

U.S. Environmental Protection Agency
Region VII
Kansas City, Kansas

INTRODUCTION

Our rivers, streams and lakes are replete with a host of organic chemicals some of which are known to be deleterious to health. These chemicals, which normally occur in low concentrations, originate primarily from domestic sewage, industrial wastes and agricultural runoff. In recent years it has become apparent that the conventional indicators of organic chemical pollution [biological oxygen demand (BOD), chemical oxygen demand (COD), total organic carbon (TOC), carbon chloroform extract (CCE), etc.] do not always reveal environmental dangers. Consider, for example, recent developments concerning methylmercury, polychlorinated biphenyls (PCB), pesticides, vinyl chloride, and phthalate ester plasticizers. The need for the determination of specific organic chemicals is obvious.

The analysis for trace amounts of organic pollutants in our air and water until recently was a difficult, if not impossible, task. Today, however, analysis for specific organic contaminants on a routine basis is possible because of the development of modern instrumentation. The most powerful analytical tool now available for the qualitative and quantitative identification of trace amounts of organic compounds is the gas chromatograph interfaced with a mass spectrometer, GC/MS.[1-3] Computerization of the GC/MS system greatly enhances the usefulness of the technique by automating the data handling requirements.

During 1971 and 1972 EPA installed several computerized GC/MS systems at various laboratories around the country including regional surveillance and analysis laboratories.[3] These instruments have been successfully applied to environmental problems in enforcement work and routine monitoring, as well as in research.

Even these modern techniques sometimes require an extremely large sample in order to determine specific compounds present in the microgram per liter (ppb) concentration range. This need for larger samples in characterizing organic pollutants resulted in the development of adsorption methods.

Activated carbon is a remarkable adsorption media for many types of organic components. It aids in the detection of low quantities of organic compounds in large volumes of water. After a large volume of water is passed through the carbon adsorption unit, the carbon containing the adsorbed organics is removed, dried and extracted with solvent. The initial extraction is generally performed using chloroform. Chloroform is sufficiently polar to desorb most nonpolar organic compounds as well as some of the more polar compounds such as phenol.

Recently, much work has been done using porous polymer resins as the adsorbing agent for the removal of organics from water.[4,5] Although the method is still (1975) largely in the developmental stage, it has many advantages over the aforementioned technique. The most frequently used resins have been the Amberlite XAD series made by the Rohm and Haas Company. In practice, the water is passed through a column of XAD resin which adsorbs the organic constituents. The organics are desorbed from the resin by elution with a small amount of organic solvent (Soxhlet extraction is not necessary) after which the extract is analyzed by gas chromatography. The method has been applied to a wide variety of organic compounds in an aqueous matrix. In general, overall recoveries are equal to or better than those in liquid-liquid extraction procedures.

This chapter deals with the application of these techniques to the analysis of drinking water for organic contaminants. The chapter has been divided into three sections which deal with the following topics: (1) analysis of drinking water (Evansville, Indiana) using carbon adsorption techniques, (2) analysis of drinking water (EPA Region VII water supplies) using resin adsorption techniques, and (3) occurrence and significance of the trihalomethanes.

EXPERIMENTAL

Sampling—Carbon Adsorption

River samples were collected using FWPCA carbon adsorption samplers (Models No. LF-1 and LF-2; see reference 13 for a description of this type of sampler). The activated carbon used was C-190 Nuchar 30 mesh from West Vaco Company. Sampling times ranged from 7 to 17 days, and sample volumes ranged from 943 to 3865 liters. After drying, the carbon was extracted with Burdick and Jackson glass-distilled chloroform for a minimum of 72 hr on a Soxhlet extractor. After extraction of the carbon the total volume was slowly reduced to below 5 ml in size by evaporation of solvent using a Kuderna-Danish evaporator. Some extracts were separated into acid, base and neutral fractions using a modified Shriner-Fuson separation technique.[6]

Sampling—Resin Adsorption

Finished water samples were collected using precleaned XAD-2 resin from Rohm and Haas according to the method of Junk and co-workers.[4] The samples were taken at a rate of about 200 ml per min over periods ranging from 7 to 14 days (2000-4000 liters). More detailed information can be found in reference 7.

Gas Chromatography—Evansville Samples

Gas chromatographic analyses were done using a Tracor MT200 instrument equipped with flame ionization, flame photometric and electron capture detectors. Generally 6 ft x 1/4 in. glass columns were used containing either 5% OV-17 on Gas-Chrom Q 60/80 mesh or 10% SE-30 on Gas-Chrom Q 60/80 mesh. In general, the following conditions were employed with flame ionization detection: temperature programming from 50° to 210° at 8° per min, detector temperature 275°, nitrogen flow 80 ml/min, sample size 0.1-10.0 μl.

Gas Chromatography/Mass Spectrometry—Evansville Samples

A Perkin Elmer model 270B GC/MS located at Indiana State University—Evansville was used extensively throughout this study. Analyses were done using a 6 ft x 1/8 in. glass column containing 5% SE-30 on 60/80 mesh Chromsorb W. The conditions employed were as follows: column flow 15 ml/min helium, temperature programming from 50° to 210°, ionizing voltage 70. Mass spectra were recorded on an oscillographic recorder with a four-element galvanometer. The total ion current was continuously recorded on a potentiometric strip-chart recorder.

GC/MS—Region VII Samples

The computerized GC/MS used in this work consists of a Varian 1400 GC interfaced to a Finnigan Model 1015 quadrupole MS with a PDP-8/E computer. The entire system was obtained from Finnigan Corporation. The software was version D developed by System Industries. At the conclusion of each run, ion abundance data are summed, normalized and plotted as a function of spectrum index number on a digital incremental plotter (Houston Instrument DP-1). Individual mass spectra for GLC peaks of interest are either plotted using the digital plotter or printed using a 30 cps printer from Digital Equipment Corporation (LA30 Decwriter). GC/MS separations were done using a 6-ft coiled glass column (i.d. 0.078 in.) packed with either 5% OV-17 on Chromsorb W 80/100 mesh or 80/100 mesh Chromsorb 101 with a helium flow of 30 ml/min. The injector temperature was 230°C. Quantitative determinations were performed on a Perkin-Elmer model 990 GC equipped with flame ionization detection.

Analysis of Grab Samples for *bis*(2-chloroisopropyl) Ether

Samples of the Ohio River and of the industrial outfall at Brandenburg, Kentucky were analyzed for *bis*(2-chloroisopropyl) ether. The 2-liter samples were extracted with two 60-ml portions of 15% ethyl ether in hexane followed by 60 ml of hexane. The samples were dried over sodium sulfate and evaporated to about 1 ml in volume. Analysis was done by gas chromatography (isothermally at 90°; 5% OV-17 column) using flame ionization and electron capture (detector temperature 300°) detection. Quantitation was achieved by comparison of peak area with that obtained from authentic standards. Complete recovery was obtained on 2 liters of distilled water spiked with 10 µg of the ether.

Volatile Organics

Volatile organics including chloroform, bromodichloromethane and dibromochloromethane were determined according to the Bellar-Lichtenberg[8] method which involved the purging of 10 ml of water with nitrogen onto a Tenax GC adsorption trap followed by heat desorption and analysis on a Chromsorb 101 column. Quantitative determinations were done by flame ionization detection. Identifications were made by GC/MS.

Identification of Organic Compounds

Initial tentative identifications were usually made solely from interpretation of mass spectra. Some interpretations were made using mass spectral computer search systems.[3] When possible, mass spectra were run on authentic standards to confirm structure identifications. In some cases retention time matching was done on two GC columns using flame ionization detection.

Halide Experiments

Stock solutions of potassium halide (1 mg/ml halide) and calcium hypochlorite (1.2 mg/ml available chlorine) were prepared in deionized-distilled water. The required volume of potassium halide solution was added to 1-liter flasks containing freshly-sampled Missouri River water. The solutions were stirred for 10 min prior to the addition of 6 ml of calcium hypochlorite solution. The mixtures were stoppered, stirred, and then allowed to stand overnight prior to analysis. Portions of the samples (about 10 ml) were analyzed for volatile organics, while the remaining portions were extracted three times with 60 ml of methylene chloride. The solvent extracts were reduced to about 1 ml in volume using Kuderna-Danish apparatus prior to gas chromatography on a 5% OV-17 column with temperature programming from 70 to 200°C at 10°C per min.

Synthesis of Dichloroiodomethane

To an aqueous solution consisting of 1500 ml water, 43 g potassium iodide and 25 g of 1,1,3,3-tetrachloroacetone in a 3000-ml round-bottomed flask equipped with stirrer was added dropwise a solution of 26 g of calcium hypochlorite in 500 ml water. The solution was stirred for 2 hr and then extracted with methylene chloride. The extract was washed with 200 ml of 0.75 N sodium thiosulfate and then concentrated to about 20 ml using a Kuderna-Danish evaporator. The concentrated extract was distilled and the product boiling at 132°C was collected.

Quality Assurance

Controls were used to assure the purity of reagents and the cleanliness of glassware. In general, quality assurance procedures as outlined in references 9 and 10 were followed.

RESULTS AND DISCUSSION

Evansville, Indiana/Ohio River

During the period from July 1971 to December 1972 the capability to analyze for organic pollutants on a routine basis was developed and maintained at the Indiana District Office, EPA, Region V.[11] A GC/MS system (Perkin-Elmer Model 270B, not computerized) located at Indiana State University—Evansville was used extensively during this time. In addition to routine analyses of samples resulting from oil spills, barge accidents and point sources, several samples were taken from the Ohio River and the Wabash River using carbon filter techniques. A study was also made of the Evansville, Indiana municipal treated water. A brief description of this work appeared in the November 1972 issues of *Environmental Science and Technology*.[12]

A typical gas chromatogram (FID) for a carbon-chloroform extract obtained from the Ohio River is shown in Figure 24.1. A gas chromatogram (FID) for an extract obtained from the Evansville tap water (Ohio River source) is shown in Figure 24.2. Compounds which were identified in these samples are shown in Table 24.1.

Of all the compounds which were detected in the Ohio River by gas chromatography, *bis*(2-chloroisopropyl) ether (or BCIPE) was by far the most abundant compound measured in every sample. The estimated concentration of BCIPE was 0.8 µg/l in raw intake water and 0.3 µg/l in tap water at Evansville in June 1972. A tap water sample taken on November 30, 1972, contained approximately 5 µg/l of the ether.

A probable source for the BCIPE is an industry at Brandenburg, Kentucky, located 148 river miles upstream (see Table 24.2). Samples taken during the Ohio River study in August 1971 indicated a daily discharge of 146 lb of BCIPE. If one assumes a river flow of 50,000 cfs, no biodegration (an authentic sample of the ether which had been incubated with raw Ohio River water at 33 mg/l BCIPE, showed no oxygen uptake after 5 days at 20°), and perfect mixing, a concentration of 0.5 µg/l would be expected downstream from the outfall.

Of particular interest at this time (1971) was the presence of the trihalomethanes (bromodichloromethane, chlorodibromomethane and bromoform) which were found in the drinking water but not in the raw Ohio River water.

Region VII

During the period from 1973 to 1975, drinking water samples from four cities in Region VII were analyzed for organic compounds using

IDENTIFICATION OF ORGANIC POLLUTANTS 405

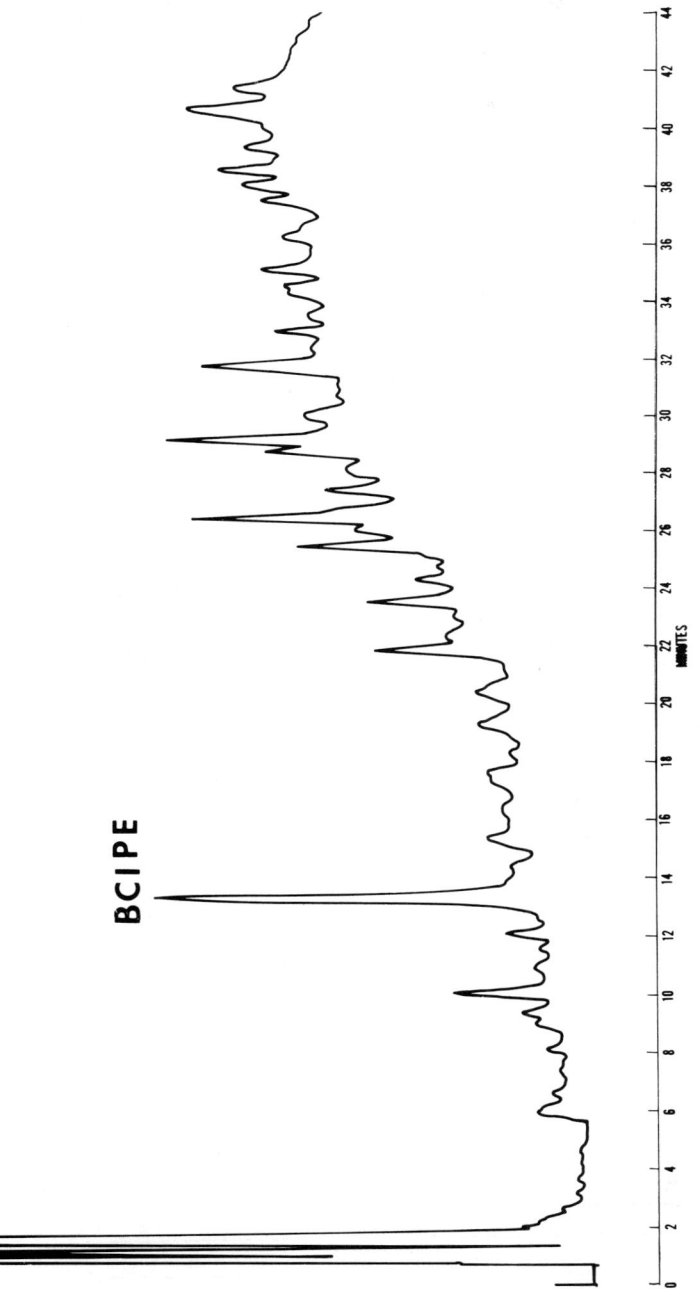

Figure 24.1 Gas chromatogram of carbon-chloroform extract, Ohio River at Evansville.

406 ORGANIC POLLUTANTS IN WATER

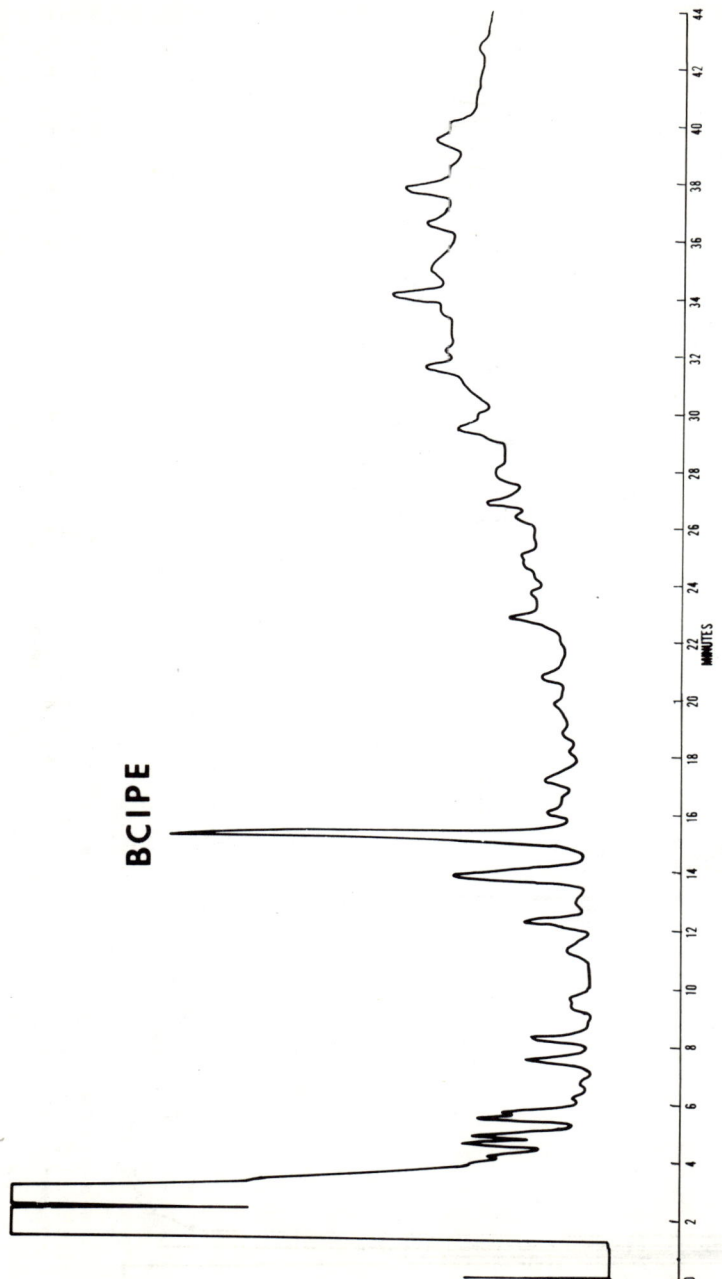

Figure 24.2 Gas chromatogram of carbon-chloroform extract, Evansville municipal water.

Table 24.1 Organic Compounds Found in the Evansville Tap Water

Bis(2-chloroisopropyl) ether	*n*-Dodecane
Bis(2-chloroethyl) ether	Toluene
Acenaphthylene	Benzene substituted with two saturated carbons
Chlorodibromomethane	
Bromodichloromethane	Benzene substituted with three saturated carbons
Bromoform	
Tetrachloroethylene	Benzene substituted with seven saturated carbons
Chlorohydroxybenzophenone	
Hexachloroethane	Indan
Hexachlorobenzene	Tetralin substituted with two saturated carbons
Bis(2-ethyoxyethyl) ether	
Styrene	Several unidentified hydrocarbons
Naphthalene	

Table 24.2 Quantitative Determination of *Bis*(2-Chloroisopropyl) Ether

Sample Location	Date and Time	*Bis*(2-chloroisopropyl) ether µg/l
Ohio River, 10 yd Upstream from Industrial Outfall	6/22/72 1230	< 0.4
Main Discharge from Industry (Ohio River Mile 643.5)	6/22/72 1233	2760
Ohio River, 10 yd Downstream from Industrial Discharge	6/22/72 1236	469
Ohio River, Raw Water Intake of Evansville's Municipal Water Supply (Ohio River Mile 791.5)	6/22/72 0930	~ 0.8[a]
Evansville's Municipal Tap Water	6/23/72 0930	~ 0.3[a]
Evansville's Municipal Tap Water	11/30/72 2030	5

[a]A larger sample was taken in order to increase the sensitivity of the measurement.

computerized GC/MS. The four cities were Kansas City, Kansas (Missouri River source), Johnson County, Kansas (Kansas River source), Jefferson City, Missouri (Missouri River source), and Kirkwood, Missouri (Meramec River source). Resin adsorption was used for sample collection and analysis for "nonvolatile" organics. The Bellar-Lichtenberg technique[8] was used for analysis of volatile organics. Compounds found in these finished waters are tabulated in Table 24.3. The gas chromatogram (total ion current profile) for the Jefferson City resin extract is shown in Figure 24.3.

Table 24.3 Organics Found in Region VII Drinking Waters

Jefferson City, Missouri
- Chloroform (92 ppb)
- Bromodichloromethane (8 ppb)
- Dichlorobromomethane
- Bromoform
- Dichloroiodomethane
- Bromochloroiodomethane
- Hexachloroethane
- Toluene
- Xylene
- Alkylbenzene, C_9H_{12}
- Anthracene
- Lasso (0.14 ppb)
- Atrazine (0.12 ppb)

Kirkwood, Missouri
- Chloroform (47 ppb)
- Bromodichloromethane
- Dibromochloromethane
- Dichloroiodomethane
- Tetrachloroethylene

Kansas City, Kansas
- Atrazine (0.06 ppb)
- Chloroform (50 ppb)
- Bromodichloromethane (4 ppb)
- Dibromochloromethane
- Dichloroiodomethane
- Chloral
- Tetrachloroethylene

Johnson County, Kansas
- Atrazine (0.06 ppb)
- Tetradeconoic Acid
- Pentadecanoic Acid
- Hexadecanoic Acid
- Octadecanoic Acid
- Chloroform (80 ppb)
- Bromodichloromethane (5 ppb)
- Dibromochloromethane
- Dacthal (0.15 ppb)

Of particular concern was the presence of the herbicides (Atrazine, Lasso and Dacthal) and the trihalomethanes (chloroform, bromodichloromethane, dibromochloromethane, bromoform, dichloroiodomethane and bromochloroiodomethane). The herbicides are present in the raw water and are not completely removed by the treatment process. The trihalomethanes are not present in the raw water but are formed during the treatment process.

IDENTIFICATION OF ORGANIC POLLUTANTS 409

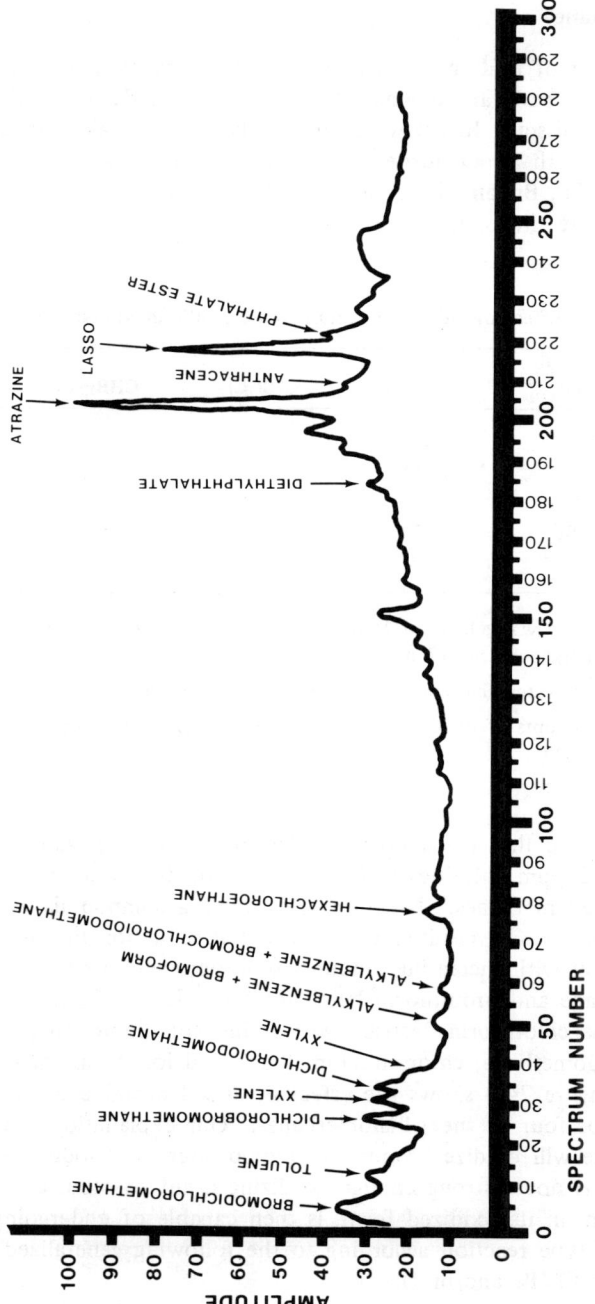

Figure 24.3 Jefferson City resin extract, computer-reconstructed gas chromatogram; OV-17 column programmed at $10°$/min from 70-240°C.

Trihalomethanes

As a result of finding six different trihalomethanes in drinking water, we decided to add various amounts of the four halides (F⁻, Cl⁻, Br⁻ and I⁻) to river (Missouri River) water and analyze for trihalomethanes after chlorination with hypochlorite.[14] The results of the addition of 1 ppm each of F⁻, Cl⁻, Br⁻ and I⁻ to Missouri River water which was subsequently chlorinated are shown in Table 24.4.

Table 24.4 Effect of Added Potassium Halide on Trihalomethane Formation

Halide Added[a]	$CHCl_3$	$CHBrCl_2$	$CHBr_2Cl$	$CHBr_3$
None.[b]	172	20	1	< 1
1 ppm F⁻	158	17	1	< 1
1 ppm Cl⁻	166	18	1	< 1
1 ppm Br⁻	21	35	30	50
1 ppm I⁻ [c]	67	15	1	< 1

[a]Potassium halide was added to Missouri River water which was then chlorinated with calcium hypochlorite (7 ppm available chlorine).
[b]Missouri River water chlorinated with calcium hypochlorite
[c]Substantial amounts of dichloroiodomethane, chlorodiiodomethane and iodoform were also formed.

It was found that the addition of KF and KCL (the water initially contained 22 ppm chloride) had little effect on the production of the trihalogenated methanes. No fluorine-containing compounds were detected. The addition of KBr resulted in a substantial reduction in chloroform concentration with increasing amounts of bromodichloromethane, dibromochloromethane and bromoform being formed. The addition of KI also resulted in a chloroform decrease with iodine-containing trihalomethanes (dichloroiodomethane, chlorodiiodomethane and iodoform) now being formed. Figure 24.4 shows the effect of added bromide ion on the formation of four of the trihalomethanes. Our explanation is that hypochlorite will oxidize bromide to hypobromite and iodide to hypoiodite, but is not a strong enough oxidizing agent to oxidize fluoride. The halogen, in its oxidized form, is then capable of undergoing the "haloform"-type reaction according to the following generalized reaction (where X = Cl, Br and/or I).

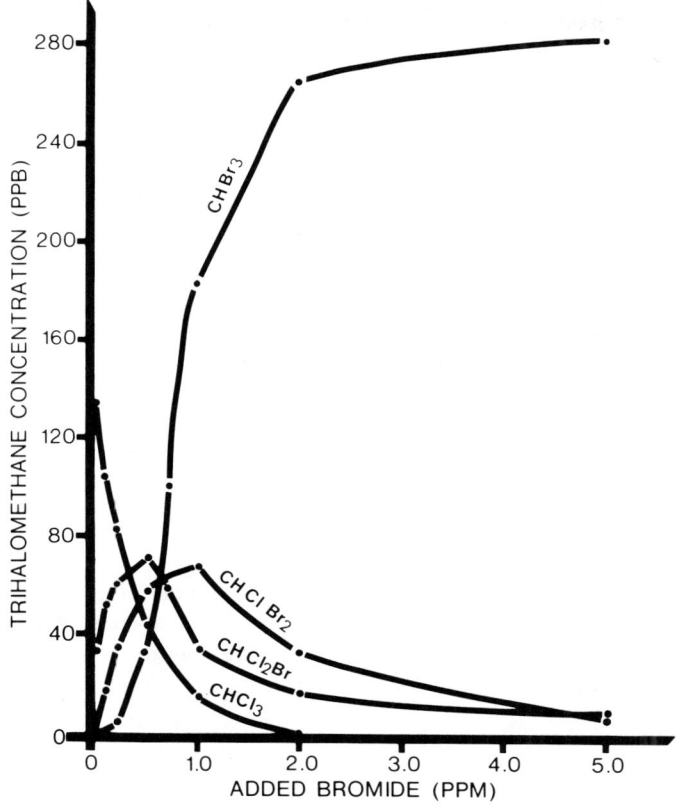

Figure 24.4 Effect of bromide ion on trihalomethane formation.

$$\underset{R-\overset{O}{\underset{\|}{C}}-CH_3}{} + 3\ OX^- \rightarrow \underset{R-\overset{O}{\underset{\|}{C}}-O^-}{} + CHX_3 + 2\ OH^-$$

Regardless of the specific nature of the precursor or precursors, this work shows that any of the ten possible trihalomethanes can be produced in the chlorination of natural water, particularly if significant amounts of iodides and bromides are present (Figure 24.5). Chlorination of Missouri River water containing 5 ppm bromide and 5 ppm iodide resulted in the formation of chloroform, bromodichloromethane, dibromochloromethane, bromoform, dichloroiodomethane, chlorodiiodomethane, iodoform, dibromoiodomethane, bromodiiodomethane and bromochloroiodomethane. These

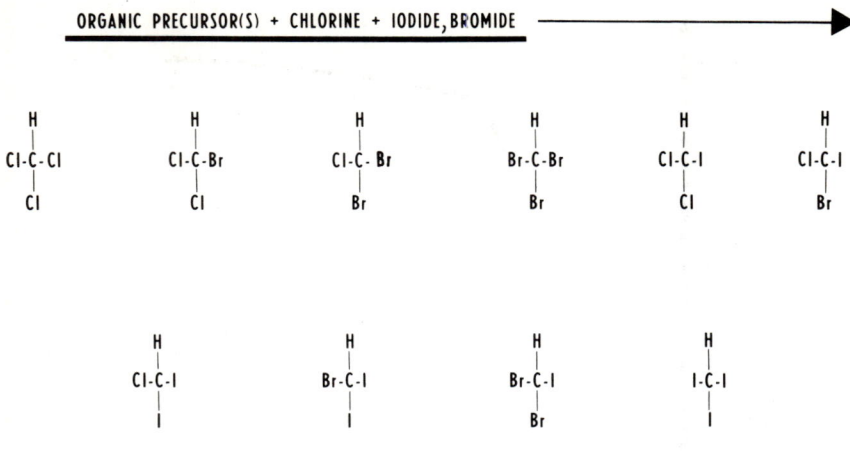

Figure 24.5 Formation of trihalomethanes during chlorination of surface water.

compounds were readily separated on an OV-17 column (Figure 24.6 shows all except the two most volatile compounds) and mass spectra were obtained for each of them (Table 24.5). The identity of dichloroiodomethane was confirmed by synthesis starting from 1,1,3,3-tetrachloroacetone.

CONCLUSIONS

To a large extent, the types of compounds found in drinking water originating from surface water reflect the types of activities occurring in the upstream areas. The Ohio River Basin is a heavily-industrialized area upstream from Evansville and consequently many industrial-type organic chemicals are found in downstream drinking water supplies. The Missouri River and the Kansas River, on the other hand, drain areas which are primarily agricultural and consequently herbicides are found in the finished drinking waters originating from those sources. However, the trihalomethanes are ubiquitous to chlorinated water which originates from surface sources (rivers, lakes). Chloroform is generally found in the highest concentration. However, if bromides or iodides are present in the raw water, the other trihalomethanes can be formed in significant concentrations upon chlorination. Obviously, because of their widespread occurrence, the long-term health effects of the compounds need to be determined.

IDENTIFICATION OF ORGANIC POLLUTANTS 413

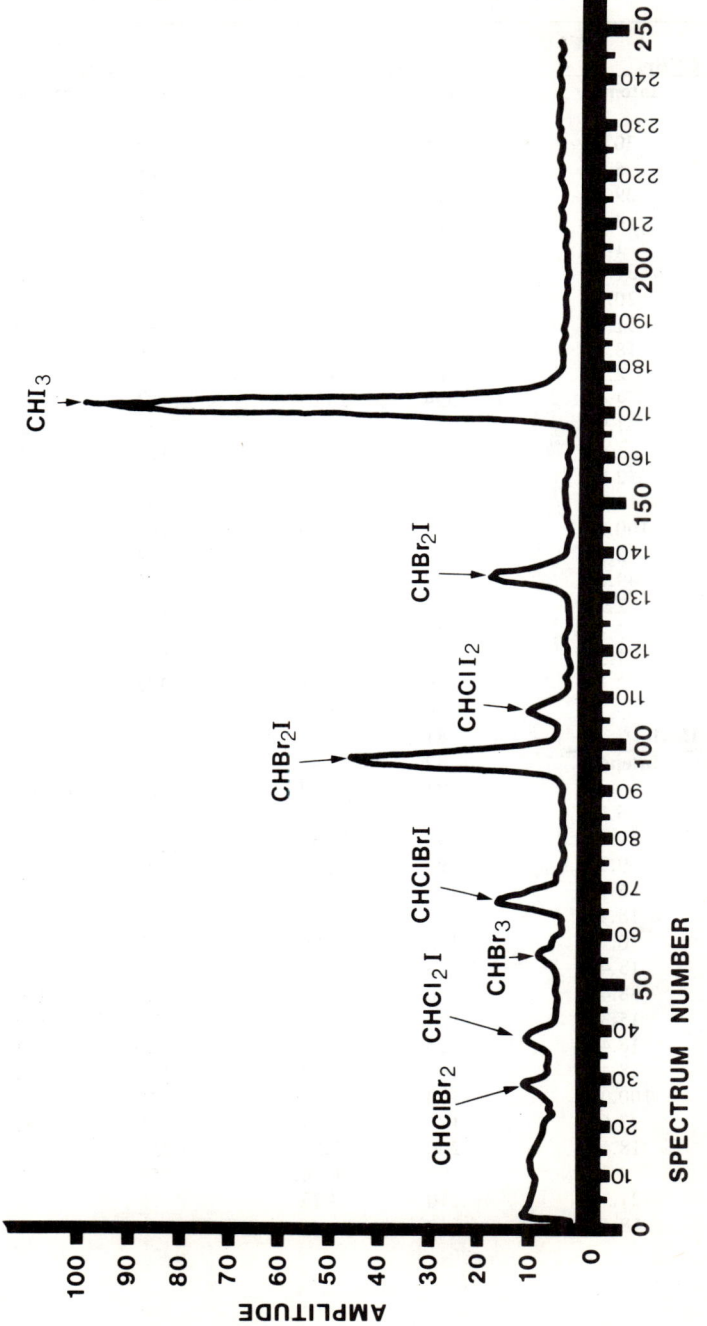

Figure 24.6 Chlorinated Missouri River water with addition of 5 ppm KBr and 5 ppm KI. Computer-reconstructed gas chromatogram on OV-17 column programmed at 10°/min from 70-240°C.

Table 24.5 Mass Spectra of Trihalomethanes

CHBr$_3$		CHI$_3$		CHClI$_2$	
Amu	Intensity	Amu	Intensity	Amu	Intensity
79	30.29	126	2.08	35	8.21
80	5.59	127	52.03	36	7.30
81	29.36	128	7.66	38	3.65
82	4.56	139	10.49	40	3.19
90	1.10	140	32.90	43	3.19
91	31.76	141	1.31	47	11.41
92	20.34	254	12.68	48	31.96
93	31.18	266	2.01	49	13.69
94	18.74	267	100.00	50	12.78
158	2.21	268	1.89	51	3.65
160	4.42	394	26.67	84	3.65
162	1.93			86	2.28
170	1.65	CHClBr$_2$		127	49.77
171	52.81	Amu	Intensity	128	9.13
172	3.94			139	7.30
173	100.00	35	3.45	140	9.13
174	2.55	36	1.29	174	5.47
175	49.12	37	1.18	175	100.00
250	2.97	47	23.90	177	33.33
252	7.39	48	27.87	197	3.65
254	7.39	50	8.63	254	9.13
256	2.35	79	24.08	302	13.69
		80	4.53	304	3.65
CHBrI$_2$		81	22.89		
Amu	Intensity	82	3.45	CHClBrI	
		91	13.17	Amu	Intensity
73	6.15	92	6.15		
79	9.23	93	12.85	47	13.63
81	9.23	94	5.50	48	26.13
91	6.15	126	3.67	50	10.22
92	18.46	127	79.21	57	5.68
93	6.15	128	4.96	79	9.09
94	18.46	129	100.00	81	9.09
127	88.46	130	2.15	127	100.00
128	15.38	131	24.94	128	9.09
139	18.46	158	1.29	129	78.40
140	28.46	160	2.59	131	22.72
219	100.00	173	2.15	175	13.63
221	97.69	175	1.29	177	9.09
254	18.46	206	2.59	254	13.63
267	9.23	208	4.96	256	9.09
346	21.53	210	3.23		
348	18.46				

Table 24.5, continued

$CHCl_2Br$		$CHBr_2I$		$CHCl_3$	
Amu	Intensity	Amu	Intensity	Amu	Intensity
35	8.91	79	17.35	35	14.55
36	3.34	80	6.42	36	4.75
37	3.34	81	18.00	37	4.78
38	1.11	82	5.14	38	1.61
47	30.08	91	19.61	41	1.64
48	19.77	92	23.46	47	34.92
49	10.02	93	20.25	48	16.07
50	6.12	94	22.18	49	12.08
79	10.02	127	54.65	50	5.27
80	1.94	128	5.14	56	2.22
81	10.02	139	10.28	69	1.03
82	4.45	140	9.00	82	6.28
83	100.00	171	54.32	83	100.00
84	3.34	172	3.85	84	5.58
85	64.34	173	100.00	85	65.58
86	1.39	175	49.18	86	1.40
87	10.30	206	3.85	87	10.73
91	3.06	208	3.85	117	1.12
93	3.06	219	7.71	118	1.52
94	1.11	221	7.71	119	1.06
126	1.11	298	6.42	120	1.43
127	6.96	300	14.46		
129	8.35	302	6.42		
131	2.22				

$CHCl_2I$			
Amu	Intensity	Amu	Intensity
35	6.21	126	1.55
36	3.49	127	34.58
37	1.94	128	3.10
47	19.81	139	2.33
48	22.73	140	1.55
50	6.79	162	2.33
82	3.88	175	11.07
83	100.00	177	3.88
85	63.27	210	12.62
87	11.07	212	7.57
		214	1.55

REFERENCES

1. Abramson, F. P. *Anal. Chem.* **44**, No. 14, 28A (1972).
2. Karasek, F. W. *Anal. Chem.* **44**, No. 4, 32A (1972).
3. Heller, S. R., J. M. McGuire and W. L. Budde. *Environ. Sci. Technol.* **9**, 210 (1975).
4. Junk, G. A., J. J. Richard, M. D. Grieser, D. Witiak, J. L. Witiak, M. D. Arguello, R. Vick, H. J. Svec, J. S. Fritz and G. V. Calder. *J. Chromatog.* **99**, 745 (1974).
5. Webb, R. G. EPA Research Report, EPA-660/4-75-003 (June 1975).
6. Eichelberger, J. W. and H. Stierli. "The Identification and Measurement of Chlorinated Hydrocarbon Pesticides in Surface Waters," U.S. Department of the Interior Publication WP-22 (November 1966).
7. Budde, W. L., ed. "Organic Analysis by Gas Chromatography/Mass Spectrometry," U.S. EPA (in press).
8. Bellar, T. A. and J. J. Lichtenberg. "The Determination of Volatile Organic Compounds at the μg/l Level in Water by Gas Chromatography," EPA-670/4-74-009, Superintendent of Documents, U.S. Government Printing Office, Washington, D. C. (November 1974).
9. Eichelberger, J. W., W. M. Middleton and W. L. Budde. "Analytical Quality Assurance for Trace Organics Analysis by GC/MS," EPA Report, EPA-660/4-75-007 (September 1975).
10. Eichelberger, J. W., L. E. Harris and W. L. Budde. *Anal. Chem.* **47**, 995 (1975).
11. Kleopfer, R. D. "Identification of Specific Organic Chemicals in the Ohio and Wabash Rivers, 1971-1972," EPA Region V Report.
12. Kleopfer, R. D. and B. J. Fairless. *Environ. Sci. Technol.* **6**, 1036 (1972).
13. *Standard Methods for the Examination of Water and Wastewater*, 13th edition (1971), p 264.
14. Bunn, W. W., B. B. Haas, E. R. Deane and R. D. Kleopfer. *Environ. Lett.* **10**, 205, No. 3 (1975).

CHAPTER 25

POSSIBLE FACTORS IN THE DRINKING WATER OF LABORATORY ANIMALS CAUSING REPRODUCTIVE FAILURE

J. D. McKinney, R. R. Maurer,
J. R. Hass and R. O. Thomas

> National Institute of Environmental Health Sciences
> Environmental Biology and Chemistry Branch
> Research Triangle Park, North Carolina

Periodically a detectable increase in reproductive failure has been noted in NIEHS laboratory animals, *i.e.,* mice and rabbits. This reproductive impairment was especially discernible in 1973 and 1974 during the month of February. The reproductive parameters affected were mating, number of embryos ovulated per female and embryonic development. An increase in abortion rate was occasionally seen. This was especially true for animals which were superovulated[1] as shown in Table 25.1 for CD-1 mice during March through November 1974, but was also detected in the same strain of mice for natural matings during November 1974 through March 1975 (Table 25.2). As indicated in these tables, the water supply is implicated as the problem since some statistically-significant differences were found in comparing the reproductive parameters of mice receiving tap water or specially-prepared purified water as their drinking water. The purified water was prepared by passing tap water through a Corning 3508-ORC organic removal cartridge and a Corning 3508-B demineralizer in tandem with a Corning "Meg-Pure" glass still and collection with all-glass tubing and bottles. Although measures have been taken to guarantee the water quality for the laboratory animals, it was considered of some importance to attempt to identify the possible causative agent(s)

Table 25.1 Superovulatory Response of Female Mice (CD-1 Strain) Drinking Tap Versus Purified Water
(Ten observations were made during March through November 1974)

	Tap Water	Purified Water	Significance Level
No. Females	96	111	
Percent Mating	56	65	NS
Percent Plugged Females with Embryos	81	89	NS
Average Number Embryos/ Females with Embryos	13.8 ± 1.4	19.9 ± 1.9	$P < 0.05$
Average Percent Blastocysts	66.0 ± 5.1	63.5 ± 3.9	NS

$\bar{X} \pm SE$ NS = nonsignificant ($P > 0.05$).

Table 25.2 Effects of Water on Reproduction in CD-1 Mice
(Natural Mating for November 1974-March 1975)

	Tap Water	Purified Water	Significance Level
No. Females	49	61	
No. Females Pregnant (%)	47 (95.9)	59 (96.7)	
No. Implantation (Average)	601 (12.8 ± 8.04)	776 (13.2 ± 0.3)	NS
No. Fetuses (Average)	542 (11.5 ± 0.4	725 (12.3 ± 0.3)	NS (P = 0.155)
No. Fetuses Resorbed and Dead (Average %/litter)	59 (9.5)± 1.2)	51 (6.7 ± 1.1)	$P < 0.05$
No. Malformed Fetuses (%)	7 (1.2)	0 (0)	
No. Litters with Malformed Fetuses (%)	5 (10.6)	0 (0)	$P < 0.05$

NS = nonsignificant ($P > 0.05$).

Statistics based on litter means ($\bar{X} \pm SE$) and analyzed with a two-sided Mann-Whitney U test.

Malformations included (4) cleft palate, (1) fused ribs, (1) meningeoencephalocele and (1) septal defect.

in the tap water. The Institute is provided water by the City of Durham which currently uses the Flat River as its source. This river provides an abundant supply of good quality water relatively free of industrial pollution and is within 12 to 15 miles from the city.

Our considerations of the possible causative factors have included both dissolved inorganics and organics and microparticulates as well as microorganisms. The concern for inorganics primarily focused on metal ions, especially the heavy metals, although the usual anions (PO_4^{-3}, SO_4^{-2} and Cl^{-1}) determined at the water treatment plant are present but not at objectionable levels. The metal analysis included both individual metal determinations using atomic absorption (AA) and total metal scans using neutron activation analysis (NAA). In both cases little or none of the more toxic heavy metals were found. The highest levels found were for zinc (2.5-170 ppb) and copper (70-790 ppb). Iron and manganese are generally present in the raw water at objectionable levels but are removed to acceptable levels (iron, 49 ppb; manganese, 6 ppb) during treatment. Reproductive studies in mice with dosages in water equivalent to 500 ppb of copper and/or zinc failed to implicate these metals as causative agents alone, although copper alone produced a slight reduction (not statistically significant) in the number of embryos per plugged female in comparison to controls.

Microparticulate matter was collected for asbestos fibers analysis by passing tap water through a teflon-impregnated glass fiber filter for 24 hr at 100 ml/min. The paper was examined microscopically for fibers before and after treatment with aqua regia. Surprisingly little was found except for very scattered pieces of what appeared to be plant detritus. Microparticulate black substance was trapped in the fibers and may be identical to particulate matter collected on a glass wool plug used in a glass column for trapping organics in the tap water. This latter plug was examined by neutron activation analysis and shown to be high in iron content when compared to a control glass wool plug.

Coliform bacteria were shown to be absent in 100 ml of tap water at the Institute and only a few colonies of other bacteria were found on a total plate count. Other bacteria-specific tests were performed on the purified water and indicated the occasional presence of several *pseudomonas* species. The kinds and numbers of bacteria found in either case were not considered objectionable or likely to account for the observed reproductive alterations.

Since there was no compelling evidence to implicate dissolved inorganics, microparticulate matter or microorganisms as the causative agents in the water, our attention turned to the dissolved organics. In view of the fact that our water supply is relatively free of industrial pollution and the recent concern over the formation of chlorinated hydrocarbon organics during chlorination of diverse water supplies in treatment processes, our initial efforts focused on determination of the volatile organics, in particular the haloforms.

In cooperation with laboratories of the EPA, qualitative analyses (some quantitative) for volatiles (VOA) were performed on samples of Institute water from the two primary animal buildings and the purified water previously described. The samples were purged according to reported procedures* using either a 5-ml stripping device (Method A—sensitivity estimated at 10 ppb) or a 500-ml stripping device (Method B—sensitivity estimated at about 0.01 ppb). All of the quantitation was done on a Perkin-Elmer 900 gas chromatograph (GC) equipped with a flame ionization detector, employing Tenax and OPN-Poracil C as GC column liquid phases. Of the 31 compounds identified by low resolution mass spectral analysis (Table 25.3—water sample taken on May 7, 1975) using a Finnigan mass spectrometer, 13 contained chlorine. More significant is the fact that of the 31 compounds identified in the tap water from each building, 11 (indicated in Table 25.3) were not present in the purified water. Of these 11 compounds, 7 contained chlorine. Chloroform was found in the purified water; however, its concentration was significantly lower (about 25 X). Although all of the halogenated compounds are of some interest from a biological point of view, our initial studies were with the haloforms, in particular chloroform, since they are clearly more concentrated and have received recent attention[2,3] for their possible chronic toxicity. Other compounds planned for study include fluorotrichloromethane, the cyanide-containing compounds, cyanogen chloride[4] and dichloroacetonitrile, as well as simple combinations of all of these.

Since the animals consuming the purified water showed significantly less reproductive alteration in some parameters and the effects appeared to be most pronounced in the month of February (generally the coldest month of the year for this area), water samples were normally collected at least biweekly from January to October 1975 in glass bottles, overfilled and stoppered with rubber to eliminate dead space, and stored under refrigeration for future analysis of chloroform. Variations in the chloroform levels with season would provide information on the range of concentrations to which animals could be exposed as well as what month or season the maximum exposure would occur. In addition, the chloroform measurement might serve as a marker for quantitating total chlorinated organics since other workers[5] in the Durham area have shown that chloroform is a product of water chlorination (a prechlorination

---

*See Chapter 6, this volume.

Table 25.3 Volatile Organic Compounds Present in Purified Water and Tap Water of NIEHS (May 1975)

		Purified Water		Tap Water Bldg. #2		Tap Water Bldg. #15	
		Identification	Conc, μg/l	Identification	Conc, μg/l	Identification	Conc, μg/l
1.	Acetaldehyde	A + B		A + B		A + B	
2.	Acetone	A + B		A + B		A + B	
3.	Benzene	B		B		B	
4.	Bromodichloromethane[a]			A + B	10.99(O)	A + B	10.99(O),11.60(O)
5.	1-Butanol	B		B		B	
6.	2-Butanone	B		B		B	
7.	Carbon Disulfide	A + B		A + B		A + B	
8.	Chlorodibromomethane[a]			A + B	2.16(T),3.40(T)	A + B	2.16(T),3.22(T)
9.	Chloroethane	B		B		B	
10.	Chloroform	A + B	3.14(T)	A + B	75.66(T),89.08(O)	A + B	80.01(T),82.12(O)
11.	Chloromethane	B		B		B	
12.	Cyanogen chloride			A + B		A + B	
13.	Dichloroacetonitrile[a]			A + B	>1.0(T), >1.0(T)	A + B	>1.0(T), >1.0(T)
14.	1,1-Dichloroethene[a]			B		B	
15.	Dichloromethane	A + B		A + B	9.0(O)	A + B	9.0(O)
16.	Diethyl ether	B		B		B	
17.	Dimethyl ether	B		B		B	
18.	Ethanol	B		B		B	
19.	Fluorotrichloromethane[a]			B		B	
20.	Methyl formate[a]			B		B	
21.	Methanol	B		B		B	
22.	3-Methylbutanal	B		B		B	
23.	2-Methylpropanal	B		B		B	
24.	2-Methyl-2-propanol	B		B		B	
25.	Nitromethane[a]			B		B	
26.	2-Pentanone			B		B	
27.	Propanal[a]			B		B	
28.	Tetrachloroethene	A + B		A + B		A + B	
29.	Tetrachloromethane (CCl$_4$)[a]			B		B	
30.	Toluene	A + B	1.02(T)	A + B	0.68(T),0.61(T)	A + B	0.62(T),0.45(T)
31.	1,1,1-Trichloroethane[a]			B		B	

[a]Denotes those compounds found only in tap water.

A = Extraction with 5-ml VOA   B = Extraction with 500-ml VOA   T = Tenax + FID   O = OPN-Poracil + FID

process is used), as it was not found in the raw water but was detectable within 30 sec after the treatment began.

Although there are several methods which can apparently meet the sensitivity requirements for analyzing tap water for chloroform, there is only one (direct aqueous injection) which can be applied to routine analysis of such large numbers of samples without extensive sample handling. Direct aqueous injection GC/MS has been previously[6] explored as a supplement to other methods for organic pollutants in water and was used by us (low-resolution Finnigan GC/MS instrument using selective ion monitoring for m/e 83 and 85 and a Porapak Q column operated isothermally at 170°C; 5 μl water injected directly with water vented) for confirmation. Since flame ionization/gas chromatography does not have the desired sensitivity,[7] it was decided to use an electron capture (EC) detector and develop our own GC conditions. Further, it was decided to use an internal standard (methylene chloride—retention time about 1/2 that of chloroform) to minimize effects of sample losses due to the high volatility and low water solubility of chloroform and the sensitivity drift of the EC detector. For these reasons, the use of an internal standard was considered an advantage over a similar recently reported[8] method.

The analyses were performed using a Tracor MT-220 gas chromatograph equipped with a ^{63}Ni electron capture detector operated at 235°C. The inlet was maintained at 230°C and the column (6 ft x 1/4 in. x 2 mm i.d. glass packed with Carbowax 400/Porasil C Durapak) temperature at 80°C. Under these conditions water is retained for approximately 30 min which permits the analysis of three samples in duplicate before the water elutes. The column flow was 50 ml/min and the detector purge 30 ml/min of nitrogen. Instrument response was determined by measuring the ratio of [peak height (p.h.) x retention time (r.t.)] for chloroform ($CHCl_3$) to the same parameter of methylene chloride (CH_2Cl_2) for a series of standard solutions containing a constant concentration of CH_2Cl_2 and various concentrations of $CHCl_3$. The ratio (p.h. x r.t.) $CHCl_3$/(p.h. x r.t.)CH_2Cl_2 was plotted as a function of [$CHCl_3/CH_2Cl_2$]. The resulting calibration curve is linear for more than an order of magnitude and covered the concentration range of interest.

The unknown samples were spiked with 3 ml of a stock CH_2Cl_2 solution to give a final [CH_2Cl_2] equal to that used in the instrument standardization. The ratio (p.h. x r.t.) $CHCl_3$/(p.h. x r.t.) CH_2Cl_2 was measured in duplicate for each water sample by direct injection. These data were averaged and then converted into [$CHCl_3$] and tabulated (Table 25.4) by date. Calculations based on previously reported[9] data on the evaporation rate of chloroform in dilute aqueous solutions

Table 25.4 Chloroform[a] Concentrations Found in NIEHS Tap Water by GC-EC from January to October, 1975 by Day and by Monthly Average

Date	($CHCl_3$), ppb	Date	($CHCl_3$), ppb	Date	($CHCl_3$), ppb
1/20/75	299	2/25/75	670	6/17/75	504
1/21/75	635	2/26/75	587	6/18/75	718
1/22/75	246	2/27/75	863	6/24/75	739
1/23/75	635	2/28/75	408	6/26/75	559
1/24/75	504	3/6/75	435	6/29/75	718
1/25/75	245	3/8/75	442	7/1/75	815
1/26/75	309	3/9/75	394	7/3/75	890
1/27/75	187	3/11/75	374	7/8/75	775
1/28/75	615	3/13/75	456	7/10/75	794
1/29/75	546	3/18/75	137	7/14/75	897
1/30/75	259	3/20/75	649	7/17/75	1100
1/31/75	406	3/25/75	546	7/22/75	1340
2/1/75	168	3/27/75	435	7/25/75	1220
2/2/75	242	4/1/75	587	7/29/75	1590
2/3/75	422	4/3/75	690	7/31/75	1220
2/4/75	204	4/10/75	162	8/5/75	1310
2/5/75	428	4/22/75	656	8/12/75	1390
2/6/75	670	4/24/75	663	8/14/75	1500
2/7/75	442	4/29/75	615	8/14/75	1340
2/8/75	153	5/1/75	697	8/28/75	1380
2/9/75	56	5/6/75	808	9/4/75	1170
2/10/75	408	5/8/75	559	9/9/75	1240
2/11/75	387	5/13/75	622	9/17/75	1170
2/12/75	401	5/16/75	532	9/18/75	1080
2/13/75	200	5/16/75	580	9/22/75	1080
2/14/75	332	5/20/75	504	9/25/75	980
2/15/75	106	5/22/75	218	9/30/75	1160
2/16/75	137	5/27/75	566	10/6/75	870
2/18/75	649	6/3/65	642	10/1/75	884;1280
2/19/75	656	6/5/75	828	10/8/75	1160
2/20/75	690	6/10/75	378	10/9/75	1050
2/21/75	746	6/11/75	690	10/21)75	1430
2/24/75	608	6/12/75	725		
January	407.2	May	565.1	September	1125.7
February	425.3	June	650.1	October	1112.3
March	429.6	July	1064.1		
April	562.2	August	1384.0		

[a]Solubility of $CHCl_3$ in water at $25°C$ is about 7420 ppm.

indicated that storage losses should not have exceeded 25% for the samples stored the longest and correspondingly lower for more recent samples even if the samples had been stored at 25°C (samples were stored at 5°C).

Although the dates in Table 25.4 indicated considerable day-to-day variations in $CHCl_3$ levels, the lowest average (407 ppb) was found for the month of January and the highest average (1384 ppb) for the month of August. Direct aqueous injection GC/MS analysis of several water samples from the month of August gave values approximately a factor of two lower than those obtained by GC/EC. A more recent sample from the month of November when compared by the two methods again differed by a factor of approximately two. The GC/MS data might be considered more reliable for quantitative purposes since this detector is less likely to suffer interferences, but there is no obvious explanation for the differences in the two methods. Applying a factor of two to the values given in Table 25.4 would reduce the range from 204 ppb to 797 ppb. However, if the average of the values given for May 1975 in Table 25.4 is divided by two (283 ppb) and compared with the average (82 ppb) of the values given in Table 25.3 for the water sample of May 7, 1975 analyzed by the VOA method of EPA, there is a factor of three to four difference which might be explained on the basis of day-to-day variations.

In an attempt to reconcile these differences, the November water sample previously compared by GC/EC (131 ppb) and GC/MS (64 ppb) in our laboratory was analyzed by the 5-ml VOA method and GC using a flame ionization detector (duplicate determinations, 135 and 148 ppb), by head space analysis using a Hall detector (average of three areas, 165 ppb) at the EPA Environmental Research Laboratory, Athens, Georgia. The values obtained were in good agreement for all methods except the GC/MS method which again was a factor of about two lower than the rest. Thus, the GC/EC analytical technique used in this work appears to be of acceptable reliability, and the differences found between methods (except for the GC/MS method) are most likely due to considerable variations in $CHCl_3$ concentration from day to day (all samples collected in the morning). A trend in $CHCl_3$ levels can be seen if weekly averages are plotted (Figure 25.1) and was shown to correlate reasonably well with temperature [lowest averages in January (43.8°F) and February (43.7°F) with the highest average in August (78.3°F)] and very well ($R = 0.92$) with chlorine feed.

The correlation with chlorine feed might be expected since chloroform in the water is a direct result of water chlorination and feed is actually determined by maintaining a fairly constant (2-2.5 ppm) chlorine residual

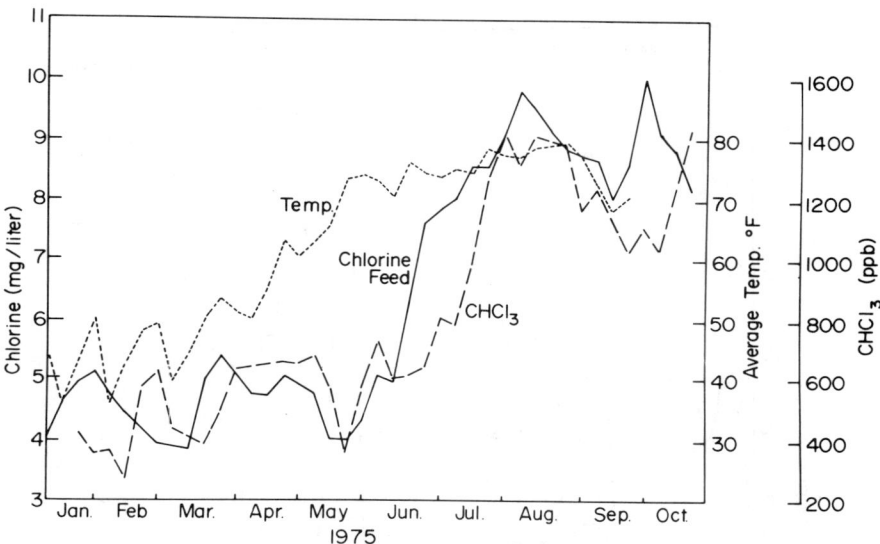

Figure 25.1 Correlation of temperature and chlorine feed with chloroform concentrations in NIEHS tap water samples.

[measured by DPD (N,N-diethyl-p-phenylenediamine) and amperometric methods]. The chlorine feed would then reflect changes both in total oxidizable dissolved organics and in rates of various oxidizing reactions (including chlorination by HOCl species). The effects of pH on the chlorination reactions are considered minimal since the pH remains fairly constant (5.8-6.4) in the plant mixing basin. The chlorine feed appears to be inversely proportional (Figure 25.2) to dissolved oxygen (measured by the Winkler method). A fairly constant oxidizing capacity would be maintained throughout the year, but a proportionately higher number of the oxidizing reactions would lead to chlorinated compounds (as measured by $CHCl_3$ determination) when chlorine feed is high. That dissolved oxygen also has an effect on the $CHCl_3$ levels found can be shown by multiple linear regression analysis of the numerical data. Such analysis of turbidity data (measured photometrically) showed that it was unrelated. The turbidity is highly variable, and extreme increases are most likely associated with heavy rainfall (Table 25.5).

In order to get a more complete picture of the dissolved trace organics in NIEHS tap water, water samples on several occasions were subjected to carbon (Calgon Corporation, Filtersorb 300) and XAD-2 resin (Rohm and Haas) adsorption techniques. The sample time was normally 24 hr

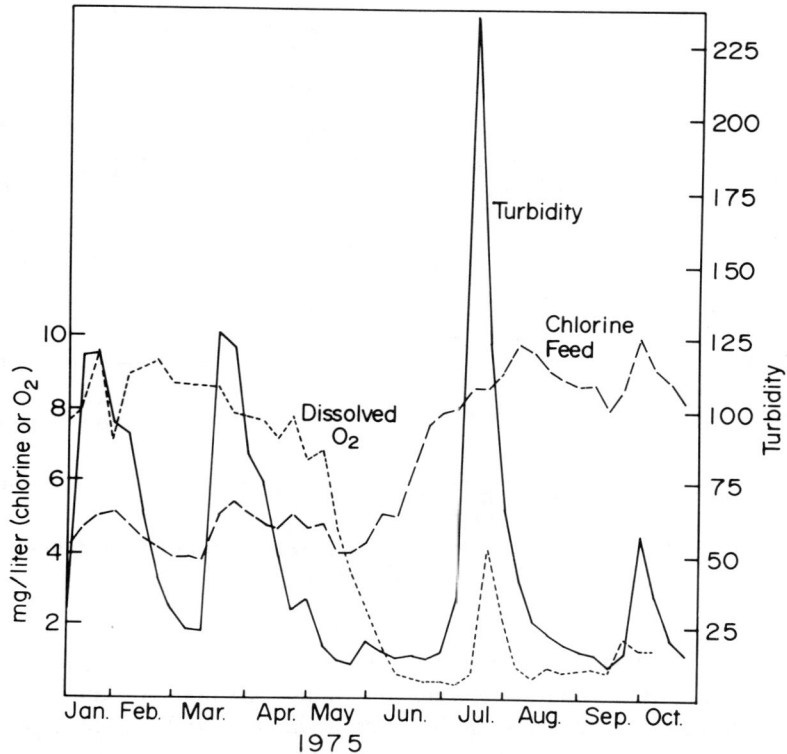

Figure 25.2 Correlation of turbidity, dissolved oxygen and chlorine feed in NIEHS tap water samples.

with a water flow through the column (1/2 in. tube filled to height of 6 cm with "clean-up" carbon or resin) equivalent to 29 gal for carbon and 39 gal for resin. The extraction and evaporation procedures were as previously described.[10,11] Control experiments (solvent blanks, etc.) were also performed. The samples were analyzed using a Finnigan GC/MS system[12] with a 2-mm x 10-ft glass column of 3% SP 2100. The GC column was held at 40° for a period of 10-11 min and then programmed at 4°/min to 300° and held for at least 5 min. Figure 25.3 shows reconstructed gas chromatograms for the dissolved trace organics from a typical comparison of resin and carbon adsorption (October 1975). Structural information (compounds were not fully identified) could be deduced from the same number (31) of halogenated organic compounds

Table 25.5 Rainfall (Variations in Water Equivalent)[a] and Temperature (°F)[b] for January-August 1975

	Rainfall	Temperature
January	6.09 (2.87)	43.8 (3.3)
February	2.85 (-0.47)	43.7 (1.5)
March	6.26 (2.82)	47.1 (-2.1)
April	1.64 (-1.43)	55.8 (-3.7)
May	3.84 (0.52)	68.2 (0.8)
June	1.66 (-2.01)	73.7 (-0.7)
July	6.74 (1.65)	75.6 (-1.9)
August	2.11 (-2.82)	78.3 (1.8)

[a]Water equivalent is in inches with ± deviations from mean in parentheses.

[b]Temperature is average degrees F with ± deviations from mean in parentheses; range from highest to lowest is 34.6°.

for both adsorption methods. However, there were only 10 compounds which appeared to be common to both. Therefore, structural information can be deduced for a larger number of compounds using both adsorption techniques than for either used alone.

The abundance of the halogenated compounds (these compounds appeared to constitute only a small percentage of the total organic content of the water) is remarkable and many appear to be of the trichloromethyl (Figure 25.4) or dichloromethyl (Figure 25.5) variety suggesting their possible formation from the haloform-type reaction during chlorination. Hexachlorocyclopentadiene was found on one occasion suggesting that others are probably not products of water chlorination. Many of the nonhalogenated compounds appeared to be characteristic of long-chain aliphatic hydrocarbons. Still others are expected water contaminants such as the phthalate esters. Interestingly, alkyl naphthalenes were also found.

The combination of chlorine feed, temperature and dissolved oxygen data may have a predictive value for determining $CHCl_3$ levels which may serve as a marker for the total concentration of chlorinated hydrocarbons resulting from water chlorination.[13] For example, multiple linear regression analyses of these data for our water system afforded an estimated equation [$CHCl_3$ (ppb)= -1156.69 + 149.39 chlorine (mg/l) + 12.99 temperature (°F) + 40.50 oxygen (mg/l)] which predicted the weekly chloroform levels within 186 ppb with one exception. This would be most characteristic of an area or water supply if it were expressed as the

428 ORGANIC POLLUTANTS IN WATER

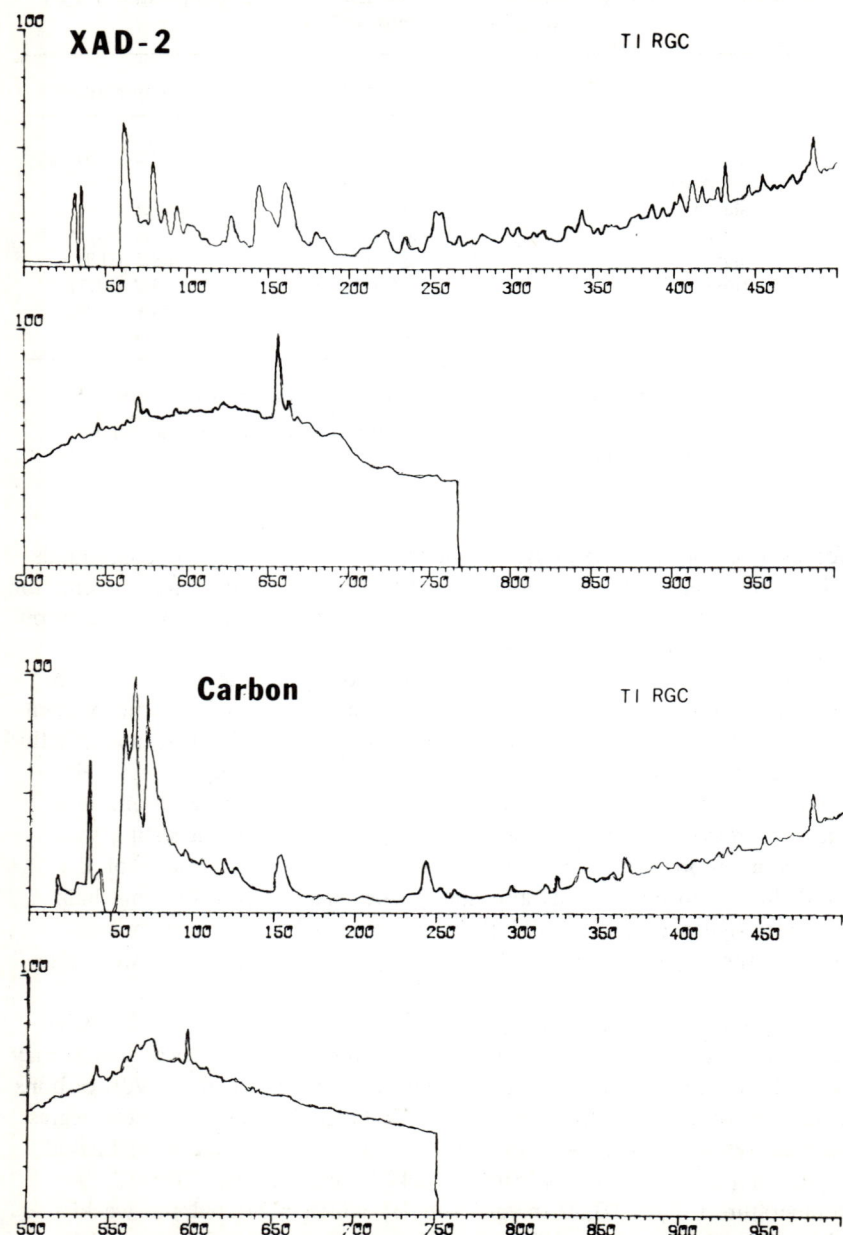

Figure 25.3 Comparison of XAD-2 resin extract and carbon chloroform extract of NIEHS tap water. Computer-reconstructed mass chromatograms using total ion current summation.

IDENTIFICATION OF ORGANIC POLLUTANTS 429

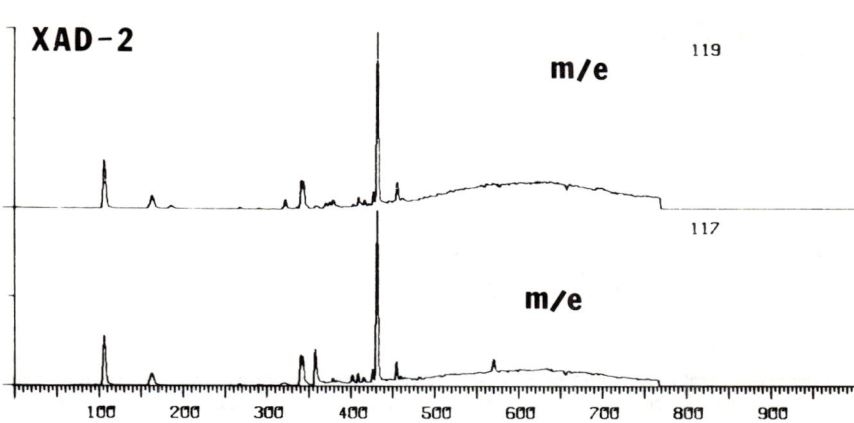

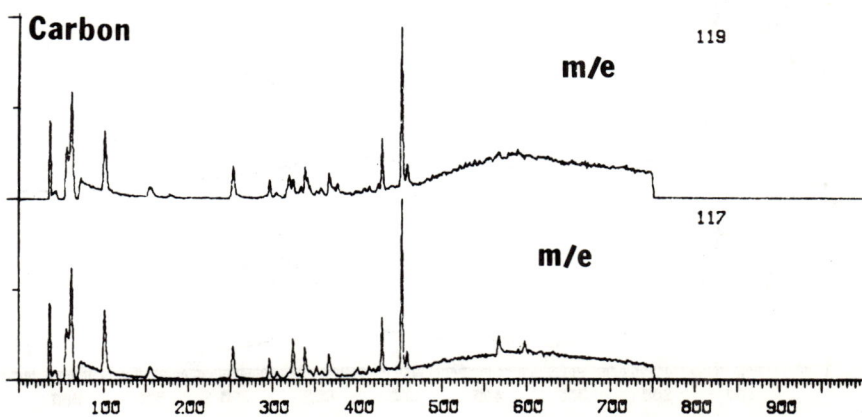

Figure 25.4 Comparison of XAD-2 resin extract and carbon chloroform extract of NIEHS tap water. Computer-reconstructed mass chromatograms using m/e 117 and 119 (CCl_3 fragment).

430 ORGANIC POLLUTANTS IN WATER

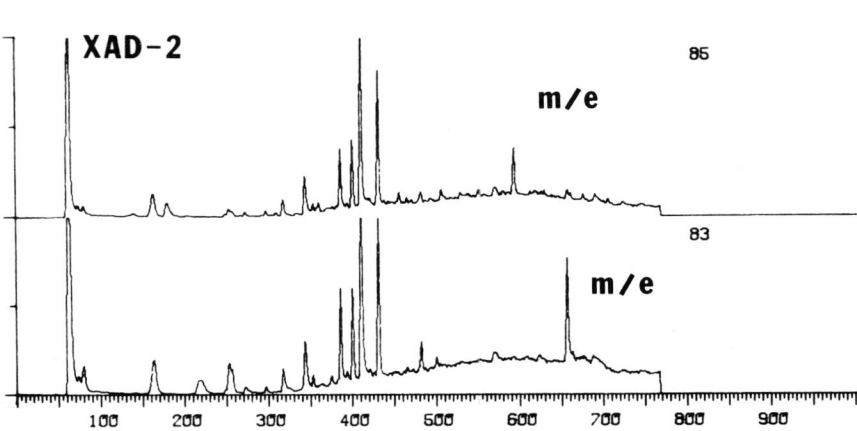

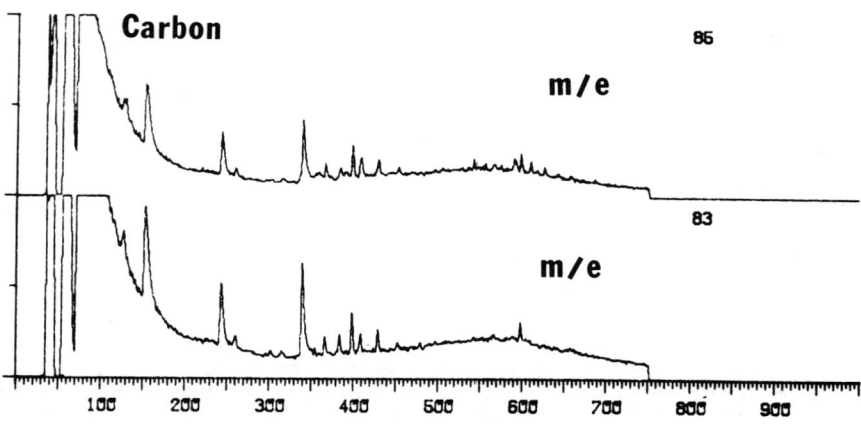

Figure 25.5 Comparison of XAD-2 resin extract and carbon chloroform extract of NIEHS tap water. Computer-reconstructed mass chromatograms using m/e 83 and 85 ($CHCl_2$ fragment).

ratio of $CHCl_3$ to total organics (TOC). This ratio may vary more for areas or water supplies which have a predominance of electrophilic chlorine addition to activated double bonds (such as the haloform reaction) which may be true of our water supply than for areas which have a predominance of electrophilic aromatic chlorination such as may be found in more industrially-contaminated waters.

The reproductive failure in laboratory animals associated with NIEHS tap water cannot be attributed to $CHCl_3$ nor possibly other chlorinated organics in the water since they are in lowest concentration when reproductive failure is highest. However, the high $CHCl_3$ concentrations found during the summer months could affect some reproductive parameters.

Preliminary mice reproductive studies with chloroform in their drinking water at concentrations of 0, 152, 760 and 3800 ppb (equivalent to 0, 40, 200 and 1000 µg/kg body weight for a 30-g mouse) showed significant decreases in at least two of the reproductive parameters normally obtained *i.e.,* dose-related decrease in embryonic development, and at the 152-ppb level decreased percentage of females mating.

However, it is possible that the reproductive failure is associated with higher total organics content found in the colder months since they may not be as rapidly degraded (temperature variation of approximately 20°C can vary chemical reaction rates three to four times) on chlorination during this time to the more volatile small molecules or may be less exhaustively chlorinated to more chemically (and biochemically) reactive intermediates. The general effects of temperature on solubility properties of organics in water should also be considered, but may be of less importance. Measurements of chemical oxygen demand (COD-related to TOC) for February and August, 1975 tap water in terms of microequivalents of dichromate per microliter (February, 14.6 ± 5.3; August 16.2 ± 2.5) of water have indicated little difference in COD. This would suggest that the difference in February and August water is not due to the amount of organics present but to the nature of the organic compounds present. On this basis, the February water may contain less chlorinated hydrocarbon compounds, and a greater percentage of the total organics present may be larger organic molecules.

Other possible causative factors include unidentified natural products, as well as the synergistic potential of various contaminants including those identified. On the other hand, the biological significance[14] of the small but finite effects on laboratory animal reproductive processes is unknown but should receive further study.

ACKNOWLEDGMENTS

We would like to thank W. Emile Coleman, Frederick C. Kopfler and Robert G. Tardiff, EPA, National Environmental Research Center, Cincinnati, Ohio for collecting and tabulating the data on volatile organics found in Table 25.3. We also thank Lawrence H. Keith and Myron Stephenson, EPA, Southeast Environmental Research Laboratory, Athens, Georgia for providing a supply of carbon and XAD-2 resin for column adsorption studies, aiding in computer mass spectral matching for compound identification, and for individual chloroform analyses by the 5-ml VOA and head space analysis methods. We further thank Joseph K. Haseman for statistical analyses of some of the data, and John Fawkes and Michael Walker for technical assistance.

REFERENCES

1. Elliott, D. S., R. R. Maurer and R. E. Staples. *Biology of Reproduction* **11**, 162-167 (1974).
2. Schwetz, B. A., B. K. J. Leong and P. J. Gehring. *Toxicol. Appl. Pharmacol.* **28**, 442-451 (1974).
3. Thompson, D. J., S. D. Warner and V. B. Robsinson. *Toxicol. Appl. Pharmacol.* **29**, 348-357 (1974).
4. Allen, L. A., N. Blezard, and A. B. Wheatland. *J. Hygiene* **46**, 184-195 (1948).
5. Christman, R. F. and F. K. Pfaender. University of North Carolina at Chapel Hill, Private Communication (1975).
6. Harris, L. E., W. L. Budde and J. W. Eichelberger. *Anal. Chem.* **46**, 1912-1917 (1974).
7. Sugar, J. W. and R. A. Conway. *J. Water Pollution Control Fed.* **40**, 1622 (1968).
8. Nicholson, A. A. and O. Meresz. *Bull. Environ. Contam. Toxicol.* **14**(4), 453-456 (1975).
9. Dilling, W. L., N. B. Tefertiller and G. J. Kallos. *Environ. Sci. Technol.* **9**(9), 833-838 (1975).
10. Grob, K. J. *J. Chromatog.* **84**, 255 (1973).
11. Junk, G. A., J. J. Richard, M. D. Grieser, D. Witiak, J. L. Witiak, M. D. Arguello, R. Vick, H. J. Svec, J. S. Fritz and G. V. Calder. *J. Chromatog.* **99**, 745-762 (1974).
12. U.S. Environmental Protection Agency. "Current Practice in GC/MS Analysis of Organics in Water," Environmental Protection Technology Series, EPA-R2-73-277 (August 1973).
13. U.S. Environmental Protection Agency. "Formation of Halogenated Organics by Chlorination of Water Supplies," Environmental Health Research Series, EPA-600/1-75-002 (March 1975).
14. Erhardt, C. L., F. G. Nelson and J. Pakter. *Amer. J. Public Health* **61**(11), 2246-2258 (1971).

CHAPTER 26

ANALYSES OF ORGANIC CONSTITUENTS IN WATER BY HIGH-RESOLUTION GAS CHROMATOGRAPHY IN COMBINATION WITH SPECIFIC DETECTION AND COMPUTER-ASSISTED MASS SPECTROMETRY

Walter Giger, Martin Reinhard,
Christian Schaffner and Fritz Zürcher

Swiss Federal Institute for Water
   Resources and Water Pollution
   Control (EAWAG)
CH-8600 Dubendorf
Switzerland

INTRODUCTION

The impact of organic constituents on the aquatic environment is of growing concern for ecological and hygienic reasons. With a few exceptions routine assessments of water quality are based on collective parameters such as total organic carbon or biological oxygen demand. But, since most biochemical reactions show a very pronounced structural dependence, studies on chemical ecology necessitate analyses for single constituents. Such investigations are hindered by two intrinsic properties of organic water constituents. First, the organic assemblages in environmental samples are of an extraordinarily high compositional complexity; and second, single components occur in trace quantities only. Therefore, very efficient enrichment, separation and detection techniques are needed.

The separation efficiency of capillary gas chromatography by far exceeds what presently is achievable by other methods. This analytical technique thus offers the best possibility to cope with highly complex mixtures of organic compounds. The limitations, however, are given by the necessary minimum volatility and thermal stability of the components amenable to gas chromatography.

This chapter reports on analyses of organic constituents in water utilizing gas chromatography with glass capillary columns. Detection and identifications are achieved by flame ionization, electron capture or computerized mass spectrometry. By applying two different enrichment procedures, namely closed-loop gaseous stripping and liquid-liquid extraction, a broad variety of water samples can be studied. Analyses of primary and secondary effluents, rain water, river water, and lake water are presented. The results are discussed with respect to both possible sources and to the fate of organics in water treatment processes and in the aquatic ecosphere.

METHODS

Enrichment

Effluents of primary and secondary sewage treatment were filtered through a glass-fritted funnel. The filtrates were then extracted by simple liquid-liquid partitioning into methylene chloride. The extracts were separated into three fractions by adsorption column chromatography on silica.[1] The nonpolar constituents eluted with pentane and methylene chloride were analyzed by glass capillary/gas chromatography after evaporative concentration. The more polar constituents were eluted with methanol, evaporated to dryness, and taken up in methylene chloride.

A rain water sample of 13 liters was concentrated to a volume of 250 ml on a rotary evaporator without removing the particulate matter. After extraction with methylene chloride the sample was subjected to an isolation procedure utilizing gel filtration and adsorption chromatography.[2] Sufficiently pure aromatic hydrocarbon concentrates are rapidly achieved by this combination of methods responding to different aspects of the physical and chemical properties of the sample.

Volatile organic constituents from river and lake water were enriched by a closed-loop gaseous stripping/adsorption/elution procedure developed by Grob.[3,4] The main advantage of this method is the high enrichment factor ($1:10^6$) which enables the detection of traces in the ng/l range. Since the gas chromatographic analyses can be performed without prior evaporation, low boiling components are quantitatively extracted as well.

It should be emphasized that no general method exists for enrichment of organics from water. The technique has to be chosen according to the type of water (*e.g.*, total organic carbon content) and considering the properties of the components to be studied (volatility, polarity, etc.).

Gas Chromatography

Gas chromatography was performed on Carlo Erba instruments equipped with glass capillary columns and Grob-type injectors.[5] The glass capillaries, coated with OV-101 and Ucon HB, were supplied by H. & G. Jaeggi, CH-9043 Trogen, Switzerland.

1-2 µl of the samples were injected without stream splitting onto the column at ambient temperature. After 30 sec the split valve was opened allowing the septum and injection port to be purged at a flow rate of 10-15 ml/min. Subsequent to the elution of the solvent, the oven temperature was raised with varying temperature programs. Hydrogen was used for carrier gas. In our GC procedure we followed the description given by Grob and Grob.[5]

The apparatus for simultaneous flame ionization and electron capture detection was kindly made available by Prof. Grob.[6] The exit of the column is split, using the platinum capillary technique developed by Etzweiler and Neuner-Jehle.[7] The low dead volume ECD was purchased from Brechbuhler AG, CH-8902 Urdorf, Switzerland.

Gas Chromatography/Mass Spectrometry

For mass spectrometric identifications and mass specific detection, a Finnigan GC/MS system (Model 1015D) combined with an on-line computer (Model 6000) was used. The glass capillary column was directly coupled to the mass spectrometer by means of a platinum capillary.[8] Helium was used for carrier gas.

In environmental samples, specified substances or groups of substances usually are of interest. In simple cases they can be detected by tracing one specific ion (mass chromatography). The structure can then be elucidated by inspection of the corresponding mass spectra. If maximum sensitivity is needed, only a small number of preselected ions are detected, each mass with an optimum signal to noise ratio (mass fragmentography). As a result, much spectroscopic information is lost and the identification has to be based on the GC retention data. The two methods can be combined by integrating certain masses of particular importance with maximum integration time during the cyclic acquisition of the spectra. This is successfully applied for the detection of chlorinated compounds by integrating mass 35 during 64 msec. Mass 35 is only formed by a few unlikely elemental combinations; its presence is therefore strongly indicative of a chlorine-containing compound.

Because of its low intensity, mass 35 is frequently overlooked in a mass spectrum. In a mass chromatogram with maximum integration time, its signal-to-noise ratio is substantially enhanced. Therefore, the sensitivity is sufficient to detect the spectra of chlorinated compounds in a GC/MS run. Moreover, the relatively high sensitivity of a quadrupole instrument in the lower mass ranges is favorable for that purpose.

Many compounds of interest, however, do not show fragments of sufficient specificity. In these cases better results can be obtained by simple combinations of mass chromatograms or simple algorithms.[9,10]

RESULTS AND DISCUSSION

Organic Constituents of Primary and Secondary Effluents

Steadily growing use of synthetic organic chemicals leads one to expect an increased load of such substances in sewage. Little is known about their behavior in sewage treatment. One suspects that many of these man-made chemicals are sufficiently refractory to survive in conventional treatment. In an attempt to characterize the organic components in effluents from primary and secondary sewage treatment, glass capillary gas chromatography has been applied.

Table 26.1 lists mean values of TOC and of the various fraction weights. A noteworthy difference in elimination rates between TOC (73%) and the methylene chloride extractable material (83%), suggests a higher amount of nonextractable polymeric constituents in the secondary effluent. The

Table 26.1 Weight Distribution of Organic Constituents in Primary and Secondary Effluents[a]

	Primary Effluent mg/l	Secondary Effluent mg/l	Elimination Rate %
Total Organic Carbon	39.3	11.1	72
Methylene Chloride Extractable	12.0	2.0	83
Pentane Eluate	0.44	0.05	89
Methylene Chloride Eluate	0.36	0.06	83
Methanol Eluate	10.3	1.7	84

[a]Data represent the mean values of three 96-hr composite samples. Total organic carbon was determined on a Beckman TOC analyzer after filtration through a glass-fritted funnel. Weights of the extractable matter and the three fractions were measured by a Cahn electrobalance. Small aliquots (20-50 μl of 1-2 ml) were transferred with a 100-μl syringe to the aluminum pan of the balance, and weighed after air-drying.

pentane and methylene chloride eluates of the silica chromatography contain hydrocarbons predominantly of petroleum origin. The bulk of the extractable material is eluted with methanol and consists of more polar compounds which are only partly amenable to gas chromatography.

In this chapter we focus on the two hydrocarbon fractions, aiming at a better understanding of the behavior of petroleum constituents in activated sludge treatment.

The gas chromatograms of Figure 26.1 characterize the aliphatic hydrocarbons in primary and secondary effluents, respectively. Before activated sludge treatment, a mixture of saturates that very much resembles the alkane distribution in No. 2 fuel oil is found. n-Alkanes are dominant with a maximum abundance in the C_{11} to C_{14} range. The pattern of branched hydrocarbons is closely related to the fingerprints often found in petroleum-derived hydrocarbon mixtures. In the secondary effluent, the distribution curve is shifted to a higher boiling range with its maximum at n-heptadecane. One possible explanation for this transformation would be the easier microbial degradation of hydrocarbons with shorter chain lengths.

However, the GC's reveal strong evidence that, at least in the C_{14} to C_{20} range, almost no biological degradation is taking place. It is a well established fact that branched and particularly isoprenoidal hydrocarbons are more slowly degraded than straight-chain alkanes. The ratios of n-heptadecane to pristane and n-octadecane to phytane can therefore be used as measures for the degree of microbial degradation.[11] As shown in Figure 26.1, these ratios are not changed by the activated sludge treatment. Hence, it can be concluded that microbial degradation of C_{14} to C_{20} alkanes is rather ineffective during this treatment.

Elimination by other mechanisms, such as gaseous stripping and adsorption, must prevail. Particularly lower-boiling and nonpolar components, such as lower alkanes, are probably stripped from the sewage during the aeration process. Adsorption on the sludge floc surfaces may be of importance as a removal mechanism for the heavier saturates.

Figure 26.2 shows the GC analyses of the aromatic hydrocarbon fractions. They mainly contain mixtures of alkylated benzenes as they are present in gasoline, diesel fuel or No. 2 fuel oil.[1] In clear contrast to the alkane fraction, the low-boiling constituents are not removed as efficiently. This may be explained by their better solubility in water, which decreases the stripping efficiency of the aeration.

p-Dichlorobenzene (peak No. 9) was found most abundant among the chlorinated benzenes of which the full series could be detected by mass chromatography and electron capture detection, respectively. In the higher boiling range a second group of components was present, but their

438 ORGANIC POLLUTANTS IN WATER

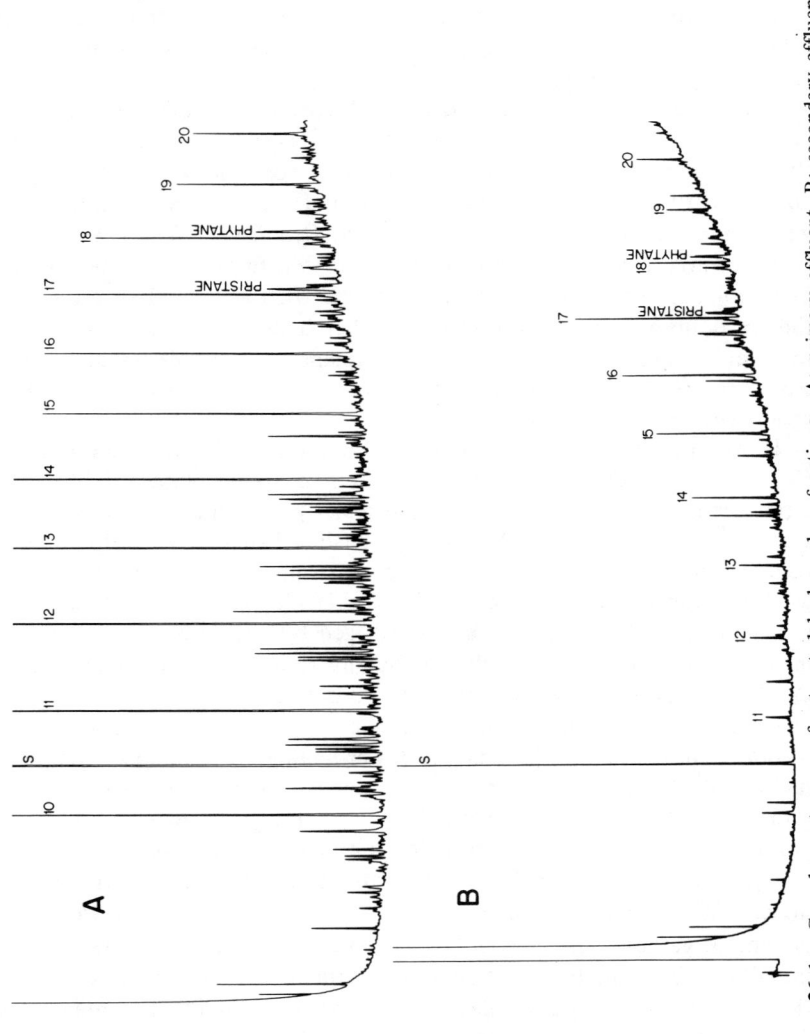

Figure 26.1 Gas chromatogram of saturated hydrocarbon fractions. A: primary effluent, B: secondary effluent. Column: OV-101 50 m × 0.36 mm, 2.8°/min from 30 to 240°C, 4 ml H_2/min, FID; 10-20: C-number of n-alkanes, S: internal standard.

IDENTIFICATION OF ORGANIC POLLUTANTS 439

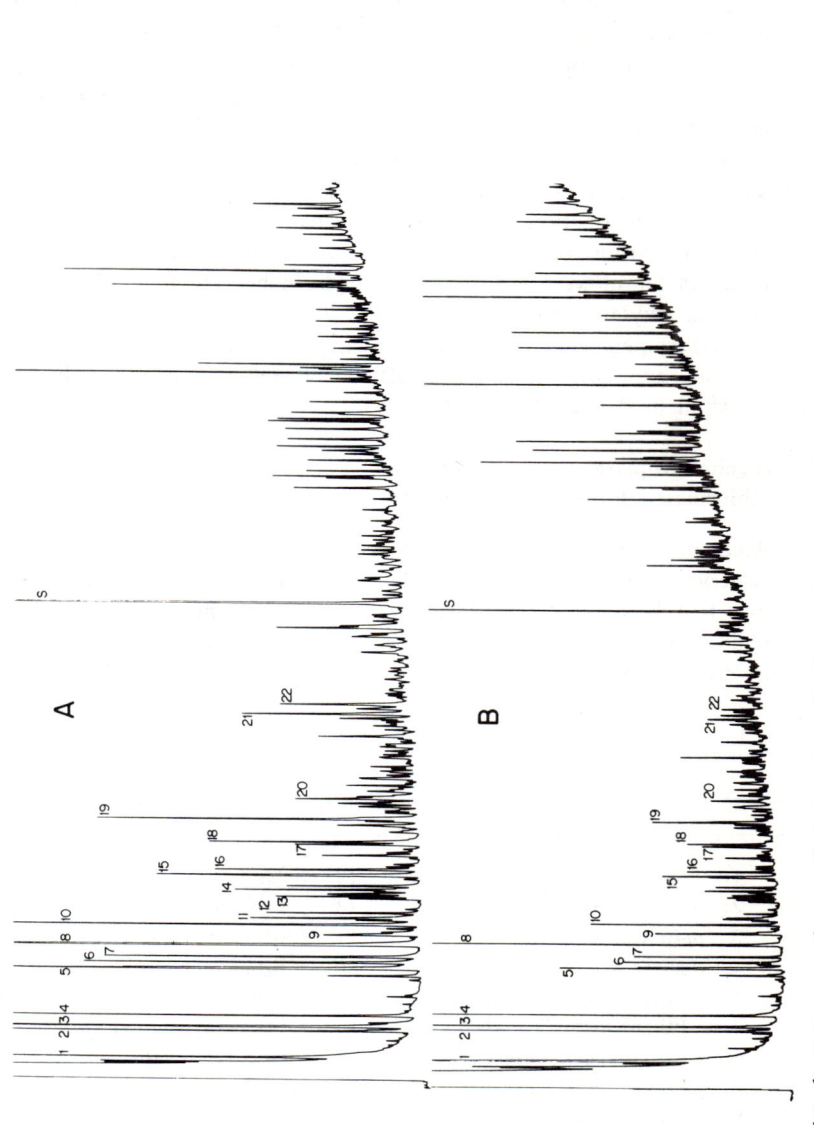

Figure 26.2 Gas chromatogram of aromatic hydrocarbon fractions. A: primary effluent; B: secondary effluent; Column: OV-101 50 m × 0.36 mm, 2.8°/min from 30 to 240°C, 4 ml H$_2$/min, FID; 1-22: identified constituents listed in Table 26.2; S: internal standard.

structures could not yet be elucidated. According to the mass spectra, these compounds are not of aromatic hydrocarbon type. Preliminary results suggest that they are probably produced by the bacteria. Their concentrations increase after percolation through activated carbon filters which contain microbial populations.[12] Concentration levels of major components are in the micro- and submicrogram-per-liter range for primary and secondary effluents, respectively.

The methanol fraction (see Table 26.1) has also been studied by capillary gas chromatography without derivatization. Among the major components found were α-terpineol, benzyl alcohol, 2-phenylethanol, phenol and alkylated phenols, together with various phthalates and adipates. α-Terpineol which is a widely-used, cheap synthetic flavor, proved to be easily eliminated showing an efficiency greater than 99%. In this case a rapid biodegradation is assumed.

Polycyclic Aromatic Hydrocarbons and Polychlorinated Biphenyls in Rain Water

Since the carcinogenic activity of 3,4-benzopyrene was discovered, many reports have been published describing polycyclic aromatic hydrocarbons (PAH) in the environment. This class of compounds has been detected in soot, tobacco smoke, petroleum, car exhaust, airborne particles, etc. PAH were also studied in various types of waters.[13]

Table 26.2 Aromatic Hydrocarbons Identified in Primary and Secondary Effluents [a]

1.	Toluene	12.	Indan
2.	Ethylbenzene	13.	1,4-Diethylbenzene
3.	m-/p-Xylene	14.	1,3-Dimethyl-5-ethylbenzene
4.	o-Xylene	15.	1,3-Dimethyl-2-ethylbenzene/ 1,3-Dimethyl-4-ethylbenzene
5.	1-Ethyl-4-methylbenzene/ 1-Ethyl-3-methylbenzene	16.	1,2-Dimethyl-4-ethylbenzene
6.	1,3,5-Trimethylbenzene	17.	1,2,4,5-Tetramethylbenzene
7.	1-Ethyl-2-methylbenzene	18.	1,2,3,5-Tetramethylbenzene
8.	1,2,4-Trimethylbenzene	19.	Tetralin
9.	1,4-Dichlorobenzene	20.	Naphthalene
10.	1,2,3-Trimethylbenzene	21.	2-Methylnaphthalene
11.	1-Isopropyl-4-methylbenzene	22.	1-Methylnaphthalene

[a]Numbers refer to Figure 26.2.

We have characterized the PAH in recent lake sediments by capillary GC/MS.[2] In the context of tracing the origin and the environmental pathways of PAH a sample of rain water was analyzed. Figure 26.3 shows the gas chromatogram of the aromatic hydrocarbons extracted from rain water. Detection was performed simultaneously by flame ionization (FID) and electron capture (ECD).

The major peaks of the FID trace are assigned to unsubstituted PAH ranging from three-ring (phenanthrene) to six-ring compounds (1,12-benzoperylene) with pyrene being most abundant. Constituents positively identified by GC/MS and co-injection of reference compounds are listed in Table 26.3. Typically, the linearly-annelated anthracene is absent, supposedly due to its lower stability. A group of peaks was attributed to methylphenanthrenes and the smallest possible naphthenolog of phenanthrene, 4,5-methylenephenanthrene, was also identified. One suspects that other alkylated and cyclo-alkylated PAH are present, but the nonselective flame ionization detection does not provide this information.

We therefore applied mass chromatographic techniques which enable the detection of these trace constituents in complex mixtures. Aromatic hydrocarbons are well suited for this method because of their intense molecular ions.[1,9] We found that all unsubstituted PAH are accompanied by smaller amounts of methylated derivatives. In the phenanthrene and pyrene/fluoranthene series the presence of dimethyl and ethyl isomers was established.

The most likely origin of the PAH is the combustion of fossil fuels. Since the sample was taken during summer, the major part is probably due to emissions of car exhausts. Washout by rain leads to an input of these pollutants into the surface waters.

Figure 26.3 also demonstrates the feasibility of dual detection. The two detection systems respond differently to various classes of compounds. The major peaks in the ECD chromatogram are due to polychlorinated biphenyls (PCB), another well-known class of environmental pollutants.

PAH and PCB resemble each other in their physical and chemical properties, and are therefore subjected to similar transport and transformation mechanisms in the environment. They have low solubilities in water and a pronounced tendency to adsorb on particulate matter. Therefore, the two classes are found in much higher concentrations in the sediments.

This analysis of rain water shows that PAH and PCB are transported through the atmosphere and enter the aquatic environment in a diffuse way. The relative importance compared to other sources (domestic and industrial wastewater, land runoff) is open to further studies.

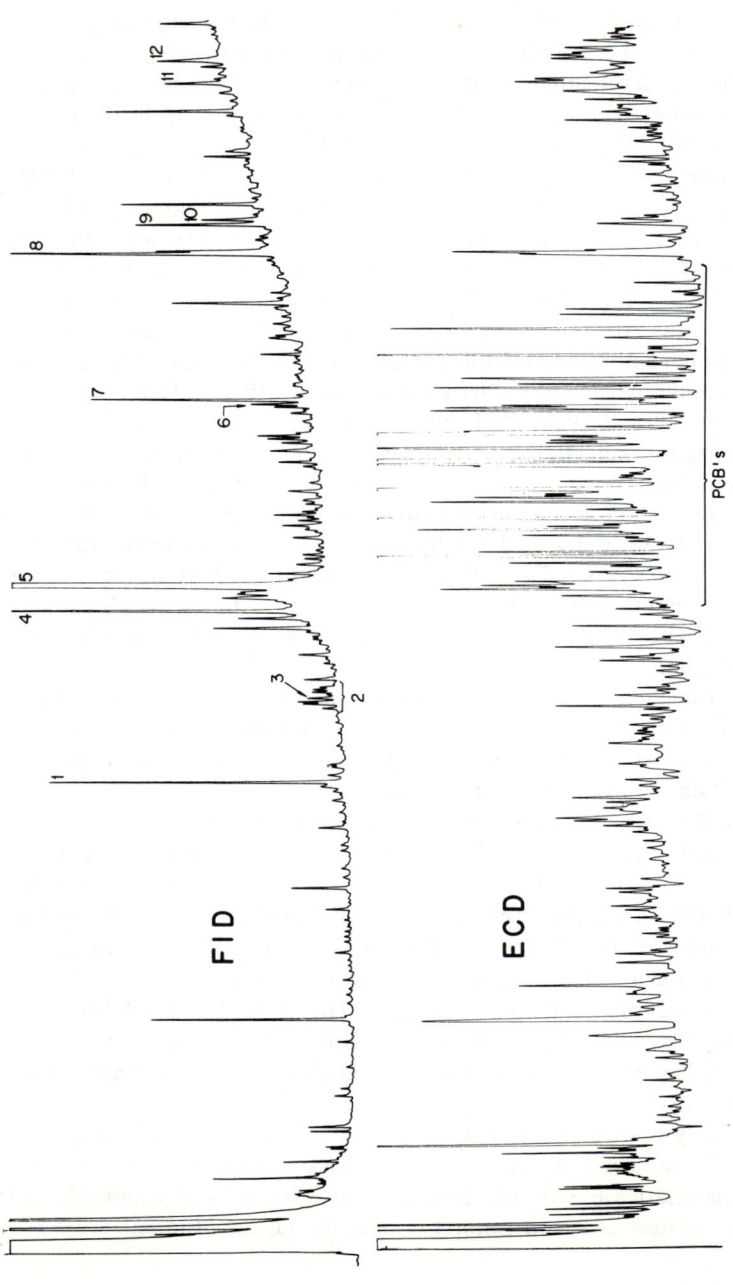

Figure 26.3 Gas chromatogram of the aromatic hydrocarbon fraction extracted from rain water. Column: OV-101 30 m × 0.35 mm, 2.5°/min from 60 to 240°C, 5 ml H$_2$/min. 1-12: identified polycyclic aromatic hydrocarbons listed in Table 26.3.

Table 26.3 Polycyclic Aromatic Hydrocarbons in Rain Water[a]

1. Phenanthrene	7. Chrysene/Triphenylene
2. Methylphenanthrenes	8. Benzofluoranthenes
3. 4,5-Methylenephenanthrene	9. 1,2-Benzopyrene
4. Fluoranthene	10. 3,4-Benzopyrene
5. Pyrene	11. Dibenzanthracenes
6. 1,2-Benzanthracene	12. 1,12-Benzoperylene

[a]Numbers refer to Figure 26.3

Organic Volatiles in River Water

Recent analytical developments[3,4] enable the detection of volatile organic compounds which are present in water as trace constituents (nanograms per liter). At this concentration level one finds organic volatiles in all natural waters. The aim of this study was to characterize the strippable part of the organic material in the river Glatt which flows partly through a densely populated area. The input and fate of a few selected components are discussed on the basis of longitudinal concentration profiles.

Figure 26.4 represents a typical gas chromatogram of the volatile organics found in the river Glatt. Enrichment was performed by the closed-loop gaseous stripping procedure. Capillary gas chromatography reveals the tremendous complexity inherent to this fraction of the organic water constituents. GC/MS and in most cases co-injection of reference samples provided the identifications listed in Table 26.4. This particular sample was taken very close to the outflow from the Greifensee, the highly eutrophic lake which feeds the Glatt. While there is no discharge of sewage into the Glatt between the Greifensee and the sample point, the Greifensee itself is subjected to a heavy sewage load. The mean residence time of water in the lake is about one year.

The identified components can be divided into the following groups:

    A: alkylated benzenes (No. 1,7-17,19)
    B: low-molecular-weight chlorinated hydrocarbons (No. 3-6,11,20)
    C: aldehydes (No. 21,24)
    D: aliphatic hydrocarbons (No. 25-30, 34)
    E: miscellaneous

Groups A and B are ubiquitous in the environment and have been found in natural and treated waters.[14-17] These components are presumably derived from human activity since they either occur in fossil

444 ORGANIC POLLUTANTS IN WATER

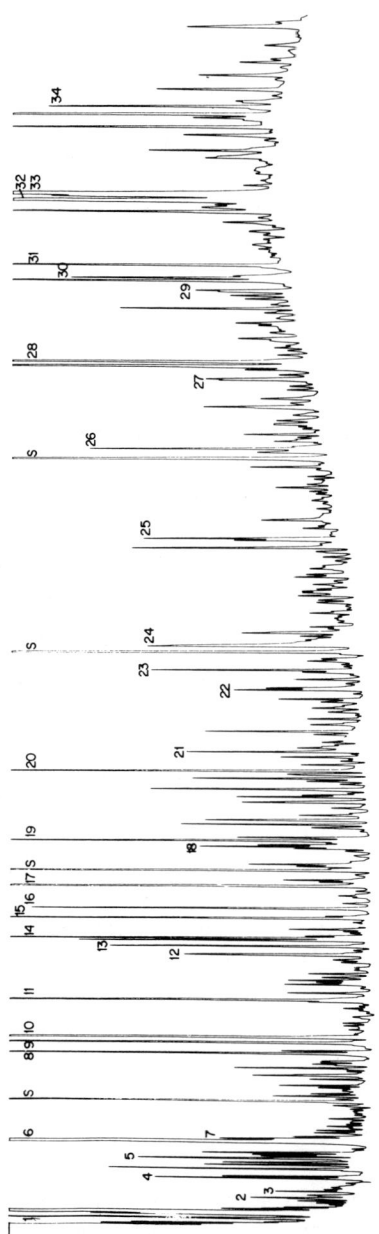

Figure 26.4 Gas chromatogram of volatiles stripped from the river Glatt. Column: Ucon HB, 50 m × 0.32 mm, 3°/min from 30 to 180°C, 2 ml H_2/min, FID. Sampling location close to the outflow from Greifensee. 1-34: identified compounds listed in Table 26.4, S: internal standards.

Table 26.4 Compounds Identified in River Water[a]

1.	Benzene	18.	Diacetone alcohol
2.	1,1-Diethoxyethane	19.	1,2,3-Trimethylbenzene
3.	cis-Dichloroethylene	20.	1,4-Dichlorobenzene
4.	Trichloroethylene	21.	n-Nonanal
5.	1,2-Dichloropropane	22.	$C_{10}H_{18}O$
6.	Tetrachloroethylene	23.	$C_{10}H_{18}O$
7.	Toluene	24.	n-Decanal
8.	Ethylbenzene	25.	n-Pentadecane
9.	p-Xylene	26.	n-Hexadecane
10.	m-Xylene	27.	Pristane
11.	o-Xylene/Chlorobenzene	28.	n-Heptadecane
12.	n-Propylbenzene	29.	Phytane
13.	1-Ethyl-3-methylbenzene	30.	n-Octadecane
14.	1-Ethyl-4-methylbenzene	31.	Tri-n-butyl phosphate
15.	1-Ethyl-2-methylbenzene	32.	n-Dibutyl phthalate
16.	1,3,5-Trimethylbenzene	33.	iso-Dibutyl phthalate
17.	1,2,4-Trimethylbenzene	34.	n-Eicosane

[a]Numbers refer to Figure 26.4.

fuels or are widely-used organic solvents. A homologous series of n-aldehydes is considered to be of biogenic origin. This series extends from C_8- to C_{11}-aldehyde with its maximum at n-decanal. The aliphatic hydrocarbons appearing in the C_{15} to C_{20} range may be either indigenous or petroleum-derived.

To get a better insight into the sources of these materials and their fate in the river, quantitative determinations were performed along the Glatt. Longitudinal concentration profiles of three components are shown in Figure 26.5. n-Decanal remains constant from the Greifensee to the Rhine. This leads to the assumption that the main source for n-decanal is the Greifensee. No change occurs in the well-aerated river, either because no elimination takes place or due to additional diffuse inputs. The other two components (tetrachloroethylene and 1,4-dichlorobenzene) clearly demonstrate the influence of civilization. The water at station 1 contains significantly lower concentrations. The concentration profile for tetrachloroethylene forms a distinct peak at station 2, suggesting a dominant point source between stations 1 and 2. This could be explained by the impact of the town of Dübendorf which is located above station 2. Evaporation processes[18] in the river are probably responsible for the

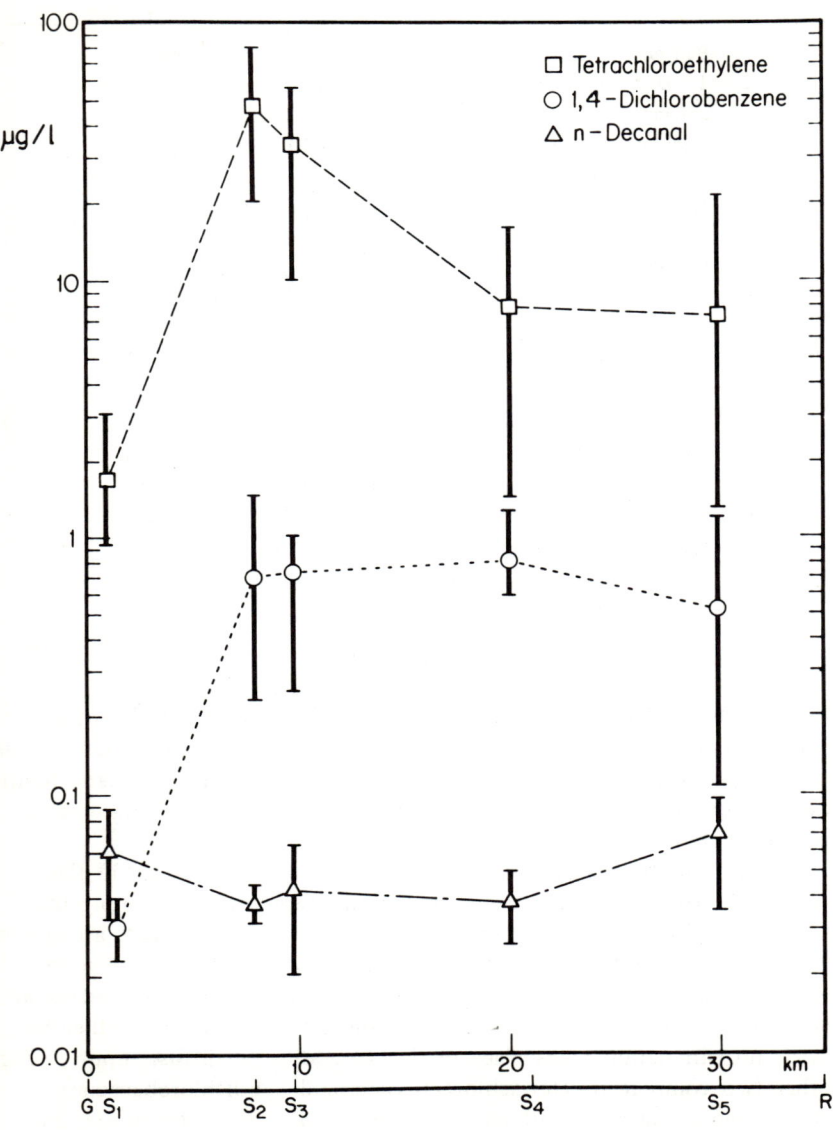

Figure 26.5 Longitudinal concentration profiles of tetrachloroethylene, 1,2-dichlorobenzene and n-decanal in the river Glatt. Mean values of 3-5 measurements and observed ranges are shown. S_{1-5}: sampling locations, G: Greifensee, R: Rhine.

concentration decrease after station 2. Dichlorobenzene, on the other hand, levels off after station 2 which might be explained by additional inputs along the river.

Chlorine-Containing Organic Volatiles
Produced by Chlorination of Lake Water

Chlorinated volatile organic compounds are produced by chlorination of water which contains organic matter. Possible health hazards of such components present in drinking water is the subject of ongoing research. Halogenated methanes rapidly formed in haloform reactions are found in microgram-per-liter quantities.[17,19] Rook has postulated that precursors of the haloform reactions are humic-type materials, based on the observation that chloroform formation correlates with color intensity. No reaction intermediates have been identified so far. To clarify the pathway for the transformation of the organic material, lake water was chlorinated in a bench-scale experiment.

Elemental chlorine was added to water from Lake Zürich (pH 8, total organic carbon 2.0 mg/l, residual chlorine 10 mg/l). After 68 hours reaction time, the free chlorine was reduced with an excess of sulfite, and the organic volatiles enriched by closed-loop stripping. The extract was subjected to GC and GC/MS analysis. A control sample of raw lake water was analyzed in the same way.

Figure 26.6 shows the gas chromatograms of the two samples detected with FID and ECD. No detailed identifications are given, but the dual detected GC's prove that a large number of components have been formed due to the impact of chlorination. Particularly in the low volatility range many halogenated compounds appear, as can be concluded from the strong responses of the ECD. GC/MS identifications are difficult because the mass spectrometer does not achieve the sensitivity of the ECD for most of the halogenated substances. The comparison of such profiles, however, provides a rapid screening procedure to assess possible changes in a water sample.

The high selectivity of mass chromatograms for mass 35 (see Methods) was used to trace chlorine-containing substances. Figure 26.7 shows such reconstructed mass chromatograms for extracts of raw lake water (A) and chlorinated lake water (B). Some of the constituents identified by mass spectrometry are listed in Table 26.5. In the absence of reference samples, the identifications are based on mass spectral evidence only.

Besides the already known haloforms, the formation of some α-chloroketones was found. These substances give evidence that other chlorine substitution reactions in α-position to carbonyl groups are going on during

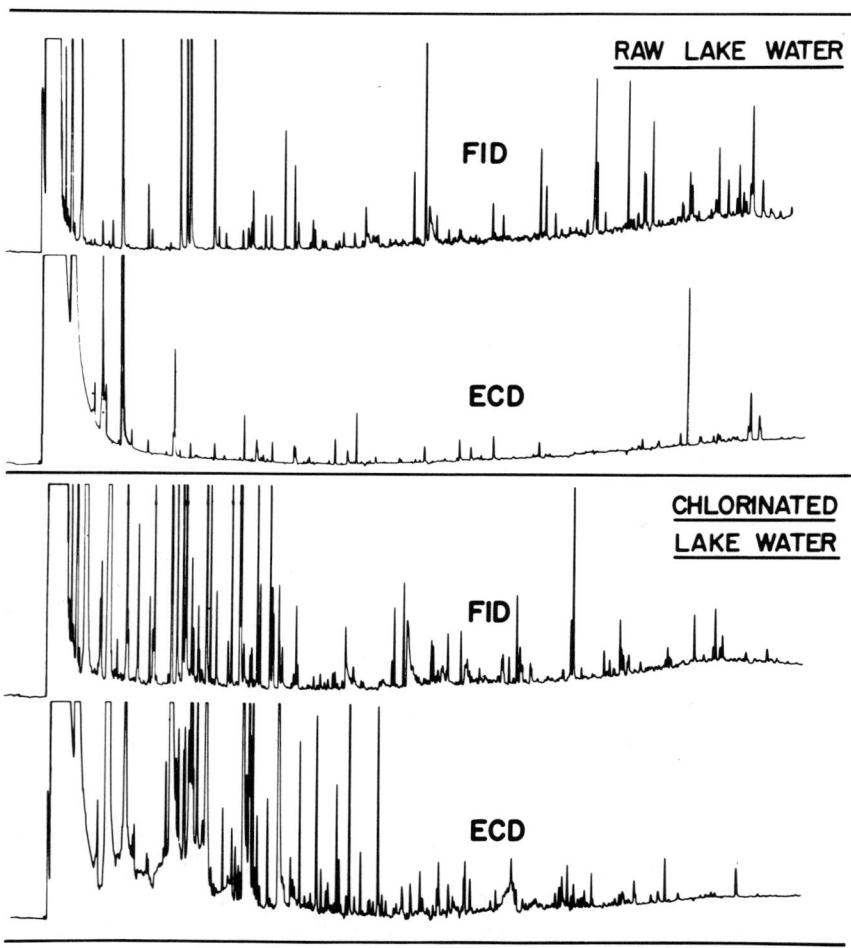

Figure 26.6 Gas chromatograms of volatiles stripped from raw and chlorinated lake water. Chlorination: 68h, 10 mg/l residual chlorine. Column: Ucon HB, 50 m x 0.32 mm, 3°/min from 30 to 180°C, 2 ml H_2/min.

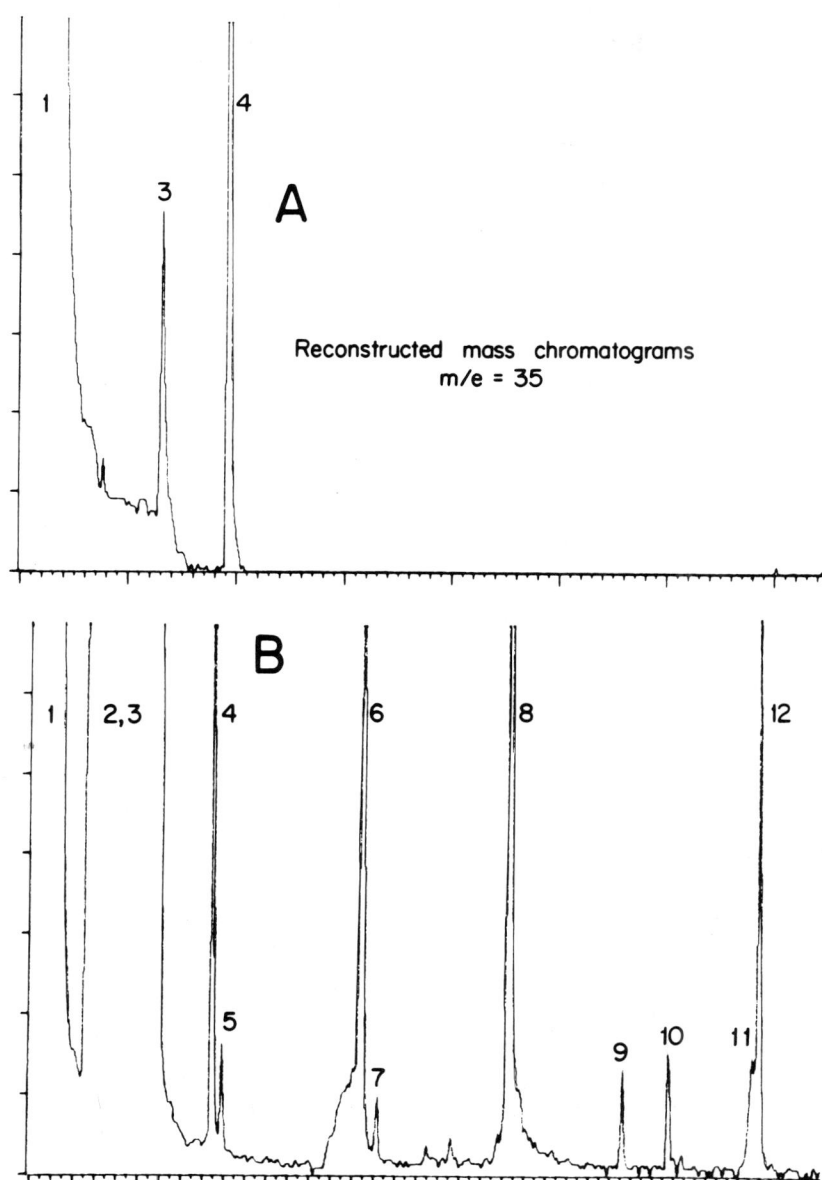

Figure 26.7 Reconstructed mass chromatograms, m/e = 35. A: raw lake water, B: chlorinated lake water. GC conditions as in Figure 26.6 but with He for carrier gas (20 psi), ion energy 70eV, emission current 400 μa. Integration times: 64 msec for mass 35, 4 msec/amu for masses 36-135, 10 msec/amu for masses 136-235.

Table 26.5 Chlorine-Containing Organic Volatiles Detected in Lake Water Before and After Chlorination[a]

Chlorinated Pollutants
1. Carbon tetrachloride
3. Trichloroethylene
4. Tetrachloroethylene

Products Formed During Chlorination
Haloforms
2. $HCCl_3$
6. $HCCl_2Br$
9. $HCClBr_2$
12. $HCCl_2I$

Chlorinated Ketones

5. $H_3C-CCl_2-\overset{O}{\underset{\|}{C}}-CH_3$

7. $H_3C-CCl_2-\overset{O}{\underset{\|}{C}}-CH_2-CH_3$

8. $H_3C-\overset{O}{\underset{\|}{C}}-CCl_3$ (+ trace of trichloroethane)

10. $H_3C-CH_2-CCl_2-\overset{O}{\underset{\|}{C}}-CH_2-CH_3$ (position of the chlorine atoms only tentatively determined)

11. Not identified chlorinated component

[a]Numbers refer to Figure 26.7.

the chlorination of lake water. It is suspected that larger molecules are the precursors which yield these ketones through cleavage reactions.

Chlorinated products of alkylated benzenes were also detected in minor quantities. The same components have been found in chlorination experiments performed in aqueous solutions of diesel fuel.[9] Such chlorinated aromatic hydrocarbons are very persistent, and may cause undesirable effects on a long-term basis.

CONCLUSIONS

Capillary gas chromatography is a method of great potential for analyses of organic constituents, particularly in combination with appropriate enrichment and preseparation techniques. Complex mixtures of volatile organic components as contained in water are unraveled by this highly efficient separation technique. Specific detectors and directly-coupled, computerized mass spectrometry further enhance the yield of information. Mass chromatographic methods enable a convenient screening for components of interest. It has to be stressed, however, that gas chromatography is capable of analyzing only a minor fraction of the organic matter in water samples. Comparisons with TOC values in the case of primary and secondary effluents show that only approximately 5% of the organics are amenable to analysis by gas chromatography. A similar relation probably holds for other types of water. Derivatization and high-pressure liquid chromatography should be applied to widen the scope of organic analyses.

For evaluating potential health hazards and ecological impacts of organic chemicals, the direct interrelation between chemical structure and biological activity has to be considered. Consequently, it may be misleading to assess environmental quality by analyses with methods not achieving full separation and identification of specific compounds.

Complete resolution of organic constituents in environmental samples is extremely difficult, if not impossible. The better our analytical methods, the more we recognize the complexity of the mixtures present. Two approaches promise to provide valuable information. First, one can concentrate on a few selected constituents which are either of high ecological relevance or represent particular classes of compounds; the pathways and fates of these constituents would be investigated quantitatively. Second, monitoring changes in distribution patterns or fingerprints can describe qualitatively the behavior of groups of components. The latter approach would be greatly improved by the development of computer methods for sophisticated processing of GC/MS data.

It is planned to follow along both these lines to get better insights into the behavior of organic constituents in water treatment processes and in the aquatic ecosphere.

ACKNOWLEDGMENTS

The authors wish to thank Professor K. Grob for his very valuable advice. The assistance of people from EAWAG's Process Engineering Group in sampling sewage effluents is gratefully acknowledged. Dr. P. Roberts kindly read and reviewed the manuscript.

This work was supported in part by the Swiss Department of Commerce (Commission of the European Communities Project Cost 64b) and the "Schweizerischer Nationalfonds zur Förderung der wissenschaftlichen Forschung."

REFERENCES

1. Giger, W., M. Reinhard, and C. Schaffner. *Vom Wasser* **43**, 343 (1974).
2. Giger, W. and C. Schaffner. *Adv. in Org. Geochem.* (in press).
3. Grob, K. *J. Chromatogr.* **84**, 255 (1973).
4. Grob, K. and F. Zürcher. *J. Chromatogr.* (in press).
5. Grob, K. and G. Grob. *Chromatographia* **5**, 3 (1972).
6. Grob, K. *Chromatographia* **8**, 423 (1975).
7. Etzweiler, F. and N. Neuner-Jehle. *Chromatographia* **6**, 503 (1973).
8. Neuner-Jehle, N., F. Etzweiler and G. Zarske. *Chromatographia* **6**, 211 (1973).
9. Reinhard, M., V. Drevenkar and W. Giger. *J. Chromatogr.* **116**, 43 (1976).
10. Clerc, J. T., M. Kutter, M. Reinhard and R. Schwarzenbach. *J. Chromatogr.* (in press).
11. Blumer, M. and J. Sass. *Science* **176**, 1120 (1972).
12. Kummert, R. and W. Giger. unpublished results.
13. Harrison, R. M., R. Perry and R. A. Wellings. *Water Research* **9**, 331 (1975).
14. Grob, K., K. Grob, Jr. and G. Grob. *J. Chromatogr.* **106**, 299 (1975).
15. Dowty, B., D. Carlisle, and J. L. Laseter. *Science* **187**, 75 (1975).
16. Rook, J. J., A. P. Meijers, A. A. Gras and A. Noordsij. *Vom Wasser* **44**, 23 (1975).
17. Bellar, T. A., J. J. Lichtenberg and R. C. Kroner. *J. Amer. Water Works Assoc.* **66**, 703 (1974).
18. Dilling, W. L., N. B. Tefertiller and G. J. Kallos. *Environ. Sci. Technol.* **9**, 883 (1975).
19. Rook, J. J. *Water Treat. Exam.* **23**, 234 (1974).

CHAPTER 27

ISOLATION AND IDENTIFICATION OF ORGANIC CONTAMINANTS IN GROUND WATER

W. J. Dunlap, D. C. Shew, M. R. Scalf and R. L. Cosby

Robert S. Kerr Environmental Research Laboratory
Ada, Oklahoma

J. M. Robertson

School of Civil Engineering and Environmental Science
University of Oklahoma
Norman, Oklahoma

Investigations were undertaken to identify organic pollutants contributed to ground water by municipal solid waste and septic tank effluents. Organic compounds were sampled in ground water from wells within a landfill and adjacent to septic laterals by adsorption techniques. Water from up-gradient control wells was also sampled to provide information on organic compounds occurring naturally in ground water in the study areas. All glass-teflon sampling systems were developed and utilized to preclude organic contamination and adsorption by sampling equipment. Activated carbon, macroreticular resins, and polyamide were employed as adsorbents during various stages of the studies. Organic fractions recovered from adsorbents were further purified by column chromatography and solubility separations as needed, and were subjected to both electron impact and chemical ionization gas chromatography/mass spectrometry (GC/MS) for identification of individual compounds. Results of this work may indicate the potential for long-term pollution of ground water by various organic chemicals.

INTRODUCTION

Historically, ground water in the United States has been a relatively minor water resource; its major role in the past has been to serve as a domestic water supply for individual homes and small communities. But now that much of our surface water has been degraded to the point that it is unfit for many uses without extensive and costly treatment, there is a growing tendency to rely more heavily on our vast supply of high-quality ground water as a primary source to meet our increasing fresh water needs.

Currently, ground water supplies about 20% of the nation's total water requirements and includes total or partial supply of over 30 of the nation's 100 largest cities, 95% of our rural population, and more than half of the water needs in agriculture. In spite of the growing dependence on ground water resources, there is an increasing tendency to dispose of unwanted wastes by various surface and subsurface waste disposal techniques without first considering the effects that these practices will have on degradation of ground water quality. When one considers the fact that it may take years or even decades for a pollutant to reach a ground water aquifer and that it may have an equally long residence time, the obvious conclusion is that we simply cannot allow this valuable water resource to become polluted because it cannot be economically recovered.

Much of our lack of basic knowledge regarding ground water pollution stems from the relatively low priority placed on ground water problems in the past. As a result, very little adequate background information has been developed on the many physical, chemical and biological changes organic pollutants undergo as they move chromatographically through an extremely complex soil matrix. The work that will be discussed here involves two projects and will describe some of our initial efforts carried out in an attempt to obtain a better understanding of the subsurface environment and its role as a pollution receptor—first, the isolation and subsequent identification of organic pollutants being leached by ground water at a solid waste disposal site; and second, the isolation and identification of some natural components of ground water preliminary to a study involving the nature, movement and fate of organic pollutants and their ultimate effect on ground water quality at a septic tank waste disposal site.

LANDFILL STUDY

Countless tons of solid waste deposited within the upper layers of the earth's crust at land disposal sites throughout the United States pose a potentially serious threat to the quality of the nation's ground water. In a recent study of ground water problems in 11 northeastern states, Miller et al. presented information on 60 cases in which landfills were pinpointed as sources of ground water pollution and noted that numerous additional cases of a similar nature were probably present in the region.[1] This situation is not unique to the northeast as indicated by the number of reports from other regions, including those of Walker,[2] Andersen and Dornbush,[3] Fuhriman and Barton,[4] Scalf et al.[5] and the California State Department of Water Resources.[6]

Among the many substances which might possibly enter ground water by leaching of solid waste are potentially hazardous organic compounds, particularly synthetic organics which may decompose slowly or be essentially nondegradable in the subsurface environment. Pollution of ground water by organic matter leached from solid waste in land disposal sites has been well documented,[3,6,7] but practically no information has previously been developed concerning the nature of the organic pollutants involved. Clearly, such information is required for realistic and comprehensive evaluation of the threat to ground water quality posed by land disposal of solid waste. Therefore, the Environmental Protection Agency, in cooperation with the University of Oklahoma, undertook an investigation which comprised an effort to provide such information by identifying specific organic pollutants contributed to ground water by a landfill.

The landfill chosen for this investigation was located at a land disposal site near Norman, Oklahoma. This site, shown in Figure 27.1, is in an area of moderately to highly permeable soil with a normally shallow water table that often lies only 2 to 5 ft below the original land surface. The site was initially operated as a dump for 38 years, but in 1960 a trench-type operation was begun in the area designated as "Landfill Site" in which layers of refuse were bulldozed into trenches and eventually covered with relatively permeable fine sand. Because of the shallow depth of ground water in the area, much of the refuse was placed below the water table, a situation common to many landfills. In 1972 a modified area fill operation was initiated in which solid waste was deposited at least 2 ft above the water table and covered at least weekly.

Ground waters from two wells, one of which was located in the southeastern part of the landfill and referred to as either well No. 3 or the test well, and a control well located about 0.7 mile from the upstream edge of the landfill, were the waters subjected to the most intensive analysis in this investigation. Well No. 3 was expected to yield ground water contaminated by solid waste since it passed through a layer of refuse 20 ft thick, whereas the control well was expected to yield water unaffected by the landfill because of its up-gradient location. Both wells

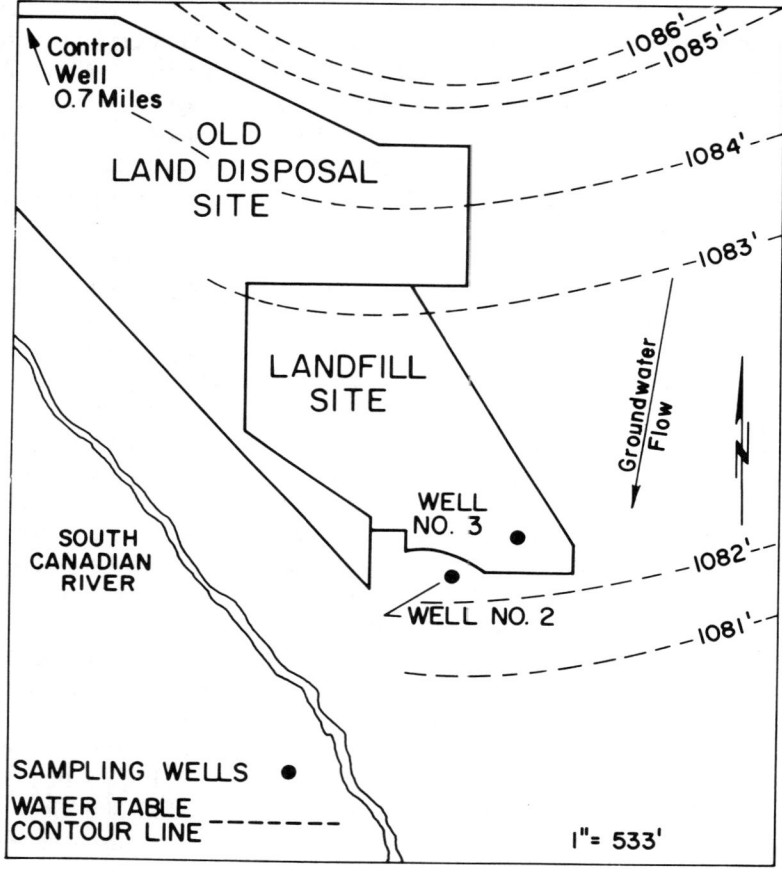

Figure 27.1 Land disposal site near Norman, Oklahoma.

were carefully drilled and cased, and were thoroughly pumped both after construction and before sampling to remove standing water and allow fresh formation water to enter the well bores. The two wells were sampled simultaneously by identical procedures so that comparison of organic matter from the control and landfill wells would clearly reveal the extent and nature of organic contamination of ground water by the landfill.

Sampling Procedures

The sampling procedure, as shown in Figure 27.2, incorporated a modified version of the low-flow carbon adsorption method[8,9] with the ground water being pumped from the saturated zone directly through columns containing activated carbon to adsorb and recover the organic compounds. Single-piece, all-glass columns, fabricated from 3-in.-diameter borosilicate tubing, were packed with 18 in. of 30-mesh activated carbon (Nuchar C-190, Plus 30, Hebert Chemical Co., St. Bernard, Ohio) held in place by solvent-washed glass wool plugs. For sampling, a packed column was placed in a vertical position at the top of the casing of each well with suitable lengths of teflon tubing attached to the bottom inlets of the columns and extending down the well shafts into the saturated zone. The ground water was then pumped up the tubing and through the carbon columns by variable-speed peristaltic-type pumps ("Masterflex" 7545 Variable Speed Drive with 7014 pump head, Cole-Parmer Instrument Co., Chicago, Illinois) attached by teflon tubing to the outlet (downstream) ends of the columns. The variable-speed pumps permitted sustained pumping of ground water from the water table through the carbon adsorption columns at accurately-controlled, constant, low-flow rates which were verified by collecting discharge water from the pumps and measuring the volumes.

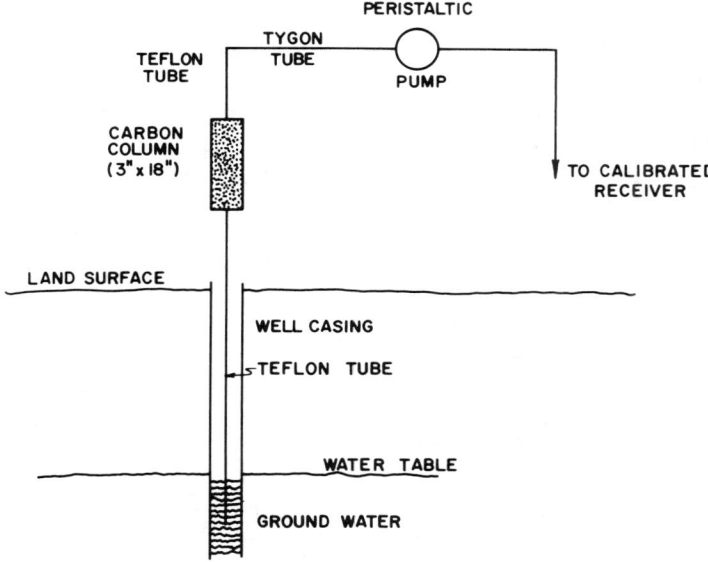

Figure 27.2 Ground water sampling system.

A total of 200 gal (757 liters) of ground water was pumped from each of the two wells through essentially identical carbon columns during a period of 126 hr at a rate of 100 ml/min. Use of the sampling systems consisting only of glass and teflon from the saturated zone to the outlet of the carbon adsorption column and placement of the pump on the downstream side of the column virtually precluded introduction of organic contaminants during the sampling operation.

Upon completion of sampling, the carbon columns containing the adsorbed organics from the water from well No. 3 and the control well were drained to remove excess water, sealed with solvent-washed aluminum foil, and transported immediately to the laboratory for processing. The carbon was dried at approximately 40°C for 48 hr under a gentle flow of clean air in a Precision-Freas mechanical convection oven (Model 845, Precision Scientific Company, Chicago, Illinois), transferred to 2200-ml modified Soxhlet extractors, and extracted for 48 hr with chloroform. These extracts were filtered through solvent-extracted glass fiber filters to remove carbon fines and then vacuum concentrated in rotary evaporators at temperatures not exceeding 27°C and pressures of about 100 mm to a final volume of 3 ml each. The crude carbon chloroform extracts (CCE's) obtained from the carbon employed in sampling well No. 3 and the control well were designated CCE3II and CCEC, respectively.

The chloroform-extracted carbon samples were dried in the Soxhlet extractors by passing a gentle stream of warm, dry air through the extraction chambers via the siphon tubes for 20 hr. The carbon was then extracted for 32 hr with pure ethanol, and the carbon alcohol extracts (CAE's) were filtered and concentrated in the same manner as the CCE's. However, it was necessary to filter the CAE's through extracted glass fiber filters when volumes of about 10 ml had been attained in order to remove precipitated material. These precipitates, together with solid material which had precipitated on the flask walls during evaporation, were dried and weighed. The filtered CAE's were then further evaporated to final volumes of 4.0 ml for CAE3II, from well No. 3 and 2.0 ml for CAEC, from the control well.

Comparison of CCE's and CAE's

Visual comparison of the various CCE's and CAE's showed CCE3II, prepared from ground water from well No. 3, to be deep yellow, while CCEC, from the control well water, was light yellow. Similarly, CAE3II was deep yellow-orange and CAEC was yellow. Both CCE3II and CAE3II were very odorous, while the control CCE's and CAE's were essentially odorless.

Aliquots of each of the concentrated CCE's and CAE's were carefully evaporated to dryness in order to determine the weights of soluble material dissolved in the concentrates. Total weights of the CCE's and CAE's were calculated from these weights and the weights of material which had precipitated during preparation of the concentrated extracts. The data obtained, presented in Table 27.1 both in terms of total weights and weights per liter of sampled water, showed CCE3II to contain approximately 25 times as much material as CCEC, and revealed CAE3II to contain about four times the material contained by CAEC, the control. Hence, the presence of much greater quantities of organic constituents in the ground water in the locale of the Norman landfill than in ground water from the same aquifer approximately one mile upstream from the landfill perimeter was clearly indicated.

Table 27.1 Weights of Carbon Chloroform and Carbon Alcohol Extracts

Source	Weight of CCE		Weight of CAE		Weight of CCE + CAE	
	Total, mg	mg/l	Total, mg	mg/l	Total, mg	mg/l
Test Well 3	304.5	0.402	1219.1	1.610	1523.6	2.013
Control Well	11.9	0.016	314.2	0.415	326.1	0.431

The various CCE's and CAE's were next compared by gas-liquid chromatography. Figures 27.3 and 27.4 show the results obtained by chromatographing, under identical conditions, aliquots of CCE3II and CCEC representing 190 ml of ground water from well No. 3 and the control wells, respectively, and aliquots of CAE3II and CAEC representing 380 ml of ground water from these wells. These chromatographic comparisons revealed that the ground water from the landfill contained a complex array of organic compounds which were readily amenable to gas chromatography and which were either not present, or were present in very much less quantity, in ground water not subject to the influence of the landfill. It was obvious that practically all of the major organic components of CCE3II and CAE3II which were sufficiently volatile for gas chromatography had been contributed to the ground water by the landfill, and identification of any of these compounds would, therefore, be helpful in elucidating the effect of the landfill on ground water.

In view of their obvious complexity, both the crude CCE3II and CAE3II extracts were further fractionated prior to GC/MS analysis. The chloroform extract was subjected to liquid-solid chromatography on a

460 ORGANIC POLLUTANTS IN WATER

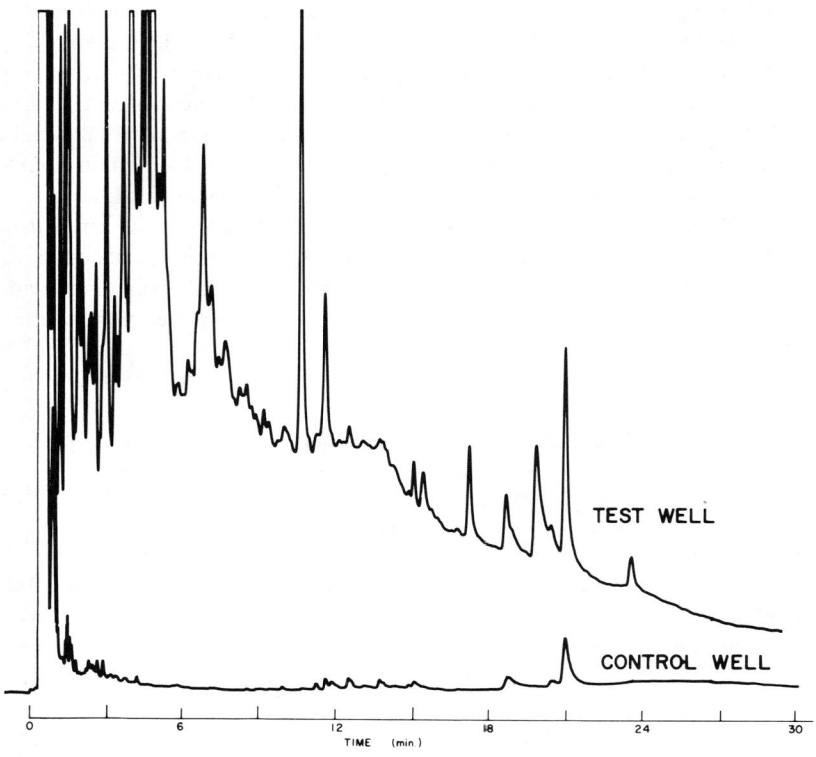

Figure 27.3 FID chromatogram of test and control well CCE.

7 x 76 mm column of 100/200 mesh Silicar CC-7 (Mallinckrodt Chemical Works, St. Louis, Missouri) using 8 ml each of hexane, benzene, chloroform-methanol (1:1) followed by 5 ml methanol. Chromatographic separation provided ten fractions which were examined by gas chromatography and those which still appeared too complex for GC/MS analysis were rechromatographed on a fresh microcolumn of Silicar CC-7 using 370 ml of solvent ranging in polarity from hexane to methanol. The fractions were examined by gas chromatography and several groups of fractions which appeared to contain very little organic matter or low levels of essentially the same components were recombined into a total of about 18 fractions suitable for GC/MS analysis.

The carbon alcohol extract from well No. 3 was separated into fractions of less complexity by a classical solubility separation scheme[10,11]

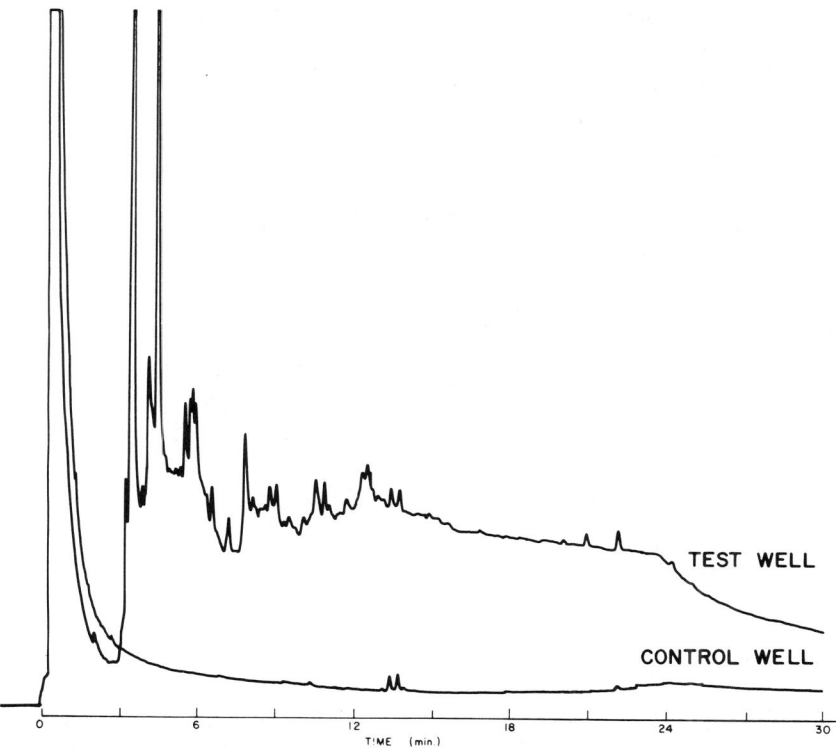

Figure 27.4 FID chromatogram of test and control well CAE.

as shown in Figure 27.5. There were five fractions obtained, namely ether insolubles, water solubles, bases, acids and neutrals, all of which were examined by gas chromatography. However, because of time limitations and the fact that the acids fraction contained, by far, the largest amount of material, most of our identification efforts were spent on this particular fraction.

Identification of Compounds in the CCE and CAE from the Landfill Well

The various fractions obtained by silica gel column chromatography of CCE3II and the acids fraction obtained by solubility separation of CAE3II were analyzed by combined GC/MS, employing computerized data acquisition and processing, to achieve identification of individual compounds

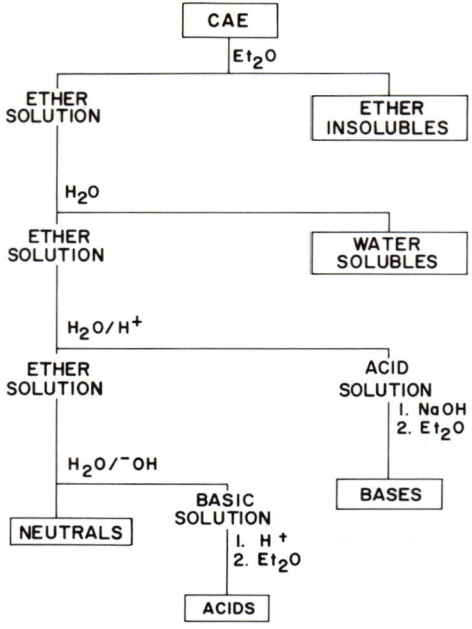

Figure 27.5 Solubility separation of CAE.

therein. A Finnigan Model 1015C GC/MS system with both electron impact and chemical ionization sources and System Industries System 150 data system was used for this work. Gas chromatography columns packed with 3% OV-1 on 100/120-mesh Gas Chrom Q or 3% Carbowax 20M on 80/100-mesh Gas Chrom Q (Applied Science Laboratories, State College, Pennsylvania) proved to be most effective for separation of the CCE3II fractions, while a column packed with 10% SP 1200/1% H_3PO_4 on 80/100-mesh Chromosorb W AW (Supelco, Inc., Bellefonte, Pennsylvania) was most useful for the CAE3II acids fraction.

Mass spectra were examined by standard interpretive procedures based on mass spectral fragmentation theory and by computerized spectral matching programs which permitted comparison of the unknown spectra with the contents of libraries of mass spectra at the National Institutes of Health, Bethesda, Maryland, and Battelle Memorial Institute, Columbus, Ohio.[1,2] Additional corroborative evidence for the structures of compounds identified on the basis of their spectra was obtained by direct comparison with standard compounds whenever such standards were available. When possible, the quantities of the identified compounds in the sampled ground water were estimated by comparing their peak heights on gas chromatograms with the peak heights produced by known quantities of standard compounds.

Most of the major substituents in the carbon chloroform and carbon alcohol extracts obtained from ground water from well No. 3 that were amenable to GC have been identified and are shown in Table 27.2. The

Table 27.2 Compounds Identified in Ground Water from Landfill Well

Compound	Estimated Concentration µg/l	Compound	Estimated Concentration µg/l
Fenchone	0.2	Tri-n-butyl Phosphate	1.7
Camphor	0.9	p-Toluenesulfonamide	–
2,6-Di-t-butylbenzoquinone	–	Methylpyridine	–
Diethyl Phthalate	4.1	N,N-Diethylformamide	–
2,6-Di-t-amylbenzoquinone	–	Triethyl Phosphate	0.3
Diisobutyl Phthalate	0.1	3-Pentanol	–
Di-n-butyl Phthalate	–	2-Methylpentan-2,4-diol	–
Butylcarbobutoxy-methyl Phthalate	–	Dipropyleneglycol Methyl Ether	–
		bis-2-Hydroxypropyl Ether	–
Butylbenzyl Phthalate	–	3-Methylcyclopentan-1,2-diol	–
Dicyclohexyl Phthalate	0.2	Triethyleneglycol Dimethyl Ether	–
Dioctyl Phthalate[a]	2.4	2-Octanol	–
p-Cresol	14.6	4-Hydroxycyclohexanol	–
o-Xylene	0.6	Acetic Acid	–
p-Xylene	0.9	Isobutyric Acid	48.7
Cyclohexanol	1.0	Butyric Acid	1.5
3-Methylcyclohexanol	–	Isovaleric Acid	0.7
2-Acetylpyrrole	–	Valeric Acid	1.1
N-Ethyl-p-toluenesulfonamide	0.1[b]	2-Ethylhexanoic Acid	4.2
N-Ethyl-o-toluenesulfonamide	–	Isomeric C_6 Acid[a]	17.1[c]
C_3 Alkylbenzenes[a]		Isomeric C_6 Acid[a]	0.2[c]
(2 compounds)	–	Isomeric C_7 Acid[a]	7.5[d]
Diacetone Alcohol	10.9	Isomeric C_8 Acid[a]	–
Butoxyethanol	–	Cyclohexanecarboxylic Acid	2.8
Ethyl Carbamate	–	Caprylic Acid	0.6
		Caproic Acid	1.1
		Heptanoic Acid	1.0

[a]General structure confirmed beyond reasonable doubt, but position of substitution or chain branching not determined because necessary standards were unavailable or compounds were not separated by GC columns employed.

[b]Determined as N,N-dimethyl-p-toluenesulfonamide.

[c]Determined as 2-ethylbutyric acid, but probably is not this compound.

[d]Determined as n-heptanoic acid.

general structures of all of the compounds shown were established beyond reasonable doubt. However, it should be noted that exact positions of substituent attachment and chain branching were not achieved for a few compounds, such as the two C_3 alkylbenzenes and some of the C_7, C_8 and C_9 isomeric acids, because of unavailability of required standards or failure of GC columns to separate closely-related compounds. In addition, there are many additional compounds at generally lower concentration levels which are still in various stages of identification and confirmation.

Discussion

The data shown in Table 27.2 clearly show that low levels of many potentially undesirable organic compounds were being contributed to ground water within and immediately under the Norman landfill by solid waste deposited in this landfill. A few of the compounds identified in this study could be leachates of natural products, including foods, or possibly end products of microbial metabolism. However, most of the identified compounds are chemicals which are commonly employed in industrial operations and, therefore, usually associated with industrial waste. Since available information indicates that the Norman landfill has never received appreciable quantities of solid waste from industrial operations, there is an obvious question concerning the source of the industrial organic chemicals leached from this landfill. It should be noted, however, that most of the compounds shown in Table 27.2 are used in the manufacture of a wide array of finished products for domestic and commercial use which ultimately will find their way into most landfills. The decomposition and/or leaching of such manufactured products deposited in the Norman landfill would appear most likely to account for the introduction of industrial organic pollutants into ground water in and near this landfill, even though it had received essentially no industrial solid waste *per se.*

Those compounds for which quantitative data were obtained during this investigation appeared to be present in ground water from well No. 3, the landfill well, only in low concentrations. However, the quantitative data resulting from this work must be considered as minimum values because of quantitative inadequacies of the procedures used, particularly the carbon adsorption method. These inadequacies result principally because: activated carbon may fail to quantitatively adsorb dissolved organic compounds from sampled water; complete recovery of adsorbed compounds from the activated carbon may not be accomplished during extraction; and, volatile sample components may be lost during drying of the activated carbon and evaporation of extracting solvents. This is illustrated by comparing the total combined weight of the landfill well CCE and

CAE, 2.013 mg/l, with the average total organic carbon content of 13.4 mg/l of water from this well. If the organic equivalent of 13.4 mg carbon/l is considered, it becomes apparent that less than 10% of the organic matter present in the sampled ground water was recovered in the combined CCE and CAE.

The history of the Norman landfill and dates of newspapers recovered from well No. 3 during drilling indicated that the refuse in the area of the fill near this well had been in place at least 3 years at the time this investigation was conducted. Based on this information, as well as the relatively low concentration of organic carbon (13.4 mg/l) in ground water from well No. 3, it appears likely that most of the readily leachable organic matter had already been removed from refuse near the test well when sampling of ground water for organic pollutants was accomplished. It is probable, therefore, that most of the compounds identified in this study are substances which were leached very slowly from the refuse in the landfill and/or substances which persisted for considerable periods of time in the aquifer in the vicinity of the refuse from which they were leached because of sorption on the earth solids comprising the aquifer. This observation implies the potential for long-term insidious pollution of ground water by undesirable organic chemicals from landfills. Slowly-decaying domestic and commercial products in landfills would appear likely to serve as reservoirs feeding low levels of industrial organic pollutants into aquifers for many years after the landfills have been closed and forgotten. Even those substances which are sorbed relatively strongly on aquifer solids could ultimately pose a pollution threat if they were resistant to biochemical and abiotic degradation in the ground water environment. Such compounds could move as zones by slow, "chromatographic" migration to finally reach wells providing water for consumption by humans or domestic animals. Because of the low levels of pollutants likely to be involved, physical properties of the polluted ground water would probably not be altered sufficiently to indicate the presence of the offending compounds. This presence could be a matter of considerable concern, however, since the health effects of chronic ingestion through water of even very low levels of compounds such as those identified in this study are largely unknown. This, coupled with the great difficulty involved in removing pollutants, particularly those which tend to adsorb significantly on aquifer solids, from a polluted aquifer, dictates the need for further investigation of this potential problem. In particular, information concerning the mobility and longevity in the ground water environment of compounds such as these are needed.

In assessing the results of this investigation, it should be clearly noted that the compounds identified included only substances readily amenable

466 ORGANIC POLLUTANTS IN WATER

to gas chromatography and represented probably less than 10% of the combined weights of the carbon chloroform and alcohol extracts. Most of the missing material was probably composed of compounds too polar and/or too high in molecular weight to yield readily to gas chromatography procedures. Characterization of this material would undoubtedly have yielded much additional information concerning the organic pollutants contributed to ground water by the Norman landfill, but the necessary analytical effort for such characterization during this study was precluded by time limitations.

SEPTIC TANK STUDY

Studies of ground water pollution problems in various parts of the United States have indicated that leachates from septic tanks pose a ubiquitous and potentially serious threat to ground water quality.[1,4,5] Consequently, EPA is currently funding and conducting jointly with Texas A&M University an investigation designed to provide information concerning the accumulating and passage of pollutants in domestic septic tank disposal fields. This information will ultimately be used for developing guidelines for septic tank siting, including maximum allowable densities of septic systems.

Field operations are being conducted at a site, shown schematically in Figure 27.6, near College Station, Texas. Two series of field studies are

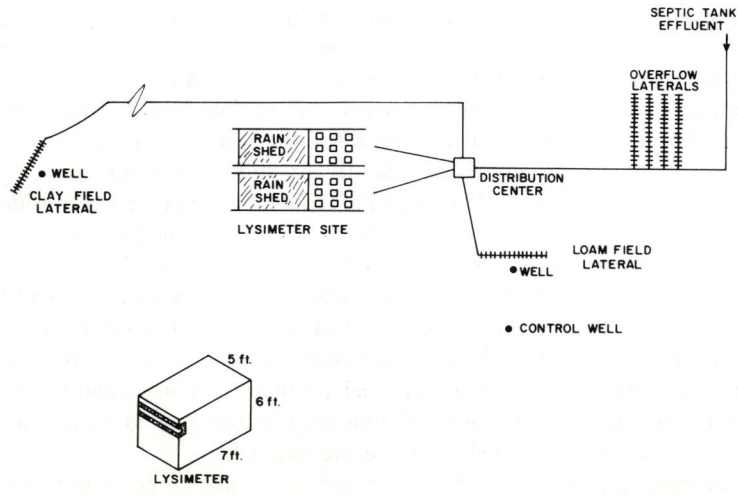

Figure 27.6 Septic tank study field site.

under way, the first of which involves the controlled application of effluent, from a large septic tank serving nine residences, to 18 drainage lysimeters each containing a large (200 ft^3) undisturbed monolith of clay, loam or sand soil. Each soil type is represented by six lysimeters, three of which are being operated under low water table (unsaturated) conditions, while the other three are being operated under high water table (partially saturated) conditions. The lysimeters are protected from natural precipitation by mobile rain shelters and are operated in a manner permitting accurate measurement of volumes and weights of water and pollutants applied, moving through, and accumulating in the soil monoliths.

In the second series of field studies, the same septic tank effluent is being applied to full-scale 50-ft septic field lateral lines installed in undisturbed clay and loam soils at the field site, as shown in Figure 27.6. The water table in the vicinity of these lateral lines is shallow, usually lying less than 10 ft below the surface. Ground water under the laterals and soil cores obtained at intervals from the surface to below the water table will be analyzed periodically for pollutants contributed by the septic tank effluent. Carefully-drilled and cased wells located adjacent to the lateral lines at both loam and clay plots and a control well approximately 35 ft from the lateral line at the loam plot, as shown in Figure 27.6, provide access to the ground water for sampling purposes.

An important phase of this investigation comprises an effort to elucidate the nature, movement and fate of organic pollutants released to the subsurface through septic tank lateral lines. Studies are currently under way to establish background data for organic compounds in lysimeter effluents, soil cores, and ground water in the aquifer underlying the field lateral lines before the media are affected by organics from the septic tank effluent. The effort that has thus far been expended in sampling and analysis of organic compounds in ground water underlying the field lateral lines is described below.

Sampling Procedures

As in the previously-described landfill studies, ground water was sampled for organic constituents by pumping directly from the saturated zone through columns containing adsorbents to concentrate and recover the organic compounds. However, the sampling procedures utilized in the present work entailed the use of columns packed with macroreticular resin and polyamide, as well as smaller, less cumbersome carbon columns in an effort to develop a more versatile and effective system for sampling organic compounds in ground water from shallow aquifers. The macroreticular resin sampling procedure was a modification of the method developed by Junk et al.[13] employing XAD-2 resin (Rohm and Haas Co.,

Philadelphia, Pennsylvania), while the reduced-scale carbon adsorption procedure was adapted from the method of Buelow, Carswell and Symons[14] employing Filtrasorb 200 (Calgon Corporation, Pittsburgh, Pennsylvania) activated carbon. The polyamide procedure was based on methods developed in our laboratories for determining diethylstilbestrol in water, employing Polyamide Woelm (ICN Pharmaceuticals, Inc., Cleveland, Ohio) as adsorbent. Polyamide resins have been much used in natural product chemistry for isolation and chromatography of phenols and other substances capable of hydrogen bond formation.[15] Because these resins usually exhibit a relatively high capacity for adsorption of selected compounds from aqueous media and little irreversible adsorption of such compounds when eluted with proper solvents, they appeared worthy of consideration for use in ground water sampling.

Prior to column preparation, the XAD-2 resin was purified by solvent extraction and stored under methanol according to the method of Junk et al.[13] Filtrasorb 200 activated carbon was boiled in a 5% solution of hydrochloric acid in organic-free water for 3 hr, washed thoroughly with organic-free water to remove all chlorides, dried at 150°C for several hours in a clean oven, and stored in glass-stoppered jars until used. This procedure, suggested by the manufacturer, greatly reduced the amount of background material found in the carbon blanks. Polyamide Woelm (hereafter called PAW) was simply suspended in distilled water about 2 hr prior to column preparation. It was washed after packing by passing successively through each column the following solvents: 100 ml water; 100 ml 2-propanol:water (1:1); 200 ml 2-propanol; 100 ml 2-propanol: water (1:1); and 200 ml water.

XAD-2 and PAW were packed from methanol and water slurries, respectively, to produce 9 x 130 mm adsorbent beds in glass columns fabricated from 12-mm o.d. borosilicate glass tubing. These columns, shown in Figure 27.7, were equipped with a 3-way teflon stopcock at each end to permit retention of fluid in the adsorbent during transport and bypassing of the sample stream during sampling. The two column sections were held together by a "Taper-Tite" teflon connector assembly (Chemplast Inc., Wayne, New Jersey), which permitted easy disassembly for packing and elution. The columns were plugged at both ends with solvent-extracted glass wool, with the short, or inlet, sections being completely filled with this material to serve as a filter protecting the adsorbent from particulate matter which might be present in the sample stream. During packing and elution the columns were inverted from the normal sampling position, with the short (sample-inlet) end replaced by a reservoir. After packing and assembly, the XAD and PAW columns were kept sealed and never allowed to become dry.

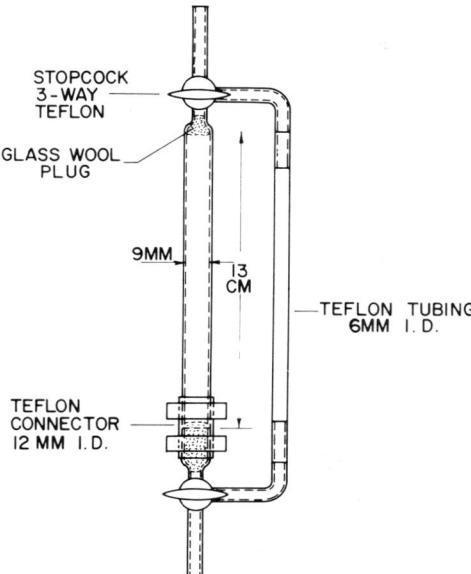

Figure 27.7 Resin adsorption column.

Carbon adsorption columns were prepared by placing 70 g of dry, purified Filtrasorb 200 in a 30-mm i.d. borosilicate glass column (Figure 27.8), thus producing a carbon bed approximately 230 mm in length. The columns were equipped with 45/30 ball and socket joints to permit disassembly for packing and removal of the carbon, which was retained in place during sampling by solvent-washed glass wool plugs. Except during actual use, the columns were kept tightly sealed by inserting 6-mm o.d. teflon plugs into 6-mm i.d. "Taper-Tite" teflon connectors attached to the column inlets and outlets.

Sampling systems were assembled by connecting two columns of a single adsorbent in series, as shown in Figure 27.9. Construction of the systems, as in the landfill studies, was such that water being sampled was in contact with only glass or teflon until it had passed through both adsorbent columns. Ground water was pumped from the water table through the columns via 6-mm o.d. teflon or glass tubing by "Masterflex" 7014 peristaltic pumps powered by 7545-10 Variable Speed Drives (Cole-Parmer Instrument Co., Chicago, Illinois).

Three separate 2-column sampling systems containing, respectively, XAD-2, PAW and carbon were employed simultaneously for sampling of ground water at the Texas A&M field site in an effort both to gain

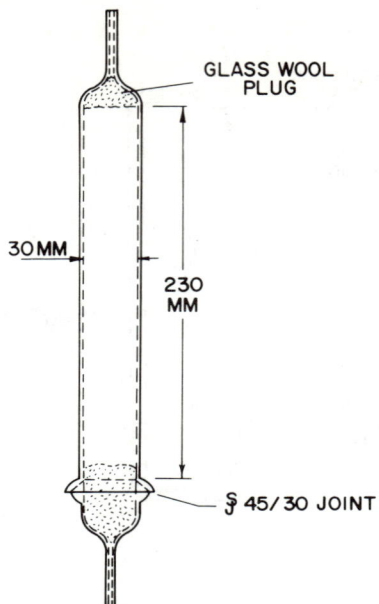

Figure 27.8 Carbon adsorption column.

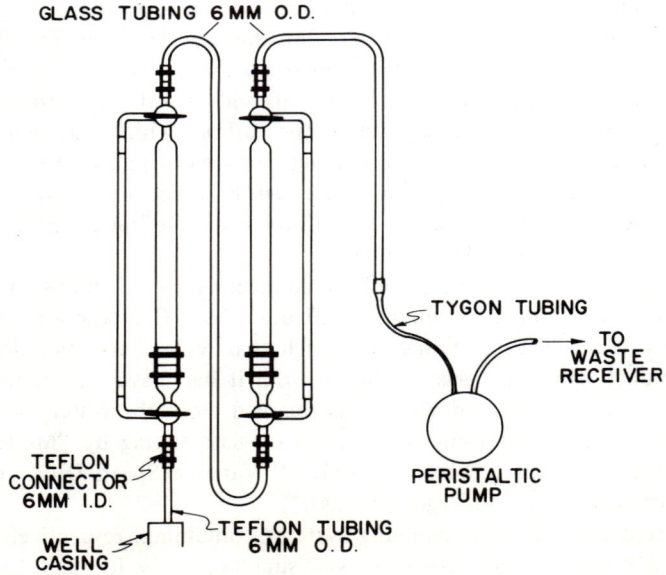

Figure 27.9 Ground water sampling system.

maximum information concerning the organic components of this ground water and to compare the effectiveness of the three adsorbents for ground water sampling. The three systems were installed at the laboratory in a specially constructed housing (approximately 60 cm high, 44 cm wide, and 34 cm deep) designed to protect the components from damage by weather and small animals in the field. After the XAD and PAW systems had been cleared of solvent and checked by pumping organic-free water through them, the housing containing the three sealed systems was transported to the field site and set up over the loam field lateral well, which had been thoroughly pumped prior to sampling to insure the presence of fresh formation water in the casing. Sampling was conducted by continuously pumping ground water through the systems for approximately 7-3/4 days. A total volume of 282 liters of water was passed through the carbon columns at an approximate flow rate of 25 ml/min. Total volumes and flow rates for the XAD-2 and PAW columns were, respectively, 145 liters at 13 ml/min and 154 liters at 13.8 ml/min. Since the peristaltic pumps for all three systems were driven by a single power unit and were hence operating at the same speed, the flow rate differences were a result of different resistances of the columns to fluid flow.

Upon completion of sampling, the columns were sealed and immediately returned to the laboratory for extracting. The XAD and PAW columns were sealed while completely filled with sample water, while the carbon columns were drained of excess liquid before sealing. Columns of each adsorbent, which had been prepared exactly as those used for sampling but which had been in contact only with organic-free water, were processed along with the sampling columns to provide blank extracts or eluates.

For extraction, the XAD-2 and PAW columns were inverted and the short inlet sections containing glass wool plugs on which considerable particulate matter had collected were replaced by 300-ml reservoirs. Each XAD-2 column was drained of excess water and extracted with 2- to 20-ml volumes of diethyl ether, with each volume of ether being allowed to stand in the column for 10 min before draining. The water layer was separated from the resulting ether-water mixture and extracted with an equal volume of ether. The combined ether extracts were dried with anhydrous sodium sulfate and reduced in volume to approximately 1 ml by the modified Micro-Snyder evaporation procedure suggested by Junk et al.[13] Volumes of the XAD ether extracts were further reduced to about 500 μl under a gentle stream of clean, dry nitrogen.

The PAW columns were eluted successively with 20 ml of 25% 2-propanol in water and 30 ml of pure 2-propanol, thus producing a propanol-water and a propanol extract for each column. These extracts were carefully reduced to 500-μl-volumes in rotary evaporators operated at about 100-mm pressure and temperatures not exceeding 25°C. Excess 2-propanol was added to the propanol-water extracts prior to the evaporation step in order to remove the water as an azeotrope (87.9% 2-propanol, 12.1% water), thus converting these fractions to solutions of pure 2-propanol.

Carbon columns were processed in essentially the same manner as the carbon columns used in the landfill study described earlier. The final volumes of the CCE's and CAE's thus produced were adjusted to approximately 500 μl.

Comparison of the Column Extracts and Identification of Some Natural Organic Constituents of Ground Water

The XAD-2-ether (XAD-E), PAW-2-propanol-water (PAW-PW), PAW-2-propanol (PAW-P), carbon chloroform (CCE), and carbon alcohol extracts (CAE) were subjected to gas chromatography on 6-ft glass columns packed with 3% OV-1 on Gas Chrom Q (Applied Science Laboratories, State College, Pennsylvania) employing flame ionization detection at relatively high sensitivity. The chromatograms obtained, as typified by the CCE chromatogram shown in the lower trace of Figure 27.10, revealed the presence in these extracts of only very low quantities of organic compounds readily amenable to GC. This would appear to indicate the presence of very little relatively nonpolar and volatile organic matter in native ground water at the field site, since such compounds, had they been present in the sampled water at the 500-ng/l level, would have produced significant GC peaks, based on the volumes of water sampled, the sensitivity of the chromatographic procedures, and the assumption of reasonable efficiency of the adsorbents.

Suitable aliquots of all the column extracts were gently evaporated to dryness by rotary evaporator and the residues were reacted with a 5:1 mixture of N,O-*bis*(trimethylsilyl)acetamide:trimethylchlorosilane (Sylon BT, Supelco, Inc., Bellefonte, Pennsylvania). Gas chromatography of the resulting reaction mixtures revealed the presence in the XAD-E's, PAW-P's, CCE's, and CAE's of organic substances which were too polar to be amenable to gas chromatography but which yielded trimethylsilyl derivatives which could be chromatographed. The probable presence of such substances in the PAW-PW extracts was also indicated. These polar compounds were present in greatest quantity in extracts of the columns occupying the first position in series in the sampling systems utilizing XAD-2 and PAW, but were found in significant quantities in extracts of both columns from the sampling system utilizing carbon.

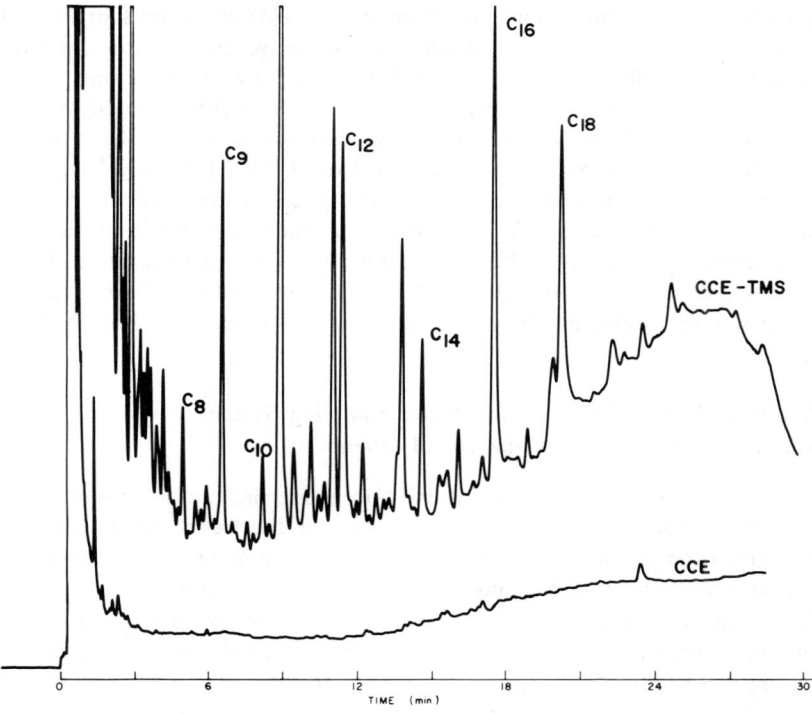

Figure 27.10 FID chromatogram of loam field lateral well CCE and CCE-TMS.

Figure 27.10, which shows results obtained by chromatographing approximately equal quantities of underivatized and derivatized CCE from the second column in the series used for sampling ground water at the field site, clearly illustrates the presence of polar substances in this CCE, and, hence, in the native ground water. The XAD-E's, PAW-P's, and CAE's yielded TMS derivatives which appeared for the most part to be different than those found in the TMS-CCE's. In particular, the TMS-PAW-P from the first column in the sampling series contained many relatively high-boiling substances which were only partially resolved by the chromatographic systems employed. Unresolved high-boiling derivatives were the principal components of the TMS-CAE's.

The silylated CCE's were analyzed by computerized GC/MS, employing the same equipment and procedures as used in the previously described landfill study, in an effort to identify individual compounds. The components indicated by carbon number in Figure 27.10 were identified and confirmed by comparison with standards as TMS derivatives of normal fatty acids, hence showing the presence in the native ground water at the field site of caprylic, nonanoic, capric, lauric, myristic, palmatic and stearic acids. Analysis of the silylated XAD-E's and PAW-P's indicated that they either did not contain the fatty acids found in the CCE's or contained them at much lower levels than the CCE's. This may have resulted from a considerable degree of dissociation of the acids at the near neutral pH of the sampled ground water and the probable inability of the resins to effectively adsorb ionic forms of organic molecules. Fatty acids have, in fact, been reported to be poorly recovered by XAD-2 unless the sampled water has been acidified previous to passing through the resin.[13]

Laboratory Study of the Ground Water Sampling System Using XAD-2, PAW and Carbon as Adsorbents

Because of the low levels of naturally-occurring organic compounds apparently present in the native ground water at the Texas A&M field site, comparison of the ground water sampling systems using the three different adsorbents was difficult in the background sampling work described above. A study was, therefore, undertaken to obtain more definitive information concerning the performance of the ground water sampling system developed in this work for recovery of organic compounds from aqueous solution when each of the three adsorbents (XAD-2, PAW, activated carbon) was used in the system.

For this work, three ground water sampling systems containing respectively XAD-2, PAW and activated carbon were used for simultaneous sampling in the laboratory of synthetic aqueous solutions of selected organic compounds. These solutions contained compounds which either were identified as being present in septic tank effluent in preliminary studies, were thought to be probable constituents of such effluent, or were of general interest as representative of particular classes of compounds found or likely to be found in polluted or natural waters. Concentrations were usually kept below 12 µg/l in an effort to insure that the organics were in true solution. Approximately 10 liters of water were sampled by each system during a single experiment, with all three systems sampling from the same region of a single 45-liter carboy to insure uniformity of sample. Controlled flow rates of 25 and 15 ml/min were used in different experiments. Processing and extraction of adsorbents was essentially as described above for the field studies, except that in one experiment XAD-2 and PAW columns were extracted with 20 ml of acetone followed by 80 ml of chloroform, as suggested by Webb in his studies of XAD resins.[16] The tentative observations recorded in this work, which is still in progress, are presented below.

1. The ground water sampling system developed for this work performed fairly effectively for recovery of low levels of organic compounds from aqueous solution when columns containing any of the three tested adsorbents were used, with each adsorbent appearing to have both advantages and disadvantages.

2. PAW was, as would be expected, particularly effective for recovery of intermediate- to high-molecular-weight compounds containing phenolic hydroxyl groups. Surprisingly, however, it appeared to perform very similarly to, if not more effectively than, XAD-2 for recovery of all those compounds tested which had a molecular weight in excess of 200. For the lower-molecular-weight compounds, XAD-2 was far superior to PAW, even when the PAW was extracted with acetone-chloroform rather than 2-propanol in order to reduce losses of volatile substances during extract processing.

3. The activated carbon used in these studies appeared much more effective than either PAW or XAD-2 for removal of carboxylic acids from solutions of pH greater than about six.

4. The washed Filtrasorb 200 appeared to differ from the Nuchar C-190 activated carbon (no longer available) used in the landfill studies in regard to strength of adsorptive bonds formed with carboxylic acids. When Nuchar C-190 was used in the earlier work, the carboxylic acids were eluted mainly by extraction with ethyl alcohol, but these compounds were mostly present in the chloroform extract of the washed Filtrasorb 200.

5. The first column in the series of two XAD-2 or PAW columns appeared to recover most of the organic matter that was removed from the sampled water by these adsorbents, but significant quantities of material were present on the second column in the series of two carbon columns.

6. Sensitive compounds such as cholesterol and diethylstilbestrol were recovered much less effectively by the carbon system than the systems employing XAD-2 and PAW, probably indicating decomposition and/or irreversible adsorption of these substances on the carbon.

7. The extraction of XAD-2 columns with acetone followed by chloroform, as suggested by Webb,[16] appeared as effective as ether extraction and was less cumbersome for the columns used in this study.

Discussion

The above work clearly shows that many problems are involved in analysis of organics in ground water. In addition to the problems normally encountered in organics analysis in surface waters, there are also severe sampling problems involved in obtaining uncontaminated and unaltered samples from the relatively inaccessible subsurface waters.

The ground water sampling systems developed for this work appear to hold considerable promise for effective sampling of organics in shallow aquifers. One problem encountered with the present systems is degassing of ground water in the columns, particularly XAD-2 and PAW, due to the pressure drop occurring across the columns as the result of the low pressures produced on the downstream side of the columns by the action of the peristaltic pumps. This does not appear to be a serious problem at present, but should be eliminated for best results. Development of a submersible, noncontaminating pump, probably of teflon construction, is needed to alleviate this problem and is a necessity for deep-aquifer sampling.

The present study indicates that the concentration of natural organic compounds in ground water at the Texas A&M field site is very low. Hence, the study of these compounds is difficult, but, conversely, the entry of any polluting organics into the aquifer should be more readily apparent because of the low background of native organics in the ground water. While chromatography of the various adsorbent extracts indicated that organic compounds in addition to fatty acids were present in the ground water at the field site, the latter compounds appear to be the predominant class of naturally-occurring organic constituents which were amenable to GC. The presence of fatty acids as natural components of ground water is not surprising, since these materials are known to persist in sedimentary deposits for extremely long periods, and the most productive ground water aquifers are most often found in those portions of the earth's crust that are sedimentary in origin.

ACKNOWLEDGMENTS

The authors wish to thank Drs. Calvin Woods, Kirk Brown and Frank Slowey of Texas A&M University for their collaborative efforts on the septic tank disposal studies.

REFERENCES

1. Miller, D. W., F. A. DeLuca and T. L. Tessier. "Ground Water Contamination in the Northeastern States," Environmental

Protection Agency Report EPA-660/2-74-056, Washington, D. C. (June 1974), pp. 211-216.
2. Walker, W. H. *J. Amer. Water Works Assoc.* **61**, 31 (1969).
3. Andersen, J. R. and J. N. Dornbush. *J. Amer. Water Works Assoc.* **59**, 457-470 (1967).
4. Fuhriman, D. K. and J. R. Barton. "Ground Water Pollution in Arizona, California, Nevada, and Utah," Environmental Protection Agency Report 16060 ERV 12/71, Washington, D.C. (December 1971), pp. 92-94.
5. Scalf, M. R., J. W. Keeley and C. J. LaFevers. "Ground Water Pollution in the South Central States," Environmental Protection Agency Report EPA-R2-73-268, Corvallis, Oregon (June 1973), pp. 100-102.
6. California State Department of Natural Resources. "Sanitary Landfill Studies," Appendix A: Summary of Selected Previous Investigations, Bulletin 147-5, Sacramento, California (July 1969).
7. Huges, G. M., R. A. Landon and R. W. Farvolden. "Hydrology of Solid Waste Disposal Sites in Northeastern Illinois, Illinois State Geological Survey, Environmental Geology Notes, Publication 45, Urbana, Illinois (April 1971).
8. Breidenbach, A. W., J. J. Lichtenberg, C. F. Henke, D. J. Smith, J. W. Eichelberger and H. Stierle. "The Identification and Measurement of Chlorinated Hydrocarbon Pesticides in Surface Waters," U.S. Department of the Interior, Federal Water Pollution Control Administration, Washington, D.C. (1966), pp. 44-50.
9. Booth, R. L., J. N. English and G. N. McDermott. *J. Amer. Water Works Assoc.* **57**, 215-220 (1965).
10. Shriner, R. L., R. C. Fuson and D. Y. Curtin. *The Systematic Identification of Organic Compounds*, 5th ed. (New York: John Wiley and Sons, Inc., 1965), pp. 67-107.
11. Breidenbach, A. W., J. J. Lichtenberg, C. F. Henke, D. J. Smith, J. W. Eichelberger and H. Stierle. "The Identification and Measurement of Chlorinated Hydrocarbon Pesticides in Surface Waters," U.S. Department of the Interior, Federal Water Pollution Control Administration, Washington, D. C. (1966), pp. 12-14.
12. Webb, R. G., A. W. Garrison, L. H. Keith and J. H. McGuire. "Current Practice in GC/MS Analysis of Organics in Water," Environmental Protection Agency Report EPA-R2-73-277, Corvallis, Oregon (1973), pp. 37-41.
13. Junk, G. A., J. J. Richard, M. D. Grieser, D. Witiak, J. L. Witiak, M. D. Arguello, R. Vick, H. J. Svec, J. S. Fritz and G. V. Calder. *J. Chromatogr.* **99**, 745-762 (1974).
14. Buelow, R. W., J. K. Carswell and J. M. Symons. *J. Amer. Water Works Assoc.* **65**, 57-72 and 195-199 (1973).
15. Endres, H. and H. Hormann. *Angew. Chem. Int. Ed.* **2**, 254-260 (1963).
16. Webb, R. G. "Isolating Organic Water Pollutants: XAD Resins, Urethane Foams, Solvent Extraction," Environmental Protection Agency Report EPA-660/4-75-003, Corvallis, Oregon (June 1975), pp. 3-8.

CHAPTER 28

CHLOROFORM-EXTRACTABLE ORGANIC COMPOUNDS IN THE INTERNATIONAL GREAT LAKES

William M. J. Strachan

Canada Centre for Inland Waters
Burlington, Ontario, Canada

INTRODUCTION

The nature of organic compounds present in natural waters is largely unknown. It is generally described in terms of either its method of isolation (*i.e.,* fulvic acid, humic acid) or by its chemical-biochemical behavior (*i.e.,* COD, BOD). Typically, dissolved organic carbon in the waters of the Great Lakes is 1-5 mg/l.[1] Unfortunately, these determinations tell us little about the chemical structure of organic material in natural waters.

At the present time, society is becoming increasingly conscious of the presence and hazards of trace amounts of toxic organic compounds which, inadvertently or otherwise, have entered the environment. Such compounds as have been noted do not share the same structural features as the bulk of the dissolved organic material but tend to be smaller, nonionic molecules. It was the intent of this study to examine the problem of the nature of such compounds already present in the Great Lakes and if possible to provide necessary background information that will undoubtedly be required in assessing impacts and trends of organic compounds in the aquatic environment.

Since the objective was to provide baseline information for analysis of relatively small, largely neutral compounds, chloroform extraction and gas chromatography were decided upon as the operational tools. About the only general instrument capable of giving information on chemical structure at the required level of less than 1 µg/l is the integrated gas chromatograph/mass spectrometer. Additionally, since a very large number of spectra can be generated per sample, a data system is essential. Two mass spectra search systems—one developed by the National Institutes of Health[2] and the other by Professor Biemann[3] and Battelle Corporation[4] were examined for their utility in identifying environmental samples from the open Great Lakes.

SAMPLES

Samples were obtained during regular cruises of the M. V. Martin Karlsen, a ship on charter at the time to the Canada Centre for Inland Waters. Sample sites are indicated in Figure 28.1. Approximately six samples for each of the four international Great Lakes were obtained during both the spring and fall of 1973. All were from offshore locations (> 2 km) away from immediate "nonpersistent" local effects. Van Dorn bottle samplers were employed and water collected from the 10-m depth to avoid contamination from the ship. A total of 12 liters (two bottle casts) were transferred to a stainless steel pressure tank and connected to a 5-10 psi nitrogen line purified by passage through molecular sieves. The samples were then pressure-filtered through large (142-mm) membrane filters of 0.22 µ porosity for sterilization. The first couple of liters were discarded as washing since the filters could not be otherwise precleaned. Approximately 10 liters were collected in chromic acid-cleaned, sterile, glass-stoppered bottles and refrigerated until extracted on shore.

Extraction was carried out in 9-liter glass vessels equipped for extraction by either heavier or lighter-than-water solvents. In this study, chloroform was employed and the arm was so constructed as to require 9 (± 0.2) liters of sample to maintain the chloroform level. Only glass or teflon parts are employed and the system is protected with an adsorption trap. The samples were made saline (0.1 M NaCl) to enhance partitioning between the two immiscible phases. One sample, obtained from Lake Ontario for another study, was examined for the influence of extraction time and acidity. While there was some increase (about 25%) in quantity of extracted material for periods longer than 24 hr, there were no qualitative differences in the chromatographic peaks. Periods of 6 hr extraction time resulted in about half the extracted material compared with 24 hr but with qualitatively similar chromatograms. Acidification was found to result in a doubling of the yield for a 24-hr extraction period, but no new chromatograph peaks. While this increase is desirable, acidification was not employed to avoid possible chemical alteration to acid sensitive compounds that might be present. Twenty-four hours was

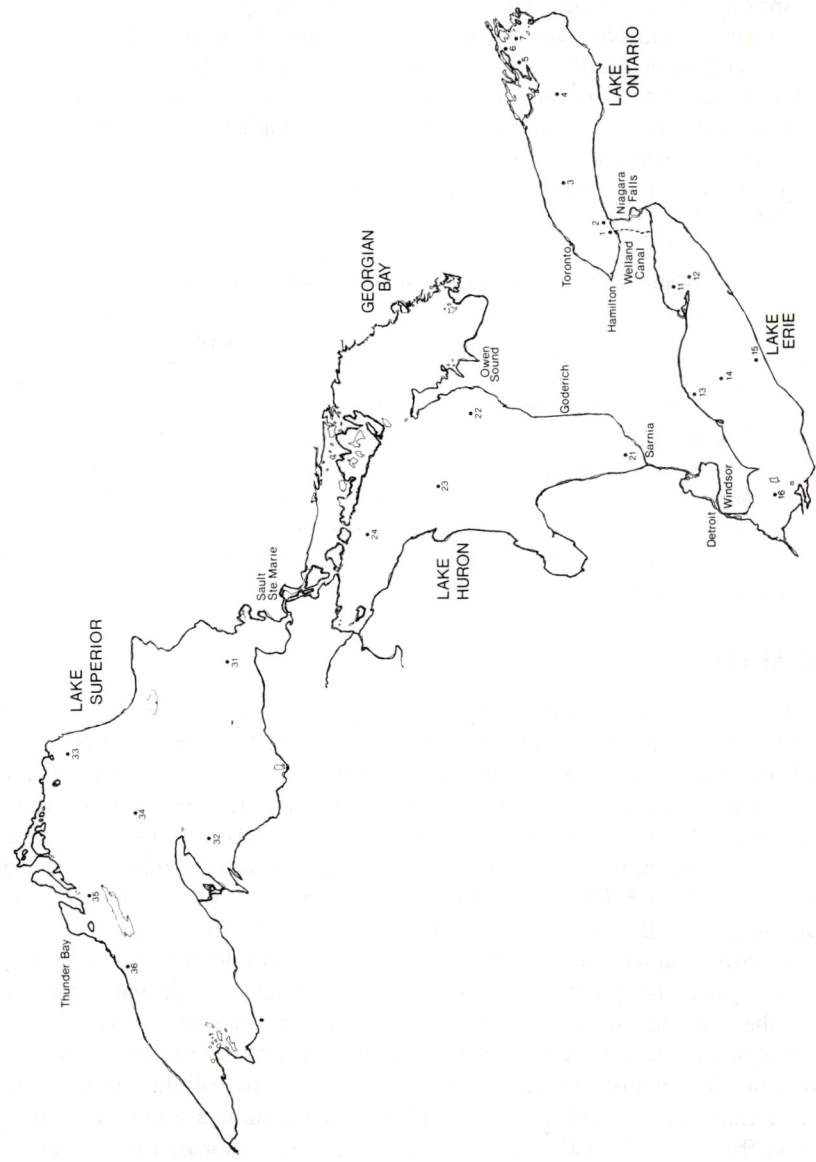

Figure 28.1 Great Lakes sampling stations.

the period of time employed, and to aid in dispersing the extracting solvent a coarse glass frit was employed. A throughput of very fine droplets of approximately 15 ml $CHCl_3$ per min was obtained.

Extracted samples were evaporated in an all-glass rotary evaporator, final evaporation occurring in small weighed ampoules at 60°C. The extracted weights were largely independent of lake or season and ranged from 0.1-0.9 mg/l with an overall mean of 0.3 mg/l (Table 28.1). This material was methylated with diazomethane and the ampoules, after evaporation of the solvent, were sealed until examined by GC and by GC/MS.

Table 28.1 Yields of Chloroform Extractable Material

Lake	Yields (mg/l)[a]	
	Spring	Autumn
Ontario	0.31 (0.18-0.93)	0.34 (0.22-0.50)
Erie	0.28 (0.18-0.53)	0.32 (0.26-0.50)
Huron	0.17 (0.11-0.21)	0.33 (0.24-0.48)
Superior	0.25 (0.16-0.44)	0.26 (0.25-0.28)

[a]Means and ranges given.

ANALYSIS

For gas chromatographic purposes, samples were taken up in 20-50 μl of benzene and 3 μl of this injected directly onto OV-1 columns in both an FID-equipped gas chromatograph and a Finnigan 1015C gas chromatograph/mass spectrometer-data system. The gas chromatographs for Ontario were programmed from 175-250°C at 4°C/min (4 min initial hold) with a carrier flow of 20 ml/min. For Superior, Huron and Erie, the program was changed to 125-250°C at 6°C/min (no initial hold) to facilitate observation of early eluting peaks. Examination of blanks (from methylation and from the chloroform) under both conditions did not show any interfering GC peaks.

One particular problem encountered and not resolved during the study was the problem of "mass defects" when data systems are employed. A data system operating a mass spectrometer looks at a very small mass window (in our instrument ± 0.05) around what the calibration program has defined as unit mass. Since perfluoro compounds are largely used for calibration, this will, *with data systems*, lead to apparently low relative intensities for higher mass/charge ratios, since the data system will not see the signal corresponding to the top of a mass peak but will be on the forward side of that peak. This can be corrected with a small offset potentiometer. For compounds such as the methyl esters of fatty acids, this correction should be approximately 0.1 amu/100 mass units. In current practice, calibration is done normally but samples are run with the offset. Without this small modification, the data system, depending on the resolution employed, can give distorted relative intensities and an example of this is shown in the spectra of methyl behenate (Figure 28.2) from sequential injections during the same GC run. The distinctly lower (half to one-eighth) intensities for the sample run in calibrate mode are indicative.

Two mass spectral search systems were available for comparing libraries of reference spectra against those from samples. The spectral search first developed by Professor Biemann compared sample and knowns by means of the two most intense mass peaks in each 14-amu window of the entire spectrum. The peak search system, developed at the National Institutes of Health in Washington, was also restricted to two masses per 14-amu window, but masses were put in sequentially at the operator's discretion. Both employ the same extensive library and the two procedures were used to attempt chemical identification of the samples.

RESULTS AND DISCUSSION

Each chromatogram from each station of a lake has minor differences from all others from the same lake, but in general there were more common than otherwise features. An illustrative chromatogram is given for each lake in Figure 28.3.

For Lake Ontario (Figure 28.3d), a number of hydrocarbons, apparently in an acyclic homologous series, were seen at every station. Comparison of the spectra, corrected for background, with the two spectral matching systems suggested both branched and normal hydrocarbons and GC retention times corresponded to the C20-C30 *n*-alkanes. Additionally, phthalate esters were prominent. While low levels of these compounds are almost automatically attributed to contamination, their source here was uncertain. The presence of triphenyl phosphate was noted in all Ontario stations during the spring cruise but nowhere else. This compound also might be a contaminant from plasticized material, but no ready explanation is available as to why it is restricted to the one cruise since processing and collection of the spring cruises on both Erie and Ontario were carried out at the same time. Semiquantitative indications of concentrations and locations of the hydrocarbons and phthalate esters observed are indicated in Table 28.2. Total hydrocarbons

484 ORGANIC POLLUTANTS IN WATER

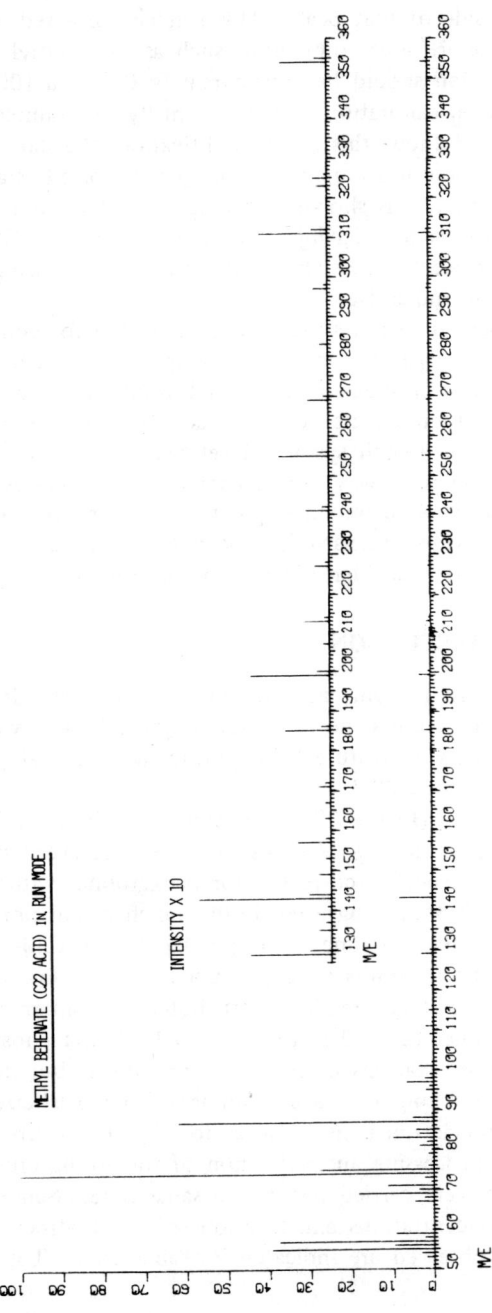

IDENTIFICATION OF ORGANIC POLLUTANTS 485

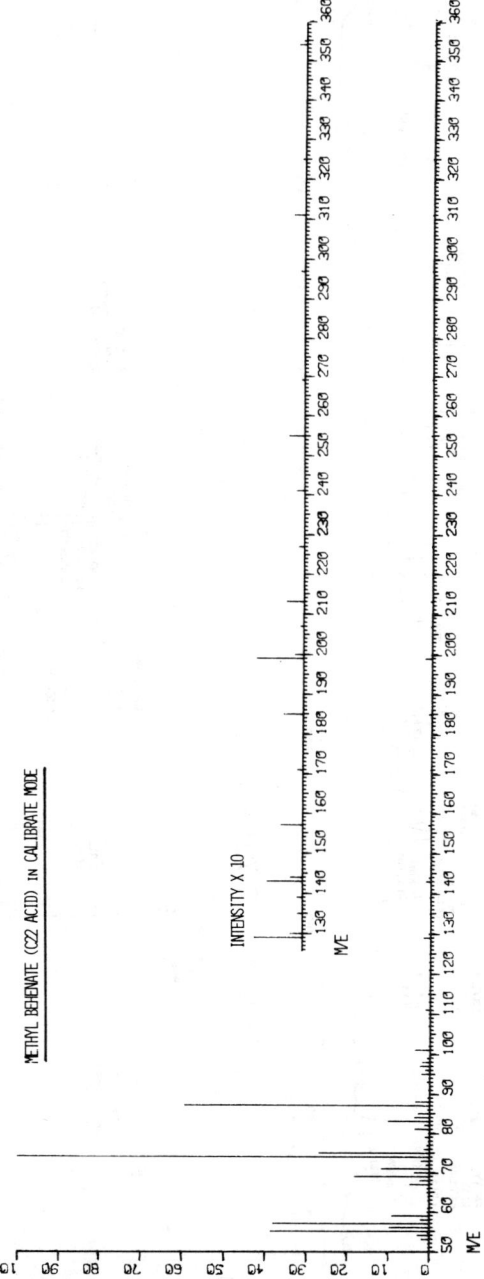

Figure 28.2 Mass spectra of methyl behenate in run and calibrate modes.

486 ORGANIC POLLUTANTS IN WATER

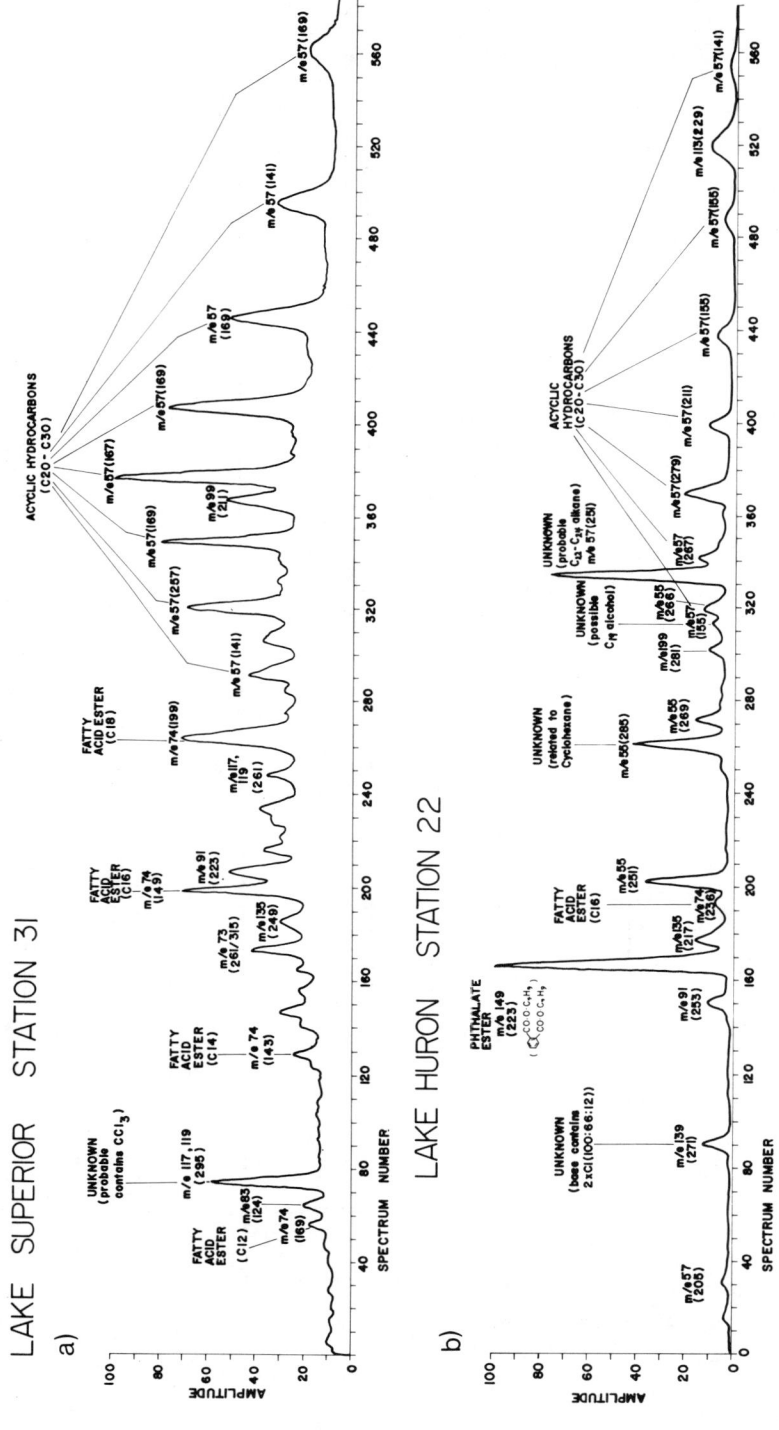

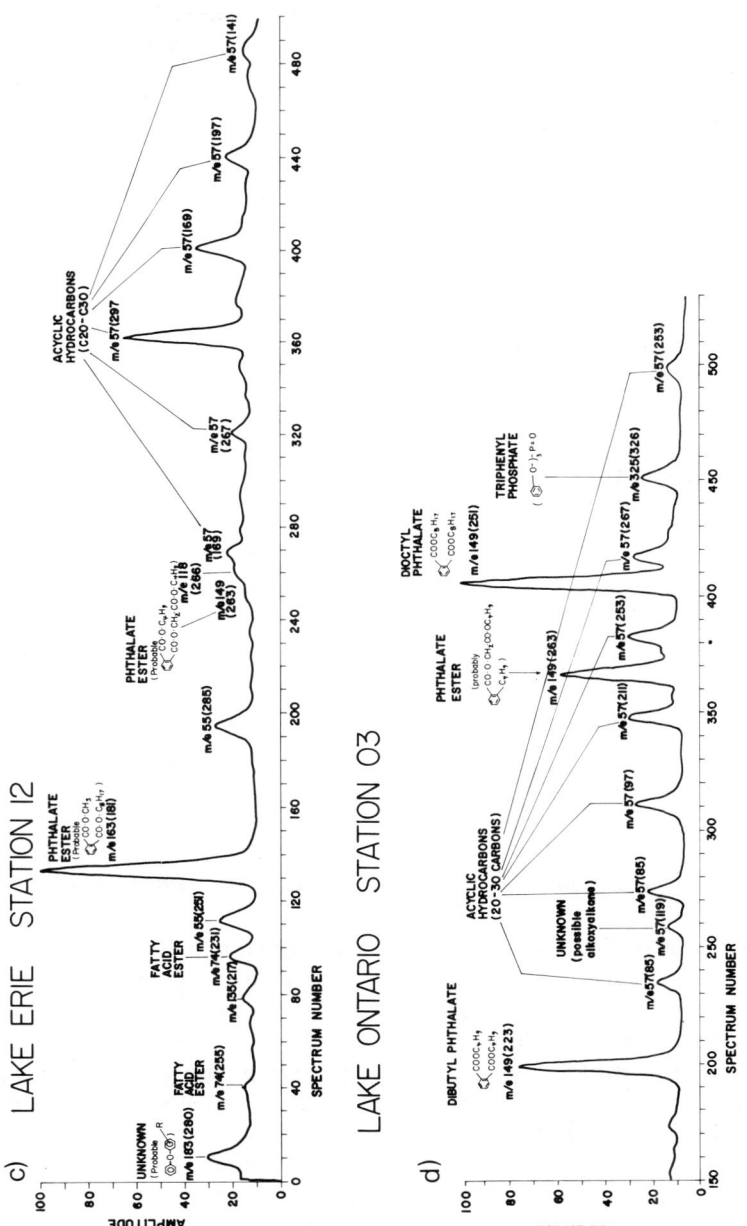

Figure 28.3 Sample GC/MS total ion chromatograms. Base peaks (and highest significant masses) given.

Table 28.2 Semiquantitative Observation of Organics[a]

		Hydrocarbons (μg/l)	Fatty Acids (μg/l)	Phthalates/Plasticizer (μg/l)
Lake Ontario	1	5	< 0.1	3
	2	4	< 0.1	20
	3	2	< 0.1	2
	4	6	< 0.1	10
	5	5	< 0.1	5
	6	6	< 0.1	(50)
	7	5	< 0.1	8
Lake Erie	11	0.4	0.4	1
	12	0.3	0.6	1
	13	0.3	0.8	0.8
	14	0.7	3	2
	15	2	0.2	0.7
	16	4	0.7	6
Lake Huron	21	7	0.3	3
	22	8	1	0.8
	23	3	0.1	3
	24	5	0.1	3
Lake Superior	31	4	0.1	< 0.1
	32	< 0.1	< 0.1	< 0.1
	33	4	2	< 0.1
	34	0.5	0	< 0.1
	35	0.5	0.1	< 0.1
	36	3	< 0.1	< 0.1

[a]Based upon average responses of fatty acid esters and saturated hydrocarbons in the flame detector gas chromatograph.

were generally observed at the low μg/l level with the phthalates somewhat above that.

In addition to the compounds mentioned, spring samples from the western end of Lake Ontario showed small amounts of fatty acids; also, stations 1 and 2 showed dehydroabietic acid in the spring and some early eluting, unidentified chlorinated hydrocarbons in the autumn (both near the detection level of 0.1 μg/l). These latter are probably related to the concentration of heavy industry, particularly chemical, along the Niagara River and the Old Welland Canal.

In the case of Lake Erie, the chromatograms frequently resembled that illustrated in Figure 28.3c. The same hydrocarbons were present as in the other lakes again with indications of branching. Levels for hydrocarbons (Table 28.2) appear to be somewhat lower than for Lake Ontario with the highest levels occurring in industrialized areas—the south shore and the Western Basin. Fatty acids were distinctly more prominent in Lake Erie than Ontario, ranging as high as 3 µg/l in the Central Basin (station 14). In addition, there were the ubiquitous phthalate esters at levels considerably lower than those for Ontario, with the exception of the industrialized Western Basin.

Several unidentified chromatographic peaks were observed at the illustrated station 12, which were also found at station 14 in mid-lake and at a number of Lake Huron stations. The one at spectrum 10 appeared to be a diaryl ether containing alkyl substituents. The spectrum of the compound at spectrum number 78 is shown in Figure 28.4 for a Lake Huron sample (22:177) where it was found in higher concentrations. The arrows indicate the masses employed here and elsewhere in the peak search mode. There are apparently no similarities to the "hits" by the two procedures but, while not noted in the illustrated structures, the simple change from mass 83 to 79 was sufficient to alter the "hits" of the peak search to nonbornane-related ones similar to that from the spectrum search. The compound corresponding to spectrum number 112 was also present in greater concentrations in a Lake Huron sample (22:202) and is illustrated for the latter in Figure 28.5. In this case, there were no apparent similarities to compounds selected by the two procedures. The compound at spectrum number 193 (illustrated for Lake Huron 22:261, in Figure 28.6) had similar "hits" by both procedures and appears to be related to a cyclohexyl structure. Also observed for Lake Erie was a very positive identification for PCB's in the autumn sample from station 16.

The chromatographs for Lake Huron (*e.g.*, Figure 28.3b) indicated that most of the compounds found in Lake Erie were also present at almost every station. The detection limit of approximately 0.1 µg/l, corresponding to a small peak on these chromatographs, represents a minimum level for the early eluting compounds since the evaporation procedure would undoubtedly lead to some loss of this material.

The C20-C30 series of hydrocarbons was present in these samples as well, but additionally there was a higher concentration from a non-homolog hydrocarbon present at both stations 21 and 22. The spectrum was typical of an *n*-alkane while its chromatographic retention time suggested 23 carbon atoms. Fatty acids (or their methyl esters) plus the phthalates were also noted. Semiquantitative levels of these are given in Table 28.2.

In addition to the particular compounds previously mentioned for Lake Erie which were also found in Lake Huron (spectra numbers 177, 202 and 261 in Figure 28.2c), the compound at spectrum number 90 and a

490 ORGANIC POLLUTANTS IN WATER

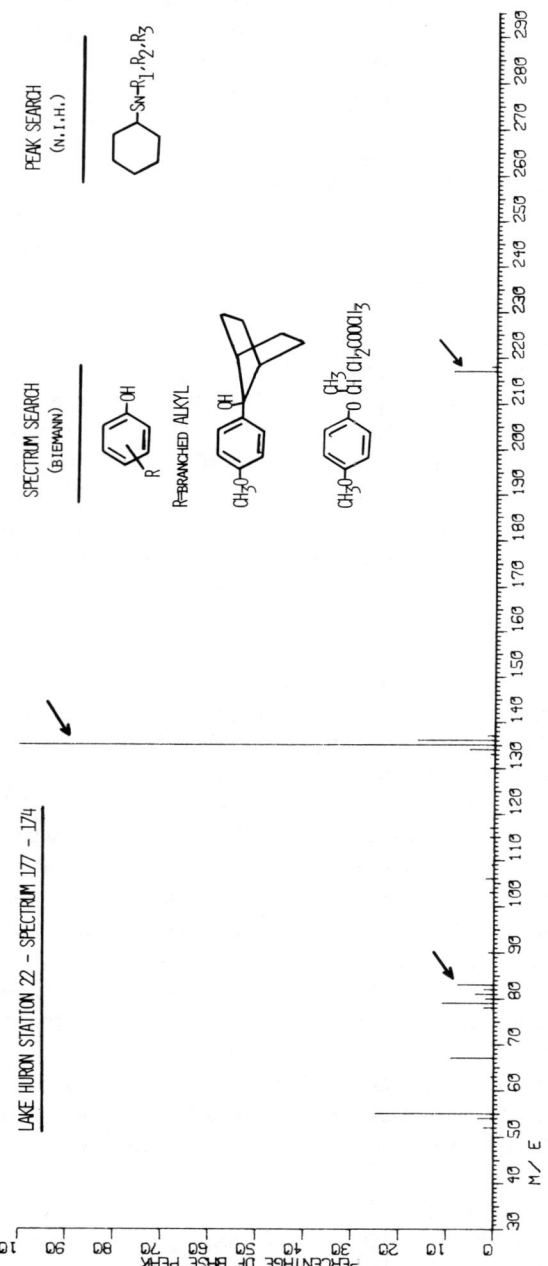

Figure 28.4 Lake Huron station 22–spectrum 177-174.

IDENTIFICATION OF ORGANIC POLLUTANTS 491

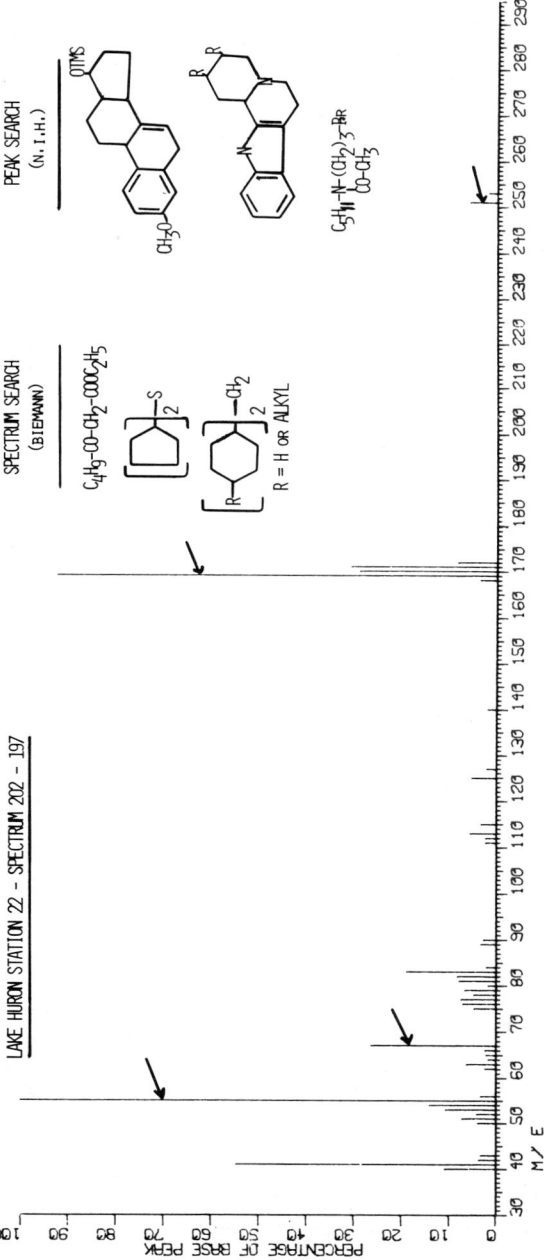

Figure 28.5 Lake Huron station 22—spectrum 202-197.

492 ORGANIC POLLUTANTS IN WATER

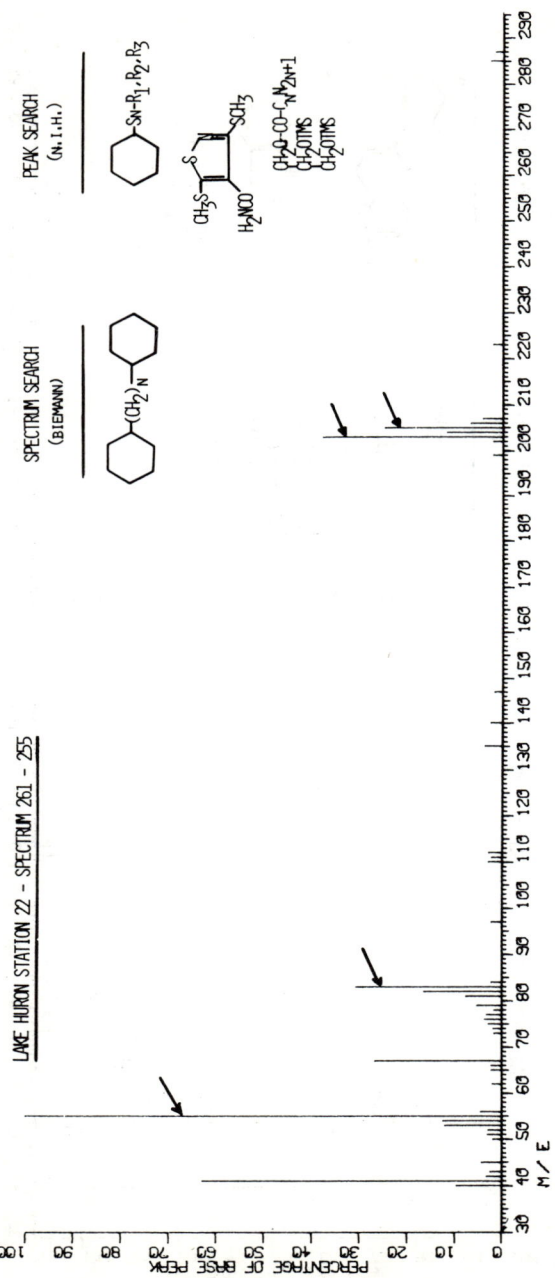

Figure 28.6 Lake Huron station 22—spectrum 261-255.

compound occurring only for station 21 at spectrum number 398 are noted here. The former (Figure 28.7) gave an almost perfect two-chlorine isotope cluster for the base peak of 139 amu and the structures arrived at by the two search methods were consistent and supported this. Spectrum number 398 (Figure 28.8) from the chromatograph at station 21, when compared by the two methods, was also consistent having strong similarities with spectra for fatty alcohol dicarboxylic esters—compounds also used commonly as plasticizers.

The total ion chromatographs for Lake Superior samples showed many of the same structures as were indicated for other lakes, especially Lake Huron. An illustration is given in Figure 28.2a. Chromatographically and spectrally, the same series of hydrocarbons was observed as in the other Great Lakes, including one apparently branched hydrocarbon which is chromatographically not the same as the one in Lake Huron samples, but elutes later. Fatty acids, identified chromatographically as the even-numbered ones from 12 to 18 carbon atoms, were also present but, interestingly, the phthalates and other plasticizers were absent in detectable amounts from all samples. Levels observed are indicated in Table 28.2.

A compound not previously mentioned but which appeared in all samples is the moderately-strong peak at spectrum number 74 of Figure 28.3d, the spectrum of which is shown in Figure 28.9. The isotope cluster at 117 indicated a trichloromethyl group and both the spectrum and peak searches turned up compounds in agreement with this. Indications at higher masses (at 199, 213 and 295) are that hexachloroethane is not the compound. The relative intensities, however, were not sufficiently accurate, as discussed earlier, to indicate the atomic composition but appearance would seem to indicate at least five. The compound at spectrum number 247, while not as strong as that at 74, is almost identical.

In summary, both the spectrum and peak search methods have been useful in identifying structures feasible for this sort of environmental sample. But the need to compensate for mass defects with data systems is apparent. Without such compensation, considerable information will be lost and its absence severely restricts the usefulness of available spectral search systems.

494 ORGANIC POLLUTANTS IN WATER

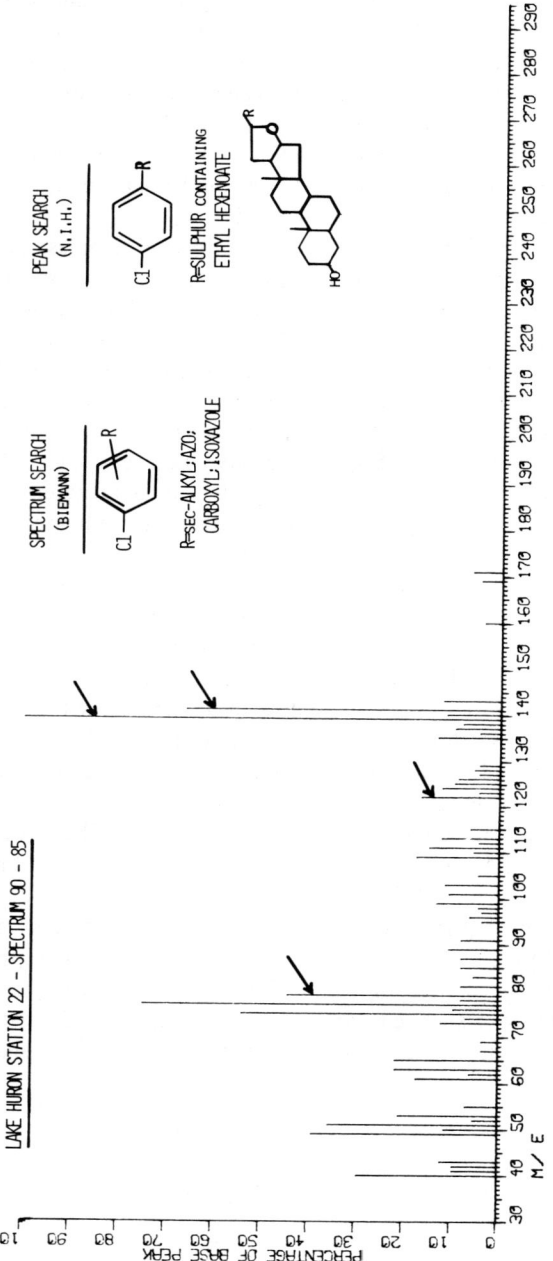

Figure 28.7 Lake Huron station 22–spectrum 90-85.

IDENTIFICATION OF ORGANIC POLLUTANTS 495

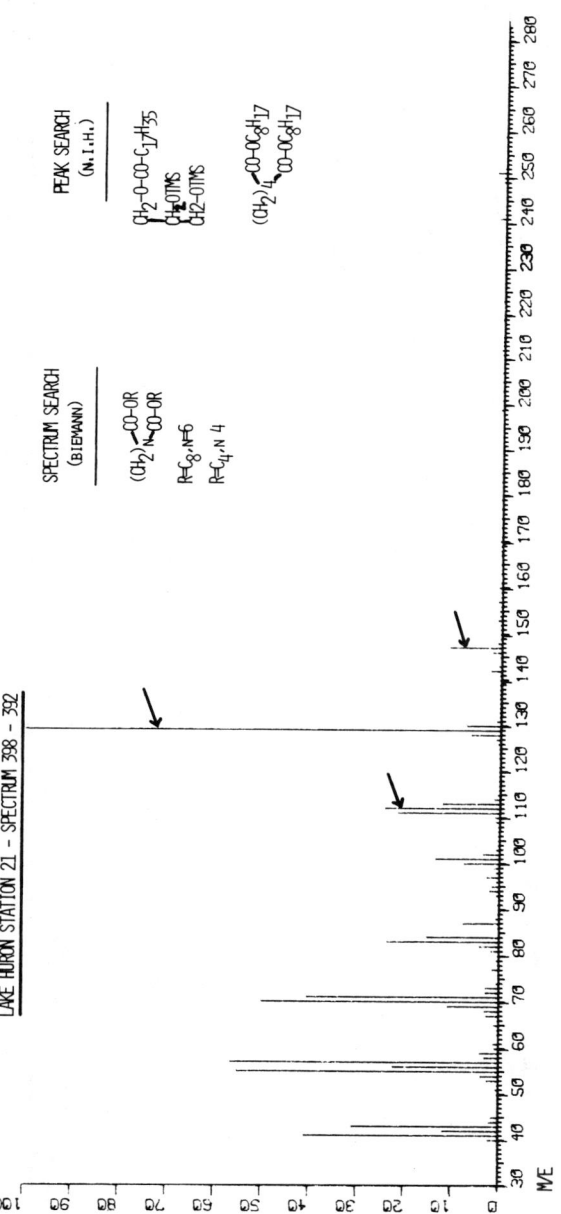

Figure 28.8 Lake Huron station 21–spectrum 398-392.

496 ORGANIC POLLUTANTS IN WATER

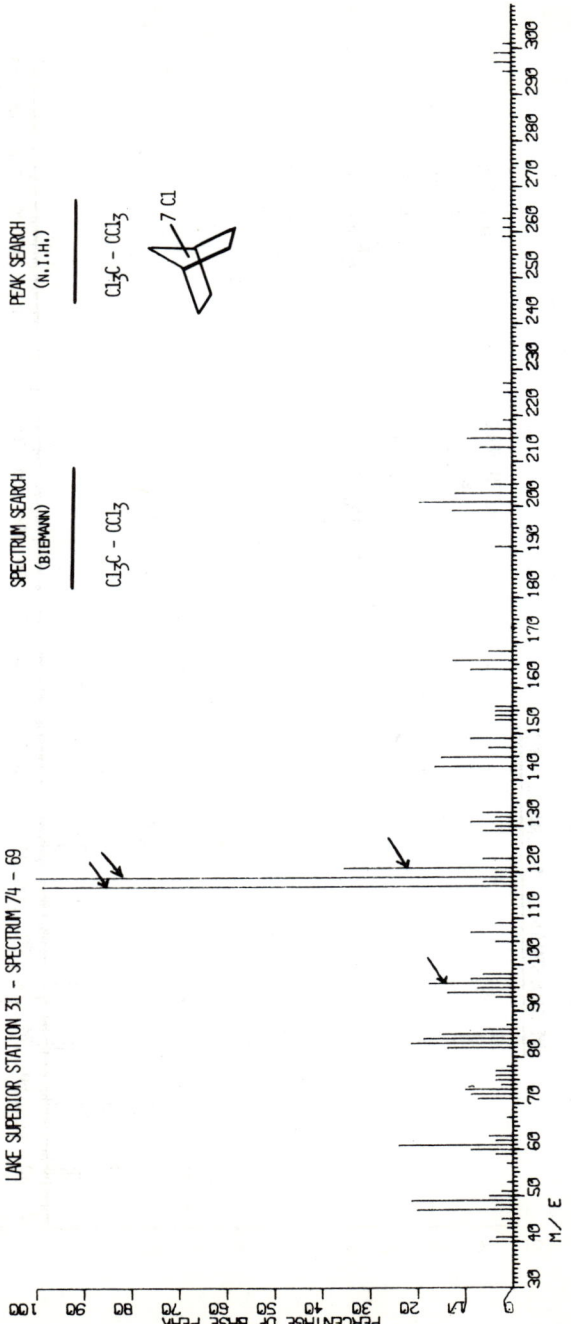

Figure 28.9 Lake Superior station 31–spectrum 74-69.

REFERENCES

1. Canada Centre for Inland Waters. Unpublished results from the STAR data file system (1970-71).
2. Heller, S. R. "Conversational Mass Spectral Retrieval System and its Use as an Aid in Structure Determination," *Anal. Chem.* **44**, 1951 (1972).
3. Hertz, H. S., R. A. Hites and K. Biemann. "Identification of Mass Spectra by Computer-Searching a File of Known Spectra," *Anal. Chem.* **43**, 681 (1971).
4. Hoyland, J. R. and M. B. Neher. "Implementation of a Computer-Based Information System for Mass Spectral Identification," U.S. EPA Publication EPA-660/2-74-048 (1974).

CHAPTER 29

USES OF WASTEWATER DISCHARGE COMPLIANCE MONITORING DATA

E. William Loy, Jr., Donald W. Brown,
J. H. Myron Stephenson

    U.S. Environmental Protection Agency
    Southeast Environmental Research Laboratory
    Athens, Georgia

John A. Little

    U.S. Environmental Protection Agency
    Region IV
    Atlanta, Georgia

INTRODUCTION

Public Law 92-500[1] of 1972 requires all point sources discharging into navigable waters of the United States to have a permit. Point sources include industries, municipal treatment plants, large agricultural feedlots, and return irrigation flows. Permits are issued under the National Pollutants Discharge Elimination System (NPDES) by the Environmental Protection Agency (EPA) or by the states when the authority has been delegated by EPA. The permit specifies, among other things, the effluent limitations on point source discharges. Therefore, the permit, ideally, should list all significant pollutants that affect the environment and specify the maximum allowable concentration in the effluent.

The EPA or state agencies that issue permits are authorized by PL 92-500 to perform periodic spot checks on the point source discharges to determine compliance with permit conditions. This surveillance program is called Compliance Monitoring. Within EPA Region IV, which comprises the eight southeastern states, the Surveillance and Analysis

(S&A) Division, located in Athens, Georgia, is responsible for performing EPA's part of sampling and analysis of effluents. Many of the states in Region IV, whether authorized to issue permits or not, have assumed much of the Compliance Monitoring program.

The Region IV S&A Division effort in Compliance Monitoring resulted in 400 inspections of dischargers by July 1, 1975. These inspections detected a permit violation rate of approximately 35%; however, many of these violations were of a minor nature. These inspections have also provided EPA with:

- first-hand knowledge of the type of waste treatment used,
- an idea of the effluent variation characteristics (3-day survey in almost all cases),
- information on waste treatment personnel,
- adequacy of laboratory and analytical techniques,
- some limited information on discharge effects on receiving waters.

The S&A Division split samples with dischargers and state laboratories in the majority of inspections. This not only provided analytical quality control information on the laboratories involved but also has helped identify analytical problems, *e.g.*, interferences with methods because of waste characteristics.

As mentioned above, all pollutants that affect the environment should be listed and concentration limitations specified in the permit. Unfortunately, in the case of specific organic pollutants, no one, including the discharger, has a complete inventory of the compounds that are present in most effluents. In many cases, the only effluent characteristics relating to organics that are often known are gross parameters such as total organic carbon, oil and grease, total phenols, chemical oxygen demand, and biological oxygen demand. It is becoming increasingly apparent that an inventory is needed of all organic compounds being discharged to the nation's surface and ground waters. Investigations such as the New Orleans drinking water study,[2] the 1975 EPA National Organics Reconnaissance Survey,[3] and the Northwest Georgia Organic Characterization Study[4] (performed by Region IV) show that the drinking water of the few cities investigated had anywhere from a few organic pollutants up to 53, as in the case of Miami, Florida drinking water. During the past three years, our laboratory has participated in ten investigations of taste and odor problems in public and private drinking water supplies. All were caused by one or more organic compounds that found their way into the water supply. Concern as to the effect of these organics in water supplies is caused not only by their potential toxic effect but also by the ever-increasing numbers of organics that are possible or probable carcinogens. In addition to health effects on humans, there are harmful effects of organics on the aquatic biota. Some are highly toxic to aquatic biota, especially pesticides, while many others have an adverse effect on the flavor of fish and other edible species.

The primary reason for the lack of knowledge on specific organics in point source discharges has been the lack of instrumentation capable of identifying trace organics (generally low part per billion concentrations in water) in a reasonable time. This situation has been greatly improved with the advent of computerized low-resolution gas chromatograph/mass spectrometer (GC/MS) systems. Scientists at the Environmental Research Laboratory, Athens, Georgia, have shown that GC/MS can be readily applied to the characterization of industrial effluents for specific organic pollutants.[5-7] Heller et al.[8] recently reported on increased capability in the computerized mass spectral matching field which should enhance the chances of identification of previously unknown organics.

The isolation of organics prior to GC/MS analysis is another time-consuming and uncertain segment of organic analysis created by the vast number of organics with wide-ranging chemical and physical properties.

In EPA Region IV, the Laboratory Services Branch, Organic Unit has the responsibility for providing the organic analytical support for the EPA compliance monitoring program. This unit has developed and adapted analytical methodology to isolate and identify as broad a spectrum of organics as possible. At present, this is primarily restricted to organics that pass through a gas chromatograph and are in a concentration range of approximately 1 $\mu g/l$ or greater.

EXPERIMENTAL

Sampling

Both grab samples and samples from automatic sampling compositors have been taken when analyzing for organics during compliance monitoring. Usually three or more samples are taken over a three-day period. Sample containers are 1-quart, wide-mouth, glass bottles with teflon-lined lids. Narrow-mouth, 4-oz glass bottles or 23-ml vials, both with screw caps containing aluminum foil liners, are used for sampling waters for volatile organic analysis. Head space is carefully eliminated in this case.

The only preservation recommended is refrigeration or icing. The 2½-gal bottle used as a receiver for the automatic compositors is packed in ice during the 24-hr sampling. The samples are then shipped to the laboratory by the fastest means and extracted or analyzed as soon as possible after receipt.

Sample Preparation

Sample preparation consists of one or a combination of the techniques shown in Table 29.1. The techniques chosen are dependent on the type of organics listed in the permit or suspected to be in the effluent.

Table 29.1 consists of organic analytical methods used by the Region IV Organic Unit. This is strictly a scheme for organics amenable to GC analysis. (The "Liquid-Liquid Extraction" method does outline a derivatization technique for acidic organics.) Organics that are thermally unstable or whose boiling point is beyond GC capability are not detected. It should be understood that the boiling points and solubilities listed are approximate boundaries as the methods definitely overlap. The breakdown is simply designed to provide the analyst with a starting point in choosing an analytical scheme for a specific compound or group of compounds. The "Direct Analysis Methods" provide a relatively rapid means for scanning a sample within the limitations specified. The "Concentration Methods" are relied on to provide a sufficient concentration of organic pollutants for identification by GC/MS. This is especially important where organics may be present that are harmful to the environment in very low concentrations. The "Direct Aqueous Injection" method, where applicable, should provide the best quantitative technique since the analyst does not have to contend with losses incurred during extraction and evaporation in the "Concentration Methods."

Instrumental Analysis

A portion of all the prepared extracts is injected on a Hewlett-Packard 5700* gas chromatograph equipped with a Hewlett-Packard 7671A automatic sampler and a flame ionization detector (FID). GC columns are chosen to best fit the extract being analyzed. This initial GC/FID scan is used to determine:

- if organics are present;
- the approximate concentration of the organics;
- the best GC conditions, and
- which sample extract has the highest concentration of organics.

This information determines if GC/MS analysis is necessary, and if so, assists in setting up optimum conditions on the GC/MS and selecting the best extract for GC/MS analysis.

The Region IV GC/MS system consists of a computerized Finnigan 1015, quadrupole mass spectrometer interfaced to a Varian 1400 gas chromatograph

---

*Trade names mentioned herein are for identification purposes only and do not imply endorsement by the U.S. Environmental Protection Agency.

Table 29.1 Organic Analysis of Water by Gas Chromatography

Concentration Methods

	Volatile Organics up to ca. 150°C		Organics 150° up to ca. 500°C	
Range of Organics				
Boiling Point				
Water Solubility	< 2%	> 2%	< 2%	> 2%
Method	Purging[9] VOA-5ml, 140 ml & 500 ml	Distillation[10]	Liquid-Liquid Extraction[11]	Water concentration[12]
Type Organics	Volatile organohalides	Alcohols, ketones, amines, aldehydes, etc.	Extractable neutrals, acids, bases	Higher molecular weight alcohols, etc.
GC Detectors[a]	GC/MS, HECD, FID	GC/MS, FID	GC/MS, FID, ECD, HECD, FPD	GC/MS, FID
GC Columns[b]	3,4,5	6,7,8,9	6,7,8,9,10	6
Minimum Detection Limit	ng/l to low µg/l	low µg/l	ng/l to low µg/l	µg/l

Direct Analysis Methods

Method	Headspace Gas Analysis[13]			
Range of Organics	Covers the whole range depending on solubility, concentration and vapor pressure. More applicable to water-insoluble, low-boiling-point compounds than to soluble high-boiling compounds.			
GC Detectors[a]	GC/MS	ECD	HECD	FID
Type Organics	Volatile general organics	Volatile organohalides	Volatile organohalides, S&N[c] compounds	Volatile general organics
GC Columns[b]	6,7,8,9,10	1,2	1,2,3,4,5	6,7,8,9,10
Minimum Detection Limit	low µg/l	ng/l	ng/l to low µg/l	low µg/l

Table 29.1, Continued

Direct Analysis Methods

	Volatile Organics up to ca. 150°C		Organics 150° up to ca. 500°C	
	< 2%	> 2%	< 2%	> 2%
Method	Direct Aqueous Injection[14]			
Range of Organics	Applicability depends only on concentration, not on solubility or boiling point.			
GC Detectors[a]	GC/MS	ECD	HECD	FID
Type Organics	General organics	Volatile organohalides	Volatile organohalides other organohalides, S&N[c] compounds	General organics
GC Columns[b]	6,7,8,9,10	1,2	2,3,5,6,8	6,7,8,9,10
Minimum Detection Limit	mg/l	low μg/l	low mg/l	low mg/l

[a] GC/MS - gas chromatograph/mass spectrometer
ECD - electron capture detector
HECD - Hall electrolytic conductivity detector
FPD - flame photometric detector
FID - flame ionization detector

[b] See Table 29.2.

[c] S&N - sulfur and nitrogen compounds

through a single-stage Gohlke jet separator. The computer system consists of a Digital Equipment Corporation PDP8E with a disc unit and a Systems Industries System 150 interface and software package. The data can be displayed in any of three modes; either on a Houston XY plotter, a Tektronix Cathode Ray Tube (CRT), or a Tektronix hard copier. Mass spectra are tentatively identified primarily by referring to the *Eight Peak Index of Mass Spectra.*[15] If the spectra are not in the Index, then the Mass Spectral Search System (MSSS) as outlined by Heller *et al.*[8] is utilized. A compound identification is not considered to be confirmed until it is compared with a standard mass spectra obtained by injecting some of the actual chemical through the GC/MS system. A second criterion for confirmation is comparison of gas chromatographic retention time of standard and sample.

Quantitation is performed by gas chromatography utilizing a flame ionization detector or a specific detector where applicable, such as the Melpar flame photometric detector or Hall electrolytic conductivity detector. Gas chromatographs utilized are either one or more of the following: a Hewlett-Packard 5700, a Tracor MT-160 or MT-220, or a Varian 1400. Recently, a Hewlett-Packard 3380 Reporting Integrator has been used for semi-automated quantitation.

The various gas chromatographic columns used in either the GC's or the GC/MS are listed in Table 29.2. Column tubing in all cases is Pyrex glass.

DISCUSSION

Three investigations in which the Organic Unit has been involved during the past year will be discussed in detail. These are presented to illustrate:

- how data gathered through Compliance Monitoring are being used to eliminate toxic pollutants from discharges;
- the need for having specific organics listed in NPDES permits to determine the source of organics found in drinking water, and
- how the analytical methods listed in the "Experimental" section are applied to environmental problems.

Organic Wastewater Characterization and Bioassay Study of a Pesticide Manufacturer

Effluent samples were collected at a plant manufacturing methyl parathion. The parameters that are limited in the effluent by the permit are listed in Table 29.3. Plant personnel made a clear point with the engineer collecting the sample that GC analysis would show a peak for ethyl parathion, but that it was not actually in the sample because the plant no longer manufactured ethyl parathion.

506 ORGANIC POLLUTANTS IN WATER

Table 29.2 Gas Chromatographic Column Information

	Packing	Column Length and Size	Oven Temperature	Use for
1.	Chromosorb 102, 80/100 mesh	3 ft x 4 mm i.d. x 1/4 in. o.d.	180°C	Volatile organohalides
2.	10% FFAP on 100/120 mesh Gas Chrom Q	6 ft x 4 mm i.d. x 1/4 in. o.d.	75°C	Volatile organohalides
3.	Chromosorb 102, 80/100 mesh	3 ft x 4 mm i.d. x 1/4 in. o.d.	Program	Volatile organohalides, organic gases and solvents
4.	Carbowax 400/Porasil C 100/120 mesh	6 ft x 4 mm i.d. x 1/4 in. o.d.	Program	Volatile organohalides
5.	0.4% Carbowax 1500/ Carbopack A	6 ft x 4 mm i.d. x 1/4 in. o.d.	Program	Volatile organohalides, organic solvents
6.	4% FFAP (or SP-1000) on 80/100 mesh Chromosorb W HP	6 ft x 2 mm i.d. x 1/4 in. o.d.	Program or isothermal	Volatile and moderately polar organics
7.	0.4% Carbowax 1500/ Carbopack A	6 ft x 2 mm i.d. x 1/4 in. o.d.	Program or isothermal	Volatile organics
8.	3% SE-30 on 80/100 mesh Chromosorb W HP	6 ft x 2 mm i.d. x 1/4 in. o.d.	Program or isothermal	General organics
9.	GP 4% Carbowax 20M/0.8% KOH on Carbopack B	6 ft x 2 mm i.d. x 1/4 in. o.d.	Program or isothermal	Volatile amines
10.	10% FFAP on 100/120 mesh Gas Chrom Q	6 ft x 2 mm i.d. x 1/4 in. o.d.	Program or isothermal	Volatile acids

Table 29.3 Comparison of Parameters Specified by EPA in an NPDES Permit with Organics Identified in the Industries' Discharge

Effluent Characteristics Listed in Permit	Organics Identified in Effluent
BOD_5	Methyl parathion
Total Kjeldahl nitrogen	Ethyl parathion
Total phosphorus (as P)	Sulfotepp
Methyl parathion	p-Nitrophenol
Ethyl parathion	Two isomers of dichlorobenzene
	Dimethyl trisulfide
	Methyl isobutyl ketone

The samples were extracted as outlined in the EPA procedure for organophosphorus pesticides in wastewater.[16]

GC analyses using a flame photometric detector in the phosphorus mode indicated the presence of ethyl parathion, methyl parathion and many unidentified compounds. GC/MS analysis of the sample confirmed that ethyl and methyl parathion were present. Sulfotepp was identified along with other peaks that gave spectra typical of organophosphates.

A bioassay study on the plant effluent was initiated because of the apparent toxicity of the effluent. These studies were conducted by the Ecology Branch of the Surveillance and Analysis Division and consisted of flow-through and static bioassays on bluegill sunfish (*Lepomis macrochirus* Raf.), static bioassays on water flea (*Daphnia magna* Strauss) and midge larva (*Paratanytarsus parthogenetics* Freeman), and flow-through on snail (*Goniobasis* sp. Raf.). The studies were performed at the plant site in a specially-designed mobile laboratory.

Organic analyses were again performed on the effluent to correlate with the bioassay data. The effluent samples were extracted for the organophosphorus pesticides,[16] acidified and re-extracted[11] for any acidic organics that might have been present. Head space analysis was also performed on the samples to detect the more volatile organics. The organics identified are listed in Table 29.3.

The bioassay study showed the plant effluent to be highly toxic to fish. The effluent had normally been passed through diatomaceous earth filters, but during the bioassay study, the diatomaceous earth was replaced with activated carbon. The sulfotepp, methyl parathion, and ethyl parathion concentrations in the effluent decreased by about 94%. Sunfish exposed to the carbon-filtered waste at 8.2% waste to dilution water ratio remained alive for 48 hr; under the same test conditions with only the diatomaceous earth filters, all of the fish died within 4 hr of exposure. The effect of the toxic effluent on the receiving stream was quite evident. A cursory examination of the benthic fauna in the receiving stream downstream from the discharge revealed a substantial reduction in numbers of invertebrate genera as compared to the large number of genera upstream from the waste discharge.

Organic residue analysis was performed on the exposed test fish and native fish taken from the receiving stream. Methyl parathion, ethyl parathion, sulfotepp, dichlorobenzene and several other compounds detected in the effluent, but not identified, were detected in all fish samples.

A meeting was held, at the request of EPA, with company representatives to discuss changes in the permit to eliminate the toxicity problem. Since this meeting, the company has agreed to comply and has submitted a schedule for meeting the new requirements. Basically, the changes involve re-engineering of the waste stream to better isolate the process wastes from cooling water and storm water runoff.

Taste and Odor Problems in Drinking Water Supplies

This is a rather frequently occurring problem that evidently has not received much attention in the past, primarily because of a lack of analytical techniques for identifying the pollutant and the source. Five of the ten taste and odor occurrences investigated by the Organic Unit have been traced to point sources, usually industrial effluents (see Table 29.4). If NPDES permits listed all significant organics present in effluents, strict limitations could be placed on those organics that have a low taste and odor threshold. This would also simplify the detective work needed to determine the source of an organic once it has been identified as the cause of the taste and odor problem.

The actual investigation of taste and odor problems is generally handled by the state and local health and environmental agencies, since they are the ones notified when a taste and odor problem occurs. The EPA Laboratory Services Branch has frquently been requested to provide analytical assistance because of the need for GC/MS analysis.

When sampling for taste and odor problems, water samples should be placed in narrow-mouth, glass bottles with no head space. This prevents loss of volatile organics.

The first decision upon receipt of taste and odor samples in the laboratory is selection of the proper analytical technique. If there is a noticeable odor associated with the sample, head space analysis[13] on the GC/MS

Table 29.4 Taste and Odor Investigations Participated in by Region IV Laboratory Services Branch from March 1973 to Date

Water System Affected	Organics Identified in Water	Probable Source
Abbeville, South Carolina	2-Ethyl-4-methyl-1,3-dioxolane	Fiberglass manufacturer
Lando, South Carolina	4-Methyl-1,3-dioxolane, methyl ethyl ketone, acetone	Waste solvent incinerator
Auburn, Alabama	Geosmin	Algal bloom
Private well in Kathleen, Florida	2-Ethyl-4-methyl-1,3-dioxolane	Polyester resin manufacturer
Evansville, Indiana	Bis-2-chloroisopropyl ether & Bis-2-chloroethyl ether	Propylene oxide manufacturer
West Columbia, South Carolina	Eucalyptol	Unknown
Private well in Orange County, Florida	t-Butyl alcohol, 2-methyl-2-butanol, Di-n-amyl ketone, 3-methyl-3-pentanol, acetone	Unknown
West Columbia, South Carolina	Acetic anhydride	Unknown
Hickory-Valdese, North Carolina	2-Ethyl-4-methyl-1,3-dioxolane	Polyester resin manufacturer
Private well in Jacksonville, North Carolina	Toluene, benzene, xylene, plus other alkyl benzenes and C_4, C_5 and C_6 alkenes and alkanes	Unknown

is a very quick and easy starting point and oftentimes results in identification of the odor-causing pollutants. Direct aqueous injection[14] on the GC/MS is another quick technique if the organic(s) causing the problem is/are present at 10 mg/l or higher, which is usually not the case. If these techniques are unsuccessful, the purging techniques[9] are excellent for isolating volatile, nonwater-soluble organics. The extraction[11] and distillation[10] techniques would be the next choice if the organics are nonvolatile or are very water-soluble.

A typical taste and odor investigation occurred in April, 1975 in the Valdese and Hickory, North Carolina water distribution systems. Water users complained about the bad taste of the water and also that they smelled worse after taking a shower than before.

The compound, 2-ethyl-4-methyl-1,3-dioxolane was suspected because of the distinctive odor that was very similar to two previous taste and odor investigations. Head space analysis[12] was not sensitive enough, even though a slight odor was noted. The distillation[10] method was used to concentrate the odor-causing compound into a smaller volume of water (700 ml to 30 ml) and then the purging[9] technique was used for final concentration and isolation. GC/MS analysis of the trapped vapors resulted in positive identification of 2-ethyl-4-methyl-1,3-dioxolane in the Hickory finished water sample at a concentration of about 1 μg/l or less. A mass spectrum of the dioxolane is shown in Figure 29.1.

The next step was to identify the source of the dioxolane. Samples were taken from two sewage treatment plants upstream from the problem area. The GC/MS analysis of head space over the samples was sufficient to identify the 2-ethyl-4-methyl-1,3-dioxolane in one of them. A quick search of the *Directory of Chemical Producers*[17] for fiberglass or resin product manufacturers, whose effluent was treated by that sewage treatment plant, listed two possible sources. GC/MS analysis of head space from one of the two sources showed that the dioxolane was present. The effluent contained 230 mg/l of the dioxolane as determined by direct aqueous injection of the sample on GC/FID. The industry took immediate steps to eliminate the problem and no reoccurrence has been reported since.

The compound, 2-ethyl-4-methyl-1,3-dioxolane, is especially troublesome as a pollutant in drinking water as normal water treatment processes fail to remove this compound. Even the addition of activated carbon has had no effect, evidently because of the compound's solubility in water.

Organic Pollutants in Miami Drinking Water

The 1975 EPA Survey of Ten Water Supplies for a Broad Range of Organics[3] included organic analyses of finished water only. The study has been completed for five of the cities, and so far the largest number of organics has been found in the Miami, Florida drinking water. There was naturally considerable interest in Region IV to determine the source of the organics.

The EPA Region IV Water Supply Branch in Atlanta, Georgia initiated action to:

- confirm the findings of the original study that was performed on one finished water sample;
- determine which organics were formed in the treatment process and which were present in the raw water;
- determine the source of any organics identified in the raw water; and
- make recommendations on how to eliminate the organics problem after answering the first three questions.

IDENTIFICATION OF ORGANIC POLLUTANTS 511

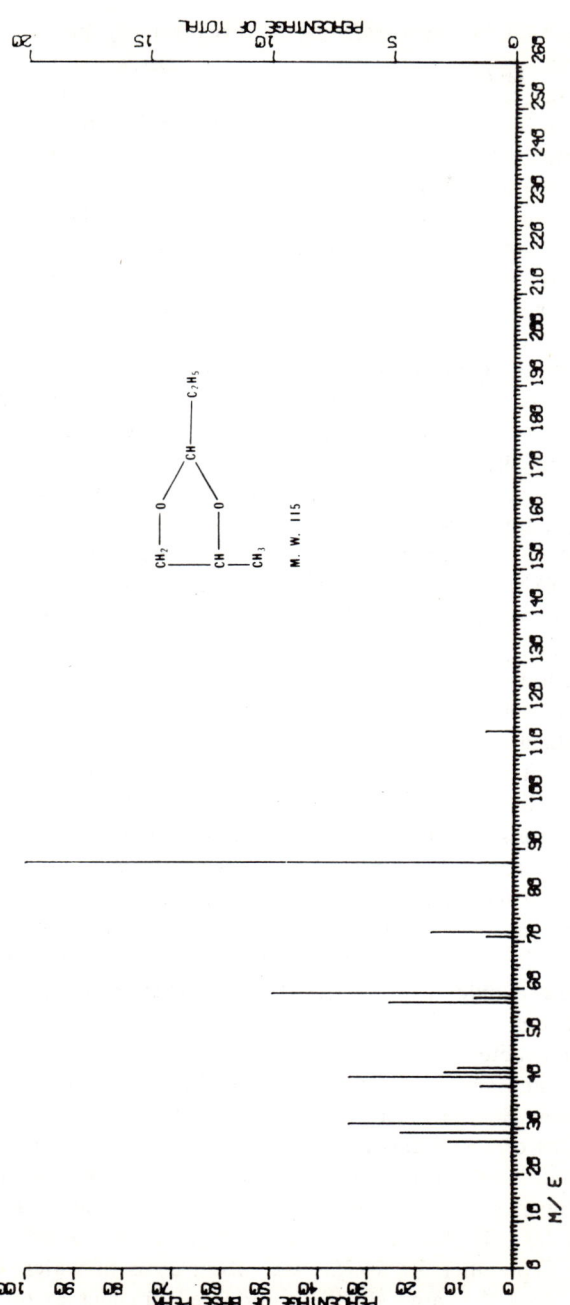

Figure 29.1 Mass spectrum of 2-ethyl-4-methyl-1,3-dioxolane.

The Organic Unit has been assisting in this effort by providing the analytical support.

The original study utilized a relatively extensive sampling and analytical scheme for finished water. It included the sampling of 6000 liters of water by activated carbon with subsequent extraction using chloroform; liquid-liquid extraction of a liter of water; and volatile organic analysis by purging[9] with the VOA-5, 140 and 500 ml. The largest number of organics was identified by means of the VOA-500 ml. Because of manpower limitations, it was decided to limit the previously-outlined Miami study to the volatile organics and use the VOA-500 ml.

The first two items in the study have been completed and the third one on determining the source or sources of organics in the raw water is still continuing. Miami's raw water supply is obtained from a shallow ground water aquifer. This complicated and limited the selection of raw water sampling sites.

Sample containers used were pint-size, narrow-mouth bottles with teflon-lined, septum-sealed screw caps. The containers were carefully filled to capacity to eliminate head space. The purging[9] technique (VOA-500 ml) was the primary analytical technique used for isolation and concentration of the volatile organics, followed by GC/MS analysis. A GC equipped with a Hall electrolytic conductivity detector was used for quantitation.

The additional work by Region IV confirmed that the Miami finished water contained a considerable number of volatile organics in the same relative concentration range determined in the original study. This work (Table 29.5) also showed that some of the organics were formed in the treatment process, thus confirming the earlier NORS work.[3] It also showed that a considerable number of organic compounds were present in Miami raw water and passed through the treatment process with little or no change in concentration. The USGS test well 628A was selected to determine the extent of contamination of the aquifer. This well is approximately 8 miles west of the Preston water treatment plant, in an uninhabited area. No halogenated organics were detected in this sample at a minimum detection limit of about 0.1 μg/l.

The next step was to try to determine the source or sources of organic pollutants in the Preston plant raw water. The well fields utilized as the sources of raw water are shown on the map in Figure 29.2. A suspected source of the organic pollutants was the sanitary landfill, also shown in the map. The seven USGS test wells located in and around the sanitary landfill plus certain wells from each of the well fields were sampled on September 30, 1975. GC/MS analysis of water supply well samples detected the same volatile chlorinated organics as those identified

Table 29.5 Organic Compounds Identified in Miami, Florida Finished and Raw Water Samples

	Sampled for:			
	NORS	Region IV		
Organics Identified	Finished Water[a] 1/29/75 $\mu g/l$	Finished Water[a] 7/7/75 $\mu g/l$	Raw Water[a] 7/7/75 $\mu g/l$	Test Well[b] 7/7/75 $\mu g/l$
Acetone	P	P	–	–
Bromodichloromethane	78	63	ND	ND
Bromoform	1.5	4.0	ND	ND
Bromotrichloroethylene	NR	T	ND	ND
Benzene	P	P	P	ND
Chlorobenzene	1.0	P	P	ND
Chloroform	311	220	0.7	ND
Carbon Tetrachloride	P	P	ND	ND
m-Dichlorobenzene	0.5	P	P	ND
o-Dichlorobenzene	1.0	P	P	ND
p-Dichlorobenzene	0.5	ND	ND	ND
Dibromochloromethane	35	37	ND	ND
1,1-Dichloroethane	P	P	P	ND
1,2-Dichloroethane	< 0.2	0.4	< 0.3	ND
1,1-Dichloroethylene	P	P	P	ND
1,2-Dichloroethylene	P	P	P	ND
Dichloroiodomethane	NR	T	ND	ND
Dichloromethane	P	P	–	–
Tetrachloroethylene	0.1	P	P	ND
Trichloroethane	P	P	P	ND
Trichloroethylene	P	P	P	ND
Vinyl chloride	5.6	8.2	5.0	ND

[a] Preston water treatment plant.
[b] U.S. Geological Survey test well #628A, water conservation district No. 3.
P – Present but not quantitated.
T – Tentative identification
NR – Not reported.
ND – Not detected.

in the raw water sampled on 7/7/75. This was also true of the samples from most of the seven sanitary landfill test wells; however, concentrations were generally lower than in the samples from the water supply well fields. The most significant results appear to be that none of these chlorinated organics were detected in test wells #1 and #2, since the underground flow of water is from the sanitary landfill toward the Preston plant well fields. However, since the concentrations were considerably

514 ORGANIC POLLUTANTS IN WATER

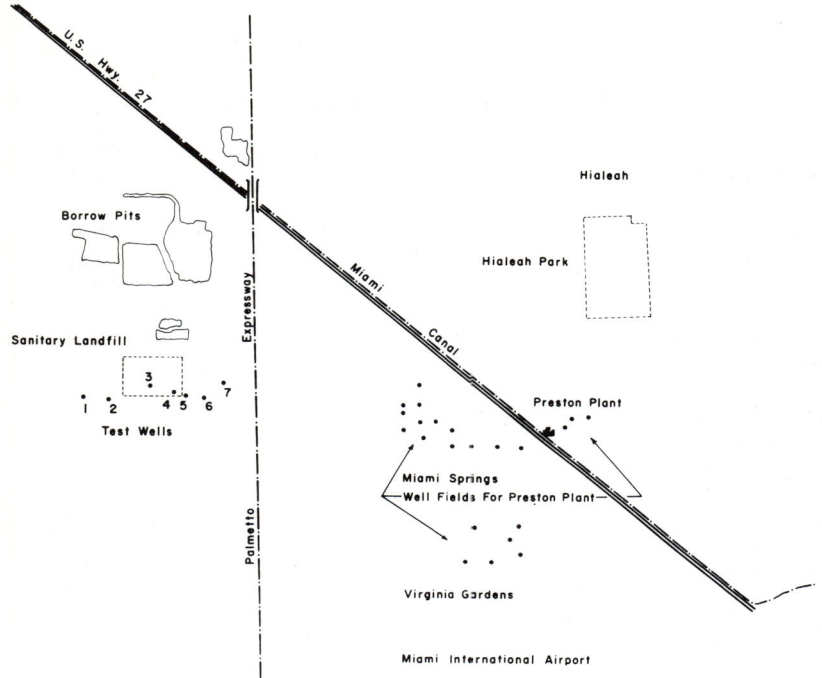

Figure 29.2 Map of Miami well fields area.

lower in the sanitary landfill than the well fields, it is still not clear that the sanitary landfill is the main source of the chlorinated pollutants.

The fact that this is an extremely shallow aquifer, criss-crossed with canals and borrow pits, indicates that, unless extensive sampling is carried out, it will be very difficult to pinpoint the source or sources of all the organic pollutants.

CONCLUSION

The above investigations along with similar investigations conducted by EPA laboratories confirm the value and necessity of specific organic analysis, as well as the need for the listing and limitations of specific organics in the NPDES permits. As more information is obtained on organics in wastewaters, there will be substantial modification of NPDES permits during the permit renewal period just around the corner. Also, as new municipal and industrial waste sources apply for permits, we will be better prepared to limit toxic, odor-causing, and/or carcinogenic organic compounds.

REFERENCES

1. "Federal Water Pollution Control Act Amendments of 1972" (Public Law 92-500). U.S. Government Printing Office, Washington, D.C. (October 18, 1972).
2. U.S. Environmental Protection Agency. "Draft Analytical Report, New Orleans Area Water Supply Study," Lower Mississippi River Facility, Slidell, Louisiana (November 1974).
3. U.S. Environmental Protection Agency. "Preliminary Assessment of Suspected Carcinogens in Drinking Water," Office of Toxic Substances, Washington, D. C. 20460 (June 1975).
4. U.S. Environmental Protection Agency. "Organic Characterization Study, Coosa River Basin—Northwest Georgia," Surveillance and Analysis Division, Athens, Georgia (September 30-October 3, 1974).
5. Alford, A. L. "Environmental Applications of Advanced Instrumental Analyses: Assistance Projects, FY 73," U.S. Environmental Protection Agency, Athens, Georgia. Publication number EPA-660/2-74-078 (August 1974).
6. Alford, A. L. "Environmental Applications of Advanced Instrumental Analyses: Assistance Projects, FY 74," U.S. Environmental Protection Agency, Athens, Georgia. Publication number EPA-660/4-75-004 (June 1975).
7. Keith, L. H. "Chemical Characterization of Industrial Wastewaters by Gas Chromatography/Mass Spectrometry," *The Science of the Total Environment* 3, 87-102 (1974).
8. Heller, S. R., J. M. McGuire and W. L. Budde. "Trace Organics by GC/MS," *Environ. Sci. Technol.* 9, 210-213 (1975).
9. "Purging a Water Sample to Determine Volatile Organic Compounds," In *Trace Organics by Gas Chromatography/Mass Spectrometry*, W. L. Budde, Ed. U.S. Environmental Protection Agency, EMSL, Cincinnati, Ohio (in press).
10. "Distillation: A Procedure for Water Soluble Volatile Organic Solvents in Effluents and Streams," In *Trace Organics by Gas Chromatography/Mass Spectrometry*, W. L. Budde, Ed. U.S. Environmental Protection Agency, EMSL, Cincinnati, Ohio (in press).
11. "Solubility Separation by pH Adjustment: A Procedure for the Extraction of Industrial Organic Chemicals in Effluents and Streams," In *Trace Organics by Gas Chromatography/Mass Spectrometry*, W. L. Budde, Ed. U.S. Environmental Protection Agency, EMSL, Cincinnati Ohio (in press).
12. "Concentration of a Water Sample: A Procedure for Concentrating Non-Volatile Water Soluble Organic Compounds," In *Trace Organics by Gas Chromatography/Mass Spectrometry*, W. L. Budde, Ed. U.S. Environmental Protection Agency, EMSL, Cincinnati, Ohio (in press).

13. "Headspace Analysis: A Quick Qualitative Technique for Volatile Organic Compounds," In *Trace Organics by Gas Chromatography/Mass Spectrometry*, W. L. Budde, Ed. U.S. Environmental Protection Agency, EMSL, Cincinnati, Ohio (in press).
14. "Direct Aqueous Injection GC/MS Analysis," In *Trace Organics by Gas Chromatography/Mass Spectrometry*, W. L. Budde, Ed. U.S. Environmental Protection Agency, EMSL, Cincinnati, Ohio (in press).
15. *Eight Peak Index of Mass Spectra*, 1st ed., Vol. I and II, Mass Spectrometry Data Centre, AWRE, Aldermaston, United Kingdom (1970).
16. U.S. Environmental Protection Agency. "Method for Organophosphorus Pesticides in Industrial Effluents," EMSL, Cincinnati, Ohio (1973).
17. Stanford Research Institute. "1975 Director of Chemical Producers, United States of America," Menlo Park, California (1975).

CHAPTER 30

GC/MS ANALYSIS OF ORGANIC COMPOUNDS IN DOMESTIC WASTEWATERS

A. W. Garrison, J. D. Pope and F. R. Allen

U.S. Environmental Protection Agency
Southeast Environmental Research Laboratory
Athens, Georgia

INTRODUCTION

In 1971 this laboratory began a program to identify extractable, volatile organic compounds in domestic wastewaters. Objectives were to develop analytical techniques for such analyses, to identify compounds being discharged into surface waters after secondary or advanced treatment, and to provide specific compound data that will help to determine waste treatment effectiveness. Knowledge of the specific compounds discharged is needed to study health effects of pollutants, to help determine the sources of compounds found in drinking water surveys,[1] and to establish effluent guidelines. Finally, some parts of the world are concerned about the possible need to renovate domestic wastewater for human consumption,[2,3] and the identification of hazardous compounds in such wastewaters is imperative for safe renovation.

Most previous studies of raw and treated sewage have been mainly nonspecific characterizations of particulate and soluble fractions.[4-7] Organics were assigned to groups such as amino acids, carbohydrates, lipids, acids, proteins, surface-active substances, tannins, fulvic acid, humic acid, etc., but relatively few specific soluble organic compounds were identified. In the past few years, with the increasing concern over trace organics in all types of water, and with the availability of adequate analytical tools, more investigators are identifying and quantifying specific soluble organics in municipal effluents. Work in Israel[8] resulted in the identification by combined gas chromatography/mass spectrometry (GC/MS) of several fatty acids and aliphatic and aromatic hydrocarbons. Considerable work of this nature has been conducted in the United Kingdom[6,9]; specific optical brighteners, amino acids, carbohydrates, steroids, pesticides and phthalates were identified. Earlier work of a less comprehensive nature resulted in the identification of volatile acids[10] and polycyclic aromatic hydrocarbons.[11]

More recently, Burlingame[12] began to use high-resolution GC/MS to analyze mixtures of organics extracted from treated municipal effluents, and has identified several specific compounds. Dunlap[13] is using accumulator columns for extraction and both electron impact and chemical ionization GC/MS to identify organics contributed to ground water by municipal solid wastes and septic tank effluents. Glaze[14] has identified more than 40 chlorinated compounds produced during wastewater chlorination. In work closely related to that reported here, Giger,[15] in Switzerland, is using capillary column gas chromatography (GC) and GC/MS to separate and identify compounds in municipal wastewaters; several aliphatic and aromatic hydrocarbons, chlorinated compounds, and alcohols have been identified.

It is generally agreed that the bulk of organic material—over 75% in most waters—is nonextractable, nonvolatile, and mostly nongas-chromatographable. Municipal wastewaters are no exception;[6] but a group in the U.S. has made significant progress in identifying these materials, including those produced by chlorination, after separation by high-pressure liquid chromatography.[16-18]

Since the principal objective of our research was to identify extractable, volatile organics in representative raw and treated domestic wastewaters, the EPA's Advanced Waste Treatment Research Laboratory (AWTRL) at the Taft Center in Cincinnati was selected as the main sample source. Their pilot-scale waste treatment plants have no industrial input, and their operation is well controlled and documented.

The activated sludge and physical-chemical pilot treatment plants take raw sewage from a common residential line (Figure 30.1). The activated sludge plant is a typical biological treatment process depending upon bacterial action for most of the organics removal; the physical-chemical system relies mostly on removal of organics by carbon adsorption. Values for total pollution parameters (Figure 30.1) are highly variable, especially those for the raw sewage, mostly due to fluctuations in dilution of the organics by relatively unpolluted water. Since the pilot plants are enclosed within the AWTRL laboratory, the temperature varies little from day to day, but the averages range from 12°C in winter to 20°C in summer.

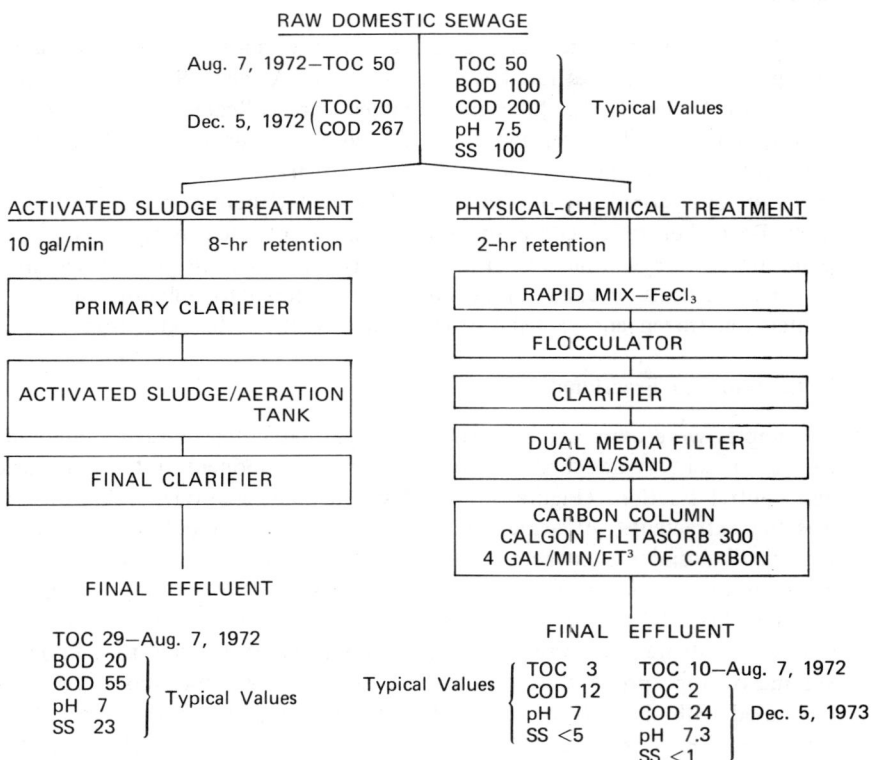

Figure 30.1 Flow diagrams of parallel sewage treatment pilot plants, with values for total pollutional parameters, at EPA's Advanced Waste Treatment Research Laboratory, Cincinnati, Ohio. Samples were taken of the raw sewage and each final effluent.

Later, samples from the Blue Plains sewage treatment plant at Washington, D.C., were included in this project at the request of the Blue Plains—EPA pilot plant manager. The EPA has several research projects at this plant. Lime-clarified raw sewage samples were collected, chlorinated and extracted by Blue Plains personnel before being shipped to ERL—Athens for analysis.

EXPERIMENTAL

Sampling

Fifteen to twenty liter samples of raw sewage, activated sludge effluent, and physical-chemical effluent were collected in August 1972 and December 1973. In August 1972, three grab samples were collected at intervals corresponding to the retention time of each treatment system. Samples were collected in glass containers, adjusted to pH 4-5 with HCl as a preservative, and stored for one to three days at 4°C before extraction.

In December 1973, composite samples were collected at each sample point for 3.5 days while the glass sample containers were in a refrigerator at 4°C. Portions of each composite were transfered to polyethylene containers, frozen and stored for about two weeks until extraction.

Extraction and Fractionation

Samples were extracted manually in 3-liter batches using a liquid-liquid extraction scheme designed to separate sample components into acid, basic and neutral fractions (Figure 30.2). Pesticide-grade methylene chloride was the extractant. All water used for reagents (HCl and NaOH solutions) was distilled and pre-extracted with methylene chloride. Before final evaporation, sample extracts were dried by passing through sodium sulfate that had been heated in a muffle furnace at 600°C for 2 hr. A gentle stream of nitrogen was used for final evaporation of extracts to a known volume of appropriate concentration for GC analyses. Extracts were combined with those from other 3-liter batches and stored in glass vials with teflon-lined septa at 4°C. The acid fractions were methylated with diazomethane in the presence of methanol.[19] A control was prepared for each sample set by taking appropriate amounts of solvents and reagents through the above extraction scheme.

One raw sewage sample was extracted into weak and strong acid fractions in addition to the usual extraction scheme. The first organic layer (Figure 30.2) was extracted with 5% sodium bicarbonate to extract strong acids, then with 5% sodium hydroxide for weak acids. Each extract was acidified, re-extracted with methylene chloride, and methylated (Figure 30.2). This fractionation was discontinued because only two compounds, phenols of very low concentration (< 1 ppb), were recovered in the weak acid fraction.

Portions (500 ml) of some samples were distilled (Figure 30.2) by directly heating, using a water-cooled condenser. The first 25 ml of distillate was collected and redistilled to collect the first 1 ml. During

IDENTIFICATION OF ORGANIC POLLUTANTS 521

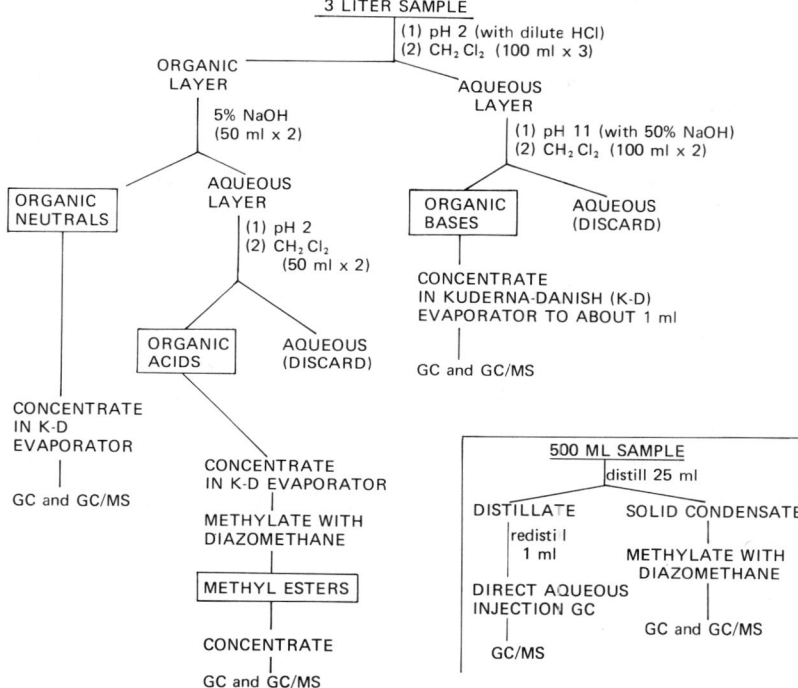

Figure 30.2 Extraction/fractionation scheme for municipal wastewater samples. Inset shows distillation scheme to concentrate samples for direct aqueous injection GC.

distillation of the raw sewage, several milligrams of waxy white material collected in the condenser. This was dissolved in methylene chloride, and methylated with diazomethane.

Chlorination

Lime-clarified raw sewage from the EPA-DC Blue Plains pilot plant in Washington, D.C., was chlorinated and extracted at that facility. Chlorination was performed in the laboratory by addition of a sodium hypochlorite solution of such a pH that the final pH of the sample at "breakpoint chlorination" was between 7 and 7.5. The residual free chlorine in the sample was 3-5 mg/l. The chlorinated sample was allowed to stand for about 30 min before extraction. Extraction followed essentially the same procedure used at ERL-Athens (Figure 30.2). The extracts were delivered to ERL-Athens for final concentration, methylation and analysis.

Ozonolysis

About 10 μl of the methylene chloride solution of the raw sewage acid fraction, previously esterified with diazomethane, was added with 2 ml of MeOH to the sample tube of an apparatus constructed after the design of Beroza.[20] After adjusting the oxygen flow to approximately 10 ml/min, the cold bath was brought into place and ozonation began. At 2 min, the indicator solution detected excess ozone, but the generation continued for another 30 sec. After the system was flushed with nitrogen, 0.5 ml of dimethyl sulfide was added to the sample tube to reduce the ozonides. To insure complete reaction, the sample tube was left in the cold bath for 5 min, then placed in an ethanol/ice bath for 1 hr. Upon removal from this bath, the sample was shaken for 5 min, and the volume adjusted for GC analysis by gentle evaporation with a nitrogen stream.

Gas Chromatography

All extracts were analyzed by GC. Typical conditions were:

GC: Varian 1400, Perkin Elmer 990 or Tracor MT220
Column: 6 ft x 1/8 in. i.d. (1/4 in. o.d.) glass
Packing: 3% SE-30 on 80/100 mesh Gas Chrom Q
Program: 50° for 2 min, then to 250° at 5°/min
Carrier Gas: helium at 50 cc/min
Sample Size: 2 μl
Detector: flame ionization

Some extracts were analyzed on a 3% SP-2100 column using similar GC conditions. This liquid phase is very similar to SE-30, but has a higher temperature limit. However, SP-2100 appeared to cause more tailing of early eluting peaks.

Aqueous distillates were analyzed by direct injection of 1 to 10 μl into a 10% FFAP GC column. (Temperature Program: 150° for 2 min, then to 180° at 4°/min; Gas Flow: 21 cc/min; Column: 6 ft x 1/4 in. glass).

Other Instrumentation

A Finnigan 1015 mass spectrometer interfaced via a Gohlke separator to a modified Varian 1400 GC was used for GC/MS analysis.[21] A System Industries System 150 interfaced the GC/MS to a Digital Equipment Corp. PDP8/e computer for data acquisition, storage and manipulation. Ionizing voltage was 70 eV.

Some low-resolution GC/MS work was done on a Varian MAT CH5/DF system interfaced to a Varian 2740 GC via a Watson-Biemann separator and to a Varian MAT SS-100 Data System. This instrument was also used at a resolution of about 5000 for determination of possible empirical formulae of the major mass spectral peaks of some unknown biodegradation products in the activated sludge effluent.

The GC conditions for GC/MS analyses were similar to those used for GC analyses described previously.

Chemical ionization mass spectrometry was performed on selected extracts with a separate computerized Finnigan 1015 mass spectrometer interfaced to a Finnigan 9500 GC using methane as a carrier/reactant gas.

Combination gas chromatograph/infrared spectroscopy instrumentation was a computerized Digilab FTS-14D/IR Fourier transform spectrophotometer equipped with the Digilab GC/IR accessory and interfaced to a Perkin-Elmer 990 GC.[22] This instrument was used to confirm the presence of the clofibrate metabolite in the acid fraction of the activated sludge effluent collected in December 1973.

After tentative identification of pollutants by GC/MS, a Perkin-Elmer PEP-1 Data System, interfaced to a Varian 1400 GC operated under the conditions described above, was used for computerized quantitation and retention time measurements.

Identification of Compounds

Sample mass spectra stored on disks from the Finnigan GC/MS runs were compared via acousticoupler connection with standard spectra in the EPA-Battelle computer files at Battelle (Columbus). Sample mass spectra were also compared through a computer terminal and acousticoupler with standard spectra in the NIH Mass Spectral Search System in Bethesda, Maryland. Later a combination of these two search systems was used— the Mass Spectral Search System, handled by the Cyphernetics Corp., Ann Arbor, Michigan. This system contains about 40,000 reference mass spectra.[23]

These computer-based searches were supplemented by manual searches of the Aldermaston *Eight Peak Index of Mass Spectra*.[24] Two publications were particularly helpful in interpreting the mass spectra of long-chain acids.[25,26] The clofibrate metabolite was first identified by matching its spectrum in the *Archives of Mass Spectral Data*.[27]

Many compound identifications were confirmed (Table 30.1) by matching their GC retention times and mass spectra with those of standards. Standards of most of the acids were obtained in kit form from Applied Science Laboratories, Inc., State College, Pennsylvania, or Supelco Inc., Bellefonte, Pennsylvania.

Table 30.1 Organic Compounds in Municipal Wastewaters

Compound by Class	Raw Wastewater		Activated Sludge Effluent		Physical-Chemical Effluent		Lime-Clarified Raw Sewage	
	8/72	12/73	8/72	12/73	8/72	12/73	Before Chlorination	After Chlorination
Normal Chain Acids								
C_3 Propionic[a]	+		+					
C_4 Butyric[a]	+		+				+	
C_5 Valeric[a]	+		+	+				
C_6 Caproic[a]	+	0.4	+			+	+	
C_7 Enanthic[a]	+			0.5		0.2		+
C_8 Caprylic[a]	+	1.1		0.1		0.1	+	+
C_9 Nonanoic[a]	+	0.2		0.3		0.2	+	+
C_{10} Capric[a]	+	2.0		0.1		0.1	+	+
C_{11} Undecanoic[a]				+				
C_{12} Lauric[a]	0.5	1.7	0.3	0.1	+	0.1	+	+
C_{13} Tridecanoic[a]						+		
C_{14} Myristic[a]	1.3	0.7	0.5	0.2	0.1	0.1	+	+
C_{15} Pentadecanoic[a]	0.3	<0.1	0.3	0.2	+	0.1	+	+
C_{16} Palmitic[a]	28.0	7.1	6.0	2.0	0.6	0.2	+	+
C_{17} Margaric[a]	0.5	<0.1	0.2	0.4	+		+	+
C_{18} Stearic[a]	32.0	6.7	10.0	2.1	0.3	0.1	+	+
C_{19} Nonadecanoic	+							
C_{20} Arachidic	0.3	+						
C_{22} Behenic[a]			0.1					
Unsaturated Acids								
C_{16} Palmitoleic[a]	0.5	3.4		0.1	0.4	0.2	+	
C_{18} Oleic[a]	7.0	7.4	1.3	0.5	0.2	0.1	+	
Branched Chain Acids								
C_4 Isobutyric[a]	+		+					
C_5 Isovaleric[a]	+		+				+	+
C_{15} Anteisopentadecanoic[b]	+				+			
C_{17} Anteisomargaric[b]	+		+					
C_{17} α-Methyl Palmitic[b]					+			
C_{19} α-Methyl Stearic[b]					+			

Table 30.1, Continued

	Concentration in Wastewater, $\mu g/l$ (+ = Present, not quantified)							
	Raw Wastewater		Activated Sludge Effluent		Physical-Chemical Effluent		Lime-Clarified Raw Sewage	
Compound by Class	8/72	12/73	8/72	12/73	8/72	12/73	Before Chlorination	After Chlorination
Oxy-Acids[b]								
C_{10} β-Hydroxy acid						+		
C_{12} β-Hydroxy acid						+		
C_{14} β-Hydroxy acid	+					+		
C_{16} β-Hydroxy acid						+		
C_{18} β-Hydroxy acid	+							
C_{17} α-Ketomargaric						+		
C_{19} α-Ketononadecanoic						+		
Miscellaneous Acids								
Benzoic[a]						+	+	+
2-Ethylhexanoic		+		+		+		
Hexahydrobenzoic							+	
Phenylacetic[a]	+	+	+	+	+		+	+
Phenylpropionic	+						+	+
Alcohols								
Benzyl alcohol							+	+
Borneol							+	
2-Butoxyethanol	+							
2-Ethyl-1-hexanol				+				
1-Pentanol				+				
2-Phenoxyethanol							+	
α-Terpineol	+	+		+		+	+	
Phthalates								
Dibutyl		+	+	+	+	+		
Diethyl				+				
Dioctyl	+		+	+	+	+		
Chlorinated Compounds								
Chlorocyclohexane								+
Chloroform[a]		+	+			+		
Dichloromethane[a]		+		+				
Hexachloroethane[a]								+
Pentachloroethane[a]								+
Pentachlorophenol[a]			0.2	+				

Table 30.1, Continued

Compound by Class	Raw Wastewater 8/72	Raw Wastewater 12/73	Activated Sludge Effluent 8/72	Activated Sludge Effluent 12/73	Physical-Chemical Effluent 8/72	Physical-Chemical Effluent 12/73	Lime-Clarified Raw Sewage Before Chlorination	Lime-Clarified Raw Sewage After Chlorination
1,1,2,2-Tetrachloroethane	+							+
1,1,1,2-Tetrachloroethane								+
Steroids								
Cholesterol		+		+				
Coprostanol		+		+				
Drugs and Drug Metabolites								
Caffeine	+	+		+			+	
2-(4-Chlorophenoxy)-2-methylpropionic acid[a] [Clofibrate metabolite]		0.8	1.0	2.0		0.1	+	+
Nicotine		+		+			+	
Salicylic acid[a]	+	+					+	
Aromatic Hydrocarbons								
Dimethylbenzene isomer							+	
Dimethylnaphthalene isomer							+	
Ethylbenzene			+					
o-Methylstyrene			+					
Toluene			+					
Xylene				+				
Miscellaneous Organics								
Acetone		+		+		+		
Benzaldehyde							+	+
m-tert-Butylphenol[a]		+						
Carvone							+	
1,1-Diethoxyethane							+	+
Dioctyl Adipate					+			
Ethyl Acetate						+		
Indene				+				
o-Phenylphenol[a]		+						
Saccharin[a]		+					+	+
Tetrahydrofuran						+		

[a]Confirmed with standard; all others by mass spectral matching.

[b]Identified only by manual interpretation of mass spectra, except that the β-hydroxy acids matched GC retention times with standards. There is considerable possibility of error in identification of the α-keto and long-branched-chain acids.

RESULTS AND DISCUSSION

Raw Sewage Components

Total Methylene Chloride Extractables

Figure 30.3 shows the FID/GC peaks observed in all fractions of the methylene chloride extract of raw sewage collected from the AWTRL in December 1973. The acid and neutral fractions contained about the same amounts of gas chromatographable organic compounds, but the basic fraction contained much less. The methylated acid fraction produced a chromatogram that was typical of this fraction in all samples, even after treatment. Palmitic (C_{16}), stearic (C_{18}) and oleic ($C_{\overline{1}8}$) acids predominated; the even-numbered straight-chain acids and palmitoleic acid ($C_{\overline{1}6}$) were at intermediate concentrations; and the odd-numbered straight-chain acids, branched-chain acids, and miscellaneous acids were at lower concentrations (see Figures 30.4 through 30.8 for other compounds in acidic fractions).

Caffeine and nicotine produced the predominant peaks in the basic fraction chromatograms; no other components have yet been identified. Caffeine, α-terpineol, and several aliphatic and aromatic hydrocarbons were found in the neutral fractions. Most neutral fractions also contained cholesterol, coprostanol, phthalates, and several components that appeared to be long-chain alcohols and/or unsaturated hydrocarbons.

Raw Sewage Condensate

During distillation of the raw sewage sample of December 1973 as a means of concentration for analysis of volatile free acids, several milligrams of a waxy white solid collected in the condenser. This material was analyzed by GC/MS (Figure 30.9) after solution in methylene chloride and methylation. Limited mass range searches of the stored mass spectral data indicated the presence of three phthalates (m/e 149), and methyl palmitate and methyl myristate (m/e 74 and 87). The acid metabolite of the drug clofibrate was the most abundant component.

The main human metabolite of the drug clofibrate [ethyl 2-(4-chlorophenoxy)-2-methylpropionate] has been shown to be 2-(4-chlorophenoxy)-2-methylpropionic acid (Figure 30.10). It is excreted in urine at 85% of the amount of the drug ingested.[28] Clofibrate is used by many older people in doses of 2 g/day to control atherosclerosis. It is not manufactured in this country, but is dispensed here.

Although the parent drug was never found in sewage, the metabolite was found in both August 1972 and December 1973 samples of the Cincinnati AWTRL pilot plant sewage, in Athens, Georgia, municipal

528 ORGANIC POLLUTANTS IN WATER

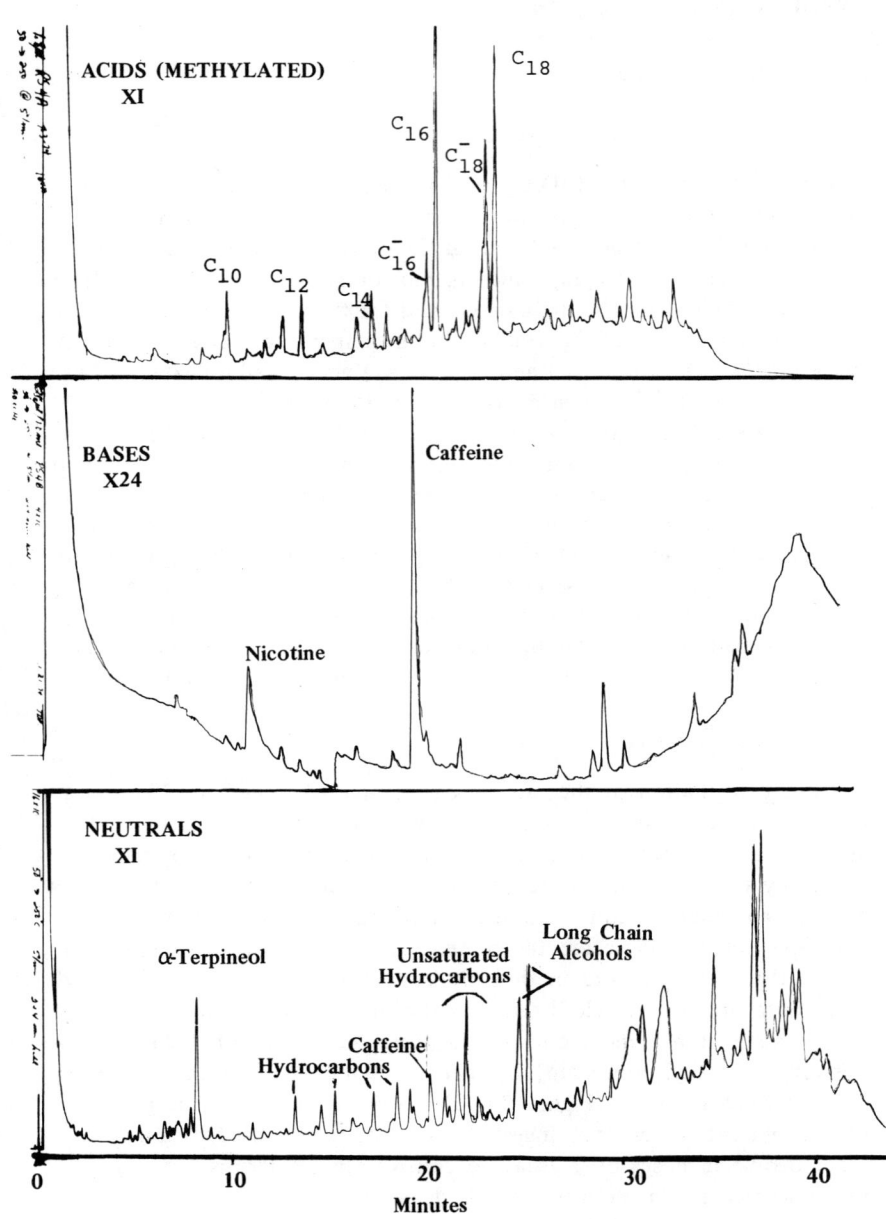

Figure 30.3 Typical methylene chloride extractables in raw domestic sewage by FID/GC.

IDENTIFICATION OF ORGANIC POLLUTANTS 529

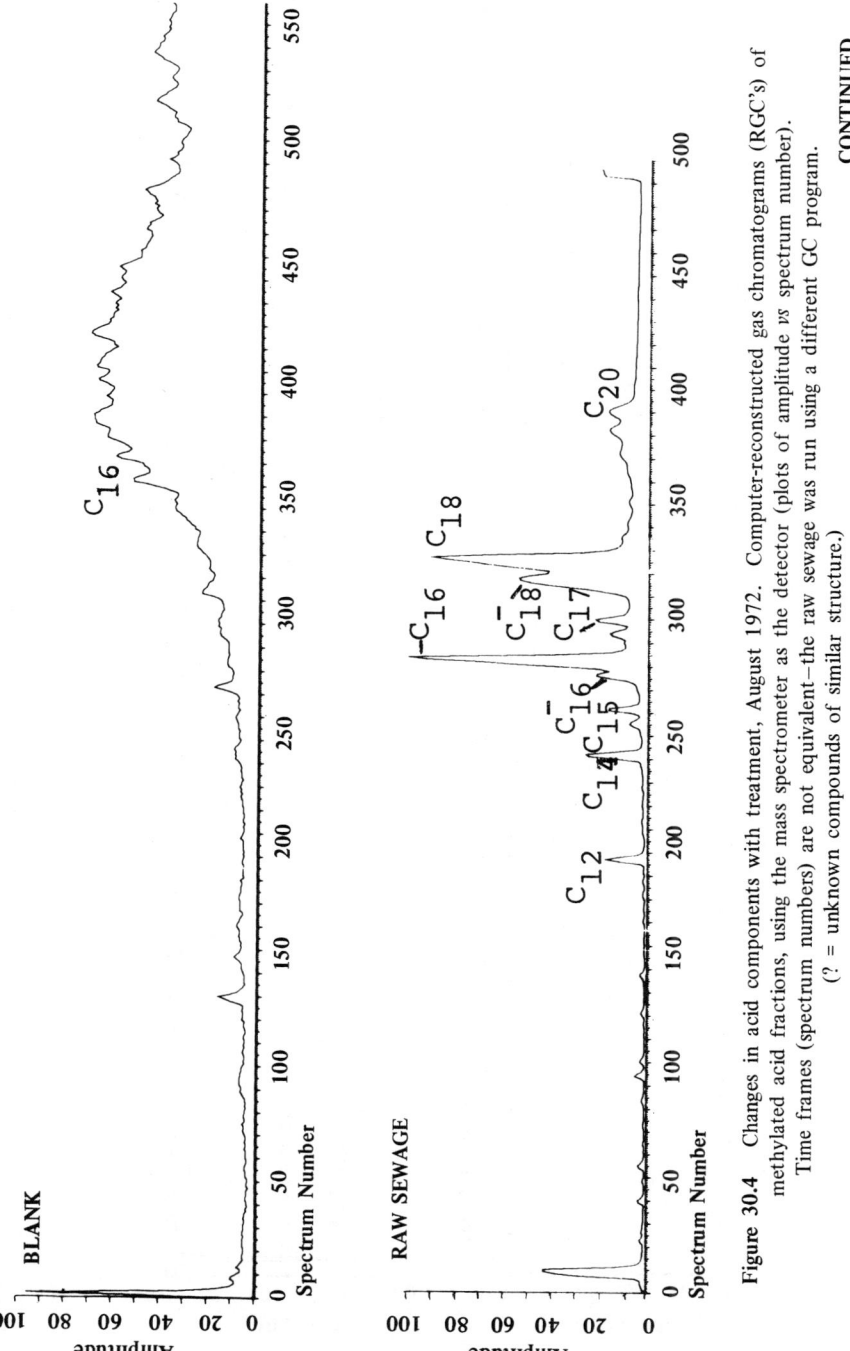

Figure 30.4 Changes in acid components with treatment, August 1972. Computer-reconstructed gas chromatograms (RGC's) of methylated acid fractions, using the mass spectrometer as the detector (plots of amplitude *vs* spectrum number). Time frames (spectrum numbers) are not equivalent—the raw sewage was run using a different GC program. (? = unknown compounds of similar structure.) CONTINUED

530 ORGANIC POLLUTANTS IN WATER

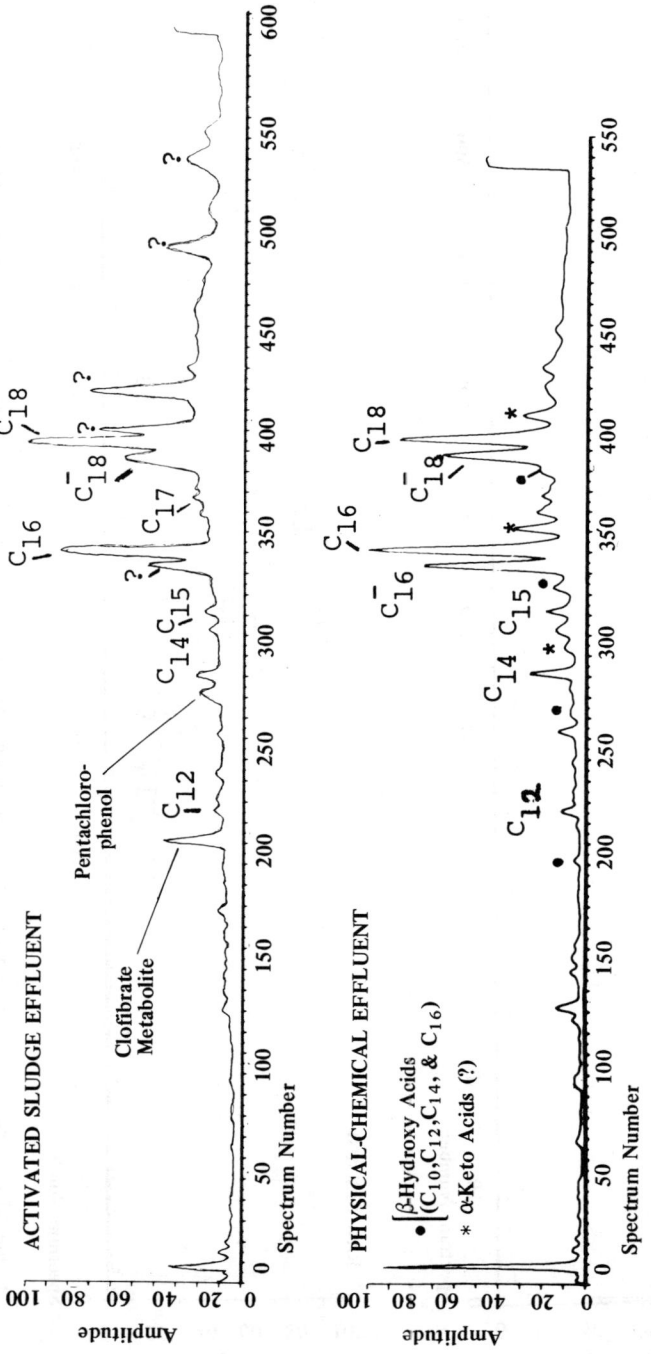

Figure 30.4, Continued

IDENTIFICATION OF ORGANIC POLLUTANTS 531

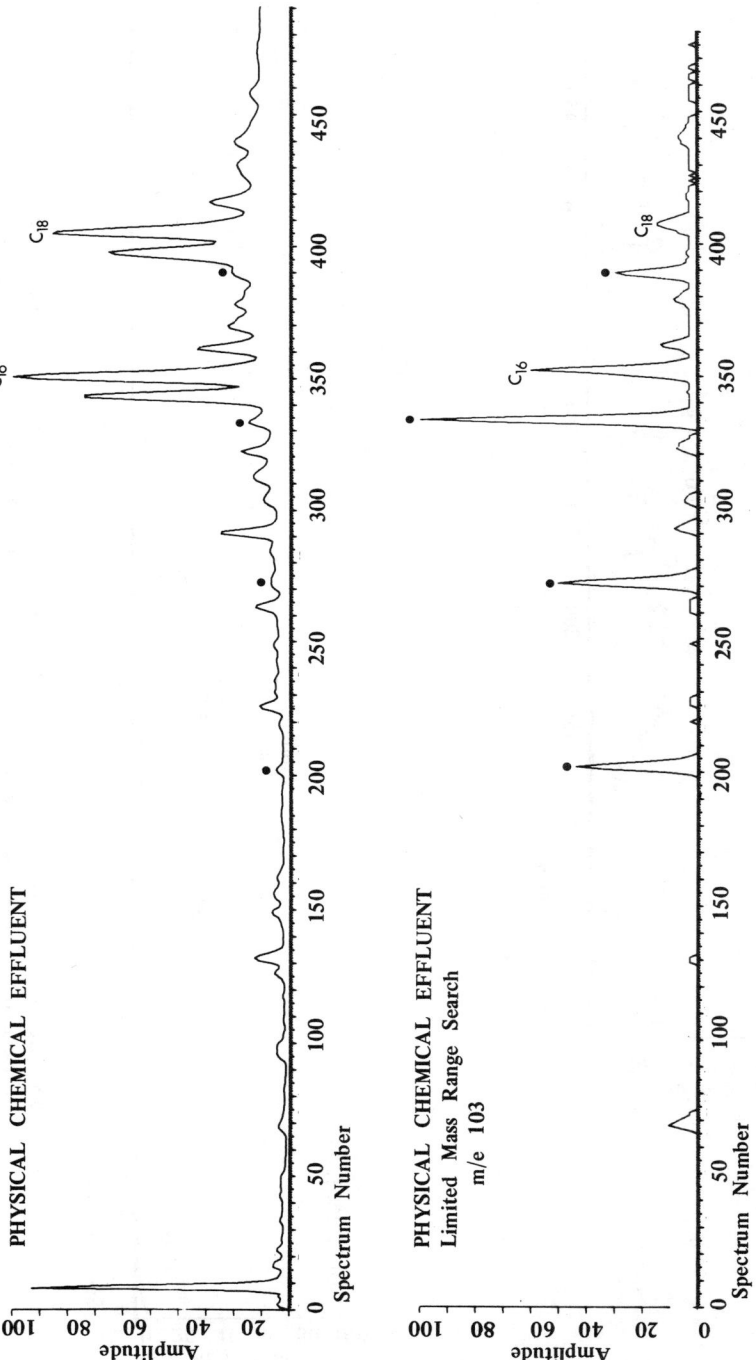

Figure 30.5 Physical-chemical effluent (as in Figure 30.4) with corresponding limited mass range search for m/e 103, indicative of long-chain β-hydroxy acids (●). (C_{16} and C_{18} fatty acids are in such abundance that their small m/e 103 peaks also give large signals.)

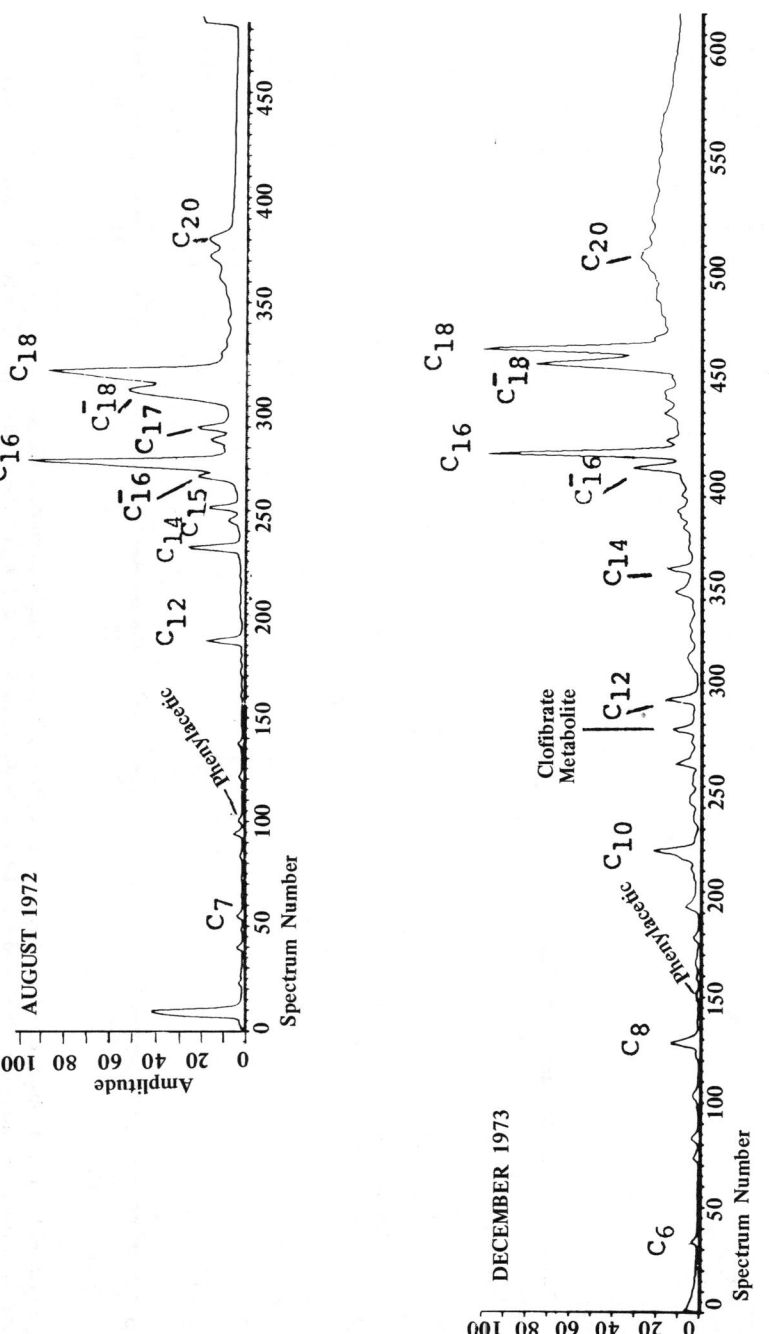

Figure 30.6 RGC's of methylated acid fractions of two raw sewage samples collected at different times. GC temperature programs were different, so time scales (spectrum numbers) are not equivalent.

IDENTIFICATION OF ORGANIC POLLUTANTS 533

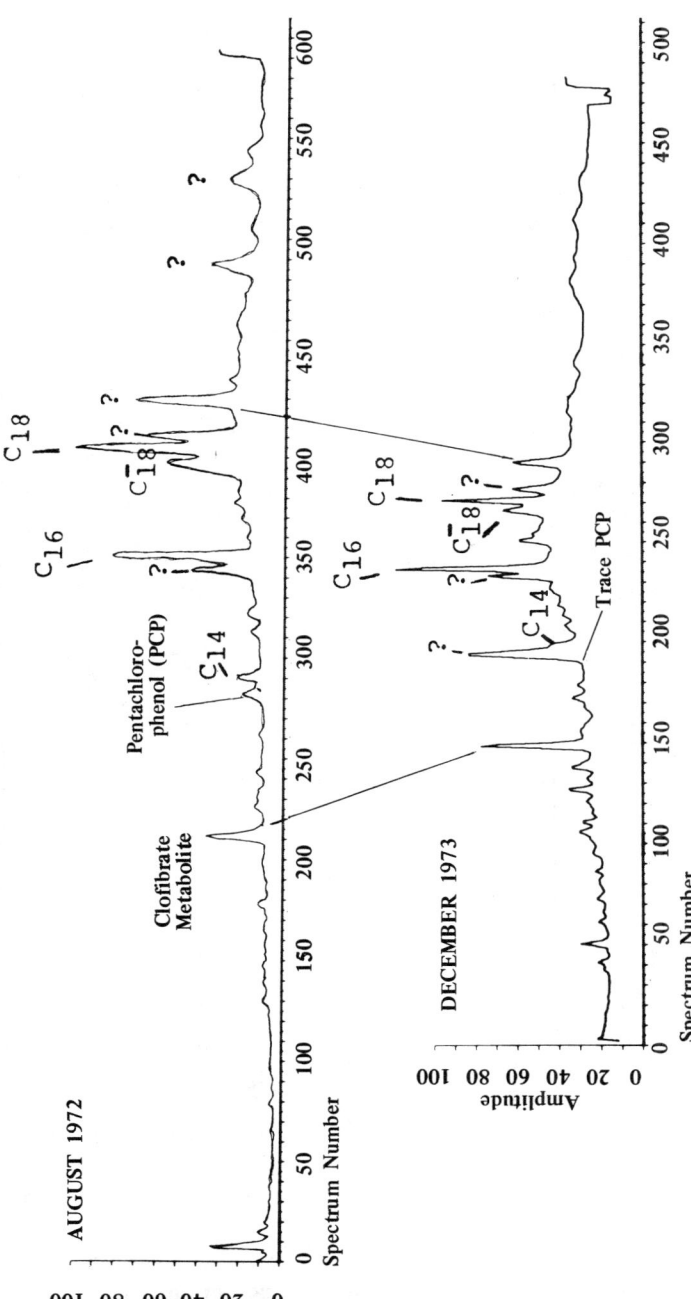

Figure 30.7 RGC's of methylated acid fractions of two activated sludge-treated effluents collected at different times. GC temperature programs were different, so time scales (spectrum numbers) are not equivalent. (? = unknown compounds of similar structure.)

534 ORGANIC POLLUTANTS IN WATER

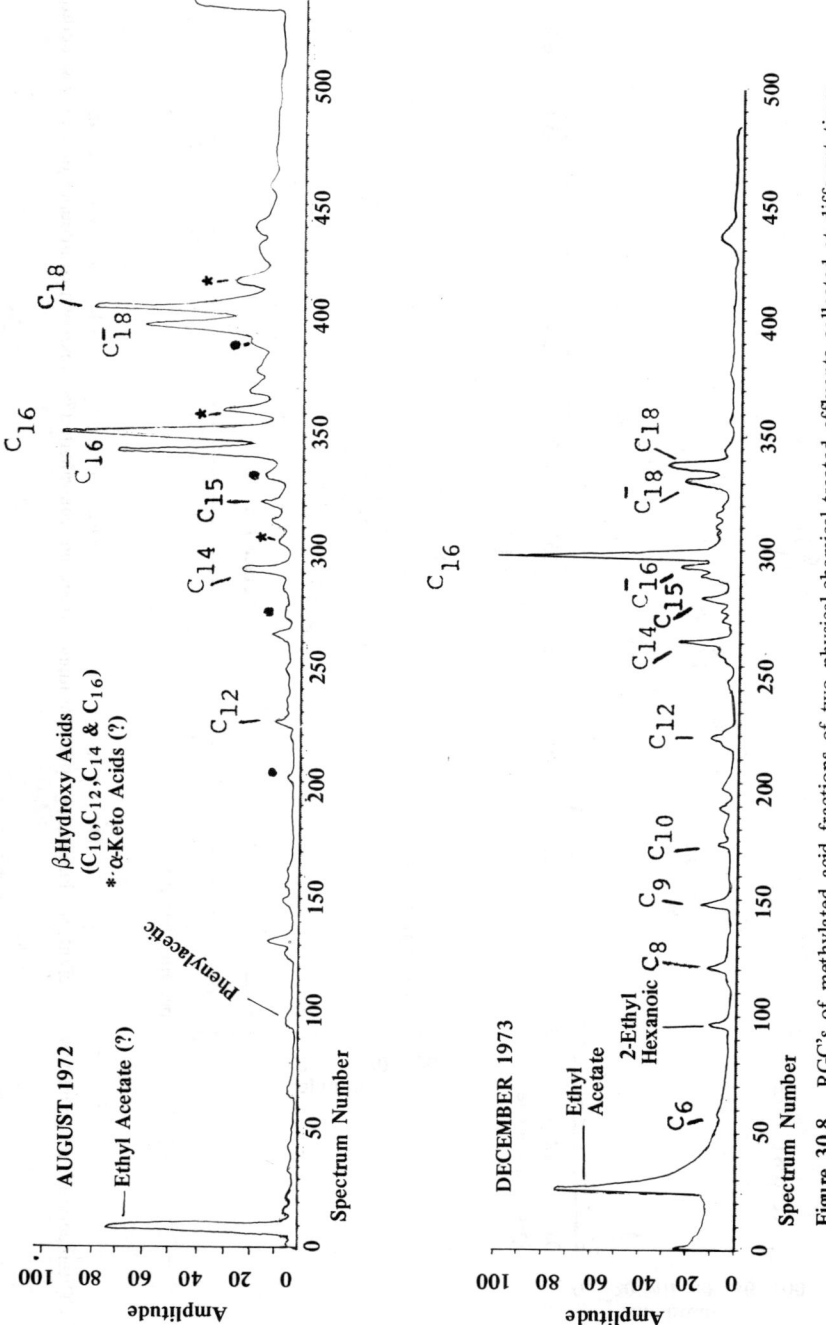

Figure 30.8 RGC's of methylated acid fractions of two physical-chemical treated effluents collected at different times. GC temperature programs were different, so time scales (spectrum numbers) are not equivalent.

IDENTIFICATION OF ORGANIC POLLUTANTS 535

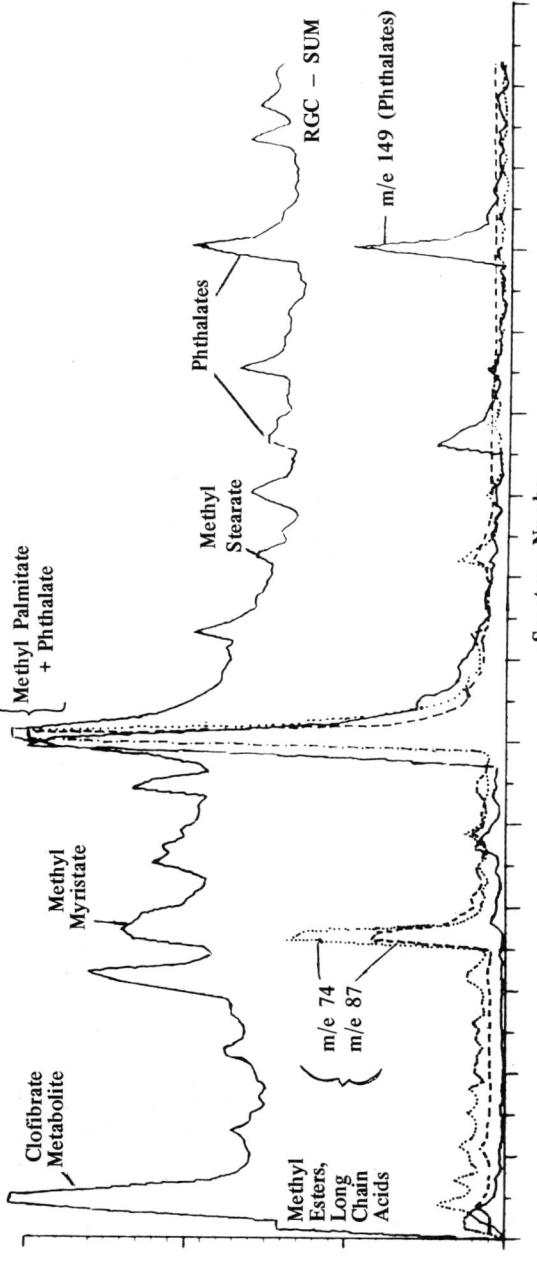

Figure 30.9 Methylated solid condensate from distillation of raw sewage—RGC and limited mass range RGC's for m/e 74, 87 and 149 using the Varian CH-5 GC/MS system (amplitude vs spectrum number).

536 ORGANIC POLLUTANTS IN WATER

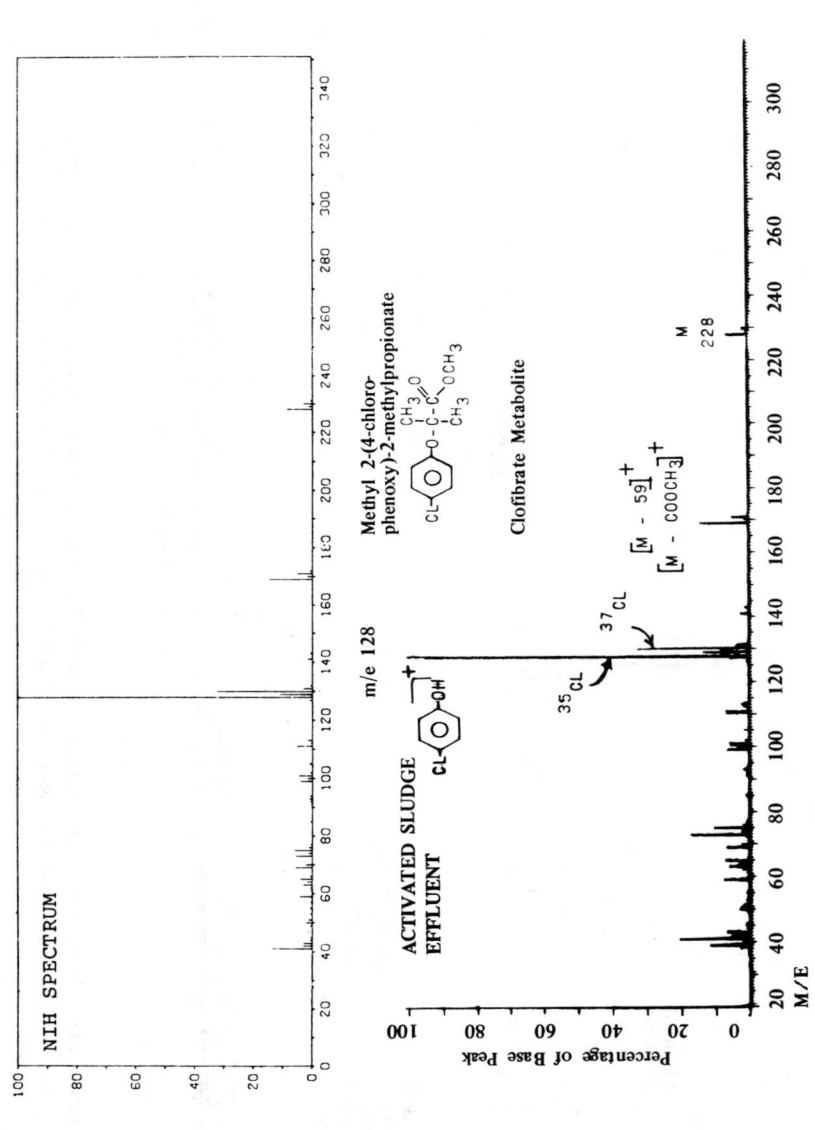

Figure 30.10 Mass spectra of the methyl ester of clofibrate metabolite. Standard (top) and the compound extracted from activated sludge effluent (bottom).

sewage, and in lime-clarified raw sewage from Washington, D.C. (see Table 30.1 for concentrations). The metabolite identification was verified by comparison of mass spectra (Figure 30.10) and infrared spectra (Figure 30.11) of a standard with those of the sample component.

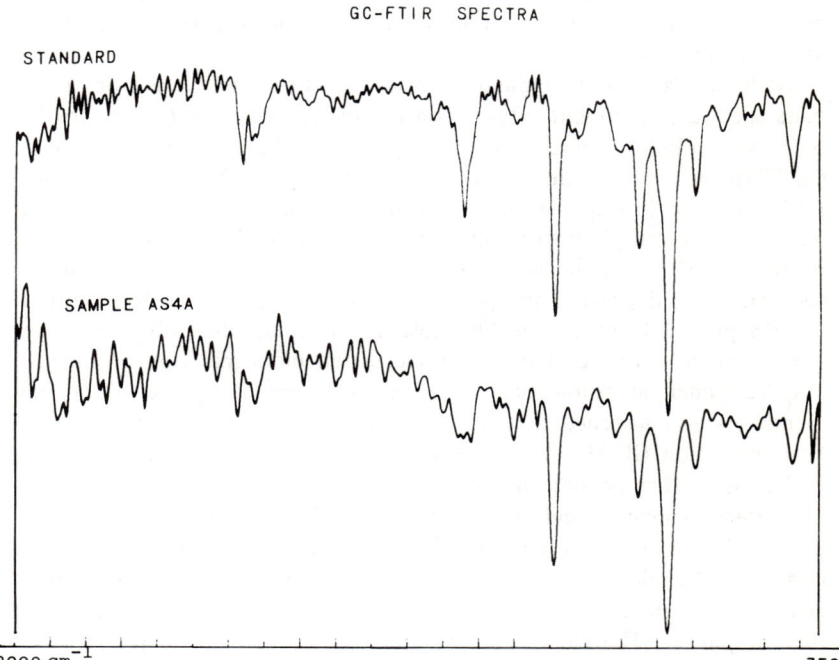

Figure 30.11 GC-Fourier Transform IR spectra of the methyl ester of the clofibrate metabolite. Standard (top) and the compound extracted from activated sludge effluent (bottom).

Effects of Physical Chemical and Activated Sludge Treatment

Treatment Effects on Acid Components

Comparison of chromatograms (Figure 30.4) showed the changes incurred during activated sludge and physical-chemical treatment of the raw sewage collected in August 1972. These chromatograms were reconstructed by the computer (RGC's) using the mass spectrometer as the detector, and were normalized on the most intense peak. Therefore, concentrations of the same components in different chromatograms were not comparable; only relative changes in components and concentrations were comparable. Measured concentrations of some components (Table 30.1) showed that there was 3- to 5-fold decrease in concentration of the main components (C_{14}, C_{16}, C_{18}, $C_{\overline{1}8}$) in raw sewage after activated sludge treatment, and a 10- to 100-fold decrease after physical-chemical treatment. These main components were present in about the same relative concentrations after treatment.

At least five new compounds with similar structures, based on their mass spectra, are present after activated sludge treatment (designated by question marks in Figure 30.4). IR spectral evidence obtained on the GC-FTIR system indicated that one of the larger unknown peaks was a LAS detergent biodegradation intermediate. This speculation correlated well with the sample history and with literature data on LAS degradation routes.[29] Although the low resolution GC/MS gave an unidentifiable spectrum, it did provide an apparent molecular weight. High-resolution GC/MS provided several possible molecular formulae for this parent ion, one of which matched that of the speculated biodegradation product, *p*-hydroxyphenyldecanoic acid. A possible formula for one of the fragment ions also matched that of a logical fragment of the same compound. High-resolution GC/MS also provided tentative structural information for two of the other possible biodegradation products.

Pentachlorophenol and the clofibrate metabolite were present only after activated sludge treatment. Their absence in the raw sewage is unexplained, but they could have been adsorbed on activated sludge particulate matter from earlier raw sewage influents and partially desorbed during this sampling. (The clofibrate metabolite was found in the raw sewage of the December 1973 series of samples, but in relatively small concentration.) Since the drug clofibrate was not found in the neutral fraction of the raw or treated sewage, the acid metabolite probably was not produced during activated sludge treatment, but as a human metabolite.

Palmitoleic acid ($C_{\overline{1}6}$) was present in relatively more abundance after physical-chemical treatment than in the raw sewage. The main changes after this treatment, however, are the formation of four β-hydroxy and three α-keto long-chain acids (Figure 30.4). The identities of these compounds have not been confirmed, but their mass spectra, especially those of the β-hydroxy acids, are fairly distinctive. A limited mass range search of the disc-stored mass spectral data for m/e 103, which is distinctive for β-hydroxy acids, indicated their presence (Figure 30.5); a homologous series was indicated by the almost equal spacing of GC peaks (time of elution). These oxygenated acids were apparently formed during physical-chemical treatment, probably on the carbon column. Biological activity has been shown to occur on columns of activated carbon used for waste treatment.[30] In this case, the occurrence of oxidation was surprising, because the effluent smelled slightly of hydrogen sulfide, which indicated anaerobic conditions.

There was little difference in composition of the raw sewage acids in samples collected 16 months apart (Figure 30.6). Except for the two unsaturated acids, concentrations of the main components were higher in the first sample than in the second sample (Table 30.1). Palmitoleic acid ($C_{\overline{16}}$) was seven times more concentrated in the second sample, and the clofibrate metabolite (about 0.8 µg/l) was found only in the second sample.

Activated sludge-treated effluents taken 16 months apart were also similar in acid composition (Figure 30.7). The clofibrate metabolite concentration was about 1 µg/l in the first sample and 2 µg/l in the second sample. Pentachlorophenol was barely detectable in the second sample. The five unknown compounds previously discussed were present in both samples, but in different ratios. A sixth unknown compound with the same mass spectral structural characteristics appeared at spectrum number 195 in the second sample. Concentrations of the main (C_{16}, C_{18} and $C_{\overline{18}}$) components were lower in the second sample; this corresponded to their lower concentrations (except for $C_{\overline{18}}$) in the second sample of raw sewage (Table 30.1).

Contrary to raw sewage and activated sludge-treated acid components, the physical-chemical effluent acids were different in composition and concentration in the two samples taken 16 months apart (Figure 30.8). Although more components were identified and quantitated in the second physical-chemical treatment sample, all components were less concentrated in the second sample than in the first. The ratios of the main components (C_{16}, $C_{\overline{16}}$, C_{18} and $C_{\overline{18}}$) were also different; C_{16} was much more predominant in the second sample. Another striking difference was the complete absence of oxygenated fatty acids in the second sample. Most of these differences could be explained. The carbon, which is usually used eight months before changing, was seven months old at the first sampling, but a different batch of carbon had been used only one month before the second sampling. The fresh carbon resulted in removal of more organics, and apparently was not old enough to allow establishment of bacterial colonies sufficient for biological metabolism of fatty acids.

Treatment Effects on Neutral Components

Many compounds were evident in chromatograms of the neutral fractions of raw and treated sewage (Figure 30.12). The total methylene

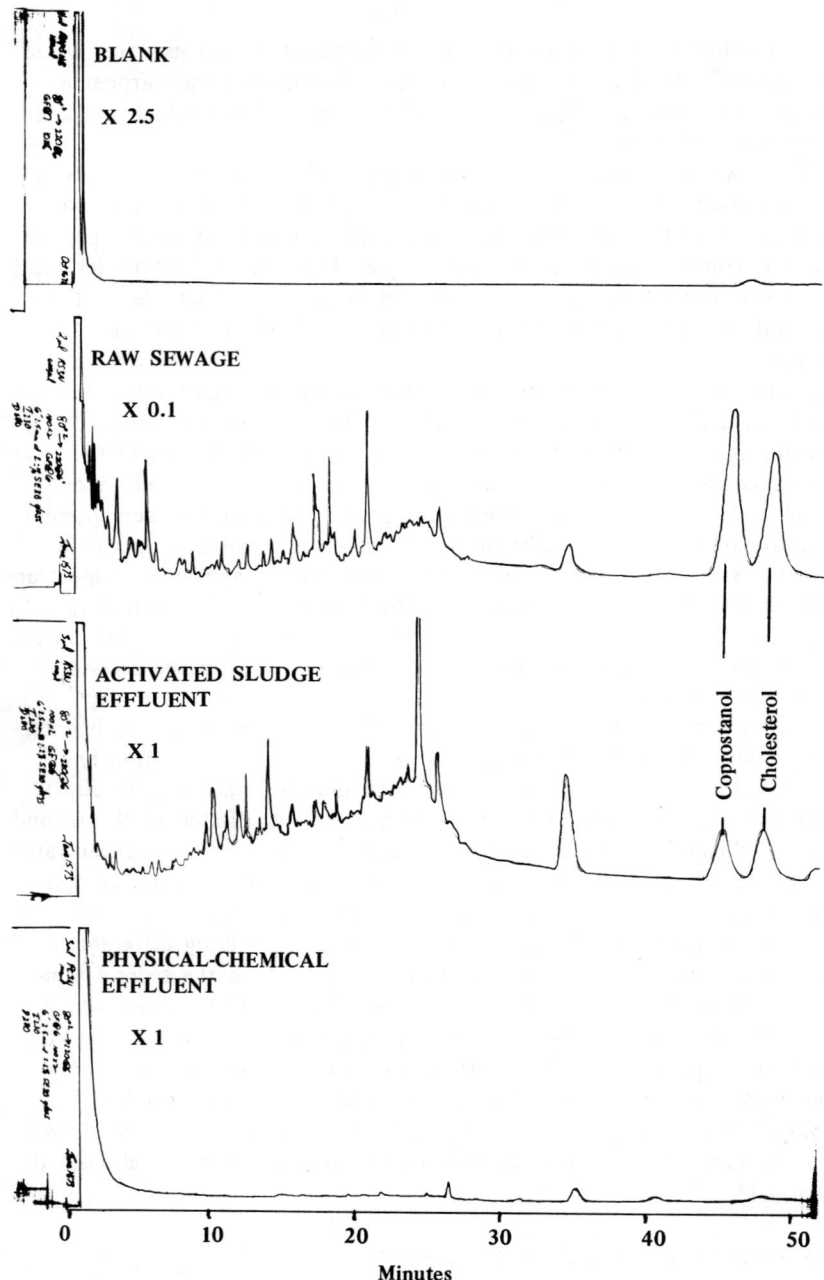

Figure 30.12 Changes in neutral components with treatment, August 1972. FID gas chromatograms.

chloride-extractable neutral organics in the raw sewage were about 10 times more concentrated (perhaps 30 times for coprostanol and cholesterol) than in the activated sludge effluent, and at least 100 times more concentrated than in the physical-chemical effluent, according to these FID/GC measurements.

Relatively few of the neutral components have been identified (Figure 30.13), and none have been quantitated. (It is only possible in these RGC's in Figure 30.13 to compare ratios of compounds before and after treatment—not concentrations of the same compound in different chromatograms.) Corresponding FID chromatograms of the December 1973 series were similar to those of the August 1972 series (Figure 30.12) in that almost no peaks showed up in the blank and few were in the physical-chemical effluent extract. The unsaturated or oxygenated hydrocarbons and α-terpineol are decreased in concentration relative to caffeine, the large unknown (*), and dioctyl phthalate after activated sludge treatment. Dibutyl phthalate predominates after physical-chemical treatment, but is probably at a very low concentration. These chromatograms were not programmed to a high enough temperature to see coprostanol and cholesterol, which were observed in other chromatograms of the same samples.

Treatment Effects on Basic Components

In general, fewer chromatographable organics were observed in the basic extracts than in the neutral or acid extracts. The raw sewage from the AWTRL contained at least 20 times the amount of methylene chloride extractable bases as did the activated sludge effluent, and at least 100 times the amount in the physical-chemical effluent (Figure 30.14). Similar conclusions were drawn from chromatograms of the second series of samples (Figure 30.15), even though the chromatograms had a high noise level and baseline rise. Both series of chromatograms showed changes in composition after treatment, but there seemed to be less change in concentration of some components after activated sludge treatment in December 1973.

Caffeine and nicotine are the only compounds identified thus far in the basic fractions. Caffeine was the principal component in the raw sewage and after activated sludge treatment (Figure 30.16); concentrations have not yet been measured. (The largest peaks in the corresponding chromatograms of Figure 30.15 are probably due to caffeine). Nicotine, the second most concentrated component in the raw sewage, was reduced in concentration relative to caffeine after activated sludge treatment. No caffeine or nicotine was observed after physical-chemical treatment (the detection limit was < 1 μg/l).

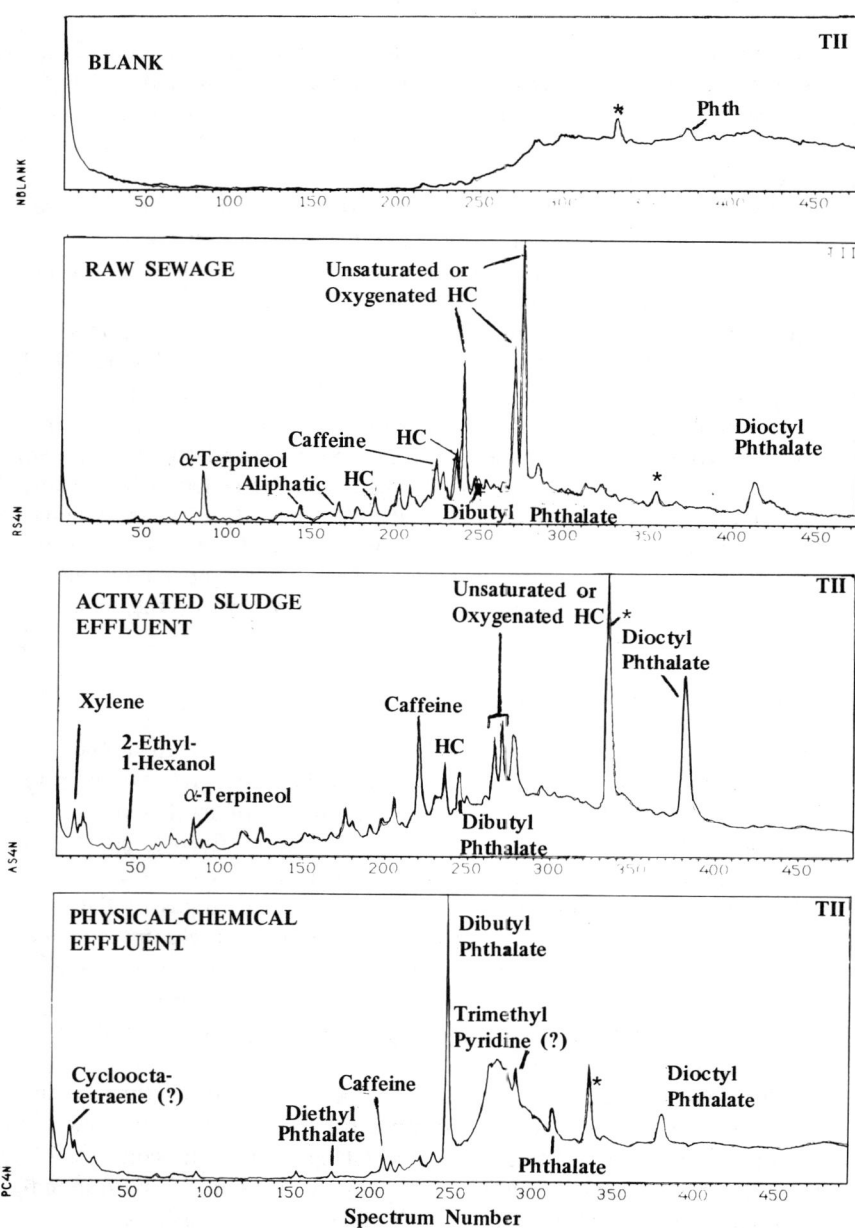

Figure 30.13 Changes in neutral components with treatment, December 1973. Computer-reconstructed gas chromatograms, using the mass spectrometer as the detector (plots of amplitude *vs* spectrum number). The asterisk (*) designates the same unknown compound in all extracts. HC = hydrocarbon.

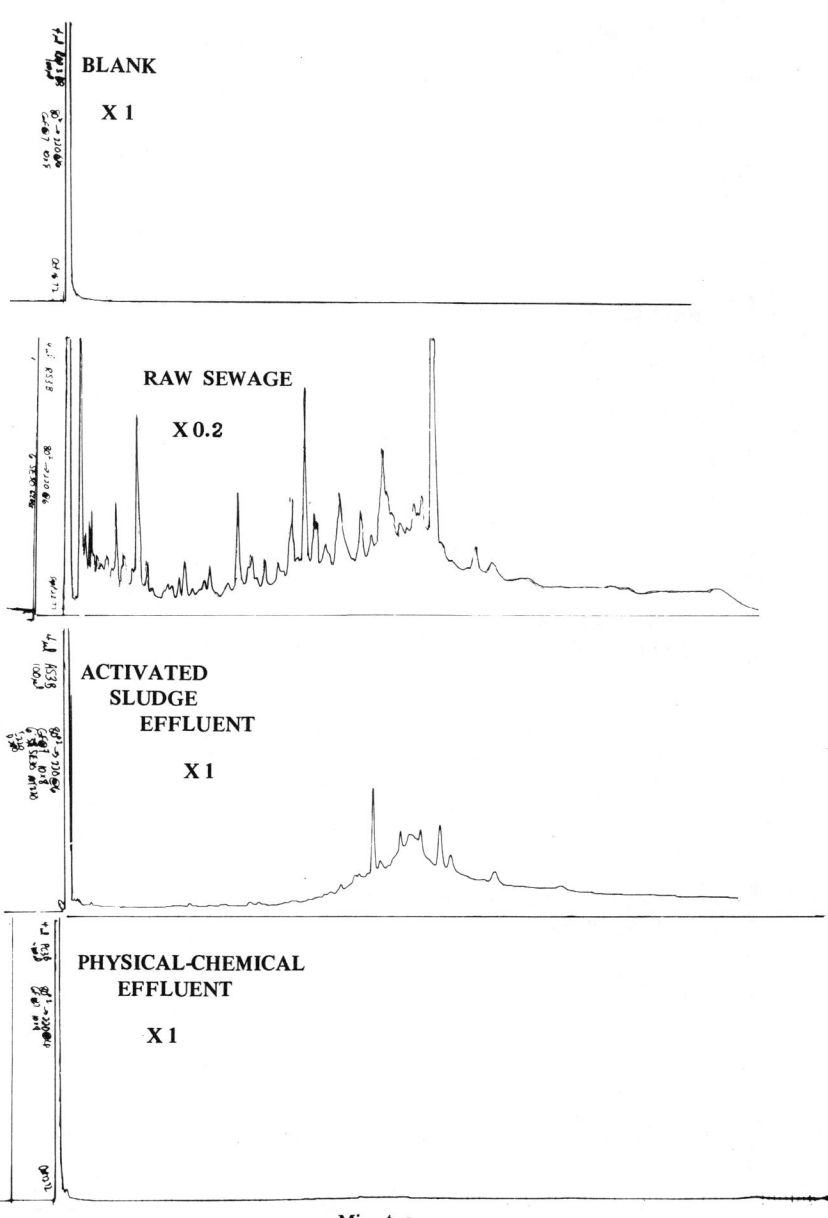

Figure 30.14 Changes in basic components with treatment, August 1972. FID gas chromatograms.

544 ORGANIC POLLUTANTS IN WATER

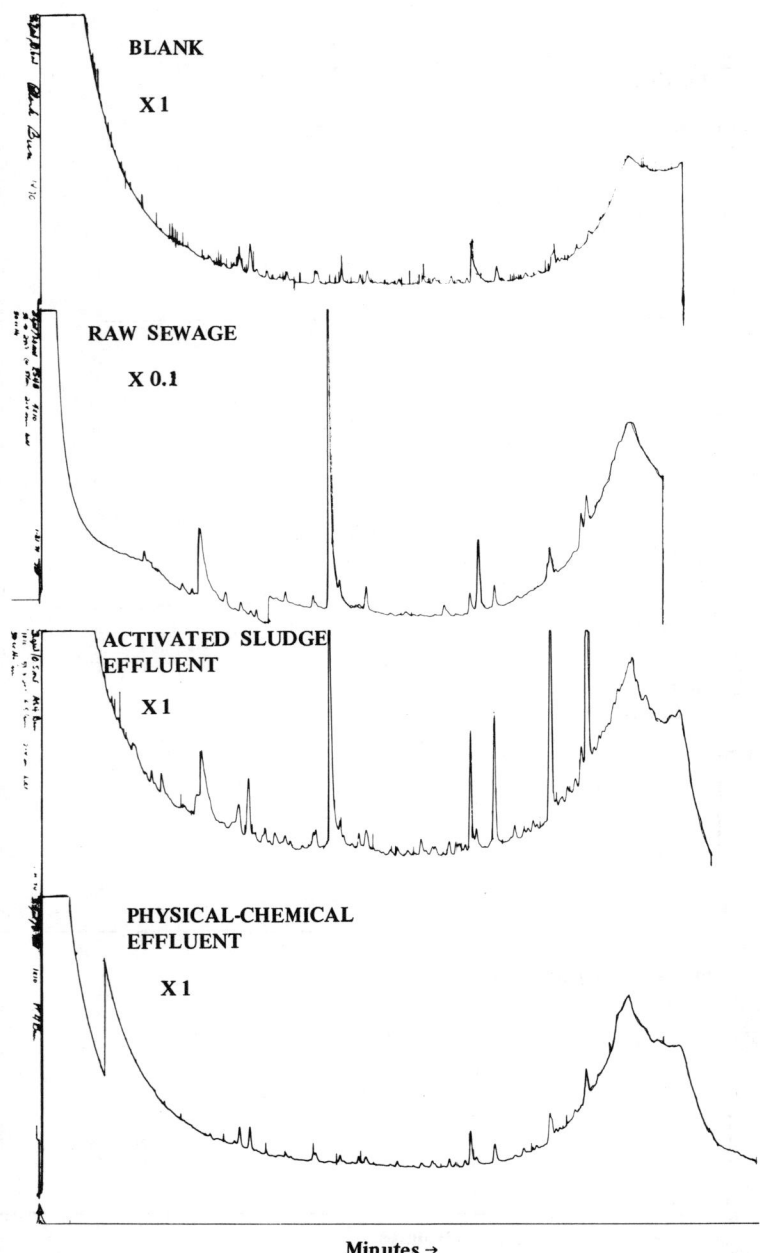

Figure 30.15 Changes in basic components with treatment, December 1973. FID gas chromatograms.

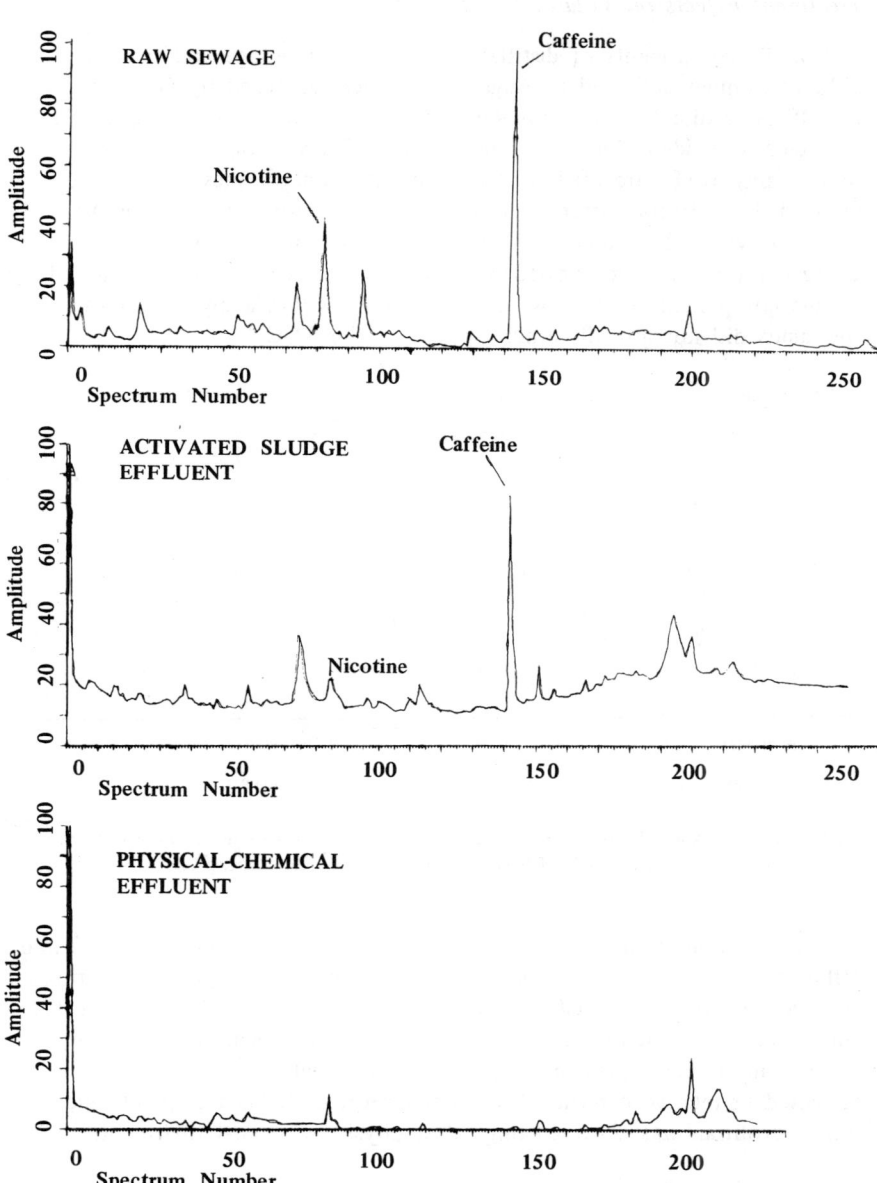

Figure 30.16 Changes in basic components with treatment, December 1973. Computer-reconstructed gas chromatograms, using the chemical ionization mass spectrometer as the detector (plots of amplitude *vs* spectrum number).

Treatment Effects on Volatile Components

Volatile components of distillate from the raw sewage and treated effluent samples collected in August 1972 were analyzed by GC and GC/MS after direct aqueous injection of a 1- to 10-μl aliquot. Six volatile acids were identified in chromatograms of raw sewage and activated sludge samples (Figure 30.17). The concentrations of these acids were reduced 5- to 10-fold after treatment, but their ratios remained about the same with valeric and isovaleric acids present in most abundance. These compounds were verified by comparison of GC retention times of sample components with those of standards. GC/MS analysis of these free acids did not give discernible peaks on the RGC.

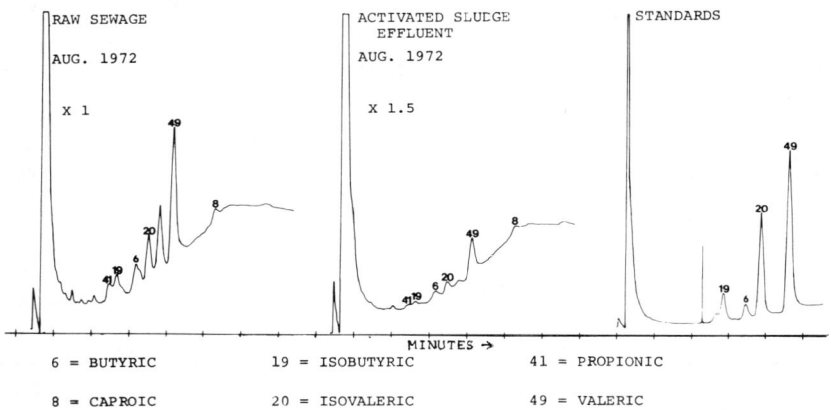

Figure 30.17 Analysis of volatile acids by direct injection of aqueous concentrates onto a 10% FFAP GC column—FID detector.

Other volatile components of the distillate from raw sewage and treated effluent samples were identified by GC/MS after direct aqueous injection and modification of the GC temperature program (Figure 30.18). Most components were neutral oxygenated or halogenated materials that were reduced in concentration or disappeared after treatment. One ketone increased in relative concentration after activated sludge treatment, and tetrahydrofuran was observed only after physical-chemical treatment.

Effects of Chlorination

Chlorination effects were determined by analysis of fractions before and after chlorination. Comparison of chromatograms of the neutral fractions

IDENTIFICATION OF ORGANIC POLLUTANTS 547

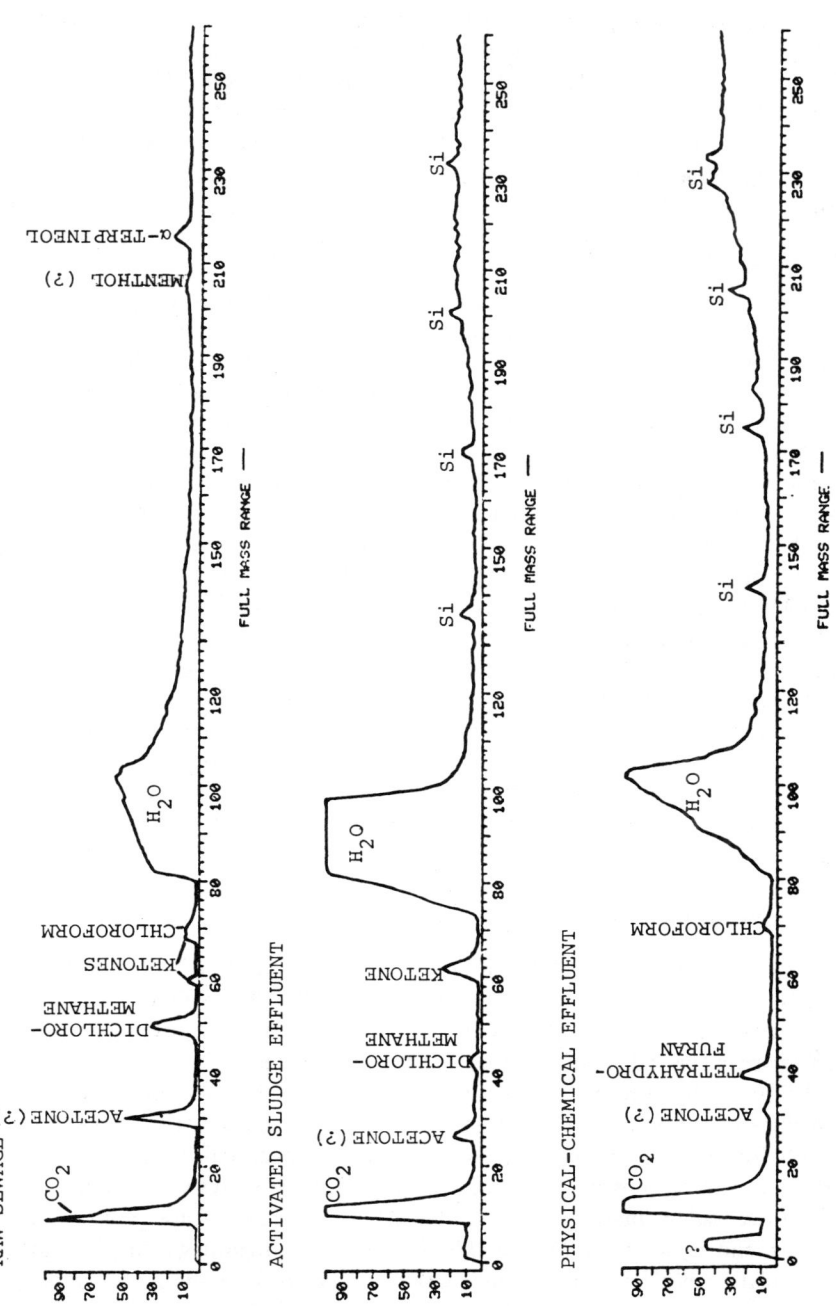

Figure 30.18 RGC's of volatile neutrals, obtained by direct injection of aqueous concentrates into the GC/MS–10% FFAP column. (Si = silicon compounds from column bleed.) Plots of spectrum number (full mass range) *vs* % total ion current.

of lime-clarified sewage from the Blue Plains waste treatment pilot plant before and after chlorination showed that some components disappeared and some new compounds were formed during chlorination (Figure 30.19).

Caffeine, α-terpineol, and the compound peak marked "x" completely disappeared during chlorination; part of the largest peak marked "*" also disappeared. Some benzyl alcohol may have been oxidized to benzaldehyde, which was present only after chlorination. Two chlorinated compounds, chlorocyclohexane and 1,1,1,2-tetrachloroethane, were present only after chlorination.

Chromatograms of the basic fractions (Figure 30.20) of the same samples showed the same type of chlorination effects observed in the neutral fractions. Caffeine and nicotine completely disappeared upon chlorination; limited mass range searches for distinctive mass spectral ions failed to detect any trace of either compound after chlorination. One or two new GC peaks were observed after chlorination. Some of the benzyl alcohol observed in the neutral fraction carried over into the basic fraction, and was gone after chlorination. This could have contributed to the benzaldehyde, which was found only in the neutral fraction after chlorination.

The most pronounced effects of chlorination were observed in chromatograms of the acid fractions of the same samples (Figure 30.21). Salicylic acid and all three mono-unsaturated fatty acids completely disappeared. Traces of the clofibrate metabolite were present in both chromatograms. Several new compounds (1,1,2,2-tetrachloroethane, pentachloroethane, hexachloroethane, and at least five unknown compounds) apparently were formed during chlorination. The presence of these neutral chlorinated compounds in the acid fraction cannot be explained at this time; they could be analytical artifacts but were not seen in any other extracts.

Effects of Ozonolysis

Ozonolysis of the AWTRL raw sewage (Figure 30.22) had some of the same effects on acids as did chlorination of the Blue Plains lime-clarified raw sewage. The mono-unsaturated $C_{\overline{1}6}$ and $C_{\overline{1}8}$ acids and several minor components (probably unsaturated) reacted completely. Several new peaks, probably aldehyde reaction products, were formed. The ozonolysis reaction was different from the chlorination reaction in that the acids were methylated before reacting with ozone, and the reaction occurred in methylene chloride solution, not in the aqueous phase. This ozonolysis technique[20] was devised as an analytical tool to detect or confirm unsaturated compounds, and it proved useful for this purpose. Although it was not designed to simulate waste treatment by ozonation, similar reaction with the unsaturated acids would be expected during such waste treatment.

IDENTIFICATION OF ORGANIC POLLUTANTS 549

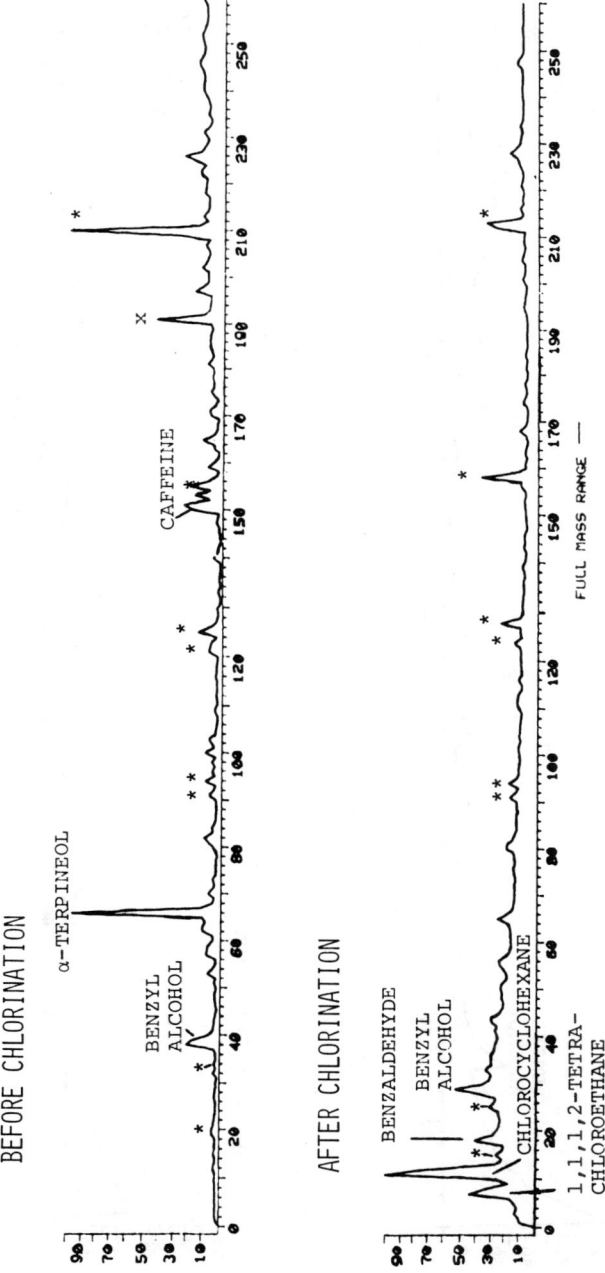

Figure 30.19 Changes in neutral components in lime-clarified raw sewage upon chlorination–RGC plots of spectrum numbers (full mass range) vs % total ion current. (* = matching peaks in the two RGC's.)

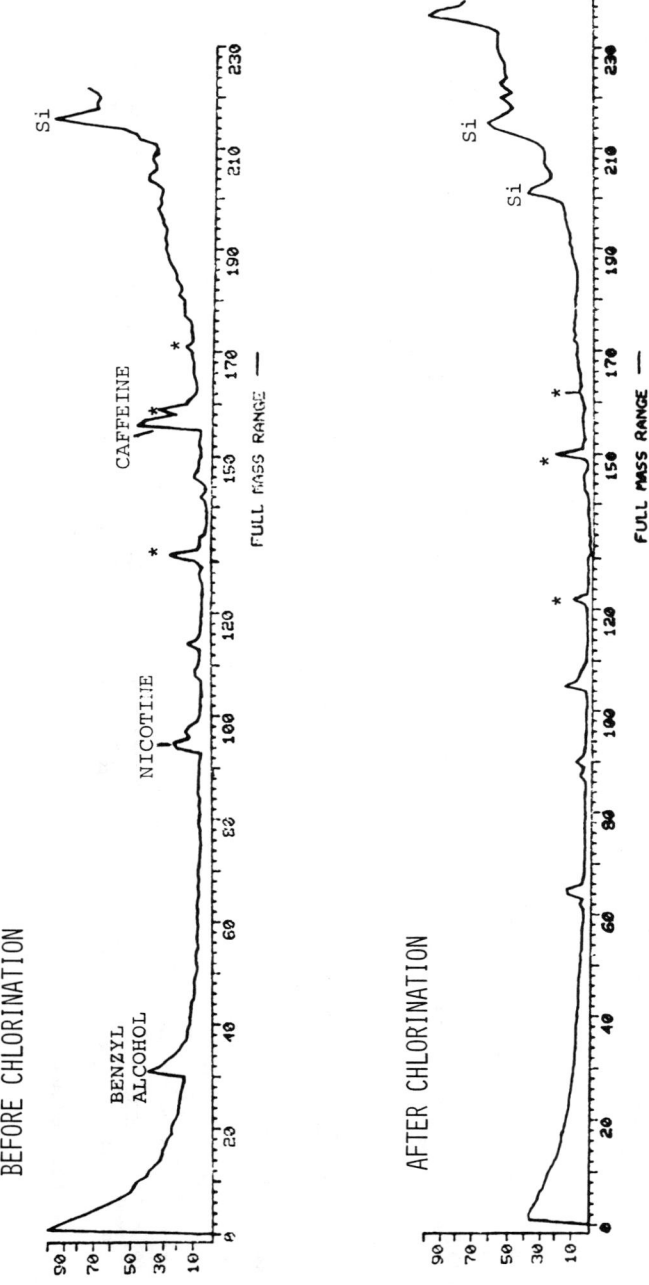

Figure 30.20 Changes in bases in lime-clarified raw sewage upon chlorination–RGC plots of spectrum number (full mass range) *vs* % total ion current. (* = matching peaks in the two RGC's.)

IDENTIFICATION OF ORGANIC POLLUTANTS 551

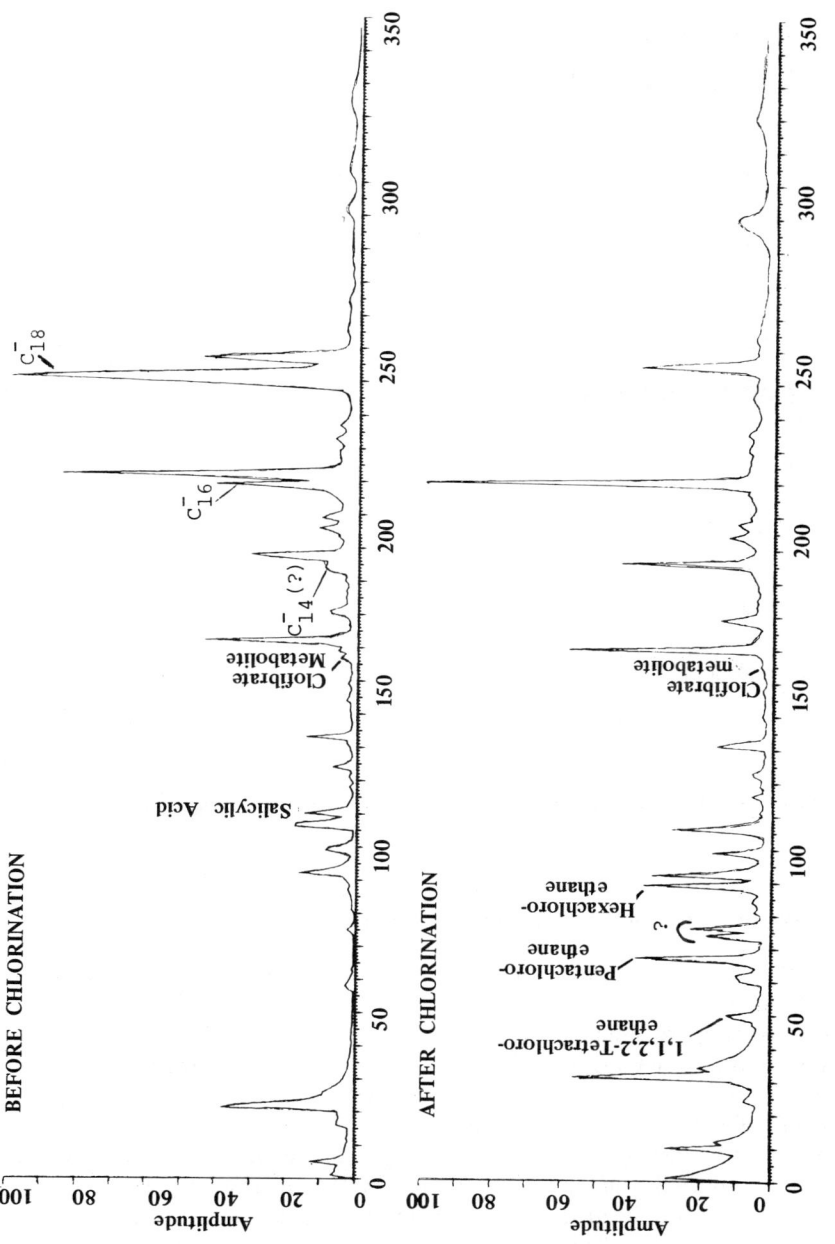

Figure 30.21 Changes in acid compounds in lime-clarified raw sewage upon chlorination—RGC's of methylated extracts.

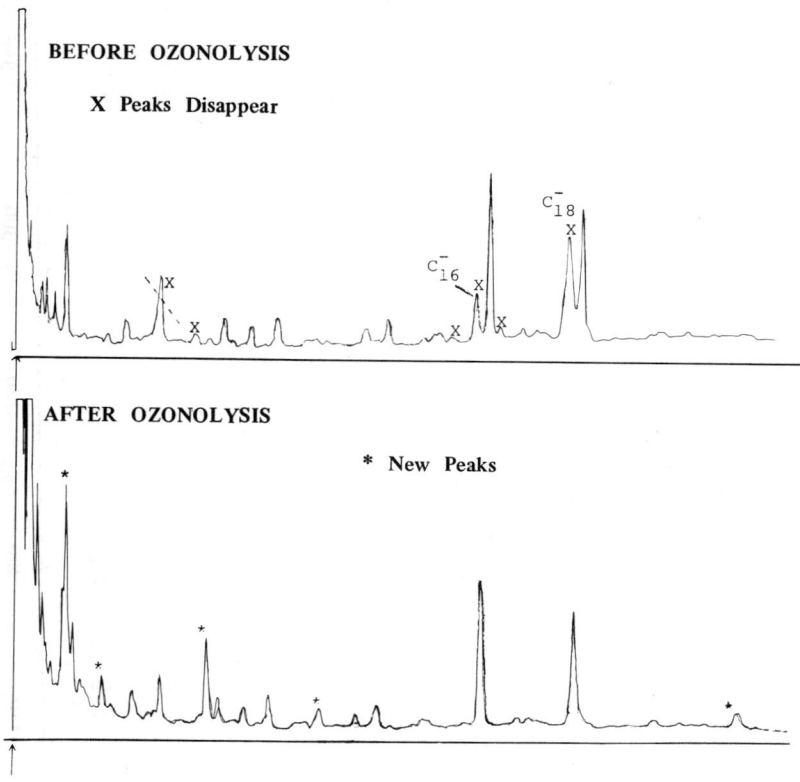

Figure 30.22 Changes in methylated acid extract of raw sewage upon ozonolysis. FID gas chromatograms.

CONCLUSIONS AND REMAINING WORK

Eighty specific volatile organic compounds were identified in raw and treated domestic wastewaters (Table 30.1), mostly by GC/MS. A series of five or more similar unknown acids of relatively high concentration were possibly detergent metabolites that were formed during activated sludge treatment of raw sewage; these need to be identified and quantitated. Only a few neutral compounds were identified in raw and treated domestic wastewaters; unsaturated and/or oxygenated hydrocarbons and many completely unknown compounds present in relatively high concentration have not been identified. While the basic fractions were less complex, several components remain unidentified.

IDENTIFICATION OF ORGANIC POLLUTANTS

Physical-chemical treatment (carbon adsorption) was more effective than activated sludge treatment in reducing the concentration of specific volatile organic pollutants, but even in physical-chemical effluents, obvious components need to be identified and quantitated, especially in the neutral fractions.

Brief analytical studies on sewage before and after chlorination indicated important changes in some components. These changes need to be better defined by identification and quantification of compounds that react with chlorine and products that are formed by chlorination.

Identification of compounds is not enough. Their quantities must be measured before the significance of changes occurring during treatment can be assessed. Concentrations of only a few of the compounds identified in this study have been measured (Table 30.1). After qualitative and quantitative analysis of components in domestic wastewaters, toxicological studies will be necessary to determine the need for improved waste treatment methods.

These volatile, methylene chloride-extractable compounds constitute only a small fraction (certainly less than 25%) of the total organic components of raw or treated domestic sewage. The next frontier in analysis of wastewaters is identification and measurement of the nonvolatile components.

ACKNOWLEDGMENTS

The authors appreciate the advice and cooperation of personnel at the EPA's Advanced Waste Treatment Research Laboratory, Cincinnati, Ohio. This research was not only condoned, but supported and encouraged by Mr. Jesse Cohen and Dr. Robert Bunch of that laboratory. Dr. James Westrick and Mr. Richard Dobbs gave valuable technical advice and helped in obtaining samples. Mr. Thomas Pressley of that Laboratory's Blue Plains (Washington, D.C.) pilot plant suggested analysis of the chlorinated samples, and collected, chlorinated and extracted these samples.

At ERL-Athens, Mssrs. Mike Carter and Alfred Thruston and Ms. Ann Alford performed the GC/MS analysis, which consumed many man-months over a three-year period. Dr. Leo Azarraga performed the GC/IR analyses. Dr. James Ryan at the EPA's pesticide analysis facility at Research Triangle Park, N. C., performed some of the high-resolution GC/MS analyses. Ms. Anne Elder did much of the drafting work.

DISCLAIMER

Mention of commercial products, trade names and companies is for informational purposes only and does not imply endorsement by the U.S. Environmental Protection Agency or the Athens Environmental Research Laboratory.

REFERENCES

1. Keith, L. H., A. W. Garrison, F. R. Allen, M. H. Carter, T. L. Floyd, J. D. Pope and A. D. Thruston, Jr. "Identification of Organic Compounds in Drinking Water from Thirteen U.S. Cities," presented at the First Chemical Congress of the North American Continent, Mexico City (December 1975), Chapter 22, this volume.
2. Ongerth, H. J., D. P. Spath, J. Crook and A. E. Greenberg. "Public Health Aspects of Organics in Water," *J. Amer. Water Works Assoc.* **65**, 495-498 (1973).
3. World Health Organization, International Reference Centre for Community Water Supply. "Health Effects Relating to Direct and Indirect Re-Use of Wastewater for Human Consumption," Report of an International Working Meeting, Amsterdam, The Netherlands, January 13-16, 1975, Technical Paper No. 7 (September 1975).
4. Bunch, R. L., E. F. Barth and M. B. Ettinger. "Organic Materials in Secondary Effluents," *J. Water Pollution Control Fed.* **33**, 122-126 (1961).
5. Hunter, J. V. and H. Heukelekian. "The Composition of Domestic Sewage Fractions," *J. Water Pollution Control Fed.* **37**, 1142-1163 (1965).
6. Painter, H. A. "Organic Compounds in Solution in Sewage Effluents," *Chem. Ind.*, 818-822 (September 1973).
7. Rebhun, M. and J. Manka. "Classification of Organics in Secondary Effluents," *Environ. Sci. Technol.* **5**, 606-609 (1971).
8. Manka, J., M. Rebhun, A. Mandelbaum and A. Bortinger. "Characterization of Organics in Secondary Effluents," *Environ. Sci. Technol.* **8**, 1017-1020 (1974).
9. Painter, H. A. Water Pollution Research Laboratory, Stevenage, England, Private Communication (October 1973).
10. Murtaugh, J. J. and R. L. Bunch. "Acidic Components of Sewage Effluents and River Water," *J. Water Pollution Control Fed.* **37**, 410-415 (1965).
11. Reichert, J., H. Kunte, K. Engelhart and J. Borneff. "Carcinogenic Substances Occurring in Water and Soil. XXVII Further Studies on the Elimination from Waste Water of Carcinogenic Polycyclic Aromatic Hydrocarbons," *J. Zbl. Bakt. Hyg. I Abt. Orig. B.* **155**, 18-40 (1971); *Chem. Abstr.* 1971, 75:154806c.
12. Burlingame, A. L., E. S. Scott, J. W. de Leeuw, B. J. Kimble and F. C. Walls. "The Molecular Nature and Extreme Complexity of Trace Organic Constituents in Southern California Municipal Wastewater Effluents," presented at the First Chemical Congress of the

North American Continent, Mexico City (December 1975), Chapter 31, this volume.
13. Dunlap, W. J., M. R. Scalf, and D. C. Shew. "Isolation and Identification of Organic Contaminants in Ground Water," presented at the First Chemical Congress of the North American Continent, Mexico City (December 1975), Chapter 27, this volume.
14. Glaze, W. H., J. E. Henderson, IV, G. Smith and O. D. Sparkman. "Analysis of New Chlorinated Organic Compounds in Municipal Wastewaters after Terminal Chlorination," presented at the First Chemical Congress of the North American Continent, Mexico City (December 1975), Chapter 16, this volume.
15. Giger, W., M. Reinhard, C. Schaffner and F. Zurcher. "Analysis of Organic Constituents in Water by High Resolution Gas Chromatography in Combination with Specific Detection and Computer Assisted Mass Spectrometry," presented at the First Chemical Congress of the North American Continent, Mexico City (December 1975), Chapter 26, this volume.
16. Pitt, W. W., Jr., R. L. Jolley and S. Katz. "Separation and Analysis of Refractory Pollutants in Water by High-Resolution Liquid Chromatography," presented at the First Chemical Congress of the North American Continent, Mexico City (December 1975), Chapter 14, this volume.
17. Jolley, R. L., G. Jones, Jr., and J. E. Thompson. "Determination of Chlorination Effects on Organic Constituents in Natural and Process Waters Using High-Pressure Liquid Chromatography," presented at the First Chemical Congress of the North American Continent, Mexico City (December 1975), Chapter 15, this volume.
18. Pitt, W. W., R. L. Jolley and S. Katz. "Automated Analysis of Individual Refractory Organics in Polluted Water," U.S. Environmental Protection Agency Res. Rep. No. EPA 660/2-74-076 (August 1974).
19. Webb, R. G., A. W. Garrison, L. H. Keith and J. M. McGuire. "Current Practice in GC/MS Analysis of Organics in Water," U.S. Environmental Protection Agency Res. Rep. No. EPA-R2-73-277 (August 1973).
20. Beroza, M. and B. Bierl. "Rapid Determination of Olefin Position in Organic Compounds in Microgram Range by Ozonolysis and Gas Chromatography," *Anal. Chem.* **39**, 1131-1135 (1967).
21. McGuire, J. M., **A.** L. Alford and M. H. Carter. "Organic Pollutant Identification Utilizing Mass Spectrometry," U.S. Environmental Protection Agency Res. Rep. No. EPA-R2-73-234 (July 1973).
22. Azarraga, L. V. and A. C. McCall. "Infrared Fourier Transform Spectrometry of Gas Chromatography Effluents," U.S. Environmental Protection Agency Res. Rep. No. EPA-660/2-73-034 (January 1974).
23. Heller, S. R., J. M. McGuire and W. L. Budde. "Trace Organics by GC/MS," *Environ. Sci. Technol.* **9**, 210-213 (1975).
24. Mass Spectrometry Data Centre. *Eight Peak Index of Mass Spectra*, 1st Ed. Atomic Weapons Research Establishment, Aldermaston, Reading, England (1970), 1493 pp.

25. Ryhage, R. and E. Stenhagen. "Mass Spectrometry of Long-Chain Esters," Chapter 9 in *Mass Spectrometry of Organic Ions*, F. McLafferty, Ed. (New York and London: Academic Press, 1963).
26. Williams, C. M., A. H. Porter and M. Greer. "Mass Spectrometry of Biologically Important Aromatic Acids," University of Florida College of Medicine and Veterans Administration Hospital, Gainesville, Florida (July 1969).
27. Stenhagen, E., S. Abrahamsson and F. W. McLafferty, Eds. *Archives of Mass Spectral Data, Vol. 3* (New York: Interscience Publishers, 1972), 193-384.
28. Fitzgerald, J. D. "Mode of Action and Clinical Results of Regelan," *Wien. Klin. Wochenschar.* **79**, 716-720 (1967); *Chem. Abstr.* 1967 67:107356f
29. Willetts, A. J. and R. B. Cain. "Microbial Metabolism of Alkylbenzene Sulphonates," *J. Biochem.* **129**, 389-403 (1972).
30. Bishop, D. F., T. P. O'Farrell, A. F. Cassel and A. P. Pinto. "Physical-Chemical Treatment of Raw Municipal Wastewater," U.S. Environmental Protection Agency Res. Rep. No. EPA-670/2-73-070 (September 1973).

CHAPTER 31

THE MOLECULAR NATURE AND EXTREME COMPLEXITY OF TRACE ORGANIC CONSTITUENTS IN SOUTHERN CALIFORNIA MUNICIPAL WASTEWATER EFFLUENTS

A. L. Burlingame, B. J. Kimble, E. S. Scott and F. C. Walls

> Space Sciences Laboratory
> University of California
> Berkeley, California

J. W. de Leeuw

> Delft University of Technology
> Department of Chemical Engineering and Chemistry
> Organic Geochemistry Unit
> Delft, The Netherlands

B. W. de Lappe and R. W. Risebrough

> Bodega Marine Laboratory
> University of California
> Bodega Bay, California

INTRODUCTION

Four billion liters of wastewaters are currently discharged daily into the ocean from treatment plants in Southern California. In contrast, the total annual surface runoff from rivers, drainage channels, and urban storm sewers averaged 400 billion liters per year between 1941 and 1970. On a daily basis, therefore, the amount of water discharged into the sea from the wastewater treatment plants is about three times greater than the total contributed by storm runoff.[1] Moreover, the wastewater constitutes approximately one-third of the total dependable water supply in Southern California and therefore represents a very significant fraction of the total water budget.

Increasing water demands in Southern California prompt the recycling of this wastewater for human consumption, industrial use, and the recharging of ground water.[2] At present 90% of the wastewaters discharged into the sea receive primary treatment only, consisting principally of the removal of flotable solids. Analysis of the effluent of six principal treatment plants in Southern California for polychlorinated biphenyls, DDT compounds, dieldrin and other pesticides by electron-capture gas chromatography[1] has indicated the presence in these effluents of a vast number of additional organic constituents. Recycling of wastewaters would require both the removal of these constituents and the development of efficient assay systems that would detect trace quantities of specific organic compounds.

Combined gas chromatography/mass spectrometry (GC/MS) has been increasingly utilized to identify organic constituents of environmental mixtures[3-6]; the recent literature is also covered in a review of the organic geochemistry of environment.[7] In very complex mixtures, however, components may not be chromatographically resolved[8] and the generation of nominal mass spectra of the chromatographic effluents poses considerable problems in their identification. The advantage of using high-resolution mass spectrometry to characterize such complex mixtures was proposed by Manka *et al.*[9]: ". . . The investigation has been somewhat limited because the mass spectrometer used in this work is a low-resolution instrument (Atlas CH_4) that can provide only nominal masses, leaving a large number of contaminants that cannot be identified unequivocally. High-resolution mass spectrometry coupled with gas chromatography would make possible identification of additional specific compounds. Once identified, those compounds could then be monitored by simple techniques. . . ."

Gas chromatography/high-resolution mass spectrometry (GC/HRMS) has been developed in this laboratory to facilitate analysis of such complex mixtures, initially using packed gas chromatographic columns,[10] and more recently utilizing glass surface-coated open tubular (SCOT) capillary columns for high-resolution separations (HRGC/HRMS).[11] The present chapter describes the application of this new technique to the analysis of organic extracts of wastewater from a Southern California municipal sewage treatment facility; it illustrates the approach used for HRGC/HRMS data set evaluation and for assessment of the molecular nature and complexity of the trace organic constituents. Also, it presents an application of an *in situ* sampling system using polyurethane foam to extract organic constituents, present in trace amounts, from large volumes of wastewaters so as to obtain quantities sufficient for analysis.[12]

IDENTIFICATION OF ORGANIC POLLUTANTS

The influent, consisting of domestic and industrial wastewater, undergoes removal of solids and flotables followed by aerobic biological treatment to give the secondary effluent. The tertiary treatment developed for the removal of viruses and of potential toxic chemical constituents so that wastewater may be used to replenish ground water at present consists of flocculation, sedimentation and filtration, followed by chlorination and adsorption on carbon (Figure 31.1).

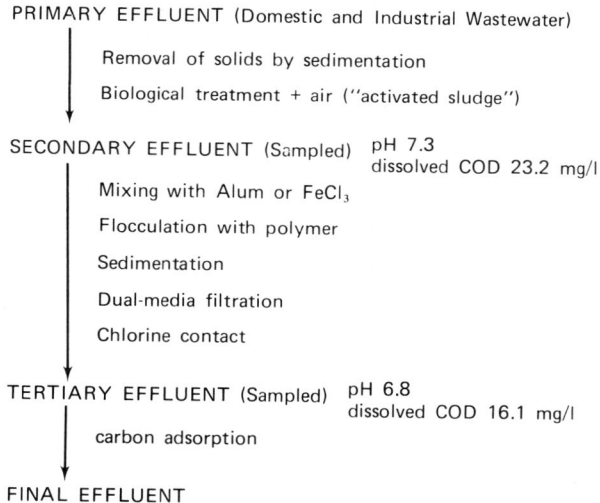

Figure 31.1 Southern California municipal wastewater secondary-tertiary sampling scheme indicating sampling points.

EXPERIMENTAL

Sample Collection and Preparation

The samples described in this paper were collected from the points in the treatment scheme indicated in Figure 31.1; they consisted of secondary effluent; tertiary effluent which had been treated by flocculation, sedimentation, filtration and chlorination but not adsorption on carbon; and of tertiary effluent which had also been treated with carbon. They were subjected to the analysis scheme outlined in Figure 31.2. Large volumes (24-36 liters) of wastewater were filtered and extracted by passage through high-density, high-ether-content polyurethane foam.[1,2] The

560 ORGANIC POLLUTANTS IN WATER

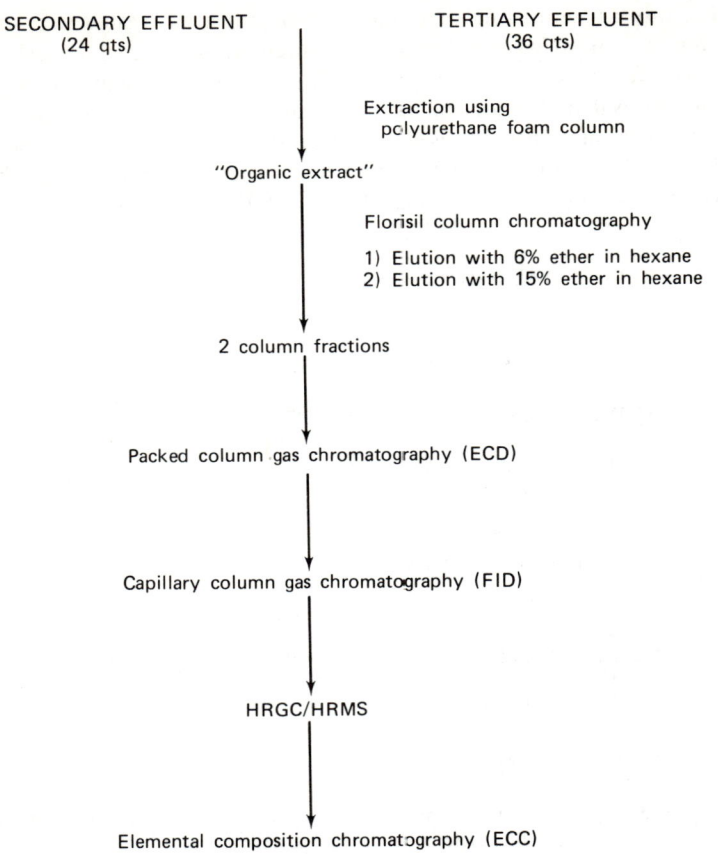

Figure 31.2 Southern California municipal wastewater sample analysis scheme.

foam plug (of length 25.4 cm and diameter 6.35 cm) was contained in a stainless steel tube fitted with a 6-liter stainless steel reservoir and a valve to control the flow rate to about 300 ml/min. After sampling, the foam column was eluted with 500 ml acetone followed by 500 ml hexane.

The eluates were combined in a 2-liter separatory funnel, and 700 ml of distilled water and 10 g kiln-fired NaCl were added before extraction.

In addition, duplicate wastewater samples were extracted with methylene chloride/hexane according to the procedures currently recommended by the Environmental Protection Agency.[13] Blank analyses with both the polyurethane foam and solvent extraction systems indicated that no peaks were present on chromatograms obtained by electron capture gas chromatography at levels greater than the equivalent of 400 picograms of p,p'-DDE in the extraction system. The hexane extract was reduced to a volume of about 10 ml and fractionated on a Florisil column,[14] eluting sequentially with 6% and 15% solutions of diethyl ether in hexane to give the 6% and 15% fractions referred to in the text. Electron capture gas chromatography (on a 2 m x 2 mm i.d. glass column packed with OV-1 on 100/200 mesh Gaschrom Q) indicated that the profiles obtained from the samples extracted with polyurethane foam were comparable to those obtained by solvent extraction with methylene chloride/hexane.

At higher dilutions the electron capture gas chromatograms exhibited relatively few, early-eluting peaks. Concentration of the extracts, however, revealed the existence in the mixture of many additional components. DDT and PCB compounds, which are readily distinguishable in extracts of primary effluents of the principal coastal wastewater treatment plants,[1] were present at concentrations below detectability (< 0.05 parts per billion) in these samples.

Two of these samples, obtained by extraction with polyurethane foam, of secondary effluent and of tertiary effluent which had been chlorinated but not treated with carbon were selected for HRGC/HRMS analysis.

Flame Ionization Gas Chromatography

Gas chromatography was carried out using a modified Perkin-Elmer model 900 gas chromatograph fitted with a flame ionization detector (FID). The column was a 60 m x 0.75 mm i.d. glass SCOT capillary[15] coated with OV-101. Helium was used as the carrier gas at a flowrate of about 6 ml/min. The injector and detector temperatures were maintained at 290° and 320°C, respectively. The column temperature was maintained isothermally at 70°C until after the solvent eluted, and was then programmed linearly at 5°C/min up to 290°C and held at this temperature until no further components eluted. Total analysis time was approximately one hour.

High-Resolution Gas Chromatography/High-Resolution Mass Spectrometry (HRGC/HRMS)

The coupling of a gas chromatograph to a fast-scanning mass spectrometer has been described previously,[10] and recent modifications now permit the use of capillary chromatographic columns as a HRGC/HRMS combination.[11] The main features of the system are summarized in Table 31.1 and result in accurate mass measurements well within 17 ppm,[11] permitting assignment of elemental compositions. The chromatographic parameters were similar to those described above.

Table 31.1 Operating Parameters for HRGC/HRMS/Computer System

GC:	modified Varian Aerograph 1400 gas chromatograph 60 m x 0.75 mm i.d. glass SCOT OV-101 column column effluent splitter: variable to FID and/or MS make-up helium inlet to separator: variable flow	
SEPARATOR:	second stage stainless steel Ryhage jet inlet jet diameter 0.010 in. outlet jet diameter 0.012 in. jet separation 0.020 in. 115 liter/sec diffusion pump	
MS:	modified AEI MS902 double focusing mass spectrometer	
	dynamic resolution	10,000
	scan rate (magnetic)	6 sec/decade
	mass range	m/e 800-60
	scan time	7.2 sec
	rescan time	2.4 sec
	cycle time	9.6 sec
DATA ACQUISITION:	LOGOS II - XDS Sigma 7 computer digitization rate 50 kHz	

Acquisition of HRGC/HRMS data is achieved using LOGOS-II,[16] an on-line, real-time software system for fast-scanning high-resolution mass spectrometry employing a Xerox Sigma 7 computer. Data processing includes the generation of plots of nominal mass spectra, calculation of elemental compositions, and construction of nominal mass chromatograms and accurate mass (elemental composition) chromatograms.

RESULTS AND DISCUSSION

Figures 31.3a through 31.6a show the capillary gas chromatograms (FID) obtained for each of the four fractions analyzed. It is apparent that the two 15% fractions (Figures 31.5a and 31.6a) give somewhat similar profiles, suggesting similar qualitative and quantitative compositions. The chromatograms of the 6% fractions (Figures 31.3a and 31.4a) indicate no such similarities, and the secondary effluent fraction (Figure 31.3a) appears to be much more concentrated than the tertiary effluent 6% fraction. Figures 31.3b through 31.6b show the total ionization chromatograms (TIC) for each fraction, obtained from the high-resolution mass spectral data set. In general the correlation between the TIC profiles

IDENTIFICATION OF ORGANIC POLLUTANTS 563

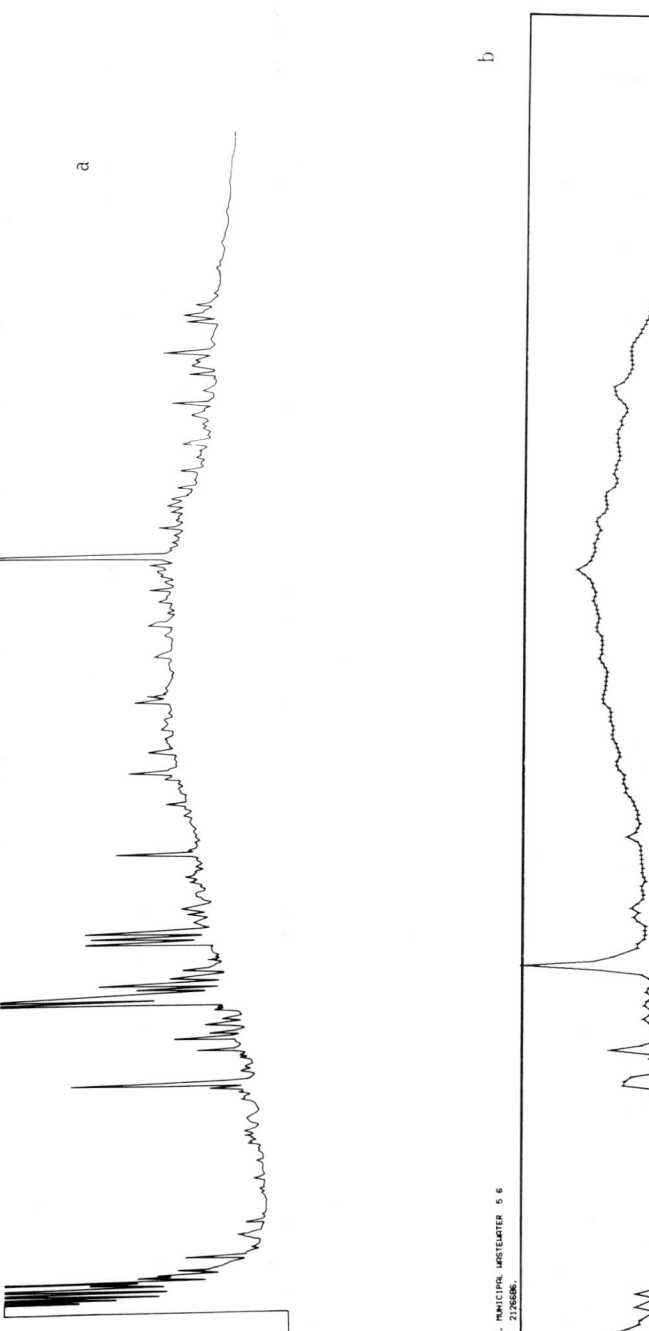

Figure 31.3 a) Flame ionization detector (FID) capillary gas chromatograms of secondary effluents, 6% fraction; b) total ionization chromatograms (TIC).

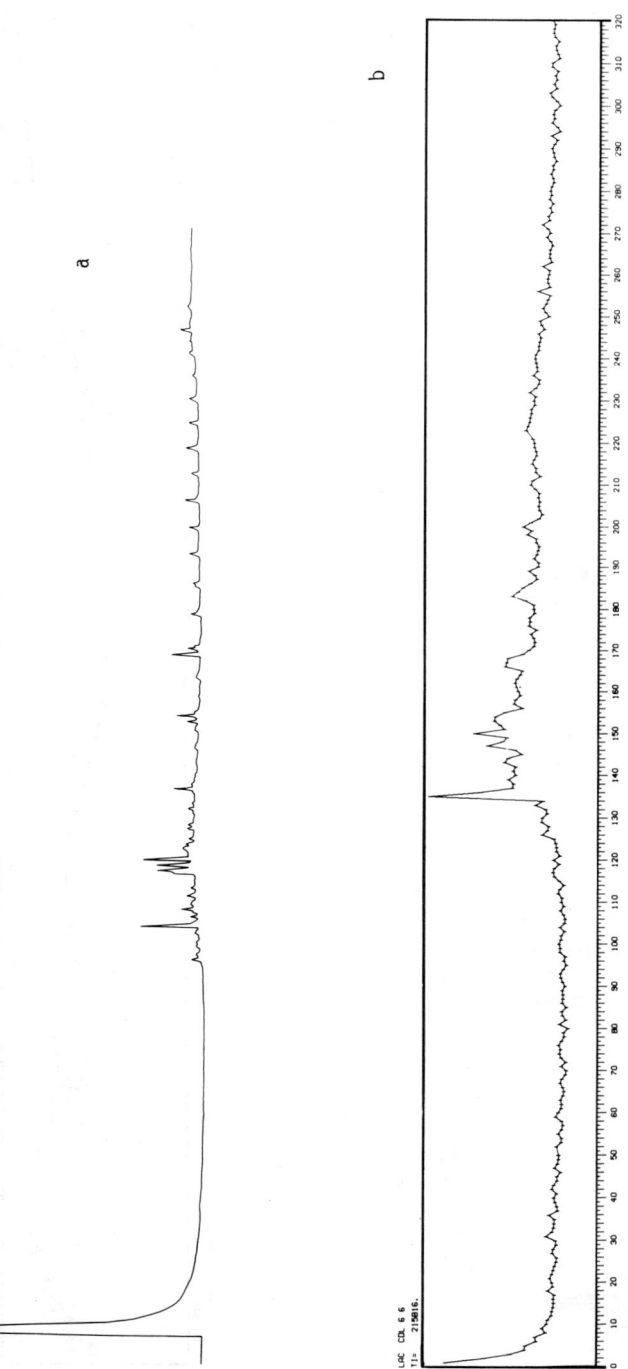

Figure 31.4 a) FID of tertiary effluents, 6% fraction; b) TIC of the same samples.

IDENTIFICATION OF ORGANIC POLLUTANTS 565

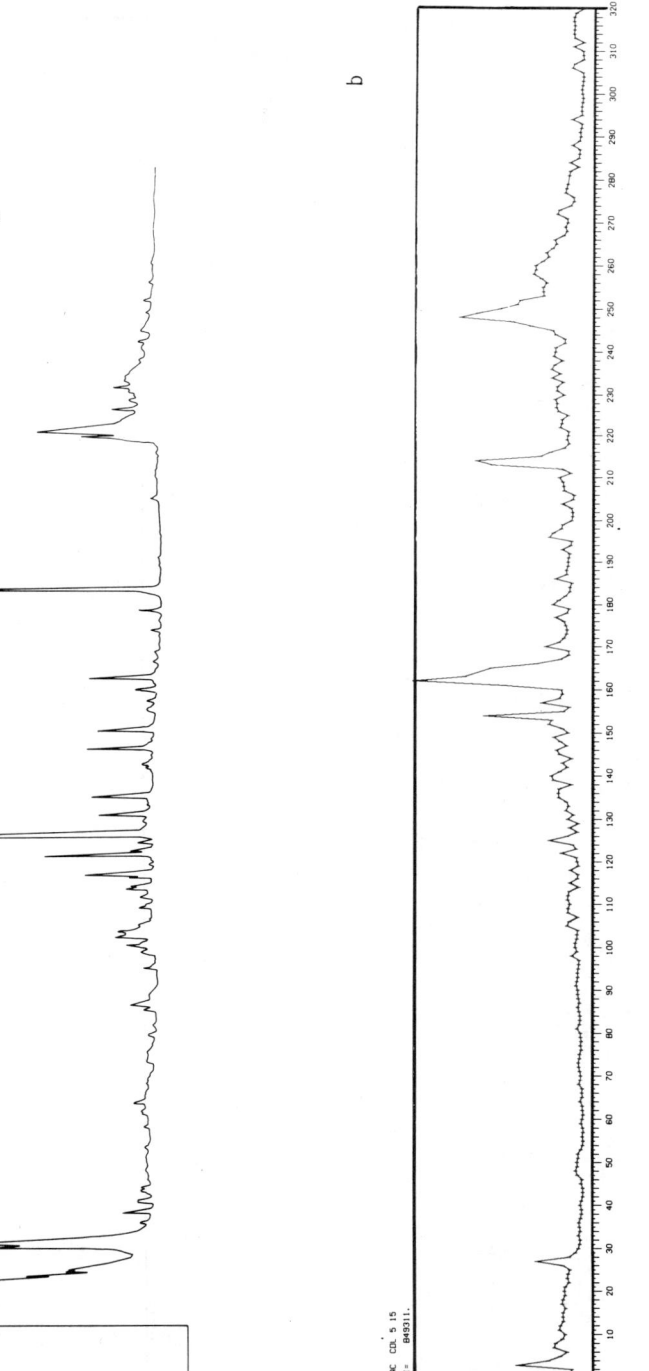

Figure 31.5 a) FID of secondary effluent, 15% fraction; b) TIC of the same samples.

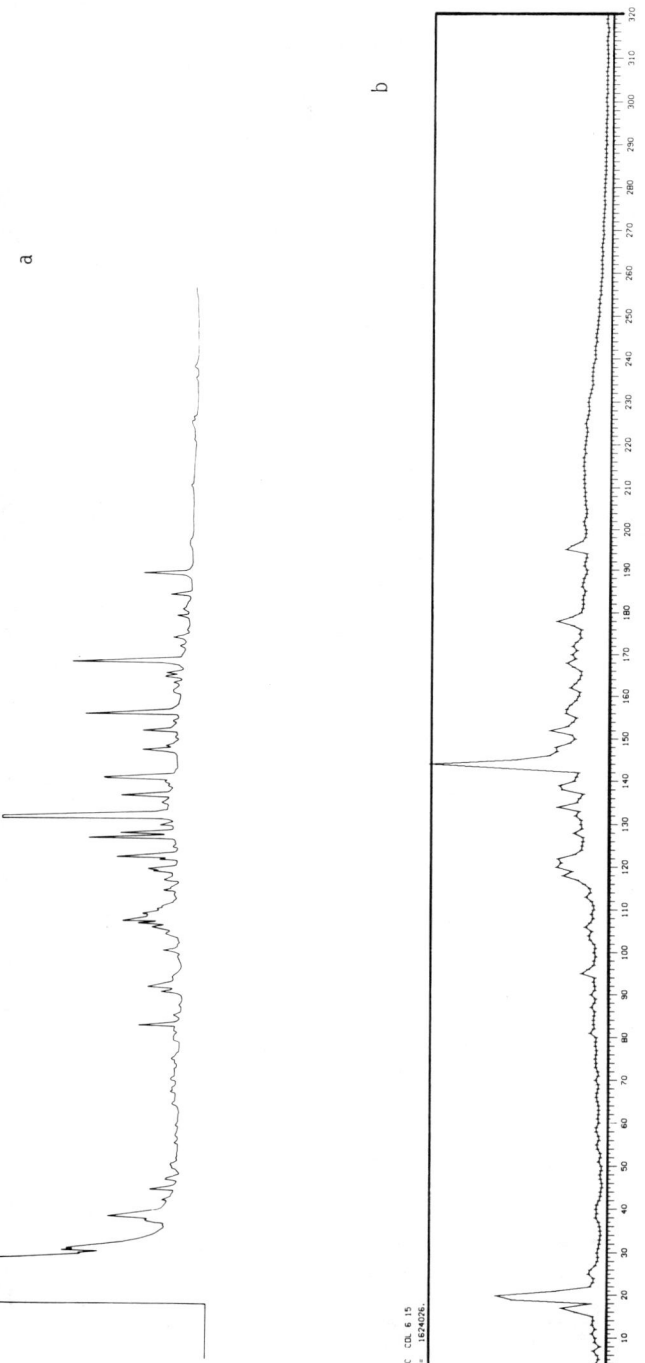

Figure 31.6 a) FID of tertiary effluent, 15% fraction; b) TIC of the same samples.

and the FID profiles is low due to both the differing relative detector responses and consideration of scan cycle time (9.6 sec) with respect to chromatographic peak elution time (~ 20 sec). The obvious "humps" in both gas and total ionization chromatograms of the secondary effluent 6% fraction (Figures 31.3a and b) are due to the complexity of the mixture and not to chromatographic artifacts (column bleed).

The availability of accurate mass measurements on all the peaks in each mass spectrum provides a very accurate and highly specific method of locating compounds of interest in such mixtures. This simply involves searching the data set for particular accurate masses (*i.e.*, specific elemental compositions) *vs* scan number (chromatographic retention time), *i.e.*, elemental composition chromatograms.

Three elemental composition chromatograms for this fraction are shown in Figure 31.7 where the accurate masses (m/e 179.9300, 181.9271 and 183.9241) correspond to the three major peaks in the isotope cluster for $C_6H_3Cl_3$, *i.e.*, $C_6H_3{}^{35}Cl_3$, $C_6H_3{}^{35}Cl_2{}^{37}Cl$ and $C_6H_3{}^{35}Cl{}^{37}Cl_2$. In all three ECC's, the relative intensities of these ions maximize at scans 26 and 32 indicating the presence of two components whose mass spectra contain ions having these elemental compositions.

Figure 31.8 shows the mass spectra of scans 26 (a) and 32 (b); these plots are of the nominal mass spectra generated from the high-resolution mass spectra by compositing the masses between, for example, 179.6000 and 180.6000 and considering this as the nominal mass 180. In these particular examples, the chlorine-containing fragments are of sufficient relative intensity to be clearly discernible in the nominal mass spectra themselves (see for example, Figure 31.10b and c). However, this is not always the case, particularly for small components eluting in the "hump" region of the gas chromatogram (see for example, Figure 31.10a).

A portion of the accurate mass measurements for scans 26 and 32 is shown in Table 31.2 which lists the measured mass and intensity (in arbitrary units), the assigned elemental composition, and the difference between the calculated and measured masses in ppm. For each scan, the elemental compositions shown are for the molecular ion (M) and (M - Cl) isotope clusters, and indicate that the two components are two of the possible three isomers of trichlorobenzene. Similar data were obtained for chlorine isotope clusters for $C_7H_5OCl_3$, $C_7H_4OCl_4$ and $C_7H_3OCl_5$; Figure 31.9a-c shows the ECC's corresponding to the ^{35}Cl isotopes only, indicating the presence of chlorinated components in HRMS 47, 68 and 87, respectively. The corresponding nominal mass spectra are shown in Figure 31.10a-c, and the corresponding accurate masses (Table 31.3) indicate that each of the three components fragments by loss of CH_3 followed by CO from the molecular ions.

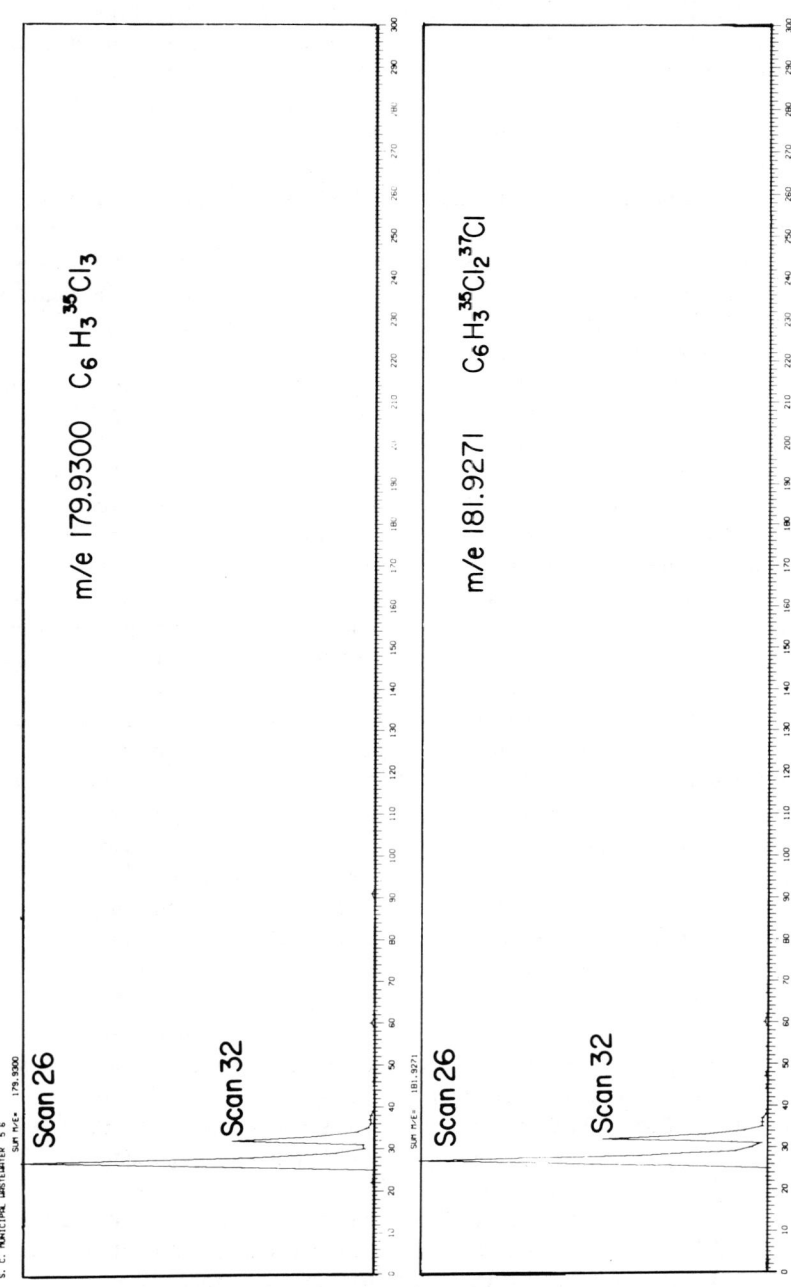

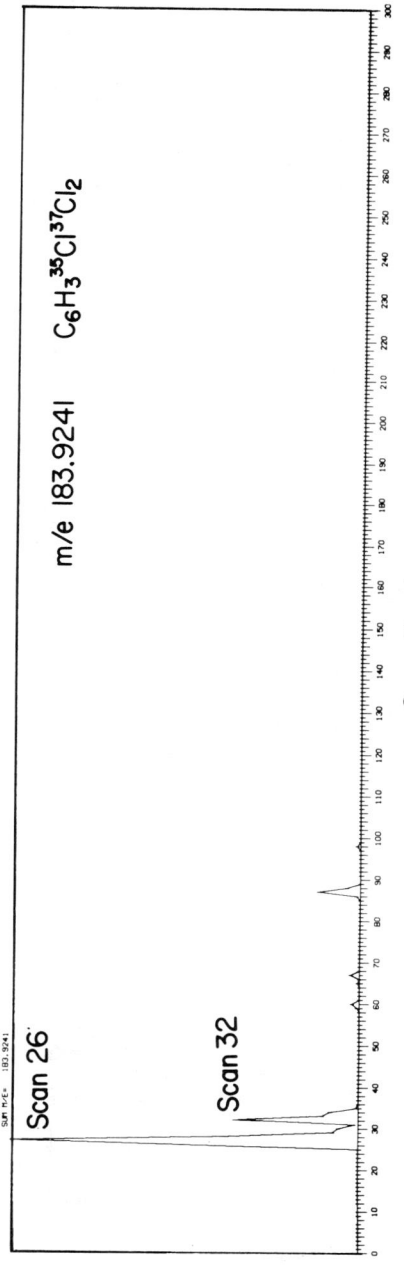

Figure 31.7 Elemental composition chromatograms for secondary effluent, 6% fraction, showing accurate masses 179.9300, 181.9271 and 183.9241, corresponding to three major peaks in the isotope cluster for $C_6H_3Cl_3$.

570 ORGANIC POLLUTANTS IN WATER

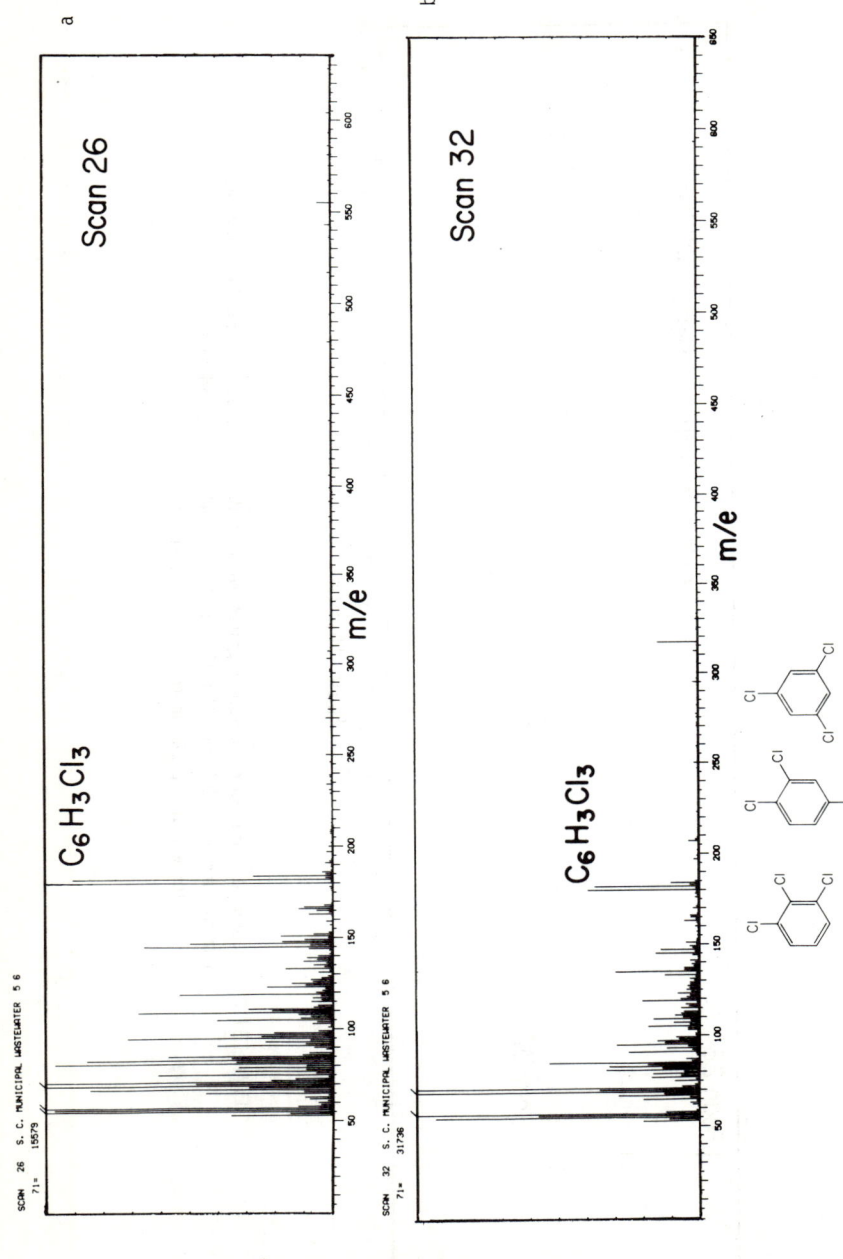

Figure 31.8 a) Mass spectrum of scan 26, secondary effluent, 6% fraction; b) mass spectrum of scan 32.

IDENTIFICATION OF ORGANIC POLLUTANTS 571

Table 31.2 Measured Accurate Masses and Relative Abundances of M and M - Cl Ion in Secondary Effluent, 6% Fraction

	Observed Mass	Intensity	Difference (ppm)	Elemental Composition			
				C	H	^{35}Cl	^{37}Cl
SCAN 26							
	179.930509	15552	-2.63	6	3	3	0
M	181.927262	14077	-0.97	6	3	2	1
	183.924736	4298	-3.27	6	3	1	2
	185.921950	549	-4.11	6	3	0	3
	144.961893	10254	-4.90	6	3	2	0
M - Cl	146.959646	7745	-9.62	6	3	1	1
	148.955451	1081	-1.13	6	3	0	2
SCAN 32							
	179.929702	12392	1.85	6	3	3	0
M	181.927650	13113	-3.10	6	3	2	1
	183.925395	3028	-6.85	6	3	1	2
	144.961214	4768	-0.22	6	3	2	0
M - Cl	146.959883	4223	-11.23	6	3	1	1
	148.957750	995	-16.57	6	3	0	2

$$M^{+\cdot} \longrightarrow M - CH_3 \longrightarrow M - CH_3 - CO + CO$$

These findings are consistent with assignment of the structures of tri-, tetra-, and penta-chloro derivatives of anisole (methyl phenyl ether):

	Aromatic substituents	Molecular composition
anisole (OCH$_3$)	3 Cl	$C_7H_5OCl_3$
	4 Cl	$C_7H_4OCl_4$
	5 Cl	$C_7H_3OCl_5$

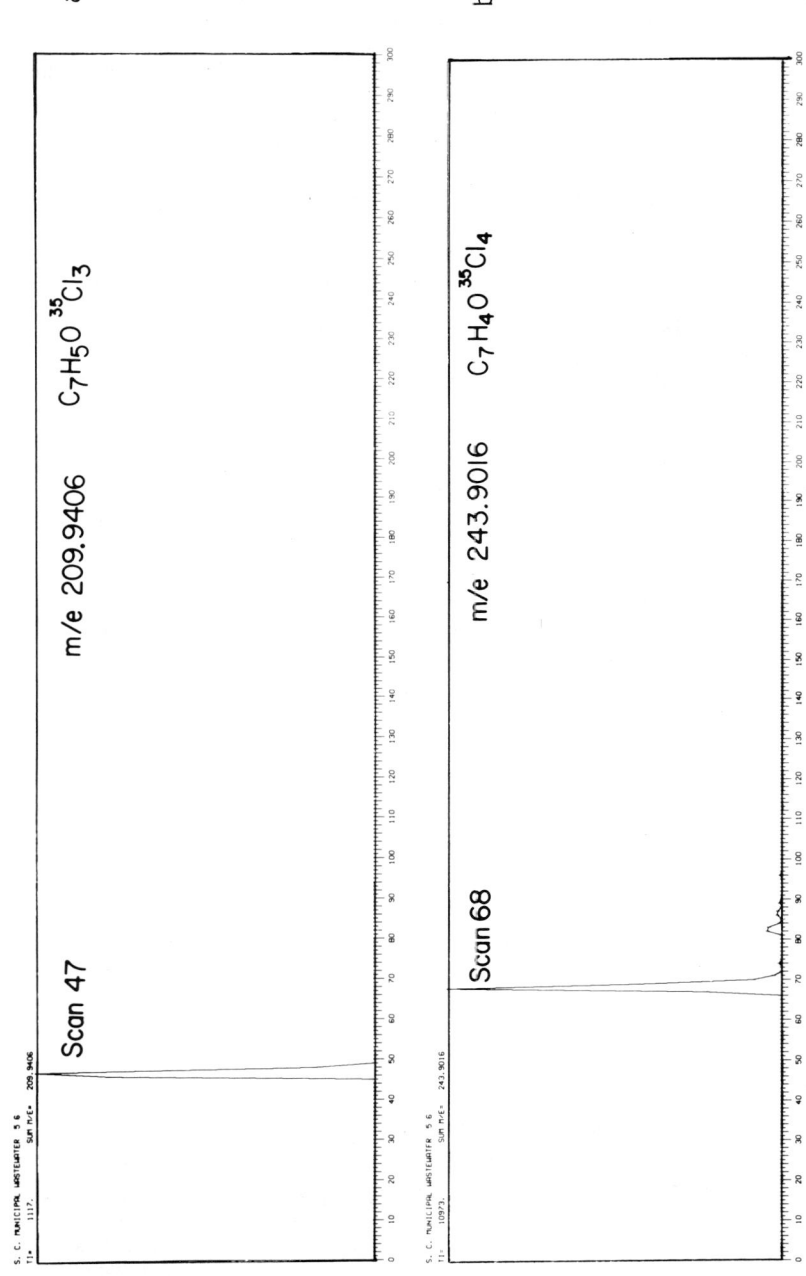

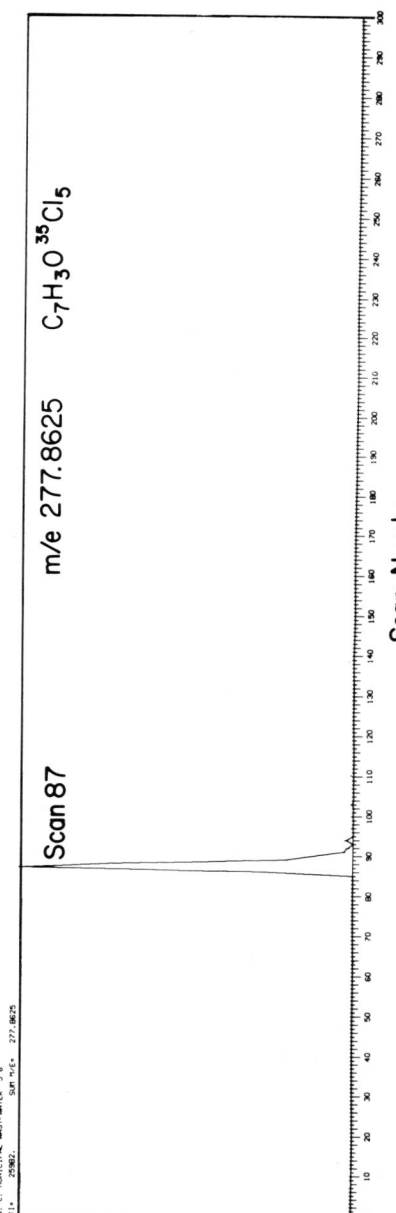

Figure 31.9 a) Elemental composition chromatogram of $C_7H_5OCl_3$, secondary effluent, 6% fraction; b) elemental composition chromatogram of $C_7H_4OCl_4$; c) elemental composition chromatogram of $C_7H_3OCl_5$.

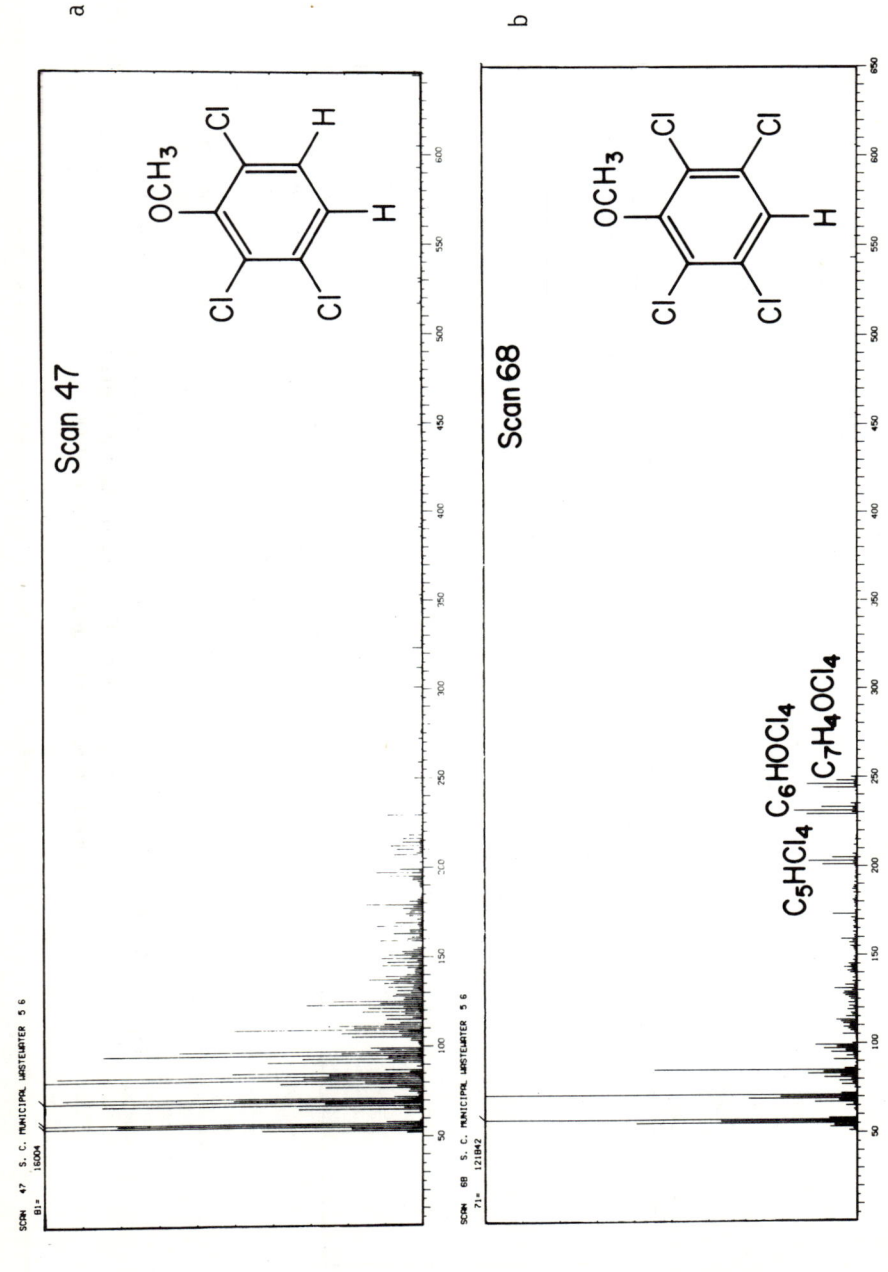

IDENTIFICATION OF ORGANIC POLLUTANTS 575

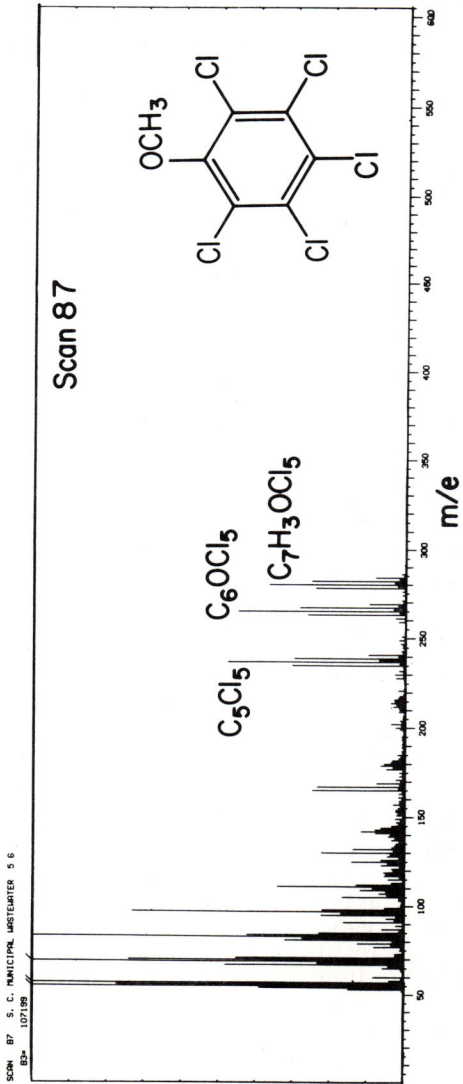

Figure 31.10 a) Low resolution mass spectrum of scan 47, secondary effluent, 6% fraction; b) low resolution mass spectrum of scan 68; c) low resolution mass spectrum of scan 87.

576 ORGANIC POLLUTANTS IN WATER

Table 31.3 Measured Accurate Masses and Relative Abundances of M, M - CH_3 and M - CH_3-CO in Secondary Effluent, 6% Fraction

	Observed Mass	Intensity	Difference (ppm)	Elemental Composition			
				C	H	O	^{35}Cl	^{37}Cl
SCAN 47								
	209.941132	1117	-2.54	7	5	1	3	0
M	211.938640	1361	-4.67	7	5	1	2	1
	213.934826	160	-0.59	7	5	1	2	2
	194.918333	2035	-6.20	6	2	1	3	0
M - CH_3	196.914319	1986	-0.74	6	2	1	2	1
	198.909322	1012	9.57	6	2	1	1	2
M - CH_3-CO	166.923376	1963	-6.99	5	2	0	3	0
	168.918402	1192	5.08	5	2	0	2	1
SCAN 68								
	243.902328	10973	-2.87	7	4	1	4	0
M	245.899728	16335	-4.27	7	4	1	3	1
	247.897216	6788	-6.00	7	4	1	2	2
	249.892952	1917	-0.70	7	4	1	1	3
	228.878746	16457	-2.59	6	1	1	4	0
	230.875367	20669	-0.72	6	1	1	3	1
M - CH_3	232.872841	11680	-2.53	6	1	1	2	2
	234.869129	2051	0.74	6	1	1	1	3
	236.868817	113	-10.40	6	1	1	0	4
	200.883076	11349	0.80	5	1	0	4	0
	202.879894	15909	1.94	5	1	0	3	1
M - CH_3-CO	204.876296	8165	5.09	5	1	0	2	2
	206.872033	1520	11.38	5	1	0	1	3
	208.872859	136	-6.80	5	1	0	0	4
SCAN 87								
	277.862422	25982	0.84	7	3	1	5	0
	279.859390	39311	1.13	7	3	1	4	1
M	281.855828	27005	3.29	7	3	1	3	2
	283.854786	8615	-3.45	7	3	1	2	3
	285.854404	1073	-12.41	7	3	1	1	4
	262.838421	28281	2.88	6	0	1	5	0
	264.833014	48338	12.14	6	0	1	4	1
M - CH_3	266.829776	30544	13.13	6	0	1	3	2
	268.827206	10550	11.62	6	0	1	2	3
	270.826086	1050	4.78	6	0	1	1	4
	234.844642	32587	-1.60	5	0	0	5	0
	236.841960	51239	-2.72	5	0	0	4	1
M - CH_3-CO	238.839334	32026	-4.05	5	0	0	3	2
	240.836925	10853	-6.26	5	0	0	2	3
	242.835470	1035	-12.37	5	0	0	1	4

It is noteworthy that in these instances the availability of elemental compositions significantly aided the interpretation of the mass spectral fragmentation pattern; attempts to identify the components from the low-resolution mass spectra alone would have been more complicated since, for example, in scan 87 the m/e 235-243 cluster (corresponding to M - 43) could also be considered to be the loss of C_3H_7 from the molecular ion. Scan 47 (Figure 31.10a) is also of interest as the data indicate the co-elution of an isomer of tetrachlorobenzene ($C_6H_2Cl_4$ cluster at m/e 214-218 and $C_6H_2Cl_3$ cluster at m/e 179-183).

In complex mixtures of this type, it is not always possible to make specific compound identifications as described above, but the availability of elemental composition chromatograms can still permit recognition of particular compound classes and an assessment of the complexity within each class. This analytical approach has been demonstrated for petroleum-derived organics isolated from refinery wastewater samples[17] and many classes of hydrocarbons and heteroaromatic components could be defined.

A similar analysis of the 6% fraction of the secondary effluent can be illustrated by reference to Figure 31.11 a-c which shows the elemental composition chromatograms of C_nH_{2n+1} fragment ions for n = 6-8 (m/e 85.1017, 99.1174 and 113.1330). These fragments are particularly characteristic of straight- and branched-chain saturated hydrocarbons and will readily locate homologous alkane series, even if the members of the series are "buried" in a chromatographically unresolved "hump."

The chromatograms clearly indicate that no such series exists in this fraction, but all show five peaks (at scans 68, 75, 79, 87 and 89) in addition to an extremely complex mixture as evidenced by the humps. Examination of the high-resolution mass spectra corresponding to these five peaks identified three n-alkanes [$C_{15}H_{32}$ (scan 68), $C_{16}H_{34}$ (small, scan 79) and $C_{17}H_{36}$ (scan 89)] and a substantial amount of a compound having the molecular formula $C_{17}H_{34}$ (monocyclic or mono-unsaturated) in scan 87 in addition to pentachloroanisole (see above). In all these scans, there were also indications of the presence of significantly unsaturated hydrocarbons, possibly alkylated aromatics.

Figure 31.12 a-c shows elemental composition chromatograms for C_nH_{2n-7} fragment ions, namely C_7H_7 (m/e 105.0704) and C_9H_{11} (m/e 119.0861), which are characteristic for alkylated benzenes. The most significant of these is the chromatogram for C_7H_7 which shows several peaks in the region between scans 70 and 100, some of which may represent homologous compounds.

Careful study of the corresponding high-resolution mass spectra suggested that these components are branched alkyl-substituted benzenes of molecular formula $C_{16}H_{26}$, $C_{17}H_{28}$, $C_{18}H_{30}$ and $C_{19}H_{32}$; however,

578 ORGANIC POLLUTANTS IN WATER

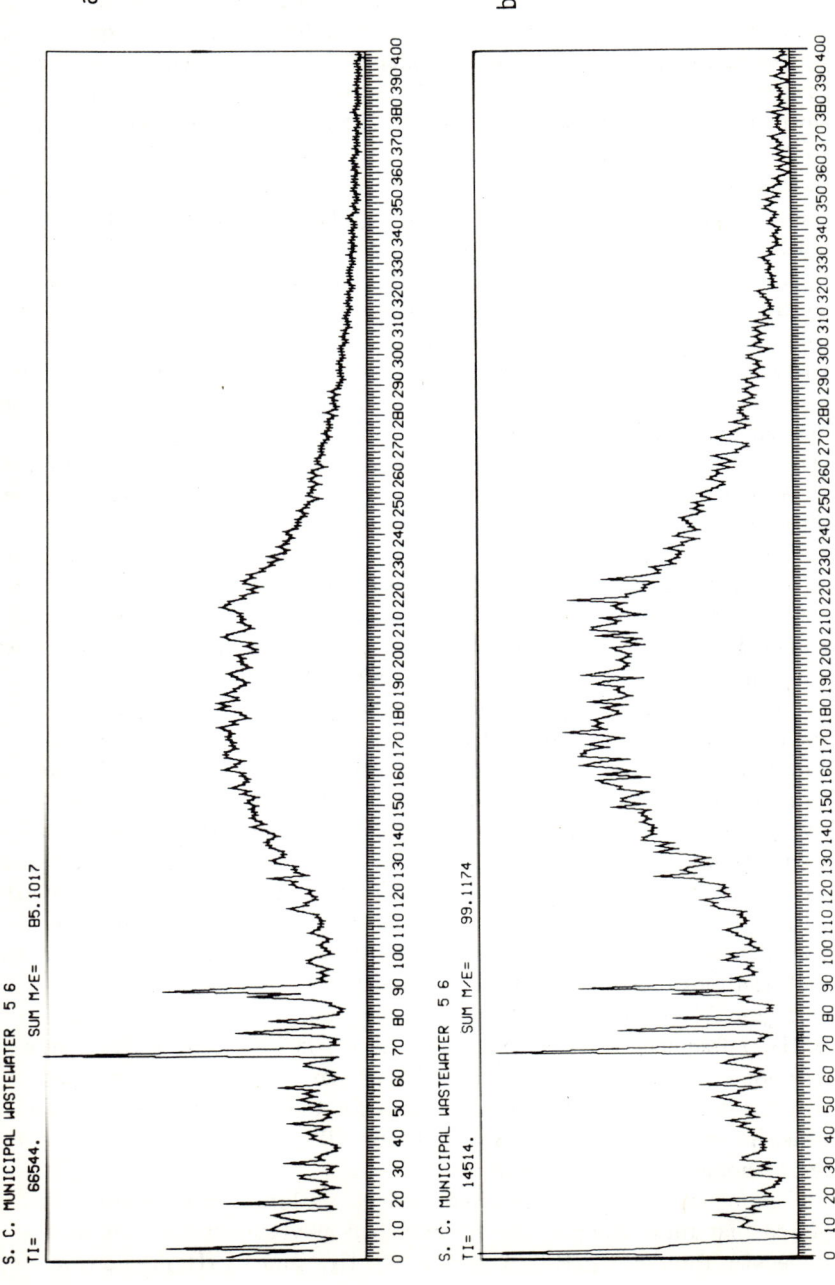

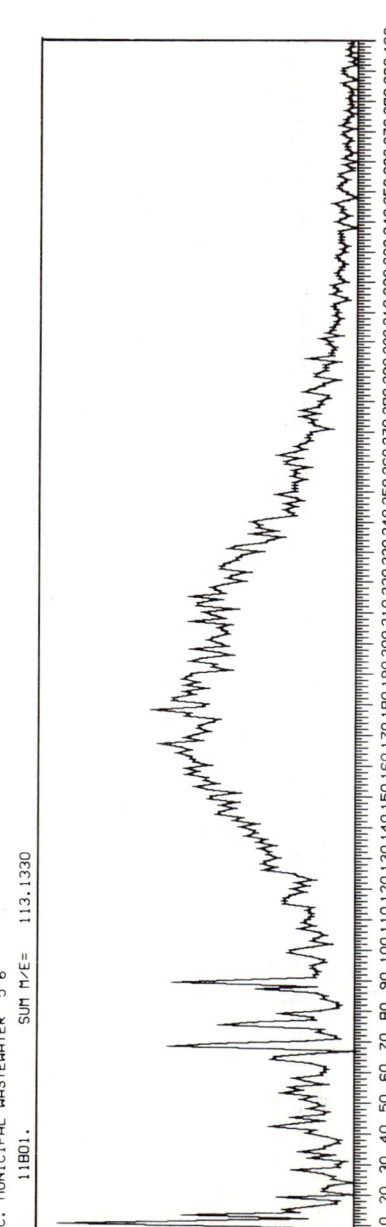

Figure 31.11 a) Elemental composition chromatogram of secondary effluent, 6% fraction, m/e 85.1017; b) elemental composition chromatogram, m/e 99.1174; c) elemental composition chromatogram, m/e 113.1330.

580 ORGANIC POLLUTANTS IN WATER

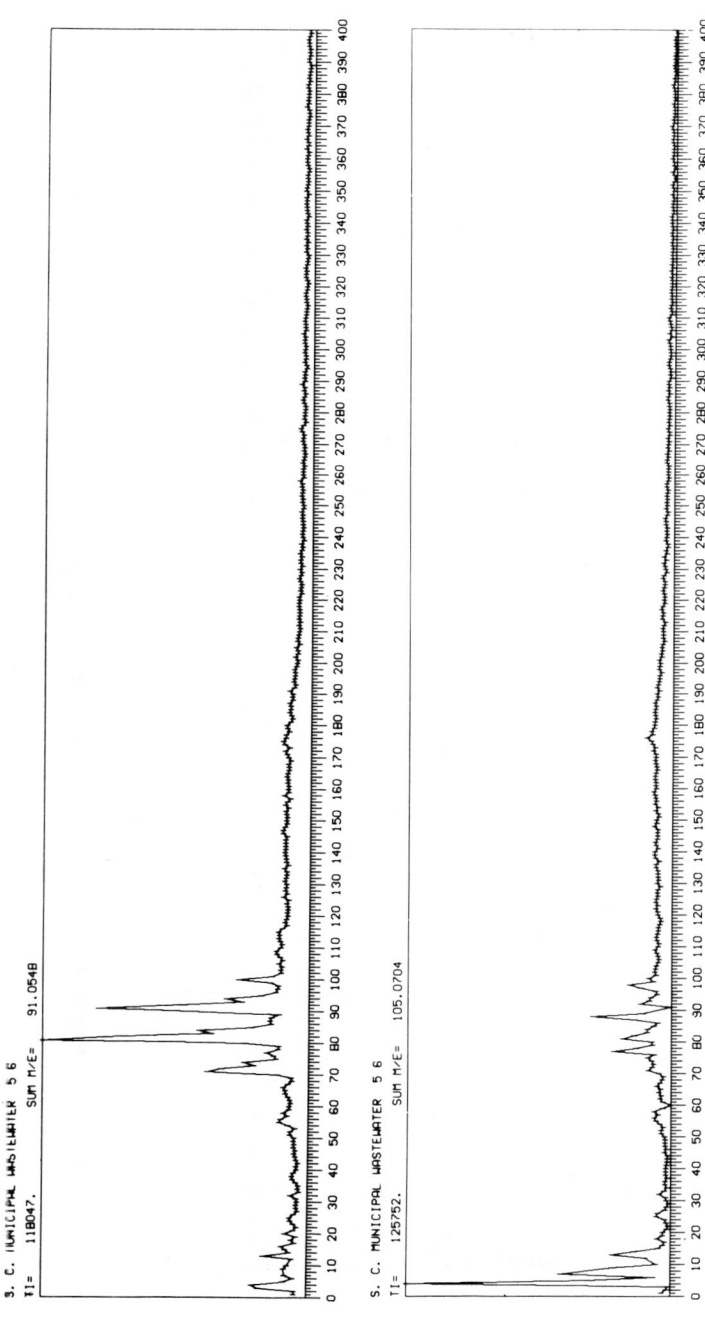

IDENTIFICATION OF ORGANIC POLLUTANTS 581

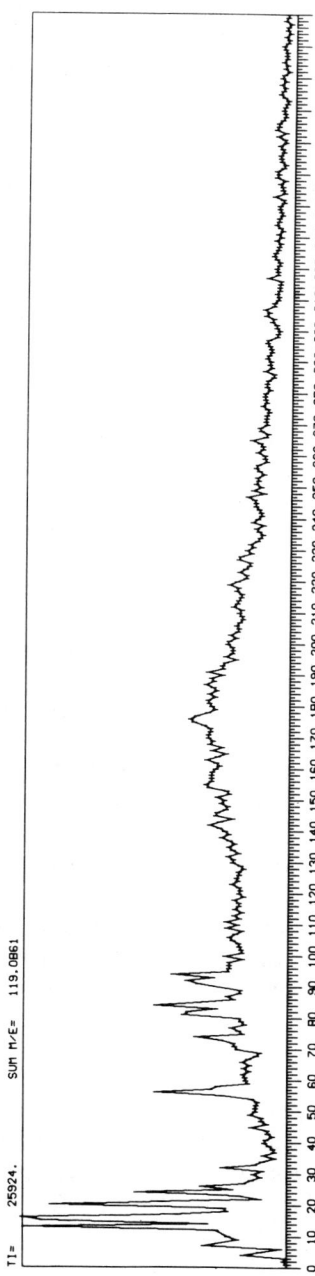

Figure 31.12 a) Elemental composition chromatogram of secondary effluent, 6% fraction, m/e 91.0548; b) elemental composition chromatogram, m/e 105.0704; c) elemental composition chromatogram, m/e 119.0861.

many other fragment ions having hydrocarbon compositions of various C/H ratios were also present in the spectra from co-eluting components making unambiguous identification difficult.

The components identified to date in the four fractions are summarized in Table 31.4, and are arranged to facilitate comparisons between the molecular composition of the secondary and tertiary effluents. The table lists the particular high-resolution mass spectrum (scan) numbers in which the components were identified, permitting some correlation of chromatographic properties, since the scan numbers can be related to relative gas chromatographic retention times.

The secondary effluent contains a series of chlorinated benzenes possessing 2 to 5 chlorine atoms per molecule and exhibiting structural isomerism in some cases. None of these compounds could be detected in the tertiary effluent despite the high specificity of elemental composition chromatography, and it was therefore considered possible that the treatment scheme might have modified these compounds by chlorination to hexachlorobenzene.[18] Specific searches in both effluents failed to produce any evidence for the presence of hexachlorobenzene, suggesting that the treatment scheme was effective in the removal of substantial amounts of chlorinated benzenes.

A series of chlorinated anisoles (methylphenyl ethers) having three to five chlorine atoms substituted on the aromatic ring were also identified in the secondary effluent, but *only* the pentachlorinated component could be found in the tertiary effluent. These identifications do correlate with the available evidence that chlorination treatment can produce more highly chlorinated components. It should be noted that pentachloroanisole is also present in the 15% fractions of both effluents, indicating incomplete separation during column chromatography using Florisil. The source of chloroanisoles in this secondary effluent is uncertain at present, although it is possible that their origin is related to that of chlorinated phenols (disinfectants). Similar compounds [chloro-derivatives of methoxy-toluene and bis(ethoxy)-benzene] together with chlorinated benzenes have been identified in samples of municipal secondary effluent which were subjected to chlorination in the laboratory.[18]

PRELIMINARY CONCLUSIONS

1. Florisil column separations of 6% and 15% ether/hexane fractions (separates PCB's, DDT's from dieldrin/aldrin, etc.) did not yield information which would not have been obtained from a mixture of these fractions using HRGC/HRMS.

Table 31.4 Summary of Comparison of Secondary and Tertiary Effluent Composition

Component Identified 6% Fractions	Spectrum Number (Related to Retention Time)	
	Secondary Effluent	Tertiary Effluent
dichlorobenzene	9	n.d.
trichlorobenzene	26,32	n.d.
tetrachlorobenzene	47,52	n.d.
pentachlorobenzene	67 (small)	n.d.
hexachlorobenzene	n.d.	n.d.
trichloroanisole	47,60	n.d.
tetrachloroanisole	68	n.d.
pentachloroanisole	87	135
alkylbenzenes		
$\quad C_{16}H_{26}$ (2?)	71,(74?)	
$\quad C_{17}H_{28}$ (2)	81,84	129 (small)
$\quad C_{18}H_{30}$	91	139,143 (small)
$\quad C_{19}H_{32}$	100	
15% Fractions		
$C_6H_{12}O$		19
$C_6H_{10}O$		25
C_9H_{14}		81
$C_{10}H_{14}O$		106
benzophenone	139	120
$C_{15}H_{24}$ (alkyl benzene)	136	123
pentachloroanisole	147	129
$C_{17}H_{24}O$	149	131
$C_{17}H_{26}O$	152	134,136
$C_{18}H_{26}O$		138,140
dibutyl phthalate	162,166,170	143,148,153
$C_{19}H_{18}O_3$	170	153
$C_{20}H_{30}O$	180	162
phthalate	190,210	172,192
dioctyl phthalate	214	195

2. The FeCl$_3$ flocculation and/or chlorine contact steps yielded only the "per" chloroanisole (penta).

3. Chlorinated benzenes observed in the secondary effluent were not detected in the tertiary effluent.

4. The electron capture detector chromatograms indicate a relatively simple mixture compared to the flame ionization detector chromatograms.

5. The organic chemical complexity of this secondary effluent is still present after flocculation, sedimentation, filtration and chlorine contact in the same order of magnitude concentration (not subjected to activated carbon step).

6. Results on the total extractable organics from these or analogous samples of wastewater are not available yet.

7. It is not possible at this time with present minimal knowledge to evaluate the effectiveness of this treatment scheme in removal of trace organics present in primary and secondary effluent.

8. It is clear that ECC (HRGC/HRMS) can be employed effectively in defining the scope of compound classes of organic substances present at various stages of treatment of municipal effluents as well as in following the qualitative or quantitative alteration of specific compounds through a given treatment *scheme* process.

ACKNOWLEDGMENTS

We gratefully acknowledge the support of the Los Angeles County Sanitation District No. 2 - 540 JTA; the NIH Biomedical Mass Spectrometry Resource Grant RR-719-02; NASA Grant NGL 05-003-003; and NSF Grant IDO 72-06412.

REFERENCES

1. Young, D. R., R. W. Risebrough, B. W. de Lappe, T. C. Heesen and D. J. McDermott. "Input of Chlorinated Hydrocarbons into the Southern California Bight from Surface Runoff and Municipal Wastewaters," *Pesticides Monitoring Journal*, in press.
2. California State Water Resources Control Board. "A State-of-the-Art" Review of Health Aspects of Wastewater Reclamation for Ground Water Recharge," Dept. of Water Resources, Dept. of Health, State of California (November 1975).
3. Jensen, S. and O. Pettersson. *Environ. Pollut.* 2, 145-155 (1971).
4. Jansson, B., S. Jensen, M. Olsson, L. Renberg, G. Sundstrom and R. Vaz. *Ambio* 4, 93-97 (1975).
5. Ten Noever de Brauw, M. C. and J. H. Koeman. *Sci. Total Envir.* 1, 427-432 (1972).
6. Webb, R. G., A. W. Garrison, L. H. Keith and J. M. McGuire. *Current Practices in GC-MS Analysis of Organics in Water,* EPA-R2-73-277, Office of Research and Monitoring, Environmental Protection Agency, U.S. Government Printing Office, Washington, D.C. (1973).

7. Eglinton, G. (Sr. Reporter). *Environmental Chemistry*, Vol. 1, The Chemical Society (London: Burlington House, 1975), 199 pp.
8. Grob, K. and G. Grob. "Glass Capillary Gas Chromatography in Water Analysis. How to Initiate Use of the Method?," presented at at the First Chemical Congress of the North American Continent, Mexico City (December 1975), Chapter 5, this volume.
9. Manka, J., M. Rebhun, A. Mandelbaum and A. Bortinger. *Environ. Sci. Technol.* **8**, 1017-1020 (1974).
10. Kimble, B. J., R. E. Cox, R. V. McPherron, R. W. Olsen, E. Roitman, F. C. Walls and A. L. Burlingame. *J. Chromatog. Sci.* **12**, 647-655 (1974).
11. Kimble, B. J., F. C. Walls, R. W. Olsen and A. L. Burlingame. "Real-Time Gas Chromatography-High Resolution Mass Spectrometry and the Analysis of Complex Organic Mixtures," presented at the 23rd Annual Conference on Mass Spectrometry and Allied Topics, Houston, Texas (May 1975).
12. de Lappe, B. W. and R. W. Risebrough. "*In Situ* Sampling for Chlorinated Hydrocarbons in Pacific Coast Waters," presented at the 1975 Pacific Conference on Chemistry and Spectroscopy, Los Angeles (October 1975).
13. *Federal Register*, Vol. 38, No. 125, Part II, June 29, 1970.
14. Food and Drug Administration Pesticide Analytical Manual, 1971.
15. German, A. L. and E. C. Horning. *J. Chromatog. Sci.* **11**, 76 (1973).
16. Burlingame, A. L., R. W. Olsen and R. V. McPherron. In *Advances in Mass Spectrometry*, Vol. 6, A. R. West, Ed. (London: Applied Science Publishers, Ltd., 1974), pp. 1053-1059.
17. Burlingame, A. L., B. J. Kimble, E. S. Scott, M. J. Stasch, F. C. Walls, and L. H. Keith. "Assessment of the Trace Organic Molecular Composition of Petroleum Refinery Wastewater Effluent by Capillary Gas Chromatography/Real-Time High-Resolution Mass Spectrometry," presented at the First Chemical Congress of the North American Continent, Mexico City (December 1975), Chapter 32, this volume.
18. Glaze, W. H. and J. E. Henderson, IV. *J. Water Pollution Control Fed.* **47**, 2511-2515 (1975).

CHAPTER 32

THE CHARACTERIZATION OF TRACE ORGANIC CONSTITUENTS IN PETROLEUM REFINERY WASTEWATER BY CAPILLARY GAS CHROMATOGRAPHY/REAL-TIME HIGH-RESOLUTION MASS SPECTROMETRY— A PRELIMINARY REPORT

A. L. Burlingame, B. J. Kimble, E. S. Scott, D. M. Wilson and M. J. Stasch

    Space Sciences Laboratory
    University of California
    Berkeley, California

J. W. de Leeuw

    Delft University of Technology
    Department of Chemical Engineering and Chemistry
    Organic Geochemistry Unit
    Delft, The Netherlands

L. H. Keith

    U.S. Environmental Protection Agency
    Southeast Environmental Research Laboratory
    Athens, Georgia

INTRODUCTION

    This chapter is intended to be a preliminary communication to present the status of an on-going research program. The program's goal is the general analytical capability to characterize extremely complex mixtures of organic substances from both biological and environmental-geochemical sources by utilizing a new powerful combination of high-resolution gas chromatography and real-time high-resolution mass spectrometry (HRGC/HRMS).

We wish to report on one aspect of a particular application of this instrumental approach which is also part of a general effort aimed at the characterization of trace organic constituents in petroleum refinery wastewater and in a particulate treatment process which is common to many plasti-petroleum refineries.[1] Starting with the primary wastewater effluent of this class of refinery, the strategy was to sample three sites prior to discharge of the wastewater into the environment. These sites were identified as: K-1, subsequent to the API separator; K-2, subsequent to the Pasveer oxidation ditch and clarifier; and K-3, subsequent to the non-aerated lagoons (Figure 32.1).

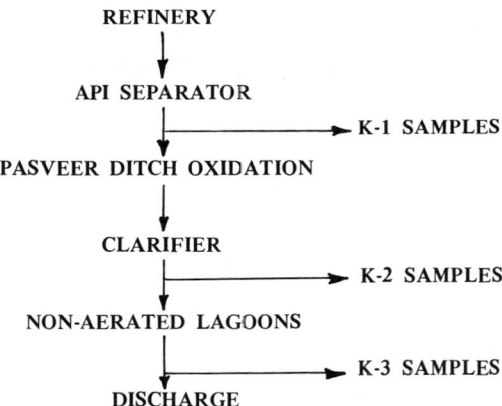

Figure 32.1 Location of sample sites in treatment system for petroleum refinery wastewater.

This communication is primarily concerned with the neutral fractions (K-1-N, K-2-N and K-3-N), since these fractions represented the bulk of the material in the primary aqueous extract effluent. Analysis of the phenolic and acidic fractions, K-1-P and K-1-A respectively, showed few major components and a wide range of minor components. The general appearance of the capillary chromatograms of K-1-N, K-2-N and K-3-N indicates an extremely complex mixture of substances distinguished by the presence of a series of normal alkanes in K-1-N and some other peaks which are not distinguishable from the unresolved chromatographic hump in K-2-N and K-3-N. A recent review in "Environmental Organic Chemistry of River and Lake Water,"[2] does not deal with organic mixtures from aqueous sources of this complexity where a chromatographic hump is the salient feature. In fact, the chromatograms of the alkanes and fatty acid methyl esters of these materials do not seem to contain the hump feature. This is also true of many ancient sediments, such as the methyl esters in Green River shales. However, a study of the organic geochemistry of sediments from contemporary aquatic environments[3] finds a chromatographic hump in the branched/cyclic alkane fraction to be a feature of their extracts. This feature is common in the case of extracts from marine sources where petroleum contamination is invoked as an explanation.[4-7] In fact, the unresolved complex mixture for a hump feature has been proposed as a characteristic useful in detecting petroleum contamination in the marine environment.[8]

To date, analyses of total organic extracts from water and sediments have not seriously attacked the complexity of the problem as indicated by mixtures that are chromatographically unresolved even under high-resolution gas chromatography conditions; they address themselves instead to discrete components which happen to persist above the envelope. However, the literature does describe techniques (developed primarily by the petroleum industry) whose analytical rigor and resolution would permit some characterization of complex chromatographically unresolved mixtures. These techniques are primarily the work of Aczell and co-workers in low-voltage high-resolution mass spectrometry, and Gallegos, Teeter and colleagues at Chevron. To our knowledge these techniques have not been applied to a problem in an environmental context with results published. It might be noted that there is a disadvantage accompanying the molecular ion advantage in the low-voltage high-resolution technique; this is the fact that the sensitivity is a factor of 10 to 100 lower, which means that scan times are in minutes and sample consumptions are in milligrams.

A general review of the microbial utilization of oil wastewater from petroleum refinery effluents has been presented by Davis.[9] Here major attention has been paid to the elimination of phenolic components in petroleum refinery wastewater, both because of the possible toxicity to the environment and also because of their ease of chlorination in municipal treatment facilities, producing odoriferously objectionable components that interfere with the reuse of this wastewater.

Under the auspices of the Environmental Protection Agency, another review has appeared which discusses the problem of refining wastewater in more current terms.[1]

There is a suggestion in the literature that microbial transformations do produce an unresolved complex mixture of hydrocarbons upon examination of their solvent extracts by gas chromatography,[10] particularly in the case of the sulfate-reducing bacterium *Desulphovibrio desulphuricans.* Cultures of this organism and other strains of this organism by Han[11] show chromatograms which maximize at n-C_{20} with a distribution from n-C_{15} to n-C_{28} alkane. There are a few minor peaks but no unresolved complex mixture. However one organism that Han studied, *Clostridium tetanomorphum H-1,* showed a distribution of hydrocarbons from n-C_{13} to n-C_{28} with a bimodal unresolved chromatographic hump which maximized at about C_{19} and C_{23} alkane.

EXPERIMENTAL

Samples were obtained from three sites (K-1, K-2 and K-3; see Figure 32.1) in the treatment scheme using teflon tubing and glass containers cooled in ice. Collections were made about every 10-15 min for 3 days, and every 12 hr the glass containers were emptied into teflon bottles which were frozen in dry ice. Later the samples were thawed, composited and refrozen giving about 5 gal of a 3-day-averaged sample from each of the three sites. One gallon of each of these composite samples was concentrated, extracted and fractionated (using 20% ethyl ether/80% methylene chloride with pH adjustments) to give phenolic (P), acidic (A), basic (B) and neutral (N) fractions for analysis.

Gas Chromatography/High-Resolution Mass Spectrometry

Phenolic and Acid Fractions

The phenol fraction was permethylated and the acid fraction was methylated. The concentrated fractions from Sites 1, 2 and 3 were subjected to preliminary gas chromatographic analysis on a 150 ft x 0.03 in. (45.7 m x 0.076 cm) i.d. glass surface-coated open tubular (SCOT) column coated with Dexsil 300 installed in a Perkin-Elmer Model 900 gas chromatograph. The column temperature was initially held at 70°C until after the solvent eluted, and was then programmed at 6.5°/min up to 280°C. GC/HRMS analysis was carried out with a 10 ft x 1/16 in. (3 m x 0.16 cm) i.d. stainless steel packed column coated with Dexsil and with a helium flow of approximately 10 ml/min. The column was programmed from 50-280°C at 8°/min. High resolution data were recorded in peak profile form using LOGOS-II[12] at a scan rate of 8 sec/decade (cyclic). The scan cycle time was approximately 22 sec. Computer processing of the resulting data involved the generation of total ion current plots, nominal and accurate mass chromatograms, low-resolution mass spectral plots, and elemental composition listings of individual scans.

Neutral Fractions

The concentrated neutral fractions from sites K-1, K-2 and K-3 (referred to as K-1-N, K-2-N and K-3-N) were subjected to preliminary gas chromatographic analysis on a 100 ft x 0.03 in. (30 m x 0.076 cm) i.d. glass SCOT column coated with OV-101 installed in a Perkin-Elmer Model 900 gas chromatograph. The column temperature was initially held at 50°C until after the solvent eluted, and was then programmed at 5°C/min up to 300°C.

High-resolution gas chromatography/high-resolution mass spectrometry analysis was carried out using the system previously described[13] but with a 30 m x 0.05 cm i.d. stainless steel capillary column coated with OV-17 (Figure 32.2). The chromatographic conditions were similar to those described above. Approximately 1/250th of the total extract of K-1-N and K-2-N and 1/50 of the total extract of K-3-N were used for this analysis without derivatization. High resolution data were recorded in peak profile form using LOGOS-II[12] at a scan rate of 6 sec/decade; this permitted acquisition of one spectrum every 11.5 sec (scan time 7.5 sec, reset time 4 sec). The resulting data were processed in the same way as described above for the phenolic and acid fractions.

Nuclear Magnetic Resonance Spectroscopy

^{13}C Fourier transform NMR spectra were obtained from all samples, which were either unaltered or made up to sufficient volume by addition of solvent (20:80 ethyl ether:methylene chloride, v:v). Spectra were also taken of the pure solvent, to which a trace amount of tetramethylsilane (TMS) was added; this spectrum provided an abscissa reference for all other spectra.

The spectrometer was a Varian XL-100-15 coupled to a Xerox Sigma 7 computer. Other operating conditions were: 8°C, 90° RF pulses at 8-sec intervals, quadrature detection, proton noise-decoupling or off-resonance decoupling to identify proton multiplicities, ~3 hr per spectrum, 0.03 ppm resolution.

Only samples K-1-A, K-1-P, K-3-A and K-3-P showed component peaks in addition to solvent peaks. However, only samples K-1-A and K-1-P were concentrated enough to identify major components, which were benzoic acid (K-1-A) and a mixture of phenol and *m*-cresol (K-1-P). The latter identification was confirmed using a solution of standards.

592 ORGANIC POLLUTANTS IN WATER

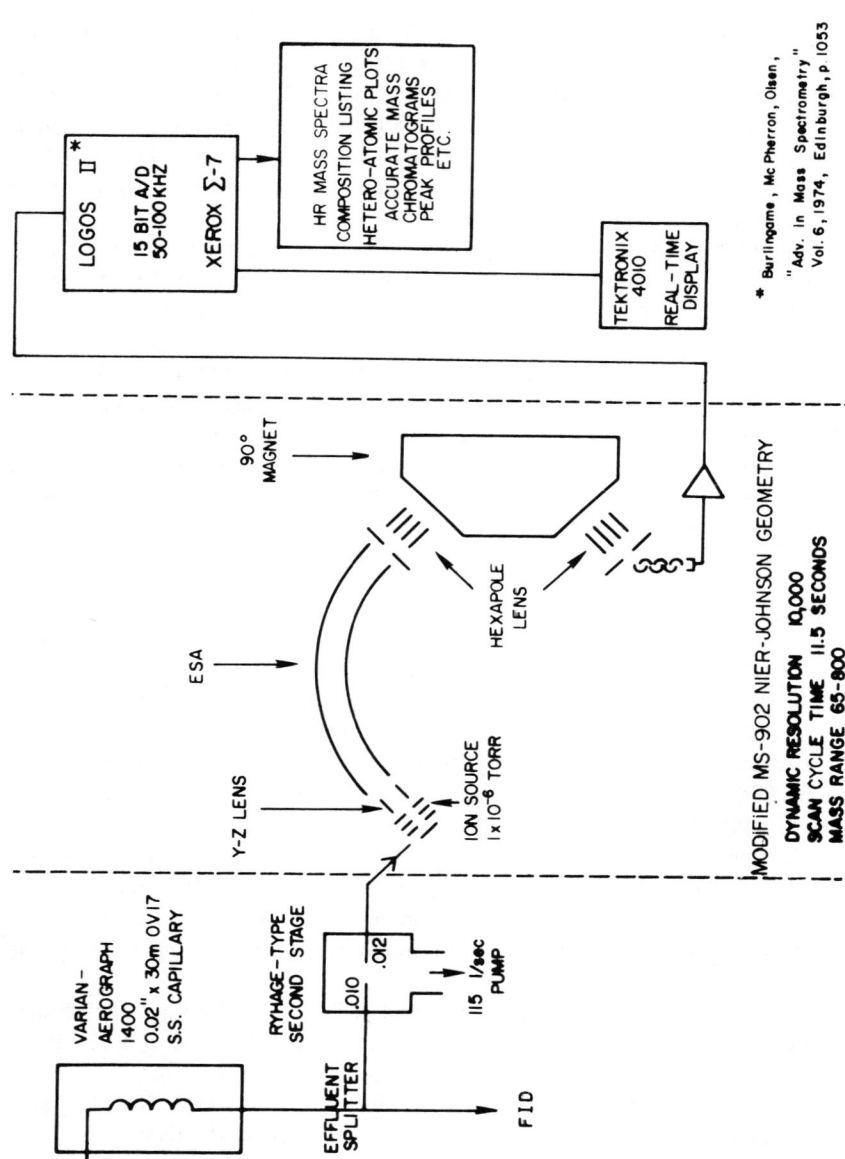

Figure 32.2 Schematic diagram of real-time capillary GC/HRMS system.

RESULTS AND DISCUSSION

Phenolic and Acidic Fractions

The three phenol fractions were analyzed as their permethylated derivatives by gas chromatography. K-1-P showed many more components than K-2-P, which was in turn more concentrated than K-3-P. Fractions K-1-P and K-2-P are illustrated in Figure 32.3 (the large peak with the long retention time that is present in both spectra is an artifact from the permethylation procedure). All three fractions showed the same major components. K-1-P was also analyzed by GC/HRMS. The total ion current plot is shown in Figure 32.4. K-1-P consists mainly of anisole and methyl anisole. Methyl benzoate, C_2 anisole, iodo anisole and iodo methyl anisole are also present. The iodines on the aromatic ring probably were formed during the permethylation procedure which uses CH_3I. These components are listed in Table 32.1. In addition to these major components, several less abundant components were identified. There is a series of methyl esters and some ethyl esters. The former may have been generated during the extraction or isolation procedures or perhaps during the permethylation procedure. There are aliphatic esters, cyclic esters, and aromatic acids of 1 and 2 rings. Several N-substituted amides are present as is a carbamate and an N-methyl-substituted amide methyl ester. These are included in Table 32.1.

The acidic fractions were analyzed as their methylated derivatives by gas chromatography (Figure 32.5). For K-1-A and K-2-A, GC/HRMS analysis showed methyl benzoate to be the main component in each fraction (Figure 32.6). GC/MS analysis made it possible to identify several minor components. They are: several saturated aliphatic normal and branched methyl esters, ethyl and propyl esters, several methyl + ethyl-substituted methyl benzoates, aromatic esters, naphthenic esters, indene esters, cyclic esters and sulfur-containing aromatic esters. These are listed in Table 32.2.

By far the largest component in the acidic fraction is methyl benzoate (by NMR, 80%) and the largest components in the phenolic fraction are phenol and cresol (by NMR, 90%). Crude oils contain relatively small quantities (up to 2%) of oxidized compounds in the form of acids and phenols.[14] Generally acid fractions consist of aliphatic acids among the compounds containing up to six carbon atoms. The six to ten carbon atom-containing compounds consist of a mixture of fatty acids and naphthenic acids. Phenol, the cresols, and the xylenols are common phenolic compounds identified in crude oils. However, crude oils usually contain less than 0.1% phenol.

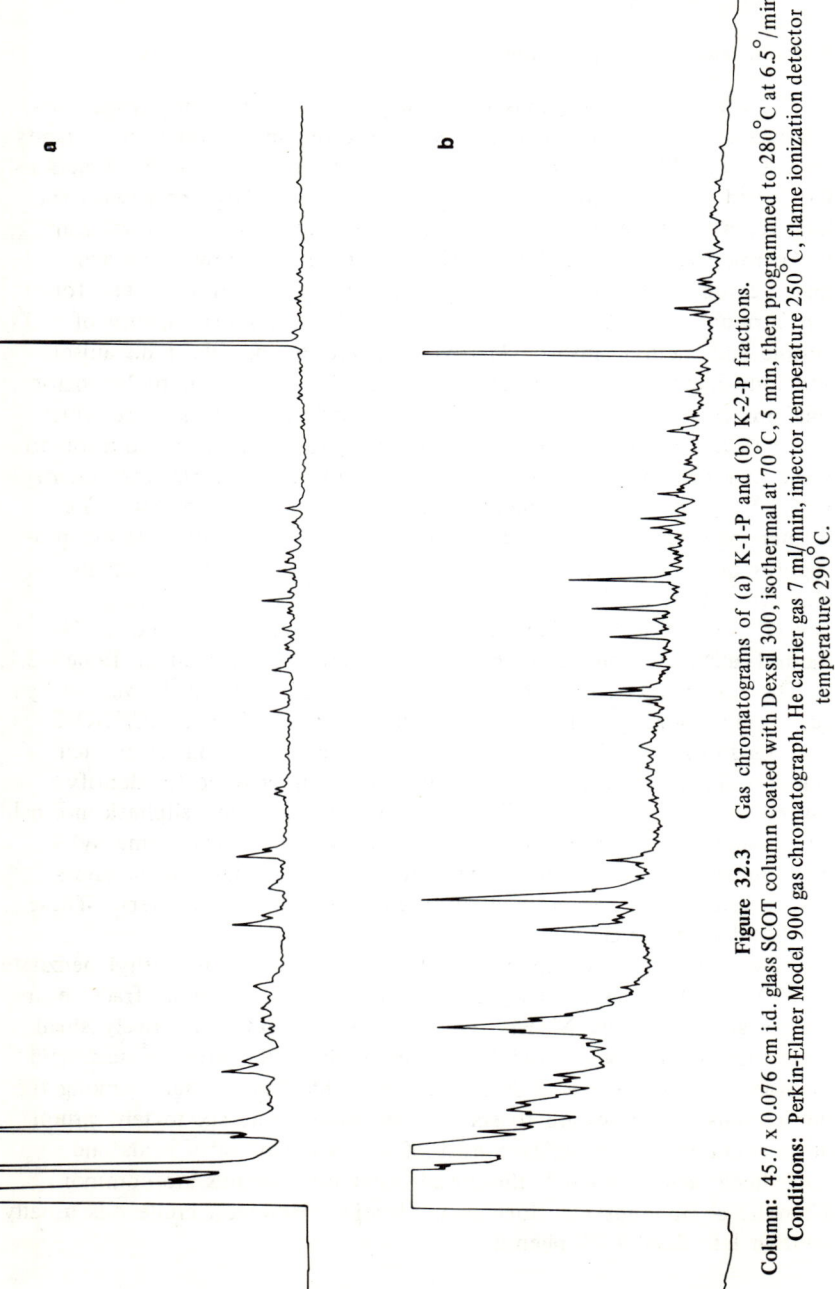

Figure 32.3 Gas chromatograms of (a) K-1-P and (b) K-2-P fractions.
Column: 45.7 × 0.076 cm i.d. glass SCOT column coated with Dexsil 300, isothermal at 70°C, 5 min, then programmed to 280°C at 6.5°/min.
Conditions: Perkin-Elmer Model 900 gas chromatograph, He carrier gas 7 ml/min, injector temperature 250°C, flame ionization detector temperature 290°C.

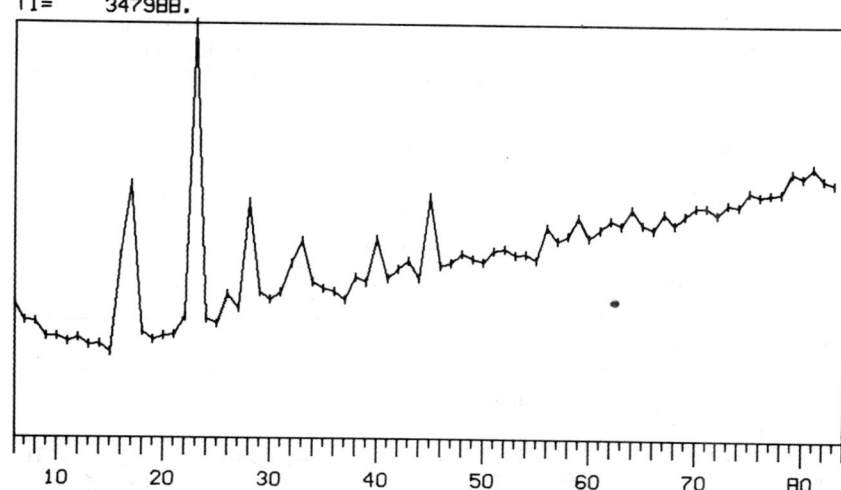

Figure 32.4 Total ion current chromatograms generated from high-resolution mass spectral data for the K-1-P fraction. Horizontal scale is scan number.

The monocyclic aromatics and homologs of benzene are characteristic for the light crude oils. Toluene, *m*-xylene, and 1,2,4-trimethylbenzene are among the major petroleum constituents.[15]

Consequently, it is suggested that methyl benzoate, phenol and cresol may be oxidation products originally derived from the monocyclic aromatics.

Neutral Fractions

The complexity of the neutral fractions is illustrated in the gas chromatograms in Figure 32.7. The K-1-N sample, however, does show a suite of discrete peaks in addition to a "hump" from a complex unresolved mixture of components, and therefore one would expect to be able to get certain useful identifications of some of the components by GC/LRMS. In contrast, the K-2-N and K-3-N fractions while very similar show few significant distinguishable components and thus represent a much more difficult analytical problem.

These same overall features are also apparent in the total ion current plots shown in Figure 32.8. The plots were generated from the high resolution data and are individually normalized to the largest value (given on the Figure as "TI" in arbitrary units) of total intensity per spectrum.

596 ORGANIC POLLUTANTS IN WATER

Table 32.1   Components Identified in Phenolic Fractions

Major Components	Formula	Scan No. in Figure
Anisole	C_7H_8O	17
Methyl anisole	$C_8H_{10}O$	23
C_2 anisole + methyl benzoate	$C_9H_{12}O$ $C_8H_8O_2$	28
Iodo anisole	C_7H_7OI	40
Iodo methyl anisole	C_8H_9OI	45

Minor Components

Esters	N-Containing	Ether	Other
$R\ CO_2\ R'$ $R' = CH_3, C_2H_5$	carbamate $C_2H_5OC-N-R$ $\quad\ \ \|\!\|$ $\quad\ \ O$ $R = (CH_3)_2, C_2H_5$	⌬—OR R'	⌬—CH$_3$ —CH$_3$
cyclohexyl-CH$_2$CO$_2$CH$_3$			
Ph-C$_n$CO$_2$R $R = CH_3, C_2H_5$ $n = 0,1$	N-substituted amides $R\ C-N-R'$ $\ \ \|\!\|\quad\ \ \backslash R'$ $\ \ O$ $R' = H, CH_3, C_2H_5$		
naphthyl-C$_n$CO$_2$CH$_3$ $R = H, C$ $n = 0,1,2,3$ R	$CH_3C-N-C-O-CH_3$ $\ \ \|\!\|\ \ \|\ \ \|\!\|$ $\ \ O\ \ CH_3\ O$		

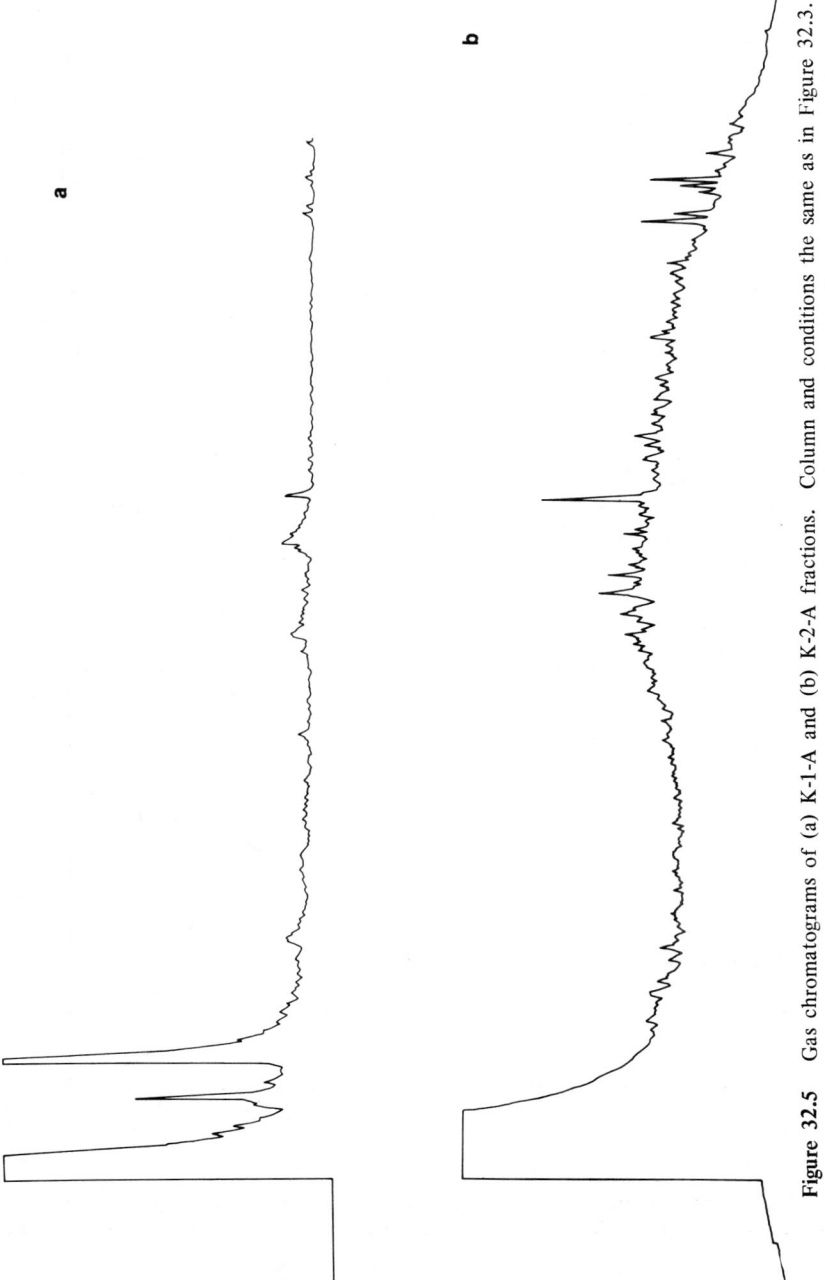

Figure 32.5 Gas chromatograms of (a) K-1-A and (b) K-2-A fractions. Column and conditions the same as in Figure 32.3.

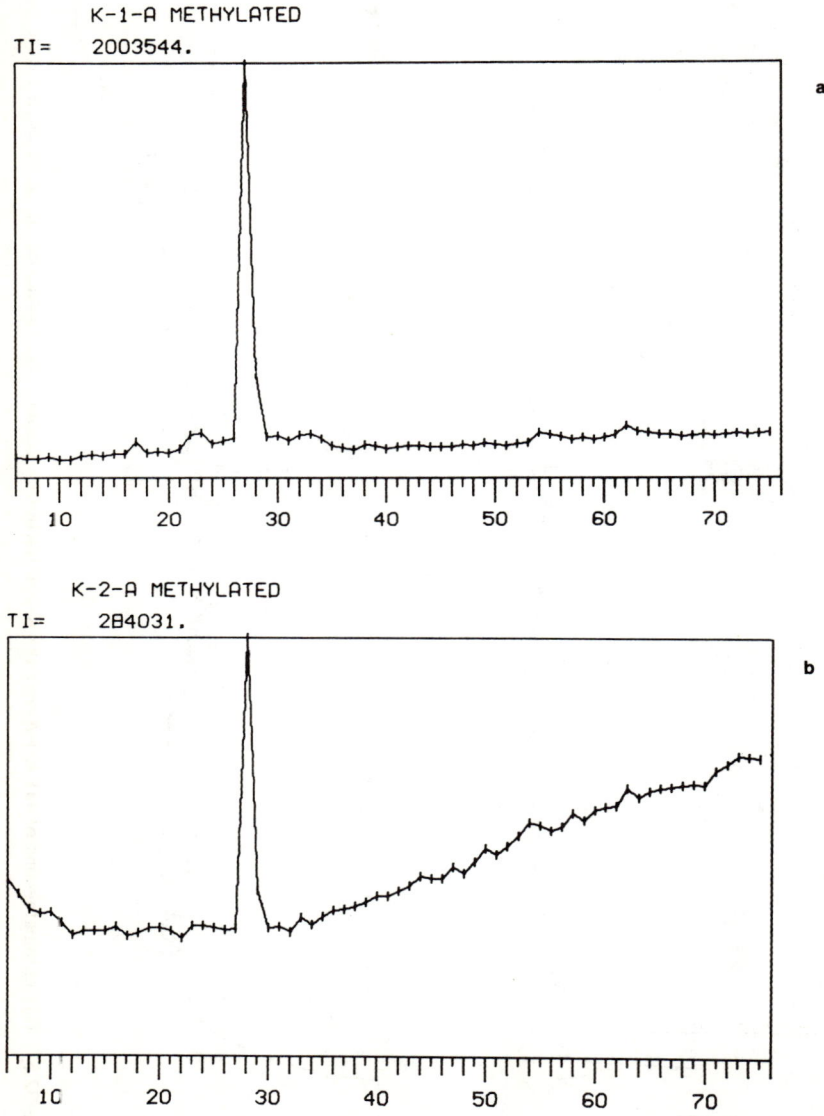

Figure 32.6 Total ion current chromatogram generated from high-resolution mass spectral data for (a) K-1-A and (b) K-2-A fractions. Horizontal scale is scan number.

Table 32.2 Minor Components Identified in Acidic Fractions

Compound Type	Formula	Present in No. of Scans
Saturated methyl esters	$C_nH_{2n}O_2$	28
Saturated ethyl esters	$C_nH_{2n}O_2$	6
Olefinic ethyl ester	$C_nH_{2n-2}O_2$	1
Saturated propyl esters	$C_nH_{2n}O_2$	1
Cyclic alkyl methyl esters	$C_nH_{2n-2}O_2$	4
Alkyl-substituted methyl benzoates	$C_nH_{2n-8}O_2$	16
Phenylalkyl methyl esters	$C_nH_{2n-8}O_2$	5
Alkyl-substituted naphthenic methyl esters	$C_nH_{2n-14}O_2$	8
Indenic methyl esters	$C_nH_{2n-12}O_2$	1
Anisole	$C_nH_{2n-6}O$	1
Sulfur-substituted aromatic methyl esters	$C_nH_{2n-6}O_2S$, $C_nH_{2n-12}O_2S$	3
Alkyl-substituted methyl sulfides	$C_nH_{2n-6}S$	1

The availability of accurate mass data permits the generation of large quantities of plots and listings used for compound identification; a small selection of this output is shown below to illustrate the types of correlations used.

Figures 32.9 and 32.10 show series of accurate mass chromatograms for C_nH_{2n+1} ions (n = 5-9) characteristic of normal and branched alkanes. From the elemental composition listings for the scans corresponding to the major peaks, these components were identified as n-alkanes, which are of particular use as internal retention index markers. From the relative intensities (maximum TI for K-1-N cf. maximum TI for K-2-N), the series is approximately three times as abundant in the K-1-N fraction. Despite the apparent lack of distinguishable component peaks in the K-2-N fraction, accurate mass chromatograms are sufficiently specific to indicate the position (*i.e.*, scan number) of clearly identifiable components.

Figures 32.11 and 32.12 show mass chromatograms for nominal mass 113 (composite of m/e 112.6000-113.6000) and the accurate masses 113.1340 (C_8H_{17}) and 113.0425 (C_6H_9S). In both fractions, the relative intensities of these ions indicate that almost all the ion current of m/e 113 is derived from C_8H_{17} and C_6H_9S ions. For K-2-N, the "hump" apparent in the nominal mass chromatogram appears to be due mainly to

600 ORGANIC POLLUTANTS IN WATER

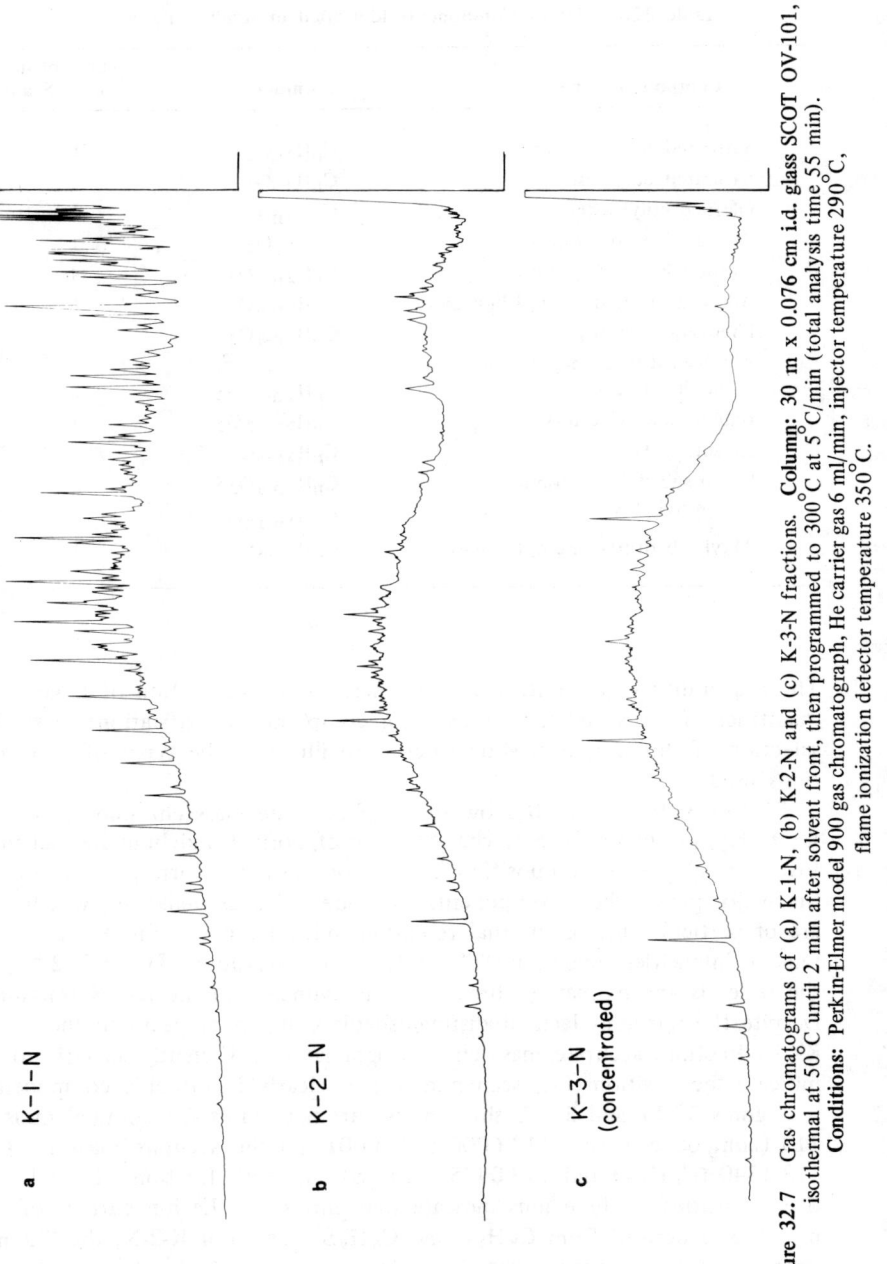

Figure 32.7 Gas chromatograms of (a) K-1-N, (b) K-2-N and (c) K-3-N fractions. **Column:** 30 m × 0.076 cm i.d. glass SCOT OV-101, isothermal at 50°C until 2 min after solvent front, then programmed to 300°C at 5°C/min (total analysis time 55 min). **Conditions:** Perkin-Elmer model 900 gas chromatograph, He carrier gas 6 ml/min, injector temperature 290°C, flame ionization detector temperature 350°C.

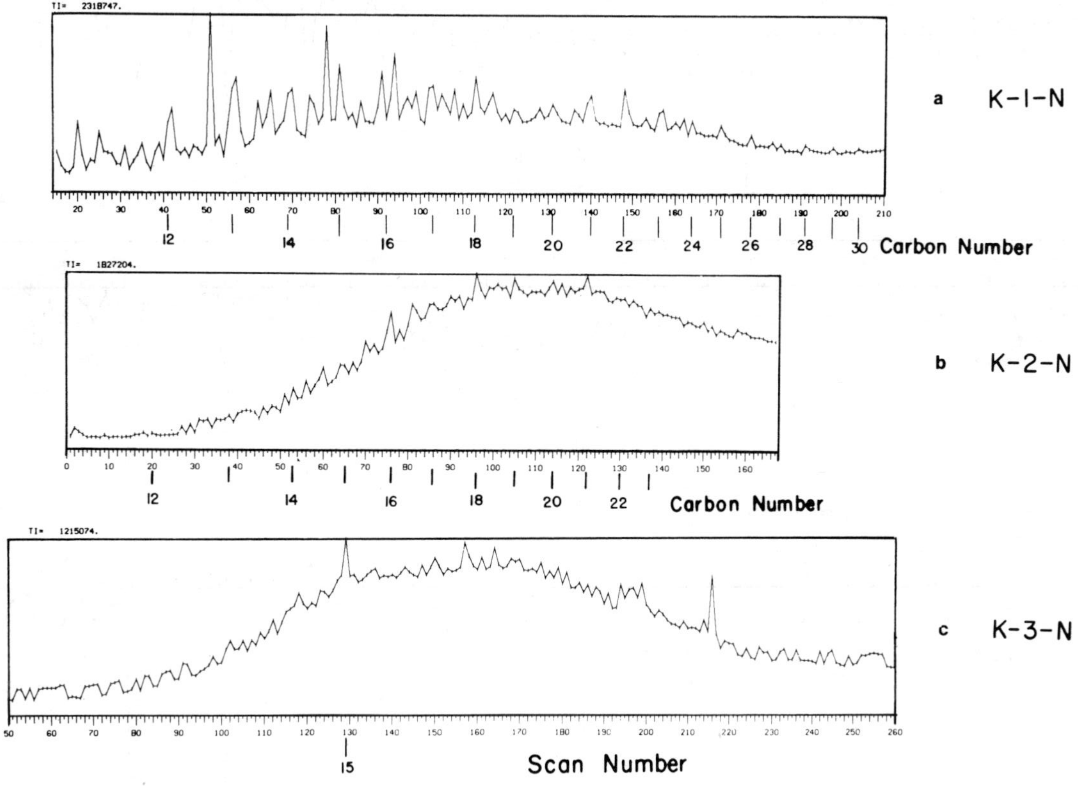

IDENTIFICATION OF ORGANIC POLLUTANTS 601

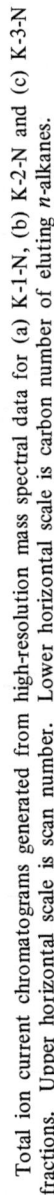

Figure 32.8 Total ion current chromatograms generated from high-resolution mass spectral data for (a) K-1-N, (b) K-2-N and (c) K-3-N fractions. Upper horizontal scale is scan number. Lower horizontal scale is carbon number of eluting n-alkanes.

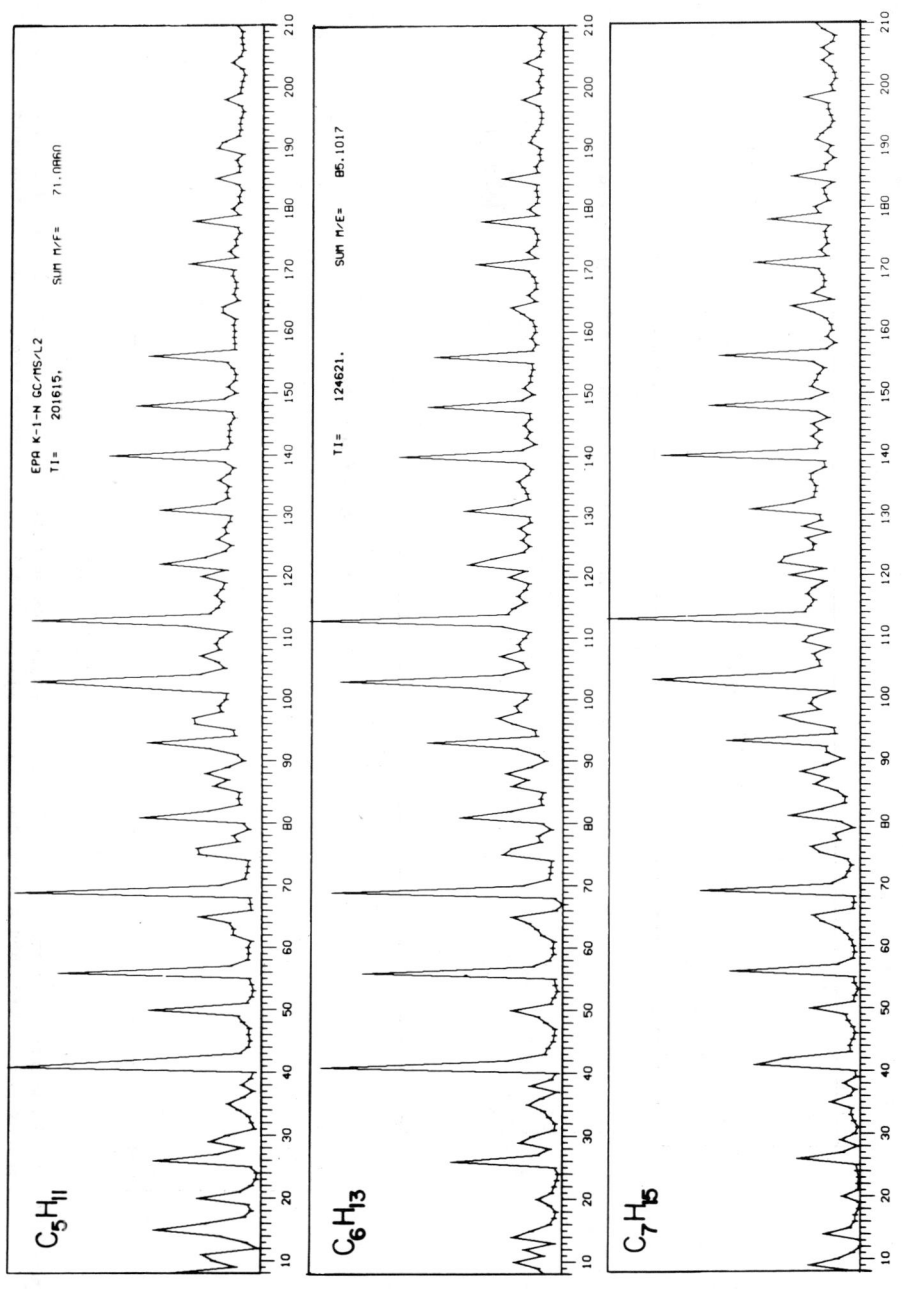

IDENTIFICATION OF ORGANIC POLLUTANTS 603

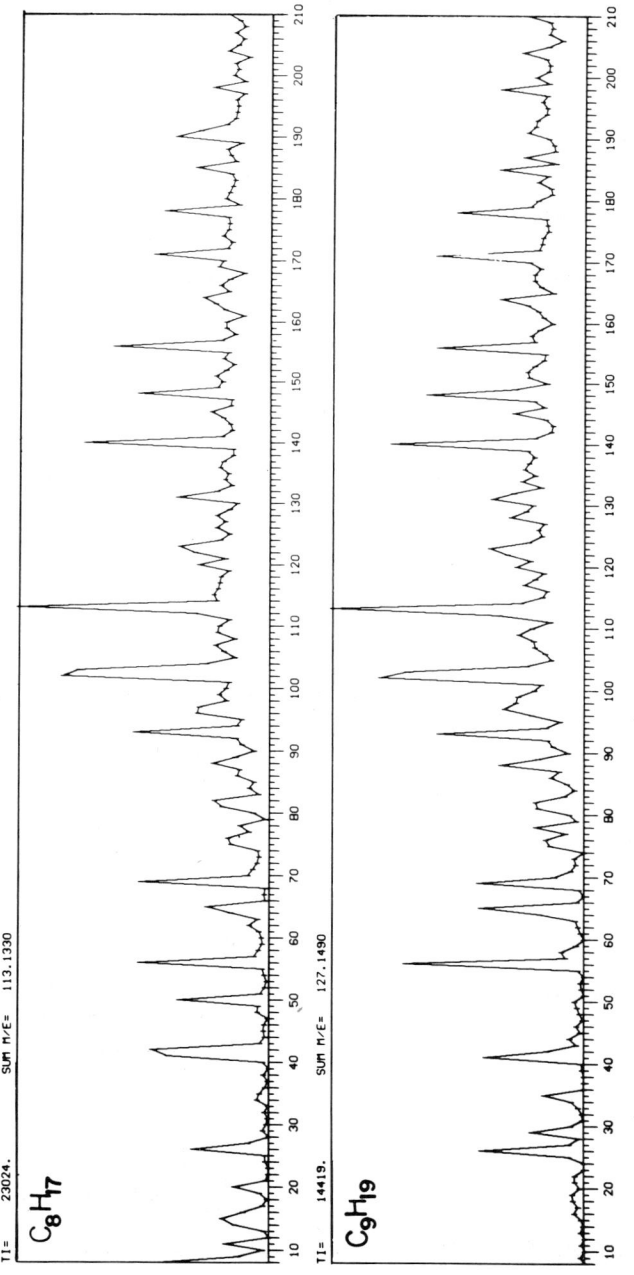

Figure 32.9 Accurate mass chromatograms for C_nH_{2n+1} (n = 5-9) ions for K-1-N fraction. Scan 113 ≡ n-C_{18}.

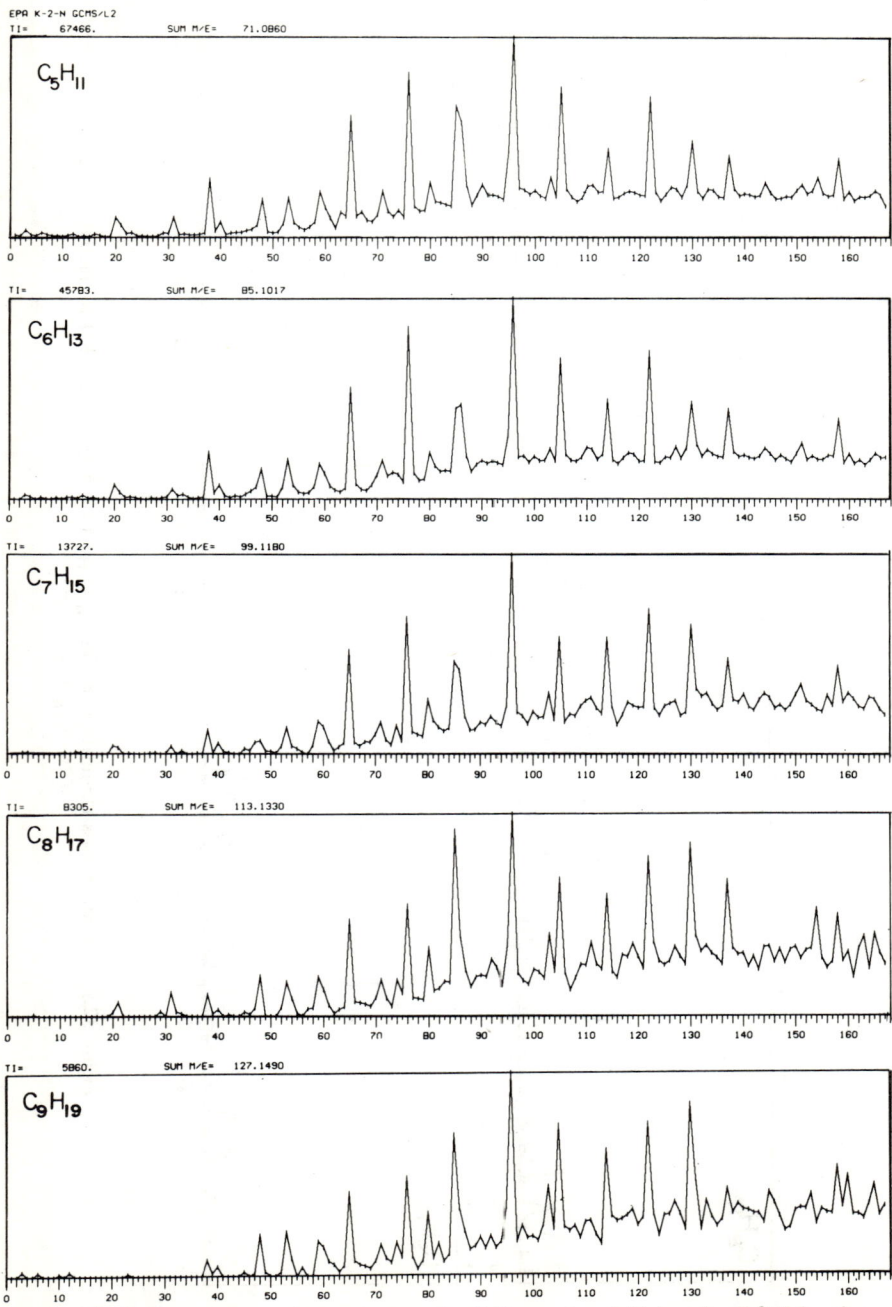

Figure 32.10 Accurate mass chromatograms for C_nH_{2n+1} (n = 5-9) ions for K-2-N fraction. Scan 96 $\equiv n\text{-}C_{18}$.

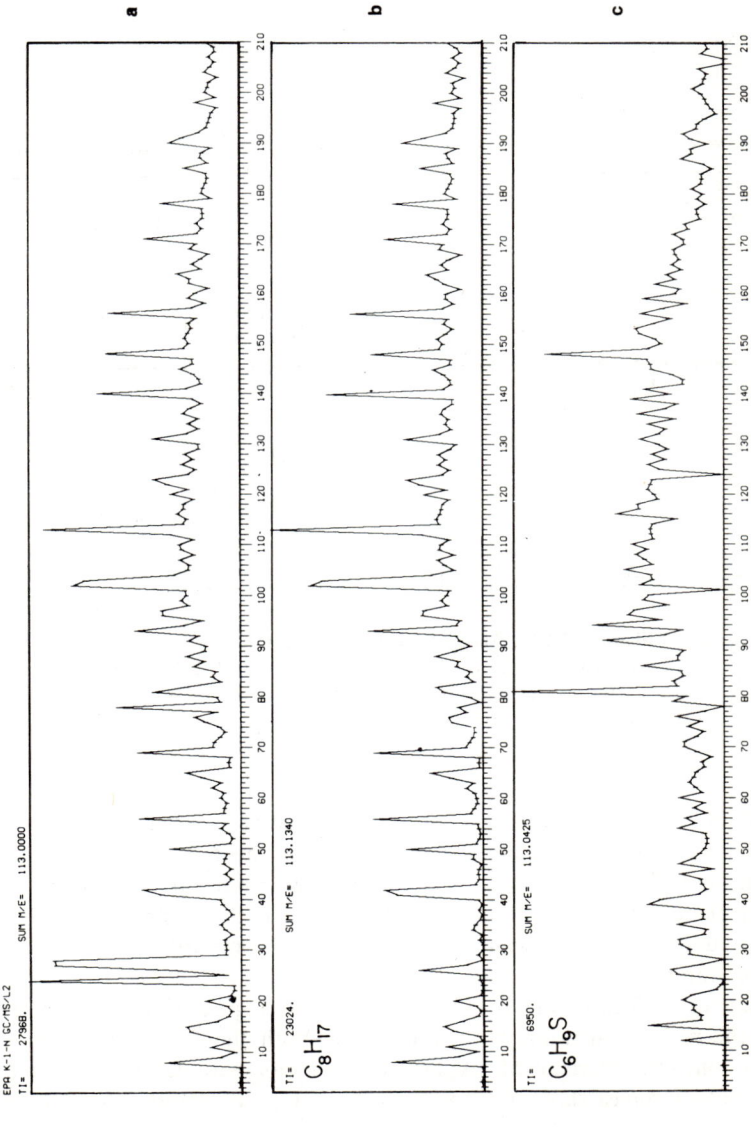

Figure 32.11 Mass chromatograms for K-1-N fraction. (a) Nominal mass chromatogram for m/e 113 (112.6-113.6); (b) accurate mass chromatogram for C_8H_{17} (113.1340); (c) accurate mass chromatogram for C_6H_9S (113.0425).

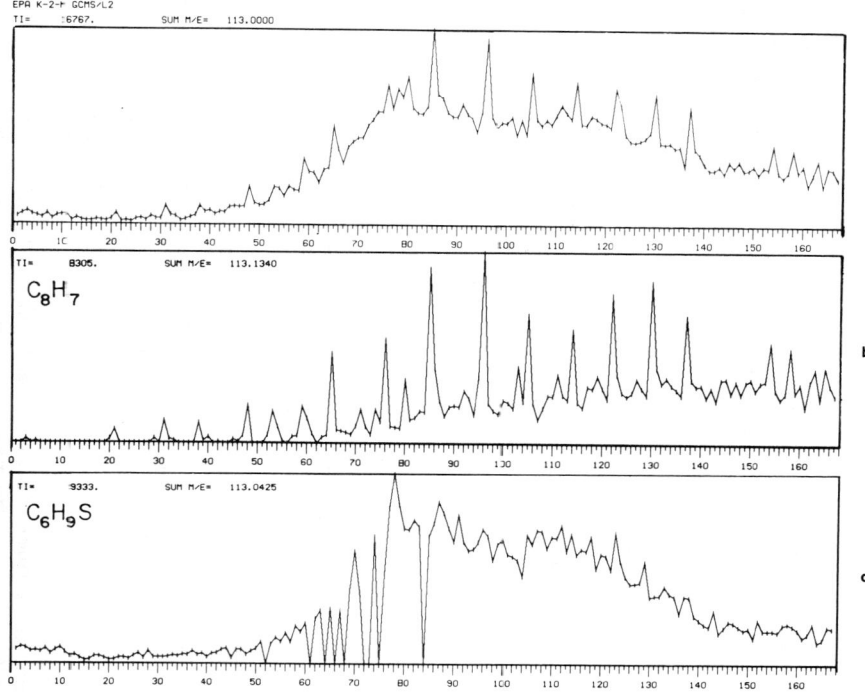

Figure 32.12 Mass chromatograms for K-2-N fraction.
(a) Nominal mass chromatogram for m/e 113 (112.6-113.6);
(b) accurate mass chromatogram for C_8H_{17} (113.1340);
(c) accurate mass chromatogram for C_6H_9S (113.0425).

the C_6H_9S ions; however, for K-1-N, the relative intensities indicate a smaller C_6H_9S "hump" and the major part of the m/e 113 ion current is due to C_8H_{17}. This type of correlation is important because if the major elemental compositions were known for each nominal mass, then this would permit an assessment of the overall composition of the unresolved hump.

Similar profiles in Figure 32.13a and b and Figure 32.14a and b show that the major contribution to m/e 105 in both the fractions is the C_8H_9 ion, characteristic of alkyl-substituted mono-aromatic compounds. Also both samples show a ratio of C_8H_9/C_9H_{12} of about 2:1 by comparison with Figures 32.13 and 32.14 (b and c) and indicate the presence of four and two C3-substituted mono-aromatic hydrocarbons, respectively, for K-1-N and K-2-N.

Figures 32.13 and 32.14 (d and e) show similar mass chromatograms for C_9H_{11} and $C_{10}H_{14}$ for C4-substituted benzenes. In Figure 32.14d, two zero values occur in the chromatogram at scans 96 and 143—a feature apparent in other chromatograms also. Investigation of m/e 119 in these scans by inspection of the actual ion current profiles revealed the presence of unresolved multiplets (see insert Figure 32.14d) of C_9H_{11} and C_8H_7O which significantly shifted the calculated center of gravity of the C_9H_{11} peak outside of the mass measurement tolerance. It is desirable to keep the tolerance on mass measurement for an accurate mass chromatogram as small as justified under fast scanning conditions[16] to preclude incorporation of other elemental compositions. However, in the analysis of such complex mixtures, the presence of unresolved multiplets occurs quite frequently, causing similar mass shifts and hence zero values in the accurate mass chromatograms. This feature does not in general detract from the overall pattern of the profiles.

Many of the chromatograms indicate that ions of a particular accurate mass contribute to the qualitative nature of the unresolved hump. This observation led to a rigorous assessment of the accuracy of measurement of the C_8H_9 accurate mass throughout the hump of the K-2-N fraction (see Figure 32.14b) to assure that no artifact of elemental compositional redundancy or data processing could be responsible for the hump in this chromatogram. The tolerance of this accurate mass chromatogram was restricted to as low as 5 ppm while preserving the overall features of the C_8H_9 envelope, and no other compositions could be determined which corresponded to this accurate mass within this small a tolerance. It was concluded, therefore, that C_8H_9 ions are genuinely contributory to the unresolved hump.

Figure 32.15a and c show the nominal mass chromatograms for the K-2-N (a) and K-1-N (c) fractions. The relative intensities indicate that for K-2-N the major ion current for m/e 125 is carried by the $C_7H_9O_2$ ion (Figure 32.15b) which may be assigned the structure

$$CH_3O-\underset{{}^+OCH_3}{\text{cyclopentadiene}}$$

The formation of this ion could occur by the loss of CH_3 followed by CO from a trimethoxybenzene component.[17] In contrast, the relative intensities in the corresponding mass chromatograms for K-1-N (Figure 32.15c and d) show that the $C_7H_9O_2$ ion contributes a relatively small proportion of the m/e 125 ion current.

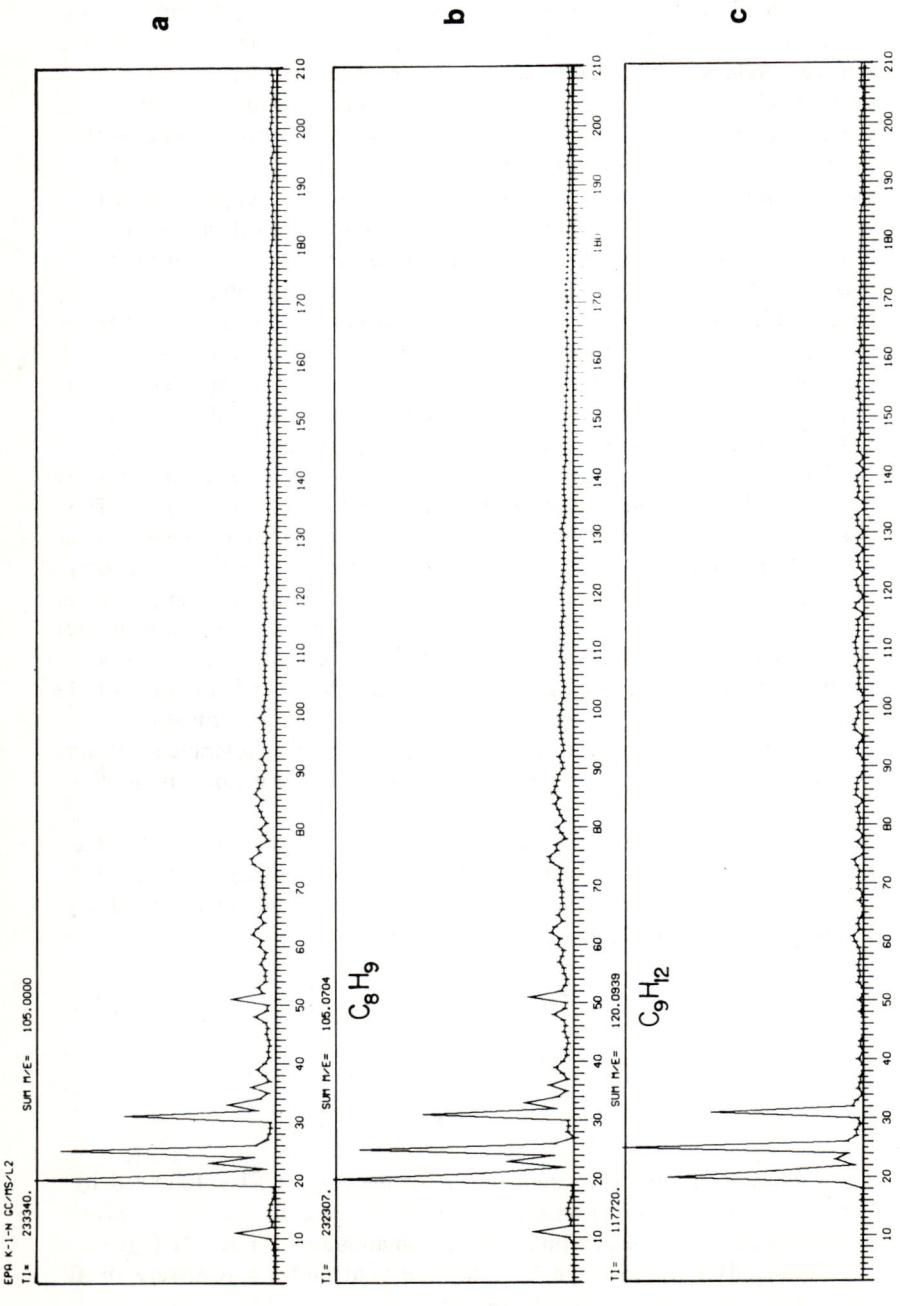

IDENTIFICATION OF ORGANIC POLLUTANTS 609

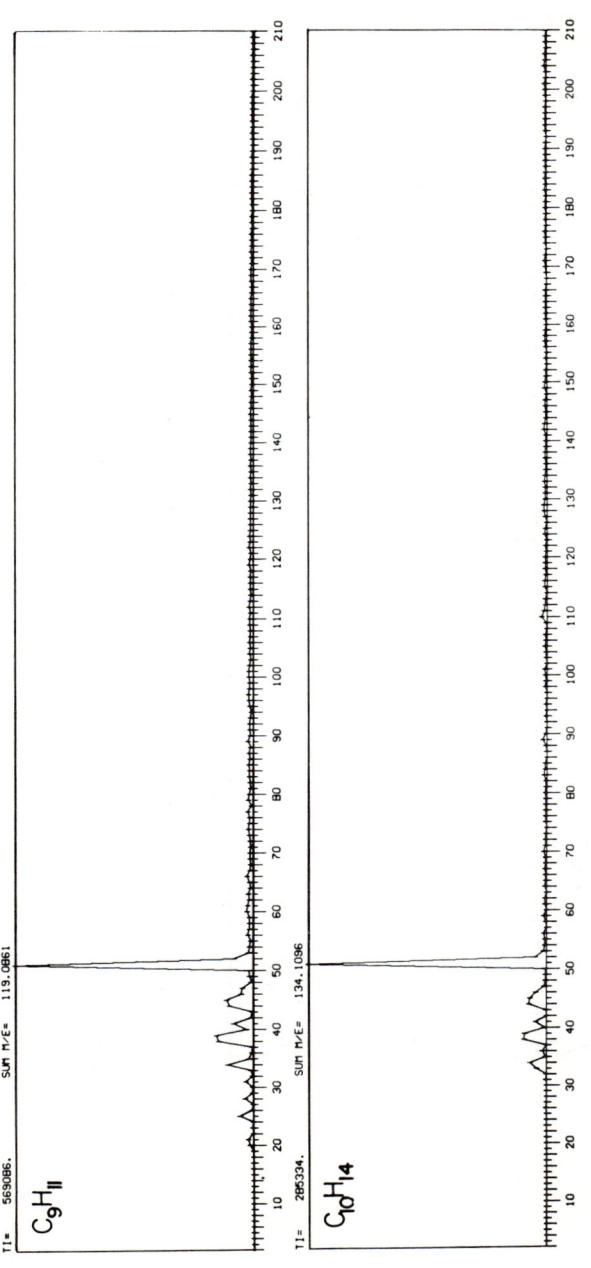

Figure 32.13 Mass chromatograms for K-1-N fraction. (a) Nominal mass chromatogram for m/e 105 (104.6-105.6); (b) accurate mass chromatogram for C_8H_9 (105.0704); (c) accurate mass chromatogram for C_9H_{12} (120.0939); (d) accurate mass chromatogram for C_9H_{11} (119.0861); (e) accurate mass chromatogram for $C_{10}H_{14}$ (134.1096).

610 ORGANIC POLLUTANTS IN WATER

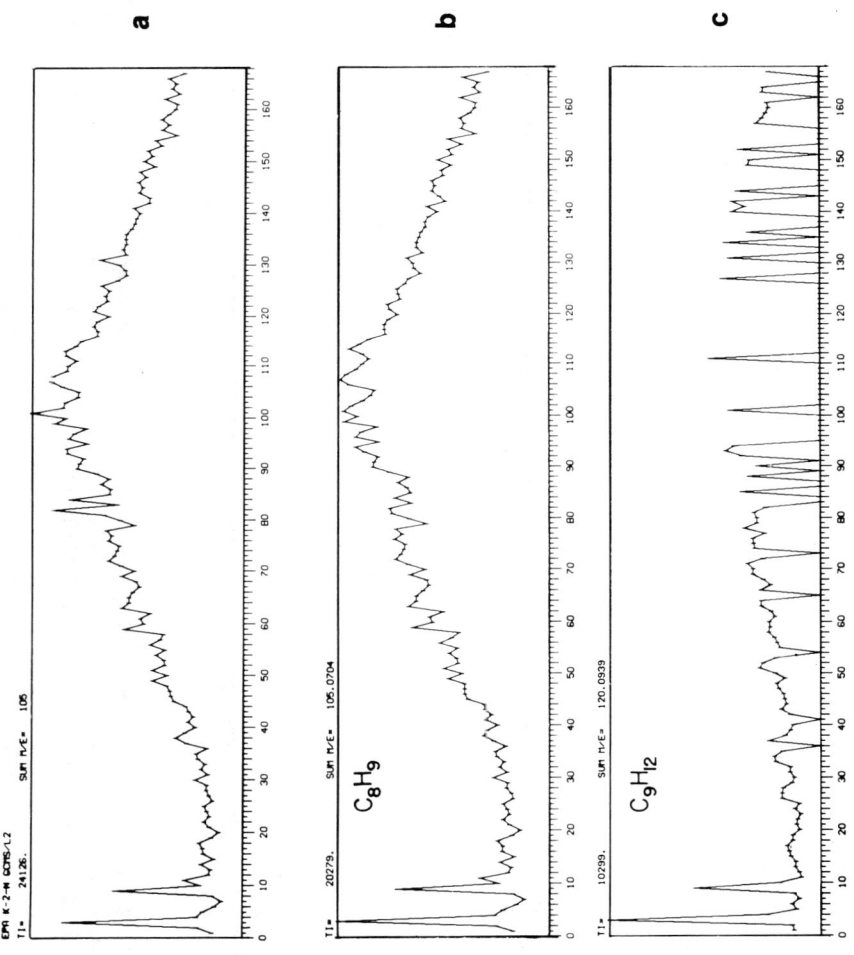

IDENTIFICATION OF ORGANIC POLLUTANTS 611

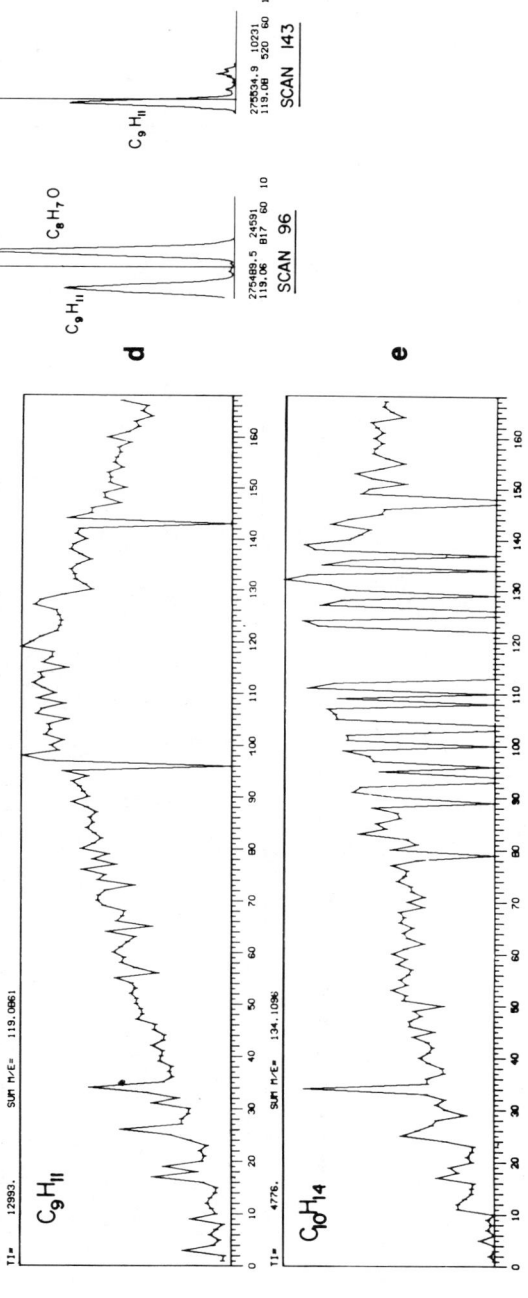

Figure 32.14 Mass chromatograms for K-2-N fraction. (a) Nominal mass chromatogram for m/e 105 (104.6-105.6); (b) accurate mass chromatogram for C_8H_9 (105.0704); (c) accurate mass chromatogram for C_9H_{12} (120.0939); (d) accurate mass chromatogram for C_9H_{11} (Inserts: ion current profiles for m/e 119 for scans 96 and 143); (e) accurate mass chromatogram for $C_{10}H_{14}$ (134.1096).

612 ORGANIC POLLUTANTS IN WATER

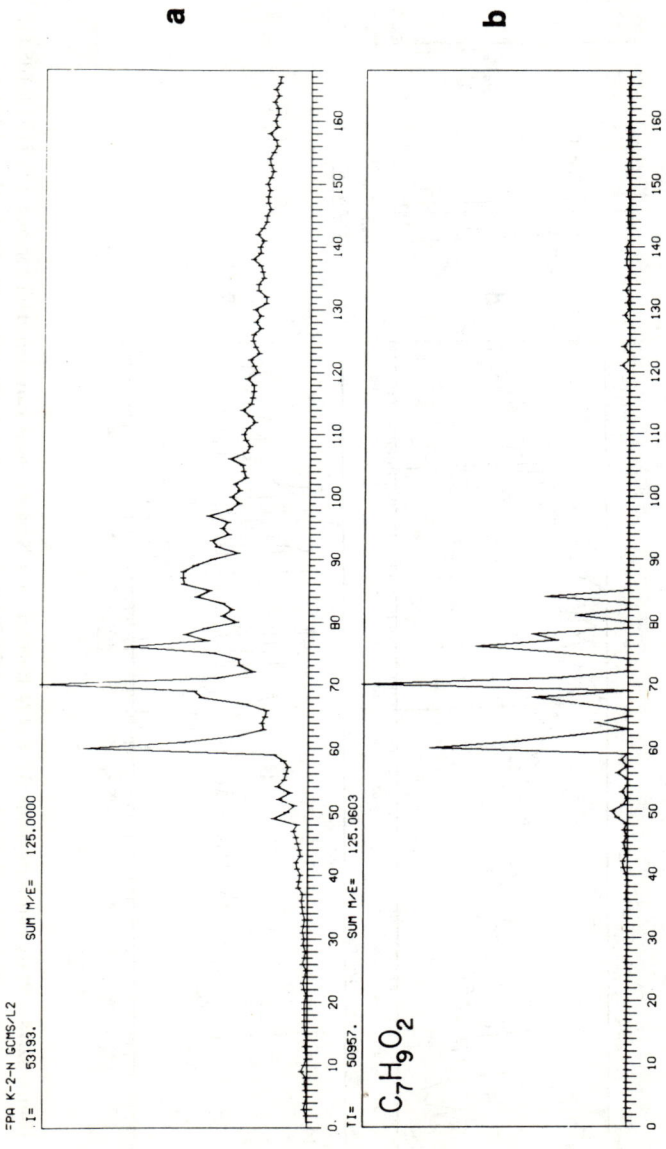

IDENTIFICATION OF ORGANIC POLLUTANTS 613

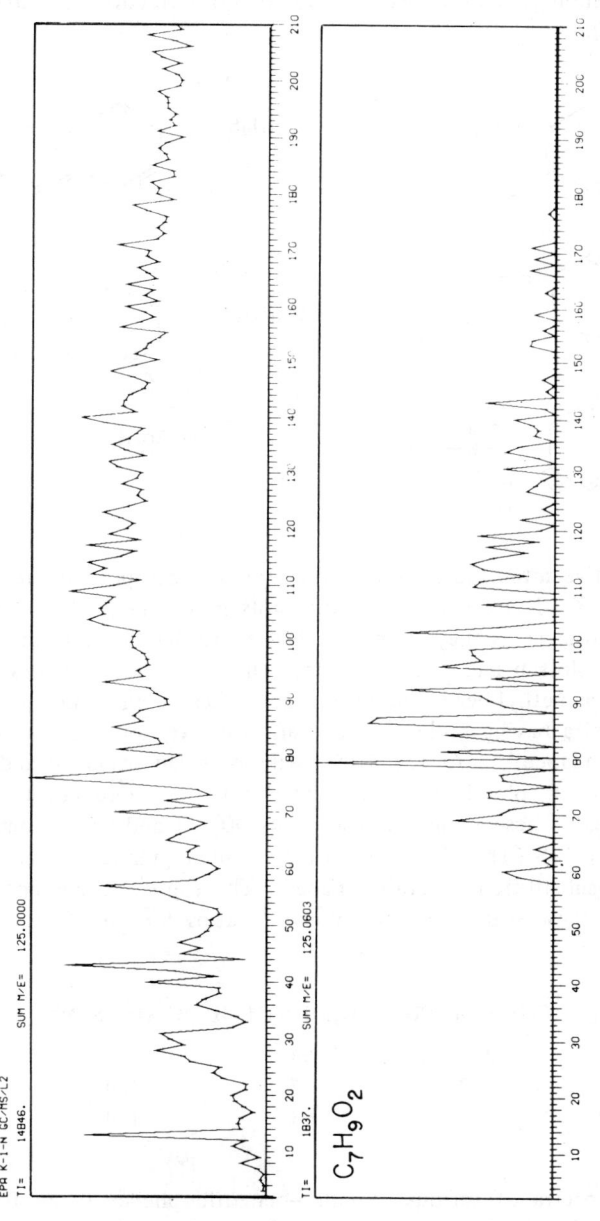

Figure 32.15 (a) Nominal mass chromatogram for m/e 125 (124.6–125.6) for K-2-N fraction; (b) accurate mass chromatogram for $C_7H_9O_2$ (125.0603); (c) as (a) for K-1-N fraction; (d) as (b) for K-1-N fraction.

Accurate mass chromatograms of ions characteristic of benzothiophene and alkylated homologues were also plotted for both fractions (Figures 32.16 and 32.17):

C_8H_6S

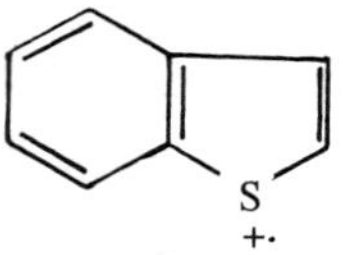

C_9H_7S

C_9H_8S

$C_{10}H_9S$

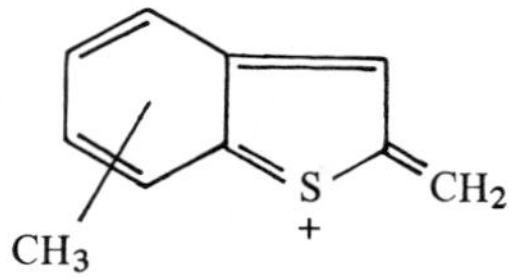

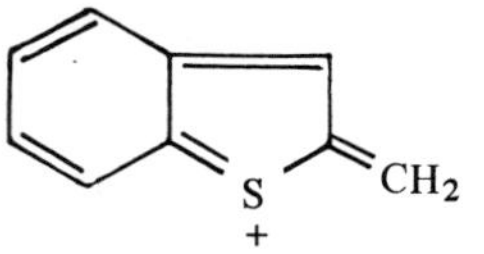

$C_{10}H_{10}S$

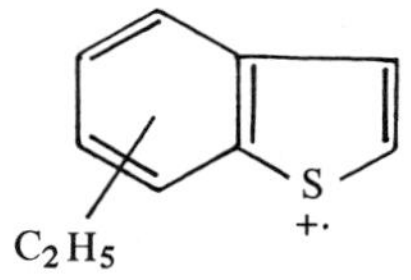

or $2(CH_3)$

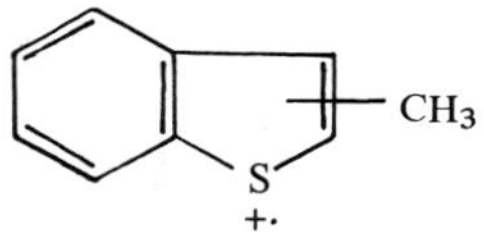

From the relative intensities, these sulfur aromatic compounds are more prominent in the K-1-N fraction. Also the plots in Figure 32.16b-e, for example, are of interest as they permit certain more detailed identifications. A comparison of plots b and c indicates that the first three peaks (scans 78-84) result from methyl-benzothiophenes since these compounds give characteristic spectra having molecular ions and (M - H) ions of approximately equal abundance.[18] As the three peaks between scans 90 and 96 are present only in plot b, they must be due to the presence of higher homologues. These same peaks at scans 90, 93 and 96 are also present in the $C_{10}H_9S$ (d) and $C_{10}H_{10}S$ (e) chromatograms, indicating that they are C_2-substituted benzothiophenes. The elemental compositional listings for these scans give the following ratios for the M, M - H and M - CH_3 ions:

Scan	C_9H_7S (M - CH_3)	$C_{10}H_9S$ (M - H)	$C_{10}H_{10}S$ (M)
90	100	34	58
93	79	77	100
96	65	98	100

Reference mass spectra of various dimethyl-benzothiophenes taken from the *Eight Peak Index of Mass Spectra*,[19] *Atlas of Mass Spectral Data*[20] and *Selected Mass Spectral Data*[21] give very similar ratios for these three ions:

C_9H_7S 27; $C_{10}H_9S$ 88; $C_{10}H_{10}S$ 100

whereas the $C_9H_7S/C_{10}H_{10}S$ ratio for ethyl-benzothiophenes[19] is about 100:40. Comparison of these ratios indicates that the component of scan 90 may be an ethyl-benzothiophene, and the components of scans 93 and 96 may be dimethyl-benzothiophenes, although the $C_9H_7S/C_{10}H_{10}S$ ratio appears to be unusually high in these spectra.

In Figure 32.18, some examples of mass spectra are shown; these were generated from the corresponding high resolution data and are normalized to the most intense mass above m/e 70. These spectra were taken from the "hump" region of K-2-N but still show noticeable features despite the relatively long scan cycle time compared to that normally used in GC/HRMS analyses.

Using mass chromatograms such as those exemplified above, together with nominal mass spectra and elemental composition listings, a variety of oxygen-, sulfur- and nitrogen-containing components have been identified in the fractions, as well as numerous aliphatic and aromatic hydrocarbons. These results are summarized in Table 32.3, together with an assessment of the relative abundance for the fractions.

For most of the compound types identified, the amount is decreased in K-2-N relative to K-1-N. However, the alkylated methyl-indeneols and compounds giving a $C_7H_9O_2$ fragment (Table 32.3b) are significantly more abundant in K-2-N, presumably reflecting the oxidative step (Figure 32.1) between the sampling points for the two fractions.

The K-3-N fraction has considerably fewer components than K-2-N, but the distribution is fairly similar. The main exception is the presence of n-C_{15} alkane in the K-3-N fraction in far greater abundance than this alkane is found in K-1-N or K-2-N. This is an unusual distribution and the pathway to the formation of n-C_{15} alkane without accompanying homologs is not presently understood. It would seem to arise as a by-product of microbial metabolism in the lagoon.

SUMMARY

In summary, the use of GC/HRMS using both packed and capillary columns can provide valuable information on the organic constituents of refinery wastewater samples, even when present in extremely complex mixtures. In many cases, the low resolution (nominal mass) spectra would not have been interpretable without the availability of the corresponding high resolution (accurate mass and composition) data, since

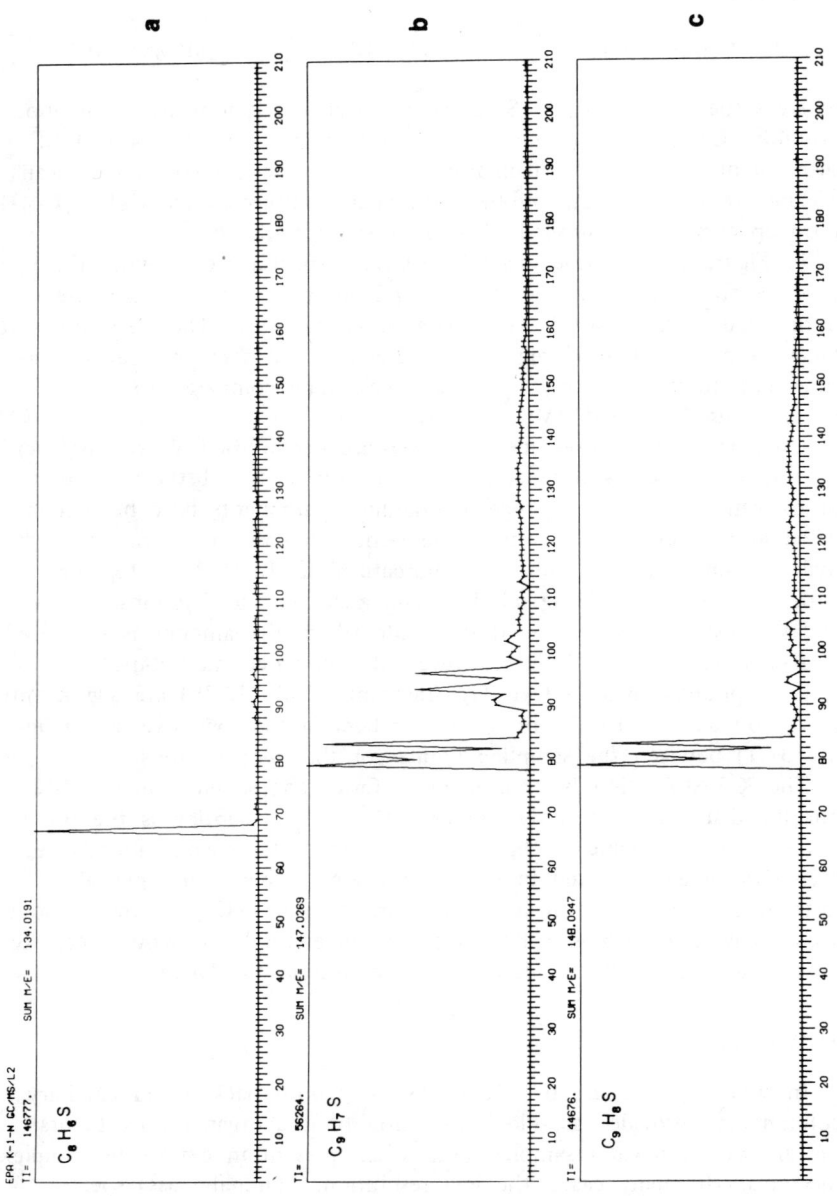

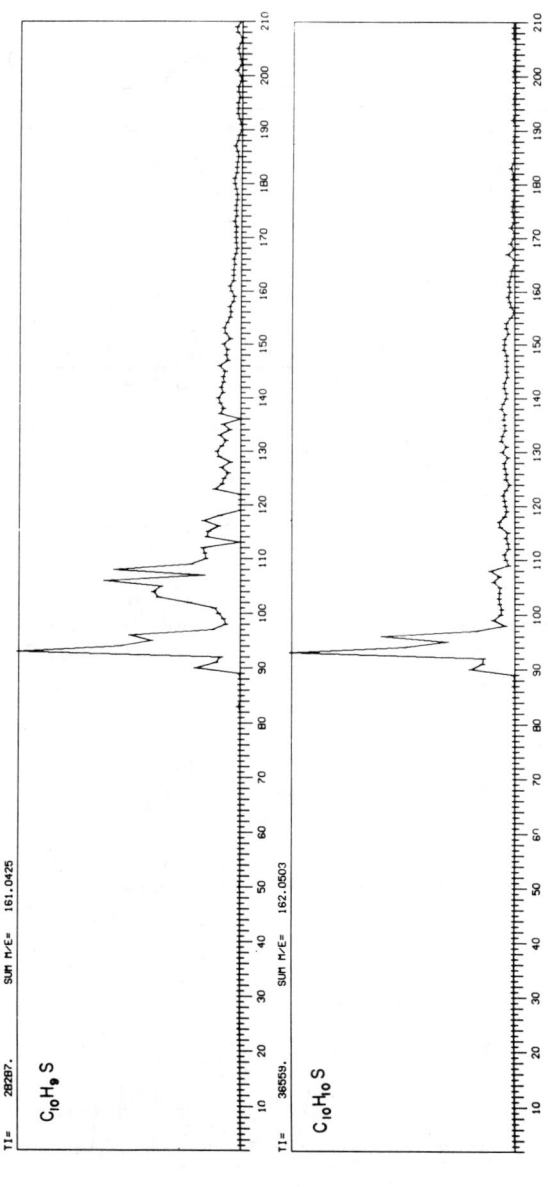

Figure 32.16 Accurate mass chromatograms for K-1-N fraction.
(a) C_8H_6 (134.0191); (b) C_9H_7S (147.0269); (c) C_9H_8S (148.0347); (d) $C_{10}H_9S$ (161.0245); (e) $C_{10}H_{10}S$ (162.0503).

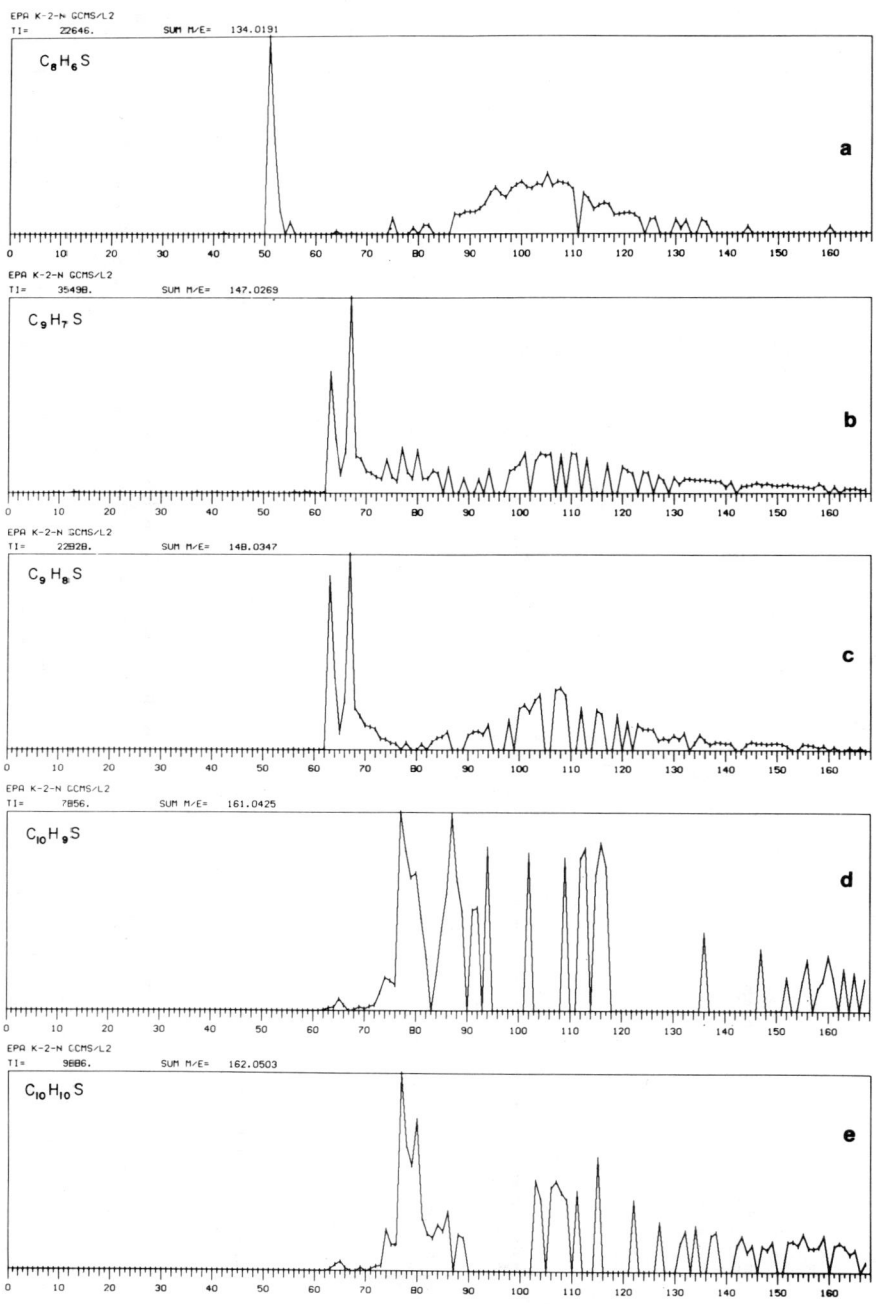

Figure 32.17 Accurate mass chromatograms for K-2-N fraction.
(a) C_8H_6S (134.0191); (b) C_7H_9S (147.0269); (c) C_9H_8S (148.0347);
(d) $C_{10}H_9S$ (161.0425); (e) $C_{10}H_{10}S$ (162.0503).

IDENTIFICATION OF ORGANIC POLLUTANTS 619

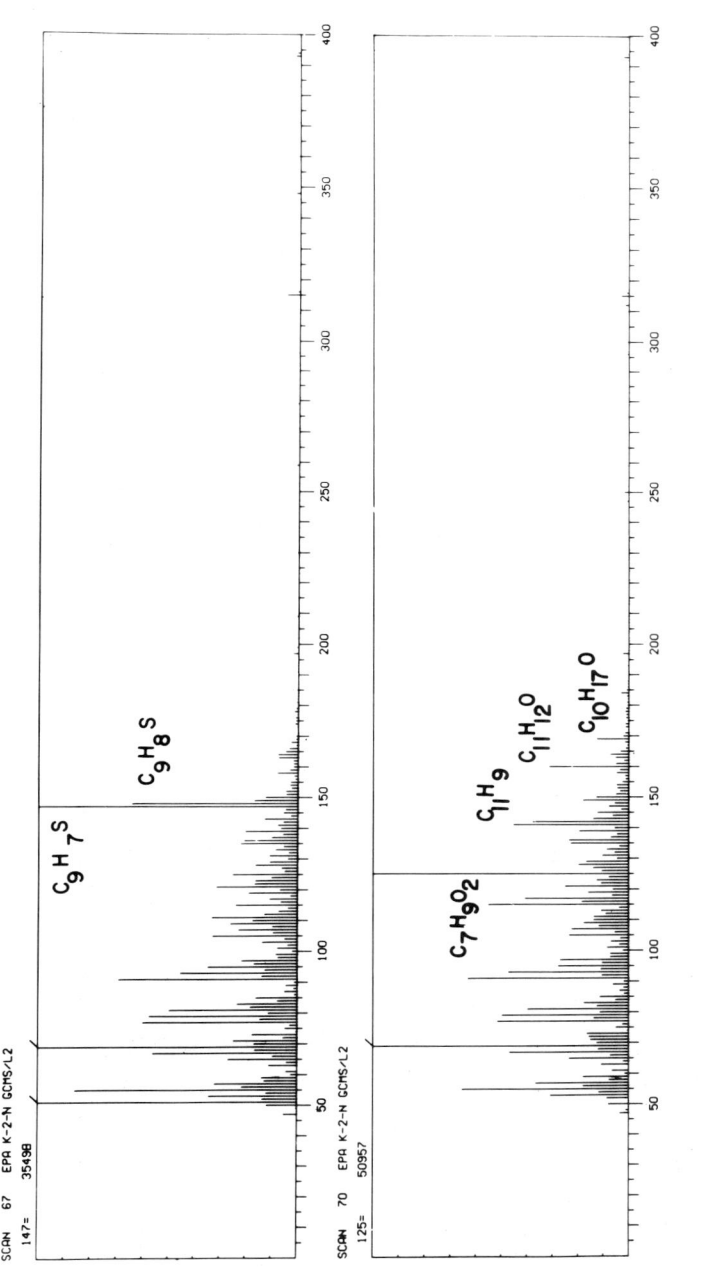

Figure 32.18 Examples of mass spectra from the "hump" region of fraction K-2-N. The spectra were generated from high-resolution (accurate mass) data and are normalized to the most intense mass above m/e 70.

CONTINUED

620 ORGANIC POLLUTANTS IN WATER

FIGURE 32.18, continued

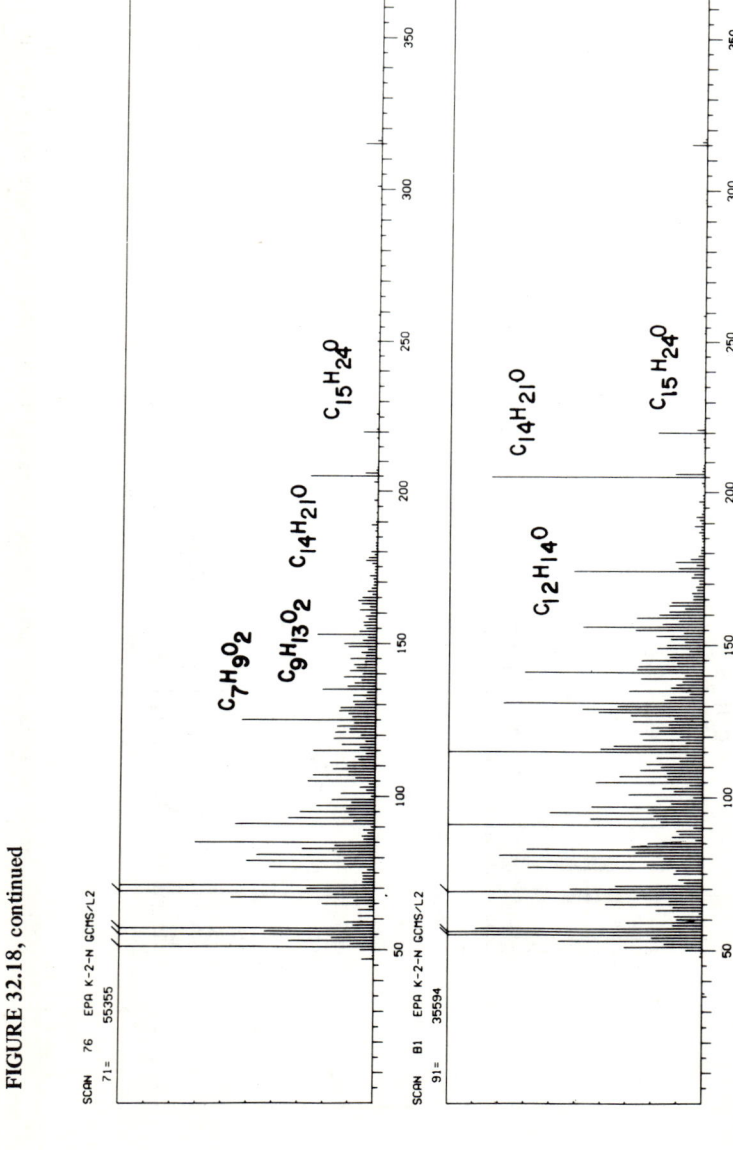

IDENTIFICATION OF ORGANIC POLLUTANTS   621

Table 32.3   Components Identified in Neutral Fractions

Compound Type	K-1-N	K-2-N	K-3-N
a) Hydrocarbons			
n-Alkanes C_nH_{2n+2}	$n = 11\text{-}33$ major constituents	$n = 12\text{-}33$	$n = 15$ (also small $n = 16\text{-}18$)
Branched alkanes C_nH_{2n+2}	series		n.d.
Mono-unsaturated or mono-cyclic alkanes C_nH_{2n}	$n = 11\text{-}28$	n.d.	n.d.
Alkyl-benzenes C_nH_{2n-6}	$n = 9$ (3 isomers) $n = 10$ (6 isomers) $n = 11$ major constituents	$n = 9$ $n = 10$ (several)	$n = 8$ (2 isomers) $n = 9$ (3 isomers) $n = 10$ (3 isomers)
Naphthalenes C_nH_{2n-12}	$n = 10\text{-}13$ (isomers present) abundant	$n = 10\text{-}14$ (isomers for $n = 11$)	$n = 12\text{-}14$ (isomers present) not very abundant
Phenanthrenes or anthracenes C_nH_{2n-18}	$n = 14\text{-}19$ ($n = 14$ relatively abundant)	$n = 14\text{-}19$ $n = 17$ relatively abundant)	$n = 15\text{-}17$ not very abundant
Pyrene or fluoroanthrene C_nH_{2n-22}	$n = 16$ (minor)	$n = 16$ (minor)	n.d.
Alkyl-biphenyls C_nH_{2n-14}	$n = 13$ (significant) $n = 14$	n.d.	$n = 12$ $n = 13$ (low abundance)
Methyl indan $C_{10}H_{12}$	present	n.d.	n.d.
b) Oxygenated Compounds			
Alkylated phenols and anisoles $C_nH_{2n-6}O$	$n = 7\text{-}12$ ($n = 8$ & 9 very abundant; isomers present) major	$n = 7\text{-}16$ (similar distribution to K-1-N)	$n = 7\text{-}13$
Alkylated methyl-indeneols $C_nH_{2n-10}O$	not detected	$n = 11\text{-}14$ ($n = 13$, 2 isomers)	n.d.

Table 32.3, continued

Compound Type	K-1-N	K-2-N	K-3-N
b) Oxygenated Compounds, continued			
Diooxygenated compounds $C_7H_9O_2$	present	present (3 components)	n.d.
c) Nitrogen Compounds			
Indoles $C_nH_{2n-9}N$	$n = 10\text{-}12$	n.d.	n.d.
Carbazoles $C_nH_{2n-15}N$	$n = 12\text{-}15$	$n = 13$ $n = 14$ (minor)	n.d.
d) Sulfur Compounds			
Thiacyclanes $C_nH_{2n}S$	$n = 6$ $n = 8$ (thiacyclopentanes)	$n = 8\text{-}11$ (several isomers; thiacyclohexanes)	$n = 6\text{-}11$ (several isomers)
Bicyclothiacyclanes C_nH_{2n-2} $C_nH_{2n-4}S$	$C_2\text{-}C_4$ substituents $n = 13$	$C_2\text{-}C_5$ substituents $n = 13$ (more abundant than K-1-N)	C_2 and C_3 substituents n.d.
Benzothiophenes $C_nH_{2n-10}S$	$n = 8\text{-}12$ (isomers present)	$n = 8\text{-}11$	$n = 9; n = 10$ isomers
Dibenzothiophene $C_nH_{2n-16}S$	$n = 12$	$n = 12$	n.d.

even with the use of capillary columns, the components of these complex mixtures could not be chromatographically resolved.

It is apparent that additional components in these fractions can be identified if additional data processing followed by detailed analysis of the output is carried out, although the incomplete fractionation of the samples is a complicating factor in these particular cases.

ACKNOWLEDGMENTS

The financial support of the Environmental Protection Agency (Grant R803019-01), the National Institutes of Health (Grant RR-719-02) and the National Aeronautics and Space Administration (Grant NGL 05-003-003) is gratefully acknowledged.

REFERENCES

1. Reid, G. W. and L. E. Streebin. "Evaluation of Waste Waters from Petroleum and Coal Processing," EPA-R-2-72-001 (Washington, D.C.: U.S. Government Printing Office, 1972).
2. Cranwell, P. A. In *Environmental Chemistry*, Vol. 1, G. Eglinton, Sr. Reporter. The Chemical Society. (London: Burlington House, 1975), pp. 22-54.
3. Eglinton, G., J. R. Maxwell and R. P. Philp. In *Advances in Organic Geochemistry 1973*, B. Tissot and F. Bienner, Eds. (Paris: Editions Technip, 1974), pp. 941-961.
4. Farrington, J. W. and G. C. Medeiros. "Evaluation of Some Methods of Analysis of Petroleum Hydrocarbons in Marine Organisms," *Proc. 1975 Conf. on Prevention and Control of Oil Pollution*, San Francisco, March 1975, pp. 115-121.
5. Blumer, M., P. C. Blokker, E. B. Cowell and D. F. Duckworth. In *A Guide to Marine Pollution*, E. D. Goldberg, Ed. (New York: Gordon and Breach Publishers, 1972), Chap. 2, pp. 19-40.
6. Farrington, J. W. and J. G. Quinn. *Estuar. and Coastal Marine Science* 1, 71-79 (1973).
7. Zafiriou, O. C. *Estuar. and Coastal Marine Science* 1, 81-87 (1973).
8. Farrington, J. W. and P. A. Meyer. In *Environmental Chemistry*, Vol. 1, G. Eglinton, Sr. Reporter, The Chemical Society (London: Burlington House, 1975), pp. 109-136.
9. Davis, J. B. In *Petroleum Microbiology*. (Amsterdam: Elsevier Publishing Co., 1967), Chap. 8.
10. Davis, J. B. *Chem. Geol.* 3, 155-160 (1968).
11. Han, J. C-Y. "Chemical Studies of Terrestrial and Extraterrestrial Life," Ph.D. Thesis, University of California, Berkeley (1970), pp. 154-181.
12. Burlingame, A. L., R. W. Olsen and R. McPherron. In *Advances in Mass Spectrometry*, Vol. 6, A. West, Ed. (London: Applied Science Publishers Ltd., 1974), pp. 1053-1059.
13. Kimble, B. J., R. E. Cox, R. V. McPherron, R. W. Olsen, E. Roitman, F. C. Walls and A. L. Burlingame. *J. Chromatog. Sci.* 12, 647-655 (1974).
14. Smith, H. M. "Crude Oil—Qualitative and Quantitative Aspects," U.S. Bureau of Mines, Rept. Invest., 1-55 (1965).
15. Nagy, B. and U. Colombo. In *Fundamental Aspects of Petroleum Geochemistry*. (Amsterdam: Elsevier Publishing Co., 1967), Chap. 3 and 4.

16. Kimble, B. J., F. C. Walls, R. W. Olsen and A. L. Burlingame. "Real-Time Gas Chromatography-High Resolution Mass Spectrometry and the Analysis of Complex Organic Mixtures," presented at the 23rd Annual Conference on Mass Spectrometry & Allied Topics, Houston, Texas (May 1975).
17. Hamming, M. C. and N. G. Foster. *Interpretation of Mass Spectra of Organic Compounds.* (London and New York: Academic Press, 1972).
18. Budzikiewicz, H., C. Djerassi and D. H. Williams. *Mass Spectrometry of Organic Compounds.* (San Francisco: Holden-Day Inc., 1967).
19. *Eight Peak Index of Mass Spectra,* Vols. 1 and 2, Mass Spectrometry Data Centre, AWRE, Aldermaston, England (1970).
20. Stenhagen, E., S. Abrahamsson and F. W. McLafferty, Eds. *Atlas of Mass Spectral Data,* Vols. 1-3 (New York: Interscience, 1969).
21. *Selected Mass Spectral Data,* Vol. I-VI, American Petroleum Institute, Research Project 44, Thermodynamics Research Center, Texas A & M University, College Station, Texas.

CHAPTER 33

IDENTIFICATION OF TWO CHLORINATED GUAIACOLS IN KRAFT BLEACHING WASTEWATERS

I. H. Rogers

>Environment Canada
>Fisheries and Marine Service
>Pacific Environment Institute
>West Vancouver, B.C.

L. H. Keith

>U.S. Environmental Protection Agency
>Southeast Environmental Research Laboratory
>Athens, Georgia

Environmental chemistry is a relatively new science which is concerned with the measurement of levels of man-made pollutants in air, water and soil and related biota. Such knowledge is required by governments in setting safe standards for the release of the many waste products of modern society and is vital to those in the biological sciences who must study complex ecological relationships. These may be disturbed from their natural balances by the presence of small concentrations of unnatural substances, many of which have potential toxic effects on biological systems. Perturbations of ecological balances often result in major adverse effects on food resources, which are ever more vitally needed as world population increases. On the other hand, more people need more jobs for their livelihood, in their quest for a better living standard, and consume more manufactured goods with correspondingly more environmental pollution. This chapter is concerned with some aspects of such a dilemma—the need to protect valuable salmon stocks in rivers used to receive the waste products of the pulp and paper industry.

British Columbia has vast areas of productive forest lands and accounts for some 69% of Canada's lumber production and 26% of the nation's woodpulp.[1] The forest industry is the mainstay of the economy of the province; 21 pulp mills now operate within its boundaries and discharge their waste products to its inland and coastal waters. The rivers involved, especially the Fraser River, are the migration pathways for adult salmon returning to their home streams to spawn and for the juvenile salmon on their way to the ocean. Two species of Pacific salmon remain in fresh water for their first year of life. Certain coastal marine receiving waters, especially estuaries, are inhabited by stocks of young salmon for the first few months of their life cycle. Unfortunately, the waste products emitted by the pulp and paper industries are acutely toxic to salmon, to various components of their foodweb and to other aquatic organisms of benefit to man, such as oysters. It is for these reasons that the Government of Canada has recently sponsored considerable research effort on the identification of toxic compounds in kraft effluents and on finding treatment methods for their removal. The Government of the United States is also interested in such information in view of the valuable salmon resources of the Pacific Northwest states and Alaska.

Previous work on the identification of toxic compounds has mainly centered on kraft pulping effluent and on whole kraft effluent. There have been few studies of bleach plant effluents, although these are known to contribute substantial amounts of toxicity and most of the color.[2,3] Tetrachloro-*o*-benzoquinone was indirectly identified as a toxic component of chlorination effluent a few years ago[4] but has never been isolated. The toxicities of di- and tetrachlorocatechol to pink salmon (*Oncorhynchus gorbuscha*) and sockeye salmon (*O. nerka*) were established in the expectation that they are present in such effluents even though this was never demonstrated.[5] In a more recent study, several low-molecular-weight compounds were identified in spent chlorination liquor. Although no toxicity bioassays were conducted, it is likely that the chloromaleic and chlorofumaric acids detected would affect salmonids.[6]

In the present study, the structures of two chlorinated guaiacols, present in low concentrations in caustic extraction effluent, are defined without isolation of the pure compounds. A preliminary report on the findings has been issued.[7] Recently another group has reported the presence of these compounds in bleach plant effluents without acknowledgment of our prior discovery.[8]

STATIC BIOASSAY TESTS

Static bioassays on all mill samples except those from mill B were conducted against sockeye salmon presmolts in fresh Cultus Lake water, as has been described elsewhere.[9] Similarly 96-hr LC50* levels of standard compounds synthesized in the laboratory were determined. Fish density ranged from 0.24 to 0.70 g/l using ten fish per test. Static bioassays on mill B effluent samples were conducted against coho salmon (*Oncorhynchus kisutch*) in a different laboratory using Cypress Creek water, fish density 0.45 g/l and five fish per test. The results are given in Table 33.1.

Table 33.1 Mortality of Coho Salmon Exposed to Mill B Effluent Samples

Sample No.	% Effluent Concentration	% Mortality Observation Time Hours				
		14	24	48	72	96
1	32	–	100	100	100	100
	18	20	80	100	100	100
	10	0	0	20	20	20
2	32	100	100	100	100	100
	18	–	100	100	100	100
	10	–	0	0	0	0
3	32	–	80	100	100	100
	18	–	0	60	80	80
	10	–	0	0	0	0
4	32	20	40	100	100	100
	10	0	0	20	20	20
	5.6	0	0	0	0	0

All mill effluent samples studied were toxic in that they failed to pass the Canadian federal standard which states that no mortality shall occur in fish exposed in a static bioassay for 96 hours in a solution composed of two parts by volume of effluent to one part of dilution water. It was not possible to compare the toxicity of samples from mill A with those from mill B, because the tests were conducted in two laboratories, using different species of salmon and with waters differing in hardness.

RECOVERY OF TOXIC FRACTIONS

Effluent samples from the caustic extraction and chlorination sewers of mill A were blended to give a pH value of 2.9, and the samples were

---

*Concentration lethal to 50% of the test organisms over a specified period of exposure.

then divided. One portion was neutralized and submitted for bioassay tests. The other was percolated through a bed of Amberlite XAD-8 resin, as described earlier.[7] The bed was regenerated with methanol and the methanol eluate concentrated and dispersed in water. After acidification to pH 2.0 the product was extracted into ether and an insoluble lignin fraction removed by filtration. Sodium fusion tests and infrared spectra indicated the presence of organochlorine functions in the lignin. The ether-soluble portion was methylated with diazomethane for analysis by combination gas chromatography/mass spectrometry (GC/MS) and by gas chromatography using both electron capture (EC) and flame ionization detectors (FID).

After passage through the Amberlite resin bed, four out of seven samples were nonacutely toxic at 100% concentration, and fish mortalities of 10%, 40% and 70% were observed in the other three. The EC traces of all seven samples contained one pair of major peaks with matching retention times. Other small peaks, varying in number from three to nine, were also present. No significant peaks were observed when the extracts were analyzed prior to methylation. The pair of peaks was also easily detectable in the FID mode but appeared small in comparison with other components of the extract. It was concluded that the peaks in question were caused by chlorinated acids or chlorinated phenols.

Samples from mill B comprised caustic extraction effluent only and were extracted into ether after acidification to pH < 4. Bioassays were conducted on neutralized samples. Four such samples were analyzed by GC as above and exhibited the identical pair of peaks in all cases.

ANALYSIS OF EXTRACTS BY GC/MS

One methylated sample from each mill was analyzed by a Finnigan 1015 quadrupole mass spectrometer interfaced with a modified Varian 1400 gas chromatograph. Separation was accomplished using a 50 ft SCOT column coated with OV-17 and temperature programmed from 150° to 220°C at 2°/min. Data collection and output were controlled by a System 150 interface and a pdp 8/e computer having associated magnetic disc and magnetic tape units. Data output from the computer was via either a Houston plotter or a Tektronix R-4012 interactive graphics terminal, with associated Digital Equipment Corpn. pdp 8/m computer and Tektronix hard copy unit.

The computer-reconstructed gas chromatogram (RGC) of the sample from mill A is shown in Figure 33.1. A total of 20 compounds was observed of which 7 were identified. In the other cases, spectra of unresolved components were obtained or the spectra were too weak for

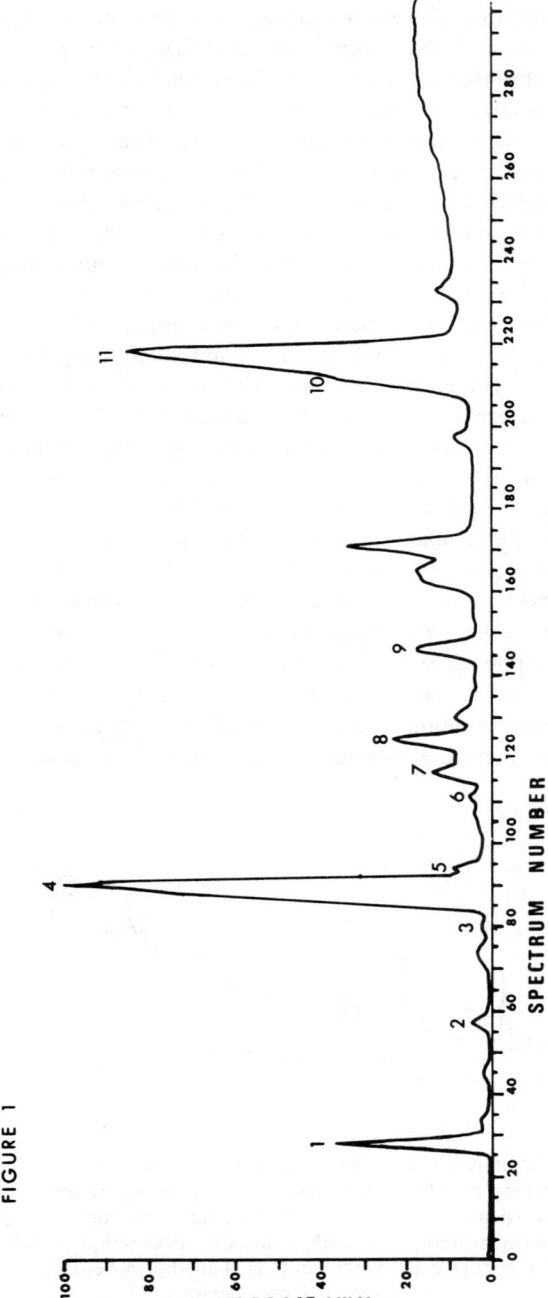

Figure 33.1 Compounds identified in mixed bleachery effluent from mill A. (1) methyl laurate; (2) methyl 2-methyldodecanoate; (3) methyl myristate; (4) methyl 2-methyltetradecanoate; (5) trichloroveratrole; (6) tetrachloroveratrole; (7,8) lignin degradation products; (9) methyl palmitate; (10) branched-chain docosane; (11) normal docosane.

positive identification. The mass spectrum of each component was printed by the computer, and the compounds identified are listed under Figure 33.1. Several saturated fatty acids of the normal series were present together with two fatty acids giving rise to very prominent ions at m/e 88 (base peak for both compounds) and m/e 101. These fatty acids are either ethyl esters of normal fatty acids, which seems unlikely, or else they possess methyl substituents on the alpha carbon atoms.[10,11] Also identified was a pair of isomeric paraffin hydrocarbons whose origin is unknown. These may have been present in some chemical formulation used as an additive within the mill. The organochlorine compound with shorter retention time gave a mass spectrum identical with that of trichloroveratrole, identified previously in an untreated sample of whole kraft effluent.[12] The second compound showed chlorine isotope distribution ratios characteristic of a tetrachloro-compound. The molecular ion was observed at m/e 274 versus m/e 240 in the trichloroveratrole; and from various similarities in the fragmentation patterns, it was concluded that the second compound is tetrachloroveratrole.

The RGC of the sample from mill B is shown in Figure 33.2. Successful identification of 12 compounds out of about 40 was accomplished by spectral matching. Again a series of saturated normal fatty acids was present, and two other fatty acids with branched chains were exhibited. These latter compounds showed prominent ions at m/e 74 (base peak) and m/e 87, typical of saturated straight-chain acids. However, they elute ahead of the corresponding saturated normal acids of similar carbon number, and are therefore ascribed to the saturated *iso* series of fatty

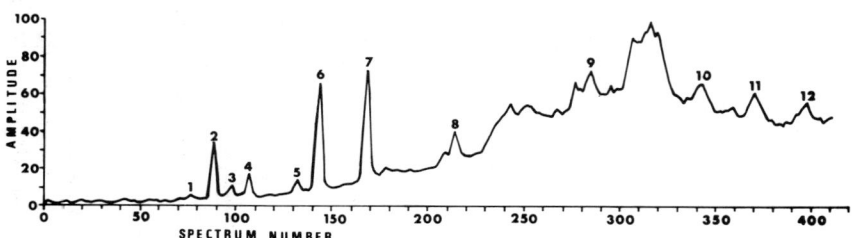

Figure 33.2 Compounds identified in caustic extraction effluent from mill B. (1) methyl myristate; (2) trichloroveratrole; (3) methyl isopentadecanoate; (4) tetrachloroveratrole; (5) methyl isopalmitate; (6) methyl palmitate; (7) methyl margarate; (8) methyl stearate; (9) methyl arachidate; (10) methyl dehydroabietate; (11) methyl behenate; (12) methyl neoabietate.

acids. Two minor peaks were identified as the resin acids dehydroabietic acid and neoabietic acid. The peaks ascribed to tri- and tetrachloroveratrole were also present.

ANALYSIS OF AN ETHYLATED EXTRACT BY GC/MS

A fresh sample of extract from mill A was ethylated with diazoethane, and the material analyzed by GC/MS as described above. Increases of 14 mass units to m/e 254 and m/e 288, respectively, were observed in the molecular ions of both the tri- and the tetrachlorinated compounds. The presence of derivatives with these molecular weights was confirmed by displaying the relevant limited mass reconstructed gas chromatograms (LMRGC's) (m/e 290 was requested instead of m/e 288 as it is the largest peak in the chlorine isotope pattern of the molecular ion). The resulting printout shown in Figure 33.3 indicates the presence of both compounds (peaks at spectra numbers 147 and 176).

Had the precursors of trichloro- and tetrachloroveratrole been chlorinated catechols, two ethyl groups would have been added in the reaction with diazoethane, so the molecular ions should have been found at m/e 268 and m/e 302, respectively. The LMRGC's corresponding to these fragments were displayed but no peaks resulted for the tetrachloro compound, and an erratic series of minor peaks appeared in the LMRGC of the trichloro compound. These were found to be due to small fragments in certain fatty acid ethyl esters and a hydrocarbon. We conclude, therefore, that the precursors of the chlorinated veratroles are chlorinated guaiacols and not chlorinated catechols.

CHLORINATION OF GUAIACOL

In an attempt to duplicate reaction conditions in the first stage of a kraft bleaching sequence, chlorine gas was slowly bubbled into an aqueous solution of guaiacol at room temperature until a pH of 1.9 was reached. After 2 hr, a stream of nitrogen gas was used to expel unreacted chlorine. Half the product was methylated with dimethyl sulfate and NaOH under previously reported conditions.[13] A portion of the methylated product was separated by preparative-scale GC on a Perkin-Elmer 900 instrument with a 10:1 splitter using a 9 ft x 1/4 in. glass column packed with 3% SE-30 on Gas Chrom Q. The first seven peaks, previously determined by GC/MS to be chlorinated monomers of veratrole, were separately collected in glass capillary tubes. The remaining peaks were dimeric reaction products and were not investigated further. From repeated injections, enough sample was separated for Fourier-Transform

632 ORGANIC POLLUTANTS IN WATER

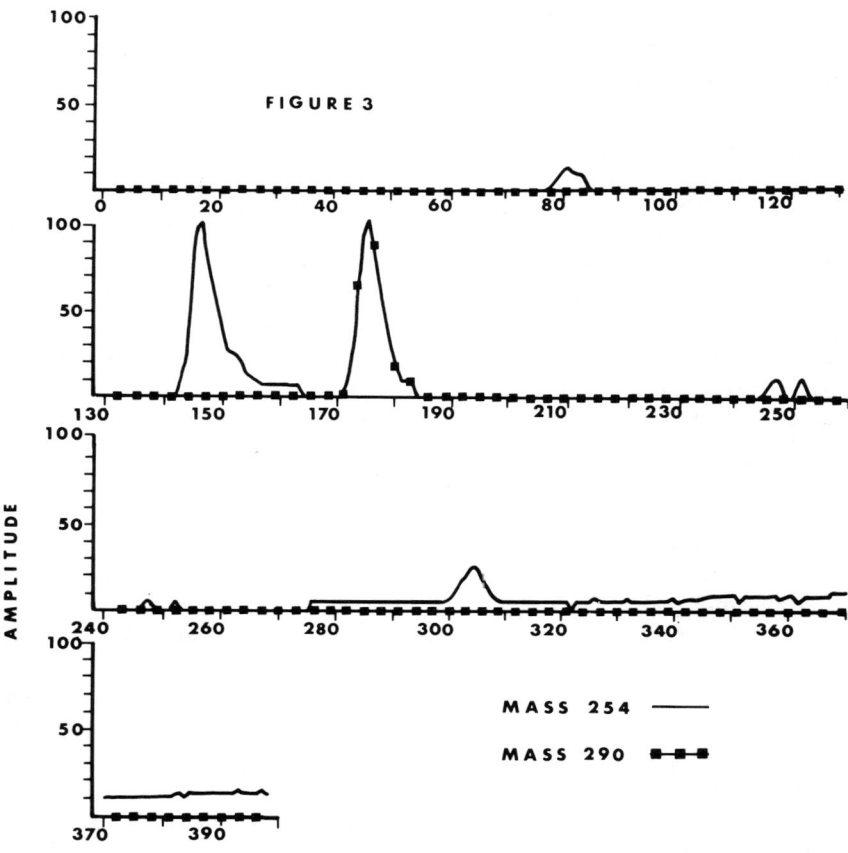

Figure 33.3 Ethylated sample from mill A. Display of parent ions at m/e 254 and m/e 290 indicating uptake of one ethyl ether residue by each compound.

Infra Red spectral measurements (FT-IR) on all seven compounds and for NMR measurements on five of them.

FT-IR spectra were recorded with a Digilab FTS-14D/IR Spectrometer equipped with the DigiLab GC/IR accessory and controller. The instrument is connected with a Perkin-Elmer 990 GC via a Wilks heated transfer line. The GC operates in the FID mode with a 14:1 splitter.

NMR spectra were recorded with a Varian HA-100 spectrometer using carbon tetrachloride as solvent and tetramethylsilane (TMS) as internal standard. Chemical shifts are expressed as τ values and are accurate to ± 1.0 Hz.

The RGC of the complex reaction products from the chlorination of guaiacol is illustrated in Figure 33.4. Peak (a) was shown by its mass spectral fragmentation pattern and molecular ion at m/e 138 to be veratrole (I), derived from unreacted starting material.

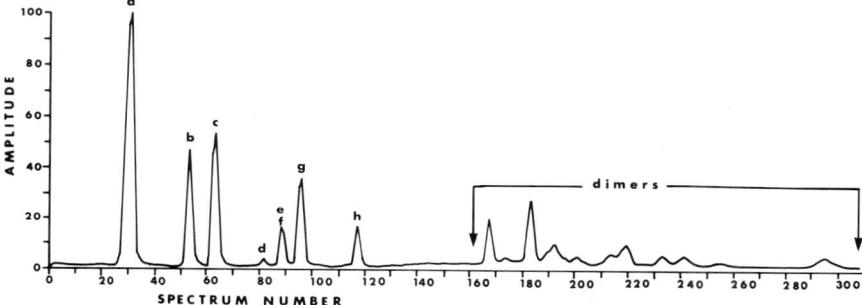

Figure 33.4 Products from chlorination of guaiacol in aqueous solution followed by methylation: (a) veratrole; (b) 3-chloroveratrole; (c) 4-chloroveratrole; (d) 3,5-dichloroveratrole; (e) 3,4-dichloroveratrole; (f) 3,6-dichloroveratrole; (g) 4,5-dichloroveratrole; (h) 3,4,5-trichloroveratrole.

Peaks (b) and (c) gave essentially identical mass spectra, with molecular ions at m/e 172 and a chlorine isotope distribution pattern indicative of only one chlorine atom. Differentiation between the two structures of monochloroveratrole was made on the basis of FT-IR spectra. Peak (c) (III) exhibited an absorption band at 880 cm^{-1} indicative of lone hydrogen C-H out-of-plane deformation. Since only 4-chloroveratrole has a lone hydrogen, this structure was assigned to peak (c) and therefore, by elimination, peak (b) is 3-chloroveratrole (II).

Peaks (d), (e + f) and (g) all gave essentially identical mass spectra with molecular ions at m/e 206 and a chlorine isotope distribution pattern indicative of two chlorine atoms.

The FT-IR spectrum of peak (g) contained an absorption band at 932 cm^{-1} that, although at higher frequency than usual, was indicative of lone hydrogen on the benzene ring. The NMR spectrum showed only one signal for the magnetically equivalent methoxyl protons (τ 6.22) and only one sharp signal for the two aromatic protons (τ 3.19). Although both 4,5-dichloroveratrole (VII) and 3,6-dichloroveratrole (VI) possess a plane of symmetry that would cause degeneracy of the methoxyl and aromatic proton signals, the relatively small downfield shift (2 Hz) of the methoxyl signal from its chemical shift in veratrole (τ 6.24) suggests that the chlorine atoms are remote from the methoxyl. This condition, together with the IR indication of lone hydrogen, is best met by the structure of 4,5-dichloroveratrole (VII).

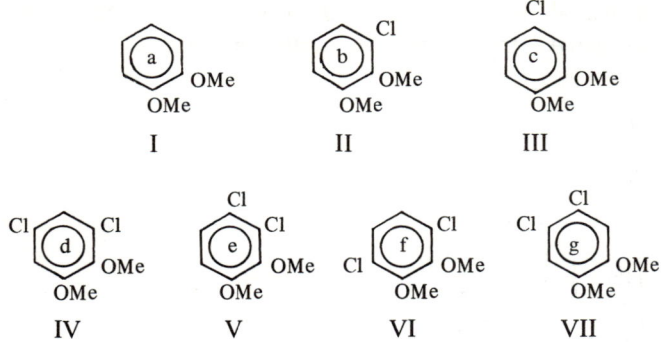

The FT-IR spectrum of peak (d) also shows an absorption band indicative of lone hydrogen on the benzene ring (898 cm^{-1}). Because 3,5-dichloroveratrole (IV) is the only other structure among the four dichloroveratrole isomers to contain lone hydrogen, this structure was assigned to peak (d). Not enough sample was available for NMR analysis.

Peak (f) was isolated in a separate experiment by column chromatography of a portion of the original reaction mixture on silica gel. This substance showed a degeneracy of the methoxyl protons in its NMR spectrum. The strongly deshielded singlet at τ 6.13 also indicated a plane of symmetry as well as *ortho*-substituted chlorine atoms. The structure of 3,6-dichloroveratrole (VI) meets these requirements. By elimination, peak (e), the only remaining dichloro isomer of veratrole, must be 3,4-dichloroveratrole (V).

The mass spectrum of peak (h) shows a molecular ion at m/e 240 and a chlorine isotope pattern characteristic of three chlorine atoms. Although there are two possible isomers of trichloroveratrole, only one (peak h) was found. The FT-IR spectrum of this compound revealed, as expected, a lone hydrogen absorption band at 930 cm^{-1}. The NMR spectrum showed two signals for the methoxyl protons (τ 6.22 and 6.16). These chemical shifts are close to the predicted values (τ 6.20 and 6.16), based on the additivity of substituent effects, for the chemical shifts of the two methoxyls in 3,4,5-trichloroveratrole (VIII). The predicted chemical shifts for the methoxyls in the other possible isomer, 3,5,6-trichloroveratrole (IX), are τ 6.11 and 6.13. Peak (h) was therefore assigned structure (VIII) on the basis of the close correlation of the methoxyl chemical shifts. Peak (h) was shown on the basis of similar retention times on a Carbowax 20M/TPA 50-ft SCOT column to be identical with the trichloroveratrole in the pulp mill wastewater extracts (Peak 5, Figure 33.1 and Peak 2, Figure 33.2). The GC/MS data were also identical.

Confirmation of the structures assigned to the various chlorinated guaiacols, by the arguments described above, must await the synthesis of pure standards corresponding to the various compounds, and the measurement of spectral data pertaining to these standards. Therefore the structures assigned must be considered tentative.

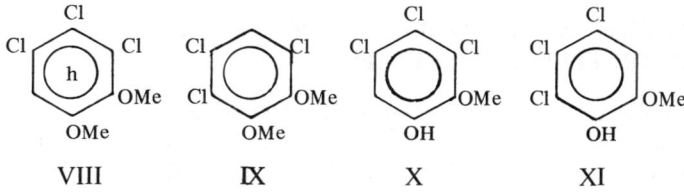

The shoulder observed at spectrum 152 in the scan for m/e 254 in Figure 33.3 indicates coexistence in the pulp mill extracts of two structural isomers of methoxyethoxytrichlorocatachol, and this was further supported by the presence of similar shoulders in LMRGC's of m/e 226 (M - C_2H_4) and m/e 211 (base peak). Furthermore, the mass spectrum observed for the shoulder (spectrum 152) was very similar to that of the more plentiful trichloro isomer (spectrum 147). On the basis of the NMR evidence discussed earlier under peak (h), the 3,5,6-trichloroveratrole system is absent, and the trichloroguaiacol is therefore considered to be a mixture of the 3,4,5-trichloro- (X) and the 4,5,6-trichloro- (XI) isomers. The spectroscopic evidence does not reveal which isomer predominates.

SYNTHESIS OF TETRACHLOROGUAIACOL AND 4,5,6-TRICHLOROGUAIACOL

Tetrachlorocatechol (10 g) was dissolved in ether (100 ml) and reacted with an ethereal solution of diazomethane (90 ml, 0.55 M) at room temperature. After standing in a fume hood overnight the solvent had evaporated, and the reaction product was partially purified by column chromatography on silicic acid (400 x 25 mm) by elution with chloroform. Tetrachloroguaiacol was separated from tetrachloroveratrole by extraction of the mixture in ethereal solution with 1 N sodium hydroxide. Both compounds were recovered pure, as determined by analytical GC. Tetrachloroguaiacol (1.07 g, 10% yield) gave white needles (mp 121.5) from 50% ethanol. The IR, NMR and mass spectra were recorded. Methylation of a sample with diazomethane yielded a product with gas chromatographic and mass spectral characteristics identical with those of peak 6 in Figure 33.1 and of peak 4 in Figure 33.2.

636 ORGANIC POLLUTANTS IN WATER

4,5,6-Trichloroguaiacol was synthesized by the chlorination of guaiacol in acetic acid solution according to a published procedure.[14] White crystals (m 110°) formed from 50% ethanol. The IR, NMR and mass spectra were recorded. Methylation of a sample with diazomethane gave a product identical with peak 5 in Figure 33.1 and peak 2 in Figure 33.2, on the basis of GC retention behavior and mass spectral fragmentation patterns.

QUANTITATIVE ANALYSIS OF CAUSTIC EXTRACTION EFFLUENT SAMPLES

Six samples of caustic extraction effluent were taken on consecutive days at mill B in December 1974. Aliquots of each sample (500 ml) were acidified to pH 4.5 with H_2SO_4 and extracted three times with ether. The combined extracts were washed, dried, evaporated and methylated with diazomethane. Standard solutions of the methyl ethers of 4,5,6-trichloro- and 3,4,5,6-tetrachloroguaiacol were prepared in ether, and quantitative GC analyses conducted on the six unknowns using an EC detector and a 6 ft x 1/4 in. o.d. glass column packed with 10% OV-1 on Gas Chrom G (80-100 mesh) at 150°C. The results are given in Table 33.2.

Table 33.2 Concentrations of Chlorinated Guaiacols in Caustic Extraction Effluents from Mill B

Sample No.	4,5,6-Trichloroguaiacol $\mu g/l$	3,4,5,6-Tetrachloroguaiacol $\mu g/l$
1	400	140
2	480	45
3	380	63
4	400	87
5	630	93
6	240	43

ORIGIN OF CHLORINATED GUAIACOLS

Two possible origins for these compounds are the chlorination of guaiacol present in evaporator condensate and used as washwater in the mill or the oxidative chlorination of residual lignin present in the brownstock. We failed to find any trace of guaiacol in a sample of brownstock taken at mill A. Evidence in favor of the origin from residual lignin is that the chlorinated guaiacols have now been found in effluent samples from three mills, widely separated geographically and using different species in their wood supply. Others have recently reported similar findings[8] involving different mills. Moreover, the various monochloro- and dichloroguaiacols and the dimeric reaction products observed when guaiacol was chlorinated *in vitro* were not found in bleach plant effluent samples.* Conversely the tetrachloroguaiacol found in the effluent was not formed in the laboratory chlorination experiment. This latter compound was isolated by Dence and Sarkanen[15] from the chlorination of spruce lignin. Lastly, sodium fusion tests and IR data indicated organochlorine functions in the lignin of mill A.

POTENTIAL SIGNIFICANCE OF RESULTS

Concentrations of trichloroguaiacol of up to 57% of the LC50 value and tetrachloroguaiacol levels of up to 32% of the LC50 value were measured simultaneously in caustic extraction effluent samples. It is unlikely that the toxicity associated with these compounds could explain more than a fraction of the overall toxicity of the caustic extraction effluent samples, some of which were killing fish at 10% concentration (see Table 33.1). Of the other compounds identified in Figures 33.1 and 33.2, only the resin acids, present in very low concentrations in mill B effluent, are known toxicants. Saturated fatty acids have been found to be nonacutely toxic to salmon.[16] It is therefore highly likely that other chemical species, not detected in this study, must contribute a large proportion of the toxic effect. These compounds are probably chlorinated lignin residues with molecular weights too high to be analyzed by gas chromatography. Such a fraction was removed during the isolation of the extract from mill A after acidification with sulfuric acid. Work on higher-molecular-weight chlorinated species is being conducted elsewhere.

It must be remembered that caustic extraction effluents account for only 10-20% by volume of the flow in the main mill sewer. Other known toxicants such as resin acids, neutral diterpenes and mill additives are present in other sewers. Low concentrations of various chlorinated species of monomeric or oligomeric structure are also present. The gross toxic effect of all these compounds on biological systems is unpredictable because the sum of the toxicities of the individual compounds may be additive, synergistic or antagonistic and the nature of these interactions is unknown.

---

*Brownlee and Strachan report finding a dichloroveratrole isomer at 7 µg/l in an Ontario mill effluent; see Chapter 35 (Ed. note).

Probably the most important consideration concerns the degree of biodegradability of these compounds, since most mills discharging to inland waters in British Columbia are equipped with aerated stabilization basins. We have analyzed a sample of whole kraft effluent discharged by mill C after biological treatment in a 5-day lagoon system. The sample was acutely toxic to fish at 65% concentration and contained no detectable resin acids. Trichloroguaiacol and tetrachloroguaiacol were found at concentrations of 98 and 70 µg/l, respectively. Over 100 compounds were detected in the GC/MS analysis and many of these were chlorinated species. Obviously, we must be concerned to find out how effectively these compounds are destroyed by biological oxidation, and what their effects are on various components of the biota in the receiving waters. A process change which could conceivably reduce the release of organochlorine compounds is the replacement of part of the chlorine in the first stage of bleaching by chlorine dioxide, provided some operational problems can be overcome. Ultimately, if oxygen bleaching is adopted in the industry, generation of such substances can be prevented. Installation of a closed bleach plant system, which eliminates discharge of these substances, would achieve the same result.

ACKNOWLEDGMENTS

Thanks are due to Professor K. Sarkanen for the gift of tetrachlorocatechol and to Professor C. Dence for the gift of samples of tetrachloroguaiacol and 4,5,6-trichloroguaiacol; to Dr. J. Servizi and Mr. R. Gordon of the International Pacific Salmon Fisheries Commission for bioassays; to Dr. L. Azarraga, Mrs. A. McCall, Messrs. M. Carter, T. Floyd and Mrs. A. Alford for obtaining the GC/MS and FT-IR spectra; to Dr. R. Bose for NMR and mass spectra; and to various colleagues for helpful discussions.

REFERENCES

1. Statistics Canada. Canadian Forestry Statistics. Catalog 25-202, Ottawa, Ontario (1972).
2. Thomas, P. CPAR Rep. 360-1, Can. Forest Serv., Ottawa, Ontario (1975).
3. Howard, T. E. and C. C. Walden. *Pulp Pap. Mag. Canada* **72**, T3 (1971).
4. Das, B. S., S. G. Reid, J. L. Betts, and K. Patrick. *J. Fish. Res. Bd. Canada* **26**, 3055 (1969).
5. Servizi, J. A., R. W. Gordon and D. W. Martens. Int. Pacific Salmon Fish. Comm. Prog. Rep. 17 (1968), 43 pp.

6. Ota, M., W. B. Durst and C. W. Dence. *Tappi* **56**, 139 (1973).
7. Rogers, I. H. and L. H. Keith. Tech. Rep. 465, Environment Canada, Fish and Marine Serv., Ottawa, Ontario (1974), 21 pp.
8. Leach, J. M. and A. N. Thakore. *J. Fish. Res. Bd. Canada* **32**, 1249 (1975).
9. Gordon, R. W. and J. A. Servizi. Int. Pacific Salmon Fish. Comm. Prog. Rep. 31 (1974), 26 pp.
10. Odham, G. and E. Stenhagen. In *Biochemical Applications of Mass Spectrometry*, G. Waller, Ed. (New York: Wiley-Interscience, 1972), Chapter 9.
11. Ryhage, R. and E. Stenhagen. In *Mass Spectrometry of Organic Ions*, F. W. McLafferty, Ed. (New York: Academic Press, 1963), Chapter 9.
12. Rogers, I. H. *Pulp Pap. Mag. Canada* **74**, T303 (1973).
13. Keith, L. H. U.S. Environmental Protection Agency Publication Number EPA-660/4-75-005 (Washington, D.C.: U.S. Government Printing Office, 1975).
14. Fort, R., J. Sleziona, and L. Denivelle. *Bull. Soc. Chim. France* **810** (1955).
15. Dence, C. and K. Sarkanen. *Tappi* **43**, 87 (1960).
16. Leach, J. M. and A. N. Thakore. *J. Fish. Res. Bd. Canada* **30**, 479 (1973).

NOTE:

We have recently confirmed the structure of the synthetic 4,5,6-trichloroguaiacol, and hence of the 3,4,5-trichloroveratrole, by the measurement of the carbon-13 NMR spectrum of the acetate derivative and the nuclear Overhauser effect in the proton NMR spectrum.

CHAPTER 34

FATE OF SELECTED ORGANIC COMPOUNDS IN THE DISCHARGE OF KRAFT PAPER MILLS INTO LAKE SUPERIOR

Michael E. Fox

    Canada Centre for Inland Waters
    Burlington, Ontario, Canada

INTRODUCTION

In 1973, studies on pulp mill wastewater discharges on Lake Superior were initiated by the International Joint Commission, Upper Lakes Reference Group. This group had the task of reporting to the International Joint Commission (IJC) on the impact of point source effluent discharges on the basically high-quality receiving waters of the Upper Great Lakes (Huron and Superior). Apart from domestic sewage, pulp and paper mill effluents constitute the major polluting discharges into the Upper Great Lakes. Pulp and paper mill effluents have been implicated in oxygen depletion, toxicity and production of taste and odor problems in the receiving water and its aquatic communities.[1,2]

The studies initiated were wide-ranging in scope and included many projects in the chemical, physical and biological fields. This chapter will report on the findings of one of those projects. Its object was to determine the fate and effective zone of persistence of dissolved organic compounds in the effluent plume.

FEASIBILITY STUDIES

A preliminary study, to determine the feasibility of unproven laboratory and field techniques, was carried out in 1973 at Marathon, Ontario

(Figure 34.1) where a kraft bleached pulp mill discharges 90,000 m^3/day of clarified effluent into Lake Superior.

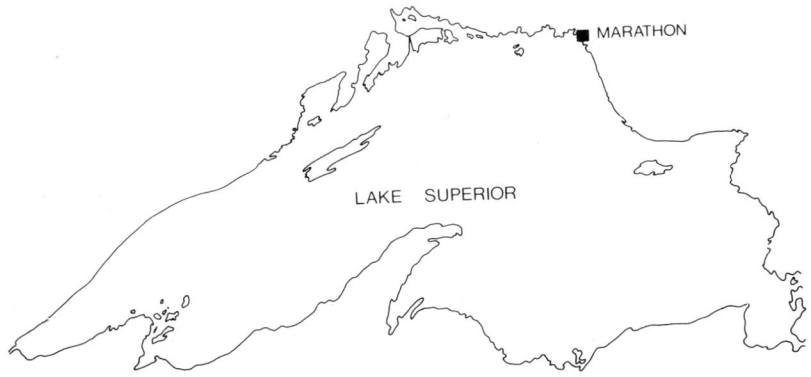

Figure 34.1 Site of 1973 studies, Marathon, Ontario.

The results of this study led to three important changes which were incorporated into the main field study at Red Rock, Ontario in 1974. Choice of study site was the most significant change. The Marathon field studies showed the site to be one of high nearshore wave energy due to its location on the open lake facing the prevailing winds. The result of this alignment was to make the mixing zone of the effluent and receiving water occur virtually at the point of discharge on a majority of occasions. The Red Rock site (Figure 34.2) where a bleached kraft and ground wood paper mill discharges 96,500 m^3/day of clarified effluent into Nipigon Bay has a much lower nearshore wave energy due to the protection afforded by offshore islands and high bluffs to the west.

At the Marathon site in 1973, drogues were used to define the movement of the effluent plume in the receiving water. This technique proved to be unsatisfactory due to low visibility, grounding by onshore winds and the nature of the plume as a thin surface layer of lower specific gravity than the underlying, colder receiving water. At Red Rock in 1974 drogues were replaced by incremental additions of Rhodamine WT, a red dye possessing high visibility at very low concentrations. Unlike Rhodamine T, Rhodamine WT is not strongly adsorbed onto particulate materials, an important characteristic for tracking pulp mill effluents.

Extraction procedures for the recovery of dissolved organic compounds used at Marathon in 1973 were based on modifications of the macroreticular resin technique used by Rogers and others.[2,3] Elution was effected by the use of acidified methanol and very good recoveries were achieved

IDENTIFICATION OF ORGANIC POLLUTANTS 643

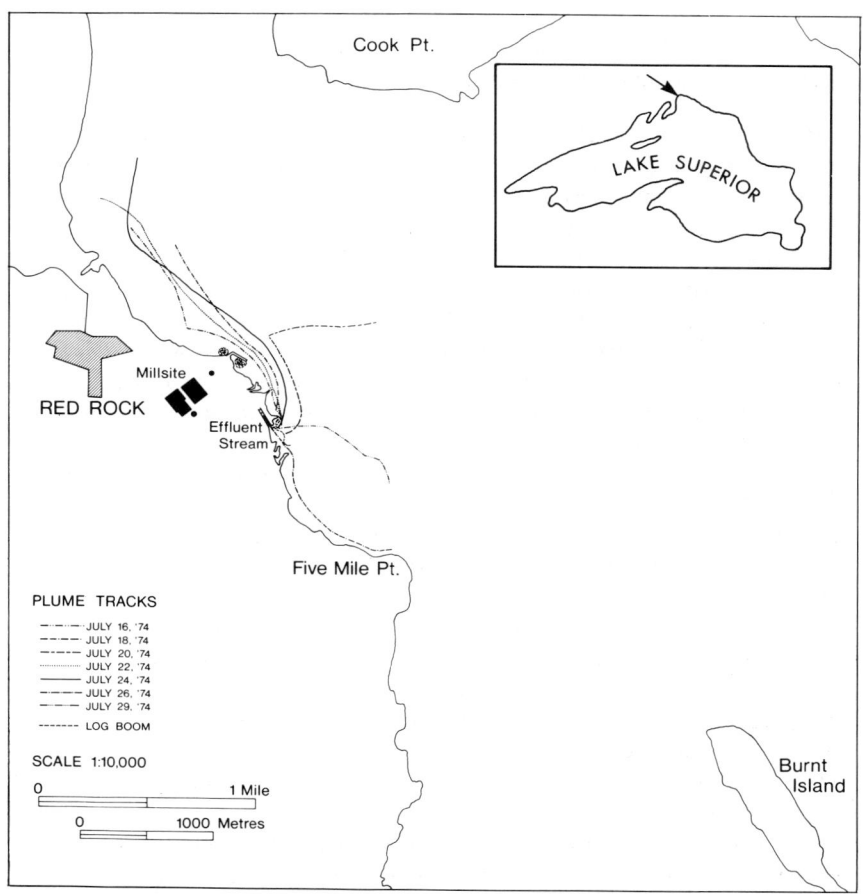

Figure 34.2 Site of 1974 studies, Red Rock, Ontario.

based on comparison with dissolved organic carbon measurements. However, when extracts eluted with methanol/HCl were analyzed by gas chromatography it was found that much of the recovered organic material was of very low volatility and interfered strongly with the gas chromatographic separations. The interfering compounds were presumed to be high-molecular-weight (> 500) lignin fragments and possibly wood sugars. Such compounds were of less interest from a toxicity and tainting point of view and were excluded from the extracts in the 1974 Red Rock Study by replacing the methanol/HCl eluant with diethyl ether. This eluant allowed good recoveries of the lower-molecular-weight compounds of interest.

Tracking and Sampling the Effluent Plume

Rhodamine WT dye dissolved in acetic acid was mixed with sufficient methanol to give a specific gravity of 1. Approximately 1 liter of the resulting dye solution was added to the Red Rock Mill effluent channel at a point immediately prior to its discharge into Nipigon Bay. After the elapse of a suitable period of time (usually 2-3 hr) the center of the dye marker patch was located using a small boat to make two or more transects of the dye patch. A continuous flow fluorometer was used to detect the point of maximum concentration of Rhodamine WT. The initial location of the dye patch was greatly facilitated by radio contact of the sampling personnel with a small aircraft engaged in aerial surveillance of the effluent plume.

The coordinates of the point were recorded, water samples were taken and a further discrete addition of dye was made to form the basis of the next sampling location further along the effluent plume. In this manner, the effluent plume was "tagged" and sampled four to six times during the course of a working day.

Since the effluent is approximately $15°C$ warmer than the receiving water it has a significantly lower specific gravity. Thus, at most times during the study period, the effluent plume and dye marker patch were largely concentrated within the top meter of the water column. Therefore all water samples and dye concentration measurements were taken at 0.5 m or by an integrated sample over the plume depth.

Effluent plume tracking and sampling, as described above, was carried out on eight different occasions at two-day intervals in July 1974. Water samples collected for extraction of dissolved organic compounds ranged in volume from 0.5 liters at the effluent discharge point and 3 liters in the plume at > 1000 m to 4.5 liters in "clean" water reference samples and river water samples. All samples were collected in clean amber glass bottles with teflon-lined caps and taken immediately to the shipboard laboratory where they were acidified to pH 2.0 with concentrated HCl.

On three of the plume tracking days, reference water samples were taken at 0.5 m in an area of the bay outside of the visible plume. On one occasion a water sample was taken near Vert Island at a distance of 20 km from the mill. This sample was considered to represent a background profile for dissolved organic compounds in Nipigon Bay.

Separate sampling and direct measurements from other vessels on the plume tracking days provided background information on chemical and physical parameters, both within and outside the plume area. A continuous record of conductivity at the effluent discharge point provided a qualitative measure of effluent composition and pinpointed unusual events such as spills.

Samples from the Nipigon River, which comprises the principal source of loading to Nipigon Bay, other than the Red Rock Mill, were collected from upstream and downstream locations on three occasions.

Surface sediment samples from the study area had been collected in an earlier phase of the study.

EXPERIMENTAL

Extraction of Organic Compounds from Water

The procedure used to recover dissolved organic compounds from water samples was essentially similar to that of Junk et al.[4] using XAD-2 macroreticular resin beads and will not be described here except where differences in procedure occurred.

The XAD-2 resin was purified by washing with distilled water followed by a double Soxhlet extraction with residue-free acetone. The purified resin was stored in amber glass bottles under "organic free" water.

The acidified water samples were passed by gravity flow at approximately 10 ml/min through 10 cm long x 2 cm diameter resin beds in glass tubes.

The loaded columns were eluted with residue-free diethyl ether, evaporated down to 1 ml and stored in glass ampoules after the removal of a 1- to 5-μl sample for preliminary gas chromatographic analysis. The samples were refrigerated for more detailed GC and GC/MS analyses at the headquarters laboratory (C.C.I.W.).

Extraction of Organic Compounds from Sediments

Wet surface sediments (50 g), which had been stored in refrigerated glass jars, were Soxhlet extracted with residue-free acetone. The acetone/water solution was evaporated to a wet slurry which was extracted with residue-free diethyl ether. The ether solution was evaporated to 1 ml and stored in refrigerated glass ampoules for further analysis.

Efficiency of Extraction Procedures

Recovery tests on the XAD-2 columns were performed using pure compounds to represent the classes of compounds found in the effluent, e.g., monoterpenes, diterpene resin acids, phenols, fatty acids. These tests indicated good recoveries ranging from 80-100%. In some cases, recoveries from the highly concentrated effluent discharge samples were believed to be low. This was thought to be due to precipitation of low-solubility free acids upon acidification rather than poor column recoveries. In any event, the recoveries obtained from the effluent plume samples were not affected by this solubility effect since the concentrations of individual compounds were relatively low.

Preparation of Extracts for GC and GC/MS Analysis

Water and sediment extracts were each divided into two aliquots. The first aliquot, to be used for direct GC, GC/MS analysis, was made up to a precise volume of 50-800 µl with acetone after evaporating the diethyl ether. The second aliquot was methylated with diazomethane and the final volume adjusted as above. The choice of final volume for each extract solution was based on the preliminary gas chromatogram run in the field.

GC Analysis of Sample Extracts

Gas chromatograms of both the methylated and unmethylated sample extracts were run on a Hewlett Packard 5700A Gas Chromatograph equipped with single flame ionization detector and a 1.8-m stainless steel column packed with 3% OV 1 on Chromosorb W-HP. The oven was programmed from 80°C to 270°C with a 4-min initial isothermal hold.

GC/MS Analysis of Sample Extracts

A limited, but representative, number of the methylated and unmethylated sample extracts were run on a Finnigan 1015 C Gas Chromatograph/Mass Spectrometer with computerized data processing. The mass spectra were produced by electron impact at 70 electron volts. In addition to mass spectra of individual compounds, the computerized data system was used to produce "Reconstructed Gas Chromatograms" from the total ion current integrated throughout the GC/MS run and "Limited Mass Range Chromatograms" from the ion current of a single selected mass or set of masses integrated throughout the GC/MS run. The "Reconstructed Gas Chromatograms" are essentially similar to a Flame Ionization Gas Chromatogram.

Identification of Components

The choice of column and temperatures allowed reasonably good separation of the nonacidic components (Figure 34.3) in the unmethylated

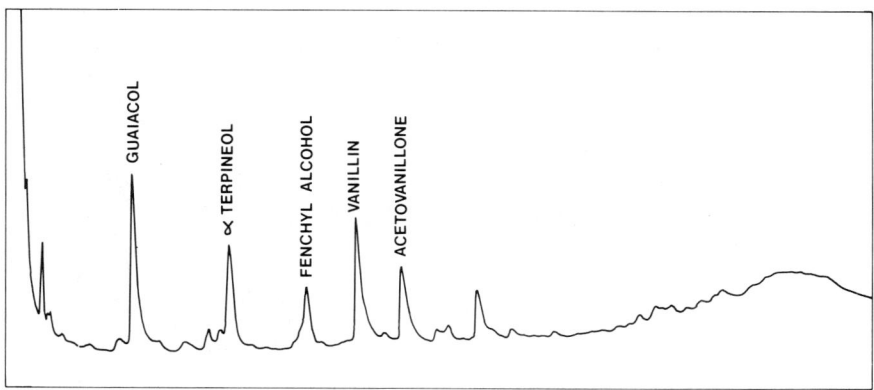

Figure 34.3 Chromatogram of extract of effluent at discharge, July 22, 1974 (unmethylated).

extracts, but was rather lacking in resolution in the methylated extracts (Figure 34.4). In particular the fatty acid methyl esters and diterpene resin acid methyl esters in the center of the chromatogram were poorly resolved. This compromise in resolution avoided the necessity of the complex separation techniques utilized by other workers in the field[2];[3] with the attendant risk of loss of trace components and could be largely overcome by the use of limited mass range chromatograms (Figure 34.5). The figure shows the sharp delineation of three typical classes of compounds in the effluent by the use of the computer to plot suitably

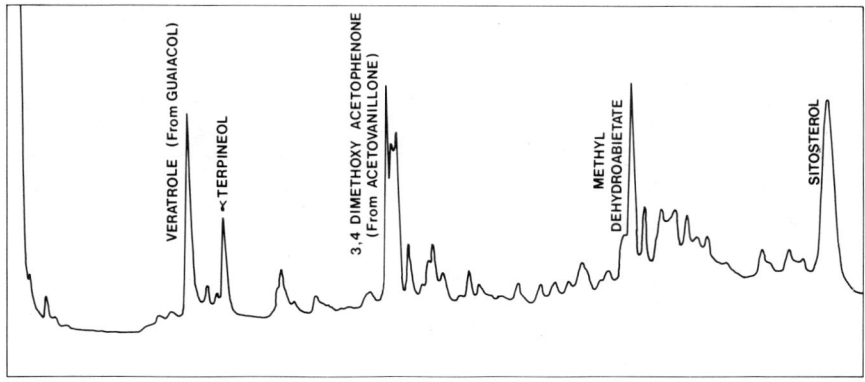

Figure 34.4 Chromatogram of extract of effluent at discharge, July 22, 1974 (methylated).

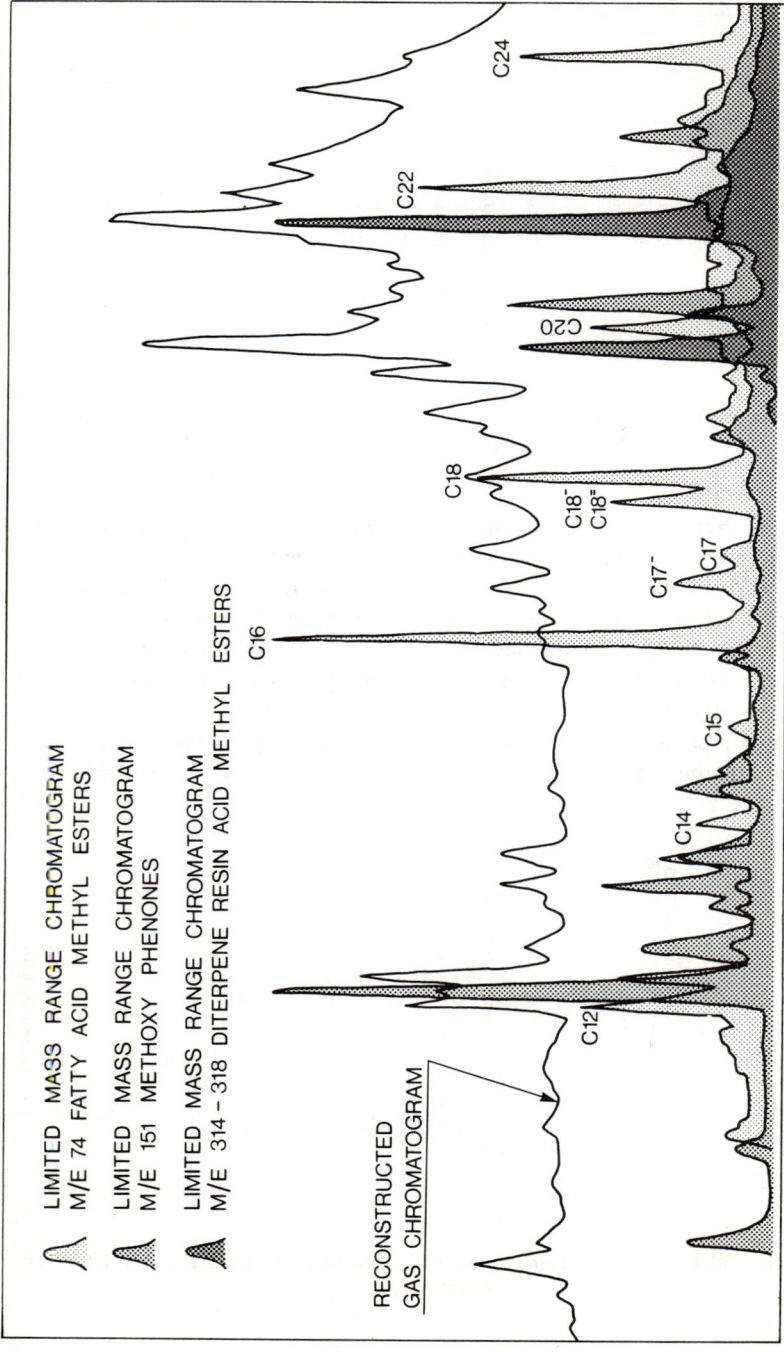

Figure 34.5 Limited mass range chromatogram of extract of effluent at discharge, July 16, 1974 (methylated).

distinctive limited mass range chromatograms. Other appropriate mass ranges such as m/e 149 for phthalic acid esters were left off the figure for the sake of clarity. Limited mass range chromatograms were used for peak identification only. Quantitation of components utilized the flame ionization chromatograms. Identification of individual mass spectra was achieved mainly through the use of the computerized mass spectral library search system of EPA-Battelle Memorial Institute.*

RESULTS

On the six plume tracking days for which data are summarized, the distance travelled by the dye marker in the effluent plume varied from 1450 to 2600 m over periods of 6.5 to 9 hr. The average distance was 2000 m in 8 hr or a velocity of 7 cm/sec. The predominant plume direction during the study period was to the northwest, parallel to the shore as can be seen in Figure 34.2. Since the plume was largely a surface structure (rarely extending below 1 meter), wind was the principal driving force.

From the 29 samples collected from the effluent discharge and effluent plume, some 85 compounds were observed, of which 35 (including all of the major ones) were identified. At distances in excess of 2000 m in the plume some 50 compounds were observable at low concentrations (Table 34.1). Many of these compounds comprise part of the natural dissolved organics of Nipigon Bay as can be seen from Table 34.2.

The dissolved organic compounds found in the Nipigon River, which constitutes the major water loading to Nipigon Bay, are essentially similar, comprising mostly fatty acids (Table 34.3).

During the course of the study, it was soon found that the effluent discharge had a characteristic dissolved organic profile. That is, the same compounds were almost always present and their relative proportions remained quite similar. Because of this observation and the complexity of the spectrum of compounds, it was decided to choose five of the eight compounds which consistently comprised the major components of the mixture for more detailed consideration. The five compounds were:

Guaiacol—a substituted phenol arising from the breakdown of lignin and one of the components known to undergo chlorination in the mill process.

α-Terpineol—a monoterpene alcohol possessing a strong taste and odor, a possible fish tainting candidate.

Acetovanillone—also a substituted phenol arising from the breakdown of lignin, a possible fish tainting candidate.

---

*Now incorporated into the **Cyphernetics MSSS system**, Ann Arbor, Michigan.

Table 34.1 List of Dissolved Organic Compounds Identified in the Red Rock Mill Effluent and Effluent Plume

Name of Compound	Major Component	Minor Component	Detected in Effluent	Detected in Effluent Plume
β-Pinene		+	+	−
Camphene		+	+	−
Guaiacol	+		+	+
Borneol		+	+	+
Fenchyl alcohol	+		+	+
Safrole[a]		+	+	+
α-Terpineol	+		+	+
Phenol		+	+	−
2-Methoxy-4-propyl phenol		+	+	−
Vanillin		+	+	−
Acetovanillone	+		+	+
3,4,5-Trichloroguaiacol		+	+	−
4-Hydroxy-3-methoxy propiophenone[a]	+		+	+
4-Hydroxy-3-methoxy phenyl acetic acid		+	+	−
Epijuvabione		+	+	−
Myristic acid		+	+	+
Pentadecanoic acid		+	+	−
Palmitic acid	+		+	+
Heptadecanoic acid		+	+	−
Stearic acid	+		+	+
Arachidic acid	+		+	+
Behenic acid	+		+	+
Lignoceric acid	+		+	+
Palmitoleic acid		+	+	+
Oleic acid	+		+	+
Linoleic acid	+		+	+
Isopimaric acid	+		+	+
Sandaracopimaric acid		+	+	+
Dehydroabietic acid	+		+	+
Abietic acid	+		+	+
6,8,11,13-Abietate-traen-18-oic acid		+	+	+
Dehydrodiconiferyl alcohol[a]		+	+	−
Sitosterol	+		+	+
Dimethyl phthalate		+	+	+
Dioctyl phthalate		+	+	+

[a]Tentative identification.

Table 34.2 List of Dissolved Organic Compounds Found Off Vert Island at 1 m (Nipigon Bay Background Profile for Dissolved Organic Compounds)

Name of Compound	Approx. Concentration µg/l
Myristic acid	5
Pentadecanoic acid	5
Palmitoleic acid	20
Palmitic acid	30
Heptadecanoic acid	20
Linoleic acid	1
Oleic acid	20
Stearic acid	10
Methyl 2-methyloctadecanoic acid	1
N-tetracosane	10
Arachidic acid	1
Dioctyl phthalate	10
Dibutyl phthalate	1
Dehydroabietic acid	<1

Table 34.3 List of Dissolved Organic Compounds Found in Nipigon River (Principal Source of Loading to Nipigon Bay Other than the Red Rock Mill)

Name of Compound	Approx. Concentration µg/l
Lauric acid	5
Myristic acid	10
Pentadecanoic acid	5
Palmitoleic acid	15
Palmitic acid	40
Heptadecanoic acid	2
Oleic acid	30
Stearic acid	20
Dehydroabietic acid	5
Arachidic acid	1
Lignoceric acid	2
Dioctyl phthalate	15
Dihexyl adipate	2

Dehydroabietic Acid—a diterpene resin acid known to be acutely toxic to fish at less than 2 ppm.[2]

Sitosterol—a phytosterol occurring a high concentration in the effluent and moderately persistent in the plume.

The concentrations of the five selected compounds in the effluent plume can be seen by referring to Tables 34.4 through 34.9. The compounds are listed in order of decreasing volatility and increasing molecular weight. The dissolved organic carbon figures give an estimate of overall disappearance of dissolved organics while the Na⁺ figures represent the disappearance of a conservative species, showing dilution effects only.

Table 34.4 Concentrations of Selected Organic Compounds, Dissolved Organic Carbon, and Sodium in the Effluent Plume
Dye Run #1, July 16, 1974

Distance from Source m	D.O.C. mg/l	Na^+ mg/l	Guaiacol µg/l	α-Terpineol µg/l	Aceto-vanillone µg/l	Dehydro-abietic Acid µg/l	Sitosterol µg/l
0	52.0	48.0	385	2000	380	1120	760
400	24.0	12.6	170	520	100	480	260
1200	9.0	3.6	25	90	18	115	60
1500	8.0	3.4	12	35	1	40	18
2000	NA[b]	3.3	3	25	<1	30	6
2000[a]	5.5	1.6	ND[c]	ND	ND	14	4

[a]Reference sample—outside plume area.
[b]Not analyzed.
[c]Not detected.

Table 34.5 Concentrations of Selected Organic Compounds, Dissolved Organic Carbon and Sodium in the Effluent Plume
Dye Run #2, July 18, 1974

Distance from Source m	D.O.C. mg/l	Na^+ mg/l	Guaiacol µg/l	α-Terpineol µg/l	Aceto-vanillone µg/l	Dehydro-abietic Acid µg/l	Sitosterol µg/l
0	72.0	76.0	144	805	425	1180	520
1000	14.0	13.3	18	120	55	155	85
1350	11.0	8.2	13	110	45	135	75
1450	17	14.3	30	220	130	380	150

Table 34.6 Concentrations of Selected Organic Compounds, Dissolved Organic Carbon and Sodium in the Effluent Plume

Dye Run #3, July 20, 1974

Distance from Source m	D.O.C. mg/l	Na$^+$ mg/l	Guaiacol µg/l	α-Terpineol µg/l	Aceto-vanillone µg/l	Dehydro-abietic Acid µg/l	Sitosterol µg/l
0	157.0	140.0	2480	2700	2260	5000	1020
150	10.0	5.5	25	50	9	290	170
1050	9.5	3.5	2	12	<1	100	18
1600	9.0	2.5	ND[a]	ND	ND	12	2
1900	6.5	1.7	ND	ND	ND	12	3

[a]Not detected.

Table 34.7 Concentrations of Selected Organic Compounds, Dissolved Organic Carbon and Sodium in the Effluent Plume

Dye Run #4, July 22, 1974

Distance from Source m	D.O.C. mg/l	Na$^+$ mg/l	Guaiacol µg/l	α-Terpineol µg/l	Aceto-vanillone µg/l	Dehydro-abietic Acid µg/l	Sitosterol µg/l
0	182.0	160.0	1145	600	1265	1225	1090
300	47.0	35.0	740	350	570	1930	480
1400	9.5	3.6	18	14	5	140	35
1900	8.0	3.4	8	7	1	60	18
2300	7.0	3.3	2	2	ND[b]	35	6
1700[a]	5.5	1.5	ND	ND	ND	18	ND

[a]Reference sample—outside plume area.
[b]Not detected.

Table 34.8 Concentrations of Selected Organic Compounds, Dissolved Organic Carbon and Sodium in the Effluent Plume

Dye Run #5, July 24, 1974

Distance from Source m	D.O.C. mg/l	Na$^+$ mg/l	Guaiacol µg/l	α-Terpineol µg/l	Aceto-vanillone µg/l	Dehydro-abietic Acid µg/l	Sitosterol µg/l
0	128.0	100.0	7040	7895	5065	9065	2400
100	65.0	53.0	595	430	510	545	520
750	14.0	8.8	60	50	60	295	85
1300	7.0	3.4	4	1	18	120	30
1900	6.5	2.7	<1	<1	1	25	5
2600	6.0	2.4	ND[b]	ND	<1	40	1
1650[a]	5.5	2.0	ND	ND	ND	35	ND

[a]Reference sample—outside plume area.
[b]Not detected.

Table 34.9 Concentrations of Selected Organic Compounds, Dissolved Organic Carbon and Sodium in the Effluent Plume

Dye Run #6, July 26, 1974

Distance from Source m	D.O.C. mg/l	Na$^+$ mg/l	Guaiacol µg/l	α-Terpineol µg/l	Aceto-vanillone µg/l	Dehydro-abietic Acid µg/l	Sitosterol µg/l
0	67.0	113.0	3400	1540	1820	4500	500
100	17.5	17.7	ND	55	25	325	65
700	7.0	3.4	12	4	2	18	6
1650	6.0	2.8	ND[a]	ND	ND	12	ND

[a]Not detected.

DISCUSSION

Guaiacol, α-Terpineol and Acetovanillone dropped to almost undetectable concentrations when Na$^+$ was < 3 µg/l, i.e., about twice the background level of 1.5 µg/l. These compounds were not detected in the reference samples taken on the same day outside of the plume at comparable distances.

They do not build up to a background level in Nipigon Bay and are not introduced via sources other than pulp and paper mills. These compounds may be removed from the plume by the processes of bacterial degradation and volatilization into the air in addition to simple dilution with the receiving water. If the figure of 100 ppb quoted for Guaiacol,[5] being the highest level not to cause fish tainting is considered to represent all three compounds and be conservatively halved to 50 ppb, then the effective zone of impact would be < 1000 m except in the case of dye run number 2 (Table 34.5). In this case the effluent plume maintained a narrow cross section close to the shoreline and little dilution with the receiving water occurred, as can be seen from the D.O.C. and Na^+ concentrations. Under these conditions a nearshore zone of impact stretching 1500 m to 5 miles point would exist.

Of the two remaining compounds, Sitosterol will be considered first as an interesting indicator of the plume chemistry, but probably of little importance from a toxicity or tainting point of view. This compound is a plant sterol of relatively high molecular weight and low volatility. Its concentration falls off somewhat more slowly than the more volatile compounds considered and is presumed to undergo bacterial degradation and possibly accretion onto particular material in addition to dilution with the receiving water. In areas outside the plume, it is generally below the detection limits of this study. Sitosterol has not previously been implicated in toxicity or tainting studies.

Dehydroabietic acid, as can be readily seen from Tables 34.4 through 34.9 and Figure 34.6, is the single major effluent component which shows a tendency to reach an equilibrium concentration in the receiving water at significant distances from the source. Dehydroabietic acid is of intermediate molecular weight and low volatility. It probably exists largely in the ionized form in the plume environment.

Disappearance due to bacterial degradation appears to be quite slow and dilution appears to be the most significant removal mechanism. A background concentration of ~ 15 µg/l is reached both in and outside of the plume area at a distance of ~ 2000 m from the effluent discharge. This background level is not maintained at very large distances from the discharge as can be seen from the trace level observed near Vert Island approximately 20 km from the discharge (Table 34.2). The data of other workers on the Red Rock project may establish the outer limits of the zone.[6] No information is available regarding possible sublethal or tainting effects at this relatively low level, although presently incomplete work on fish tainting with a range of dilutions of the Red Rock effluent may establish this. Dehydroabietic was also detected in the Nipigon River (Table 34.3) at up to 15 µg/l and averaging 5 µg/l. This river was used

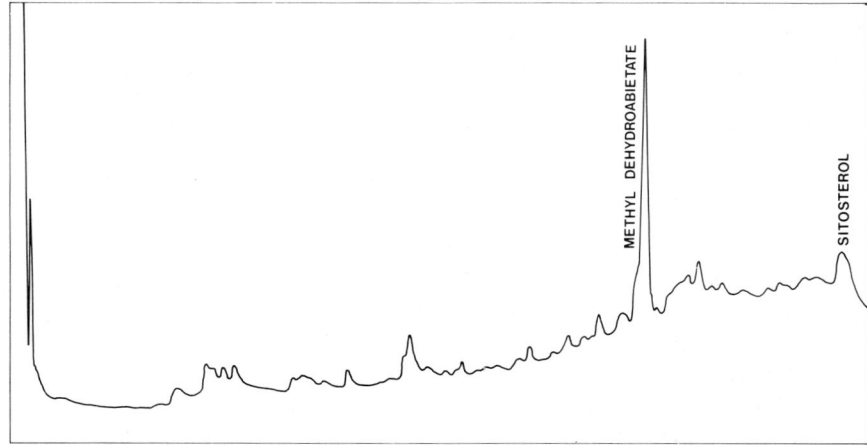

Figure 34.6 Chromatogram of extract of effluent plume 2000 m from discharge, July 22, 1974 (methylated).

for logging drives to the mill for many years until quite recently, and no doubt many logs lie on the bottom of Lake Helen on the lower part of the river. Leaching of dehydroabietic acid from the sunken logs may account for the levels observed in the river. This is supported by the fact that no dehydroabietic acid was detected in the Jackfish River nearby, a small river not used for log drives.

If the flow of the Nipigon River is considered to be approximately 27,000 m^3/min and that of the Red Rock Mill effluent 66 m^3/min, then the loading factor from the Nipigon River at 5 µg/l dehydroabietic acid is more than one half that of the mill effluent at 3.7 mg/l.* Thus the Nipigon River is a significant source of dehydroabietic acid into Nipigon Bay.

Since dehydroabietic acid appeared to be moderately persistent in the water column, analyses were made on a series of surface sediment samples from Nipigon Bay in the Red Rock area to determine if accumulation in the sediment was observed. The results showed a considerable accumulation of dehydroabietic acid in the sediments with a clearly defined distribution of high concentration from the effluent discharge running south and southeast along the deep channel toward Nipigon Strait (Figure 34.7). Levels of dehydroabietic acid in the sediment near the discharge are almost 100 times the background levels in Nipigon Bay, and in the deep

---

*Average value for dye runs 1-6.

IDENTIFICATION OF ORGANIC POLLUTANTS 657

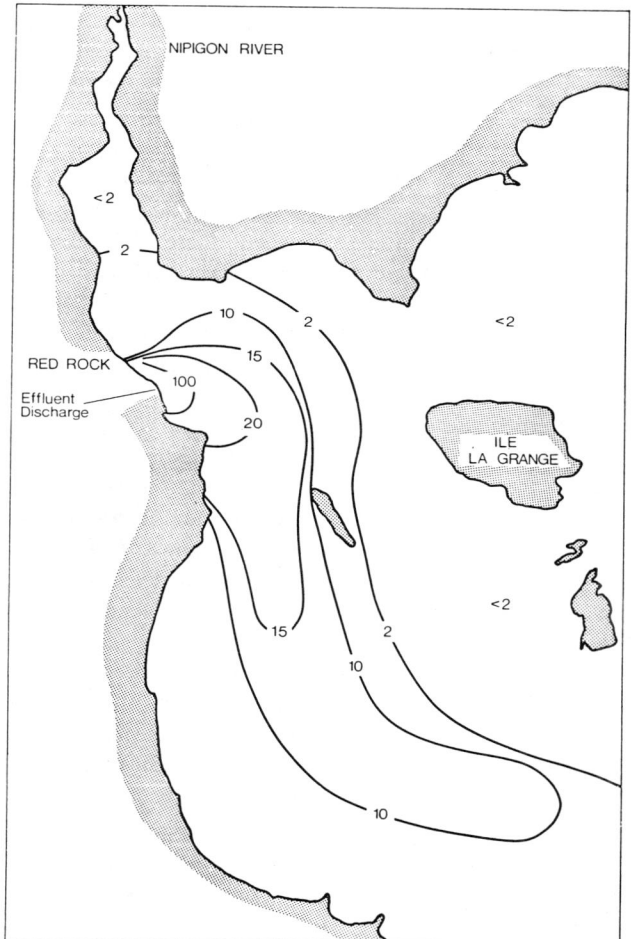

Figure 34.7 Dehydroabietic acid in surface sediment, µg/g dry weight.

channel 15 km from the discharge are almost 10 times the background level. Interestingly, the levels in the sediment near the mouth of the Nipigon River are close to those of the Nipigon Bay background levels of less than 2 µg/g. Thus the sediment loading clearly originates from the mill effluent and not from the Nipigon River.

Although a wide range of fatty acids were detected in the effluent and effluent plume, some of which are known to be toxic to fish at the low ppm level, they have not been considered in this discussion. This is because, at the extreme dilutions encountered in the plume, the levels of these fatty acids are not readily distinguishable from those of the background water samples (Tables 34.2 and 34.3). The natural production of these acids in both lakes and rivers is felt to be such that the greatly diluted effluent beyond a few hundred meters does not exert a significant effect.

Other materials noted in the effluent plume, receiving waters and tributary rivers (Tables 34.1, 34.2 and 34.3) were phthalic acid esters. No satisfactory interpretation for the presence of these ubiquitous materials is offered. Certainly phthalic acid is a component of pulp mill effluents and would account for the dimethyl phthalate observed. This source would not, however, account for the dioctyl phthalate which was only observed in the mill effluent on one occasion, albeit at a very high concentration. The presence of dioctyl phthalate in the Nipigon River and the sample off Vert Island could be accounted for by uptake from the atmosphere as has been suggested by many researchers, or by laboratory contamination of the samples. The latter explanation is not considered to be likely in view of the precautions taken and the production of blanks free from this component. Other compounds identified but not quantified, such as monoterpene hydrocarbons and trichloroguaiacol, might be expected to cause fish tainting and toxicity problems at much higher concentrations than those observed.[7]

A high percentage of the total dissolved organic carbon present in the effluent occurred as compounds not observed in this study due to the selective nature of the extraction and analysis. These compounds included high-molecular-weight and polymeric compounds such as large lignin breakdown fragments, wood sugars, starch, pectin and some very low-molecular-weight compounds such as organosulfur compounds and methanol. With the exception of the organosulfur compounds, these compounds are not thought to have significant detrimental effects on the biota of the receiving waters.

CONCLUSIONS

Of the 35 identified organic compounds which are dispersed by the effluent plume into Nipigon Bay, only dehydroabietic acid exhibits a degree of persistence in the water column. It is not known whether the concentrations detected could give rise to sublethal effects on the biota such as the fish tainting which has occurred in Nipigon Bay. Dissipation of organic compounds by dilution with the receiving water appears to be the principal short-term mechanism.

Surface sediment samples show a pattern of elevated dehydroabietic acid levels which clearly originate from the effluent discharge. The movement of this material can be seen to follow the deep channel toward Nipigon Strait and was measured at almost 10 times the Nipigon Bay background level at a distance of 15 km from the effluent discharge.

ACKNOWLEDGMENTS

The author would like to thank Dr. W. M. J. Strachan and Dr. B. Brownlee for much helpful discussion and Mr. R. G. Sandilands for providing surface sediment samples.

REFERENCES

1. Thomas, N. A. "Assessment of Fish Flesh Tainting Substances," *Biological Methods for the Assessment of Water Quality,* ASTM STP 528, ASTM, Philadelphia, 178 (1973).
2. Rogers, I. H. "Isolation and Chemical Identification of Toxic Components of Kraft Mill Wastes," *Pulp Paper Mag. Can.* 74(9), 303 (1973).
3. Leach, J. M. and A. N. Thakore. "Identification of the Constituents of Kraft Pulping Effluent that are Toxic to Juvenile Coho Salmon (*Oncorhynchus kisutch*)," *J. Fish. Res. Board Can.* 30, 479 (1973).
4. Junk, G. A. *et al. J. Chromatog.* 99, 745 (1974).
5. Shumway, D. F. and J. R. Palensky. "Impairment of Flavour of Fish," Oregon State University, Department of Fisheries and Wildlife. Project 18050 D.D.M. Corvallis (February 1973).
6. Brownlee, B. and W. M. J. Strachan. "Persistent Organic Compounds from a Pulp Mill in a Near-Shore Freshwater Environment," presented at the First Chemical Congress of the North American Continent, Mexico City (December 1975), Chapter 35, this volume.
7. Rogers, I. H. and L. H. Keith. "Organochlorine Compounds in Kraft Bleaching Wastes. Identification of Two Chlorinated Guaiacols," Fisheries and Marine Service Technical Report No. 465, Environment Canada, Ottawa, Ontario (1974).

CHAPTER 35

PERSISTENT ORGANIC COMPOUNDS FROM A PULP MILL IN A NEAR-SHORE FRESHWATER ENVIRONMENT

B. Brownlee and W. M. J. Strachan

    Canada Centre for Inland Waters
    Burlington, Ontario

STUDY SITE

The study area was a 25 km^2 section of the northwest corner of Nipigon Bay which is on the north shore of Lake Superior. A kraft pulp mill located at the town of Red Rock is the major industry in the area. The effluent from this mill enters the receiving waters of Nipigon Bay as a point source. The study was conducted during July and August of 1974 as part of a group effort by environmental chemists and fisheries biologists.

PURPOSE

Our objective was to determine if any organic compounds present in the mill effluent could be detected at significant distances (up to 5 km) from the effluent outfall and therefore might be considered to be persistent. Procedurally this involved first, identifying as many constituents as possible in the mill effluent and second, determining if any of these compounds could be detected in water, seston or sediment samples taken according to a predetermined pattern of sampling sites. Conversely, those compounds which were not detectable outside the immediate mill area could be considered to be relatively nonpersistent.

The sampling pattern chosen is shown in Figure 35.1. Effluent plume movement was a major consideration as we were concerned primarily with

662 ORGANIC POLLUTANTS IN WATER

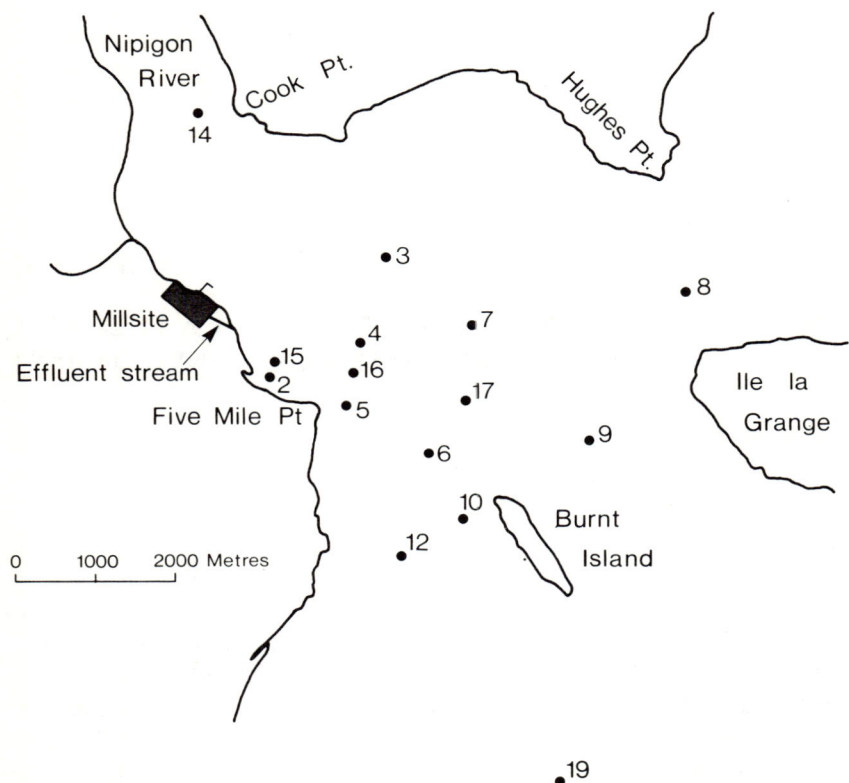

Figure 35.1 Location of sample stations with corresponding station numbers.

compounds detectable outside the area influenced by the plume. Local reports were that predominant plume movement is in a southeasterly direction extending as far as Five Mile Point. This predominance was taken into account when the sampling pattern was chosen.

Fish tainting has been a problem for the commercial fishery in this area.[1] Another study conducted at the same time as ours was directed toward the identification of suspect compounds in fish taken from the mill area.[2] We hoped to be able to correlate our results with this study.

METHODS

Each station was sampled at least once for water and seston. Sampling was limited to two or three stations per day. Water samples of about 35 liters were collected from a depth of 5 m (less in shallow water) by aspiration using a teflon and stainless steel apparatus. The water sample was pressure-filtered through precombusted (450°C), 142-mm Nucleopore® glass fiber filters (two required) in a teflon and stainless steel filtration apparatus at 10 psi or less of nitrogen which was purified by passing through molecular sieves. A 9-liter portion of filtrate was subjected to solvent extraction the same day. The glass fiber filters with retained particulate material (seston) were placed in clean containers and stored frozen. A limited number of sediment samples were collected by Shipek sampler and stored frozen.

The water sample was made saline (0.1 N) with precombusted (450°C) sodium chloride and extracted with 700 ml of glass-distilled chloroform in a continuous liquid-liquid extractor for 22 hr. The extract was reduced to small volume and stored temporarily in a clean, sealed ampoule. Later, the extracts were taken to dryness at 60°C, the yield of extractable material determined, and the residue methylated with diazomethane in ether-methanol. After removal of solvent at 60°C, the methylated extracts were stored at 4°C in sealed ampoules until analysis.

The seston filters were thawed and homogenized with 150 ml of glass-distilled chloroform-methanol (2:1, v/v). After heating to reflux for 1 hr, the mixture was filtered; two drops of 6 N hydrochloric acid were added to the filtrate (extract), and the filtrate was taken to dryness, methylated, and stored as for the water samples. Sediment samples were freeze-dried and extracted for 24 hr with glass-distilled chloroform in a Soxhlet extractor. The extracts were treated the same as the water extracts.

The ampoules were opened, the residues taken up in a small volume of glass-distilled benzene, and an aliquot analyzed by gas chromatography (GC). The conditions used were: 6 ft by 1/8 in. stainless steel column packed with 3% OV-1 on Chromosorb W-HP, operated at 125-250°C with programming at 6°C/min, with a nitrogen carrier flow of 20 ml/min, and a flame ionization detector (FID). An aliquot was also examined by gas chromatography/mass spectrometry (GC/MS) using the same chromatographic conditions except that a glass column was used and helium was the carrier gas. The gas chromatograph is coupled *via* a glass jet separator to a computer-operated Finnigan model 1015 S/L quadropole mass spectrometer. Our operating procedure is described in greater detail by Strachan.[3] The electron impact mass spectral data thus obtained were used to identify components in the extracts by conducting Biemann-type spectrum searches of a computer-based library of mass spectra. Having identified a compound in one or more extracts we could then examine to see whether this compound was present in the other extracts on the basis of GC retention time in the total ion current chromatogram from

GC/MS data, and characteristic fragment and molecular ions in the mass spectra. The corresponding peak in the FID gas chromatogram could then be located and a semiquantitative value determined for the level of this compound in the original sample on the basis of GC response (peak height) relative to external standards. These levels represent minimum values only, and in the case of more volatile materials the actual levels may be considerably higher than observed due to loss on evaporation. Under our conditions, the approximate detection limits for a compound are 0.1 µg/l in water, 10 µg/g dry weight of seston, and 0.1 µg/g dry weight of sediment. The detection limits are partly governed by the complexity of the chromatograms and the location of a particular peak in a chromatogram.

RESULTS

Table 35.1 lists the compounds identified in the mill effluent and the concentrations observed. Several other compounds remain unidentified

Table 35.1 Compounds Identified in Derivatized Chloroform Extracts of Mill Effluent

Similarity Index[a]	Compounds	Concentration in Total Effluent µg/l
	Dichloroveratrole[b]	7
0.54	3,4-Dimethoxyacetophenone	6
	3,4,5-Trichloroveratrole[c]	2.5
0.87	Methyl Palmitate	50[d]
0.73	Ethyl Palmitate	
0.50	Methyl Linolenate	60
0.86	Methyl Linolelaidate	150
0.59	Methyl Stearate	9
0.73	Methyl Sandaracopimarate	170
0.64	Methyl Isopimarate	380
0.61	Methyl Dehydroabietate	1300
0.74	Methyl Abietate	1500
0.44	Methyl Neoabietate	40
0.51	Methyl 9,10-Dihydroxystearate	3
	Methyl 7-Ketodehydroabietate[b]	25
0.60	Dioctyl Phthalate	15

[a]Explained in text.
[b]Tentative assignment (see text).
[c]Structural assignment based on similarity of mass spectrum to published spectrum of 3,4,5-trichloroveratrole (Rogers and Keith, 1974).[4]
[d]For methyl and ethyl palmitate combined.

as yet. With the exception of dioctyl phthalate, all are derivatives of phenols, fatty acids or resin acids. The similarity index is a measure of the confidence that can be placed in the identification; a perfect match would give a similarity index of 1.00. In the Biemann-type search, the comparison is done on the two most intense peaks in each 14-amu segment of the mass spectrum.

One complicating feature of our results was the presence of ethyl esters of fatty and resin acids in the derivatized extracts. These are most likely artifacts of the extraction procedure, the source of the ethyl groups being the ethanol added as a preservative to the chloroform used in the extractions. We have further confirmed the identity of ethyl palmitate and ethyl dehydroabietate (the latter found in derivatized extracts from some water samples) by comparison of mass spectra with those of authentic materials prepared by esterification of the parent acids with ethanol/hydrochloric acid.

Some of the structural assignments require elaboration. 3,4,5-Trichloroveratrole was assigned by comparison of our mass spectrum with the published spectrum of this compound which has been found (as the isomeric guaiacols) in kraft bleachery effluent.[4] The spectra are qualitatively very similar with some differences in relative intensities. Both show an isotope cluster at m/e 240, 242, 244 for the molecular ion ($C_8H_7Cl_3O_2$), another cluster at m/e 225, 227, 229 for M-15 (-CH_3), and the base peak at m/e 162.

The structure for dichloroveratrole is assigned without specifying the isomer and is done on the basis of spectral similarity to 3,4,5-trichloroveratrole, with important fragment ions being shifted lower by 34 amu for dichloroveratrole. While 3,4,5-trichloroveratrole has an isotope cluster at m/e 240, 242, 244 for the molecular ion ($C_8H_7Cl_3O_2$) and the base peak at m/e 162, dichloroveratrole has an isotope cluster at m/e 206, 208, 210 for the molecular ion ($C_8H_8Cl_2O_2$) and the base peak at m/e 128. Similar loss patterns can be observed in both spectra for M-15 (-CH_3), M-43 (-CH_3, -CO), and M-78 (-CH_3, -CO, -Cl) the base peak.

While ethyl dehydroabietate did not appear in extracts of effluent samples, it did appear in extracts of some water samples and its structural assignment will be discussed at this point. As mentioned above, the mass spectra of ethyl dehydroabietate from extracts were the same as those of an authentic sample. The spectrum of this compound has a molecular ion peak for $C_{22}H_{32}O_2$ at m/e 328 and the base peak at m/e 239. It is very similar to methyl dehydroabietate which has a molecular ion peak for $C_{21}H_{30}O_2$ at m/e 314 and the base peak at m/e 239. Significant losses are M-15, M-73 and M-89 for ethyl dehydroabietate, and M-15, M-59 and M-75 for methyl dehydroabietate, in agreement with the proposed fragmentation scheme for esters of dehydroabietic acid.[5]

The mass spectrum of the compound we have tentatively assigned as methyl ketodehydroabietate has a molecular ion peak at m/e 328 ($C_{21}H_{28}O_3$) and a base peak at m/e 253, both 14 amu higher than for methyl dehydroabietate. This increment corresponds to introduction of a keto function or an extra methylene group on the dehydroabietane skeleton and the keto function is much more likely. The keto function was assigned to the 7-position as this position, being adjacent to the aromatic ring, is the most likely site of oxidative attack on the dehydroabietane skeleton. This parent material also appeared in the extracts of some water samples as the ethyl ester. When present as the methyl ester, its quantitation in the presence of dioctyl phthalate was complicated by the fact that the two nearly co-elute on an OV-1 column.

With respect to the matter of persistent compounds as operationally defined earlier, many of the compounds listed in Table 35.1 (as derivatives) were found only in the effluent. We can conclude that either they are not persistent, *e.g.*, abietic acid, or are present in the effluent in such small quantities as to be undetectable in the water at even small distances from the mill, *e.g.*, 3,4,5-trichloroguaiacol isomers (3,4,5-trichloroveratrole after methylation). The C_{18} unsaturated fatty acids, linolenic and linolelaidic, were identified in the effluent as the methyl esters. Peaks having the same retention times appeared in chromatograms for several samples, but we found them difficult to confirm by their mass spectra as they gave variable base peaks and highest masses, with molecular ions being only occasionally observed.

Three compounds might be described as slightly or moderately persistent. They are: acetovanillone (3,4-dimethoxyacetophenone after methylation, see Reference 6), sandaracopimaric acid and 7-ketodehydroabietic acid. The locations and levels at which these compounds were observed in water samples are summarized in Table 35.2.

Three compounds were more widely distributed. They are: palmitic acid, dehydroabietic acid and dioctyl phthalate. The locations and levels of these three compounds in water, seston and sediment samples are summarized in Table 35.3. Some palmitic acid is contributed by the effluent but, as it is a natural product as well, no inferences regarding its persistence can be reasonably drawn from its wide distribution. The frequent observation of phthalate esters in the environment and their notoriety as accidental contaminants leads to a good deal of caution in interpreting the finding of these materials in our samples. We have taken precautions to protect against contamination of our samples from identifiable sources of phthalate esters such as plastics, and have detected dioctyl phthalate in moderate quantities in mill effluent (see also Reference 6). We conclude that the mill effluent is at least one source of dioctyl phthalate in

Table 35.2 Levels of Acetovanillone (I),[a] Sandaracopimaric Acid (II)[a] and 7-Ketodehydroabietic Acid (III)[a] in Water Samples

Distance from Source km	Station[b]	Concentration, µg/l		
		I	II	III
0	Effluent	6	170	25
1.0	2	0.3	2.2	0.4
1.0	15			9
1.7	16			0.6
1.8	5			Trace
4.7	9			0.1

[a]These are the presumed parent compounds actually present in the water; they were identified as the methylated derivatives in the chloroform extracts.
[b]See Figure 35.1 for location of sampling stations.

Nipigon Bay and that this material is quite widely distributed (persistent) throughout the study area.

Dehydroabietic acid is added to the receiving waters of Nipigon Bay in large quantities in the mill effluent, the only other identifiable source being the Nipigon River (see also Reference 6). The loadings from these two sources are probably of the same order because the flow of the Nipigon River is about 400 times that of the mill effluent.

DISCUSSION

Lack of mention of a compound in a sample is not meant to imply its total absence, rather that it was not present in identifiable and quantifiable amounts. It should also be stressed that we were only observing those compounds sufficiently volatile to elute from a GC column and that these constitute only a fraction of the total extractable material. Acetovanillone was found in only one water sample indicating only slight persistence for this compound. Fox[6] has observed much higher levels of this compound in water samples taken from the effluent and plume during the same study. This may be due to the use of ether as the extraction solvent. In our case use of the higher-boiling chloroform may have resulted in considerable underestimation of the amounts of this material due to evaporation losses.

7-Ketodehydroabietic acid appears to be moderately persistent. An alternative explanation is that it is also produced in the water column from dehydroabietic acid.

668 ORGANIC POLLUTANTS IN WATER

Table 35.3 Levels of Palmitic Acid (IV),[a] Dehydroabietic Acid (V),[a] and Dioctyl Phthalate (VI) in Water, Seston and Sediment Samples

Distance from Source, km		0	1.0	1.0	1.6	1.7	1.8	2.2	2.5	3.0	3.1	3.6	3.9	4.7	5.7	3.0	6.8
Station[b]		Effluent	2	15	4	16	5	3	7	6	17	12	10	9	8	14	19
Concentration in Water, μg/l	IV	50	0.5	1.5	0.1	1	0.2	0.2	0.1	0.5	0.4	0.4	0.7	0.2	0.6	0.9	0.2
	V	1300	15	100		10	0.4		0.04	0.3	0.7	0.1	0.6	2	0.08	0.2	0.1
	VI	15	1	55	0.1	2						1.5	1	0.6	0.3	0.5	0.3
Concentration in Seston, μg/g Dry Weight	IV		550	500	750	700	1000	1500	150		1000	500	1300	2000	2000	400	
	V			60		40	Trace		7			35					Trace
	VI																
Concentration in Sediment, μg/g Dry Weight	IV		100											0.3	1	3	
	V		150											0.1	2	0.3	
	VI														0.7		

[a]These are the presumed parent compounds actually present in the samples; they were identified as the methyl and ethyl esters in the derivatized chloroform extracts.

[b]See Figure 35.1 for location of sampling stations.

Dehydroabietic acid was found in most of the water samples, all sediment samples, and a few seston samples. The levels observed in water samples 1 km from the source were 10-100 times higher than levels observed in water samples from 2 km and beyond. Without accurate physical data to determine dilution factors at sites within the study area it is impossible to determine how much of the decrease in concentration (disappearance) is due to dilution and how much is due to removal processes (biodegradation, sedimentation). However, these observations can be made: (a) the mill effluent is a significant source of dehydroabietic acid to the receiving waters, (b) some dehydroabietic acid enters the bay in particulate form as indicated by its occurrence in some seston samples, (c) dehydroabietic acid is found in surface sediment—in significant amounts 1 km from the effluent outfall (see also Reference 6)—and (d) dehydroabietic acid is quite persistent in the water column.

We found no evidence for accumulation of mill-derived compounds from the water column into the seston. However, our detection limits for seston are about 100,000 times higher than for water and so only strongly-accumulated substances would be observed to partition from water into seston. As yet none of the persistent compounds observed by us can be implicated in fish tainting. This awaits further study.[2]

The listing of compounds in Table 35.1 does not include all compounds observed or identified in this study. Some have been difficult to identify accurately and others have been identified but were found in samples other than the effluent. Examples of the former case are several hydrocarbons. Examples of the latter case are the C_{14}, C_{22}, and C_{24} fatty acids found in several water and seston samples, and hexachlorobenzene found in a water sample from station 9.

REFERENCES

1. Chatterjee, R. M. "Fish Tainting in the Upper Great Lakes," Water Resources Branch, Ontario Ministry of the Environment (1974).
2. Kaiser, K. L. E. Canada Centre for Inland Waters, Environment Canada, studies in progress.
3. Strachan, W. M. J. "Chloroform Extractable Organic Compounds in the International Great Lakes," presented at the First Chemical Congress of the North American Continent, Mexico City (December 1975), Chapter 28, this volume.
4. Rogers, I. H. and L. H. Keith. "Organochlorine Compounds in Kraft Bleaching Wastes. Identification of Two Chlorinated Guaiacols," Fisheries and Marine Service, Environment Canada, Technical Report No. 465 (1974).

5. Enzell, C. R. and I. Wahlberg. "Mass Spectrometric Studies of Diterpenes," *Acta Chem. Scand.* **23**, 871 (1969).
6. Fox, M. E. "Fate of Selected Organic Compounds in the Discharge of Kraft Paper Mills into Lake Superior," presented at the First Chemical Congress of the North American Continent, Mexico City (December 1975), Chapter 34, this volume.

CHAPTER 36

GC/MS ANALYSES OF ORGANIC COMPOUNDS IN TREATED KRAFT PAPER MILL WASTEWATERS

Lawrence H. Keith

    U.S. Environmental Protection Agency
    Southeast Environmental Research Laboratory
    Athens, Georgia

"The need to know what chemicals may escape into the environment and at what levels they may be harmful leads rather quickly to a realization that until one can identify these compounds with certainty and measure their presence in selected compartments of the environment, effective control of these chemicals is essentially impossible."[1] Although much is known about the identities of specific chemical compounds in various types of trees and, to a lesser degree, in the wastewaters of paper mills, little is known about what happens to these chemicals as they pass through wastewater treatment systems and enter receiving waters. To our knowledge this study represents the first attempt to characterize a wastewater chemically, trace the dissolved volatile organics through a treatment system, and correlate this information with the traditional collective pollution parameter measurements (BOD, TOC). The results detailed in this chapter were gathered over a six-year period and portions of them have been presented previously.[2-4]

By tracing the chemicals through the treatment system one can identify which compounds are being effectively removed and which are resistant to the treatment in use. Any new chemicals produced during treatment are readily apparent. Once identifications are made, the approximate concentration of each compound can be calculated at each stage of the treatment. By comparing two or more different types of treatment, their effects on the individual compounds, or on the various classes of compounds in the wastewaters, can be ascertained. This knowledge will be particularly useful for advanced wastewater treatment and control studies, especially those involving wastewater recycling. A build-up in concentration of compounds resistant to the proposed treatment could be detrimental to closed-loop systems. If the segregated wastewater streams are analyzed before they are combined for treatment, the main sources of compounds resistant to treatment can be identified and singled out for more economical, specialized treatment, if desired or needed.

Knowledge of the specific chemical composition of treated wastewaters is also basic to the evaluation of the environmental impact of these wastewaters and to the problem of analyzing and controlling their discharge. Once the "refractory" compounds are identified, studies involving their fate and ecological effects can commence. Acute and, possibly more significant, chronic effects of these chemicals on aquatic life can be determined.

PAPER MILL TREATMENT FACILITIES

Two mills, having similar processes but different waste treatments, were used in this study. The first, Paper Mill "A," daily produced about 1400 tons of containerboard in March 1972 when the samples were taken. Approximately 13 million gallons of water passed through the treatment system daily. Treatment consists of a primary clarifier, trickling biofilter, secondary clarifier, and five aerated lagoons with a total retention time of 2 to 2½ days (Figure 36.1). The trickling biofilter is 30 ft high and 80 ft in diameter, and is packed with a vinyl core material having 97% void space. It handles an average daily load of 56,000 lb BOD and provides 50% BOD removal. The total BOD reduction through the whole system is reported to be in excess of 70%.

The second mill, Interstate Paper Corp. at Riceboro, Georgia, daily produced about 540 tons of containerboard in March 1972 when the samples were taken. Approximately 5.5 million gallons of water passed through the treatment system daily. Treatment consists of lime addition at 0.1%, a 40-min flocculation period followed by gravity clarification, and a 650-acre (3-6 month retention) stabilization lagoon (Figure 36.2). The highly alkaline effluent (pH 12) first undergoes partial neutralization to pH 10 by surface absorption of atmospheric carbon dioxide, causing precipitation of nearly all the remaining calcium salts in the inlet section of the stabilization basin. Lime treatment removes about 90% of the color from the effluent. Overall BOD reduction is reported to be 93%, with a concentration of about 6 mg/l in the lagoon effluent.[5,6]

IDENTIFICATION OF ORGANIC POLLUTANTS 673

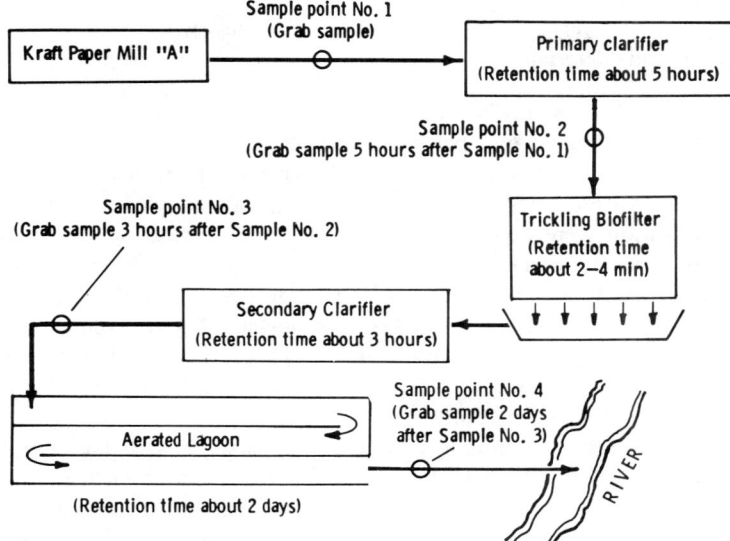

Figure 36.1 Waste treatment system diagram of Mill A.

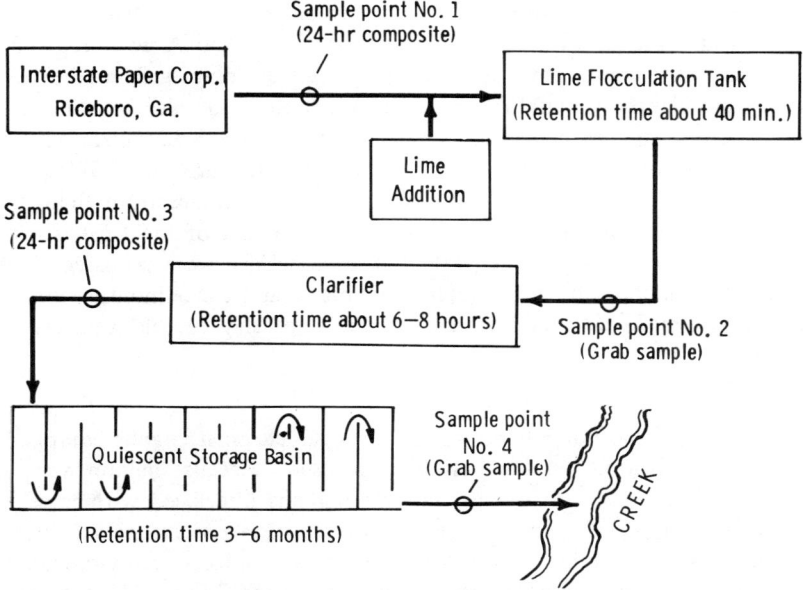

Figure 36.2 Waste treatment system diagram of the Interstate Paper Corporation mill at Riceboro, Georgia.

EXPERIMENTAL

Sampling and Materials

Burdick and Jackson "distilled in glass" solvents were used for all extractions. Samples were placed in plastic containers and immediately frozen. They were stored at -10°C and thawed for use as necessary.

Both mills were sampled twice. In 1972, grab samples for chemical characterization were taken at various stages of the treatment system and time delays were programmed so that a "slug" of the effluent would be followed through the treatment facilities. However, quantitative analysis indicated that the slug was missed at the outfall of Mill A and, of course, the chemical composition of Interstate's discharge at sampling points 1, 2 and 3 bore no relationship to that at point 4 because of the long retention time in the lagoon.

During the second sampling period, in January 1974, 3-day composites were collected from the raw effluent and the outfalls of both mills by automatic sampling devices. Samples were not taken at intermediate points in the treatment system of either mill because automatic sampling devices were not available there. All samples were collected during the same week. Samples of the raw wastewaters and of the outfalls, collected over 24-hr periods, were frozen each day.

A 3-day composite composed of equal volumes from the Monday, Tuesday and Wednesday raw wastewater samples of Mill A was prepared. A second 3-day composite was composed of equal volumes from the Wednesday, Thursday and Friday treated wastewater samples of Mill A. This provided samples of a slug of wastewater, with a 3-day averaged composition, both before and after it had passed through the 2.5-day treatment system. Three-day composites of equal volumes from the Monday, Tuesday and Wednesday raw wastewater and of the treated wastewater samples of the Interstate Mill at Riceboro were prepared. Because the stabilization pond retention time is at least 3 months, no attempt was made to obtain samples of the same slug of this wastewater.

Instrumentation

Both Varian 1400 and Perkin-Elmer 900 gas chromatographs equipped with flame ionization detectors (FID) were used. Because helium was used as a carrier gas with our gas chromatographs interfaced with mass spectrometers (GC/MS), we used helium as a carrier gas routinely in developing chromatographic conditions. Generally, a commercially-prepared (Perkin-Elmer) 50-ft support-coated open tubular (SCOT) capillary column coated with carbowax 20 M/terephthalic acid (K 20 M/TPA) was used for separation. The optimum carrier gas flow for our GC/MS systems operating under a vacuum and using a Gohlke jet separator is 16 to 18 ml/min. If a helium flow of about half this amount is used for optimizing conditions with an auxiliary GC (operating under atmospheric pressure), the other chromatographic variables (temperature program rate, initial temperature, initial hold) hold true when the same column is transferred to the GC/MS system.

From 1968 to 1971, a Perkin-Elmer/Hitachi RMU-7 double focusing mass spectrometer connected to a Perkin-Elmer 900 GC through a Watson-Biemann separator was used. Now, a Finnigan 1015 quadrupole mass spectrometer connected to an all-glass, single-stage Gohlke jet separator is used. A System Industries interface permits a Digital Equipment Corporation (DEC) PDP8/e computer to control the mass spectrometer during calibration and data acquisition; to accept data from the mass spectrometer; and to control a Houston plotter during data reduction. Programs, raw data, and reduced data were stored on either two DEC tape units or a Diablo disc. Output of the reduced data was achieved under computer control *via* the plotter, the teletypewriter, or an acousti-coupler. The coupling device connected the PDP8/e to the CDC 6400 computer at Battelle Laboratories, Columbus, Ohio, and permitted semiautomatic spectrum identification by a matching program, fully described elsewhere.[7,8]

ANALYTICAL PROCEDURES

Chloroform vs Carbon Chloroform Extracts

Adsorption of organics on granulated carbon followed by extraction with chloroform provides a large amount of sample with which to work. Accordingly, 1800 gal of the Mill A aerated lagoon effluent was passed through a bank of eight carbon filters in parallel. The carbon was dried and extracted with chloroform. The resulting extract was concentrated in a Kuderna-Danish apparatus and chromatographed under conditions identical to those used for a 1-liter chloroform extract. Figure 36.3 shows a comparison of the two chromatograms. Most of the same peaks are present in both chromatograms. Although the relative intensity differs somewhat, the chromatograms are similar enough that, from a qualitative aspect, the two methods of sample concentration were essentially equivalent.

However, the carbon adsorption method has numerous disadvantages:

- It is more time-consuming than it is to simply "grab" a sample for solvent extraction.

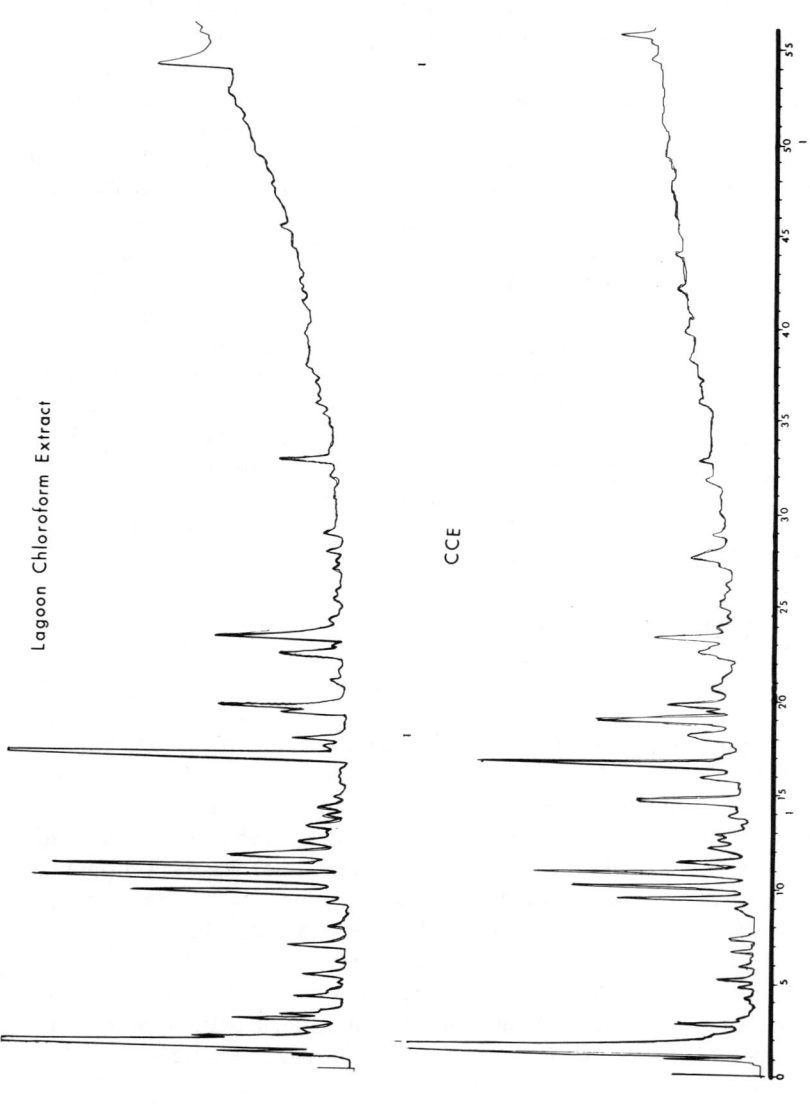

Figure 36.3 Gas chromatograms of the extracts of Mill A lagoon effluent from (top) carbon chloroform extract (CCE) and (bottom) liquid-liquid chloroform extract.

- Some compounds may be only partially desorbed from the carbon.
- A greater possibility of chemical change (isomerization, hydrolysis, etc.) exists when carbon with its large, active surface area is used.
- Extent of adsorption varies with size of carbon particles, contact time with carbon (flow rate), and turbidity of the water.
- Carbons from various sources differ markedly in their adsorptive ability.
- A carbon column can provide a medium for bacterial growth, causing biological degradation of the adsorbed organics.

We therefore elected to obtain the rest of our samples by the simpler technique of extraction with chloroform followed by Kuderna-Danish concentration.

Comparison of Methylation Techniques

Paper mill wastewater extracts contain two types of extractable volatile compounds: neutrals (predominantly terpenes and their derivatives), and acidic compounds that are converted to their methyl derivatives to facilitate gas chromatographic (GC) separation. After the pH of the wastewater is raised to 11 with sodium hydroxide and the neutrals are extracted, the aqueous solution can be methylated directly with dimethyl sulfate and sodium hydroxide, followed by extraction of the methyl derivatives with chloroform. This method has the advantage of being an *in situ* procedure; however, it is complex and time consuming. Alternatively, after extraction of the neutrals at a high pH, the solution can be made strongly acidic with concentrated hydrochloric acid, and the free acids and phenols extracted with chloroform. Methyl derivatives of the acids and phenols can then be made in a separate step using diazomethane, on-column GC methylation techniques, or several other common methylation procedures. An evaluation of each method was made using representative standard compounds and also using samples of kraft mill wastewater.

Duplicate samples consisting of 10 mg each of guaiacol (I), vanillin (II), palmitic acid (III) and dehydroabietic acid (IV) were dissolved in 600 ml of water made to pH 11 with sodium hydroxide. Each solution was placed in a separatory funnel, made acidic with concentrated hydrochloric acid, and extracted with three 133-ml portions of chloroform. After each combined extract was evaporated to near dryness in a Kuderna-Danish concentrator, 2 ml of 10% methanol in ether was added and the samples were methylated by the standard procedure with diazomethane. Chloroform was then added to each to bring the volumes to 10.0 ml. Duplicate injections into a gas chromatograph were made and the areas of the peaks were compared to the peak areas from standards of veratrole

(V), veratraldehyde (VI), methyl palmitate (VII) and methyl dehydroabietate (VIII) (10 mg each) in 10 ml of chloroform (1000 ppm). Calculated average percent conversion and recovery values were as follows: I → V, 18%; II → VI, 47%; III → VII, 70%; IV → VIII, 96%.

I: R = H   II: R = H   III: R = H   IV: R = H
V: R = CH$_3$ VI: R = CH$_3$ VII: R = CH$_3$ VIII: R = CH$_3$

The same model compounds were methylated with dimethyl sulfate. The reaction with dimethyl sulfate is dependent on temperature control, addition time, and reaction time, and on how vigorously the reaction mixture is stirred. The procedure we use[9] is a variation of that described by Bicho et al.[10,11] To determine the average yield and recovery of methylated model compounds from this reaction, four mixtures of I through IV were subjected to the same reaction and work-up procedures. After methylation, the products were extracted with chloroform, dried over sodium sulfate, and concentrated to less than 10 ml in a Kuderna-Danish apparatus. The volume was adjusted to 10.0 ml with chloroform. Comparison of GC peak areas with peak areas of standards V through VIII provided the average percent methylation and recovery: I → V, 90%; II → VI, 72%; III → VII, 58%; IV → VIII, 60%.

A 500-ml portion of the Mill A wastewater from sample point 2 (primary clarifier effluent) was made alkaline to pH 11 with sodium hydroxide and extracted with chloroform to remove the neutral compounds. Methylation of the aqueous layer with dimethyl sulfate was followed by re-extraction with chloroform to remove the methylated organics. After concentration in a Kuderna-Danish apparatus to 0.5 ml, 1.2 μl of the extract was chromatographed on a 50-ft SCOT column coated with Carbowax 20 M/TPA and programmed from 100 to 200°C at 4°/min with an initial 2-min hold at 100°. The chromatogram is shown in Figure 36.4A. A control sample was prepared exactly the same way using 500 ml of distilled water; its chromatogram is shown in Figure 36.5A.

IDENTIFICATION OF ORGANIC POLLUTANTS 679

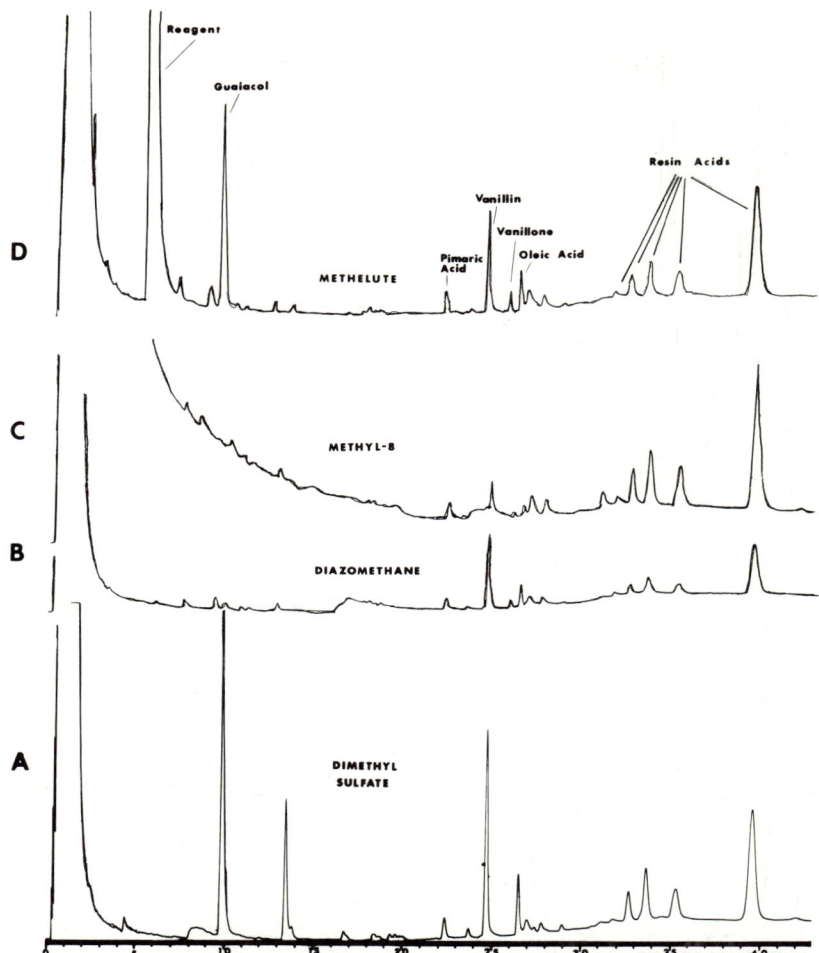

Figure 36.4 Gas chromatograms of the methyl derivatives from Mill A wastewater using (A) dimethyl sulfate, (B) diazomethane, (C) Methyl-8 and (D) MethElute as methylating reagents.

A 1-liter portion of the same wastewater was extracted with chloroform at pH 11 to remove neutral compounds and then made acidic to pH < 2 with concentrated hydrochloric acid. The aqueous layer was re-extracted with four 200-ml portions of chloroform to remove the acids and phenols. The extracts were combined and divided into two equal portions, each representing 500 ml of the wastewater extract.

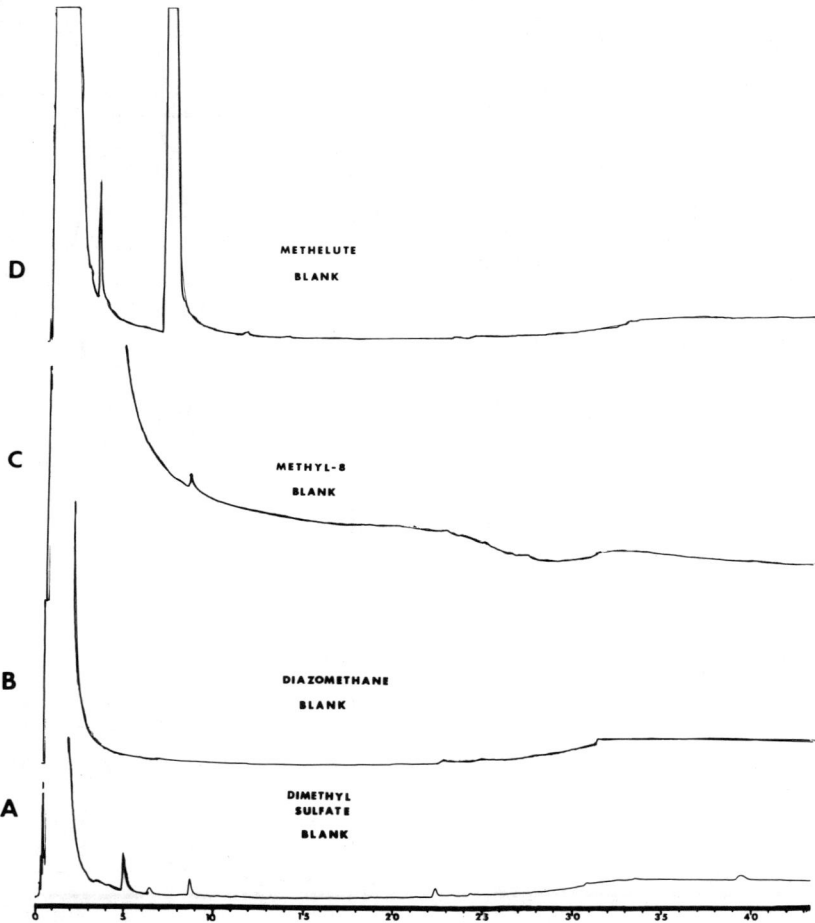

Figure 36.5 Gas chromatograms of the methyl derivatives of control samples treated with (A) dimethyl sulfate, (B) diazomethane, (C) Methyl-8 and (D) MethElute.

One portion was concentrated to near dryness in a Kuderna-Danish apparatus and methylated with diazomethane as previously described; the final volume was adjusted to 0.5 ml. The chromatogram of 1.2 µl of this sample under conditions identical to those of the previous sample is shown in Figure 36.4B. A control sample was prepared from 400 ml of chloroform subjected to the same procedure; the chromatogram of 1.2 µl of the control is shown in Figure 36.5B.

The second portion of the chloroform extract was concentrated in a Kuderna-Danish apparatus to near dryness and MethElute (trimethylanilinum hydroxide in methanol; Pierce Chemical Co.), which methylates the sample on-column, was added to bring the volume to 0.5 ml. The chromatogram of 1.2 µl of this sample under conditions identical to those of the previous two samples is shown in Figure 36.4D. A control sample was prepared from 400 ml of chloroform subjected to the same procedures. The chromatogram of 1.2 µl of the control is shown in Figure 36.5D.

Another 500-ml portion of the wastewater, after extraction of neutral compounds at pH 11, was made acidic and extracted with four 100-ml portions of chloroform. The extracts were combined, dried and concentrated to near dryness as before in a Kuderna-Danish apparatus. Enough Methyl-8 (DMF dimethyl acetal in pyridene; Pierce Chemical Co.) was added to bring the volume to 0.5 ml and the solution was heated at 60°C for 15 min in a reacti-vial. The chromatogram of 1.2 µl of this sample, under conditions identical to those of the other three, is shown in Figure 36.4C. A control sample was prepared from 400 ml of chloroform subjected to the same procedure. The chromatogram of 1.2 µl of the control is shown in Figure 36.5C.

GC/MS Techniques

After a sample is injected into the computer GC/MS system, the mass spectrometer automatically scans its preset mass range every 4-5 sec. When each run is complete, the computer plots a reconstructed gas chromatogram (RGC). The RGC peaks are all normalized to the amplitude of the largest peak, arbitrarily plotted at 100. Each point on the spectrum number scale under the RGC represents a complete mass spectrum. Figure 36.6A shows the RGC of the methylated acids and phenols from the lagoon wastewater of Mill A.

Specialized techniques of MS or data reduction can be used to detect a specific material or class of materials in a mixture. The most common technique is the generation of the limited mass reconstructed gas chromatogram (LMRGC). For example, in Figure 36.6A, the RGC shows as peaks those spectra that contain significant numbers of any ion fragments of m/e 33 to 450. Below the RGC is the LMRGC in which the computer was instructed to respond only to those spectra that contain the m/e 149 fragment (Figure 36.6B) due to protonated phthalic anhydride found in the spectra of phthalate esters. The LMRGC, therefore, indicates that two of the sample peaks, 44 and 61, are phthalates. The numbers 44 and 61 refer to the compounds listed in Table 36.1.

Similar characteristic fragments exist for other classes of compounds. The most useful ones in this study were m/e 74 and 87, which are

682 ORGANIC POLLUTANTS IN WATER

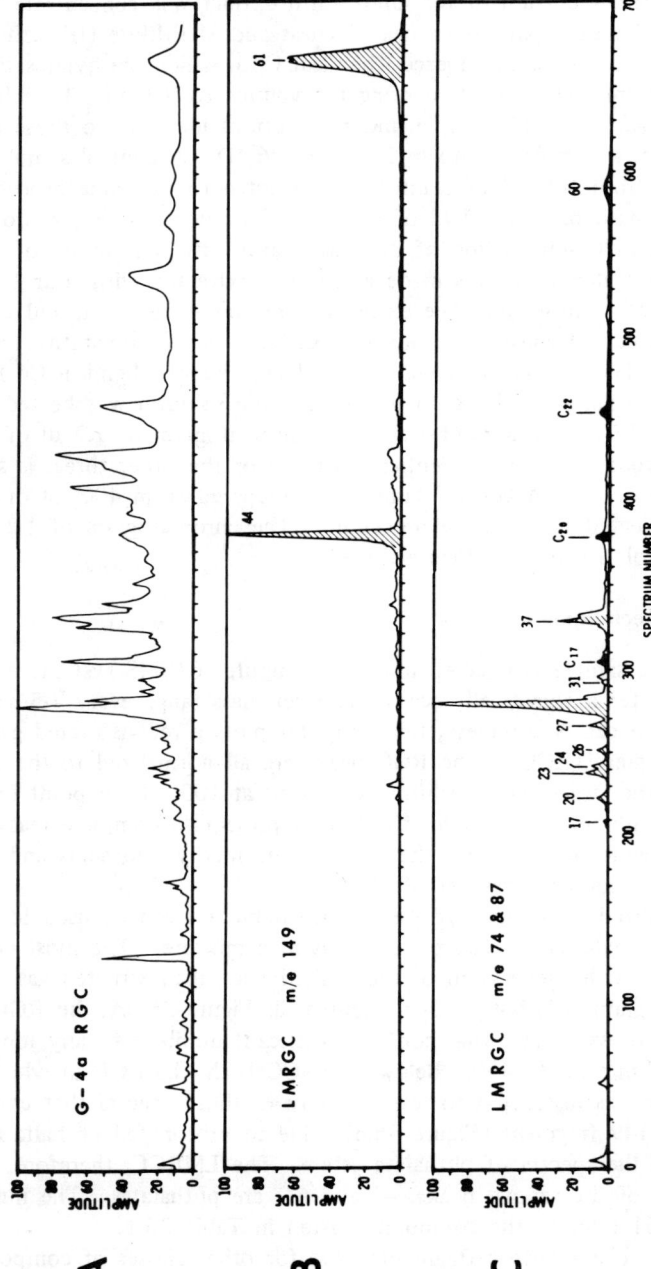

Figure 36.6 Mill A lagoon extract (A) RGC, (B) LMRGC m/e 149 search for phthalates and (C) LMRGC m/e 74 and 87 search for fatty acid methyl esters.

IDENTIFICATION OF ORGANIC POLLUTANTS 683

Table 36.1 Acids and Phenols Identified in Both Kraft Paper Mill Effluents with Approximate Concentrations

Peak No.	Compound Identified as Methyl Derivative (Parent Compound)	Confirmed By	Approximate Concentration in mg/l Mill A Sample Points				Approximate Concentration in mg/l Interstate Sample Points			
			1	2	3	4	1	2	3	4
1	Furfuryl methyl ether (furfuryl)	GC	0.010	0.005	—	—	0.010	—	—	—
2	Anisole (Phenol)		0.008	—	—	—	0.005	0.005	—	—
3	1,3-Dimethoxy-2-propanol		—	—	—	—	—	0.010	—	—
4	Methyl trisulfide		—	0.005	—	—	—	—	—	—
5	Unidentified–apparent MW=103		0.035	0.035	0.040	0.035	0.025	0.055	0.020	0.015
6	Benzaldehyde	GC	0.002	—	—	—	0.005	—	—	—
7	Dimethylsulfoxide	GC	—	—	0.055	—	0.010	0.080	0.020	0.020
8	Ethyl carbamate	GC	0.010	0.015	0.025	0.020	—	0.080	0.020	0.010
9	Borneol	MS,GC	—	—	—	—	0.010	0.020	0.060	—
10	α-Terpineol	MS,GC	0.020	0.002	0.001	—	0.005	0.020	0.115	0.002
11	Veratrole (Guaiacol)	MS,GC	2.700	2.400	0.360	0.170	2.200	3.200	3.700	0.035
12	o-Nitrotoluene	MS,GC	—	—	—	0.015	—	—	—	—
13	Methyl o-hydroxybenzoate (o-hydroxybenzoic acid)		0.010	0.005	—	—	—	—	—	—
14	Methyl mandelate (mandelic acid)		—	—	—	0.025	—	—	—	—
15	Dimethylsulfone	GC	0.240	0.400	0.400	0.055	0.240	0.540	0.400	0.130
16	Unidentified aromatic MW=166		0.130	0.090	0.050	0.005	0.050	0.030	0.160	—
17	Methyl isomyristate (C_{14} fatty acid)	GC	—	—	—	0.003	—	—	—	0.001
18	p-Methoxybenzaldehyde (p-hydroxybenzaldehyde)		0.055	0.050	0.020	0.025	0.090	0.240	0.190	0.005
19	Eugenol		0.025	0.025	0.025	—	—	—	—	—
20	Methyl myristate (C_{14} fatty acid)	MS,GC	—	—	—	0.020	—	—	—	0.010
21	Unidentified aromatic, MW=196		—	—	0.045	0.035	—	—	—	—
22	Unidentified nonaromatic, MW=106		—	—	—	—	0.170	0.175	0.245	0.065

Table 36.1, Continued

Peak No.	Compound Identified as Methyl Derivative (Parent Compound)	Confirmed By	Approximate Concentration in mg/l Mill A Sample Points				Approximate Concentration in mg/l Interstate Sample Points			
			1	2	3	4	1	2	3	4
23	Methyl anteisopentadecanoate (C_{15} fatty acid)	MS,GC	—	—	0.010	0.020	—	—	—	0.005
24	Methyl 10-methyltetradecanoate (C_{15} fatty acid)		—	—	—	0.030	—	—	—	0.002
25	p-Methoxyacetophenone (p-hydroxyacetophenone)	GC	0.020	0.025	0.030	—	0.060	0.130	0.080	0.010
26	Methyl pentadecanoate (C_{15} fatty acid)	MS,GC	—	—	—	0.020	—	—	—	0.010
27	Methyl isopalmitate (C_{16} fatty acid)	MS,GC	—	—	0.015	0.030	—	—	—	0.003
28	Methyl palmitate (C_{16} fatty acid)	MS,GC	0.070	0.190	0.180	0.430	0.140	0.160	0.035	0.017
29	Methyl palmitelaidate (C_{16} trans-9-unsaturated fatty acid)	MS,GC	0.005	0.020	0.025	0.125	—	—	—	0.010
30	Ethyl palmitate (C_{16} fatty acid)	MS,GC	—	—	—	0.030	—	—	—	—
31	Methyl anteisomargarate (C_{17} fatty acid)	MS,GC	0.020	0.025	0.040	0.050	0.030	0.120	0.040	0.001
32	Methyl 3,4-dimethoxyphenylacetate (Homovanillic acid)		0.050	0.075	0.090	0.080	0.090	0.365	0.150	0.003
33	Veratraldehyde (Vanillin)	MS,GC	1.500	1.700	0.450	0.410	2.100	4.400	2.600	0.070
34	2-Methylthiobenzothiazole (2-Mercaptobenzothiazole)	GC	0.035	0.025	0.030	0.025	—	—	—	—
35	Methyl vanillate (Vanillic acid)	GC	—	—	0.005	0.005	—	—	—	—
36	Acetoveratrone (Acetovanillone)	GC	0.420	0.490	0.450	0.370	0.820	1.600	0.860	0.120
37	Methyl stearate (C_{18} fatty acid)	MS,GC	0.025	0.065	0.045	0.110	0.100	0.100	0.020	—
38	Methyl oleate (C_{18} cis-9-unsaturated fatty acid)	MS,GC	0.470	0.600	0.430	0.400	0.570	0.510	0.120	0.080
39	3,4,5-Trimethoxybenzaldehyde (Syringaldehyde)	MS,GC	0.070	0.070	0.070	0.040	0.070	0.100	0.070	0.020

IDENTIFICATION OF ORGANIC POLLUTANTS 685

#	Compound	Method								
40	Methyl linoleate (C$_{18}$ cis,cis-9, 12-diunsaturated fatty acid)	MS,GC	0.350	0.470	0.230	0.100	0.450	0.920	0.160	—
41	3,4-Dimethoxypropiophenone (3-Methoxy-4-hydroxypropiophenone)	GC	0.060	0.080	0.025	—	—	—	—	—
42	3,4,5-Trimethoxyacetophenone (Acetosyringone)	GC	0.055	0.060	0.050	0.050	0.090	0.200	0.130	0.005
43	Methyl arachidate		0.030	0.035	0.030	0.025	0.035	0.025	—	—
44	a Dihexyl phthalate									
45	Unidentified resin acid "A" methyl ester		—	—	—	—	0.070	0.075	0.130	0.020
46	Unidentified resin acid "B" methyl ester		0.030	0.055	0.095	0.100	—	—	—	—
47	Unidentified resin acid "C" methyl ester		0.100	0.055	0.010	—	—	—	—	—
48	Unidentified resin acid "D" methyl ester		—	—	—	—	—	—	—	0.145
49	Unidentified unsaturated fatty acid methyl ester		—	—	0.005	0.035	—	—	—	—
50	Methyl pimerate (resin acid)	MS,GC	0.245	0.475	0.570	0.800	1.270	0.825	0.610	0.500
51	Methyl sandaracopimerate (resin acid)	MS,GC	0.050	0.060	0.125	0.045	0.340	0.245	0.275	0.110
52	Unidentified resin acid "E" methyl ester		0.025	—	—	—	—	—	—	—
53	Methyl-13-abieten-18-oate (resin acid)	MS,GC	—	0.430	0.800	1.400	0.100	0.035	0.050	1.050
54	Unidentified unsaturated fatty acid methyl ester similar to araconidate		0.035	—	—	—	0.075	—	—	—
55	Methyl isopimerate (resin acid)	MS,GC	0.430	0.660	0.770	0.780	4.400	1.700	1.200	0.800
56	Methyl abietate (resin acid)	GC	0.370	0.430	0.420	0.050	3.300	1.500	1.900	—
57	Methyl dehydroabietate (resin acid)	MS,GC	1.500	1.300	4.000	1.000	3.300	2.700	3.600	3.900
58	Methyl 6,8,11,13-abietatetraen-18-oate (resin acid)	MS,GC	0.065	0.070	0.170	0.095	0.160	0.115	0.280	0.180
59	Methyl neoabietate (resin acid)	MS,GC	0.105	0.125	—	—	1.300	1.200	0.450	—
60	Methyl lignocerate (C$_{24}$ fatty acid)	MS,GC	—	—	0.055	0.030	—	—	—	—
61	a Dioctyl phthalate		—	—	—	—	—	—	—	—
	Total		9.380	10.692	10.246	7.123	21.690	21.480	17.700	7.243

characteristic of long-chain saturated fatty acid methyl esters.[1,2] We have found that using the m/e 74 and 87 characteristic fragments together results in a more reliable indication of long-chain saturated fatty acid methyl esters than either one by itself. The LMRGC corresponding to these fragments for the lagoon wastewater from Mill A is shown in Figure 36.6C. The numbers above the shaded peaks correspond to the compounds in Table 36.1.

To minimize the time spent identifying the hundreds of spectra resulting from the GC/MS data output from complex mixtures such as these extracts, we first executed computer matching of the spectra. This was followed by a manual comparison of the computer-matched spectra with the unknown mass spectra. Widespread use of GC/MS spectra-matching in pollutant identification requires easy access to a central spectra library,[7] rapid matching, and an indication of the similarity of the unknown spectra to the reference spectrum for each match. An EPA program that provides such information was developed using the algorithm of a matching program described in the literature.[13] This rapid-matching program* was developed jointly by Battelle and the Environmental Research Laboratory—Athens, and utilizes a CDC 6400 time-shared computer.[14]

RESULTS

Methylation Evaluation

Comparison of the chromatograms in Figure 36.4 shows that dimethyl sulfate and MethElute are better reagents for methylating the paper mill samples than diazomethane or Methyl-8. Phenols, especially guaiacol, were not methylated well with diazomethane or Methyl-8. Although extraction of guaiacol and other phenols with chloroform is not 100%, incomplete extraction can be eliminated as a major contributor to the small peak height of the methylated guaiacol (veratrole) in the diazomethane sample because the samples for diazomethane and MethElute treatment came from the same extract, which was divided into two portions, and the MethElute sample shows a large veratrole peak.

The dimethyl sulfate appears to be equivalent to MethElute with respect to methylation of the resin and fatty acids. However, the larger phenolic peaks in the dimethyl sulfate sample and the absence of the MethElute reagent peak caused us to make use of dimethyl sulfate for all remaining methylations involved in analysis of these wastewaters.

*The mass spectral files and program now reside in the Cyphernetics Corp. computer, Ann Arbor, Mich. They are commercially available on the Mass Spectral Search System (MSSS).

Identification of Terpenes

Table 36.2 lists the individual compounds found in the neutral fraction of the wastewater extracts. Most of these are terpenes. The corresponding gas chromatograms of Mill A extract are shown in Figure 36.7 with peak numbers that correspond to the compounds in Table 36.2. Vertical displays of gas chromatograms of extracts taken at consecutive points in a waste treatment system, with the compounds identified by name or code, are called "chemical profiles." The compounds present and their relative concentrations are shown as a function of the treatment. Such a display allows one to see at a glance the effectiveness (or lack of it) of each step in the treatment process with respect to an individual compound, to a class of compounds, or to all chromatographable compounds in the wastewater, whether they are identified or not. This information, in conjunction with the traditional collective pollution parameters (*e.g.*, BOD and TOC), provides a much better understanding and evaluation of the effluent composition and its possible environmental effects than the chemical profiles or collective parameters alone.

Two of the neutral extracts from Mill A were analyzed by a Digilab Fourier Transform infrared spectrophotometer interfaced with a gas chromatograph (GC/IR). Spectra of the eluting compounds are obtained "on the fly" as they are with the GC/MS system. These analyses were helpful in confirming the functional groups of several unidentified compounds. Compounds 67, 68 and 70 were all shown to be ketones. Computer matching of the mass spectra of these same peaks had led to tentative identifications of 1-cyclohexenyl methyl ketone, 4-nonyne, and 3-cyclohexen-1-yl methyl ketone, respectively. The presence of ketone carbonyl absorption in the infrared spectra eliminates the possibility that one is 4-nonyne and confirms the methyl alkyl ketone structures of all three compounds. Four of the GC/IR identifications (camphor, borneol, 2-acetylthiophene, and 2-propionylthiophene) were confirmed by comparing the sample spectra with standards. The two terpenes had previously been confirmed by GC/MS, but the two thiophene isomers were confirmed only by GC/IR.

In low concentrations these neutral compounds are probably not significant pollutants. The chemical profile, supported by Table 36.2, shows that the treatment systems are able to reduce the concentration of most of these chemicals by large factors. An exception may be the terpene ketones. Several ketones either are not reduced in concentration or are increased in concentration in short-term biological oxidation systems. Ketones 65, 67 and 68 were not reduced in concentration in the Mill A wastewater and two (camphor and fenchone) were increased. The

Table 36.2 Neutral Volatiles Identified in Both Kraft Paper Mill Effluents with Approximate Concentrations

Peak No.	Compound Identified as Methyl Derivative (Parent Compound)	Confirmed By	Approximate Concentration in mg/l Mill A Sample Points				Approximate Concentration in mg/l Interstate Sample Points			
			1	2	3	4	1	2	3	4
4	Methyl trisulfide	MS,GC	0.003	0.004	0.008	0.001	—	—	—	—
62	Fenchone		0.007	0.007	0.015	0.015	0.015	0.002	0.001	<0.001
63	Hexachloroethane		—	—	—	—	—	—	—	<0.001
64	Sabinene		—	—	—	—	—	—	—	0.003
65	Unidentified terpene ketone	MS,GC,IR	0.055	0.050	0.055	0.045	0.040	0.010	0.015	0.004
66	Camphor		0.045	0.045	0.060	0.090	0.090	0.015	0.020	0.035
67	Unidentified terpene ketone		0.045	0.040	0.050	0.045	0.045	0.006	0.020	0.008
68	Unidentified terpene ketone		0.020	0.020	0.025	0.020	0.020	0.003	0.008	0.002
69	Fenchyl alcohol	GC	0.065	0.065	0.035	0.010	0.105	0.040	0.060	—
70	Unidentified terpene ketone		0.025	0.025	0.020	0.004	0.015	0.005	0.009	—
71	Terpene-4-ol	MS,GC	0.050	0.045	0.040	0.010	0.030	0.010	0.015	<0.001
72	2-Formylthiophene		0.010	—	—	—	—	—	—	—
73	Methyl chavicol	MS,GC	0.045	0.040	0.030	—	0.030	0.010	0.020	—
9	Borneol	MS,GC,IR	0.275	0.200	0.155	0.090	0.470	0.200	0.260	0.080
10	α-Terpineol	MS,GC	0.645	0.700	0.625	—	0.490	0.215	0.280	—
11	Veratrole	MS,GC	0.020	0.015	0.015	0.008	0.015	0.004	0.008	—
74	2-Acetylthiophene	GC,IR	0.025	0.025	0.030	0.025	0.012	0.002	0.004	<0.001
75	Myrtenol		0.010	0.010	0.008	—	0.008	—	—	—
76	2-Propionylthiophene	GC,IR	0.025	0.025	0.025	0.010	0.020	0.005	0.010	<0.001
77	Anethole		0.007	—	—	—	—	—	—	—
78	Benzyl alcohol		0.013	0.012	0.008	—	0.025	0.004	0.007	—
79	Methyl eugenol	GC	0.002	0.001	—	—	—	—	—	—
80	Unidentified terpene alcohol		0.006	0.007	0.010	—	0.030	0.015	0.015	0.008
81	Unidentified aromatic similar to methyl isoeugenol (MW=178)		0.006	0.008	0.009	—	0.009	0.006	0.008	—
30	Ethyl palmitate	MS,GC	0.006	—	—	—	—	—	—	—
82	Unidentified monounsaturated C_{19} fatty acid methyl ester		0.038	—	—	—	—	—	—	—
83	Unidentified diunsaturated C_{19} fatty acid methyl ester		0.016	—	—	—	—	—	—	—
84	Unidentified phthalate diester		—	—	—	—	—	—	—	—
Total			1.464	1.344	1.223	0.711	1.469	0.552	0.760	0.145

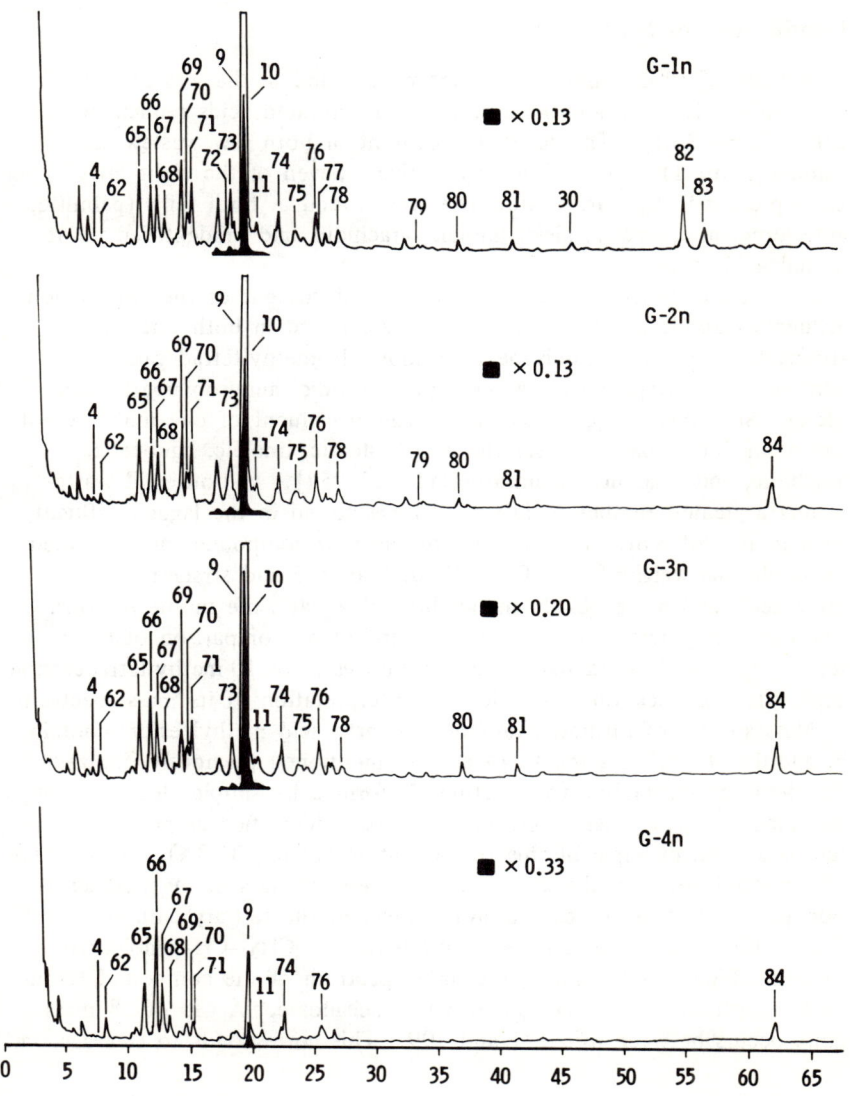

Figure 36.7 Chemical profile of neutral volatile compounds from Mill A extracts of sample points 1-4.

observation of an increase in the concentrations of camphor and fenchone in some instances has been verified in an independent report.[15]

Identification of Fatty Acids

A total of 17 different fatty acids were found in the wastewaters of both mills; 15 were identified and two unsaturated acids remain unidentified (Table 36.1). The fatty acid content of both raw wastewaters (sample point 1) were qualitatively similar. Seven of the eight fatty acids present in the raw wastewaters were found at both mills (palmitic, anteisomargaric, stearic, oleic, linoleic, arachidic, and unidentified unsaturated acid #54).

More dissimilarity existed in the fatty acid content of the two lagoon effluents (sample point 4). Ten acids were found in both effluents (isomyristic, myristic, anteisopentadecanoic, 10-methyltetradecanoic, pentadecanoic, isopalmitic, palmitic, palmitelaidic, anteisomargaric, and oleic). Six others were found in the lagoon effluent of one mill but not the other [ethyl palmitate (as the ester), stearic, linoleic, lignoceric, arachidic, and unidentified unsaturated acid #54]. In both mill wastewaters a greater number of fatty acids was found in the lagoon effluent than in the influent. The majority of the new compounds are saturated low-molecular-weight (C-14, C-15, C-16) branched and straight-chain fatty acids and are probably metabolites of aquatic life in the lagoons. The only fatty acid identification not verified by comparison of GC retention time with a standard is the methyl ester of 10-methyltetradecanoic acid. Its identification rests solely on interpretation of its mass spectrum.

Mass spectra of saturated unbranched fatty acid methyl esters contain, in addition to m/e fragments 74 and 87 mentioned previously, ionized fragments corresponding to structure IX formed by simple cleavage along the chain.[12] The mass spectrum of the peak identified as methyl palmitate is a good example of this fragmentation (Figure 36.8A). Introduction of a methyl group in the carbon chain causes changes in the fragmentation pattern because of easy cleavage alpha to the tertiary carbon atom.[16] This results in a gap of 28 mass units when the CH_3-CH- moiety is cleaved. Figure 36.8B shows the mass spectrum of the compound tentatively identified as methyl 10-methyltetradecanoate. A gap of 28 mass units occurs between m/e 171 and 199. This corresponds to cleavage on

$$\left[CH_3O-\underset{\underset{O}{\|}}{C}-(CH_2)_{\overline{n}} \right]^+$$
$$IX$$

IDENTIFICATION OF ORGANIC POLLUTANTS 691

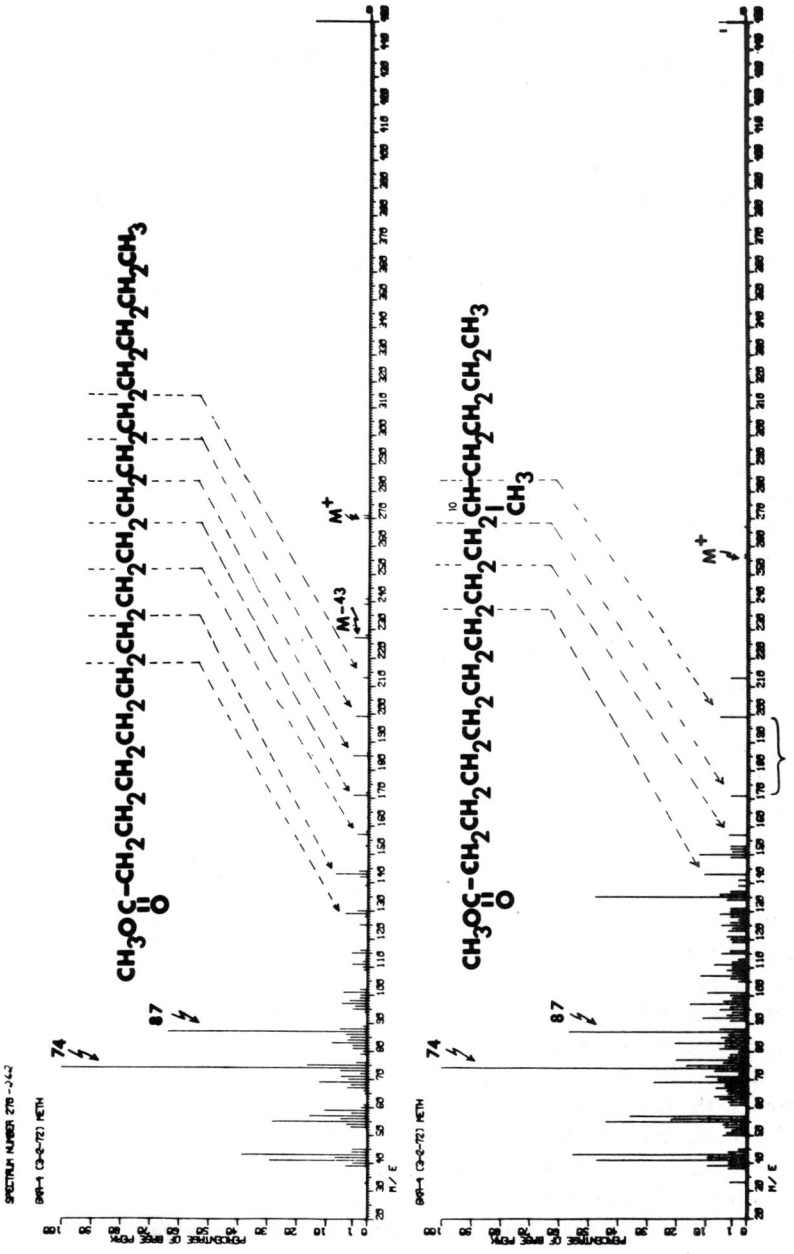

Figure 36.8 Mass spectra of (A) methyl palmitate and (B) methyl 10-methyltetradecanoate.

each side of the number 10 carbon. The mass spectrum of methyl 10-methyltetradecanoate is not in the mass spectral data base; however, all of the first five matches were fatty acid methyl esters.

The computer unexpectedly matched two mass spectra with ethyl palmitate (ethyl hexadecanoate). The base peak (m/e 88) and second largest fragment (m/e 101) are shifted 14 mass units higher than their counterparts in the mass spectra of saturated fatty acid methyl esters. These fragments would correspond to X and XI, respectively. However,

$$\left[\begin{array}{c} C_2H_5O-C=CH_2 \\ | \\ OH \end{array} \right]^+ \qquad \left[\begin{array}{c} C_2H_5O-C-CH_2-CH_2 \\ \| \\ O \end{array} \right]^+$$

$$X \qquad\qquad\qquad XI$$

they could also be rationalized as coming from structures XII and XIII corresponding to the compound methyl 2-methylhexadecanoate. The mass spectrum of the 2-methylhexadecanoate was not in the computer data bank. The mass spectrum of a standard of ethyl palmitate closely resembled the mass spectra of the sample peaks, and the GC retention time of ethyl palmitate was identical to the peaks in question. Spiking the samples with ethyl palmitate increased the size of this peak in the sample extracts, confirming the identity.

$$\left[\begin{array}{c} CH_3O-C=CH \\ | \quad | \\ OH \quad CH_3 \end{array} \right]^+ \qquad \left[\begin{array}{c} CH_3O-C-CH-CH_2 \\ \| \quad | \\ O \quad CH_3 \end{array} \right]^+$$

$$XII \qquad\qquad\qquad XIII$$

Fatty acids are not generally considered to be very toxic although the California "Water Quality Criteria" lists the minimum lethal dose (MLD) for several of them at 5.0 mg/l for fish.[17] Recent work by Leach and Thakore[18] has indicated that the sodium salts of unsaturated fatty acids are more toxic to fish than those of saturated fatty acids. Furthermore, among the C-18 unsaturated fatty acids, the toxicity of the sodium salts increased with increasing unsaturation in the order oleic < linoleic < linolenic (oleic has one, linoleic has two, and linolenic has three double bonds).

Little or no work has been reported on the analyses of fatty acids in treated paper mill effluents. We have found no previous report of odd-numbered carbon and branched fatty acids in treated wastewaters. The presence of 10-methyltetradecanoic acid and ethyl palmitate, although in small amounts, was unexpected.

Identification of Resin Acids

A total of thirteen different resin acids (Table 36.1) was found in the wastewaters of both mills in the 1972 samples; eight were identified and confirmed and five were not identified. The resin acid contents of the finished wastewaters were qualitatively similar for the two mills. Seven resin acids were common to the raw wastewater extracts of both mills and six of these were found in both finished wastewater extracts. Differences were primarily found among the small, unidentified resin acid GC peaks.

In addition to the isopimaric, abietic, dehydroabietic and sandaracopimaric acids found in a recent study by Leach and Thakore,[18] we identified pimaric, 6,8,11,13-abietatetraen-18-oic, neoabietic, 13-abieten-18-oic and several unidentified resin acids. To our knowledge neoabietic, 13-abieten-18-oic, and 6,8,11,13-abietatetraen-18-oic acids have not been reported before in kraft pulp mill wastewaters.* All of these resin acids (as their methyl esters) were initially identified by comparison of their mass spectra with published mass spectra.[19] They were all confirmed by comparison of gas chromatographic retention times and GC mass spectra from our instrument with standards obtained from Dr. Duane F. Zinkel, USDA Forest Products Laboratory, Madison, Wisconsin.

In general, most compounds decreased significantly in concentration as the wastewater passed through the treatment system. One exception appeared to be the resin acid content. Some of the peaks appeared to increase in concentration. This apparently was the result of large variances in the concentration of the organics in the mill wastewaters with time and our failure to have precisely sampled the "slug" of effluent we tried to follow through the treatment systems.

Resin acids have long been known to be toxic in various degrees to aquatic life. In addition, resin acid salts or "soaps" are responsible for much of the foaming in kraft mill effluents. The foam causes additional expense in the treatment of these effluents because defoamers must be added. If the foam is discharged to the receiving waters it can float long distances before it is dispersed or degraded.[4]

Rogers[20] has recently reported that bioassays using sockeye salmon have shown the lethal effects of a mixture of resin acids isolated from Douglas fir oleoresin to occur at concentrations as low as 2.0 mg/l. Others have found resin acid soaps to be toxic to minnows at 1.0 mg/l,[21] or to the water flea, *Daphnia pulex*, at 2.0 mg/l.[22] Resin acids have been reported to be toxic to various fish at concentrations of 1 mg/l to 5 mg/l.[23,24]

---
*Fox has now also reported identifying 6,8,11,13-abietatetraen-18-oic acid (Chapter 34 this volume). Brownlee and Strachan have now also reported identifying neoabietic acid (Chapter 35 this volume).

Identification of Phenols

Eleven different phenols were identified in the kraft mill wastewaters (Table 36.1). Based upon the data obtained in 1974, some correlation can be made between the percent removal of the phenols in the two waste treatment systems and the complexity or degree of substitution of the phenol molecules. Table 36.3 lists the percent removal of these phenols and shows their structures. The analytical data for the calculations were obtained from 3-day composite samples quantitated with the aid of the Perkin-Elmer PEP-1 computer system. The phenols are generally more resistant to treatment as the complexity of the molecule increases. This is especially true with Mill A, which has only biological treatment, with the single exception of p-hydroxybenzaldehyde, which might be produced in the aerated lagoons by biological degradation of lignin or vanillin.

Phenols have long been associated with kraft paper mill wastewaters and many studies of their effects have been made. Concentrations of phenol that have been reported as lethal or damaging to fish range from 0.079 to 1900 mg/l but the most reliable information relating to pure phenol under carefully standardized conditions indicates that the 24-, 48- and 96-hour TLm concentrations are in the general range of 10-20 mg/l at 20°C.[25] Probably a greater problem caused by the phenolic constituents of kraft mill wastewaters is the impairment of the flavor of fish, shrimp and other edible aquatic life. One recent investigation[26] showed that the flavor of cooked coho salmon was impaired after exposure of the fish to untreated kraft pulp mill effluent for 72-96 hours at concentrations of 1-2% or more by volume. No flavor impairment was noted when these fish were exposed to 2-9% by volume of biologically-treated effluent. These results are in agreement with our findings that, in general, the phenolic constituents of kraft paper mill wastewaters are reduced in concentration from 50 to 98% by biological treatment. The chemical profiles of the acids and phenols from both mills are shown in Figures 36.9 and 36.10. The peak numbers correspond to the compounds listed in Table 36.1. The compositions of the raw effluents from the two mills were similar, but the treated effluent samples from the two mills exhibited more differences.

Computer-Assisted Gas Chromatographic Analyses

Based on the compound identifications from the 1972 analyses, a computer-assisted gas chromatographic analysis was used to analyze the 1974 wastewater samples. GC/MS was used to verify compound identifications. Once compounds were identified, computer-assisted GC

Table 36.3 Percent Removal of Phenols Versus Their Structural Complexity

Parent Compound	Identified as	Structure	% Removed Mill A	% Removed Inter-state
Guaiacol	Veratrole	OH, OCH₃ on benzene	96%	98%
p-Hydroxybenzaldehyde	p-Methoxybenzaldehyde	OH, CHO (para) on benzene	21%	75%
Vanillin	Veratraldehyde	OH, OCH₃, CHO on benzene	65%	93%
Acetovanillone	Acetoveratrone	OH, OCH₃, COCH₃ on benzene	77%	90%
Homovanillic Acid	Methyl Homovanillate	OH, OCH₃, CH₂COOH on benzene	50%	?
Syringaldehyde	3,4,5-Trimethoxy-benzaldehyde	CH₃O, OH, OCH₃, CHO on benzene	62%	—
Acetosyringone	3,4,5-Trimethoxy-acetophenone	CH₃O, OH, OCH₃, COCH₃ on benzene	57%	67%

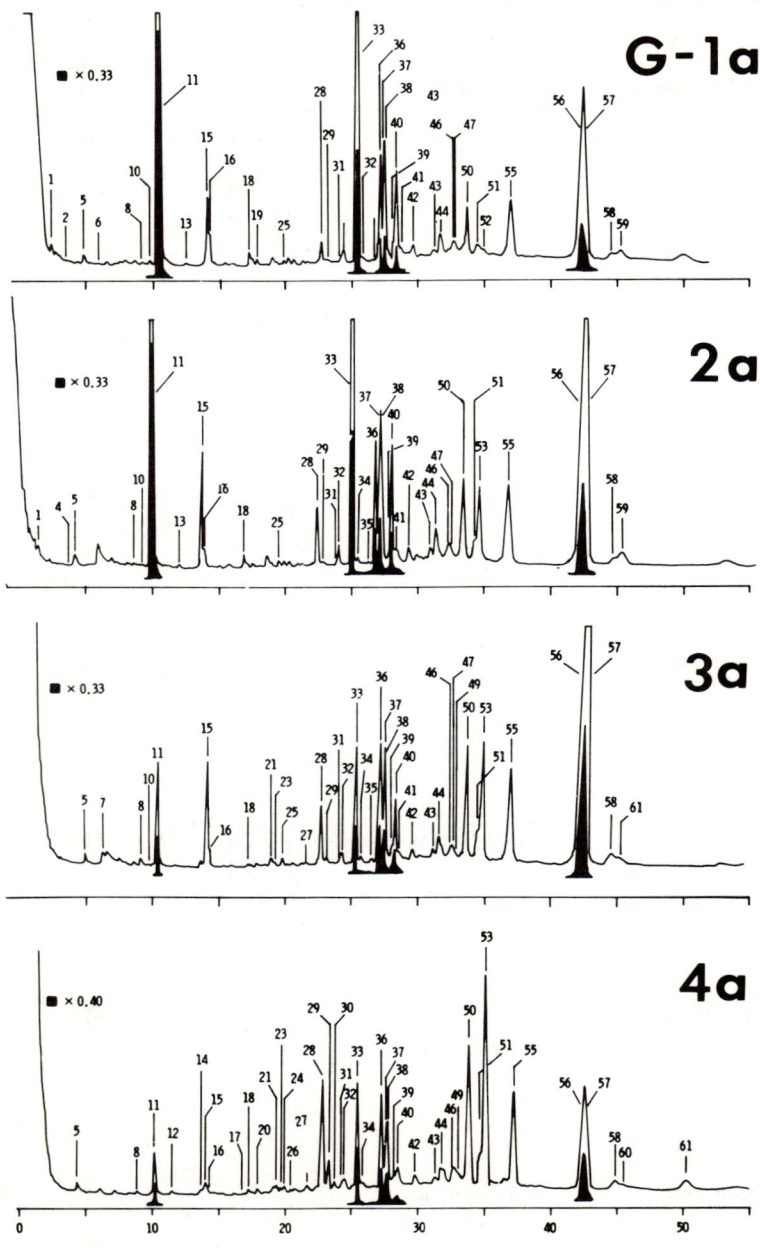

Figure 36.9 Chemical profile of acids and phenols from Mill A extracts of sample points 1-4.

IDENTIFICATION OF ORGANIC POLLUTANTS 697

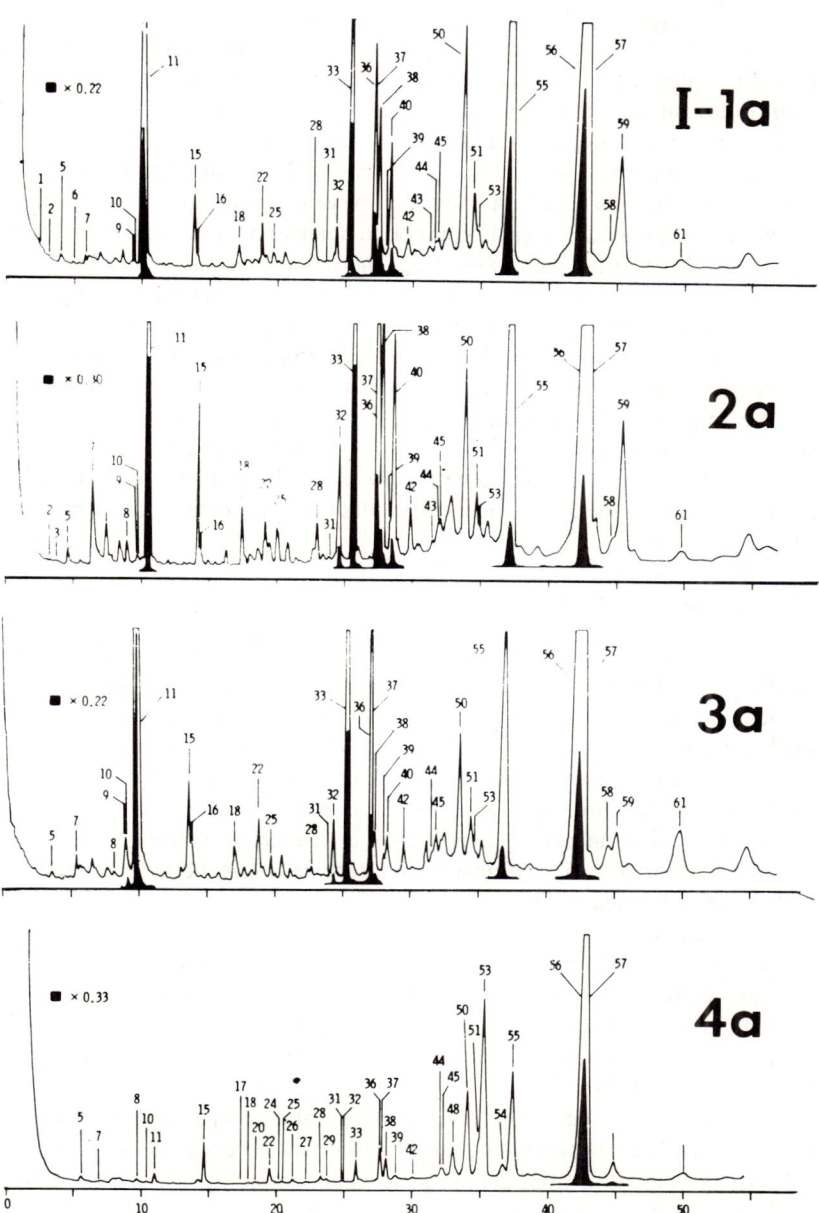

Figure 36.10 Chemical profile of acids and phenols from Interstate Mill extracts of sample points 1-4.

analysis was sufficient for subsequent qualitative and quantitative analyses. Chloroform extracts of 1 liter of each sample were concentrated to 1.0 ml. A 300-μl aliquot of each extract was spiked with 30 μl of a standard solution (11 mg/ml) of acenaphthene, the internal standard, bringing its concentration in the extract to 1.0 mg/ml. A standard solution of acenaphthene, veratrole, methyl palmitate, and methyl dehydroabietate was prepared and chromatographed using the PEP-1 computer system. From the known concentrations and the measured peak areas, response factors were calculated for each compound relative to that of acenaphthene (veratrole, 1.49; methyl palmitate, 2.05; methyl dehydroabietate, 2.32). Concentrations of the various phenols and fatty and resin acid methyl esters in the samples were then calculated by computer, assuming that all compounds in a given chemical class have similar response factors.

Computer analyses of the 3-day composite extracts from Mill A and from the Interstate mill are shown in Figures 36.11 and 36.12, respectively. Peak number designations, notation of the presence or absence of the compound in the corresponding 1972 sample, and the sum of all the concentration values were manually added to the data printed out by the computer.

Collective Pollution Parameters

Comparison of collective pollution parameters (Table 36.4) indicates that BOD removal is around 70-90% in both mill wastewaters during treatment. TOC reduction in the treated wastewaters of both mills probably varies between 60-90%. Because of large amounts of suspended solids in the raw effluent of Mill A, the TOC values may be erroneous. Reduction in total chromatographable acidic materials is about 65-80% in both mills. The decrease of only 24% in the acidic material for the 1972 sampling of Mill A is probably not representative because of faulty sampling, as explained earlier.

If treatment effectiveness is considered with respect to classes of compounds, the phenols appear to be the most susceptible to treatment. Reduction in the chromatographable phenolic content of each mill was very consistent, ranging in Mill A wastewaters from a 73% to a 77% reduction. Chromatographable phenols were reduced in the Interstate wastewaters by 94% to 95%.

Resin acid reductions in the treated wastewaters ranged from about 50% to 90%. The apparent increase in resin acid content of Mill A in 1972 may be due to a nonrepresentative grab sample in which the "slug" of effluent being sampled was apparently missed.

Fatty acid content of the Interstate wastewaters was decreased by about 85% to 90%, but the overall fatty acid concentrations increased

IDENTIFICATION OF ORGANIC POLLUTANTS 699

```
RUN      1  MILL A RAW EFFLUENT-ACIDS AND PHENOLS
INST  4  , METHOD  50  , FILE  37  3:
```

TIME	AREA	RRT	RF	C	NAME	PEAK	IN '72
10.14	3.1384	1.000,	1.4900,	.5051,	VERATROLE:	11	YES
13.66	1.7237	1.347,	2.0500,	.3817,	DIMETHYLSULFONE:	15	YES
16.70	.2877	1.646,	1.4900,	.0463,	METHOXYBENZALDEHYDE:	18	YES
18.48	.4908	1.822,	1.0000,	.0530,	:	—	--
19.10	9.2566	1.883,	1.0000,	1.0000,	ACENAPHTHENE:	A	--
20.08	.3557	1.980,	1.0000,	.0382,	:	—	--
22.10	1.3317	2.184,	2.0500,	.2949,	PALMITATE:	28	YES
		2.220,	2.0500,		PALMITELAIDATE:	29	YES
23.75	.4753	2.350,	1.4900,	.0765,	ME HOMOVANILLATE:	32	YES
24.73	5.2457	2.450,	1.4900,	.8443,	VERATRALDEHYDE:	33	YES
26.51	3.0875	2.629,	1.4900,	.4969,	VERATRONE:	36	YES
26.82	3.5064	2.661,	2.0500,	.7765,	STEARATE AND OLEATE:	37,38	YES,YES
27.47	.9110	2.726,	1.4900,	.1466,	ME SYRINGALDEHYDE:	39	YES
27.66	.7588	2.746,	2.0500,	.1680,	LINOLEATE:	40	YES
27.94	.2840	2.774,	1.4900,	.0457,	3,4-DMP:	41	YES
28.98	.9289	2.879,	1.4900,	.1495,	3,4,5-TMA:	42	YES
30.59	.5729	3.037,	1.0000,	.0618,		--	--
31.13	.5133	3.087,	2.3200,	.1286,	ME RESIN ACID:	—	NO
31.82	.4960	3.152,	2.3200,	.1243,	ME RESIN ACIDS B,C:	46,47	YES,YES
32.86	4.0798	3.250,	2.3200,	1.0225,	PIMERATE:	50	YES
34.10	6.7065	3.366,	2.3200,	1.6808,	SAND-P AND 13-AB-18-:	51,53	YES,NO
36.20	3.8577	3.563,	2.3200,	.9668,	ISOPIMERATE:	55	YES
41.92	26.4614	4.100,	2.3200,	6.6320,	AB- AND DEHYDROAB-:	56,57	YES,YES
44.11	.3482	4.305,	2.3200,	.0872,	6,8,11,13-AB-:	58	YES
44.59	.5035	4.350,	2.3200,	.1262,	NEOABIETATE:	59	YES
45.81	.4653	4.464,	2.0500,	.1030,	LIGNOCERATE:	60	NO

```
                            TOTAL:  14.9834 MG/L
```

```
RUN      MILL A TREATED EFFLUENT-ACIDS AND PHENOLS
INST  4  , METHOD  50  , FILE  36  3:
```

TIME	AREA	RRT	RF	C	NAME	PEAK	IN '72
10.12	.1291	1.000,	1.4900,	.0193,	VERATROLE:	11	YES
13.64	.1732	1.347,	2.0500,	.0356,	DIMETHYLSULFONE:	15	YES
16.70	.2452	1.650,	1.4900,	.0366,	METHOXYBENZALDEHYDE:	18	YES
17.42	.1129	1.721,	2.0500,	.0232,	MYRISTATE:	20	YES
18.39	.1112	1.817,	1.4900,	.0166,	AROMATIC M W 168:	--	--
18.71	.2002	1.848,	2.0500,	.0412,	ANTEISO C-15 :	23	YES
19.11	9.9590	1.888,	1.0000,	1.0000,	ACENAPHTHENE:	A	--
19.79	.1403	1.955,	2.0500,	.0288,	PENTADECANOATE:	26	YES
20.12	.1804	1.988,	1.0000,	.0181,	:	--	--
21.03	.1711	2.079,	2.0500,	.0352,	ISOPALMITATE:	27	YES
22.11	1.8876	2.187,	2.0500,	.3885,	PALMITATE:	28	YES
22.57	1.1136	2.234,	2.0500,	.2292,	PALMITELAIDATE:	29	YES
23.66	.2549	2.343,	1.4900,	.0381,	ME HOMOVANILLATE:	32	YES
24.44	.2631	2.421,	2.0500,	.0541,	MARGARATE:	M	NO
24.72	1.9480	2.450,	1.4900,	.2914,	VERATRALDEHYDE:	33	YES
26.48	.7781	2.626,	1.4900,	.1164,	VERATRONE:	36	YES
26.93	2.2860	2.671,	2.0500,	.4705,	STEARATE AND OLEATE:	37,38	YES
27.44	.3729	2.723,	1.4900,	.0558,	ME SYRINGALDEHYDE:	39	YES
27.64	.1832	2.743,	2.0500,	.0377,	LINOLEATE:	40	YES
27.92	.1779	2.771,	1.4900,	.0266,	PHTHALATE:	--	--
28.55	.3314	2.834,	1.0000,	.0332,	:	--	--
28.98	.4289	2.877,	1.4900,	.0641,	3,4,5-TMA:	42	YES
29.54	.2222	2.934,	1.0000,	.0223,	:	--	--
30.53	.6486	3.031,	2.0500,	.1335,	ARACHIDATE:	43	YES
31.04	.4631	3.081,	1.4900,	.0692,	DIHEXYLPHTHALATE:	44	YES
31.78	.2201	3.154,	2.3200,	.0512,	RESIN AND FATTY ACID:	46,49	YES
32.78	.8428	3.252,	2.3200,	.1963,	PIMERATE:	50	YES
33.96	1.4073	3.368,	2.3200,	.3278,	SAND-P AND 13-AB-18-:	51,53	YES
35.48	.2199	3.516,	2.0500,	.0452,	UNSAT FATTY ACID:	54	NO
36.06	.7912	3.573,	2.3200,	.1843,	ISOPIMERATE:	55	YES
41.43	2.5728	4.100,	2.3200,	.5993,	AB- AND DEHYDROAB-:	56,57	YES
44.17	.2192	4.368,	2.3200,	.0510,	NEOABIETATE:	59	NO
45.48	.2818	4.496,	2.0500,	.0580,	LIGNOCERATE:	60	YES
49.59	.1314	*******,	1.0000,	.0131,	!		

```
                            TOTAL:   3.8114 MG/L
NOTE: 10-METHYLTETRADECANOATE UNDER ACENAPHTHENE IS ESTIMATED AT
      A CONCENTRATION OF 0.03 MG/L.

   FATTY ACID METHYL ESTER AT 31.78 MIN. IS UNSATURATED.
```

Figure 36.11 Computer analysis of Mill A acids and phenols from (top) raw wastewater and (bottom) fully-treated effluent.

700 ORGANIC POLLUTANTS IN WATER

```
RUN           INTERSTATE RAW EFFLUENT-ACIDS AND PHENOLS
INST   4  , METHOD   50  , FILE   35   3:
```

TIME	AREA	RRT	RF	C	NAME	PEAK	IN '72
10.10	8.3184	1.000,	1.4900,	1.3637,	VERATROLE:	11	YES
11.89	.4989	1.177,	1.0000,	.0548,	:	--	--
13.65	2.7822	1.351,	2.0500,	.6275,	DIMETHYLSULFONE:	15	YES
15.60	.2228	1.544,	1.0000,	.0245,	:	--	--
16.66	.5923	1.649,	1.4900,	.0971,	METHOXYBENZALDEHYDE:	18	YES
17.39	.3927	1.721,	1.0000,	.0432,	:	--	--
17.85	.5308	1.767,	1.0000,	.0584,	:	--	--
18.36	1.6764	1.817,	1.0000,	.1844,	:	--	--
18.72	.6656	1.853,	1.4900,	.1091,	AROMATIC MW 182:	--	NO
19.07	9.0883	1.888,	1.0000,	1.0000,	ACENAPHTHENE:	A	--
20.05	.4136	1.985,	1.0000,	.0455,	:	--	--
22.05	.8961	2.186,	2.0500,	.2021,	PALMITATE:	28	YES
22.80	.2664	2.262,	1.0000,	.0293,	:	--	--
23.30	.3619	2.312,	2.0500,	.0816,	ANTEISOMARGARATE:	31	YES
23.68	.7918	2.351,	1.4900,	.1298,	ME HOMOVANILLATE:	32	YES
24.37	.2071	2.420,	2.0500,	.0467,	MARGARATE:	M	NO
24.66	7.7353	2.450,	1.4900,	1.2681,	VERATRALDEHYDE:	33	YES
26.41	3.8972	2.626,	1.4900,	.6389,	VERATRONE:	36	YES
26.71	2.2447	2.656,	2.0500,	.5063,	STEARATE AND OLEATE:	37, 38	YES, YES
27.57	1.9567	2.743,	1.4900,	.4413,	MOSTLY LINOLEATE:	40	YES
28.88	.4292	2.875,	1.4900,	.0703,	3,4,5-TMA:	42	YES
29.37	.2510	2.925,	1.0000,	.0276,	:	--	--
29.96	.2554	2.984,	2.0500,	.0576,	ARACHIDATE:	43	YES
30.47	.4545	3.031,	2.3200,	.1160,	ME RESIN ACID:	--	NO
31.01	.5225	3.081,	2.3200,	.1333,	ME RESIN ACID:	45	YES
31.73	1.1206	3.148,	2.3200,	.2860,	ME RESIN ACID MW 314:	--	NO
32.78	7.8243	3.246,	2.3200,	1.9973,	PIMERATE:	50	YES
33.60	2.1625	3.322,	2.3200,	.5520,	SANDARACOPIMERATE:	51	YES
33.98	.9214	3.357,	2.3200,	.2352,	13-ABIETEN-18-OATE:	53	YES
34.54	.6673	3.409,	1.0000,	.0734,	:	--	--
36.23	24.6361	3.566,	2.3200,	6.2886,	ISOPIMERATE:	55	YES
38.77	.6480	3.802,	1.0000,	.0933,	:	--	--
41.98	50.9977	4.100,	2.3200,	13.0176,	AB- AND DEHYDROAB-:	56, 57	YES, YES
42.80	.7900	4.176,	1.0000,	.0869,	:	--	--
44.62	6.8304	4.345,	2.3200,	1.7436,	6,8,11,13-AB, NEOAB-:	58, 59	YES, YES
45.92	.3696	4.465,	2.0500,	.0833,	LIGNOCERATE:	60	NO

TOTAL: 30.7714 MG/L

```
RUN           INTERSTATE TREATED EFFL-ACIDS AND PHENOLS
INST   4  , METHOD   50  , FILE   41   3:
```

TIME	AREA	RRT	RF	C	NAME	PEAK	IN '72
10.12	.1315	1.000,	1.4900,	.0220,	VERATROLE:	11	YES
		1.350,	2.0500,		DIMETHYLSULFONE:	15	YES
16.65	.1423	1.645,	1.4900,	.0239,	METHOXYBENZALDEHYDE:	18	YES
18.41	.1368	1.819,	1.0000,	.0154,	:	--	--
19.06	8.8748	1.883,	1.0000,	1.0000,	ACENAPHTHENE:	A	--
20.07	.1581	1.983,	1.0000,	.0178,	:	--	--
22.07	.2320	2.186,	2.0500,	.0535,	PALMITATE:	28	YES
22.74	.2390	2.254,	1.0000,	.0269,	:	--	--
24.41	.2475	2.423,	2.0500,	.0571,	MARGARATE:	M	NO
24.67	.5476	2.450,	1.4900,	.0919,	VERATRALDEHYDE:	33	YES
26.45	.3995	2.630,	1.4900,	.0670,	VERATRONE:	36	YES
26.89	.3413	2.675,	2.0500,	.0788,	STEARATE AND OLEATE:	37, 38	NO, YES
		2.720,	1.4900,		ME SYRINGALDEHYDE:	39	YES
28.96	.1394	2.885,	1.4900,	.0234,	3,4,5-TMA:	42	YES
29.51	.2972	2.941,	1.0000,	.0334,	:	--	--
30.02	.2021	2.991,	1.0000,	.0227,	:	--	--
31.06	1.6762	3.093,	2.3200,	.4381,	ME RESIN ACID:	45	YES
31.72	1.3443	3.157,	2.3200,	.3514,	ME RESIN ACID:	48	YES
32.74	2.9344	3.257,	2.3200,	.7670,	PIMERATE:	50	YES
33.91	3.6459	3.371,	2.3200,	.9530,	SAND-P AND 13-AB-18-:	51, 53	YES, YES
35.98	2.8126	3.572,	2.3200,	.7352,	ISOPIMERATE:	55	YES
41.39	11.7420	4.100,	2.3200,	3.0694,	AB- AND DEHYDROAB-:	56, 57	NO, YES
43.54	.2298	4.309,	2.3200,	.0600,	6,8,11,13-AB, NEOAB-:	58, 59	YES, NO
		4.460,	2.0500,		LIGNOCERATE:	60	NO

TOTAL: 6.9079 MG/L

Figure 36.12 Computer analysis of Interstate Mill acids and phenols from (top) raw wastewater and (bottom) fully-treated effluent.

Table 36.4 Collective Pollution Parameter Measurements and Total Concentrations of the Volatile Components in the Acid-Phenol Extracts

Concentrations (mg/l)

	Mill A						Interstate Paper at Riceboro, Georgia					
	1972			1974			1972			1974		
	Raw	Outfall	% Change	Raw	Outfall	% Change	Raw	Outfall	% Change	Raw	Outfall	% Change
BOD$_5$	323	88	-73	320	45	-86	438	70	-84	440	52	-88
TOC	240	230	-4	350	350	0	470	200	-57	490	85	-83
Total GC Organics	9.38	7.12	-24	14.98	3.81	-75	21.69	7.24	-67	30.77	6.91	-78
Total Phenols	4.96	1.15	-77	2.31	0.62	-73	5.53	0.27	-95	3.57	0.23	-94
Total Fatty Acids	1.01	1.46	+45	1.36	1.59	+17	1.40	0.14	-90	1.34	0.19	-86
Total Resin Acids	2.04	4.27	+109	10.77	1.39	-87	14.24	6.71	-53	24.37	6.37	-74

*Related to concentrations of total gas chromatographable organic material.

in both the 1972 and the 1974 samples of Mill A treated wastewaters. This is probably due to the production of fatty acids by the microbiota in the aerated lagoons. An increase in number of branched and odd-carbon fatty acids was also noted.

To maintain the proper perspective with respect to a mass balance, the total volatile acidic material as well as the three major classes of compounds comprising it is presented as a percentage of the TOC in Table 36.5. The total of all the volatile acidic components in the wastewater extracts was less than 10% of the total organic carbon content of

Table 36.5 Volatile Acidic Components as Percentages of TOC

	Percentage of TOC			
	Mill A - 1974		Interstate - 1974	
Component	Raw Effluent	Outfall	Raw Effluent	Outfall
Total Acidic Volatiles	4.28	1.09	6.28	8.13
Total Phenols	0.66	0.18	0.73	0.27
Total Fatty Acids	0.39	0.45	0.27	0.22
Total Resin Acids	3.08	0.40	4.97	7.49

the wastewaters. The neutral components were less than 1% of the TOC. The sum of all the volatile components therefore is still less than 10% of the TOC. Yet this relatively small amount of dissolved organic material probably represents the bulk of problem-causing compounds—both toxic and taste- and odor-causing compounds. The dark brown color of kraft pulp mill wastewaters is believed to be due to partially degraded lignin molecules. These high-molecular-weight nonvolatile compounds form a significant portion of the balance of the TOC. Other contributors to the TOC are the carbohydrates and, to a lesser extent, tannins and various other highly polar or nonvolatile compounds.

The increase in the proportion of the TOC represented by total volatiles in the Interstate wastewaters is a reflection of the greater decrease in the nonvolatile portion of the organic content as compared to the volatile content. Lime flocculation probably removed a large amount of the nonvolatile organics.

The major difference in the volatile organic content of wastewaters from the two mills was the fatty acid content. Whereas the fatty acids decreased significantly during the Interstate treatment, they increased during treatment in Mill A. Reductions in BOD, TOC, total GC peak areas, and phenolic and resin acid content were similar in the wastewaters from the two mills.

DISCUSSION

The development of GC/MS has been the single most important factor in our ability to identify specific organic pollutants in water, whether it be industrial wastewaters, as in this study, environmental receiving waters, or drinking water. Computer-assisted GC/MS has greatly reduced the amount of time required to make these identifications. From another viewpoint, within the time period of a given study it makes the identification process much more efficient and results in a larger number of organic chemical identifications.

With the results of this study it will now be possible to obtain routinely by GC analysis alone much pertinent data concerning the quantities of specific chemicals being discharged into environmental waters from other kraft paper mills. The efficiency of various types and combinations of existing treatment facilities can be measured and compared in their ability to remove undesirable compounds from kraft paper mill wastewaters. The methods of analysis are sensitive—the GC/MS techniques and extraction/concentration methods we employed permitted identification of some compounds at concentrations of less than 1 part per billion in the wastewaters.

Although these results primarily describe the analysis of only two Georgia mills and two different sampling periods, earlier analyses of the same mills,[2-4] while not as complete, support these data. Furthermore, an independent study by Hrutfiord[15] also supports some of these results. A number of general conclusions can be drawn from this study:

- Raw effluents from kraft pulp mills producing paper from the same raw materials contain essentially the same volatile organics.
- The volatile organic composition of kraft wastewaters remains relatively constant over the years as long as the raw materials remain the same and the process is not changed.
- The relative quantities of the volatile organics in pulp mill wastewaters can fluctuate rapidly.
- Because of rapid fluctuations in the quantities of organics in the wastewaters, composite samples are needed for representative quantitative results.
- When peaks in the GC of a pulp mill wastewater have been identified, routine GC analyses in a mill laboratory are feasible.

Conclusions can also be reached from a comparison of the effects of biological treatment (Mill A's trickling filter and aerated lagoon) and chemical/biological treatment (Interstate's lime flocculation/stabilization lagoon) on the volatile organic compounds in these wastewaters. All values used in these comparisons came only from the January 1974 composite samples.

- BOD reductions were in the range of 85-90% with both treatments.
- TOC reductions were in the range of 80-85% with chemical/biological treatment and 60-65% with only biological treatment.
- Total volatile acids and phenols were reduced by 75-80% in both treatments.
- Total resin acids were reduced in concentration by 87% with biological treatment but were still present at about 1.4 mg/l in the treated effluent. Total resin acids were reduced in concentration by 74% with chemical/biological tratment but were still present at about 6.4 mg/l in the treated effluent.
- Total phenols were reduced in concentration by 73% with biological treatment and were present at 0.6 mg/l in the treated effluent. They were reduced by 94% with chemical/biological treatment and were present at 0.2 mg/l in the treated effluent.
- Total fatty acids increased 17% in concentration with biological treatment and were present at 1.6 mg/l in the treated effluent. They decreased by 86% with chemical/biological treatment and were present at 0.2 mg/l in the treated effluent.
- Biologically treated wastewaters from both mills contained a greater *number* of fatty acids than did the raw wastewaters. The new fatty acids were branched and odd-numbered carbon compounds.
- Total volatile materials ranged from 1-10% of the TOC.
- Terpenes were reduced by 90% or more by both treatments.

Only one difference was noted in the compositions of the two effluents. The total fatty acid content was decreased in Interstate's treated effluent but not in that of Mill A. The fatty acids in the oxidation ponds of Mill A may be biological metabolites from aquatic life in these ponds.

This study should raise many questions and hopefully stimulate further work that will result in an accumulation of data from many other mills with other types of wastewater treatment. Although we relied heavily on GC/MS for compound identification, when the identifications are once made and GC retention times are established, GC/MS is not always necessary. Much useful identification work can be accomplished with GC and standards of the compounds identified in this report. Other wastewater treatment systems should be evaluated using the techniques developed for this study. In particular, the effectiveness of activated sludge and of activated carbon filters in reducing the concentrations of the organics identified in these wastewaters would be useful information. The following questions should be answered:

- Does the chemical composition of kraft paper mill wastewaters remain fairly consistent across large geographic areas or only within smaller geographic areas such as the Southeast?

- How does the composition of these wastewaters vary with changes in both types and amounts of wood and changes in the pulping process?
- Does the chemical composition of these wastewaters vary with season?
- What types of compounds are likely to build up in concentration if the wastewater is recycled?
- Can the in-plant source of the more refractory compounds be determined so that one or more small-volume wastewater streams can be selected for special intensive treatment *before* they are combined with other in-plant wastewaters?

Chronic and acute toxicity data are lacking for most of these compounds. Now that they have been identified as environmental contaminants they can be examined in more detail for their ecological effects. Taste and odor thresholds will hopefully be determined for the phenolics, where this information is lacking. This kind of supplemental data, when added to a computerized Distribution Register of Organic Pollutants in Water (acronym: Water DROP) presently being developed by EPA, will be an invaluable aid to many other laboratories and programs concerned with pollution studies.

ACKNOWLEDGMENTS

All sample preparations and gas chromatographic separations were done by Terry Floyd. Mass spectral data were provided by Ann Alford and Mike Carter. Infrared data were provided by Leo Azarraga and Ann McCall.

Charles Davis, Lloyd Chapman and William J. Verross provided valuable assistance and wastewater samples from the Interstate Paper Corporation at Riceboro, Georgia. Officials at Mill A, who prefer to remain anonymous, were also very helpful in providing both samples and information. Without the cooperation of the staff from these two mills this study would not have been possible.

Duane F. Zinkel (USDA Forest Products Laboratory, Madison, Wisconsin) kindly supplied us with reference standards of the resin acids.

REFERENCES

1. National Academy of Sciences. Report of the Committee for the Working Conference on Principles of Protocols for Evaluating Chemicals in the Environment, "Principles for Evaluating Chemicals in the Environment," Washington, D. C. (1975), p. 299.
2. Keith, L. H. "Identification of Organic Contaminants Remaining in a Treated Kraft Paper Mill Effluent," 157th National Meeting of the American Chemical Society, Division of Water, Air and Waste Chemistry (1969), pp. 76-81.

3. Keith, L. H., A. W. Garrison, M. M. Walker, A. L. Alford and A. D. Thruston, Jr. "The Role of Nuclear Magnetic Resonance Spectroscopy and Mass Spectrometry in Water Pollution Analysis," presented at the 158th National Meeting of the American Chemical Society, Division of Water, Air and Waste Chemistry (1969), pp. 3-6.
4. Garrison, A. W., L. H. Keith and M. M. Walker. "The Use of Mass Spectrometry in the Identification of Organic Contaminants in Water from the Kraft Paper Mill Industry," presented at the 18th Annual Conference on Mass Spectrometry and Allied Topics (1970), pp. B205-213
5. Davis, C. L. "Color Removal from Kraft Pulping Effluent by Lime Addition," U.S. Environmental Protection Agency, Washington, D.C., Publication Number 12040 ENC (December 1971), p. iii.
6. U.S. Environmental Protection Agency. "Color Removal from Kraft Pulping Effluent by Lime Addition," Technology Transfer Capsule Report 2 (1972).
7. McGuire, J. M., A. L. Alford and M. H. Carter. "Organic Pollutant Identification Utilizing Mass Spectrometry," U.S. Environmental Protection Agency, Washington, D. C., Publication Number EPA-R2-73-234 (July 1973), pp. 10-13.
8. Hoyland, J. R. and M. B. Neher. "Implementation of a Computer-Based Information System for Mass Spectral Identification," U.S. Environmental Protection Agency, Washington, D.C., Publication Number EPA-660/2-74-048 (June 1974), pp. 5-33.
9. Keith, L. H. "Analysis of Organic Compounds in Two Kraft Mill Wastewaters," U.S. Environmental Protection Agency, Washington, D.C., Publication Number EPA-660/4-75-005 (June 1975), p. 99.
10. Bicho, J. G., E. Zavarin and D. L. Brink. "Oxidative Degradation of Wood II," *TAPPI* **49**, 218-226 (1966).
11. Brink, D. L., Y. T. Wu, H. P. Noveau, J. G. Bicho and M. M. Merriman. "Oxidative Degradation of Wood IV," *TAPPI* **55**, 719-721 (1972).
12. Ryhag, R. and E. Stenhagen. "Mass Spectrometry of Long-Chain Esters," in *Mass Spectrometry of Organic Ions*, F. W. McLafferty, Ed. (New York: Academic Press, 1963), Chapter 9, pp. 399-443.
13. Hertz, H. S., R. A. Hites and K. Biemann. *Anal. Chem.* **43**, 681 (1971).
14. Hoyland, J. R. and M. B. Neher. "Implementation of a Computer-Based Information System for Mass Spectral Identification," U.S. Environmental Protection Agency, Washington, D.C. Publication Number EPA-660/2-74-048 (June 1974), p. 1.
15. Hrutfiord, B. F., T. S. Friberg, D. F. Wilson and J. R. Wilson. "Organic Compounds in Pulp Mill Lagoon Discharge," U.S. Environmental Protection Agency, Washington, D.C., Publication Number EPA-660/2-75-028 (June 1975), p. 15.
16. Odham, G. and E. Stenhagen. "Fatty Acids," in *Biochemical Applications of Mass Spectrometry*, G. Waller, Ed. (New York: Wiley-Interscience, 1972), pp. 211-228.
17. McKee, J. E. and H. W. Wolfe. In "Water Quality Criteria," California State Water Resources Control Board, Sacramento, Publication Number 3-A, Second Edition (1963), p. 187.

18. Leach, J. M. and A. N. Thakore. "Identification of the Constituents of Kraft Pulping Effluent that are Toxic to Juvenile Coho Salmon (*Oncorkynchur kisutch*)," *J. Fish. Res. Brd. Canada* **30**, 479-483 (1973).
19. Zinkel, D. F., L. C. Zank and M. F. Wesolowski. "Diterpene Resin Acids," U.S. Department of Agriculture Forest Service, Madison, Wisconsin, Forest Products Laboratory, pp. C1-C32 (1971).
20. Rogers, I. H. "Secondary Treatment of Kraft Mill Effluents: Isolation and Identification of Fish-Toxic Compounds and Their Sublethal Effects," *Pulp and Paper Magazine of Canada* **74**, T303-T308 (1973).
21. Van Horn, W. M., J. B. Anderson and M. Katz. "The Effect of Kraft Paper Mill Waters on Fish Life," *TAPPI* **33**, 209-212 (1950).
22. Maenpaa, R., P. Hynninen and J. Tikka. "On the Occurrence of Abietic and Pimaric Acid Type Resin Acids in the Effluents of Sulphite and Sulphate Pulp Mills," *Pap. ja pun.* **50**, 143-150 (1968).
23. Hagman, N. "Resin Acids in Fish Mortality," *Finnish Paper and Timber J.* **18**, 32-34 and 40-41 (1938).
24. Ebeling, G. "Recent Results of the Chemical Investigation of the Effect of Wastewaters from Cellulose Plants on Fish," *Vom Wasser* **5**, 192-200 (1931); *Chem. Abstr.* **36**, 2262.
25. McKee, J. E. and H. W. Wolf. "Water Quality Criteria. California State Water Resources Control Board," Sacramento, Publication Number 3-A, Second Edition (1963), p. 238.
26. Shumway, D. C. and G. G. Chadwick. "Influence of Kraft Mill Effluent on the Flavor of Salmon Flesh," *Wat. Res.* **5**, 997-1003 (1971).

NOTE

This information is published in part in *Environmental Science & Technology*, June, 1976.

INDEX

AAF
  See 2-acetylaminofluorene
AAF, N-hydroxy-   53,54
6,8,11,13-abietatetraen-18-oic acid
  650,685,693
13-abieten-18-oic acid   685,693
abietic acid   650,664,666,685,693
acenaphthene   698
acetaldehyde   94
acetic anhydride   509
acetone   252,354
acetosyringone   685
acetovanillone   649,652-654,666,
  667,684
2-acetylaminofluorene   53-54,65-66
2-acetylthiophene   687-688
acids   9,300,463
activated carbon   4,6,148,271-
  272,354,370,400-401,425,453,
  457,464,468,471,474-475,512,
  518,539,553,675
activated sludge   518,520,536-
  539,541,546,553
Alachlor   146,337,345-349,354-
  355,365,408
alachlor chlorine homolog (ACH)
  346,348-349,355,365
alcohols   300,463,525
aldehydes   445
aliphatic esters   593
aliphatic hydrocarbons   7-8,445
n-alkanes   483,588-589,621
alkyl benzenes   577,583,621
alkyl biphenyls   621
alkyl halides   91,96
alkyl-substituted methyl benzoates
  599

alkyl-substituted methyl sulfides
  599
alkyl-substituted naphthenic
  methyl esters   599
alkylated chlorobiphenyls   21
alkylated methyl-indeneols   615
amines   255-256,300
ammonia   240-241
anion exchange resin   217
anisole   593,599,621
anthracene   210,621
aqueous injection   422,424,502,
  509,546
Arizona, Tucson   331,362
Aroclor 1016
  See polychlorinated biphenyls
aromatic acids   525,593
aromatic hydrocarbons   8,440,
  442,463,526,577
asbestos   419
atrazine   138,143-148,261,337,
  341-343,349,354-355,362,365,
  368-369,408
atrazine, deethyl-   349,365
automatic compositors   501

bacterial mutagenesis   53,71
benzaldehyde   393,548
benzene   252,313,509,513
1,2-benzocyclo-octa-1,3,5-triene
  32
benzoic acid   591
benzo(ghi)perylene   209
1,12-benzoperylene   441
benzophenone   583
benzo(a)pyrene   51,63,208-209

benzoquinones   463
benzothiophene   614,622
benzyl alcohol   440,548,688
bioaccumulation   208,297,299
bicyclothiacyclanes   622
bioassay   507,627-628
bioconcentration factor   299,
   301-302
biological treatment   687,694,
   703-704
biomagnification   151
biphenyl   24
bladex
   See cyanazine
borneol   687-688
branched chain acids   524
British Columbia   626,638
2-bromobutane   101
bromochloroiodomethane   408,
   411
bromodichloromethane   89,105,
   123,317,321,354,356,362,402,
   404,407-408,410-411,421,513
bromodiiodomethane   411
bromoform   105,318,354,356,
   362,365,386,388,404,407-408,
   410-411,513
1-bromonaphthalene   44
bromotrichloroethylene   513
bromotrichloromethane   323
butachlor   365

CAE's
   See carbon alcohol extracts
caffeine   526-527,541,548
calcium hypochlorite   403
camphor   687-688
Canadian waters   194-195,
   198-200
cancer   51,71,205
capillary columns • 75,77-78,
   81-82,186-187,194,200,204,
   335-336,434-436
captan   261
carbazoles   622
carbohydrate analyzer   217
carbohydrates   283
Carbon-14   6

carbon alcohol extracts (CAE)
   4,6,458-459,461-462,465
carbon chloroform extracts (CCE)
   4,6-7,458-460,465
carbon tetrachloride   105,421
carcinogenic metabolite   64
carcinogens   49,51-53,55,64,71,
   205,255,500
carcinogens, presumptive   52-53,
   58,63-65,67,70-71
CCE's
   See carbon chloroform extracts
cerate oxidimetry   217,235
Charles River   205-206,210-212
chemical derivatization   159
chemical ionization mass spectrometry (CIMS)   170,189,200,
   204,235,337,340,346,523
chemical mutagenesis   54
chemical profiles   687,689,696-697
chloralkylene-12   21,24
chloral   351-352,362
chloral hydrate   351-354
chloramine   236
chlordane   11,301,334
chlordene   334
chlorinated acetones   251
chlorinated alcohols   251
chlorinated aliphatics   251
chlorinated anisoles   582-583
chlorinated aromatics   251,364
chlorinated aromatic acids   243
chlorinated benzenes   582-583
chlorinated catechols   631
chlorinated compounds   311
chlorinated dibenzofurans   37
chlorinated guaiacols   626,631,
   636-637
chlorinated hydrocarbons   69,248
chlorinated nucleosides   243
chlorinated organics   251,253
chlorinated pesticides   8,11
chlorinated phenols   243,582
chlorinated phthalates   251
chlorinated purines   243
chlorinated pyrimidines   243
chlorinated veratroles   631
chlorination   89,105,223,233-234,
   236,240-242,247-249,306,313,
   349,352,381,395,411-412,420,

425,427,431,447,450,518,521,
546,548,553,559,582,589,626-
627,633,636-638
chlorine 96,110,240,242,271
chlorobenzene 313,318,362,513
chlorobinaphthyl 45
chlorobiphenylols 37
chlorocyclohexane 525,548
5-chlorocytosine 242
chlorodibromomethane 91,105,
 317,321,362,365,404,407,421
chlorodiiodomethane 410-411
4-chloro-2,6-dimethoxyterphenyl
 43
chloroethane 421
1,2-*bis*(2-chloroethoxy)ethane 395
bis(2-chloroethyl) ether 7,356,
 365,377-378,386,389-390,395,
 407,509
chloroform 89-92,105,123,158,
 242,251-253,313,314,317,320-
 321,331,352-354,356,386,388-
 389,394,402,408,411-412,420-
 424,431,513,525
bis(2-chloroisopropyl) ether 356,
 365,402,404,407,509
α-chloroketones 447
chloromethane 421
1-chloronaphthalene 44-45
chloro-organics 240,242-243
chlorophenols 10
chloropicrin 173-174,312-313,
 317,320-321,323-324,362
chloroplatinates 9
chlorotoluene 313,320,323,362
4-chloroterphenyl 42
5-chlorouracil 242
3-chloroveratrole 633
4-chloroveratrole 633
chloroxylene 320,323-324
cholesterol 475,526-527,541
chrysene 207-209
^{36}Cl analysis 236
clofibrate metabolite 523,527,
 536-539,548
coal liquefaction 216,224
coal pyrolysis 224-226,229-230
coal tar 51
compliance monitoring 499-501,
 505

composite sampling 141
computer analysis 699-700
computer matching 675,686
computerized quantitation 523
concentration methods 502
continuous dialysis 129
copper 271
coprostanol 526-527,541
coulometric detectors 11
m-cresol 591,595
cyanazine (bladex) 356,365
cyanogen chloride 420-421
cyclic esters 593,599
1-cyclohexenyl methyl ketone
 687
cytosine 242

D.C., Washington 519,537
2,4-D
 See 2,4-dichlorophenoxyacetic
  acid
dacthal 408
DDD 151
DDE 138,143-148,151,301,357,
 561
DDT 35,69,151,301,558,561
decafluorotriphenylphosphine 165
n-decanal 445-446
dechlorination 381
deethylatrazine (desethylatrazine)
 See atrazine, deethyl-
dehalogenation 37
dehydroabietic acid 488,631,
 650,652-659,664,666-669,677,
 685,693
Delaware River 376
detergents 8,283
dialysis 125-129,133
diazoethane 631
diazomethane 482,679-680,686
dibenzothiophene 622
dibromochloromethane 357,386,
 388,402,410-411,513
1,2-dibromoethane 107
dibromoiodomethane 411
2,6-di-*t*-butyl-*p*-benzoquinone
 357,362,365
dibufylnitrosamine (DBN) 257
dibutyl phthalate 224,541,583

dichloramine 240-242
dichloroacetonitrile 420-421
dichlorobenzenes 313,394,437,
  445-447,507-508,513
2,4'-dichlorobiphenyl 24
dichlorobromomethane 354,386,
  388,408
1,1-dichloroethane 421
dichloroguaiacol 637
dichloroiodomethane 313,318,
  323,354,365,403,408,410-412,
  513
dichloromethane 421,525
2,4-dichloro-6-methoxyterphenyl
  43
dichloronaphthalene 44
1,2-dichloronaphthalene 44
1,8-dichloronaphthalene 44
dichloronitromethane 320
2,4-dichlorophenoxyacetic acid
  (2,4-D) 148
2,4-dichloroterphenyl 42
dichloroveratrole 664-665
3,4-dichloroveratrole 633-634
3,5-dichloroveratrole 633-634
3,6-dichloroveratrole 633-634
4,5-dichloroveratrole 633-634
dieldrin 138,143-148,151,334,
  342,354,357,365,558
diethylamine 255
diethylnitrosamine (DEN) 257
diethylstilbestrol 475
dihydrocarvone 354
9,10-dihydroxystearic acid 664
dimeric quaterphenyls 37
3,4-dimethoxyacetophenone 664
dimethyl benzothiophenes
  614-615
dimethyl phthalate 394,658
dimethyl sulfate 678-680,686
dimethyl trisulfide 507
dimethylamine 255,259
2,6-dimethyl-4-heptanone 323
dimethylnitrosamine (DMN)
  255,257,259-261
1,2'-dinaphthyl 8
2,4-dinitrotoluene (DNT) 271
dioctyl adipate 526
dioctyl phthalate 224,541,583,
  658,665-666,668

dipropylnitrosamine (DPN) 257
drinking water 89,96,105,147,151,
  164,233,261-262,311,313,326,
  329,348,351,354-355,362,368,
  371,375-377,384-386,393-394,
  400,404,408,431,447,500,510
Dübendorf 445

earthquakes 9
ecological effects 705
EDTA 271
elemental composition chromatograms 567,577,582
endrin 9,177,301,334,354,357
environmental carcinogenesis 69
environmental monitoring 157
epijuvabione 650
esters 300
ethanol 94,107
ethers 300
ethylbenzene 252
ethylbenzothiophene 615
ethyl dehydroabietate 665
2-ethyl-1-hexanol 383,389
2-ethyl-4-methyl-1,3-dioxolane
  509-510
ethyl parathion 505,507-508
ethyl palmitate 665,692
explosives 265,267-268,271-272,
  274,276,278
extracted ion current profile
  (EICP) 166

fatty acids 474,476,483,488-489,
  493,518,524,527,531,548,589,
  630-631,637,647,650-651,657-
  658,664-666,669,683-686,690,
  692,698,701-702,704
fenchone 687
fenchyl alcohol 688
field ionization 170
field polarographs 266,272,275-
  276,278
fish tainting 662
flame retardants 16,21,24,30
Flat River 419
Florida, Miami 97,305-306,311-
  312,314,321,331,362,510,512,514

fluoranthene 210,621
fluorotrichloromethane 420-421
foam 693
2-formylthiophene 688
fractional purging 95,97
Fraser River 626
fulvic acid 283-284,286,289-291, 294

gas chromatography/high resolution mass spectrometry (GC/HRMS) 340,346,558,593,615
gas chromatography/infrared spectroscopy (GC/IR) 337,340, 346,348,523,632-634,687
gas chromatography/mass spectrometry (GC/MS) 157-158, 164,166,170,179-180,182,185, 194,206,208,212,253,297,306, 309-310,313,322,326,332,334-335,337,340,343,371,376-377, 382,384-385,389,393-395,399-402,404,422,424,435,443,447, 453,460-461,473,482,500,502, 505,507-508,510,512,522-523, 538,546,552,558,628,631,635, 638,645-646,663-664,674,681, 686-687,694,703
Georgia, Athens 527
Georgia, Riceboro 672
geosmin 11,509
Glatt River 443-444
glycol-1-palmitate 224
Great Lakes 479-480,493,641
Greifensee River 443,445-446
ground water 110,146,453-455, 457-459,465,467,469,472-476, 512,518,558-559
guaiacols 631,636-637,649,652-655,677,686

Hall electrolytic conductivity detector 505,512
haloforms
  See halomethanes
halogenated compounds 105,314
halomethanes 97,101,106,109-110,394

HCE 151
headspace analysis 113,133,508, 510
n-heptadecane 437
herbicides 364
2,2',4,4',5,5'-hexabromobiphenyl 29
hexabromocyclododecane 32
hexachlorobenzene 301,407,582, 669
hexachlorocyclopentadiene 427
hexachloroethane 358,363,365, 386,407-408,525,548,688
hexadecane 158
high performance liquid chromatography (HPLC) 257,261
high pressure liquid chromatography (HPLC) 224,234,236,242,248, 299
high resolution gas chromatography/ high resolution mass spectrometry (HRGC/HRMS) 558,561-562, 582,587,591
high resolution liquid chromatography (HRLC) 215-216,231
HMX
  See 1,3,5,7-tetranitro-1,3,5,7-tetraazacyclooctane
hollow-fiber membranes 129
homovanillic acid 684
Houston Ship Channel 61
humic acid 242,284,286,288-289, 291,294
humin 284,286,291-292,294
β-humus 284,286,289
hydrocarbons 300,437,483,488-489,493,577,582,669
β-hydroxy acids 531,538
p-hydroxybenzaldehyde 694
p-hydroxyphenyldecanoic acids 538
hymathomelanic acids 283
hypochlorous acid 240-241

Illinois, Cairo 11
indene 526
indene esters 593,599
indeneols 621
Indiana, Evansville 404,406-407

indoles 622
infrared spectroscopy (IR)
    5,8-10,178
inositol 224
internal standard 101,107
iodide 271
iodoform 410-411
2-iodopropane 101
ionic strength 115,117
Iowa, Ottumwa 305-306,314,
    321,331,362
Iowa cities, pesticides in 142,
    144-145,147
Iowa rivers, pesticides in 142,
    145,150-151
isocratic elution 227
isopimaric acid 650,664,685,693
isopropylbiphenyl isomers 24
isotopic dilution 241
isotopic exchange 236
isovaleric acid 546

Jackfish River 656

Kanawa River 11
Kansas, Kansas City 408
Kansas River 408,412
Kentucky, Brandenburg 402,404
7-ketodehydroabietic acid
    664,666-667

lagoons 672,694
Lake Erie 483,488-489
Lake Helen 656
Lake Huron 489,493
Lake Marion 218
Lake Ontario 480,483,488-489
Lake Superior 299,493,641-642,
    661
landfills 455,459,464-465,467,
    469,512-514
Lasso
    See Alachlor
Lazo
    See Alachlor
leaching 455,464
lindane 354,363

lignins 283,628,637,694
lime 672,702-703
liquid chromatography 12,78
liquid-liquid extraction 7,502,
    512,520
liquid-liquid extraction, continuous
    380,390,392
Louisiana, New Orleans 5,9,88,
    144-145,147,261,305,329-332,
    335,339,342,348-350,354-355,
    368-370
Louisiana, Terrebonne Parish
    331,351,362
lyophilization 216
lysimeters 467

macroreticular resin
    See XAD resins
mannitol 224
Maryland, Baltimore 257,259-262
Massachusetts, Boston 212,260
Massachusetts, Lawrence 331,362
Massachusetts, Waltham 261
mass chromatography (MC) 166,
    325,390,428-430,435-437,441,
    447,562,599,606-607,611,613-
    615
mass defects 482
mass spectral search system (MSSS)
    160,162,180-181,505,523
mass transfer 126
membrane/mass spectrometric
    response 120-124
membrane/mass spectrometry
    119,123,133
membrane probe 124
membrane separations 118
Meramec River 408
2-mercaptobenzothiazole 684
metabolic activation 53
metallurgical industry 281
metals 281-283,288,292,294-295,
    299,419
MethElute 679-681,686
methoxychloroterphenyls 43
methoxyclor 301
methoxyethoxytrichlorocatachol
    635
methoxynaphthalene 44

Methyl-8   679-681,686
methyl anisole   593
methyl benzoate   593,595
methyl benzothiophenes   614
methyl dehydroabietate   665,678, 698
methyl docosanoate (behenate)   483,485
methyl indan   621
methyl isobutyl ketone   507
methyl 2-methylhexadecanoate   692
methyl palmitate   678,698
methyl parathion   505,507-508
methyl trisulfide   683,688
methylene chloride   159
4,5-methylenephenanthrene   441
o-methylinositol   224
10-methyltetradecanoic acid   690
Mississippi River   234,236,239, 241-242,261,329-330,350
Missouri, Jefferson City   408
Missouri, Kirkwood   408
Missouri, St. Louis   8
Missouri, St. Louis county   129
Missouri River   403,408,410-413
monochloramine   240-241
monocyclic alkanes   621
mutagenesis, bacterial   53,71
mutagenesis, chemical   54
mutagenic activity   59-60,64,69
mutagenic activity ratios   55,58-59,70
mutagenic analysis   69
mutagens   66
mutagens, urinary   70

1-naphthol   45
naphthalenes   427,621
naphthenic esters   593
National Organics Reconnaissance Survey (NORS)   87,105,305-306,320,330,332,334,342,352, 354,370
natural products   464
natural waters   218,221,223
neoabietic acid   631,664,685,693
New York, New York   331,351, 362
Niagara River   488

nicotine   526-527,541,548
Nipigon Bay   642,644-645,649, 651,655,657-659,661,667
Nipigon River   645,649,651, 655-658,667
o-nitrochlorobenzene   8
nitrogen dioxide   256
nitroglycerine (NG)   266,268,270
nitromethane   313
p-nitrophenol   507
N-nitrosamines   255,257,259-261, 350
N-nitroso amino acids   255
N-nitroso atrazine   349-350
N-nitroso pyrrolidine   255
N-nitroso sarcosinate   257
nitrosyl radical   256
North Carolina, Durham   419
North Carolina, Hickory   509
North Carolina, Valdese   509
North Dakota, Grand Forks   331,351,362
nuclear magnetic resonance (NMR) spectroscopy   178,591,632-635

octabromobiphenyl   29
n-octadecane   437
Ohio, Cincinnati   305-306,314, 321,331,351,362,518,527
Ohio River   11,404-405,407,412
oil   187,189,593
Oklahoma, Norman   455-456,459
Ontario, Marathon   641
Ontario, Red Rock   642
ordnance waste   266
organohalides   129,132
oxidation   236
oxy-acids   525
ozonolysis   522,548,552

PAH's
   See polycyclic aromatic hydrocarbons
palmitic acid   668,677
palmitoleic acid   538-539
partition coefficients   299,301
PBB's
   See polybromonated biphenyls

PCB's
    See polychlorinated biphenyls
PCN's
    See polychlorinated naphthalenes
PCT's
    See polychlorinated terphenyls
Pennsylvania, Philadelphia 305-306,314,321,331,351,362,376, 386,395
pentabromochlorocyclohexane 30,32
pentachloroanisole 577,**582-583**
pentachloroethane 525,548
pentachlorophenol 525,538-539
pentachlorophenyl methyl ether 334
pentane 106-107,109
perfluorotributylamine 308
permethylation 593
persistence 16,297-298
pervaporation 119,123,133
pesticides 11,15,50,157,159, 299-300,364,558
petroleum 208
petroleum industry 281,588-589, 615
phenanthrene 209-211,441,621
phenols 6,9,227-228,300,440, 591,593,595,650,665,683,686, 694-695,698,701-702,704
phenylalkyl methyl esters 599
2-phenylethanol 440
phosphates 463
photochemical degradation 36
photochemical excitation 36
photographic industry 281
photolysis of PBB 37
photolysis of PCB 36
photolysis of PCN 43
photolysis of PCT 41
phthalates 129,166,427,463,483, 488-489,525,527,681
phthalic anhydride 681
phytane 437
pimeric acid 685,693
plasticizers 80,364,493
PNA's (polynuclear aromatics)
    See polycyclic aromatic hydrocarbons
polarogram 267-268,271

polarography 274
polybrominated biphenyls 24,35
polychlorinated biphenyls (PCB's) 11,21,24,35-37,69-70,129, 354,441,489,558,561
polychlorinated naphthalenes (PCN's) 35,43
polychlorinated terphenyls (PCT's) 35,41
polycyclic aromatic hydrocarbons (PAH's) 7-8,150,159,186-187, 205-206,208-210,354,440-441, 443
polyethylene membranes 127
polyurethane foam 7,558-561
potassium ferrocyanide 95
pristane 437
probability based matching program (PBM) 181,383-384,395
propachlor 146
propazine 359
2-propionylthiophene 687-688
1,2-propyleneglycoldinitrate (PGDN) 266-267,271-276,278
proteins 283
proximally active metabolites 67
pyrene 208-211,441,621
pyridine bases 9
pyrolysis-gas chromatography 12

quality index (QI) 174-175

radioactivity 235
rain water 434,441-443
RDX
    See 1,3,5-trinitro-1,3,5-hexahydrotriazine
recovery efficiency 257-258
recycling 558,672
reproductive failure 417,431
resin acids 637,665,685,693, 698,701-702,704
resin sorption 138-140,150,152
resin traps 118
retention volume 299-300
Rhine River 17,445-446

INDEX 717

saccharin 526
salicylic acid 526,548
salmon 626-627,637,693-694
salmonella *typhimurium* 54
sandaracopimaric acid 650,664, 666-667,685,693
β-santalene 354
Schylkill River 376
sea water 262
selected ion monitoring 120
Self-Training Interpretive and Retrieval System (STIRS) 383-384,395
separation efficiency 83
septic tank 466-467
sewage 10,216,218,233-234,236, 240-243,249-251,253,255,261-262,282-284,286,434,436-437, 510,517-522,527,537-539,541, 546,548,552,558,641
shale oil 63
simazine 359
similarity index (SI) 172,174
sitosterol 196,198-200,650, 652-655
sludge 284,293-295
sodium thiosulfate 403
solvent extraction 159
specific ion monitoring (SIM) 179
splitless injection 79,81
steranes 204
sterols 189,194
stigmasterol 196
stripping efficiency 115,117
sulfotepp 507,508
superchlorination 249,253
syringaldehyde 684

tannins 283
Tennessee, Memphis 234
Tennessee, Oak Ridge 234
terpenes 463,650,677,683,687, 704
terpene alcohols 650,688
terpene ketones 650,688
α-terpineol 440,527,541,548, 649,652-654,688
2,3′,4′-5-tetrabromobiphenyl 37
tetrachloroacetone 363,403,412

tetrachlorobenzene 577
tetrachlorocatechol 635
2,3,7,8-tetrachloro-*p*-dibenzodioxin (TCDD) 150
1,1,1,2-tetrachloroethane 359, 526,548
1,1,2,2-tetrachloroethane 388, 526,548
tetrachloroethane 421
tetrachloroethylene 323,359, 363,388-389,395,407-408,445-446,513
tetrachloroguaiacol 635,637-638
3,4,5,6-tetrachloroguaiacol 636
tetrachloromethane
  See carbon tetrachloride
tetrachlorophenol 251
1,1,3,3-tetrachloro-2-propanone
  See tetrachloroacetone
tetrachloroveratrole 630,635
1,3,5,7-tetranitro-1,3,5,7-tetraaza-cyclooctane (HMX) 266-267, 270-271
Texas, Denton 249-250,252
Texas, Fort Worth 284
textile industry 281
Thermal Energy Analysis (TEA) 256-257,261,350
thermal extraction 91-92
thin layer chromatography (TLC) 43
thiacyclanes 622
*ortho*tolidene 234
toluene 252,313,323,509
total organic carbon (TOC) 702
total organic-bound chlorine (TOCl) 248,252-253
toxicity 297-298,627
trichloroacetaldehyde
  See chloral
trichloroacetone 386,388,394
trichlorobenzene 567
1,1,1-trichloroethane 91,388,421
1,1,2-trichloroethane 323,359,365
trichloroguaiacol 658
3,4,5-trichloroguaiacol 635,650, 666
4,5,6-trichloroguaiacol 635-636
trichloromethane
  See chloroform

trichloronitromethane
  See chloropicrin
2,4,5-trichlorophenoxyacetic acid
  (2,4,5-T)   149
1,1,1-trichloropropane   359
1,2,3-trichloropropane   359
2,4,5-trichloroterphenyl   42
3,4,5-trichloroveratrole   630,633-634,664-666
3,4,6-trichloroveratrole   635
3,5,6-trichloroveratrole   634
trickling biofilter   672
tridecane   343
trimethyl isocyanurate   350,359
1,3,5-trimethyl-2,4,6-trioxohexa-hydrotriazine
  See trimethyl isocyanurate
2,4,7-trinitrofluorenone   8
1,3,5-trinitro-1,3,5-hexahydrotriazine
  (RDX)   266-268,270-271
2,4,6-trinitrotoluene (TNT)   266, 270-271
triphenyl phosphate   483
triplet excited state   37,41
triterpanes   204

ultraviolet-analyzer   217
ultraviolet spectroscopy   9
unsymmetrical dimethyl hydrazine   257
uracil   242
urea complexes   7

valeric acid   546
vanillic acid   684
vanillin   677,684,694
veratraldehyde   678
veratrole   631,633,677,698
vinyl chloride   51,97,315,513
volatile organic analysis (VOA)   89,107,114,307,310,312,320, 331,352,420,424,512

Wabash River   404
Washington, Seattle   305-307, 313-314,321-322,324,331,351,362
waste treatment   673
Water DROP   705
water solubilities   210
Watts Bar Lake   218
wide-bore glass capillary columns   82-83

XAD resins   7,136-137,139,247-249,253,331,334,354-355,364, 368,377-378,381-382,389,392-394,400-401,425,428-430,467-468,471,474-475,645
xenobiotics   15
X-ray diffraction   9
xylene   509

zinc   271
zone of persistence   641